# Nanohybrids in Environmental & Biomedical Applications

## Monograph Series in Physical Sciences

**Exchange Bias**

From Thin Film to Nanogranular and Bulk Systems

*Surender Kumar Sharma*

**Fundamentals of Charged Particle Transport in Gases and Condensed Matter**

*Robert Robson, Ronald White, Malte Hildebrandt*

**Nanohybrids in Environmental & Biomedical Applications**

*Surender Kumar Sharma*

For more information about this series, please visit:

[https://www.crcpress.com/Monograph-Series-in-Physical-Sciences/book-series/MPHYSCI]

# Nanohybrids in Environmental & Biomedical Applications

Edited by
Surender Kumar Sharma
Federal University of Maranhao, Sao Luis, Brazil

CRC Press is an imprint of the
Taylor & Francis Group, an **informa** business

CRC Press
Taylor & Francis Group
6000 Broken Sound Parkway NW, Suite 300
Boca Raton, FL 33487-2742

First issued in paperback 2021



CRC Press is an imprint of Taylor & Francis Group, an Informa business


ISBN 13: 978-0-367-77942-9 (pbk)
ISBN 13: 978-0-8153-6762-8 (hbk)



**Library of Congress Cataloging-in-Publication Data**

Names: Sharma, Surender K. (Professor of physics), editor.
Title: Nanohybrids in environmental & biomedical applications/edited by
Surender Kumar Sharma. Other titles: Nanohybrids in environmental and biomedical applications |
Monograph series in physical sciences.
Description: Boca Raton, FL: CRC Press, Taylor & Francis Group, [2019] |
Series: Monograph series in physical sciences
Identifiers: LCCN 2018048957| ISBN 9780815367628 (hardback; alk. paper) |
ISBN 0815367627 (hardback; alk. paper)
Subjects: LCSH: Nanostructured materials. | Heterostructures. | Biomedical
materials. | Composite materials.
Classification: LCC TA418.9.N35 N2528 2019 | DDC 620.1/15–dc23
LC record available at https://lccn.loc.gov/2018048957

**Visit the Taylor & Francis Website at**
**http://www.taylorandfrancis.com**

**and the CRC Press Website at**
**http://www.crcpress.com**

# Contents

# Preface

This book will serve as an overview of nanohybrid materials and their prospective applications in the field of biomedical research and development and environmental industry for a wide audience: from beginners and graduate-level students up to advanced specialists. It provides an extensive overview of the synthesis of multifunctional nanomaterials presenting key examples of their safe utilization for biomedical and environmental applications.

This contributed volume shares up-to-date advancements in the area of important nanohybrid materials with in-depth investigation of the cutting-edge developments on this particular subject. The first chapter serves as preparatory material for researchers working in nanotechnology.

It provides a forum for the critical evaluation of multifunctional nanohybrids for benign exploitation in the biomedical and environment sectors that are at the forefront of research in nanoscience and nanotechnology. It also presents highlights from the extensive literature on the topic, including the latest research in this field from the agriculture sector.

**Surender Kumar Sharma**
*Sao Luis, Brazil*

# Acknowledgments

Since the beginning of my task in selecting the title of this book and bringing it into life, I have always experienced a special source of inspiration, guidance and shower of blessings from my lovely family members. They have all shaped me into who I am today. I am unable to find words to express my gratitude to them for their unconditional love and support throughout my life.

A heartfelt thank you to my wife Dr. Sonia Malik for her time and advice in accomplishing this task. I am really blessed to have a wonderful son Aadit, whose smiling face always gives me the strength to pass through tough times.

I am grateful to the Taylor & Francis, CRC Press Team with whom I have had the pleasure to work during this project. I appreciate them for their technical support, consistent help, and suggestions in editing this book. Last but not least, I acknowledge all the authors who have contributed to this book for sharing their experiences and knowledge on their respective subjects.

# *Editor*

**Surender K. Sharma** obtained his Ph.D. degree from H. P. University, Shimla, India. After spending several years in research/teaching positions in Brazil, France, Czech Republic, India, and Mexico working in the area of nanomagnetism and functional nanomaterials, he joined the Federal University of Maranhão, Brazil, as a faculty member in Physics. He is actively involved in research, teaching, and supervising students. He was awarded FAPEMA Senior Researcher grants and has published more than 74 research articles in reputed journals, 5 books as a single author, and 5 book chapters, and has been active in professional organizations.

# Contributors

**Aman Akash**
Department of Chemistry
Savitribai Phule Pune University
Ganeshkhind, India

**Kanwal Akhtar**
Department of Physics
University of Agriculture, Faisalabad
Faisalabad, Pakistan

**Khuram Ali**
Department of Physics
University of Agriculture, Faisalabad
Faisalabad, Pakistan

**Balaprasad Ankamwar**
Department of Chemistry
Savitribai Phule Pune University
Ganeshkhind, India

**Hafeez Anwar**
Magnetic Materials Laboratory
Department of Physics
University of Agriculture, Faisalabad
Faisalabad, Pakistan

**Iram Arif**
Department of Physics
University of Agriculture, Faisalabad
Faisalabad, Pakistan

**Damayanti Bagchi**
Department of Chemical, Biological and Macromolecular Sciences
S. N. Bose National Centre for Basic Sciences
Kolkata, India

**Eduardus Budi Nursanto**
Department of Chemical Engineering
Universitas Pertamina
Jakarta Selatan, Indonesia

**Allah Ditta**
Department of Environmental Sciences
Shaheed Benazir Bhutto University
Pakhtunkhwa, Pakistan
and
School of Biological Sciences
The University of Western Australia
Perth, Australia

**S. Del Sol Fernández**
Instituto Politécnico Nacional
Centro de Investigación en Ciencia Aplicada y Tecnología Avanzada
Ciudad de México, México

**Saee Gharpure**
Department of Chemistry
Savitribai Phule Pune University
Ganeshkhind, India

**Debanjan Guin**
Department of Chemistry
Institute of Science
Banaras Hindu University
Varanasi, India

**Lienda Handojo**
Department of Chemical Engineering
Institut Teknologi Bandung
Bandung, Indonesia

**Antonius Indarto**
Department of Chemical Engineering
Institut Teknologi Bandung
Bandung, Indonesia

**Carlos Jacinto da Silva**
Grupo de Nano-Fotônica e Imagens
Instituto de Física
Universidade Federal de Alagoas
Maceió-AL, Brazil

**Uswa Javeed**
Department of Physics
University of Agriculture, Faisalabad
Faisalabad, Pakistan

**Yasir Javed**
Department of Physics
University of Agriculture, Faisalabad
Faisalabad, Pakistan

**George Z. Kyzas**
Hephaestus Advanced Laboratory
Eastern Macedonia and Thrace Institute of Technology
Kavala, Greece

**Efstathios V. Liakos**
Hephaestus Advanced Laboratory
Eastern Macedonia and Thrace Institute of Technology
Kavala, Greece

**Jaise Mariya George**
School of Chemical Sciences
Mahatma Gandhi University
Kerala, India

**Beena Mathew**
School of Chemical Sciences
Mahatma Gandhi University
Kerala, India

**Athanasios C. Mitropoulos**
Hephaestus Advanced Laboratory
Eastern Macedonia and Thrace Institute of Technology
Kavala, Greece

**Oscar F. Odio**
Instituto Politécnico Nacional
Centro de Investigación en Ciencia Aplicada y Tecnología Avanzada
Ciudad de México, México

**Samir Kumar Pal**
Department of Chemical, Biological and Macromolecular Sciences
S. N. Bose National Centre for Basic Sciences
Kolkata, India

**Ragam N. Priyanka**
School of Chemical Sciences
Mahatma Gandhi University
Kerala, India

**E. Ramón-Gallegos**
Instituto Politécnico Nacional Laboratorio de Citopatología Ambiental
Ciudad de México, México

**Edilso Reguera**
Instituto Politécnico Nacional
Centro de Investigación en Ciencia Aplicada y Tecnología Avanzada
Ciudad de México, México

**Uéslen Rocha Silva**
Grupo de Nano-Fotônica e Imagens
Instituto de Física
Universidade Federal de Alagoas
Maceió-AL, Brazil

**Ilias T. Sarafis**
Hephaestus Advanced Laboratory
Eastern Macedonia and Thrace Institute of Technology
Kavala, Greece

**Naveed A. Shad**
Department of Physics
Government College University
Faisalabad, Pakistan

**Syedda Shaher Bano**
Department of Physics
University of Agriculture, Faisalabad
Faisalabad, Pakistan

**Navadeep Shrivastava**
Institute of Physics
Federal University of Goiás
Goiânia, Brazil

**Chandra Shekhar Pati Tripathi**
Department of Chemistry
Institute of Science
Banaras Hindu University
Varanasi, India

**Herlys Viltres**
Instituto Politécnico Nacional
Centro de Investigación en Ciencia Aplicada y Tecnología Avanzada
Ciudad de México, México

**Erving Clayton Ximendes**
Grupo de Nano-Fotonica e Imagens Instituto de Fisica
Universidade Federal de Alagoas
Maceio-AL, Brazil

# Part I

# Biological Applications

# 1

# *Controlled Wet Chemical Synthesis of Multifunctional Nanomaterials: Current Status and Future Possibility*

**Navadeep Shrivastava and Surender Kumar Sharma**

**CONTENTS**



## 1.1 Introduction

The term "nano" means "dwarf" in Greek. The present scientific and technological achievements in day-to-day human life are continuously being replaced through the use of nanoscience and nanotechnology. The Royal Society of Chemistry defines nanotechnology as "the design, characterization, production and application of structures, devices and systems by controlling shape and size at nanometer scale" (RSRAE, 2004). In fact, nano-based products are being considered as part of the next industrial revolution and are delivering huge impacts on society, the economy and life in general. Nanoscaled materials have the potential to be used in a wide spectrum of areas as in medicine, information technologies, biotechnologies, energy production and storage, material technologies, manufacturing, instrumentation, environmental applications and security. Basically, nanotechnology is not a core industry, but an enabled tool and technology that, combined with other technologies, has the potential to impact most other industries in various ways (Sarveena et al., 2017; Shrivastava et al., 2017). The many promising application areas of nanotechnology have boosted the public funding for research and development rapidly with the support of billions of dollars by several agencies. The basic rationale is that nanomaterials, typically of size 1–100 nm in at least one dimension, feature exceptional structural and functional properties, very different to those in bulk materials or discrete molecules. Nanostructured materials have a significant commercial impact due to their unique properties such as finite size and surface effects. Finite size effects are related to the manifestation of so-called quantum size effects, which arises when the size of the system is commensurable with the de-Broglie wavelengths of the electrons, phonons or excitons propagating in them. Surface effects can be related, in the simplest case, to the symmetry breaking of the crystal structure at the boundary of each particle, but can also be due to different chemical and magnetic structures of internal "core" surface "shell" parts of nanoparticles (Liz-Marzán & Kamat, 2003).

The next generations of nanoparticles are about hybridization of two characteristics that mainly come due to individual characteristics of distinct entities. The design and synthesis of materials that simultaneously consists of more than one functional part, called multifunctional or bifunctional material (Shrivastava et al., 2017). Bifunctional nanoparticles have exhibited potentially promising physiochemical properties, which can revolutionize and transform the landscape of the bio-clinical industry to next-generation advance devices. In particular, colloidal nanoparticles have been extensively investigated as probes in biomedical/devices industries due to their unique size-dependent electronic, optical and magnetic properties among all possible building blocks. Thus, hybrid nanoparticles with combined magnetic and optical properties are much more powerful and can be used in a broad range of applications.

There are also some potential negative environmental and health aspects that may follow the nanotechnology. Engineered nanomaterials can penetrate the skin, lungs and intestinal tract with unknown effects to human health as nanoparticles can travel around in the body and reach, for example, the brain (Oberdörster et al., 2005). The new engineered nanoparticles have novel properties not previously known and it is likely that exactly because of these novel properties they will cause impacts on ecosystems and organisms. Nanoparticles can cause other effects if they react with other substances or even carry other substances into organisms, soil or groundwater. In this chapter, we have focused on the synthesis of nanomaterials with special attention on hybrid nanoparticles and their current status and future possibility in biomedical and environmental applications.

## 1.2 Classification of Nanomaterials

Nanomaterials are mainly classified into two categories: (a) Organic nanomaterials, and (b) Inorganic nanomaterials. The central focus of this chapter is to present various synthesis methodologies of the modern-day inorganic nanomaterials; hence we are highlighting inorganic nanomaterials in detail in the next subsection.

### 1.2.1 Organic Nanomaterials

This era of nanoscience has an unbiased growing interest in all kinds of nanomaterial for the ease of day-to-day applications. Organic nanomaterials have shown their positive impacts in the past; hence researchers are highly interested in discovering novelty applications of such materials for biomedicals, sensors, energy storage, catalysis, and environment (Virlan et al., 2016). There is focus on the growing evidence of organic structured based nanomaterials (natural and synthetic) for regeneration of bone, cartilage, wound healing, skin or dental tissue in biomedical applications. Mostly, nanoforms of chitosan, silk fibroin, synthetic polymers, poly acrylic acids (PAAs), Poly (methyl methacrylate) (PMMA) or their combinations are in use but researchers are working on several other organic molecules and composites to reduce to the nano level because of the progress in synthetic chemistry. Novel methods in investigation and manipulation of individual molecules and small ensembles of molecules have produced major advances in the field of organic nanomaterials. The new visuals of optoelectronic characteristics of organic molecules using single-molecule spectroscopy (SMS) and scanning probe microscopy (SPM) have encouraged chemists toward novel molecular and supramolecular designs (Grimsdale & Mullen, 2005). Biodegradable polyesters, such as poly(lactic acid), poly(glycolic acid), their co-polymer PLGA and poly(e-caprolactone), have been of particular interest for drug delivery applications (Soppimath et al., 2001). Several organic films/nanosheets are of interest in the case of radiation detection. As the major focus of this book is on inorganic nanomaterials-based hybrid structures, hence we have confined our discussion on the detailed synthesis procedures and case studies of these nanohybrids for biomedical applications and environmental sciences.

### 1.2.2 Inorganic Nanomaterials

In this section, we discuss the general synthesis schemes of inorganic nanomaterials. A bottom-up wet chemistry approach has been adopted throughout the book chapter while discussing hybrid nanomaterials. This procedure promotes the preparation of nanoparticles by assembling the individual atoms or molecules in the presence of stabilizer or protecting agents. They prevent the agglomeration of nanoparticles by either steric or electrostatic repulsion among nanoparticles, as agglomeration is a main key issue in the nanoregime. Several chemical methods have been reported for the synthesis of nanomaterials via a bottom-up approach, such as co-precipitation, thermal decomposition, hydrothermal, microemulsion, etc. Wet chemical synthesis permits the manipulation of matter at the molecular level. Additionally, by understanding the relationship between how matter is assembled on an atomic and a molecular level and the material's macroscopic properties, molecular synthetic chemistry can be designed to prepare novel starting components. Better control of particle size, shape and size distribution can be achieved in particle synthesis.

## 1.3 General Inorganic Synthesis Schemes

There are several standard bottom-up or wet chemical methods to prepare functional nanomaterials. Before discussing the synthesis procedures and protocols to prepare different bifunctional nanomaterials, it is important to understand the general synthesis procedures for hydro (solvo) thermal, microemulsion, co-precipitation, polyol and microwave-assisted approaches, to provide a better idea of the synthesis protocols.

### 1.3.1 Hydro (Solvo) Thermal Synthesis

The heterogeneous/homogeneous chemical reaction in water (hydro) or non-aqueous medium under constant pressure and temperature is known as hydro (solvo) thermal synthesis (see Figure 1.1a). Usually, the reaction takes place in an autoclave (a sealed thick-walled steel vessel with a Teflon cup) at high temperatures (150–220°C) and high vapor pressure (>1 bar), allowing the subsequent growth of single crystals or crystallization of substances from the solution. Reaction conditions (temperature, pressure, pH, reaction timing, etc.) and solvent (reaction medium) choice are the key issues to tune the size, shape, phase composition, crystallinity, etc. of nanomaterials (Gai et al., 2014). The solvent selection varies from water to different organics, depending upon the need and specification of the reaction, although water still remains the most widely used solvent. In hydro (solvo) thermal synthesis, some organic additives or surfactants with specific functional groups, e.g., oleic acid (OA), polyethylenimine (PEI) or cetyltrimethylammonium bromide (CTAB) are generally added along with the reaction precursors to achieve simultaneous control over the crystalline phases, sizes and morphologies as well as the surface functional groups for the resulting nanoparticles (Ye et al., 2018). The broad range of nanomaterials can be synthesized with the hydro (solvo) thermal method, using optimized reaction conditions such as temperature, pressure and pH.

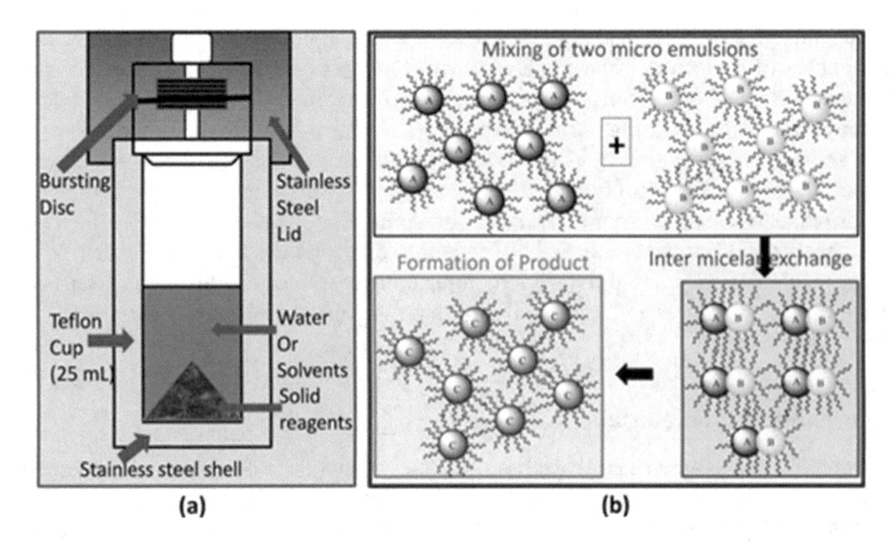


**FIGURE 1.1** Schematic explanation of synthesis mechanism for the formation of nanoparticles by (a) hydrothermal and (b) microemulsion method.

### 1.3.2 Microemulsion Synthesis

A microemulsion is a thermodynamically stable dispersion of two immiscible liquids (e.g., water and oil) with the aid of a surfactant. Small droplets of one liquid are stabilized in the other liquid by an interfacial film of surfactant molecules. In water-in-oil microemulsions, the aqueous phase forms droplets (~1–50 nm in diameter) in a continuous hydrocarbon phase. Consequently, this system can impose kinetic and thermodynamic constraints on particle formation, such as a nanoreactor. The surfactant-stabilized nanoreactor provides confinement that limits particle nucleation and growth. By mixing two identical water-in-oil emulsions containing the desired reactants, the droplets will collide, coalesce and split, and induce the formation of precipitates (see Figure 1.1b). Adding a solvent like ethanol to the microemulsion allows the extraction of the precipitate by filtering or centrifuging the mixture. The main advantage of the reverse micelle or emulsion method is better control over the nanoparticles' size by varying the nature and amount of surfactant and co-surfactant, the oil phase or the other reacting conditions. The working window for synthesis in microemulsions is usually quite narrow and the yield of nanoparticles is low compared to other methods, such as hydrothermal and co-precipitation methods. Furthermore, because large amounts of solvent are necessary to synthesize appreciable amounts of material, microemulsion is not a very efficient process and is rather difficult to scale-up (Liu et al., 2008; Kharissova et al., 2013).

### 1.3.3 Co-Precipitation Method

Generally, co-precipitation can be defined as "the simultaneous precipitation of more than one substance from homogeneous solution," which results in the formation of a crystal structure of a single phase (e.g., $Fe_3O_4$). Co-precipitation is probably the most convenient and efficient chemical pathway to synthesize a broad range of nanomaterials, including magnetic nanoparticles, due to relatively mild reaction conditions, low cost of required equipment, simple protocols and short reaction times (Laurent et al., 2008). Therefore, this method is not only a preferred route to synthesize magnetic iron oxide nanoparticles, but it is also commonly used to prepare a broad range of rare earth ion ($RE^{3+}$)-doped luminescent nanomaterials, such as alkaline-earth tungstates: $MWO_4$:$RE^{3+}$ ($M^{2+}$: Ca, Sr and Ba) or rare earth fluorides: $NaYF_4$:$RE^{3+}$, $LaF_3$:$RE^{3+}$. The great advantage of the co-precipitation method is the obtainment of a large number of nanoparticles (Mascolo et al., 2013). However, polydispersity and wide particle size distribution are usually obtained with this method due to the kinetic factors that control the growth of the crystal. Generally, two processes are involved in the growth and formation of the particles: rapid nucleation (aggregation of nanometric building blocks such as pre-nucleation clusters), which occurs when the concentration of the species reaches critical supersaturation. The other one is the slow growth of the nuclei by diffusion of the solutes to the surface of the crystal. Therefore, to produce monodisperse nanoparticles, controlling these processes is very important. In a supersaturated solution when the nuclei form at the same time, subsequent growth of these nuclei results in the formation of particles with very narrow size distribution. Co-precipitation methods are popular because these methods allow for the production of large quantities of nanoparticles with a moderate degree of particle uniformity. The major disadvantage of co-precipitation is the lack of fine particle size control as size is primarily determined by kinetic factors. These reactions are typically performed in aqueous environments and utilize a base to initiate the reaction.

### 1.3.4 Polyol Synthesis

Polyol synthesis (a special kind of sol–gel synthesis) is an excellent method for the synthesis of nanoparticles from metallic salts by using a poly-alcohol, which acts as amphiprotic solvents (Ghosh Chaudhuri & Paria, 2012), as well as complexing, reducing and surfactant agents, depending on the studied system. The poly-alcohols used in this process are ethylene glycol (EG), diethylene glycol (DEG), 1,2-propanediol, tetraethylene glycol and glycerol. Owing to the high boiling point and high dielectric constant of these solvents, they offer a wide range of reaction temperature. Hydrolysis and reduction reactions can be performed in these liquids, allowing the production of a wide variety of size and shape-controlled inorganic nanoparticles from the nano- to the microregime. In this method (Dong et al., 2015a), the precursor compound is suspended in the liquid polyol and heated to the boiling point of the polyol at constant stirring. During this process, the precursor gets dissolved in the polyol leading to the formation of an intermediate followed by reduction to form nanoparticles. This method allows preparation of biocompatible nanoparticles by coating hydrophilic polyol over the nanoparticles, producing particles with higher crystallinity that leads to a better saturation magnetization and narrow particle size distribution as compared to the traditional methods. This process is heralded for its self-seeding mechanism and lack of required "hard" or "soft" templating materials, making it an ideal process for industrial scale-up owing to the low cost of processing.

### 1.3.5 Microwave-Assisted Method

In recent years, a microwave-assisted synthesis method has been widely used over conventional heating methods to produce nanomaterials with higher yield (Dallinger and Kappe, 2007). A specially designed microwave synthesis reactor allows exquisite control of the reaction temperature, stirring rate and pressure inside the reaction vessel, which is the unique feature of microwave apparatus over conventional heating principles. The heating effect in the microwave arises from the interaction of the electric field component of the microwave with charged particles in the material through both conduction and polarization. The use of microwave heating as a non-classical energy source has been shown to dramatically reduce reaction times, increase product yields and enhance purity or material properties compared to conventionally processed experiments. In conventional methods, the heating of chemical reactions has been achieved using mantles, oil baths, hot plates and reflux set-ups, where the highest reaction temperature achievable is dictated by the boiling point of the solvent used. This traditional form of heating is rather slow and has low efficiency for transferring energy to a reaction mixture as it depends on convective currents and on the thermal conductivity of various compounds or materials that have to be penetrated. This often results in the temperature of the reaction vessel being higher than that of the reaction solution. In contrast, microwave irradiation produces efficient volumetric heating by raising the temperature uniformly throughout the whole liquid volume by direct coupling of the microwave energy to the molecules that are present in the reaction mixture (Gawande et al., 2014).

A single mode microwave reactor closed-vessel microwave synthesis purchased from Synth-wave has been utilized and discussed here: It consists of a single reaction chamber (SRC) which is a large, pressurized stainless steel reaction chamber into which all reactions mixture are placed and prepared simultaneously. The pressurized chamber in SRC serves as the reaction vessel and the microwave cavity and the reaction vessel enables the intensity and distribution of the microwave energy to be optimized with the shape and size of the reaction vessel. There are several benefits of using SRC (Nishioka et al., 2013). Some of them are: (i) it can run multiple parameters or reactions simultaneously; (ii) since all reactions are in the same vessel, nearly any reaction type can be processed simultaneously; (iii) a large range of temperature and pressure; (iv) several stoichiometry combinations and element changes can be modified for inorganic hydrothermal syntheses in a single run, saving weeks of research labor; and (v) control of every reaction is made possible by direct pressure and temperature control.

### 1.3.6 Thermal Decomposition Synthesis

This is one of the most rated and well-established methods for the synthesis of monodispersed nanoparticles with uniform shape, tailored size and single crystal structure. In this method, organometallic precursors in the presence of organic solvents (such as 1-octadecane [ODE]), and a surfactant (e.g., OA and oleylamine) form organometallic complexes which are rapidly broken down in a hot solvent-containing surfactant at high temperature (usually the boiling points of the solvents). In general, the synthetic process is conducted at elevated temperatures (250–330°C) in an oxygen-free and anhydrous environment, wherein the precursors decompose to form the nucleus for a particle to grow on (Zhang et al., 2005; Unni et al., 2017). The size and shape of the nanoparticles can be controlled by varying the molar ratios of precursor and surfactants and solvents. The most commonly used precursors include lanthanide trifluoroacetate, iron acetoacetonate, iron pentacarbonyl, lanthanide oleates, iron chlorides, lanthanide acetates and lanthanide chlorides. The use of a surfactant can form a protective layer around the nanoparticle, and plays an important role as a dispersant. The surfactants usually contain a functional group to cap the surface of the nanoparticle, for example, up conversion nanoparticles for controlling their growth and a long hydrocarbon chain to assist their dispersion in organic solvents.

## 1.4 Illustrations of Magnetic and Non-Magnetic Nanomaterials

### 1.4.1 Synthesis of Magnetic Nanoparticles

The preparation methods of iron oxide nanoparticles are quite simple and easy to handle. They can be obtained by chemical, physical and biological methods (Bhargava et al., 2013; Xu & Sun, 2013). There are several methods and the majority of researchers are interested in exploring the cheap, green and non-toxic protocols and its characteristics for biomedicals, sensors, devices and environmental applications. Among these methods, co-precipitation and thermal decomposition methods are utilized more and these are well understood mechanisms. The initial solution consists of iron sources dissolved in an aqueous solution. Typical iron sources for synthesis are: $FeCl_3$, $FeCl_2$, $Fe(acac)_3$, (acac = acetylacetonate) or $Fe(CO)_5$. In the co-precipitation synthesis developed by Massart (1981), these precursors ($FeCl_3$, $FeCl_2$, $Fe(acac)_3$) are able to easily dissociate in solution, forming Fe cations leading the reaction in a basic environment to form an iron oxide. Here, the addition of a base to an aqueous solution of ferrous ($Fe^{2+}$) and ferric ($Fe^{3+}$) ions in a 1:2 stoichiometry produced a black precipitate of spherical magnetite ($Fe_3O_4$) in the absence of oxygen (Massart, 1981). The quality of the final product depends on rapid bursting followed by a particle growth phase, explained by Lamer (Lamer & Dinegar, 1950; Vreeland et al., 2015) (see Figure 1.2). For example, $Fe_3O_4$ is formed according to the reaction:

$$Fe_2 + 2Fe_3 + 8OH \rightarrow Fe_3O_4 + 4H_2O.$$

In this reaction, nuclear bursting and particle growth phases are overlapping little or there is no overlap. If these phases are not well separated, control of particle size will be severely hindered. The type and properties of the iron oxide nanoparticles produced can be controlled by varying the reaction conditions. One of the most important conditions to control during the synthesis of iron oxides is the oxygen content of the reaction mixture (Laurent et al., 2008). Synthesis of $Fe_3O_4$ is typically performed in oxygen controlled environments due to the inherent instability of $Fe_3O_4$ in oxygen. Moreover, $Fe^{2+}$ is oxidized to

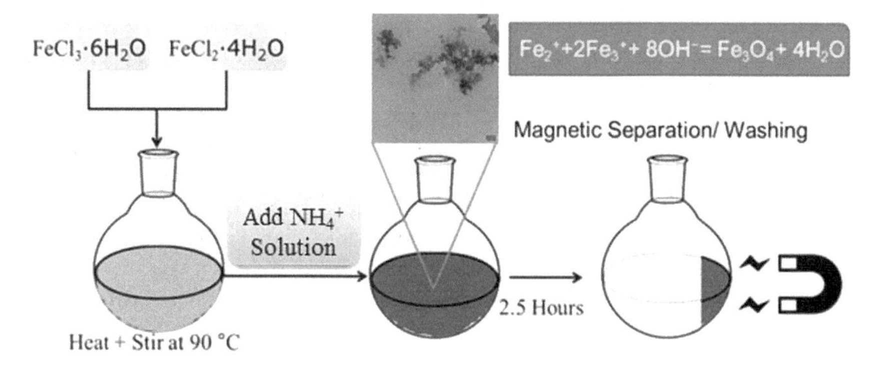


**FIGURE 1.2** Synthesis of iron oxide nanoparticles.

$Fe^{3+}$ in aqueous solutions, altering the $Fe^{2+}/Fe^{3+}$ ratio in the reaction mixture, resulting in the potential formation of unwanted species of iron oxides. Thus, nitrogen or argon gas are typically used to purge the reaction mixture. Control over particle size, morphology and uniformity is achieved by altering such factors as the $Fe^{2+}/Fe^{3+}$ ratio, temperature, time and stir speed (Jeong et al., 2004; Zhu et al., 2011).

As stated in Section 1.3.6, the thermal decomposition method is mainly used to obtain precise growth and control over nanoparticles. Recently, Unni and co-workers produced iron oxide nanoparticles with a diminished magnetic dead layer by the controlled and careful addition of molecular oxygen during high temperature synthesis (Unni et al., 2017), which is a step ahead of previous thermal decomposition syntheses (Hufschmid et al., 2015; Lassenberger et al., 2017). The conventional thermal decomposition method of organometallic precursors of iron oxide nanoparticle synthesis with no oxygen and *in situ* condition, in the past, provided excellent control over shape–size tunability but it was not sufficient to compare with the magnetic characteristics of the bulk data. This conventional method gives a "magnetically dead layer" (Luigjes et al., 2011), crystal defects and mixed iron oxide phases (magnetite/maghemite phases). The dead layer is determined experimentally or through modeling where the magnetic diameter is significantly smaller than the physical diameter. In their novel synthesis, Unni and co-workers (2017) first prepared iron oxide nanoparticles by preparing the iron oleate precursors using Park and co-workers' method (Park et al., 2004) which utilizes iron chloride hexahydrate [$Fe(Cl_3)·6H_2O$], water, sodium oleate, hexane and ethanol at around 60°C (see Figure 1.3). Next, iron oxide was produced by the heating-up thermal decomposition route in the absence of oxygen. This iron oxide includes iron oleate with OA dissolved in trioctylamine as the non-reacting solvent in a three-neck reactor in the presence of a controlled environment of $N_2$ gas, stirring at 340°C for 1 h at a constant increment of temperature. Further, hexane and acetone were used to wash nanoparticles. In the next crucial step, they prepared magnetic nanoparticles by the extended LaMer thermal decomposition route in the absence of oxygen. The main steps consist of (a) 48.3 mmol of docosane heated to 350°C for 50–60 min at a ramp rate of

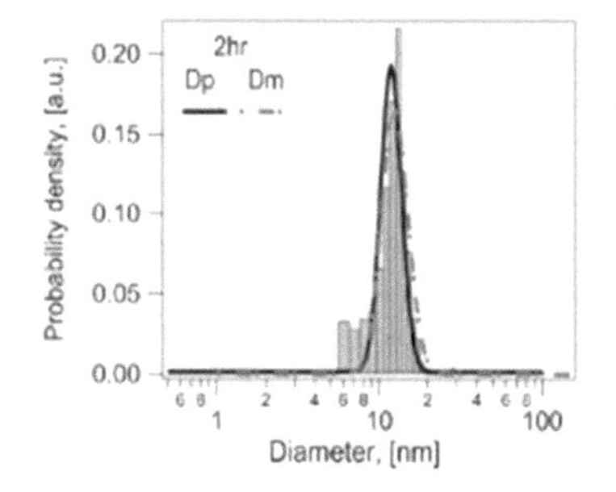


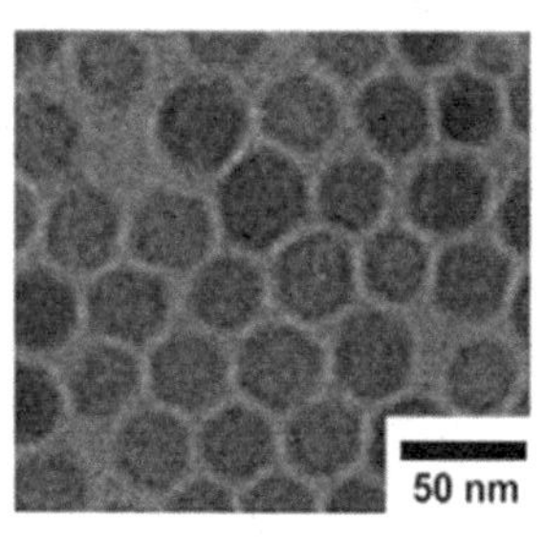


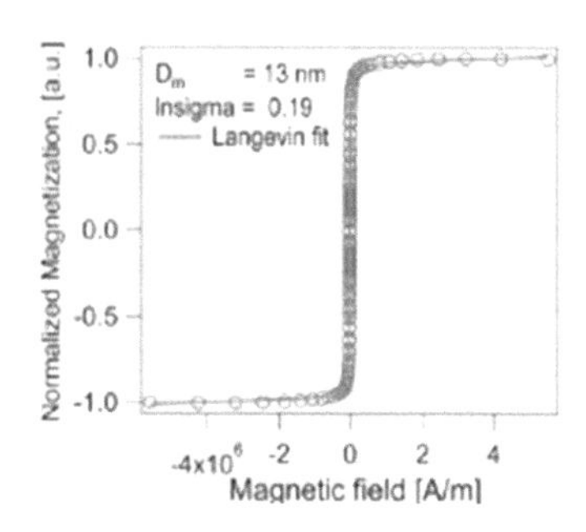


**FIGURE 1.3** **(See color insert.)** Controlled growth of both physical and magnetic diameter for iron oxide nanoparticles obtained using the Extended LaMer mechanism based synthesis thermal decomposition synthesis in the presence of molecular oxygen for 2 h. Reused with permission from Unni et al., 2017, American Chemical Society.

7–8°C/min in a 100 mL three-neck reaction flask under controlled inert gas atmosphere and (b) the controlled addition (using a syringe pump) of 30 mL of iron oleate precursor (0.63 M Fe) mixed with 55 mL of 1-ODE at 350°C. The resulting solution was stirred at a constant rate for 5 h. After this, the mixture was cooled to room temperature, and iron oxide nanoparticles obtained at the end of the reaction were purified by suspending 5 mL of the black waxy liquid in 10–20 mL of hexane. The particles were precipitated using 20–40 mL of acetone by centrifuging in an Eppendorf. Postsynthesis oxidation of iron oxide nanoparticles was used in order to improve the magnetic properties. Nanoparticles suspended in organic solvents were thermally treated at 120°C with and without bubbling air for particles synthesized by an Extended LaMer mechanism.

***Oxidation by Bubbling in Air:*** Approximately 15 mg/mL of iron oxide nanoparticles were suspended in 2 mL of OA and placed in a heating block from Fisher for 4 h at 120°C with air being bubbled in the samples. There was significant color change of the sample from dark brown to light brown, and high rates of OA boil off were observed.

***Oxidation by Solvent Transfer without Bubbling in Air:*** Approximately 5 mg of as-synthesized iron oxide nanoparticles were suspended in 2 mL of hexane and sonicated for 10 min. Further, 2 mL of 1-ODE was added to the vial and the particles were left in an oven for 2 h at 120°C. The particles remained well-suspended in 1-ODE, and color change was observed during oxidation.

The difference between Unni and co-workers and other researchers is the use of the oxygen molecule during synthesis (Unni et al., 2017). They also prepared iron oxide nanoparticles using the same general method, but instead of using a perfectly inert atmosphere of argon gas, 20% oxygen gas and 80% argon gas at a constant rate was supplied in a controlled way (see Figure 1.3).

The studies reported here demonstrate the critical role that molecular oxygen plays in determining the magnetic properties of iron oxide nanoparticles obtained through thermal decomposition syntheses. Experiments carried out in "inert" atmospheres using argon and nitrogen as inert blanket and carrier gases demonstrated that the resulting particles possessed magnetic diameter distributions that were smaller and broader than the corresponding physical diameter distributions. Further, experiments with postsynthesis oxidation demonstrated that while an improvement in magnetic diameter distributions can be obtained with such methods, this improvement is difficult to control and limited for larger particles. The results demonstrate that the safe and judicious addition of molecular oxygen as a species in the thermal decomposition synthesis leads to nanoparticles with practically equal physical and magnetic diameter distributions, magnetic properties that resemble those expected for bulk magnetite and improved application-relevant properties such as thermal energy dissipation rate and resolution for magnetic particle imaging (see Figure 1.3).

In another work, magnetic cobalt ferrite ($CoFe_2O4$) nanoparticles have been successfully synthesized by thermal decomposition of Fe (III) and Co (II) acetylacetonate compounds in organic solvent in the presence of OA/OLA as surfactants and 1,2-hexadecanediol (HDD) or octadecanol (OCD-ol) as accelerating agent (Lu et al., 2015). Lu and co-workers studied the influence of surfactant concentration (dioctyl ether) and reaction time on the morphology of the nanoparticles. In the case of using dioctyl ether, the volume of the solvent was reduced to 20 mL but the concentration of reaction agents was kept constant. Hachani and his colleagues (Hachani et al., 2015) synthesized iron oxide nanoparticles of low polydispersity through a simple polyol synthesis in high pressure and high temperature conditions (see Figure 1.4). The control of the size and morphology of the nanoparticles was studied by varying the solvent used, the amount of iron precursor and the reaction time; this process yields nanoparticles with a narrow particle size distribution in a simple, reproducible and cost-effective manner without the need for an inert atmosphere. They used triethylene glycol (TREG), diethylene glycol (DEG) and tetraethylene glycol (TEG) as solvents. $Fe(acac)_3$ was used as precursor. A polyol and $Fe(acac)_3$ was mixed to obtain a red dispersion that was then placed into a 45 ml capacity Teflon liner and the latter was assembled with the autoclave jacket and placed into an oven at 300°C and maximum working pressure of 115 bar for 2 h. The resulting black dispersion was washed with acetone and centrifuged. This procedure yielded iron oxide nanoparticles coated with polyols (see Figure 1.4).

In a work by Mameli and co-workers (Mameli et al., 2016), a thermal decomposition method was utilized to prepare Zn-substituted cobalt ferrite ($Zn_xCo_{1-x}Fe_2O_4$) nanoparticles (see Figure 1.5). They studied complex magnetic properties' dependence on the properties of the material with the variation of Zn-amount with constraints of the same particle size and distribution, crystallite size and capping agent.

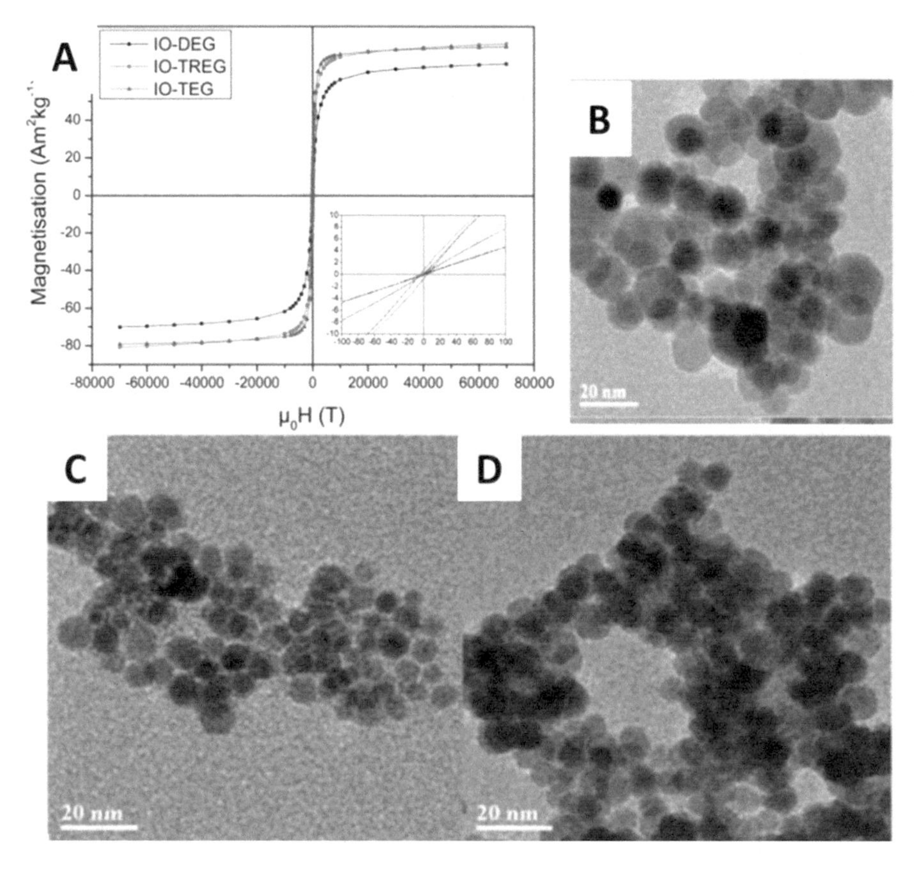


**FIGURE 1.4 (See color insert.)** (A) Magnetization curves of iron oxide NPs obtained with different polyols (DEG, TREG and TEG). (B) TEM images of iron oxide nanoparticles synthesized using TEG. (C) and (D) TEM images of iron oxide nanoparticles synthesized using different reaction times in tri (EG): (C) 1 h, (D) 2 h. (Reproduced from Hachani et al., 2015. With permission.)

Magnetic properties at 300 K were engaged and correlated to specific absorption rate (SAR) values by combining results originally observed from dc/ac magnetometry and $^{57}Fe$ Mössbauer spectroscopy in different time spans of experimental acquisitions (see Figure 1.5). Synthesis was carried out using acetylacetonate precursors of Fe, Co and Zn; 1,2-hexadecanediol, oleic acid, oleylamine and dibenzylether in a calculated amount. During syntheses, much attention was given to the control temperature that was elevated from room temperature to 200°C for 2 h and then to 280°C, for 1 h.

## 1.4.2 Synthesis of Semiconducting or Quantum Dot Nanoparticles

Quantum dots (QDs) have emerged as one of the most exciting fluorescent nanoparticles (in the order of 2–10 nanometers containing approximately 200–10,000 atoms) with a potential for diagnostic and therapeutic application in the field of nanomedicine. In general, QDs are produced using atoms from group II and VI of the periodic table, e.g., cadmium–selenide (CdSe), cadmium tellurium (CdTe), zinc–selenium (ZnSe), group III–V elements, e.g., indium phosphate (InP), indium arsenate (InAs), gallium arsenate (GaAs), gallium nitride (GaN) or group IV–VI elements, e.g., lead–selenium (PbSe). The most commonly used QDs are CdSe or CdTe with a passivation shell made of ZnS which protects the core from oxidation and increases the photoluminescence quantum yield. The surface of the QD is further coated with solubilization ligands, making them water soluble for their use in biology and environmental sciences. QDs have emerged as one of the most exciting nanoparticles with a potential for diagnostic and therapeutic application in the field of nanomedicine (LaRocque et al., 2009). The current fluorophores

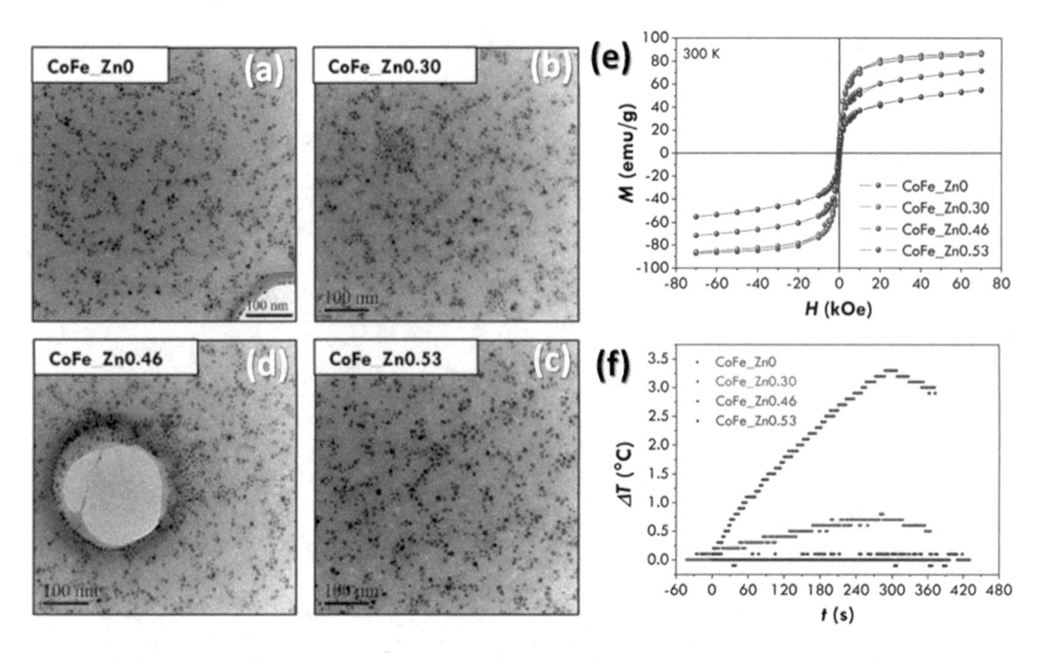


**FIGURE 1.5 (See color insert.)** (a-d) Morphological properties/TEM images of $Zn_xCo_{1-x}Fe_2O_4$ (with x = 0, 0.30, 0.46, 0.53) samples using the JEM-2010 UHR, (e) Magnetization versus magnetic field curves of $Zn_xCo_{1-x}Fe_2O_4$ (with x = 0, 0.30, 0.46, 0.53) samples measured at 300 K. (f) Heating curves of $Zn_xCo_{1-x}Fe_2O_4$ (with x = 0, 0.30, 0.46, 0.53) samples at 25°C, obtained under a magnetic field of 183 kHz and 17 kA $m^{-1}$. (Reused with permission from open access reference Mameli et al., 2016.)

such as organic dyes, fluorescent proteins and lanthanide chelates suffer the problems of instability, photobleaching and sensitivity to environmental conditions such as pH variations. The unique optical and spectroscopic properties of QDs offer a compelling alternative to traditional fluorophores due to their high quantum yield, high molar extinction coefficient (~600,000 $M^{-1}$ $cm^{-1}$, roughly an order of magnitude higher than even the strongly absorbing Rhodamine, exceptional resistance to photobleaching as well as to photo and chemical degradation (Huo, 2007; Zrazhevskiy et al., 2010). In addition, the intensity of fluorescence produced by the QDs is 10–20 times brighter than the organic dyes. Conventional dyes suffer from narrow excitation spectra. This requires excitation by light of specific wavelength, which varies between particular dyes. In addition, they have broad emission spectra. This means the spectra of different dyes may overlap to a large extent limiting the number of fluorescent probes that may be used to tag different biological molecules.

An ideal strategy for the synthesis of semiconductor nanoparticle should have the following characteristics: (i) control of the size and shape, (ii) high monodispersity and good crystallinity, (iii) high luminescence quantum yield and (iv) reproducibility. Chemical methods allow the preparation of a broad range of semiconductor nanoparticles and core–shell nanostructures with excellent control over size, shape and dispersity. This method also enables the functionalization of surfaces of semiconductor nanoparticles with various ligands, and the design of higher order nanostructures through self-assembly. Colloidal synthesis offers a versatile route, where nanoparticles are grown using the chemical reaction of their precursors in an appropriate solvent. It follows the usual nucleation and crystal growth mechanism where the latter process is controlled through the use of an appropriate surfactant or capping agent (sometimes the solvent molecule itself serves as capping agent). Control over the size, shape, dispersity and crystallinity can be achieved by adjusting the reaction conditions such as time, temperature, the concentration of the precursor, chemical nature of the capping ligands and surfactants.

A breakthrough in producing monodispersed QDs of cadmium chalcogenide such as CdS, CdSe and CdTe (II–VI semiconductors) was reported by Bawendi and co-workers in 1993 (Murray et al., 1993). They were able to synthesize highly crystalline QDs using organometallic precursors in the presence of a coordinating solvent, trioctylphosphine oxide (TOPO). The reaction was carried out at elevated

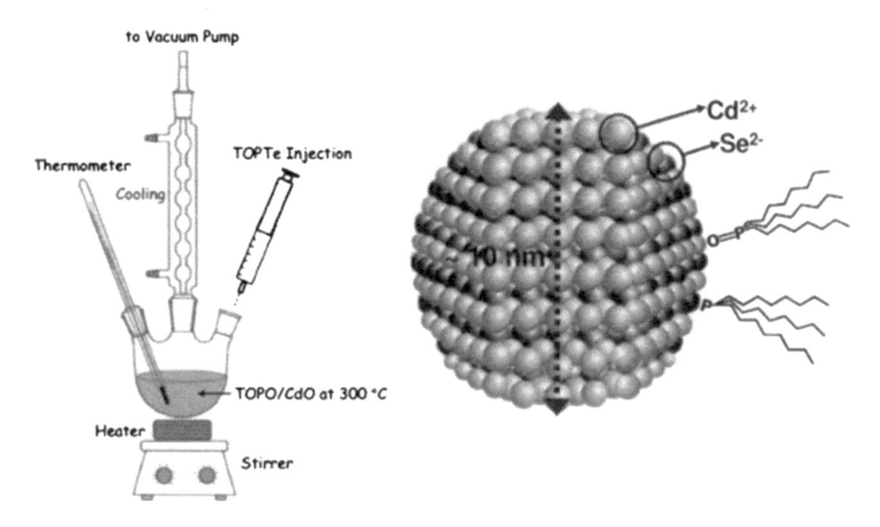


**FIGURE 1.6 (See color insert.)** Figure presents an experimental set-up for QD synthesis by high temperature organometallic method, along with a schematic illustration of CdSe QD capped with TOPO (bound to $Cd^{2+}$) and TOP (bound to $Se^{2+}$). (Adapted from Kuno, 2008.)

temperatures (~360°C) under vacuum, using selenium precursor (TOPSe, selenium coordinated to trioctylphosphine) which was injected rapidly into a solution of dimethylcadmium in TOPO (see Figure 1.6). This resulted in burst nucleation followed by crystal growth leading to the formation of crystalline QDs. TOPO served as the main capping agent and the QDs produced possessed narrow size distribution ($\sigma_r < 10\%$), having excellent solubility in nonpolar solvents. Low surface defects, compared to other synthetic methods is another advantage of the high temperature organometallic synthesis. Excellent control over crystal growth was achieved by controlling the precursor concentration, temperature, reaction duration etc. Further, the organic capping agent, which is electrostatically bound to the QD surface, could be replaced by functionalization with a thiol. It is possible to introduce photoactive or electroactive groups on to the surface of a nanoparticle by following this method (place exchange reaction). Numerous attempts have been made to replace dimethyl cadmium with other organometallic precursors due to its extreme toxicity, pyrophoric nature, expensiveness and instability (Hambrock et al., 2001; Peng & Peng, 2001). A marked improvement in the above synthesis strategy was achieved by Peng and co-workers by using a nonpyrophoric and stable cadmium precursor, cadmium oxide, instead of dimethylcadmium (Peng & Peng, 2001).

A significant reduction in reaction temperature (~300°C) as well as in size distribution ($\sigma_r < 5\%$) was observed for the resulting nanocrystals. This was attributed to the thermal stability as well as slow decomposition of the cadmium oxide compared to dimethylcadmium, which led to a slow and homogeneous nucleation and crystal growth (avoiding Ostwald's ripening) (Chen et al., 2008). Several other synthesis protocols were reported by varying the cadmium precursor (e.g. cadmium acetate, cadmium stearate), reaction medium (e. g. 1-ODE, olive oil) and capping agents such as amines, fatty acids, phosphonic acids, etc. (Chen et al., 2008). The design and fabrication of anisotropic semiconductor nanostructures such as nanorods (quantum rods), nanowires (quantum wires), dipods, tripods and tetrapods have drawn attention in recent years. These nanostructures have a lot of potential, particularly for fabricating nanoelectronic, photovoltaic and energy storage devices due to their shape-dependent properties. For example, CdSe nanorods possess polarized emission related to their cylindrical symmetry which makes them a novel photonic material (for, e.g., laser systems) (Alivisatos, 1996).

Recently, Ramasamy et al. (2018), demonstrated a two-step approach to synthesize highly monodisperse InP QDs (<10%) absorbing from 490 to 650 nm (size range of ~1.9 to 4.5 nm). Initially, InP QDs with an average size of 1.9 nm were synthesized and later used as "seeds" to grow larger QDs (see Figure 1.7) by continuously injecting Zn(In)–P complexes as monomers source.

A very nice protocol regarding the synthesis of chiral QDs has been developed by Gun'ko and co-workers (Moloney et al., 2015). They reported the procedure for the synthesis of chiral optically active

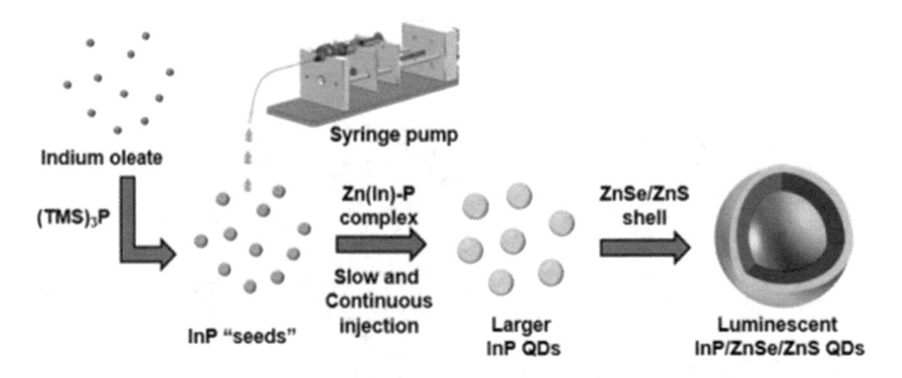


**FIGURE 1.7** Schematic of "Seed-Mediated" synthesis of larger InP QDs and InP/ZnSe/ZnS QDs. (From Ramasamy et al., 2018. With permission.)

QD nanostructures and their quality control using spectroscopic studies and transmission electron microscopy imaging. They closely examined various synthetic routes for the preparation of chiral CdS, CdSe, CdTe and doped ZnS QDs, as well as of chiral CdS nanotetrapods. Most of these nanoparticles can be produced by a very fast (70 s) microwave-induced heating of the corresponding precursors in the presence of d- or l-chiral stabilizing coating ligands (stabilizers), which are crucial to generating optically active chiral QDs. Alternatively, chiral QDs can also be produced via the conventional hot injection technique, followed by a phase transfer in the presence of an appropriate chiral stabilizer.

Cui's group reported the first example of Ag-doped ZnInSe QDs, which not only have extremely high stability but also have tunable emission in a broad emission scope (Wang et al., 2014a). Due to their extremely high stability, good biocompatibility, low cytotoxicity and tunable emission, as-prepared Ag:ZnInSe QDs are expected to be applicable in optical coding, white light emitting diode (LED) and bioimaging. Nitrates of Zn and Ag were used as precursors in 3-mercaptopropionic acid (MPA). It was observed that pH plays a very important role and can be managed by NaOH solution. NaHSe solution was injected into the mixture after the solution was a controlled atmosphere of $N_2$ gas for 30 min. Then, the solution was refluxed at 100°C.

### 1.4.3 Synthesis of Upconverting Rare Earth Nanoparticles

The major focus of biomedical applications is in upconverting nanoparticles where a near infrared (NIR) laser is used to excite the nanoparticles. NIR downconverting nanoparticles are in focus nowadays. A wide variety of synthesis protocols are on trend for nanoparticles such as co-precipitation, thermal decomposition, hydro(solvo)thermal synthesis, microwave-assisted, combustion synthesis and urea homogeneous precipitation with their competitive advantages/disadvantages (Wang & Liu, 2009; Niu et al., 2014; Zhou et al., 2015). Two different protocols can also be combined for preparation. Postsynthesis processes are used as well for the development of upconversion nanoparticles with controlled particle size, chemical composition or surface functionalization. The most convenient and highly cited method from the beginning for producing ultra-small particles is co-precipitation, which has been used by several groups for the synthesis of nanoparticles (5–10 nm) with better size distribution. While having no need for costly equipment and complicated procedures, this method has a fast growth rate and requires post-annealing processes. In a typical process, a solution of lanthanide salts is injected into a solution of the host material. Examples of hosts used are $LaF_3$, $NaYF_4$, $LuPO_4$ and $YbPO_4$ (Haase & Schäfer, 2011). Particle size and growth rate can be regulated by using capping ligands or chelating agents such as polyvinylpyrrolidone (PVP), polyethylenimine (PEI) or ethylenediaminetetraacetic acid (EDTA) (Dong et al., 2015b). In particular, PEI provides a good platform for direct surface functionalization of nanoparticles with bioligands. However, a heat treatment step is required for post processing, which is one of the disadvantages of this method.

Thermal decomposition is another synthetic method that produces highly monodispersed particles with no need for post-annealing processes. In a typical procedure, metal precursors, mostly triuoroacetate, are thermolyzed together with OA and ODE (Wang et al., 2014b). OA acts as the primary solvent that prevents agglomeration of particles and ODE acts as the high boiling solvent (350°C). The advantages of oleate-based synthesis include narrow particle size distribution, high luminescence efficiency and high phase purity of the particles. Although this method produces high quality particles, it requires expensive air sensitive precursors, results in mostly hydrophobic particles and has toxic by-products (Wang et al., 2010; Wang et al., 2011a,b). Examples of the reported hosts for this method are $LaF_3$, $NaYF_4$ and GdOF (Eliseeva & Bunzli, 2009; Park et al., 2009).

The hydro (solvo) thermal method is a good method for preparation of highly crystalline materials at a much lower temperature and with no need for post annealing. The process is performed in an autoclave reaction vessel that maintains the required temperature and pressure for a desired period of time. By keeping solvents in temperatures and pressure above their critical point, the autoclave enhances the solubility of solid and increases the reaction rate between solids. Since the reaction happens in a closed cylinder, growth monitoring is virtually impossible. Particle size and morphology can be manipulated in this method by using polyol- or micelle-mediation. Some of the popular hosts reported for this method include $LaF_3$, $NaYF_4$, $La_2(MoO_4)_3$ and $YVO_4$ (Liu et al., 2009; Chen et al., 2017; Shrivastava et al., 2018). In contrast to time-consuming solvo(hydro)thermal methods, microwave-assisted synthesis is fast and provides good control over size and shape tunability. The particle sizes and shapes depend on the reaction temperature, which is governed by the ratio of the metal ions to solutions involved (Shrivastava et al., 2017, 2018).

Recently, You and co-workers prepared uniform lanthanide (Ln+)-doped $NaREF_4$ (RE=rare earth) nanocrystals (You et al., 2018). They reported this synthesis via a facile solid–liquid thermal decomposition (SLTD) method by directly employing $NaHF_2$ powder as a fluoride and sodium precursor. The proposed SLTD strategy is easy to perform, time-saving and cost-effective, making it ideal for scale-up syntheses. Particularly, over 63 g of β-$NaGdF_4$:Yb,Er@$NaYF_4$ core–shell NCs with narrow size variation (<7%) were synthesized via a one-pot reaction. This sodium precursor, which decomposes to NaF and HF at an elevated temperature and may react with RE compounds in the liquid phase (OA and ODE) for the nucleation and growth of $Ln^{3+}$-doped $NaREF_4$ nanoparticles. The crystalline structure, particle size and morphology can be well tuned by adjusting the reaction temperature or the added amounts of sodium acetate (NaAc). The synthesis scheme and quality of the product have been presented in Figure 1.8.

### 1.4.4 Synthesis of Plasmonic Nanoparticles

Plasmonic nanoparticles are metal particles whose electron density can couple with electromagnetic waves (in most cases, infrared or visible light) of wavelengths far larger than the size of the particles; unlike in the usual case where there is a limit of the wavelength that can be coupled to a piece of metal depending on its size (Eustis and El-Sayed, 2006). This is due to the properties of the metal–dielectric interface between the particles and the surrounding medium. The plasmonic effect can occur only when the light frequency is lower than or equal to the plasma frequency of the nanoparticles. For the light of a higher frequency than the resonance frequency, the destructive interference of electron oscillation will occur inside the particle and there will be no plasmonic effect. The reason why plasmonic nanoparticles are important in biomedicals or environmental applications is that they have a very strong optical absorption at the resonance wavelength. Some other characteristics of plasmonic nanoparticles (Au, Ag, Pt) can be stated as: they are chemically inert and therefore samples labeled with Au nanoparticles are very stable and have a very long self life and lifetime; they have a very strong plasmonic resonance and, hence strong absorption, making them ideal labels for detection and the third is that they can be easily functionalized so that they can specifically bind to the target biomarkers (Merkel et al., 2009). The most straightforward approach for the synthesis of plasmonic nanoparticles is the direct one-pot-one-step synthesis. Depending on the reaction conditions (e.g., temperature, solvent, additives), certain morphologies can be obtained as spheres, stars, rods, plates and prisms. Usually poly vinyl pyrrolidone (PVP) is utilized as a reducing and a capping agent for plasmonic nanoparticles, playing an important role in the nucleation, the growth and the final shape of nanoparticles due to its interaction with metal ions and surfaces as well as the preferable attachment to distinct crystal facets of certain metals (e.g., silver) (Pastoriza-Santos & Liz-Marzán, 2002; Behera & Ram, 2013). The most common strategy to synthesize

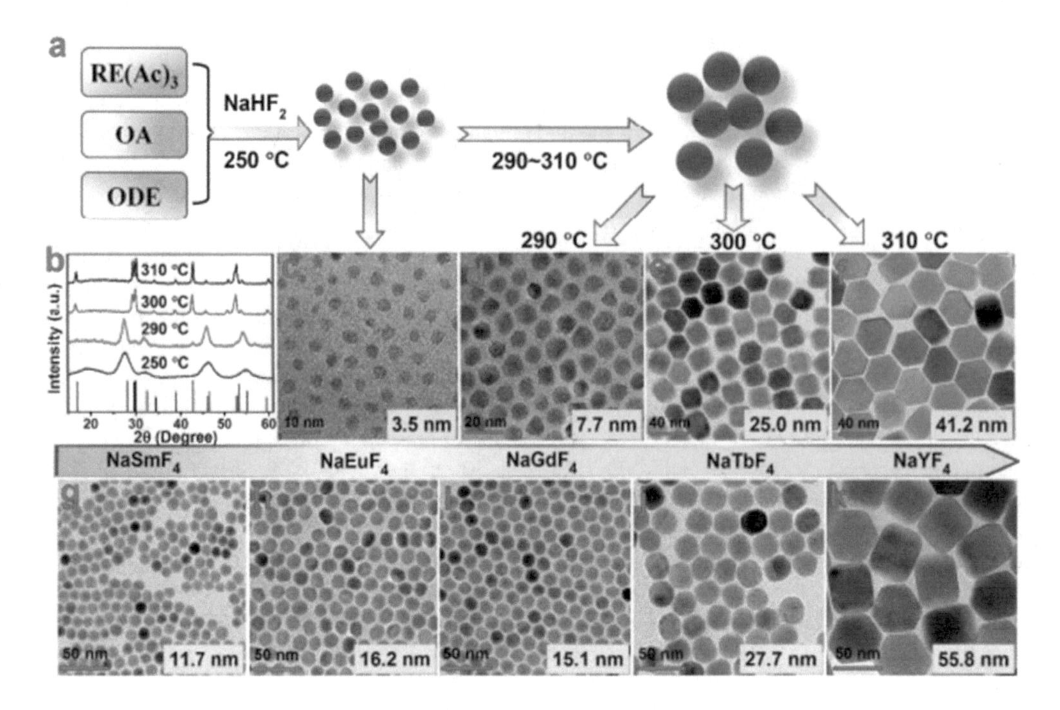


**FIGURE 1.8 (See color insert.)** (a) Schematic procedure for the synthesis of $Ln^{3+}$-doped $NaREF_4$ NPs. (b) XRD of $NaGdF_4$:Yb,Er NPs synthesized at different temperature. The black and red vertical lines in (b) represent the standard pattern of β-$NaGdF_4$ (JCPDS No. 027-0699) and α-$NaGdF_4$ (JCPDS No. 027-0697), respectively. (c) TEM image of α-$NaGdF_4$:Yb,Er NPs synthesized at 250°C. (d–f) TEM images of $NaGdF_4$:Yb,Er NCs synthesized at 290, 300, and 310°C, respectively. (g–k) TEM images of the as-prepared β-$NaSmF_4$, β-$NaEuF_4$, β-$NaGdF_4$, β-$NaTbF_4$ and β-$NaYF_4$ NOs, respectively. (Reproduced with permission from You et al., 2018.)

metal nanoparticles is to dissolve and reduce metal salt in the presence of molecules that can prevent the nanoparticles from aggregating. The concentration ratio of reduction agent to metal salt affects the growth kinetics of the nanoparticles, and thus affects their size. Often the reduction agent also stabilizes the nanoparticles, thus filling a dual role. As an illustration, gold nanoparticles of sizes at the nanoscale (10–200 nm) can show the surface plasmon, particularly, localized surface plasmon resonances (LSPRs) upon illumination (Figure 1.9). These are usually coherent (particular range of light illumination) with free conduction band electrons. (Abadeer & Murphy, 2016).

**Synthesis of Au Nanoparticles:** One example is the method first described by Turkevich et al. in 1951 where gold chloride ($HAuCl_4$) was reduced by sodium citrate generating fairly monodispersed gold nanoparticles suspended in an aqueous solution. This method was later improved by Frens who produced gold nanoparticles with a diameter of ~10–150 nm (Frens, 1973). In this procedure, citrate acts both as a reducing agent and electrostatic stabilizer where the size of the particles can be controlled by varying the ratio between the gold salt and the citrate. The reaction temperature is also important for the formation of the metal nanoparticles since both the reaction kinetics and oxidation potential are dependent upon the temperature. The excess of citrate anions forms a complex multilayer around the particles which prevent aggregation and gives the particles a net negative surface (Martinsson, 2014).

A more efficient method to control the size and shape of gold nanospheres was proposed in 2000 by Natal and co-workers using a seed growth process (Brown et al., 2000). Small seed nanoparticles are first synthesized and further grown via the reduction of the metal salt in the presence of surfactants to stabilize the nanoparticles. Seeds are generated using a first reducing agent (sodium borohydride or sodium citrate) whereas the growth is done with a second milder reducing agent (often ascorbic acid). The ascorbic acid reduces the metal salt in an intermediate state which becomes metallic gold ($Au^0$) only after a catalyzed reduction occurs at the surface of the seeds. Even if the work of Natan et al. gave good spherical gold nanoparticles, gold nanorods were also obtained during the synthesis, affecting strongly

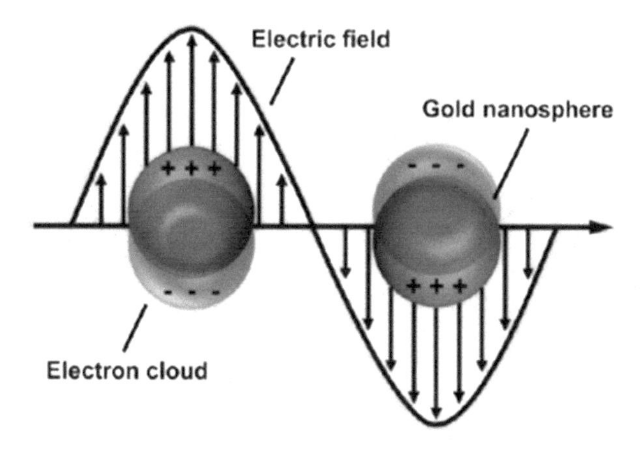


**FIGURE 1.9** Schematic of the localized surface plasmon resonance on gold nanospheres. Upon illumination at resonant wavelengths, conduction band electrons on the gold nanoparticle surface are delocalized and undergo collective oscillation. (From Abadeer & Murphy, 2016. With permission.)

the resulting batch monodispersity. In 2011, Bastús and colleagues successfully synthesized citrate-stabilized monodisperse gold nanospheres via seed growth (Bastús, et al., 2011). The factors influencing the size distributions of nanoparticles are the initial seeds concentration, the number of growth steps and of secondary nucleation. CTAB (hexadecyl-trimethyl-ammonium bromide) was used as a surfactant during the seed growth and enables better control of the final morphology of the gold nanoparticles. This new process gave gold nanoparticles a more spherical appearance while forming fewer gold nanorods. However, this method leads to the nanoparticles that are not easily functionalizable a posteriori, which can be disadvantageous. Indeed, the CTAB is strongly anchored and attached onto the surface of gold nanoparticles and ligand exchange with another molecule having an affinity for the gold surface is not straightforward, requiring multiple steps and specific synthesis conditions. In 2016, Malassis et al. demonstrated that it is possible to obtain rather spherical monodisperse gold nanoparticles of different sizes through the reduction of gold salt using ascorbic acid by tuning the pH of the mixture (Malassis et al., 2016). The real advantage of this synthesis was its easy implementation. Indeed, the synthesis method only involves one step as compared to the multiple synthesis steps necessary in the seed growth process. Moreover, ascorbic acid is very easily exchanged by other stabilizing agents allowing multiple post-functionalizations of the obtained gold nanoparticles.

Recently, Xiahou et al. have reported high quality gold nanoparticles with size ranging from 5.5 to 17.8 nm in high yield via a simple, one-step seeded growth method (see Figure 1.10) with the assistance of Tris-base (TB) (Xiahou et al., 2018). The sizes of these small Au nanoparticles can be finely controlled by varying the gold precursors ($HuCl_4$) and/or seed amount according to an established relationship between the nanoparticles' diameter and the particle number of the seeds (or the mass of gold precursor in some cases).

A controlled preparation of Au nanoparticles in the size range of 6 to 22 nm (see Figure 1.11) was explored by Suchomel et al. (2018). The Au nanoparticles were prepared through the reduction of

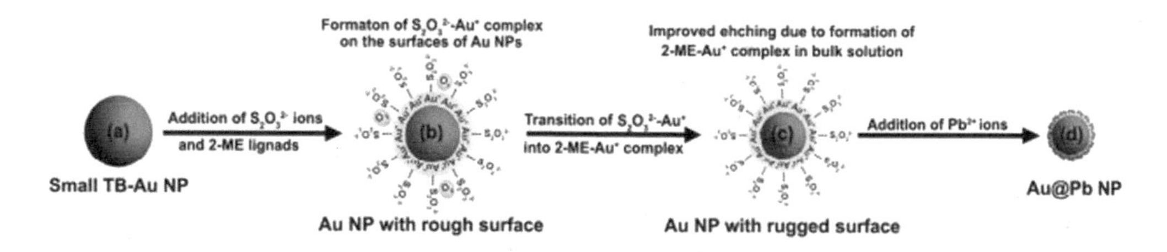


**FIGURE 1.10** Representation of the detailed test principle of the 2-ME/$S_2O_3^{2-}$-TB-$Au_{5.5}$ NP sensor for the detection of $Pb^{2+}$ ions. (From Xiahou et al., 2018. With permission.)

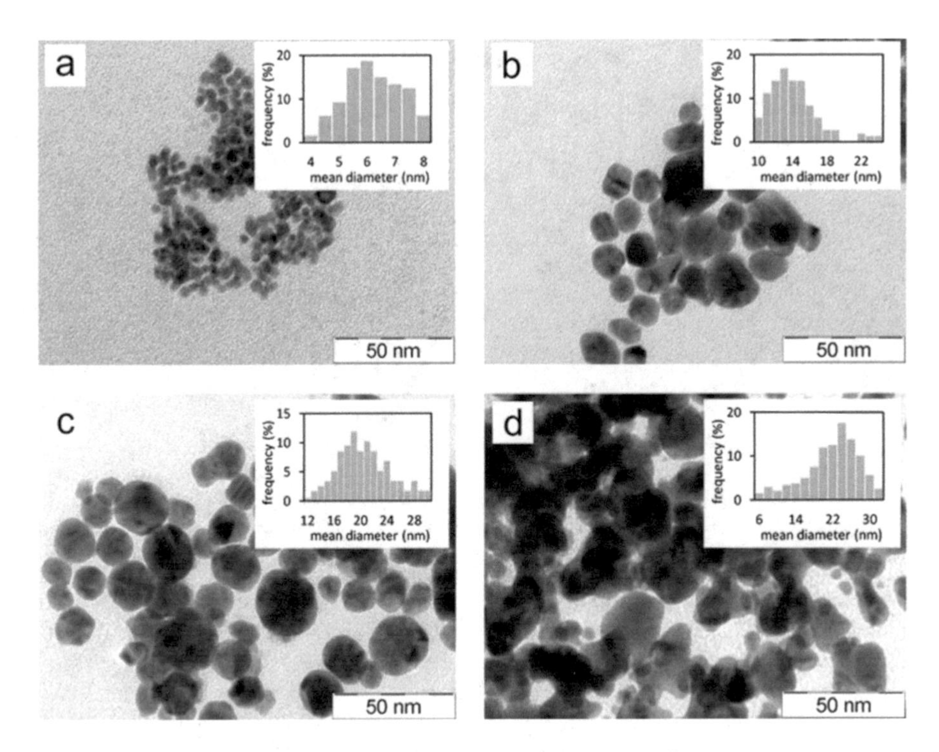


**FIGURE 1.11** TEM images and particle size distribution histograms of gold nanoparticles prepared via reduction of tetrachloroauric acid by maltose at alkali pH at various concentrations of Tween 80: 10 mmol/L (a), 1 mmol/L (b), 0.1 mmol/L (c), 0.01 mmol/L (d). (From Suchomel et al., 2018. With permission.)

tetrachloroauric acid using maltose in the presence of nonionic surfactant Tween 80 at various concentrations to control the size of the resulting Au nanoparticles. With increasing concentration of Tween 80, a decrease in the size of produced Au nanoparticles was observed, along with a significant decrease in their size distribution.

**Synthesis of Ag and/or Pt Nanoparticles:** The most fundamental reactions to prepare silver nanoparticles can be divided into two groups where $AgNO_3$ is reduced into products at a high temperature in (i) oleylamine (OAm), that is, cis-1-amino-9-octadecene and (ii) ethylene glycol (EG). These two solvents (reaction media) are helpful in understanding the nucleation and growth of Ag nanoparticles (Sun, 2013). Surfactants can be added to OAm and EG. When $AgNO_3$ is dissolved into OAm/EG (solvents), it releases $Ag^+$ ions which further associates with solvent or surfactant molecules (Deng et al., 2009). The solvent's capability for reducing silver precursors depends on the reaction temperature; hence, nucleation of Ag nanoparticles can be tuned finally by tuning temperature; there is obviously not only a single factor in preparing Ag nanoparticles. Liz-Marzán and co-workers have successfully synthesized silver nanoplates in boiled N,N-dimethylformamide (DMF) containing $AgNO_3$ and poly vinyl pyrrolidone (PVP) (Pastoriza-Santos & Liz-Marzán, 2002). They could demonstrate that tuning reaction precursor ($AgNO_3$), hence $Ag^+$ ions with respect to PVP has an impact on the morphology of nanoparticles as particles were showing changes from isotropic spheres to anisotropic nanowires and nanoplates on increasing $Ag^+$ ions. The aqueous-based colloidal synthesis using $C_nTABr$, a cationic surfactant, has attracted much attention as a simple and facile method (see Figure 1.12) that fabricates size- and shape-tunable metal nanoparticles (Zheng et al., 2014; Seo et al., 2018). Unlike other surfactants including carbonyl groups, CnTABr binds weakly to metal surfaces making it easy to regulate particle shape and preserve catalytically active sites (Zheng et al., 2014). Recently, a study was published showing a correlation between the surfactant concentration and particle size of synthesized Pt nanoparticles using cationic surfactants ($C_nTABr$), a $K_2PtCl_4$ precursor and a $NaBH_4$ reducing agent (Seo et al., 2018). A solution mixture of $C_nTABr$ and $K_2PtCl_4$ heated to 50°C in an oil heat bath under stirring was formed cloudy to transparent. This cloudy compound is known as the real precursor formed by coulombic interaction between

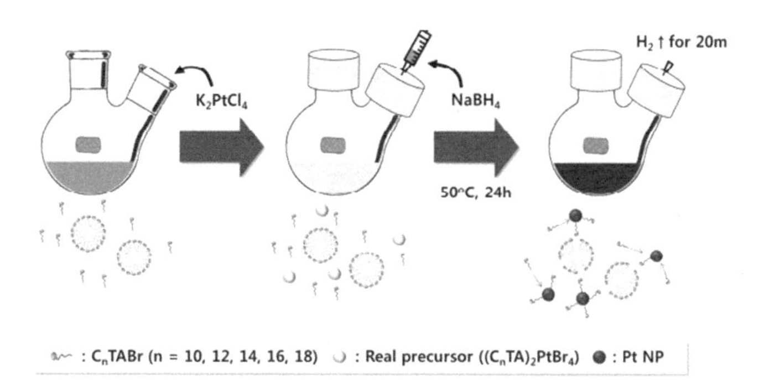


**FIGURE 1.12** **(See color insert.)** Schematic flowchart of the synthetic procedure for the Pt nanoparticles (NPs). (From Seo et al., 2018. With permission.)

$[C_nTA]^-$ and $[PtBr_4]^{2-}$ forming $[C_nTA]_2[PtBr_4]$ (Sau & Murphy, 2004). Subsequently, an ice-cooled aqueous solution of $NaBH_4$ was added as a reducing agent (Zhuang et al., 2008). The aqueous solution was kept at 50°C during the reaction for 24 h while releasing hydrogen gas in a controlled way.

Direct synthesis of Pt nanoparticles can be obtained by reducing a hexachloroplatinic complex in EG at room temperature, but elemental platinum is obtained at 110°C. The product can be obtained after a few hours of reaction and was found to be composed of colloidal particles with a size smaller than 10 nm. The PVP can be used to avoid sintering in colloidal platinum dispersions.

## 1.5 Preparations and Illustrations of Hybrid Nanomaterials

Currently, the major research area regarding preparation of hybrid nanomaterials can be categorized into three different sections: (i) inorganic–organic, (ii) inorganic–inorganic, and (iii) organic frameworks. Hybrid nanoparticles stand for their structure, characteristics, and unusual joining of nanoentities. Some major classes of functional nanoplatforms, such as (i) magnetic nanoparticles, (ii) nanophotonic materials and (iii) plasmonic nanoparticles have been extensively studied and widely used in a variety of fields. The next generation of nanoparticles is about hybridization of two characteristics that mainly come due to individual characteristics of distinct entities. Bifunctional nanoparticles have exhibited potentially promising physio-chemical properties, which can revolutionize and transform the landscape from bio-clinical industry to next-generation advance devices. In particular, colloidal nanoparticles have been extensively investigated as probes in biomedical/devices industries due to their unique size-dependent electronic, optical and magnetic properties among all possible building blocks. Thus, hybrid nanoparticles with combined magnetic and optical properties are much more powerful and can be used in a broad range of applications. Moreover, they offer new modalities that neither luminescence nor magnetic nanoparticles exhibit. For example, the demonstration of real field dependent luminescence phenomenon called magneto-luminescence, and magnetomotive-photoacoustic phenomenon. This novel phenomenon shows remarkable potential in several other important applications, such as high accuracy and secure communication, aircraft guidance and radiation field detection and modulation of the magnetic and optical fields. Furthermore, these nanocomposites can be used as nanoblocks to build various nanoelectronic and photonic devices by applying an external magnetic field to manipulate or arrange the magnetic nanoparticles and optical signals.

There are several kinds of strategies to prepare nanoparticles and a few of the key terms that can be mentioned here are core–shell engineering, composites, multi-shell nanoparticles, yolk–shell nanoparticles, porous silica-based hybrid nanoparticles, polymeric hybrid nanoparticles, doped nanohybrids, dumbles, tetrapods, Janus, and core–satellite. The core–shell structure is the most common type of bifunctional nanosystem. In this type of nanoarchitecture, generally, a core (e.g. magnetic) is combined with (fluorescent) molecules, metal nanoparticles or semiconductor nanocrystals, etc. and therefore exhibits unique optical, magnetic and electrical properties. However, different components in these kinds of bifunctional nanoparticles interact with each other, which generally weaken or reduce the functionality of each unit. Recent advances in the synthesis of various functional superparamagnetic iron oxide nanoparticles with heterostructures offer promising solutions to this problem by creating anisotropic nanoparticles. They consist of functional units with different chemistry, polarity or other physico-chemical properties on asymmetric sides without sacrificing their own properties. The superior properties of these heterostructures rely on synergistically enhanced magnetism and synthesis strategies for precise control of particle size, morphology and chemical composition along with functionalization with fluorescent material (e.g., dumbbell nanoparticles). The desire to design more sophisticated nanoarchitecture with unusual properties is definitely needed. This kind of integration offers exciting opportunities for discovering new materials, processes and phenomena. At the nanoscale, these kinds of bifunctional materials have their own advantage: (i) tuned and controlled size depending on application, (ii) nanoparticles manipulated by external force (magnetic field) "action at a distance," and (iii) multimode applications, e.g., MRI contrast agents and multimodal imaging sensors.

### 1.5.1 Magnetic-Luminescent Nanoparticles

Since magnetic-luminescent bifunctional nanoparticles may provide new and promising two-in-one characteristics, there are also some challenges to overcome in their fabrication: (i) the complexity in the preparation, which frequently involves multi-step synthesis and many purification stages leading their production which are quite technical and time-consuming; (ii) the risk of quenching of the luminophore on the surface of the particle by the magnetic core and (iii) the risk of quenching each other by the incorrect number of fluorescent molecules attached to the surface of the particle. To resolve these challenges, several techniques have been proposed to integrate the luminescence and magnetic nanomaterials. The fabrications of these are commonly achieved as hybrid conjugates of magnetic and luminescent entities, core–shell or composite structures using coating, layer-by-layer deposition and optical materials (QDs) on magnetic nanoparticles or vice-versa. Another method is the functionalization of magnetic iron oxide with fluorescent dyes and luminescent complexes, called the cross-linking attachment of molecules (Khan et al., 2014; Shrivastava, 2017). Moreover, the use of $SiO_2$, ZnS or C as spacers between iron oxide and luminescent materials is of great importance to the performance of nanohybrids.

Magnetic-luminescent multifunctional molecularly imprinted polymer (MIP) nanospheres containing luminescent $LaVO_4:Eu^{3+}$ nanocrystals (NCs), magnetic $Fe_3O_4$ nanoparticles, and the recognition sites of template organophosphate pesticides were successfully prepared via a facile and versatile bottom-up self-assembly strategy (Ma et al., 2012). Interestingly, both the favorable photoluminescent $LaVO_4:Eu^{3+}$ nanoparticles and magnetic $Fe_3O_4$ nanoparticles are incorporated within the final composite nanospheres, thus rendering them susceptible to both magnetic guidance and photoluminescent detection. The whole synthesis process has been summarized in Figure 1.13.

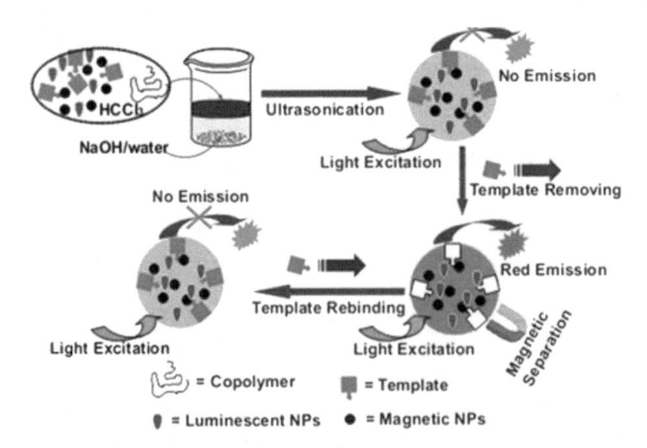


**FIGURE 1.13 (See color insert.)** Scheme for the fabrication of multifunctional MIP composite nanospheres and the luminescence response via the removal and rebinding of template analytes. (From Ma et al., 2012. With permission.)

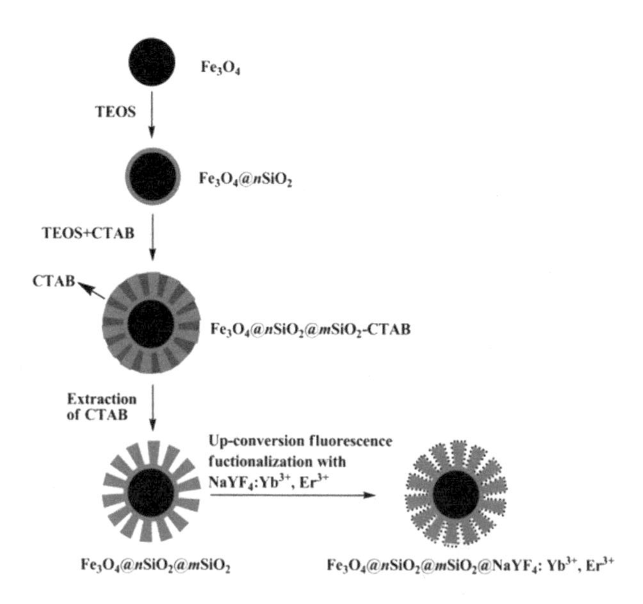


**FIGURE 1.14** The formation process of multifunctional $Fe_3O_4$@n$SiO_2$@m$SiO_2$@$NaYF_4$:$Yb^{3+}$,$Er^{3+}$ nanocomposites.

Acar's group prepared nanohybrids composed of luminescent QDs and SPIO iron oxides by the ligand exchange mechanism in a simple and versatile extraction method (Kas et al., 2010). In this method, aqueous QDs (CdS or CdTe) coated with carboxylated ligands exchange the fatty acid (lauric acid) coating of SPIOs in a water–chloroform extraction process. QDs form a coating around SPIOs and transfer them into the aqueous phase in high efficiency.

The synthesis (using a facile two-step sol–gel process), characterization and application in controlled drug release are reported for monodisperse core–shell-structured $Fe_3O_4$@n$SiO_2$@m$SiO_2$@$NaYF_4$:$Yb^{3+}$,$Er^{3+}$/$Tm^{3+}$ nanocomposites with mesoporous, upconversion luminescent and magnetic properties (Gai et al., 2010). The synthesis protocol can be understood from Figure 1.14. Authors have proposed a novel and facile strategy for the fabrication of multifunctional nanocomposites using silica-coated $Fe_3O_4$ spheres as the core, ordered mesoporous silica as the shell, and are further functionalized by the deposition of $NaYF_4$: $Yb^{3+}$, $Er^{3+}$/$Tm^{3+}$ upconversion phosphors.

## 1.5.2 Magnetic-Plasmonic Nanoparticles

As we discussed in previous sections, multifunctional nanohybrid designs have been proven to be an important way to fulfill the several biomedical requirements and environmental applications. In this section, we discuss two efficient nanoheaters combined into a multifunctional single nanohybrid (Espinosa et al., 2015). This nanohybrid consists of a multi-core iron oxide nanoparticle optimized for magnetic hyperthermia, and a gold branched shell with tunable plasmonic properties in the NIR region, for photothermal therapy which impressively enhanced heat generation, in suspension or *in vivo* in tumors, opening up exciting new therapeutic perspectives, at low excitation doses (Espinosa et al., 2015). Gold nanoparticles such as gold nanoshells, nanostars and nanorods are often preferred due to their strong absorption cross-sections. We are all aware of the application of iron oxide nanoparticles. This combination of systems is called a magneto-plasmonic nanoparticle, which is subjected to external magnetic and laser stimulations. Indeed, gold-based nanoparticles exhibit strong LSPRs in visible and NIR spectral regions; this property has been explored in a plethora of exciting applications ranging from molecular dark-field, reflectance and photoacoustic imaging to photothermal therapy and drug release (Sun, 2013). Similarly, superparamagnetic nanoparticles are of great interest because of their applications in various cell and molecular separation assays, magneto-motive optical and ultrasound imaging, MRI and hyperthermia cancer treatment.

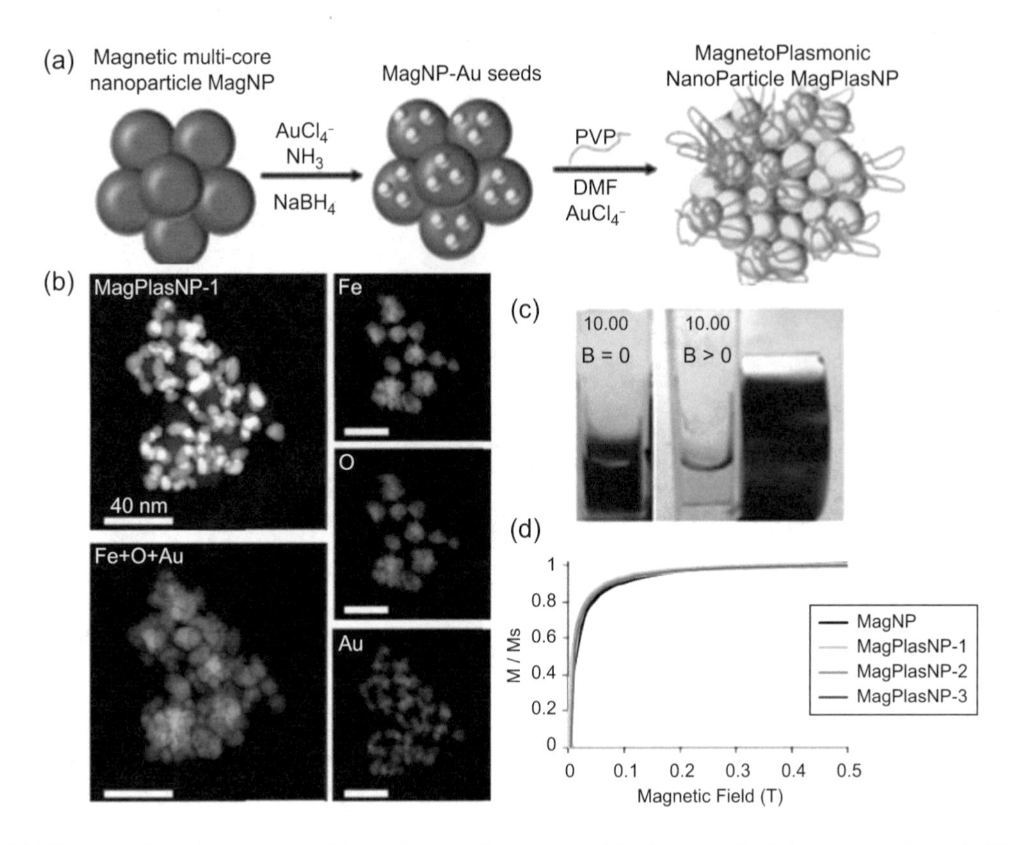


**FIGURE 1.15 (See color insert.)** (a) The seeding-growth process used for the synthesis of the magneto-plasmonic NPs (MagPlasNP) of different ratio Au/Fe. (b) Elemental analysis of different MagPlasNP-1. HAADF image, EELS Fe map (green), EELS O map (blue), EELS Au map (red) and overlaid maps. Superparamagnetic behavior of the nanohybrid: (c) Attraction of MagPlasNP-2 by a permanent magnet and (d) Magnetization curves of MagNP and MagPlasNP-1,2 and 3. (From Espinosa et al., 2015. With permission.)

Espinosa et al. prepared this kind of nanoparticle using a two-step polyol synthesis process (see Figure 1.15). First, maghemite multi-core nanoparticle (MagNP) synthesis was performed using $FeCl_3.6H_2O$ and $FeCl_2.4H_2O$ precursors in a liquid mixture of NMDEA and DEG with 1:1 (v/v) ratio. NaOH was dissolved in polyols (Hugounenq et al., 2012), which was later mixed in the first solution and stirred. Then, the temperature was elevated to 220°C slowly and the solution was stirred for 12 h at this temperature. After cooling, it was cleaned and washed to get magnetic nanoparticles. Fe-nitrate, dissolved in water, was mixed with these nanoparticles. Then, this mixture was heated to 80°C and stirred. Again, they were cleaned with appropriate agents. In this way, multi-core magnetic nanoparticles were created.

In the second step of the preparation of materials, first seeding the surface of MagNPs with gold (as MagNP-seeds) was allowed, and further growth of gold nanoparticles over MagNPs was performed. In order to deposit seeds on the surface of the MagNPs, their surface was functionalized with citrate anions at 80°C. Under vigorous stirring, in 100 ml of Millipore water alkalinized with ammonia, citrate treated MagNPs and 680 μl of a 35 mM $HAuCl_4$ aqueous solution were mixed. The mixture was heated at 40°C, followed by magnetic separation and redispersion of the NPs in 100 ml of water. 1 ml of a freshly prepared 0.1 M $NaBH_4$ solution was added to the previous suspension under vigorous stirring. Nanoparticles were magnetically separated and washed. Finally the nanoparticles were redispersed in a solution of PVP in ethanol. The schematic scheme, quality of materials, magnetic separation and evidence of superparamagnetic nanoparticles are highlighted in Figure 1.15. Recent advances in nanoparticle research allow the synthesis of various composite nanoparticles with core/shell and dumbbell structures that offer a promising solution to this problem encountered with single-component magnetic nanoparticles. For example, Au–$Fe_3O_4$ dumbbell nanoparticles were applied by conjugating anticancer drug cisplatin (cis-diamminedichloroplatinum) on Au and Herceptin antibodies on $Fe_3O_4$ with the number of cisplatin molecules and antibody molecules being controlled by the size of the Au and $Fe_3O_4$ components.

### 1.5.3 Other Hybrids

Research on the hybrid characteristics of nanoparticles cannot be limited only to the previous discussions. The accessibility of hybrid nanoparticles are increasing and are the major focus of material scientists. In this section, we discuss a few more examples to illustrate the synthesis of so-called novel nanohybrids.

(1) Magnetic nanoparticles, ferrites, oxyhydroxide and oxyfluoride compounds are of great industrial and scientific importance. Magnetites are known for their multi-field applications from energy storage to catalysis, to environmental, to biomedical due to their unique electric and magnetic properties. Doped iron oxides, particularly the spinel-structured $TMFe_2O_4$, (TM = transition metal) have shown tunable magnetic properties and attracted research interest for magnetic storage and catalysis. Five kinds of RE ions (Gd, Sm, Nd, Y and Lu) were selected to be doped into magnetite to investigate the influence of RE doping on the magnetic properties of magnetite (Wang et al., 2010). Here, the concentrations of the RE ions are 50%. These materials can be prepared using standard synthesis protocols by choosing appropriate precursors. A new class of fluorescent and superparamagnetic bifunctional nanocrystal (Tb-doped $\gamma$-$Fe_2O_3$) has been successfully prepared by a facile, non-hydrolytic modified thermal decomposition method (Zhang et al., 2009). The synthesized Tb-doped $\gamma$-$Fe_2O_3$ nanocrystals were found to be monodispersed (after being amine-functionalized), non-toxic (postsynthesis), superparamagnetic featuring green emission. The nanocrystals were amine-functionalized to render them water-dispersible and provided ease of further functionalization by other biomolecules. It should be noted here that incorporation of RE ions into iron oxide is not straight forward.

(2) Ramasamy and co-workers were able to dope Fe into an upconverting $\beta$-NaGdF4:$Yb^{3+}$/$Er^{3+}$ nanosystem, successfully (Ramasamy et al., 2013). Fe doping was increased up to 40% by replacing Gd ions in the matrix. The synthesis was carried out by a modified procedure obtained from the literature using OA as the coordinating ligand and 1-ODE as the non-coordinating solvent (Li et al., 2008; Ramasamy et al., 2013). A layer of silica was coated onto the surface of upconversion nanoparticles to form a core–shell structure in order to make the nanoparticles water soluble. These core–shell nanoparticles were covalently linked with folic acid (FA) and successfully used for fluorescence imaging of HeLa cells.

(3) Dai et al. (2017) showed that upconverting NaGdF4:$Yb^{3+}$/$Er^{3+}$ nanosystem, when combined with Au nanorods, in a single unit, showed no change (decrease) in the emission intensity under the influence of increasing magnetic field up to 6 Tesla indicating that the plasmon-enhanced upconversion luminescence is independent of the magnetic field.

(4) Interestingly, organic $\pi$-conjugated polymers have been combined with metallic or semiconducting nanoparticles or nanocrystals as stabilizing ligands to afford more stable, processable, soluble, rigid nanoparticles to enhance their optoelectronic properties by means of energy or charge transfer (Lahav et al., 2006; Bronstein & Shifrina, 2011; Park & Advincula, 2011). Hybrid nanocomposites consisting of semiconducting QDs and conjugated polymers have in recent years attracted much attention for several applications in sensing and energy-related studies. Nanoparticle-cored dendrimers or dendrimer-encapsulated nanoparticles (metals: Au, Ag, iron oxide etc.) are another class of hybrid nanomaterials that are metal-containing dendrimers (Gopidas et al., 2003; Jeong et al., 2004). Several groups demonstrated that, as the surface groups of the dendritic ligand become close-packed at the periphery of the hybrid material, these ligands offer a relatively low degree of surface passivation even at low loadings of the ligand onto the metal surface. These are also hybrid nanoparticles (Guo and Peng, 2003; Scott et al., 2005).

## 1.6 Postsynthesis and Surface Passivation

The postsynthesis process involves several important steps. The most important is the cleaning of nanoparticles from the solutions. There are several ways to wash nanoparticles as magnetic separation, centrifugation, decantation, etc. After washing nanoparticles, it is now important to change the surface characteristics of these nanoparticles as it plays a vital role in their overall property. The surface atoms are chemically more active compared to the bulk atoms because they have fewer adjacent coordinate atoms and unsaturated sites or more dangling bonds. Therefore, the imperfection of the particle surface induces additional electronic states in the band gap.

After the surface modification step, nanoparticles are conjugated with vulnerable biomolecules, which tend to lose their functionality and three-dimensional structure in extreme conditions. To passivate the nanoparticles' surface, an organic stabilizer is often used to form a monolayer of ligands on the particle surface to cap. The types and structure of the ligands such as the bonding elements and strength (S-, P- or N-based covalent or ionic bonding), carbon-chain length and structure are important factors for the passivation of nanoparticle's surface. This organic capping layer can not only sterically stabilize the QDs in the solution, but most importantly, it can passivate the surface in nanoparticles, thus decreasing surface defects and improving surface characteristics (Burda et al., 2005; García-Calvo et al., 2018). Therefore, nanoparticles compatible with a water-based environment having a pH close to the physiological value (pH 7.4) and hydrophilic coatings with uniform functional groups are preferred. The modification step itself or the final result must not induce agglomeration of the individual nanoparticles. Repulsion between the nanoparticles may be evoked by introducing either negatively or positively charged groups favoring the even distribution of the separate particles. The coating should be sufficiently transparent at excitation and emission wavelengths and stable in storage and assay conditions. The thickness of the coating layer becomes extremely relevant in the applications based on resonance energy transfer from the nanoparticles' donors to acceptor molecules, and a very thin monolayer is preferred in this case.

Although the organic capping layer improves/stabilizes the surface originated characteristics of nanoparticles, it still cannot provide perfect passivation and remove all of the surface dangling bonds due to the large structure difference between the ligands and the nanoparticles. An ideal passivation method is to use inorganic materials (ZnS, $SiO_2$, carbon dots) with the same crystal structure and a close lattice parameter to cap the nanoparticles. A universal route to functionalize an inorganic particle is to grow an amorphous silica shell around the particles (Guerrero-Martínez et al., 2010). Several postsynthesis methods such as ligand exchange, ligand oxidation, layer-by-layer coating, and coating by amphiphilic molecules and polymers have been recently summarized in several review articles (Erathodiyil & Ying, 2011; Jiang et al., 2013).

### 1.6.1 Bioconjugation

In this section, we discuss conjugating nanoparticles with biomolecules (biocompatible molecules). This is a most important postsynthesis process and is used frequently for biomedical and sometimes in environmental applications. Both parties, the biomolecule and the nanoparticle, must bear suitable functional groups that react with each other, forming a covalent bond. Physical adsorption of biomolecules on the surface of the nanoparticle may also be exploited, but the stability of the conjugate in assay and storage conditions cannot be assured, which renders the covalent bioconjugation a more favorable method. A suitable ligand candidate should allow attachment of a diversity of biomolecules including nucleic acids, proteins (avidin/streptavidin, albumin, adaptor proteins and antibodies), polysaccharides and peptides. This acts like organic–organic hybrid systems or inorganic–organic nanohybrids. Biomolecules are commonly conjugated to QD surfaces through the following approaches (Shan et al., 2008):

(i) Covalent cross-linking: link the carboxyl groups on the nanoparticle surface to amines, examples: (1-ethyl-3-(3-dimethylaminopropyl) carbodiimide (EDC), N-hydroxysulfosuccinimide (NHS)

(ii) Thiolated peptides (cysteine residues) can be directly conjugated to the nanoparticle surface through the affinity to shell layer
(iii) Adsorption or non-covalent self-assembly using engineered proteins (positively charged domain, polyhistidine (HIS) residues

Owing to the large surface area-to-volume ratio, several biomolecules over varying types can be attached to a single nanoparticle where each of these biomolecules provides the desired function which grants single nanoparticles multi-functionality (Klostranec & Chan, 2006).

The major challenge in bioconjugation is to maintain the functionality of the biomolecule. In addition, the orientation of the attached molecule is critical to ensure access to the analyte recognition site. Favorable binding orientation affects the assay performance significantly (Zhang & Meyerhoff, 2006). After the conjugation step, the excess biomolecules should be removed thoroughly without losing the nanoparticle conjugates, and the success of the bioconjugation needs to be confirmed using nuclear magnetic resonance (NMR) and Fourier transformation infrared (FTIR) techniques. Several works of bioconjugation with different types of nanoparticles can be found in review articles (Sperling & Parak, 2010; Sapsford et al., 2011; Massey and Russ, 2017).

## 1.7 Conclusion and the Future of (Hybrid) Nanostructures

To summarize, we have discussed standard wet chemical synthesis protocols for nanohybrids with numerous illustrative examples from simple magnetic iron oxide to more complex systems consisting of bifunctional nanosystems. Indeed, synthesis of hybrid nanoparticles is a challenging task and requires thorough investigation. The major challenges include the optimization of the application of target ligands in appropriate ligand density that will improve the acceptance of nanohybrids in biomedical and environmental purpose in a facile and economical manner.

There are several questions that need to be addressed around the synthesis of novel materials for selected and multimodal nanomaterials: How does a nanohybrid crystal grow in solution? What properties do they have and why are they size/shape-dependent? How can we engineer hybrid nanostructures (metal-metal, metal-dielectric, metal semiconductor)? How can we control the location of one component in relation to other component(s) in the nano-entity regime? How can nanocrystals be used as building blocks to engineer novel materials? What is the current knowledge base of the risks of nanoparticles to human health and the environment? Though eminently fundamental, this research is required for the design of nanomaterials with tailored properties that can be used for practical applications in the field of catalysis, optics and biology. In particular, proper design of the hybrid nanoparticle should permit control over the interaction of the material components to combine different confinement-induced properties, create new ones or introduce new functionalization. Combining material components of different nature in the same nanoparticle is a new challenge in nanoscience and offers a wide range of new and largely unexplored possibilities for developing novel materials.

## Acknowledgments

NS and SKS jointly thank CNPq, FAPEMA and CAPES for financial support.

## REFERENCES

Abadeer N. S., Murphy C. J., *J. Phys. Chem. C*, 2016, 120, 4691–4716.
Alivisatos A. P., *Science*, 1996, 271, 933–937.
Bastús N. G., Comenge J., Puntes V., *Langmuir*, 2011, 27, 11098–11105.
Behera M., Ram S., *Int. Nano Lett.*, 2013, 3, 17.
Bhargava A., Jain N., Bharathi M. L., Akhtar Mohd. S., Yun Y.S., Panwar J. *J. Nanopart. Res.*, 2013, 15.

Bronstein L. M., Shifrina Z. B., *Chem. Rev.*, 2011, 111, 5301–5344.
Brown K. R., Walter D. G., Natan M. J., *Chem. Mater.*, 2000, 12, 306–313.
Burda C., Chen X., Narayanan R., El-Sayed M. A., *Chem. Rev.*, 2005, 105, 4, 1025–1102.
Chen H., Zhang P., Cui H., Qin W., Zhao D., *Nanoscale Res. Lett.*, 2017, 12, 548.
Chen O., Chen X., Yang Y., Lynch J., Wu H., Zhuang J., Cao Y. C., *Angew. Chem.*, 2008, 120, 8766–8769.
Dai G., Zhong Z., Wu X., Zhan S., Hu S., Hu P., Hu J., Wu S., Han J., Liu Y., *Nanotechnology*, 2017, 18, 155702.
Dallinger D., Kappe C. O., *Chem. Rev.*, 2007, 107, 2563–2591.
Deng Z., Mansuipur M., Muscat A. J., *J. Phys. Chem. C*, 2009, 113, 867–873.
Dong H., Chen Y. C., Feldmann C., *Green Chem.*, 2015a, 17, 4107–4132.
Dong H., Du S., Zheng X., Lyu G., Sun L., Li L., Zhang P., Zhang C., Yan C., *Chem. Rev.*, 2015b, 115, 10725–10815.
Eliseeva S., Bunzli J., *Chem. Soc. Rev.*, 2009, 39, 189–227.
Erathodiyil N., Ying J. Y., *Acc. Chem. Res.*, 2011, 44, 925–935.
Espinosa A., Bugnet M., Radtke G., Neveu S., Botton G. A., Wilhelm C. and Abou-Hassan A., *Nanoscale*, 2015, 7, 18872–18877.
Eustis S., El-Sayed M. A., *R. Soc. Chem.*, 2006, 35, 209–217.
Frens G., *Nature Physical Science*, 1973, 241, 20–22.
Gai S., Yang P., Li C., Wang W., Dai Y., Niu N., Lin J., *Adv. Func. Mat.*, 2010, 20, 7, 1166–1172.
Gai S., Li C., Yang P., Lin J., *Chem. Rev.*, 2014, 114, 2343–2389.
García-Calvo J., Calvo-Gredilla P., Ibáñez-Llorente M., Romero D. C., Cuevas J. V., García-Herbosa G., Avella M., Torroba T., *J. Mater. Chem. A*, 2018, 6, 4416–4423.
Gawande M. B., Shelke S. N., Zboril R., Varma R. S., *Acc. Chem. Res.*, 2014, 47, 1338–1348.
Ghosh Chaudhuri R. G., Paria S., *Chem. Rev.*, 2012, 112, 2373–2433.
Gopidas K. R., Whitesell J. K., Fox M. A., *J. Am. Chem. Soc.*, 2003, 125, 6491–6502.
Grimsdale A. C., Müllen K., *Angew. Chem. Int. Ed.*, 2005, 44, 5592–5629.
Guerrero-Martínez A., Pérez-Juste J., Liz-Marzán L. M., *Adv. Mater.*, 2010, 22, 1182–1195.
Guo W. Z., Peng X. G., *C. R. Chim.*, 2003, 6, 989–997.
Haase M., Schäfer H., *Angew. Chem. Int. Ed.*, 2011, 50, 5808–5829.
Hachani R., Lowdell M., Birchall M., Hervault A., Mertz D., Begin-Coline S. and Thanh N. T. K., *Nanoscale*, 2015, 8, 3278–3287.
Hambrock J., Birkner A., Fischer R. A., *J. Mater. Chem.*, 2001, 11, 3197–3201.
Hufschmid R., Arami H., Ferguson R. M., Gonzales M., Teeman E., Brush L. N., Browning N. D., Krishnan K. M., *Nanoscale*, 2015,, 7, 11142–11154.
Hugounenq P., Levy M., Alloyeau D., Lartigue L., Dubois E., Cabuil V., Ricolleau C., Roux S., Wilhelm C., Gazeau F., Bazzi R., *J. Phys. Chem. C.*, 2012, 116, 15702–15712.
Huo Q., *Colloids Surf. B Biointerfaces*, 2007, 59, 1–10.
Jeong J.R., Lee S., Kim J., Shin S., *Phys. Stat. Sol. (B)*, 2004, 241, 1593–1596.
Jiang S., Win K. Y., Liu S., Teng C. P., Zheng Y., Han M. Y., *Nanoscale*, 2013, 5, 3127.
Kas R., Sevinc E., Topal U., Acar H. Y., *J. Phys. Chem. C*, 2010, 114, 7758–7766.
Khan L. U., Brito H. F., Hölsä J., Pirota K. R., Muraca D., Felinto M. C. F. C., Teotonio E. E. S., Malta O. L., *Inorg. Chem.*, 2014, 53, 12902–12910.
Kharissova O. V., Kharisov B. I., Jiménez-Pérez V. M., Muñoz F. B., Ortiz M. U., *RSC Adv.*, 2013, 3, 22648.
Klostranec J. M., Chan W. C. W., *Adv. Mater.*, 2006, 18, 1953–1964.
Kuno M., *Phys. Chem. Chem. Phys.*, 2008, 10, 620–639.
Lahav M., Weiss E. A., Xu Q. B., Whitesides G. M., *Nano Lett.*, 2006, 6, 2166–2171.
LaMer V. K., Dinegar R. H., *J. Am. Chem. Soc.*, 1950, 72, 4847–4854.
LaRocque J., Bharali D. J., Mousa S. A., *Mol. Biotechnol.*, 2009, 42, 358–366.
Lassenberger A., Scheberl A., Stadlbauer A., Stiglbauer A., Helbich T., Reimhult E. *ACS Appl Mater Interfaces* 2017, 9(4), 3343–3353.
Laurent S., Forge D., Port M., Roch A., Robic C., Vander Elst L. V., Muller R. N., *Chem. Rev.*, 2008, 108, 2064–2110.
Li Z. Q., Zhang Y., Jiang S., *Adv. Mater.*, 2008, 20, 4765–4769.
Liu B., Xie W., Wang D., Huang W., Yu M., Yao A., *Mater. Lett.*, 2008, 62, 3014–3017.
Liu X., Zhao J., Sun Y., Song K., Yu Y., Du C., Kong X., Zhang H., *Chem. Comm.*, 2009, 43, 6628–6630.
Liz-Marzán L. M., Kamat P. V. *Nanoscale Materials*, 2003.

Lu L. T., Dung N. T., Tung L. D., Thanh C. T., Quy O. K., Chuc N. V., Maenosono S., Thanh N. T. K., *Nanoscale*, 2015, 7, 19596–19610.
Luigjes B., Woudenberg S. M. C., de Groot R., Meeldijk J. D., Torres Galvis H. M., de Jong K. P., Philipse A. P., Erné B. H., *J. Phys. Chem. C*, 2011, 115, 14598–14605.
Ma Y., Li H., Wang L., *J. Mater. Chem.*, 2012, 22, 18761–18767.
Malassis L., Dreyfus R., Murphy R. J., Hough L. A., Donnio B., Murray C. B., *RSC Adv.*, 2016, 6, 33092–33100.
Mameli V., Musinu A., Ardu A., Ennas G., Peddis D., Niznansky D., Sangregorio C., Innocenti C., Thanh N. T. K., Cannas C., *Nanoscale*, 2016, 8, 10124–10137.
Martinsson E., 2014, Nanoplasmonic Sensing Using Metal Nanoparticle. Dissertation, 1624, Linköping University, Sweden.
Mascolo M. C., Pei Y., Ring T. A., *Materials*, 2013, 6, 5549–5567.
Massart R., *IEEE Trans. Magn.*, 1981, 17, 1247–1248.
Massey M., Russ A. W., *Chemoselective and Bioorthogonal Ligation Reactions: Concepts and Applications*, Volume 2, W. Russ Algar et al. (ed.), WILEY-VCH Verlag GmbH & Co. KGaA, 2017.
Merkel T. J., Herlihy K. P., Nunes J., Orgel R. M., Rolland J. P., DeSimone J. M., *Langmuir*, 2009, 26, 13086–13096.
Moloney M. P., Govan J., Loudon A., Mukhina M., Gun'ko Y. K., *Nat. Proto*, 2015, 10, 4, 558.
Murray C.B., Norris D.J., Bawendi M.G., J. *Am. Chem. Soc.* 1993, 115, 8706.
Nishioka M., Miyakawa M., Daino Y., Kataoka H., Koda H., Sato K., Suzuki T. M., *Ind. Eng. Chem. Res.*, 2013, 52, 4683–4687.
Niu J., Wang X., Lv J., Li Y., Tang B., *Trends Anal. Chem.*, 2014, 58, 112–119.
Oberdörster G., Oberdörster E., Oberdörster J., *Environ. Health Perspect.*, 2005, 113, 823–839.
Park J., An K. J., Hwang Y. S., Park J. G., Noh H. J., Kim J. Y., Park J. H., Hwang N. M., Hyeon T., *Nat. Mater.*, 2004,, 3, 891–895.
Park Y., Advincula R. C., *Chem. Mater.*, 2011, 23, 4273–4294.
Park Y. I., et al., *Adv. Mater.*, 2009, 21, 4467–4471.
Pastoriza-Santos I., Liz-Marzán L. M., *Nano Lett.*, 2002, 2, 903–905.
Peng Z. A., Peng X., *J. Am. Chem. Soc.*, 2001, 123, 183–184.
Ramasamy P., Chandra P., Rhee S. W., Kim J., *Nanoscale*, 2013, 5, 8711.
Ramasamy P., Ko K.-J., Kang J.-W., and Lee J.-S., *Chem. Mater.*, 2018, 30 (11), 3643–3647.
RSRAE (The Royal Academy of Engineering, Nanoscience and Nanotechnologies): opportunities and uncertainties. RS Policy documents. The Royal Society and Royal Academy of Engineering London, 19/04, July 2004.
Sapsford K. E., Tyner K. M., Dair B. J., Deschamps J. R., Medintz I. L., *Anal. Chem.*, 2011, 83, 4453–4488.
Sarveena S. N., Singh M., Sharma S. K., *Complex Magnetic Nanostructures: Synthesis, Assembly and Applications*, Springer International Publishing, Cham, 2017, pp. 225–280.
Sau T. K., Murphy C. J., *J. Am. Chem. Soc.*, 2004, 126, 8648–8649.
Scott R. W. J., Wilson O. M., Crooks R. M., *J. Phys. Chem. B*, 2005, 109, 692–704.
Seo J., Lee S., Koo B., Jung W.C., *CrystEngComm*, 2018, 20, 2010–2015.
Shan Y., Wang L., Shi Y., Zhang H., Li H., Liu H. i., Yang B., Li T., Fang X., Li W., *Talanta*, 2008, 75, 1008–1014.
Shrivastava N., *Development of Magnetic-Luminescent Bifiunctional Nanoparticles and Their Application in Radiation Detection*, Federal University of Maranhao, 2017, p. 195.
Shrivastava N., Rocha U., Muraca D., Jacinto C., Moreno S., Vargas J. M., Sharma S. K., *AIP Adv.*, 2018, 8, 056710.
Shrivastava N., Sarveena M. S., Sharma S. K., *Exchange Bias: From Thin Film to Nanogranular & Bulk Systems*, CRC Publishers, Boca Raton, 2017, pp. 1–35.
Soppimath K. S., Aminabhavi T. M., Kulkarni A. R., Rudzinski W. E., *J. Control. Release*, 2001, 70, 1–20.
Sperling R. A., Parak W. J., *Phil. Trans. R. Soc. A*, 2010, 368, 1333–1383.
Suchomel P., Kvitek L., Prucek R., Panacek A., Halder A., Vajda S., Zboril R., *Sci. Rep.*, 2018, 8, 4589.
Sun Y., *Chem. Soc. Rev.*, 2013, 42, 2497–2511.
Turkevich J., Stevenson P. C., Hillier J., *Discuss. Faraday Soc.*, 1951, 11, 55–75.
Unni M., Uhl A. M., Savliwala S., Savitzky B. H., Dhavalikar R., Garraud N., Arnold D. P., Kourkoutis L. F., Andrew J. S., Rinaldi C., *ACS Nano*, 2017, 11, 2284–2303.

Virlan M. J. R., Miricescu D., Radulescu R., Sabliov C. M., Totan A., Calenic B., Greabu M., *Molecules*, 2016, 21, 207.
Vreeland E. C. et al., *Chem. Mater.*, 2015, 27, 6059–6066.
Wang C., Xu S., Shao Y., Wang Z., Xu Q. and Cui Y., *J. Mater. Chem. C*, 2014a, 2, 5111.
Wang F., Banerjee D., Liu Y., Chen X., Liu X., *Analyst*, 2010, 135, 1839–1854.
Wang F., Deng R., Liu X., *Nat. Protoc.*, 2014b, 9, 1634–1644.
Wang F., Deng R., Wang J., Wang Q., Han Y., Zhu H., Chen X., Liu X., *Nat. Mater.*, 2011a, 10, 968–973.
Wang F., Liu X., *Chem. Soc. Rev.*, 2009, 38, 976–989.
Wang G., Peng Q., Li Y., *Acc. Chem. Res.*, 2011b, 44, 322–332.
Xiahou Y., Zhang P., Wang J., Huang L., Xia H., *J. Mater. Chem. C*, 2018, 6, 637–645.
Xu C., Sun S., *Adv. Drug Deliv. Rev.*, 2013, 65, 732–743.
Ye N., Yan T., Jiang Z., Wu W., Fang T., *Ceram. Int.*, 2018, 44, 4521–4537.
You W., Tu D., Zheng W., Shang X., Song X., Zhou S., Liu Y., Li R., Chen X., *Nanoscale*, 2018, 10, 11477–11484.
Zhang H., Meyerhoff M. E., *Anal. Chem.*, 2006, 78, 609–616.
Zhang Y., Das G. K., Xu R., Tan T. T. Y., *J. Mater. Chem.*, 2009, 19, 3696–3703.
Zhang Y.-W., Sun X., Si R., You L.-P., Yan C.-H., *J. Am. Chem. Soc.*, 2005, 127, 3260–3261.
Zheng Y., Zhong X., Li Z., Xia Y., *Part. Part. Syst. Charact.*, 2014, 31, 266–273.
Zhou J., Liu Q., Feng W., Sun Y., Li F., *Chem. Rev.*, 2015, 115, 395–465.
Zhu P., et al., *Adv. Mater. Res.*, 2011, 287, 77–80.
Zhuang J., Wu H., Yang Y., Cao Y. C., *Angew. Chem. Int. Ed.*, 2008, 47, 2208–2212.
Zrazhevskiy P., Sena M., Gao X., *Chem. Soc. Rev.*, 2010, 39, 4326–4354.

# 2

# *Probing Crucial Interfacial Dynamics of Nanohybrids for Emerging Functionalities*

**Damayanti Bagchi and Samir Kumar Pal**

**CONTENTS**

## 2.1 Introduction

Hybrid materials are composites of two constituents at the molecular level.[1] The interactions at the microscopic level between the two constituents impart new characteristics in between the two original phases or can sometime provide altered properties in the hybrid. In the case of hybrid materials, the primary interactions occur mainly at the nanoscale region ($10^{-9}$ m), and thus they often termed as nanohybrid materials or nanohybrids.[2] The molecular cross-talking at the interface between the two constituents is often able to generate novel properties of the nanohybrids and thus can improve their activities compared to their bulk counterpart.[3] The molecular level interaction can be classified into two segments such as non-covalent (electrostatic, van der Waal, etc.) interaction and covalent conjugation.[4] The hybrids formed through non-covalent interaction are often termed as class-I nanohybrids and the covalently attached hybrids are classified in the class-II section. However, depending upon the nature and strength of interaction, the functionalities of the nanohybrids get altered. In most cases, nanohybrids impart new effectivity and thus nanohybrids are considered as a new class of functional materials.[5]

The two parent counterparts of nanohybrids can be divided into various sections. In general, nanohybrids consist of an inorganic and an organic building block assembled at the nanoscale.[6] However, sometimes one organic material interacts with another organic dye at the nanoscale imparting an alteration in the dye characteristics. Organic nanoparticles mainly focuses on the crystalline synthesis product into the finest particulate dispersion possible, with greater water stability, which is essential for pharmaceutical activity.[7] These organic nanomaterials are mostly sized between 10 and 500 nm.[8] The charge transport or energy transfer processes at the interfacial position present in organic nanomaterials often improves its functionalities. The other important classification is the inorganic–organic nanohybrids, which mainly interact through covalent conjugation and surface attachment and thus can be included as class-II nanohybrids.[9] Hybrid organic–inorganic materials have been developed in the past 30 years or so as intimate combinations of these dissimilar materials that are not constrained by the classical material compromises.[10] The quantum leap in behavior often possible with hybrids arises principally from two sources: the reduction of the domain size of the inorganic phase to 100 nm or below (often much less), and generation of enormous interfacial areas, which enable numerous covalent bonds or other compatibilization between the phases. In some hybrid materials systems, the individual component has not been detected as a distinct phase. The realm of quantum effects and tuneable electronic, optical, and magnetic properties are also readily accessible.[10]

Nanohybrids mainly possess great interfacial area which imparts new properties with improvements in functionalities.[11] Thus molecular interactions at the inorganic–organic interface, often termed as the nano–bio interface are of immense importance and need to be evaluated in detail.[12] The available microscopic techniques, namely electron microscopy, such as scanning electron microscopy (SEM),[13,14] transmission electron microscopy (TEM),[15] or scanning tunneling microscopy (STM),[16] could provide information on the inorganic parts. High-resolution microscopic images can provide the lattice fringes and also the crystal interplanar distance. X-ray diffraction (XRD) could also provide the crystal plane diffraction pattern of the inorganic counterpart. Fourier-transform infrared spectroscopy (FTIR) or nuclear paramagnetic resonance spectroscopy (NMR) can analyse the perturbation in the organic part. Thus, all the above-mentioned methods are able to only observe the nano–bio interface in an indirect manner which might be of severe importance but cannot be sufficient enough to completely understand the interfacial processes.

The quantum mechanical phenomena present at the interfaces of the nanohybrids also show the necessity to study the interfacial processes using experimental tools and to fully understand the interfacial dynamics at the nano-junction; electronic spectroscopy, including steady-state and excited state spectroscopy, could be considered as one of the best tools. Ultrafast spectroscopy is a spectroscopic technique that uses ultrashort pulse lasers for the study of dynamics on extremely short time scales (attoseconds to nanoseconds).[17] Different methods are used to examine the dynamics of charge carriers, atoms, and molecules. Time-correlated single photon counting (TCSPC) is used to analyse the relaxation of molecules from an excited state to a lower energy state. Since various molecules in a sample will emit photons at different times following their simultaneous excitation, the decay must be thought of as having a certain rate rather than occurring at a specific time after excitation. By observing how long individual molecules take to emit their photons and then combining all these data points, an intensity vs. time graph can be generated that displays the exponential decay curve typical to these processes. The interface between the inorganic and organic parts can be visualized using the ultrafast TCSPC technique. Various dynamical relaxation techniques such as solvation dynamics,[17,18] anisotropy,[19] and various energy transfer mechanism as Förster resonance energy transfer (FRET),[20] nano-surface energy transfer (NSET)[21] can be observed and can be directly correlated with nano-bio interfacial interactions. Finally, applications for hybrid materials have expanded from their original base of enhanced mechanical properties and abrasion resistance. The current generation of materials shows novel optical, catalytic and biological properties.

In the following description, we first addressed interfacial dynamics at organic nanomaterial based two hybrid systems, for better drug delivery applications. Next, we move on to inorganic–organic hybrids and subdivided by the nature of components: inorganic nanomaterials are divided as metal and metal-oxides and the organic counterpart is classified in common biomolecules (as protein, DNA) and organic ligands (dye, drug, or small ligands). Each of the sections starts with a brief overview of the work, synthesis methodologies for preparing hybrid materials followed by general characterization including common microscopic and spectroscopic tools. The interfacial dynamics are evaluated using ultrafast spectroscopic methods and excited state relaxation methods. The molecular cross talking is confirmed using FRET, NSET, and an electron transfer mechanism. Finally, the nanohybrids are shown to improve overall functionalities including drug delivery applications and various active biological functions, such as anticancer and antioxidant processes.

## 2.2 Methodologies: Spectroscopic Tools Used to Probe the Interface of Nanohybrids

### 2.2.1 Steady-State Spectroscopy

Absorbance measurements were performed in a UV-2600 spectrophotometer (Shimadzu).[22] Fluorescence measurements were performed in a Fluorolog fluorimeter (Jobin Yvon Horiba).[23,24] Dynamic light scattering (DLS) measurements are done with a Nano S Malvern instrument employing a 4 mW He-Ne laser ($\lambda_{ex}$ 632 nm) and equipped with a thermostated sample chamber.[25] TEM samples were prepared by dropping sample stock solutions onto a 300-mesh carbon-coated copper grid and dried overnight in air. Particle sizes were determined from micrographs recorded using a FEI TecnaiTF–20 field-emission high-resolution transmission electron microscope operating at 200 kV.[26] A JASCO FTIR–6300 spectrometer was used for the FTIR to confirm the covalent attachment. For FTIR measurements, powdered samples were mixed with KBr powder and pelletized.[20] Background correction was made by using KBr pellet reference. Raman scattering measurements were performed in a back scattering geometry using a micro-Raman setup consisting of a spectrometer (model LabRAM HR, JobinYvon) and a Peltier-cooled charge-coupled device (CCD) detector. An air-cooled argon ion laser with a wavelength of 488 nm was used as the excitation light source.[27] Raman spectra of all samples have been recorded at room temperature in the frequency range 50–4000 $cm^{-1}$. Magnetic measurements were performed in a Lake Shore vibrating sample magnetometer (VSM) with an electromagnet that can produce field up to 1.6 T. XRD patterns were obtained by employing a scanning rate of 0.02° $s^{-1}$ in the $2\theta$ range from 5° to 80° by a PANalytical XPERT–PRO diffractometer equipped with Cu K$\alpha$ radiation (at 40 mA and 40 kV).[28] Thermo gravimetric analysis (TGA) was performed under a nitrogen atmosphere. The samples were heated from 30 to 600°C at a rate of 10°C $min^{-1}$ by using a PerkinElmer TGA-50H.[29]

### 2.2.2 Time-Resolved Spectroscopy

All the picosecond-resolved fluorescence transients were measured by using commercially available TCSPC setup with MCP-PMT from Edinburgh Instruments, U.K. (instrument response function (IRF) of ~75 ps) using 375 nm, 409 nm, 445 nm, 510 nm, and 633 nm excitation laser sources.[30,31] Details of the time-resolved fluorescence setup have been discussed in our previous reports.[32] Time-resolved emission spectra (TRES) were constructed following the methods described earlier[33] to determine the time-dependent fluorescence Stokes shifts. For the fluorescence anisotropy measurements, the emission polarizer was adjusted to be parallel and perpendicular to that of the excitation and the corresponding fluorescence transients are collected as $I_{para}$ and $I_{per}$, respectively. The time-resolved anisotropy is calculated using the magnitude of G, the grating factor of the emission monochromator of the TCSPC system, was found using a long tail matching technique.[34] FRET distance between donor–acceptor ($r$) was calculated from the equation $r^6 = [R_0^6(1-E)]/E$, where $E$ is the energy transfer efficiency between donor and acceptor and following the procedure published elsewhere.[35–38]

## 2.3 Crucial Charge Transfer Dynamics at the Organic–Organic Interface of Organic Nanohybrids

Nanoparticles of organic materials could have interesting applications in the pharmaceutical field as the microemulsions can be regarded as a potential drug delivery agent.[39] Appropriate localization of a drug and its structure-functional integrity in a delivery agent essentially dictates the efficacy of the vehicle and the medicinal activity of the drug.[40] Hence, the photo-induced dynamical processes present in a drug into delivery agent provide significant information regarding the therapeutic efficacy of the composite system. In the next section, we will provide an overview of how the photoinduced dynamics of organic nanohybrids represent the knowledge of drug localization and effectiveness of the nano-system.

### 2.3.1 Essential Dynamics of an Effective Phototherapeutic Drug in a Nanoscopic Delivery Vehicle

The family of linear furocoumarins compounds commonly known as Psoralen (PSO) provide an active dermal phototherapeutic effects in the presence of UVA irradiation.[41,42] The photosensitizer PSO will absorb the UV radiation and conduct an array of photochemical-redox, and/or radical reactions. Moreover, PSO intercalates with DNA upon irradiation with UVA and subsequently, it forms an adduct with pyrimidine base of the opposite strands of DNA resulting in cross-linking in DNA strands.[43] These inhibit cell division and thus PSO followed by UV radiation is widely used for the treatment of psoriasis. However, most naturally occurring PSO is insoluble in water with very low penetration capability within the phospholipid membrane of the cell to reach their cellular targets.[44] Topical delivery of PSO using liposomal formulations has attracted considerable interest in recent decades because of the improved therapeutic action.[45] However, classical liposomes are of little use as they cannot penetrate the cellular or bacterial membrane. Conversely, several research works indicate that ethosomes, which is in a class of liposomes containing some amount of ethanol in the core could be a better tool for sub-dermal delivery of macromolecules.[46]

We herein explored the photoinduced dynamics of PSO encapsulated in the ethosome drug delivery vehicle. In the present study, ethosomes of 110 nm vesicular size were prepared and PSO was entrapped in the ethosomal formulation. The spherical droplets with an average hydrodynamic diameter of 110 nm were measured from the DLS instrument (inset of Figure 2.1a). Absorption spectra of PSO in a water–ethanol mixture and in ethosome are shown in Figure 2.1a, which shows that PSO in a water–ethanol mixture (red) has three peaks at 244 nm, 294 nm, and 340 nm. PSO encapsulated in ethosome (blue) also shows three peaks with a base line upliftment due to scattering of colloidal ethosome.[47] The transitions at 294 nm and 340 nm are reported to be resulting from $n \rightarrow \pi^*$ transition of non-bonding electrons on the C-2 carbonyl group in PSO and $\pi \rightarrow \pi^*$ transition of $\pi$ electrons of PSO ring system respectively. It was estimated that 88.7 μM PSO is associated with 3.4 μM ethosome. In other words, 26 PSO are found to be attached with each ethosome vesicle. Emission spectra of PSO in a water–ethanol mixture (red) and PSO–ethosomes (blue) (Figure 2.1b) show that the fluorescence intensity of PSO is quenched upon encapsulation of PSO in the ethosomes. The emission quenching of PSO in a nonpolar medium, cyclohexane (green) is also shown in the Figure 2.1b. The quantum yield of PSO in aqueous ethanol, ethosomes, and cyclohexane are estimated to be $1.0 \times 10^{-2}$, $6.0 \times 10^{-3}$, and $2.0 \times 10^{-4}$ respectively. Studies on

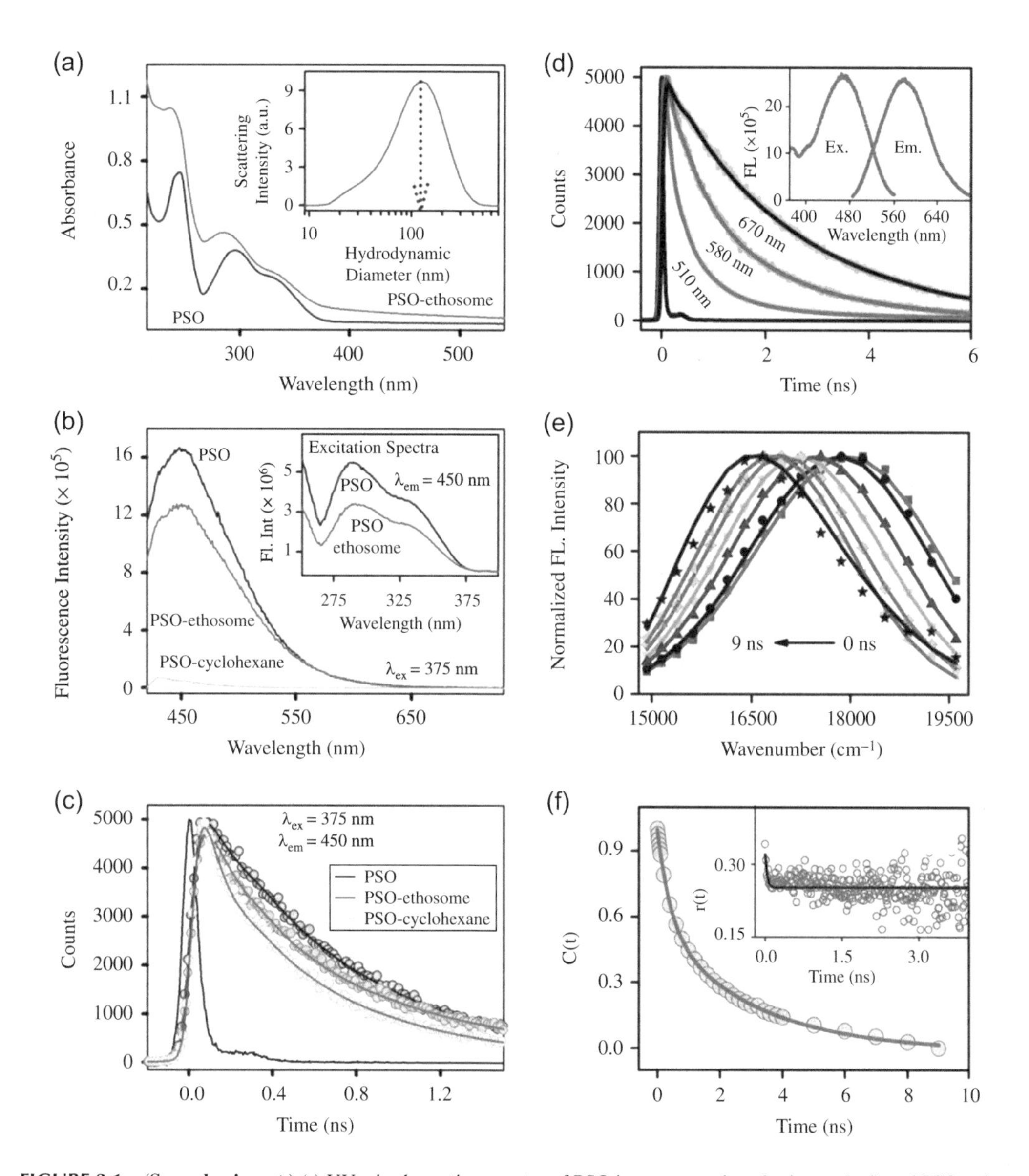


**FIGURE 2.1** **(See color insert.)** (a) UV–vis absorption spectra of PSO in a water–ethanol mixture (red) and PSO–ethosomes (blue). The inset shows the hydrodynamic diameter of the ethosomes measured by DLS. (b) Room-temperature emission spectra of PSO (red), PSO–ethosomes (blue), and PSO in cyclohexane (green). The excitation wavelength is 375 nm. The inset shows the excitation spectra of PSO (red) and PSO–ethosome (blue). (c) Fluorescence transients of PSO in a water–ethanol mixture (red), PSO–ethosomes (blue), and PSO in cyclohexane (green). (d) Fluorescence transients at different wavelengths for DCM–ethosomes. The inset shows excitation and emission spectra of DCM. (e) TRES. (f) Decay of the solvation correlation function, C (t), with time. The inset shows temporal decay of fluorescence anisotropy, r(t), of DCM–ethosomes. (Reproduced from Bagchi et al.[41] With permission.)

spectroscopic properties of PSO have reported that the energies of lowest excited states (singlet and triplet) of PSO are strongly dependent upon solvent polarity. With the increase in polarity of the solvent the energy level ordering of PSO is assumed to be changed and thereby fluorescence becomes predominant compared to non-radiative inter system crossing. Upon encapsulation into ethosome, the polarity of the solvent around PSO decreases which results in the quenching of fluorescence intensity without changing the emission maxima. Figure 2.1c shows picosecond-resolved emission transients of PSO in various mediums including ethosomes. The numerical fitting of the transients reveals the average time constant of 0.62 ns for aqueous ethanol and 0.34 ns for cyclohexane. In the ethosome the average time constant of 0.50 ns is close to that of the aqueous ethanol revealing the location of PSO to be at the interface.[48]

In order to compare the microenvironment of the ethosomes with liposomes, we have used a well-known fluorescent solvation probe 4-(dicyanomethylene)-2-methyl-6-(p-dimethylaminostyryl)-4H-pyran DCM, widely used in the characterization of various liposomes.[49] The inset of Figure 2.1d represents excitation and emission spectra of DCM in ethosome. The emission peak of DCM in ethosome is blue shifted to 570 nm compared to DCM in the buffer. This corresponds to the binding of DCM to the hydrophobic core of ethosome. Figure 2.1d shows fluorescence transients of DCM encapsulated in the ethosomes at three characteristic detection wavelengths (510 nm, 580 nm, and 670 nm), which fall in the blue, peak, and red end of the emission spectrum of the probe in the ethosomes. In the excited state of DCM, intra-molecular charge transfer takes place which gives rise to a very high dipole moment (23.6 Debye) compared to the ground state. In the blue edge of the spectrum (510 nm), the observation is consistent with the fact that DCM undergoes solvation relaxation inside signal is seen to decay faster compared to the red edge (670 nm), where a rise component is apparent in ethosomes. Using the fitting parameters of the fluorescence decays and the steady-state emission spectrum, the TRES and the solvent correlation function ($C(t)$) of DCM in the ethosomes were constructed (Figure 2.1e,f). The associated dynamical Stokes shift ($\Delta\nu$) of DCM in the ethosomes is calculated to be 1458 $cm^{-1}$. $C(t)$ decay of DCM is fitted with biexponential function with two time components of 430 ps (44%) and 3.04 ns (56%). The time constants are consistent with the reported values of the solvation relaxation time of DCM in dipalmitoyl–phosphatidylcholine (DPPC) liposomes. The temporal decay of fluorescence anisotropy $r(t)$ of DCM in ethosome is shown in the inset of Figure 2.1f. The decay transient can be fitted mono-exponentially with a time constant of 38 ps (29%) with a significant part (71%) that persists within our experimental time window of 20 ns (we show up to 4 ns). The faster time constant is consistent with the tumbling motion of DCM in the ethosomes.[50]

The co-localization of a model cationic drug crystal violet (CV) with the hydrophobic DCM in the ethosome was investigated using a FRET strategy. Figure 2.2a shows the spectral overlap of energy donor DCM emission with the absorption spectrum of acceptor CV. A significant quenching in steady-state emission of DCM in the presence of CV in the ethosome is observed, where DCM:CV concentration in the ethosomal mixture is 1:1 (inset of Figure 2.2b).

The picosecond-resolved fluorescence decay profile of DCM in the absence (green) and in the presence (blue) of CV was monitored at 620 nm upon excitation at 375 nm (Figure 2.3b). The shorter component in decay profile suggests an excited state energy transfer process. A FRET efficiency of ~75% with an average donor–acceptor distance of 5.8 nm was calculated for the DCM–CV pair. It is also observed that the distance between DCM and CV is independent of detection wavelength by detecting fluorescence transients at different wavelengths (data not shown). This indicates the homogeneous distribution of DCM in ethosomal mixture. It is reported that the maximum thickness of the phospholipid bilayer is 7.3 nm.[51] Therefore the possible location of CV could be at the polar interface of the ethosome. After having an idea about the location of CV with respect to DCM, we exploit the spectral overlap of CV with PSO in order to study the localization of PSO in the ethosomes. A considerable spectral overlap of PSO emission with the absorption spectrum of CV is shown in Figure 2.2c, where the concentration of PSO: CV is 1:10, indicating the possibility of FRET from PSO to CV in the ethosome. A significant quenching in steady-state emission of PSO in presence of CV in the ethosome is observed (inset of Figure 2.2d). Picosecond-resolved fluorescence decay profile of PSO in ethosome in the absence (red) and in the presence (cyan) of CV was monitored at 450 nm upon excitation at 375 nm (Figure 2.2d). The excited state lifetime ($\tau_{av}$) of PSO reduces upon interaction with CV. We have estimated the FRET efficiency to be ~84% and donor (PSO)-acceptor (CV) distance around 2.2 nm. The distance is almost independent of detection wavelength as there is no indication of change in lifetime by changing the detection wavelength. PSO–ethosome has been evaluated as an antimicrobial agent followed by a drug delivery system to inhibit the growth of bacterial biofilms of gram-negative *E. coli*.[52] For photodynamic therapy experiments, we have added PSO–ethosome (85 μM PSO) to bacterial cultures in the presence and absence of UV-A. The inhibition of bacterial growth after photodynamic treatment is clearly visible. The colony forming units (CFU) indicates insignificant antibacterial activity of PSO–ethosome in dark. In the case of PSO–ethosome treated samples with UVA irradiation, the bacterial growth is inhibited sharply

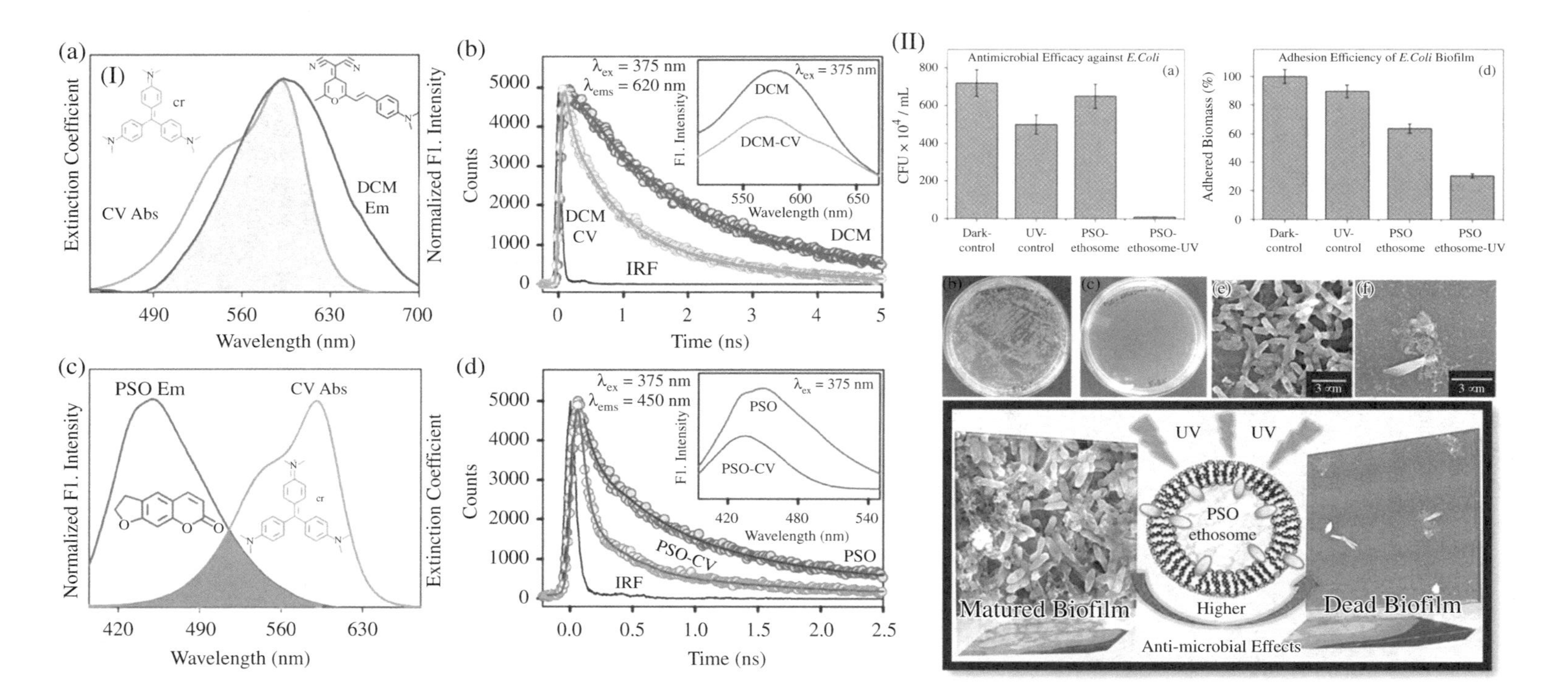


**FIGURE 2.2** **(See color insert.)** (I) (a) Overlap of DCM–ethosome emission and CV absorption in water. (b) Picosecond-resolved fluorescence transients of DCM–ethosomes (excited at 375 nm) in the absence (green) and presence (blue) of CV collected at 620 nm. The inset shows steady-state emission spectra of DCM–ethosomes in the absence (green) and presence (blue) of CV. (c) Overlap of PSO–ethosome emission and CV absorption in water. (d) Picosecond-resolved fluorescence transients of PSO–ethosome (excited at 375 nm) in the absence (red) and presence (cyan) of CV collected at 450 nm. The inset shows steady-state emission spectra of PSO–ethosomes in the absence (red) and presence (cyan) of CV. (II) (a) Antibacterial activity of PSO–ethosomes against *E. coli* in the absence and presence of UVA. (b, c) Images of PSO–ethosome treated E. coli culture plates before and after UVA irradiation, respectively. (d) Adhesion efficiency of PSO–ethosome treated *E. coli* biofilms in the absence and presence UVA irradiation. SEM images of an *E. coli* biofilm (e) without treatment and (f) treated with PSO–ethosomes followed by UVA illumination for 30 min. The lower panel depicts a schematic representation of the processes. (Reproduced from Bagchi et al.[41] With permission.)

indicating immense photoinduced antimicrobial activity of PSO–ethosome. The maximum inhibition of *E. coli* is obtained for PSO–ethosome treated samples where a 95% decrease in CFU is observed after photodynamic treatment (Figure 2.2(II)a). The pictures of *E. coli* cultures treated with PSO–ethosome in the absence and presence of UV-A light also clearly suggest bacterial growth inhibition. The inhibition of bacterial growth after photodynamic treatment is clearly visible. Next, the nanomaterial is tested for biofilm inhibition assay. Total adhered biomass of the biofilms is monitored through quantitative assay using crystal violet stain. There is a decrease in biomass (~ 30%) for *E. coli* treated with ethosome containing PSO. The bacterial biomass could be further reduced to ~60% when PSO–ethosome treated bacteria were exposed to UVA light for 30 min (Figure 2.2(II)d and 2.5d). The structural and morphological changes of the biofilms are observed through SEM images which show clear destruction and inhibition in biofilm formation (Figure 2.2(II)e,f). Hence, to our understanding, the studies could pave the way in designing novel high potential therapeutic drugs with improved pharmacological efficacy to treat multidrug resistant bacteria-induced diseases.

### 2.3.2 Dynamical Study on Photo-Triggered Destabilization of Nanoscopic Vehicles by a Photochromic Material

The efficacy of photoresponsive destabilization to phosphatidylcholine liposome, which is used as potential drug delivery vehicles, was investigated by using a synthesized photochromic dye dihydroindolizine (DHI).[53,54] The structural conversion of DHI from closed to open isomer (see Figure 2.3) can fluctuate or defect the liposomal membrane by mechanical stress and leads to photoresponsive destabilization of the liposome. The rate of isomerization in the liposome is evaluated and shown in Figure 2.3b. The consequence of different isomerization of DHI on liposome stability was monitored by steady-state, time-resolved fluorescence and polarization-gated fluorescence spectroscopy by labeling the liposome with fluorescent probe ANS. The small red-shift (5 nm) in emission maximum of ANS-PC-DHI upon UVA irradiation with a decrease in the fluorescence intensity corroborates that ANS is now experiencing higher polarity upon closed-to-open transition of DHI (inset of Figure 2.3c).[24] Furthermore, the faster fluorescence decay of ANS-PC-DHI upon UVA irradiation leading to the enhanced internal rotation of the fluorophore relative to the liposome also indicates a progressive release of restriction on the probe might be due to the increase in the mobility of solvating species (see Figure 2.3g,h). To understand the fusion phenomenon, FRET techniques were used where ANS was encapsulated in a group of liposomes, and doxorubicin (DOX) was encapsulated in another group of liposomes both having DHI. The decrease in average lifetime of ANS-PC-DHI upon UVA irradiation due to energy transfer from ANS to DOX, which otherwise does not depict any quenching when simple DOX is diluted in the medium containing ANS-PC-DHI, indicates that UVA-irradiated DHI-sensitized liposome could not lead to rearrangement of bilayer and total membrane perturbation; rather, it leads to the fusion of the liposome (see Figure 2.3e,f).

Based on the photoresponsive properties and microstructural change investigated earlier, DHI liposomes could be considered for photoresponsive drug delivery systems. Figure 2.4a shows a burst release which occurred upon UVA irradiation at the 30 min, followed by a slower sustained release up to 1 h, whereas spontaneous release of DOX was observed in the group without UVA irradiation. The therapeutic efficacy of the drug-loaded liposome was further evaluated against cervical cancer cell line HeLa by exposing the cells directly to the PC-DHI-DOX in presence or absence of UVA (Figure 2.4b). MTT assay studies reveal an enhanced cellular uptake of DOX leading to a significant reduction in cell viability (40%) of HeLa, followed by photoresponsive destabilization of the liposome. Qualitative analysis of the confocal fluorescence micrographs clearly showed that under UVA exposure, cellular uptake of DOX in the HeLa cells treated with PC-DHI-DOX was significantly higher with respect to the other three systems as evident from the variation in intensity of the red color in Figure 2.4c) The flow cytometry-based quantitative analysis (see Figure 2.5d) revealed that the percentage of cell populations that underwent DOX uptake was higher (98.3%) in the case of cells treated with PC-DHI-DOX under UVA exposure in comparison to PC-DHI-DOX treated cells in dark condition (81.8%). Cells treated with PC-DOX showed no significant changes upon UVA irradiation. These findings altogether indicate higher DOX delivery by UVA trigger PC-DHI liposome. These results presented in this study indicated that a DHI-encapsulated liposome could serve as a safe and promising drug delivery vehicle.

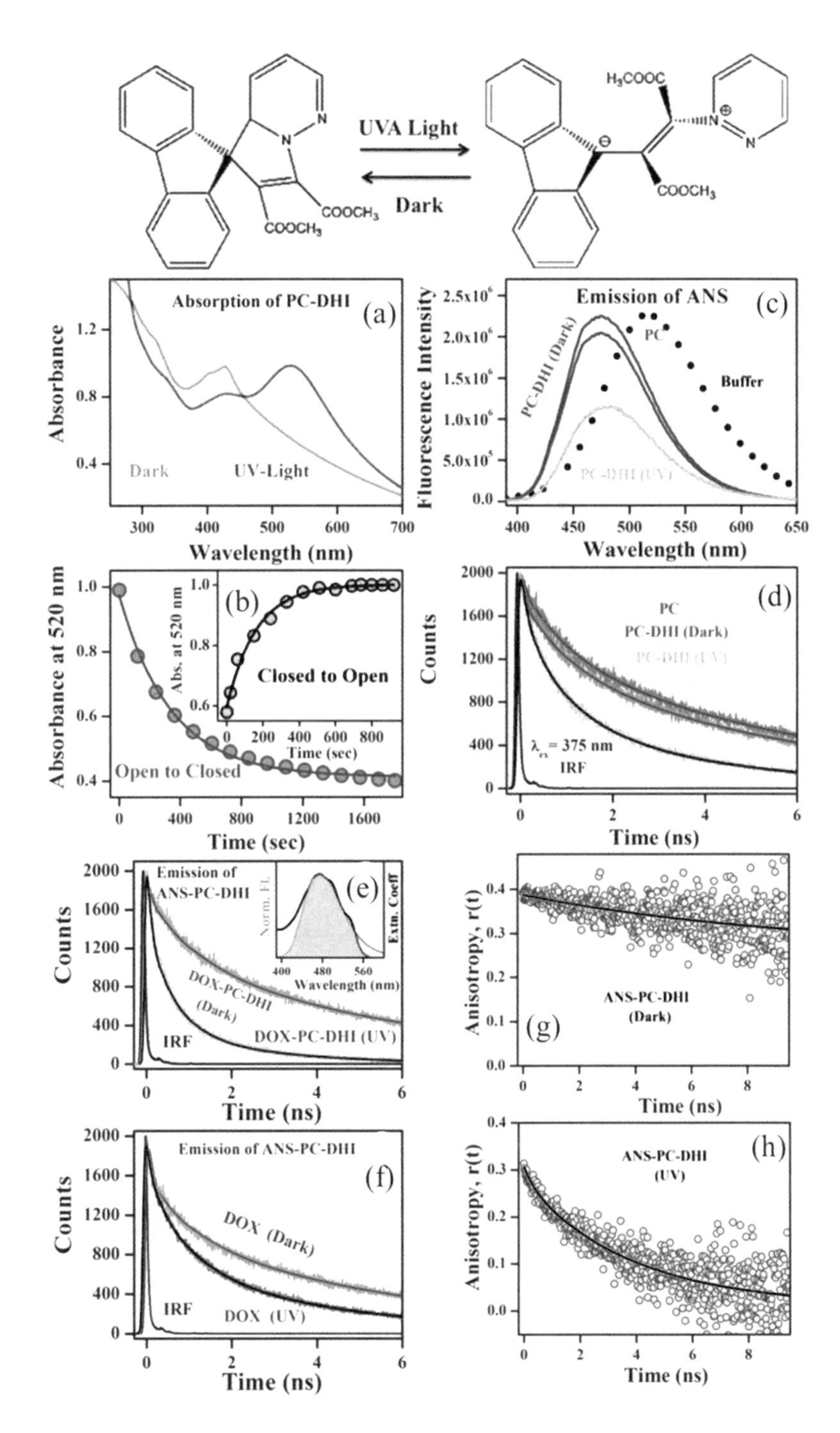


**FIGURE 2.3 (See color insert.)** Structures of the closed and open isomers of DHI. (a) Absorption spectra of DHI in liposome: closed and open isomers. (b) Kinetics of the open to closed transition of DHI in liposome (SD=±0.008, $n$=3). Inset shows the corresponding closed to open conversion rate (SD=±0.003, $n$=3). (c) Steady-state emission spectra of ANS in buffer, ANS bound to PC and PC-DHI in presence and absence of UVA light. (d) Time-resolved transients of ANS, ANS bound to PC and ANS bound to PC-DHI in the presence and absence of UVA light. Picosecond-resolved transients of the donor-acceptor in the absence and presence of UVA light, (e) donor is (ANS-PC-DHI) and acceptor is (DOX-PC-DHI) and (f) donor is (ANS-PC-DHI) and acceptor is free DOX. Insets depict the corresponding spectral overlap between donor (ANS-PC-DHI complex) emission and acceptor (DOX-PC-DHI) absorbance. Time-resolved anisotropy of ANS bound to PC-DHI in presence of (g) dark and (h) UVA light. (Reproduced with permission from P. Singh et al.[53], © 2017 Elsevier, B.V. All rights reserved.)

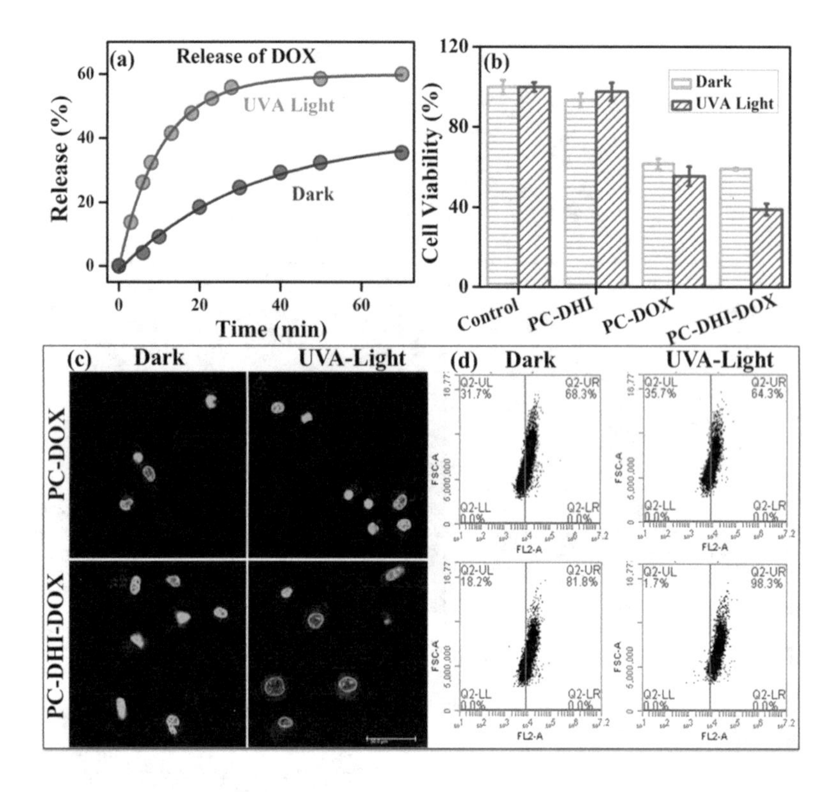


**FIGURE 2.4** (a) Release profile of DOX from PC-DHI in the absence (SD = ±0.4, $n=3$) and presence of UVA light (SD = ±0.9, $n=3$). (b) Cytotoxicity assay in HeLa cells with PC-DHI, PC-DOX, and PC-DHI-DOX with MTT as an indicator dye in the presence and absence of UV light. (c) Confocal microscopy images of HeLa cells treated with PC-DOX liposome and PC-DHI-DOX liposome both in the absence and presence of UVA light. (d) Flow cytometry of HeLa cells treated with PC-DOX liposome and PC-DHI-DOX liposome both in the absence and presence of UVA light. (Reproduced with permission from Singh et al.[53], © 2017 Elsevier, B.V. All rights reserved.)

## 2.4 Key Interfacial Dynamics at Metal–Organic Interfaces in Metallic Nanohybrids

### 2.4.1 Dynamical Evaluation of Protein-Assisted Metal Nanoparticles Synthesis Strategy

Biomolecule-assisted synthesis of metal nanoparticles can be divided into two categories.[55] One uses multi-domain protein cages (templates) and the other relies on the self-assembly of biomolecules such as small peptides, DNA, and denatured protein. In the present discussion, the protein-assisted synthesis route to prepare highly crystalline 3–5 nm gold nanoparticles (Au NPs) is explained.[55,56] The synthesis strategy relies on systematic thermal denaturation of a number of proteins and protein mixture from *E. coli* in the absence of any reducing agent (see Figure 2.5 Scheme). By using UV–vis spectra, Au NPs synthesized using human serum albumin (HSA), BSA, SC, and *E. coli* extract protein reveal surface plasmon resonance (SPR) bands at 530, 531, 540, and 600 nm, respectively (see Figure 2.5 (I) a). On the basis of Mie theory, the average diameter of Au NPs (spherical), can be approximately estimated average diameters to be in the range of 3.4, 3.5, and 3.7 nm for HSA, BSA, and SC, respectively.

Figure 2.5I (a) shows the UV–vis absorption spectra obtained at different time intervals after mixing with aqueous $AuCl_4^-$ solution with HSA in phosphate buffer at 76°C temperature. The formation of Au nanoparticles in the colloidal solution was monitored from their absorption spectra as the small noble metal particles reveal absorption band in the UV–vis spectral region due to SPR. The sharp absorption band peaking at 530 nm indicates a relatively high monodispersity, both in size and shape of the Au particles, consistent with the

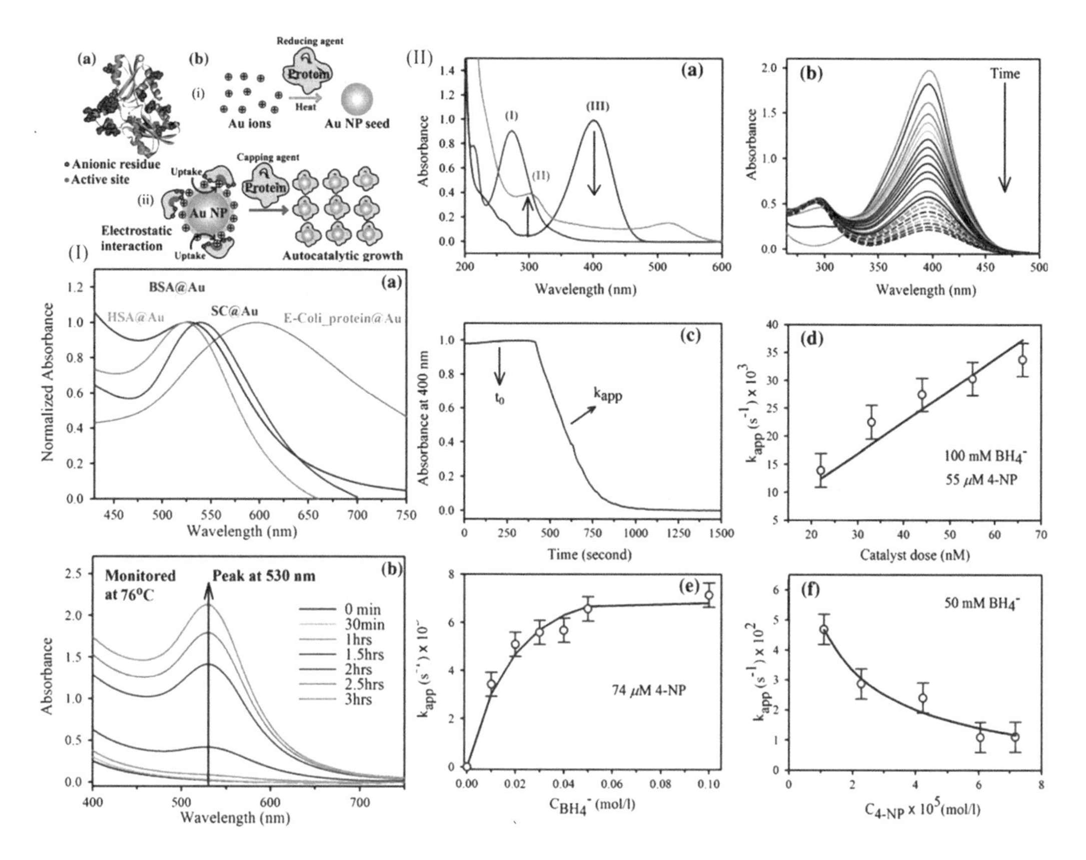


**FIGURE 2.5** Schematic representation of nanoparticle synthesis (a) Molecular structure of subtilisin Carlsberg. (b) Generalized two-step [(i) and (ii)] mechanism for solution-phase Au nanoparticle synthesis (I) (a) UV–vis spectra of Au NPs conjugated with various proteins, HSA, BSA, SC, and E. coli extract, respectively. (b) Time-resolved UV–vis spectra for one of the representative proteins (HSA) at 76°C. (II) (a) Absorption spectra of 4-NPA (I) in absence of NaBH4, (II) in the presence of $NaBH_4$ at 0 min, and (III) in the presence of Au@protein nanobioconjugates. Conditions: [4-NP-] = $5.5 \times 10^{-5}$ M; [Au NPs] = $2.2 \times 10^{-7}$ M; [$NaBH_4$] = 0.1 M. (b) Concentration versus time plot (monitored at 400 nm) for 4-NPA reduction by $NaBH_4$. Conditions: [4-NP] = $5.5 \times 10^{-5}$ M; [$NaBH_4$] = 0.1 M. (c) Typical time dependence of the absorption of 4-NPA at 400 nm. (d) Plot of apparent rate constant ($k_{app}$) versus catalyst dose for 4-NPA reduction by $NaBH_4$ in the presence of Au@protein solution as catalyst. Conditions: [4-NP] = $5.5 \times 10^{-5}$ M; [$NaBH_4$] = 0.1 M. (e, f) Dependence of the apparent rate constant $k_{app}$ on the concentration of $BH_4^{-1}$ e and 4-NP f. The solid lines are the fit of the Langmuir–Hinshelwood model. The surface area of Au nanoparticles is 0.0106 $m^2$/L in both cases. (From Goswami et al.[55] With permission.)

HRTEM images. Long term stability of the Au NP in aqueous solutions (for several months) indicates that the HSA serves as capping agent. In the case of the other proteins, a similar trend was observed as that of the HSA. The protein-capped as-prepared Au nanoparticles are found to serve as an effective catalyst to activate the reduction of 4-nitrophenol in the presence of $NaBH_4$. The kinetic data obtained by monitoring the reduction of 4-nitrophenol by UV–vis spectroscopy, revealing the efficient catalytic activity of the nanoparticles, have been explained in terms of the Langmuir–Hinshelwood (L–H) model (see Figure 2.5II). The fitted data in Figure 2.5II f clearly demonstrate that the reduction of 4-NP can be described by the L–H model with good accuracy. The methodology and the details of the protein chemistry presented here may find relevance in the protein-assisted synthesis of inorganic metal nanostructures in general.

### 2.4.2 Sensitization of an Endogenous Photosensitizer: Electronic Spectroscopy of Riboflavin in the Proximity of Metal Nanoparticles

Riboflavin (Rf) or vitamin $B_2$ is an essential micronutrient and substantially present in dietary products, and is alternatively used as a potential photodynamic therapeutic (PDT) agent.[57,58] Synthesis of gold nanoparticle–riboflavin (Au–Rf) conjugate provides a novel nanohybrids system.[57] Figure 2.6a depicts the lattice fringes of Rf–Au NP distinctly show an interplanar distance of ~0.31 nm corresponding to the interspace between two (222) planes. The average particle size for an Rf–Au NP is found to be ~29.2 ± 0.1 nm. Raman spectroscopic studies provide an insight into the nature of the attachment of Rf to the surfaces of the metal (Au NPs). Rf attached to Au NPs show that the characteristic Rf peak is shifted from 1346 to 1358 $cm^{-1}$ and is also broadened (see Figure 2.6b). Moreover, the peak at 1563 $cm^{-1}$ for Au NPs is moved to 1540 $cm^{-1}$ in the presence of Rf molecules. This suggests a covalent bond formation between Rf and Au NP. Time-resolved fluorescence studies on the nanohybrids clearly depict the phenomenon of NSET is predominant in the riboflavin-gold. Figure 2.6c illustrates a large spectral

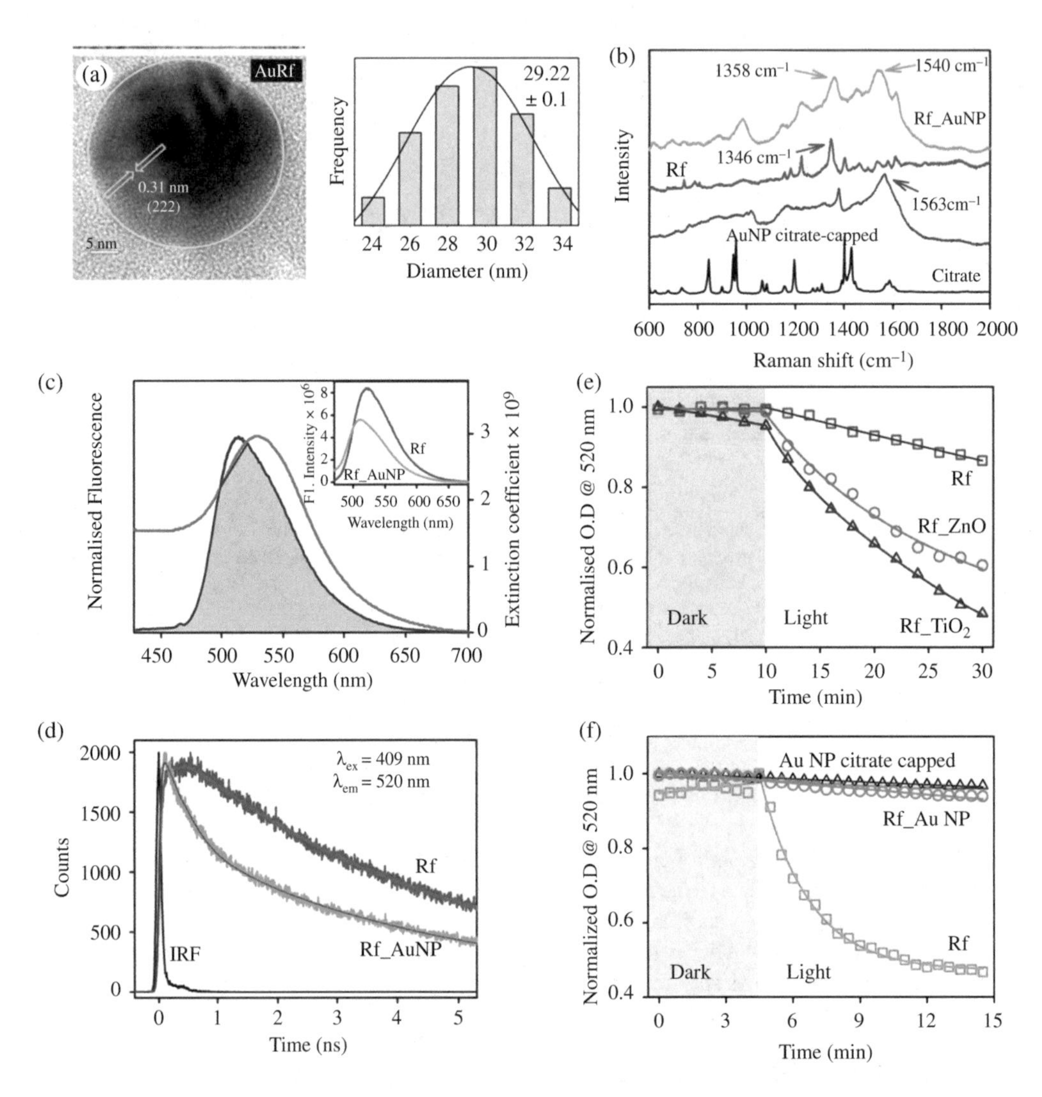


**FIGURE 2.6 (See color insert.)** (a) HRTEM images of Au nanoparticles. The right panel shows the corresponding particle size distribution of the NPs. (b) Raman spectra of Rf and Rf–Au NPs in the solid phase. (c) Spectral overlap of donor Rf (dark green) and acceptor Au NPs (pink). The inset shows the steady-state emission quenching of Rf after complexation with Au NPs. (d) The picosecond-resolved fluorescence transients of Rf and Rf–Au NPs (excitation at 409 nm). All the spectra were taken in aqueous medium. Absorption kinetics (at 520 nm) of DPPH in the presence of (e) Rf–ZnO, Rf–$TiO_2$ in ethanol and (f) Rf–Au NPs in ethanol:water (1:1) mixture under blue light irradiation. Part (f) shows the degradation kinetics under dark only up to 5 min for the representation as antioxidant activity under dark for a longer time window of 10 min shows no significant activity. (Reproduced with permission from Chaudhuri et al.,[57] © 2015, American Chemical Society.)

overlap of Rf emission and surface plasmon absorbance of Au NP which indicates a fair possibility of energy transfer from Rf to Au NP.

The fluorescence intensity of the Rf is quenched and also the peak is shifted by 12 nm when it is attached to Au NPs in an aqueous medium as shown in Figure 2.6c, inset. The fluorescence decay time of the Rf in water without Au NP is given by a single exponential with the average lifetime value of 4.76 ns. However, there is a quenching of decay time in the Rf–Au NP and the time scales are fitted by biexponential decay (see Figure 2.6d). The associated fluorescence transient of Rf–Au NPs is one faster component of 0.4 ns (48%) due to the energy transfer from Rf to Au NPs and a longer component of 4.65 ns (52%) consistent with the decay of Rf only. FRET is employed for the determination of the donor–acceptor distance. The energy transfer efficiency and the overlap integral [J(λ)] are calculated to be 53.7% and $1.36 \times 10^{20}$ $M^{-1}$ $cm^{-1}$ $nm^4$ respectively. The Förster radius ($r_0$) is found to be 34 nm. The calculated donor–acceptor separation (r) is 28 nm. As the estimated donor–acceptor distance exceeds 100 Å, the phenomenon of the Au NP-based surface energy transfer (SET) process is a convenient spectroscopic ruler for long distance measurement, which follows $1/d^4$ distance dependence.

NP-induced lifetime modification serves as a ruler to unravel the distance range well beyond 10 nm. In order to approve the NSET formulism,[59] the distance between the donor Rf and the acceptor Au NP is determined to be 16.71 nm ($d_0$ = 17.36 nm) using NSET equations. Herein, NSET from the donor Rf to the acceptor Au NPs as the calculated donor–acceptor distance is in consonance with the size of the Au NPs (radius 15 nm). In order to measure the photo-antioxidant activity of the nanohybrids, the well-known 2,2-diphenyl-1-picrylhydrazyl (DPPH) assay was provided under blue light irradiation ($\lambda_{max}$ = 450 nm). The control Rf in Figure 2.6f shows enhanced activity over the Rf–Au. Hence, it is worth emphasizing that the energy transfers from Rf to Au NP result in the reduced antioxidant activity of the Rf–AuNP nanohybrids, as shown in Figure 2.6f. This study unravels a mechanistic pathway of drug sensitization with metal NPs to modulate the photoinduced antioxidant property of the important vitamin as well as drug molecule.

## 2.5 Ultrafast Charge Transfer Dynamics at Semiconductor Nanoparticle–Organic Molecule Interface

### 2.5.1 Protein-Capped Mn-Doped ZnS Nanoparticle: A Multifunctional, UV-Durable Bio-Nanocomposite

The design of synthetic NPs capable of recognizing given chemical entities in a specific and predictable manner is of great fundamental and practical importance.[60] Herein, we report a simple, fast, water-soluble, and green phosphine free colloidal synthesis route for the preparation of multifunctional enzyme-capped ZnS bio-nanocomposites (BNCs) with/without transitional metal-ion doping (see Figure 2.7I).[60] Figure 2.7(I) a-h shows a set of TEM images of CHT and Cys-capped, with/without Mn-doped ZnS NPs. It has to be noted that the shape of the NPs in the protein matrix is relatively quasi-spherical compared to that of the Cys-capped NPs. The observation could be consistent with the fact that the NPs in the protein matrix are associated with a number of sulfur-containing Cys residues from various locations of a protein which essentially direct the shape of the NPs to be quasi-spherical. Also, plenty of free Cys residues in the solution for the Cys-capped NPs lead to uniform growth of the NPs and make the shape to be spherical. The corresponding high-resolution TEM (HRTEM) images clearly demonstrate lattice fringes with an observed d-spacing of ~0.31 nm and ~0.23 nm for CHT and Cys-capped NPs, respectively, which are in good agreement with the high-crystallinity in the materials with zinc-blended structures. The particle sizes are estimated by fitting our experimental TEM data on 100 particles which provides the average diameter of 3 and 2.7 nm for CHT and Cys-capped NPs (left insets, Figure 2.7(I) b, d), respectively. It is noticeable that CHT-capped NPs are fairly monodispersed in the protein matrix while for Cys-ZnS:Mn, some of the particles are agglomerated up to 10 nm. The room temperature photoluminescence (PL) spectra (see Figure 2.7 (II) a, b) of doped and undoped ZnS NPs have been recorded at an excitation wavelength of 300 nm (4.13 eV). Cys-capped undoped ZnS NPs show one broad emission band centered at ~420 nm, which is attributed to defect-state recombinations, possibly

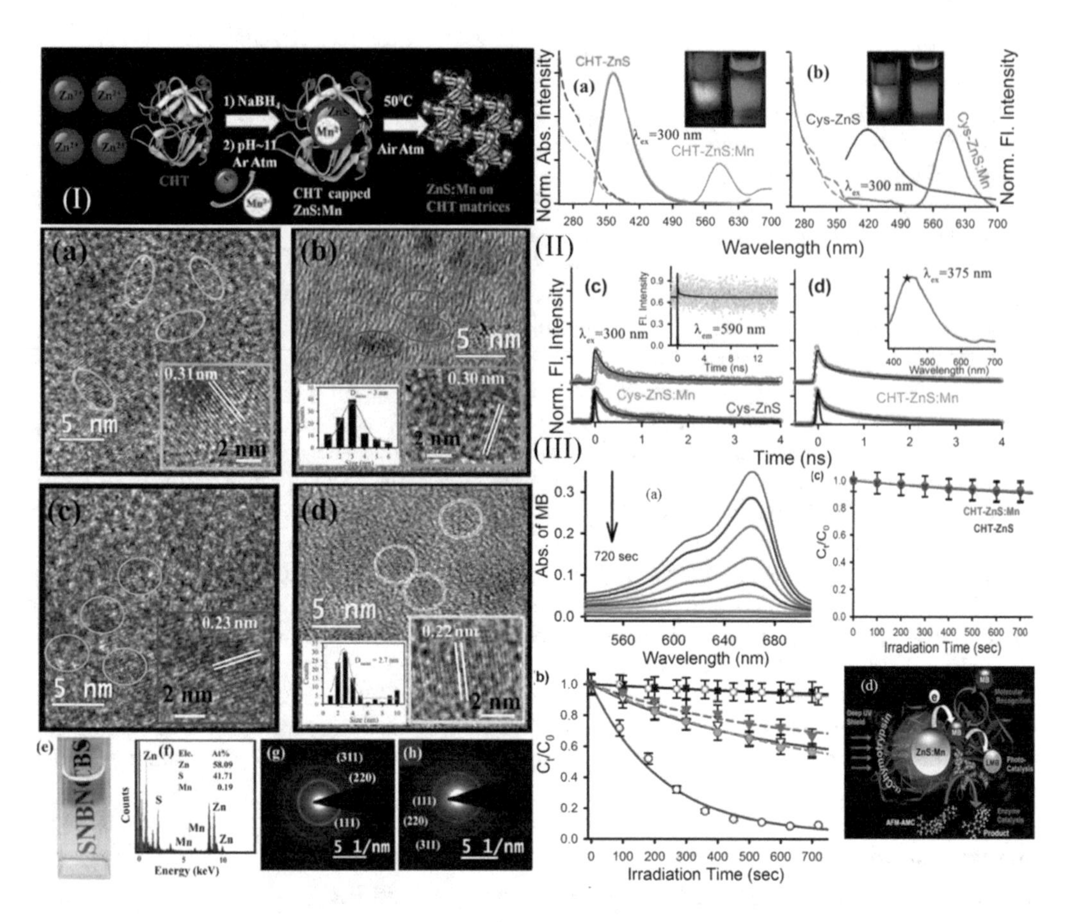


**FIGURE 2.7** **(See color insert.)** (I) Synthetic strategy of enzyme-mediated Mn-doped ZnS BNCs. TEM and HRTEM images (inset) of (a) CHT-ZnS, (b) CHT-ZnS:Mn, (c) Cys-ZnS, (d) Cys-ZnS:Mn NPs. Inset left of panels (b) and (d) represent the size distribution analysis of CHT-ZnS NPs and Cys-ZnS:Mn NPs, respectively. (e) Optically transparent solution of CHT-ZnS:Mn BNCs under daylight. (f) EDAX analysis and atomic percentages elements, (g) and (h) SAED analysis of CHT-ZnS and CHT-ZnS:Mn BNCs, respectively. (II) Optical absorption and steady-state emission spectra of (a) CHT-ZnS and CHT-ZnS:Mn BNCs and (b) Cys-ZnS, Cys-ZnS:Mn NPs, respectively. Inset of (a) and (b) shows PL photos of the corresponding solutions upon 300 nm excitation. (c) The picosecond-resolved fluorescence transients of Cys-ZnS and Cys-ZnS:Mn NPs (excitation at 300 nm) collected at 420 nm and inset shows fluorescence transient of Cys-ZnS:Mn NPs collected at 590 nm. (d) The picosecond-resolved fluorescence transients of CHT-ZnS:Mn NPs (excitation at 375 nm) collected at 460 nm (green) and 590 nm (red). The inset shows PL spectra upon 375 nm excitation. (III) (a) Time-dependent UV–vis spectral changes of MB in the presence of CHT-ZnS BNCs under UV-light irradiation. (b) Plot of relative concentration ($C_t/C_0$) versus irradiation time for the degradation of MB (monitored at 655 nm) is shown. The degradation is performed in the presence of BNCs: CHT-ZnS (empty circle), CHT-ZnS:Mn (filled circle), Cys-ZnS (empty triangle), Cys-ZnS:Mn (filled triangle), only CHT (empty square), no catalysts (crossed). (c) Plot of $C_t/C_0$ versus irradiation time in the presence of CHT-ZnS (filled triangle) and CHT-ZnS:Mn (filled circle) upon selective excitation with a 350 nm high-pass filter. (Reproduced with permission from Makhal et al.[60] © 2012, American Chemical Society.)

at the surface. Since an excess of the cations have been used in the synthesis procedure, it was expected that sulfur vacancies at the surface would give rise to Zn dangling bonds that form shallow donor levels. Thus, the recombination is mainly between these shallow donor levels and the valence band.[61] Upon Mn incorporation in nanocrystal samples, blue ZnS emission is quenched whereas an orange emission band develops at ~590 nm (see Figure 2.7(II) b), corresponding to the spin-forbidden $^4T_1$-$^6A_1$ Mn d-d transition in a tetrahedral site.[61] The insets of Figures 2.7(II)a,b show PL photographs from the undoped (blue) and doped (orange) solutions upon 300 nm excitation. In the CHT-capped BNCs, NP associated proteins show a strong emission band centered at 367 nm (see Figure 2.4a) which possibly augments ZnS PL band at 420 nm. In the picosecond-resolved emission study (see Figure 2.7(II) c, d), the excited state population of charge carriers in Cys-ZnS:Mn NPs are monitored at 420 nm followed by excitation at 300 nm. It is to be noted that Cys-ZnS and Cys-ZnS:Mn sample solutions show almost the same decay pattern (time constants) when both the decays are monitored at 420 nm.

This phenomenon reveals that the ZnS PL quenching upon Mn-doping is either static in nature or may be too fast to be resolved in our TCSPC instrument with 60 ps IRF. Upon below band-edge excitation (with 375 nm, i.e., 3.3 eV), no Mn emission peak is noticeable in the doped NPs (see Figure 2.7(II) d, inset). The picosecond-resolved fluorescence decays (excitation at 375 nm) monitored at 460 (to avoid Raman scattering at 428 nm) and 590 nm are shown in Figure 2.7(II) d which exhibits similar time constants of ZnS. The observation suggests that the below-band gap excitation is not sufficient to excite the doped material (Mn) via energy transfer from the host's conduction band to the Mn state. Considering that the excitation process generates an electron–hole pair across the band gap (3.9 eV) of the ZnS nanocrystal host, the present results make it obvious that there is a more efficient excitonic energy transfer from the host to the doped Mn site compared to that of the defect states in these materials; revealing a strong coupling between the Mn d-levels and the host states. The energy transfer is unlikely to occur directly from the semiconductor trap (defect) states to the low-lying Mn d-states. This observation demonstrates that the trap states are not in a direct coupling with the Mn d-states and Mn-doping do not affect the trap state lifetimes of the excited state electrons at the host ZnS surface. The enzymes α-Chymotrypsin (CHT), associated with the NPs, are demonstrated as an efficient host for the organic dye Methylene Blue (MB), revealing the molecular recognition of such dye molecules by the BNCs. An effective hosting of MB in the close proximity of ZnS NPs (with ~3 nm size) leads to photocatalysis of the dyes which has further been investigated with doped semiconductors (see Figure 2.7(III) a, b, c). The NP-associated enzyme α-CHT is found to be active toward a substrate (Ala-Ala-Phe-7-amido-4-methyl-coumarin), hence leading to significant enzyme catalysis. Irradiation-induced luminescence enhancement (IILE) measurements on the BNCs clearly interpret the role of surface capping agents which protect against deep UV damaging of ZnS NPs.

### 2.5.2 Modulation of Defect-Mediated Energy Transfer Dynamics from ZnO Nanoparticles for Photocatalytic Applications

For biomedical applications of nanotechnology, the utilization of the ZnO nanoparticles for the efficient degradation of bilirubin (BR) through photocatalysis is described in this section.[62] BR is a water-insoluble by-product of heme catabolism that can cause jaundice when its excretion is impaired.[63] The photocatalytic degradation of BR activated by ZnO nanoparticles through a non-radiative energy transfer pathway can be influenced by the surface defect states (mainly the oxygen vacancies) of the catalyst nanoparticles. These were modulated by applying a simple annealing in an oxygen-rich atmosphere. Figure 2.8(I) a shows a typical TEM image of the fairly monodispersed ZnO nanoparticles annealed at 250°C. As shown in Figure 2.8(I) b, the lattice spacing of 0.26 nm indicates the (002) plane of the wurtzite structure of ZnO nanoparticles.[64] The polycrystalline nature of the nanoparticles is confirmed by the corresponding selected area electron diffraction (SAED) pattern (see Figure 2.8(I) c). The mean size for the ZnO nanoparticles annealed at 250°C was obtained to be 5.5 nm. Maximum mean particle size of 5.9 nm was obtained in the case of the NPs annealed at 350°C. The room temperature PL spectra of annealed ZnO nanoparticles are shown in Figure 2.8(II) a. All the NP samples show a small UV emission at approx. 355 nm, which can be attributed to the near band-edge transitions in the ZnO nanoparticles, and a large and broad green–yellow emission centered at around 530 nm, which can be attributed to the oxygen vacancy defect states (mostly present at the surface of the NP) mainly due to the transition of a photoexcited electron from the conduction band of ZnO to a deep-level trap state ($VO^{++}$) as the origin of the green luminescence.[65,66] The significant spectral overlap between the emission spectrum of the donor ZnO and the absorption spectrum of the acceptor species BR is shown in Figure 2.8(II) b. The fluorescence decay kinetics of the as-synthesized and annealed ZnO nanoparticles in the presence and in the absence of BR were then studied by using the picosecond-resolved TCSPC technique. The emissions from ZnO nanoparticles were detected at 540 nm with a laser excitation wavelength of 375 nm. As shown in Figure 2.8(III) a-f, upon addition of BR into the system, a faster fluorescence decay component was observed, attributed to the energy transfer process between ZnO and BR via FRET.

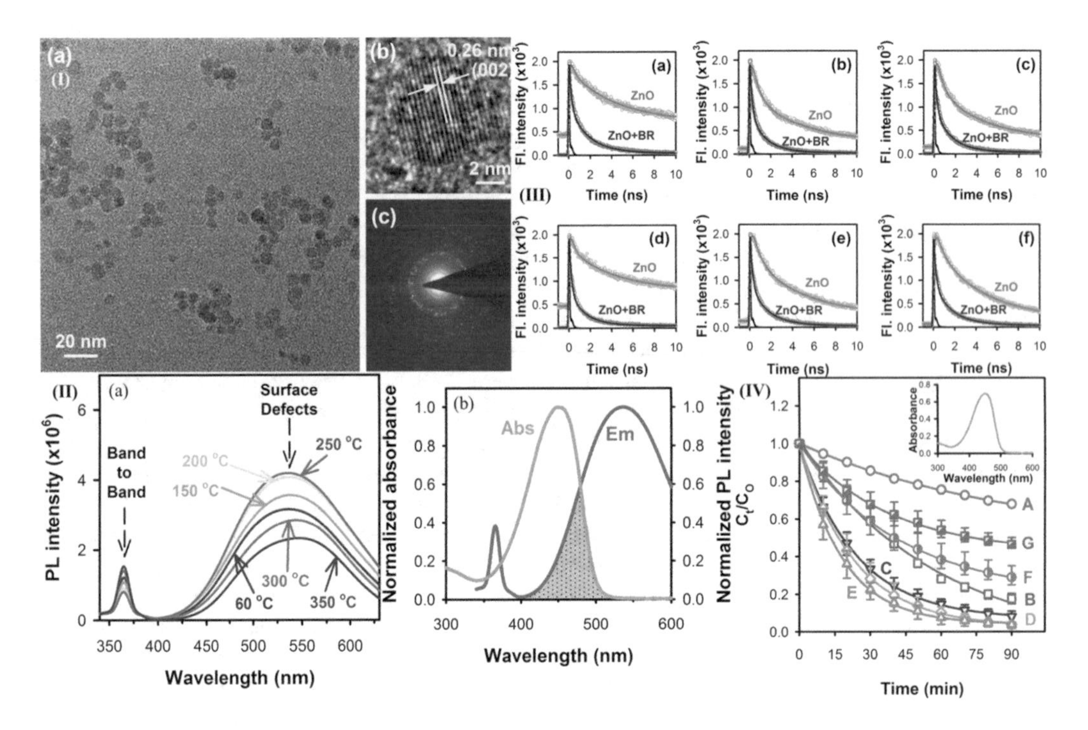


**FIGURE 2.8 (See color insert.)** (I) (a) Transmission electron micrograph, (b) HRTEM image of a single ZnO nanoparticle and (c) SAED pattern of the ZnO nanoparticles annealed at 250°C. (II) (a) Room temperature photoluminescence (PL) spectra of the ZnO nanoparticles annealed at various temperatures in air for 1 h (excitation wavelength: 320 nm) (b) Spectral overlap (shaded area) between the defect-mediated ZnO nanoparticle emission and the BR absorption. (III) The picosecond-resolved fluorescence transients of (a) as-synthesized (hydrolyzed at 60°C), (b) 150°C, (c) 200°C, (d) 250°C, (e) 300°C and (f) 350°C annealed ZnO nanoparticles in the presence (blue) and in the absence (red) of BR. The fluorescence decay was monitored at 540 nm with an excitation wavelength of 375 nm. (IV) Relative concentration ($C_t/C_0$) of BR with varying UV irradiation time during photocatalytic degradation (monitored for peak absorbance at 450 nm) (A) in the absence and in the presence of (B) as-synthesized particles, (C) annealed at 150°C, (D) annealed at 200°C, (E) annealed at 250°C, (F) annealed at 300°C and (G) annealed at 350°C particles. The inset shows the UV/Vis optical absorption spectrum of BR solution. (Reproduced with permission from Bora et al.[62])

After evaluating, the mechanism of the energy transfer process between the ZnO nanoparticles and the BR molecules adsorbed at the surface using time-resolved fluorescence spectroscopy, the photocatalytic degradation of BR has been obtained (see Figure 2.8(IV)). Upon inclusion of the as-synthesized ZnO nanoparticles (hydrolyzed at ≈60°C) in BR, almost 50% degradation was observed to occur within 40 min of UV irradiation, leading to an approximately 70% faster photocatalytic reduction of BR compared to the degradation of BR in 40 minutes when no catalyst was used. The reduction in the BR concentration increased further when annealed ZnO nanoparticles (up to 250°C) were used as the photocatalysts because of their higher concentrations of surface defects. The rates of the photocatalytic degradation of BR were found to follow a first-order exponential equation with a maximum photocatalytic activity for the ZnO nanoparticles annealed at 250°C. However, when the surface defects were reduced by annealing the ZnO nanoparticles at temperatures above 250°C, a significant drop in the catalytic activity was observed, which suggests the vital role of the surface defects in the photocatalytic degradation of BR. It should be noted that the FRET process between the ZnO nanoparticles and the BR molecules does not interfere with the normal phototherapy process of BR under UV irradiation. Hence, the photoproducts formed after the photocatalysis of BR in the presence of ZnO nanoparticles should mainly contain the structural (Z-lumirubin) and configurational ((Z,E)-BR) isomers of water-insoluble BR, which are the usual photoproducts of the BR phototherapy. In addition, the presence of methylvinylmaleimide (MVM) as an outcome of the photocatalytic degradation of BR in the presence of ZnO nanoparticles has also been evidenced in our previous study. A correlation of photocatalytic degradation and time-correlated single photon counting studies revealed that the defect-engineered ZnO nanoparticles that were obtained through post-annealing treatments led to an efficient decomposition of BR molecules that was enabled by FRET.

### 2.5.3 Direct Observation of Key Photoinduced Dynamics in a Potential Nano-Delivery Vehicle of Cancer Drugs

In recent times, significant advances in the use of zinc oxide (ZnO) NPs as delivery vehicles for cancer drugs have been achieved.[67,68] The current section will explore the key photoinduced dynamics in ZnO NPs upon complexation with a model cancer drug protoporphyrin IX (PP). A typical HRTEM image of ZnO NPs is shown in Figure 2.9(I) a. The lattice fringe of the ZnO NP shows an interplanar distance of 0.26 nm, corresponding to the spacing between two (002) planes. The average particle size is estimated by fitting our experimental TEM data over 60 particles and it is found to be ~5 nm. Nanohybrids have been characterized by FTIR, Raman scattering and UV–vis absorption spectroscopy. As characterized by UV–vis spectroscopy, the Soret band peak of the drug PP resides at 405 nm while the Q-band peaks are observed in the range between 500 nm and 650 nm. The FTIR technique is used to investigate the binding mode of the carboxylate group of PP on the ZnO surface as the attachment is very crucial for the precise and safe delivery of the drug. For free PP, the stretching frequencies of the carboxylic group are at 1695 $cm^{-1}$ and 1406 $cm^{-1}$ for antisymmetric and symmetric stretching vibrations, respectively, as shown in Figure 2.9(II) a. When PP is attached to ZnO, the stretching frequencies of the carboxylic group are located at 1604 $cm^{-1}$ and 1418 $cm^{-1}$ for antisymmetric and symmetric stretching vibrations, respectively. The shifting of the stretching frequencies clearly indicates the formation of a covalent bond between the drug PP and the carrier ZnO NPs. To further investigate the binding between the drug and delivery vehicle, Raman spectra were collected from PP, ZnO NPs and PP–ZnO nanohybrids as shown in Figure 2.9(II) b. The Raman spectrum of PP does not show any peak in the wavenumber range of 300–600 $cm^{-1}$. However, four vibration peaks at 328, 378, 438, and 577 $cm^{-1}$ are observed in the Raman spectrum of ZnO NPs, indicating the presence of a wurtzite structure.

After the binding of PP on the ZnO surface, the characteristic bands of ZnO are all present but slightly blue, shifted, and broadened, which is indicative of their good retention of the crystal structure and shape. The strong peak at 438 $cm^{-1}$ is assigned to the nonpolar optical phonon, E2 mode of the ZnO NPs at high frequency, which is associated with oxygen deficiency.

The room temperature PL spectrum of a ZnO NP is comprised of two emission bands upon excitation above the band-edge ($\lambda_{ex} = 300$ nm) as shown in Figure 2.9(III) a. The narrow UV band centered at 363 nm in the emission spectra of ZnO NPs is due to the band gap emission. The broad emission in the blue–green region is due to defect centers located near the surface. The broad emission is composed of two bands: one arises from the doubly charged vacancy center at 555 nm (P2) and the other arises from the singly charged vacancy center located at 500 nm (P1). The spectral overlap of the donor ZnO NP emission with that of the PP absorption is shown in Figure 2.9(III) b. The fluorescence decay profile of the donor ZnO NPs in the presence and absence of the acceptor PP was obtained upon excitation of a 375 nm laser and monitored at 500 nm (P1) and 555 nm (P2). The excited state lifetime of the ZnO NPs quenches in the PP–ZnO nanohybrid compared to that of bare ZnO NPs. Picosecond-resolved FRET from the defect-mediated emission of ZnO NPs to PP has been used to study the formation of the nanohybrid at the molecular level.

Picosecond-resolved fluorescence studies of PP–ZnO nanohybrids reveal efficient electron migration from photoexcited PP to ZnO, eventually enhancing reactive oxygen species (ROS) activity (see Figure 2.10 (I) a-b). The dichlorofluorescin (DCFH) oxidation[69] and no oxidation of luminol in PP/PP–ZnO nanohybrids upon green light illumination unravel that the nature of ROS is essentially singlet oxygen rather than superoxide anions. Direct evidence of the role of electron transfer as a key player in enhanced ROS generation from the nanohybrids. The nanohybrid is employed for model photodynamic therapy application in a light sensitized bacteriological culture experiment. The maximum inhibition of bacterial growth is obtained for PP–ZnO treated samples where a 65% decrease in CFU is observed after photodynamic treatment. The results clearly indicate the enhanced ROS generation in the presence of PP–ZnO nanohybrids compared to that of PP only as the presence of ZnO NPs in the proximity of PP drugs facilitates the charge separation which is evident from our picosecond-resolved fluorescence studies.

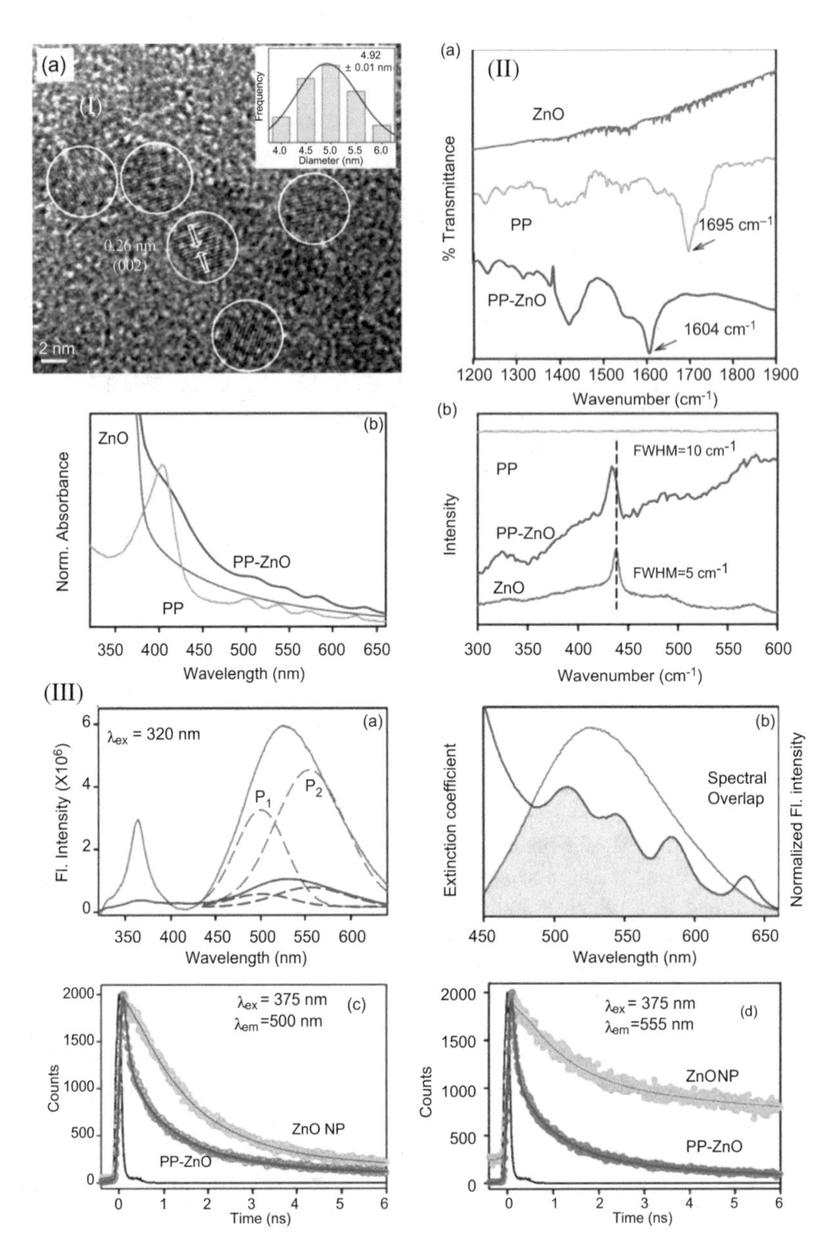


**FIGURE 2.9** **(See color insert.)** (I) (a) HRTEM images of ZnO NPs. The inset shows the size distribution of the ZnO NPs. (b) UV-Vis absorption of ZnO NPs (green), PP (red) and PP–ZnO (blue) in the DMSO-ethanol mixture. (II) (a) FTIR and (b) Raman spectra of PP (red), ZnO NPs (green) and PP–ZnO composites (blue). (III) (a) Room temperature PL spectra of ZnO NPs (green) and PP–ZnO composites (blue) are shown. The excitation wavelength was at 300 nm. The broad emission band is composed of two components, P1 (500 nm) and P2 (555 nm). (b) Shows the overlap of the ZnO NP emission and PP absorption. The picosecond-resolved fluorescence transients of ZnO NPs (excitation at 375 nm) in the absence (green) and in the presence of PP (blue) collected at (c) 500 nm and (d) 555 nm are shown. (Reproduced with permission from Sardar et al.[67])

### 2.5.4 Photoinduced Dynamics and Toxicity of a Cancer Drug in Proximity of Inorganic Nanoparticles under Visible Light

Rose bengal (RB; 4,5,6,7-tetrachloro-20,40,50,70-tetraiodofluorescein disodium) is a water-soluble, anionic, xanthene photosensitizer, which generates singlet oxygen ($^1O_2$) from oxygen molecules ($O_2$) when irradiated with green light.[70] RB is considered as a propitious sensitizer in PDT of tumors with minimal

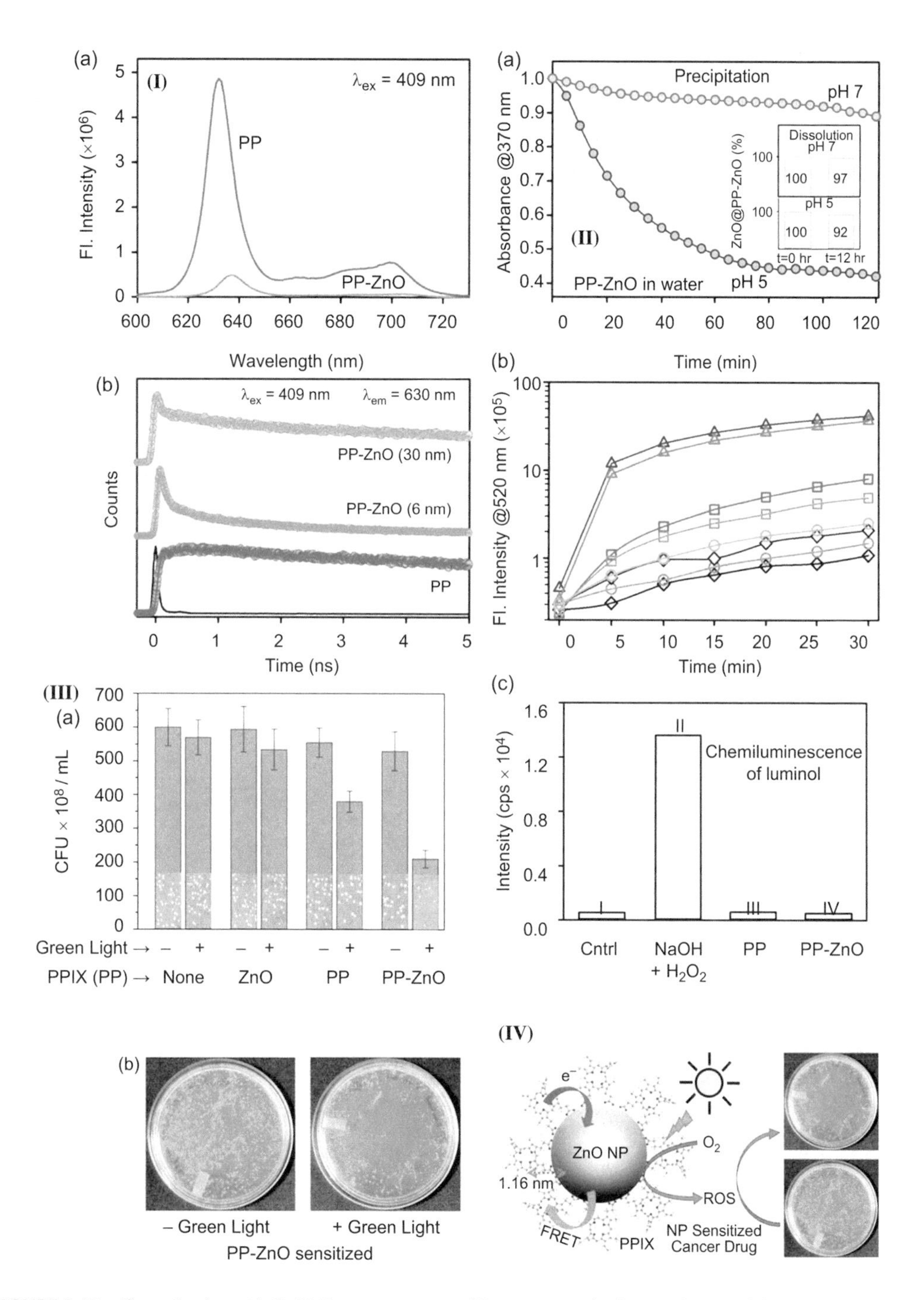


**FIGURE 2.10** **(See color insert.)** (I) (a) Room-temperature PL spectra (excitation at 409 nm) of free PP (red) and PP–ZnO (green). (b) Fluorescence decay profiles of PP (red), PP–ZnO (5 nm) (blue) and PP–ZnO (30 nm) (cyan) at 630 nm (excitation at 409 nm). (II) (a) Stability of the nanohybrid from absorbance at 370 nm of PP– ZnO dispersed in water at pH 7 and pH 5. Inset shows the dissolution of PP–ZnO nanohybrids in water at pH 5 and pH 7 after 12 hours. (b) The DCFH oxidation with time in the presence of PP–ZnO (blue at pH 5 and pink at pH 7), PP (red at pH 5 and dark yellow at pH 7), ZnO (green at pH 5 and orange at pH 7) and DCFH only (black at pH 5 and dark red at pH 7) under green light irradiation. The excitation was at 488 nm. (c) Chemiluminescence of luminol after green light irradiation for 15 minutes in the presence of (I) control, (II) $NaOH+H_2O_2$, (III) PP and (IV) PP–ZnO. (III) (a) Antibacterial activity of PP–ZnO, PP and ZnO in the presence and absence of green light. (b) Images of E. coli. in PP–ZnO sensitized plates before and after green light irradiation. (IV) Schematic representation of the entire process. (Reproduced with permission from Sardar et al.[67])

side effects.[71] There are considerable reports in the literature that describe the use of RB as a photodynamic sensitizer for cancer chemotherapy.[72] Constructive photosensitization predominantly depends on the physical and chemical characteristics of the PS, such as chemical purity, charge, solubility, distinct localization in tumor cells, sufficiently long residence time, and minimal time interval between the drug administration and its accumulation in neoplastic cells.[73] The intracellular localization and uptake of a photosensitizer in cells are vital to the photodynamic process as the photoinduced cellular damage occurs proximally with the oxidizing species formed by the excited molecules.[74] The present report illustrates the sensitization of RB with ZnO NPs leading to an increase in the photodynamic activity of the drug.[75] Herein, in the present work, we synthesized nanohybrids of RB with ZnO NPs of approximately 24 nm in size. Figure 2.11(I)a shows the sizes of the NPs and crystallinity confirmed using high-resolution transmission electron microscopy (HRTEM). The absorption spectra suggest the presence of two distinct peaks in the nanohybrids – one for RB, another for ZnO NPs. The thermo-gravimetric analysis shows better thermal stability of RB in nanohybrids compared to the bare dye (see Figure 2.11(I)c).

Picosecond-resolved fluorescence experiments on the nanohybrids were performed to understand the efficient electron transfer from photoexcited RB to ZnO NPs (see Figure 2.11(II)b), which eventually upgrades the ROS activity in the RB–ZnO nanohybrids. Picosecond-resolved FRET from ZnO NPs to RB was used

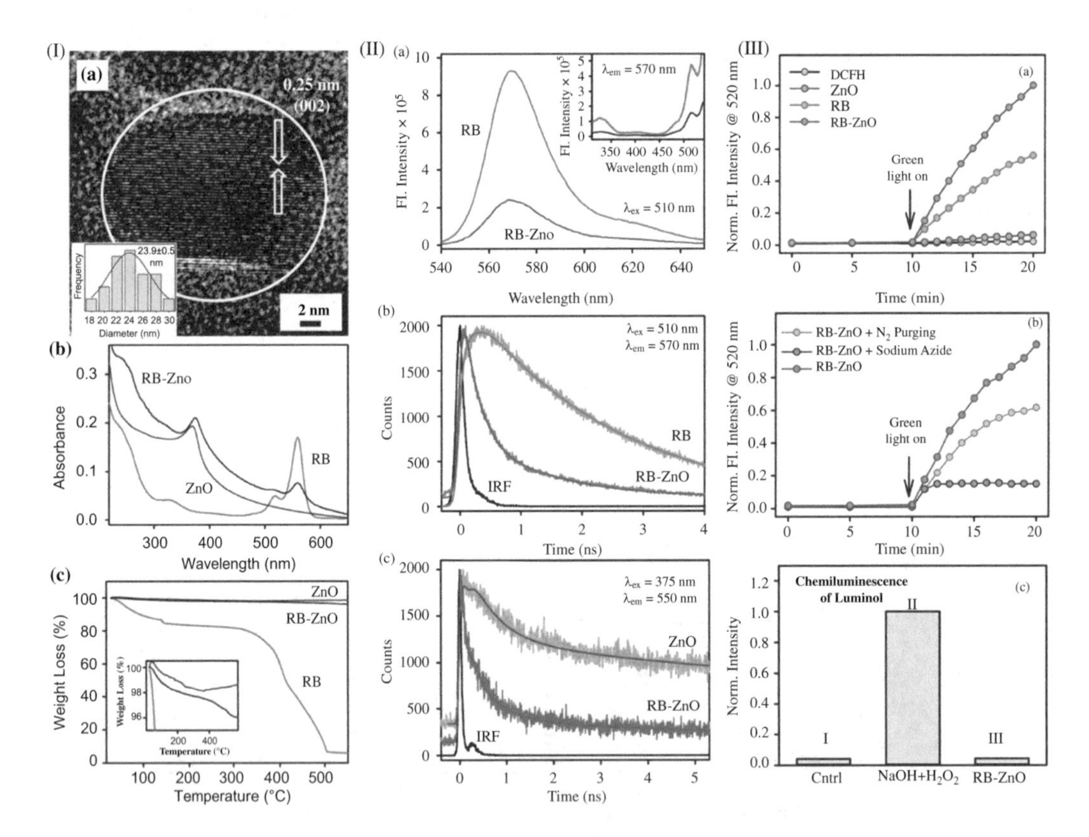


**FIGURE 2.11 (See color insert.)** I (a) HRTEM image of ZnO NPs. Inset: The size distribution of ZnO NPs. (b) Absorption spectra of RB–ZnO (blue), RB (red), and ZnO (green). (c) TGA profile of RB–ZnO (blue), RB (red), and ZnO (green). II (a) Room temperature PL spectra of RB (red) and RB–ZnO (30 nm; blue) upon excitation at 510 nm. Inset: The excitation spectra of RB (red) and RB–ZnO (30 nm; blue) at detection wavelength 570 nm. (b) Fluorescence decay profiles of RB (red) and RB–ZnO (30 nm; blue) upon excitation at 510 nm and detection wavelength at 570 nm. (c) Fluorescence decay profiles of ZnO (5 nm; green) and RB–ZnO (5 nm; blue) upon excitation at 375 nm and at a detection wavelength at 550 nm. III a) DCFH oxidation with respect to time with the addition of RB–ZnO (blue), RB (red), ZnO (green), and control DCFH (black) under dark with subsequent green-light irradiation. b) DCFH oxidation with respect to time with RB–ZnO addition in an atmosphere of purged nitrogen (pink), sodium azide (green), and a control (blue) under dark with subsequent green-light irradiation c) Chemiluminescence of luminol prior to green-light illumination for 15 min for the control (I), NaOH+H2O2 (II) and RB–ZnO (III). (Reproduced with permission from Chaudhuri et al.,[70] © 2015, John Wiley and Sons.)

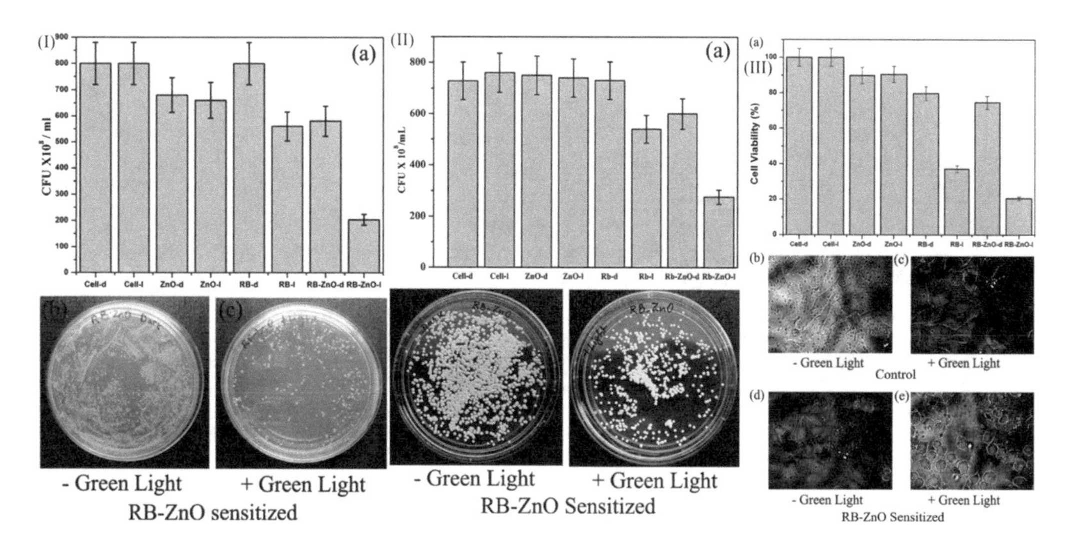


**FIGURE 2.12** **(See color insert.)** (I) (a) Antibacterial activity of RB–ZnO, RB, and ZnO in the presence (with suffix -l) and absence (with suffix -d) of green light. Images of RB–ZnO treated *E. coli* plates in the absence (b) and presence (c) of green light. (II) a) Antifungal activity of RB–ZnO, RB, and ZnO in the presence (with suffix -l) and absence (with suffix -d) of green light. Images of RB–ZnO treated *C. albicans* plates in the absence (b) and presence (c) of green light. (III) (a) *In vitro* cytotoxicity assay in HeLa cells with RB–ZnO, RB, and ZnO with MTT as an indicator dye in the presence (with suffix -l) and absence (with suffix -d) of green light. Images of control and RB–ZnO treated HeLa cells in the absence (b, d) and the presence (c, e) of green light. (Reproduced with permission from Chaudhuri et al.,[70] © 2015, John Wiley and Sons.)

to understand the nanohybrids formation at the molecular level (see Figure 2.11(II)c). The ROS formation was monitored by DCFH oxidation. Enhanced ROS generation was observed in the presence of the RB–ZnO nanohybrids compared with that of free RB upon green light illumination (see Figure 2.11(III)). The nanohybrid was used as a model photodynamic therapeutic agent in bacterial (see Figure 2.12(I)), fungal (see Figure 2.12(II)), and HeLa cell lines (see Figure 2.12(III)). The enhanced ROS generation, as a result of charge separation of the drug in the proximity of semiconductor NPs, is responsible for the augmentation of the drug activity. The present study will therefore be helpful for the design efficient photodynamic drugs.

## 2.6 Dynamical Evaluation of Surface Modulated Spinel Metal Oxide Nanoparticles: Role of Ligand Functionalization

### 2.6.1 A Novel Nanohybrid for Cancer Theranostics: Folate-Sensitized $Fe_2O_3$ Nanoparticles

A novel nanohybrid has been synthesized and thoroughly characterized for both the diagnosis and therapy of colorectal cancer.[76] A facile and cost-effective synthesis of folic acid (FA)[77]-templated $Fe_2O_3$ nanoparticles has been established using a hydrothermal method. Figure 2.13(I)a and b depict the field emission scanning electron microscopy (FESEM) image of the as-prepared $Fe_2O_3$ and FA–$Fe_2O_3$, respectively. $Fe_2O_3$ has a branched network-like morphology having a segment length of 300–500 nm and width of 50–70 nm. FA–$Fe_2O_3$ has a nano-egg-like morphology of length 150–200 nm and width 60–110 nm. Figure 2.13(I)c,d represent the corresponding HRTEM images of $Fe_2O_3$ and FA–$Fe_2O_3$, while insets show the corresponding HRTEM images at higher magnification. For both samples, the fringe distance was found to be 2.52 Å which corresponds to the spacing between the (110) planes of a-Fe2O3.[78] The powder XRD pattern of FA–$Fe_2O_3$ and $Fe_2O_3$ is shown in Figure 2.13(I)e. The XRD pattern and fringe distance from the HRTEM image imply that the crystal structure of α-$Fe_2O_3$ remains intact after FA template synthesis. The FTIR spectra of FA, $Fe_2O_3$, and FA–$Fe_2O_3$ are shown in Figure 2.13(II)a. Two sharp peaks at 472 $cm^{-1}$ and 550 $cm^{-1}$ are the characteristics of Fe–O stretching in

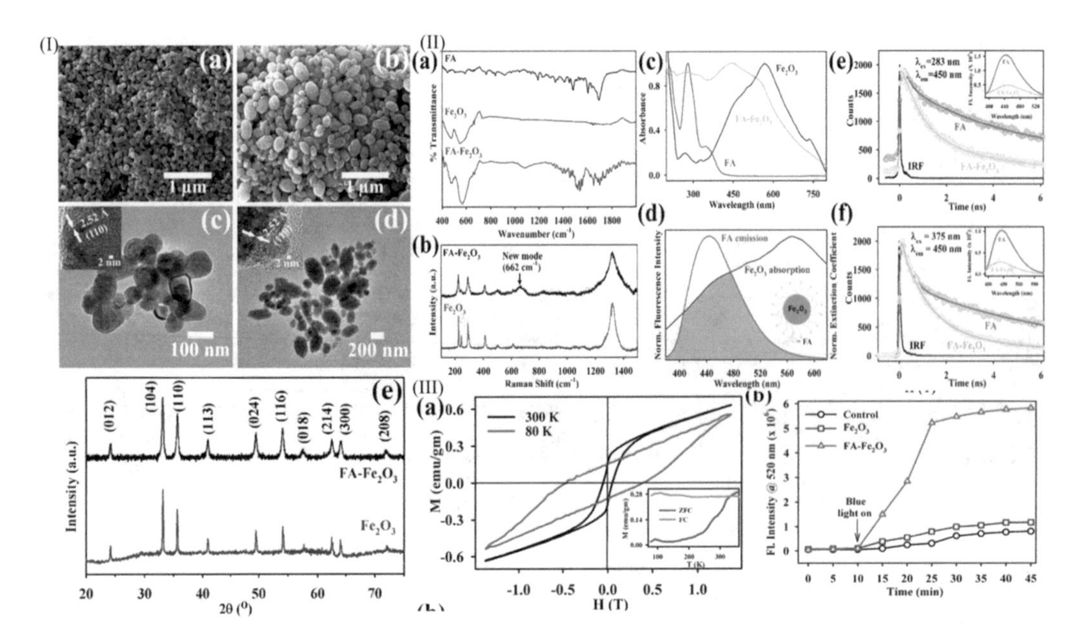


**FIGURE 2.13 (See color insert.)** (I) FESEM image of (a) $Fe_2O_3$ and (b) FA– $Fe_2O_3$ NPs. HRTEM image of (c) $Fe_2O_3$ and (d) FA–$Fe_2O_3$ NPs. Inset of (c) and (d) shows the corresponding HRTEM images at higher magnification. (g) XRD pattern of $Fe_2O_3$ and FA– $Fe_2O_3$ NPs. (II) (a) FTIR spectra of FA, $Fe_2O_3$, and FA–$Fe_2O_3$. (b) Raman spectra of FA–$Fe_2O_3$ and $Fe_2O_3$. (c) Absorption spectra of FA, $Fe_2O_3$, and FA–$Fe_2O_3$. (d) Spectral overlap between the emission of FA and absorption of $Fe_2O_3$. (e) and (f) are the picosecond-resolved fluorescence transient spectra of FA in the absence and presence of $Fe_2O_3$, recorded at 450 nm upon excitation at 283 nm and 375 nm, respectively. Insets show the corresponding steady-state emission spectra. (III) (a) Field-dependent magnetization (M vs. H) of FA– $Fe_2O_3$ at 300 K and at 80 K. Bottom right inset shows temperature dependence of $M_{ZFC}$ and $M_{FC}$ curves measured at H = 1.6 T. (b) DCFH oxidation with respect to time in addition of $Fe_2O_3$, FA–$Fe_2O_3$ and control under dark with subsequent blue light irradiation. (Reproduced with permission from Nandi et al.[76])

the $Fe_2O_3$ spectrum. The lower frequency absorption remains unchanged but the higher one gets shifted to a higher frequency of 570 $cm^{-1}$ in FA–$Fe_2O_3$, which indicates a change in size during FA templated synthesis of $Fe_2O_3$. The sharp peak at 1699 $cm^{-1}$ corresponds to the C=O stretching of the carboxylic acid group of FA. In FA–$Fe_2O_3$, the C=O stretching perturbed into two peaks may be due to one of the free carboxylic acids of FA in FA–$Fe_2O_3$ and another one is covalently attached to $Fe_2O_3$. The differences in UV–vis absorption spectra of $Fe_2O_3$ and FA– $Fe_2O_3$ are clearly depicted in Figure 2.13(II)c. FA has two well-known absorption bands at 360 and 280 nm. Three possible electronic transitions are observed in the case of the $Fe_2O_3$ absorption spectra, absorption arising in the range of 200–400 nm is mainly due to ligand-to-metal charge transfer transition (LMCT) from oxygen to iron, the absorption band at 445 nm is attributed mainly to pair excitation. In the case of FA–$Fe_2O_3$ the LMCT band is in the wavelength range 200–400 nm and the absorption band at 445 nm gets triggered; surprisingly the other one at 568 nm become sedate due to the FA functionalization of $Fe_2O_3$. The significant spectral overlap of the emission of FA with the absorption of $Fe_2O_3$ is shown in Figure 2.13(II)d. Steady-state fluorescence quenching of FA was observed in the presence of $Fe_2O_3$ upon excitation at 283 nm and 375 nm as shown in the insets of Figure 2.13(II)e and f, respectively.[76]

Picosecond-resolved fluorescent transients were recorded for both FA and FA–$Fe_2O_3$ at 450 nm upon excitation at 283 nm and 375 nm as shown in Figure 2.13(II)e and f, respectively, and the well-known FRET strategy was used to confirm the molecular level attachment of FA with $Fe_2O_3$. The fluorescence lifetime of FA is found to be quenched in the presence of $Fe_2O_3$ NPs, which suggest a non-radiative energy transfer process from the donor (FA) to the acceptor ($Fe_2O_3$). The donor–acceptor distances ($r_{DA}$) have been calculated to be 1.11 nm and 1.13 nm for the excitation at 283 nm and 375 nm, respectively. Measurements of magnetic properties of FA–$Fe_2O_3$ using a VSM display magnetic field dependent

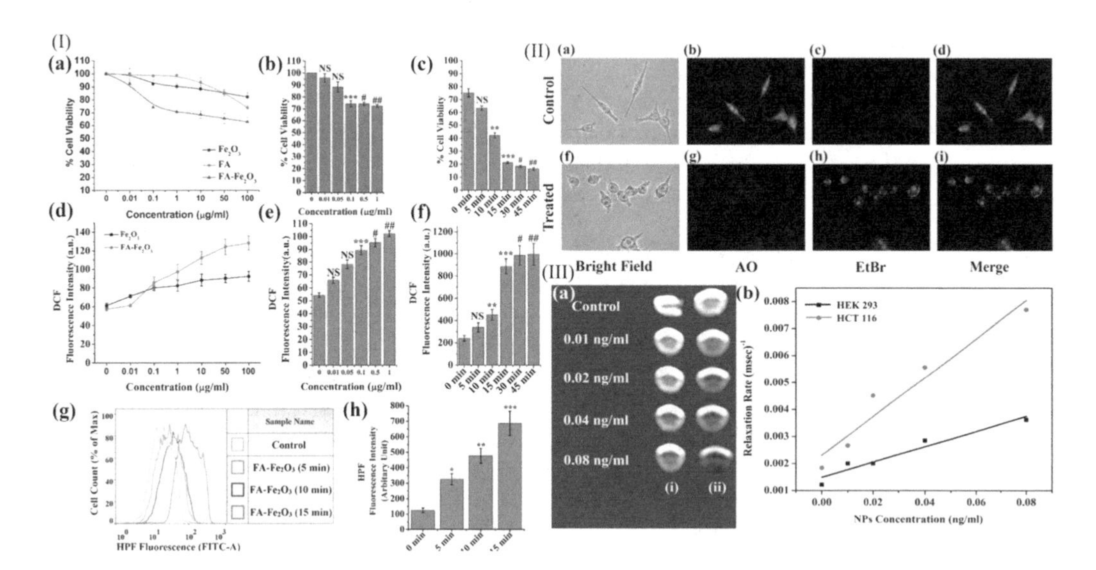


**FIGURE 2.14 (See color insert.)** (I) HCT 116 cells were treated with the requisite amount of drugs for 24 h prior to various experiments. (a) MTT assay quantified cell viability with different concentrations of FA, $Fe_2O_3$ and FA–$Fe_2O_3$ in the absence of blue light. (b) The same at different concentrations of only FA– $Fe_2O_3$. (c) Light-induced cytotoxicity after treatment with 0.1 mg $ml^{-1}$ of FA– $Fe_2O_3$ followed by blue light irradiation for different time durations (0–45 min). (d) $Fe_2O_3$ and FA– $Fe_2O_3$ treated dose-dependent (0–100 mg $ml^{-1}$) intracellular ROS level in terms of DCF fluorescence intensity without blue light treatment. (e) FA– $Fe_2O_3$ treated dose-dependent (0–1 mg $ml^{-1}$) intracellular ROS level in terms of DCF fluorescence intensity without blue light treatment. (f) Intracellular ROS measurement after treatment with 0.1 mg $ml^{-1}$ FA– $Fe_2O_3$ followed by time-dependent (0–45 min) blue light exposure. (g) Intracellular hydroxyl radical (OH) determination by HPF staining after treatment with FA– $Fe_2O_3$ followed by time-dependent (0–15 min) blue light exposure, using flow cytometer. (h) Flow Jo analyzed data of hydroxyl radical (OH) determination. (II) Morphological changes of FA–$Fe_2O_3$ (0.1 mg ml1 for 24 h) treated HCT 116 cells followed by blue light exposure (15 min). (a) and (f) control and treated cells under bright field. (b) and (g) AO stained control and treated cells. (c) and (h) EtBr staining of control and treated cells. (d) and (i) Merged image of AO and EtBr stained cells. (III) (a) T2 weighted MRI phantom images of (i) HEK 293 and (ii) HCT 116 cells treated with different concentrations of FA– $Fe_2O_3$. (b) Relaxivity study of FA–$Fe_2O_3$ incubated in HEK 293 and HCT 116 cell lines. (Reproduced with permission from Nandi et al.[76])

magnetization (M vs. H) at 80 K and 300 K, and also hysteresis features with coercivity 0.43 T and 0.053 T, respectively (see Figure 2.13(III)a). Blue light induced ROS generation (Figure 2.13(III)b) is attributed mainly to the absorption band of $Fe_2O_3$ at 445 nm, which gets triggered after folic acid functionalization in FA–$Fe_2O_3$. The nanohybrid (FA–$Fe_2O_3$) is a combination of two nontoxic ingredients, FA and $Fe_2O_3$, showing remarkable PDT activity in human colorectal carcinoma cell lines (HCT 116)[79] via the generation of intracellular ROS. The light-induced enhanced ROS activity of the nanohybrid causes significant nuclear DNA damage, as confirmed by the comet assay. Assessment of p53, Bax, Bcl2, cytochrome c (cyt c) protein expression and caspase 9/3 activity provides vivid evidence for cell death via an apoptotic pathway (Figure 2.14(I)). *In vitro* magnetic resonance imaging (MRI) experiments in folate receptor (FR)-overexpressed cancer cells (HCT 116) and FR-deficient human embryonic kidney cells (HEK 293) reveal the target specificity of the nanohybrid toward cancer cells, and are thus pronounced MRI contrasting agents for the diagnosis of colorectal cancer (Figure 2.14(III)).

### 2.6.2 Interfacial Dynamics at the Surface Modulated Spinel Metal Oxide Nanoparticles

#### *2.6.2.1 Rational Surface Modification of $Mn_3O_4$ Nanoparticles to Induce Multiple Photoluminescence and Room Temperature Ferromagnetism*

The size reduction or facet-specific reactivity of nanocrystals can further be modulated by employing surface chemistry based approaches.[80,81] The effects of capping ligands to regulate the surface properties of $Mn_3O_4$ NPs followed by the subsequent appearance of novel optical/magnetic properties have recently studied thoroughly by our group. Besides being of fundamental scientific interest, such an understanding is important for optimizing NP properties. So, it would be of great interest to develop approaches to control the surface chemistry to gain a better understanding on the origin of surface-induced optical and magnetic properties, as subtle differences in ligand functional groups or the structural position of the functional groups can dramatically change the optical or magnetic responses.[82]

A series of surface modification studies carried out by systematic variation of the nature of surface-protecting ligands to define how NP–ligand interactions modify the electronic properties of the NPs that ultimately govern multiple photoluminescence states.[80] The $Mn_3O_4$ NPs synthesized using the ultrasonic-assisted method revealed no characteristic absorption signature, whereas upon ligand functionalization, $Mn_3O_4$ NPs exhibit distinct features depending upon the type of ligand functional groups used. The absorption spectra of functionalized $Mn_3O_4$ NPs where functional groups of the ligands are chosen as –OH (the hydroxyl group of glycerol), –OH and $-NH_2$ (the hydroxyl and amine groups of ethanol amine), $-NH_2$ (the amine group of guanidine), –COO (the carboxylate group of succinate), –COO and $-NH_2$ (the carboxylate and amine groups of glycine), and –COO and –SH (the carboxylate and thiol groups of thioglycolate) are provided in Figure 2.15. In all cases, a characteristic absorption band between 300 and 360 nm has been observed which can be depicted as high energy LMCT band originated by the interaction between the ligand functional groups and the $Mn^{2+}/Mn^{3+}$ ions present on the surface of the nanoparticles. The presence of an α-hydroxy carboxylate moiety in the capping ligand appears to be necessary to activate the Jahn–Teller (J–T) splitting of $Mn^{3+}$ ions in the NPs which corresponds to the d–d transitions and is subsequently the deciding factor for inducing diverse optical responses.

The tartrate-modified $Mn_3O_4$ (T–$Mn_3O_4$) exhibits optimal optical characteristics and thus have been chosen to further study the surface chemistry. The water-soluble spherical T–$Mn_3O_4$ NPs of an average size of 3 nm exhibit two absorption peaks at 315 and 430 nm, a shoulder descending into lower energies around 565 nm and a broad band at 752 nm (Figure 2.15i). The observed peak at 315 nm could be assigned to the possible high energy LMCT processes involving tartrate–$Mn^{2+}/Mn^{3+}$ interactions. The other bands at 430, 565, and 752 nm are attributed to d–d transitions of $Mn^{3+}$ due to J–T effect in a high-spin octahedral environment and the observed bands correspond to the transitions of $^5B_{1g}$ to $^5E_g$, $^5B_{1g}$ to $^5B_{2g}$, and $^5B_{1g}$ to $^5A_{1g}$, respectively. The synthesized T–$Mn_3O_4$ shows high fluorescence from both the LMCT and d–d transition band in alkaline conditions where the absorption peak at 430 nm and the lower energy shoulder at 565 nm (both originate due to d–d transitions involving $Mn^{3+}$) are significantly perturbed and blue-shifted to 385 and 440 nm, respectively, although the LMCT band at 315 nm and another d–d band at 758 nm remain almost unaffected. Multiple PL of T–$Mn_3O_4$ NPs starting from blue, cyan, green, to the near-infrared region (PL maximum at 417, 473, 515, and 834 nm) of the spectra against excitation at four different wavelengths (315, 370, 440, and 760 nm, respectively) originated predominantly from the LMCT excited states and ligand field excited states of the metal d orbitals. In order to confirm the origin of multiple PL states, picosecond-resolved fluorescence decay transients of T–Mn3O4 NPs have been collected at three different peak maxima of 410, 470, and 515 nm using three different excitation sources of 293, 375, and 445 nm wavelengths, respectively. The observed differences in the excited-state lifetime of T–$Mn_3O_4$ NPs at 410 nm PL compared to the lifetimes at 470 and 515 nm PL suggest the difference in the origin of the PL. The obtained difference in lifetime values of 470 nm PL upon excitation using 293 (4.04 ns) and 375 (1.13 ns) nm sources clearly distinguish between the origin of the two excitations suggesting that the LMCT excited states are responsible for the PL at 417 nm, whereas the J–T excited states lead to the PL maxima at 470, 515, and 834 nm.

The effect of surface functionalization on the magnetic behavior of $Mn_3O_4$ nanodots was recently studied and it was shown that depending on the ligand structure the magnetic behavior varies to a great extent.[83] The paramagnetic $Mn_3O_4$ nanocrystals demonstrate a distinctly different magnetization response at 300 K upon ligand functionalization revealing a different degree of ferromagnetism.[84] While the room temperature ferromagnetism can be activated by functionalization with glycerol and guanidine, it can be further enhanced by succinate and tartrate and could be correlated with the crystal field splitting energy (CFSE) of $Mn^{3+}$ ions upon interaction with the ligand fields and also on the field strength of the functional ligands (see Figure 2.16e). According to ligand field theory, transition metal

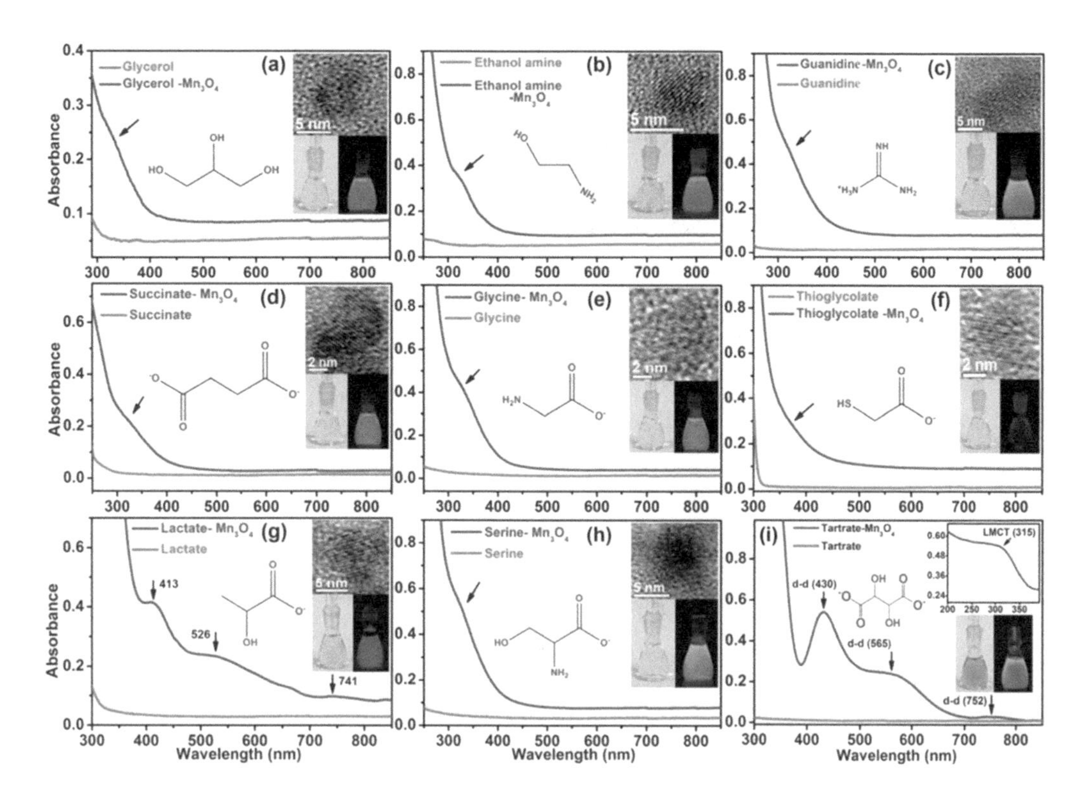


**FIGURE 2.15** **(See color insert.)** (a)–(i) represents the UV–vis absorption spectra of ligand functionalized $Mn_3O_4$ NPs in aqueous solution at pH~7. Different combinations of ligand functional groups have been employed in order to activate the J–T splitting of $Mn^{3+}$ ions in the NPs surface and to bring out optimal optical responses from the functionalized NPs. (a) –OH (hydroxyl) group of glycerol (b) –OH & $–NH_2$ (hydroxyl and amine) groups of ethanol amine, (c) $–NH_2$ group of guanidine, (d) –COO– (carboxylate) group of succinate, (e) –COO– and $–NH_2$ groups of glycine, (f) –COO– and –SH (carboxylate and thiol) groups of thioglycolate, (g) –COO– and –OH (at α position) groups of lactate, (h) –COO– and –OH (at β position) groups of serine and (i) –COO– and –OH (two α hydroxyl groups) groups of tartrate have been used respectively, to functionalize the as-prepared $Mn_3O_4$ NPs. Upper inset of Figure (a)–(h) show the corresponding HRTEM image of various ligand functionalized $Mn_3O_4$NPs. Photographs under visible (left) and UV light (right) of various ligand functionalized $Mn_3O_4$ NPs have been shown in the lower inset. (Reproduced with permission from Giri et al.[80])

ions having a larger d orbital splitting energy due to ligand coordination should have a smaller spin–orbit coupling. Any decrease in the spin–orbit coupling of surface magnetic cations results in a smaller surface magnetic anisotropy and subsequently, the coercivity of the NPs will be reduced. It is also well known that σ donor ligands result in larger CFSEs than π donors. Thus, tartrate functionalized NPs show higher coercivity due to the presence of both σ donors (–OH) and π donors (–COO-). The presence of $Mn^{3+}$ ions at the surface of the functionalized NPs and substantial interactions of carboxylate and hydroxyl groups with the nanoparticle surface (in the case of T–$Mn_3O_4$ NPs), suggest surface modification induced alteration of the $Mn^{3+}$ ($3d^4$)-ligand anion p orbital hybridization strength and the density of a midgap state with strong O 2p character have been correlated to the observed ferromagnetism.

The T–$Mn_3O_4$ NPs were finally evaluated for their photocatalytic properties by following the decomposition rate of MB, a commonly used textile dye. The photodegradation curve of MB in the presence of T–$Mn_3O_4$ NPs has been found to follow a first-order exponential equation with a total photodegradation of 48% within 40 minutes of UV irradiation (see Figure 2.16i). The addition of $H_2O_2$ into the system enhances the degradation rate significantly (85% decomposition in 40 min), which can be corroborated with the fact that $H_2O_2$ can act as the electron acceptor to reduce the recombination rate of a photo-generated electron–hole pair as well as a source of OH radicals. The small average diameter, favorable surface functionalization (by tartrate) for cationic MB attachment and various PL states are responsible for accelerating the photocatalytic activity of $Mn_3O_4$ NPs.

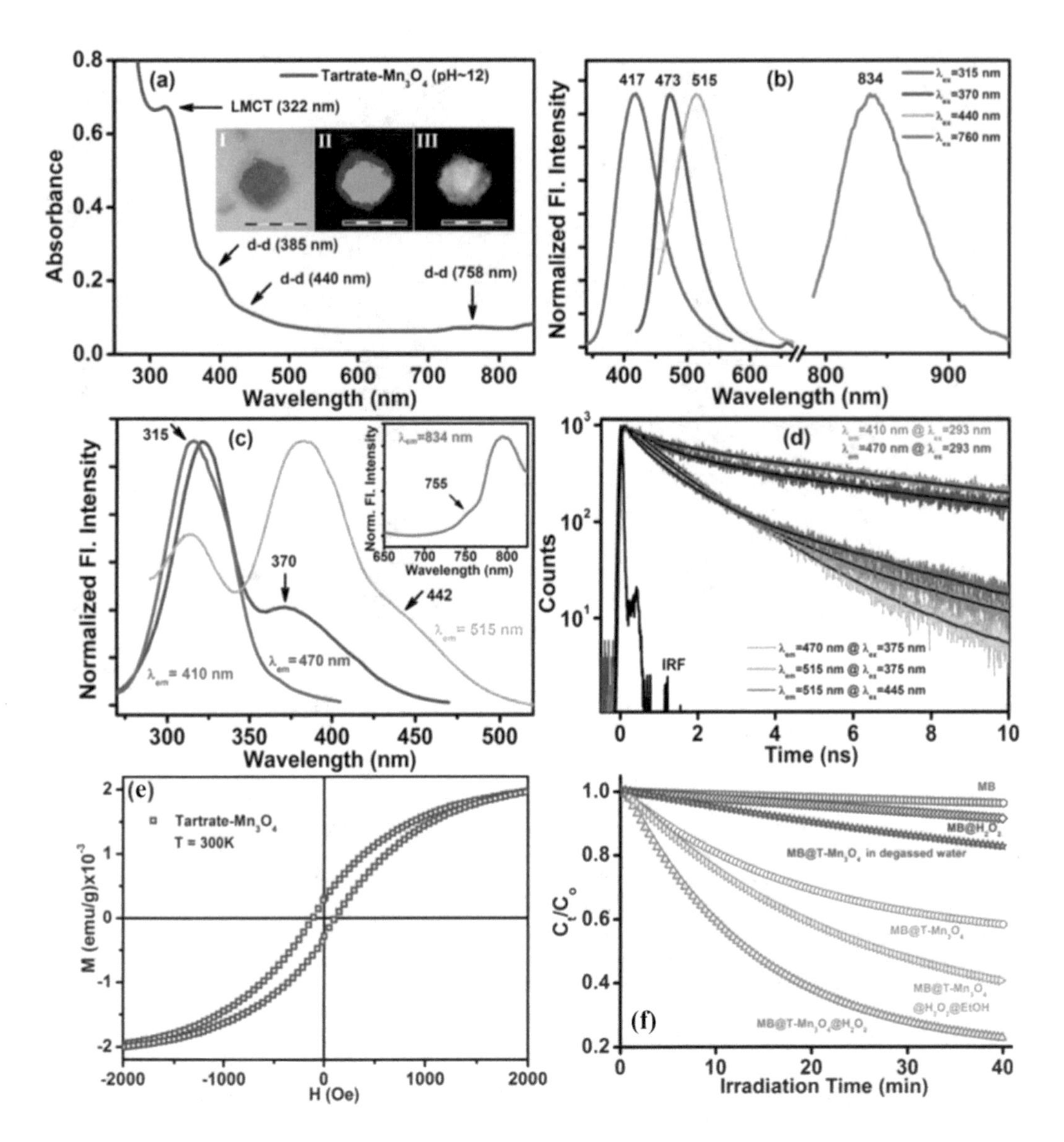


**FIGURE 2.16** **(See color insert.)** (a) UV–vis absorption spectrum of T–$Mn_3O_4$ NPs after treatment (at pH ~12 and 70°C for 12 hrs). Inset shows the fluorescence microscopic images of the same under irradiation of white light (bright field, I) and light of two different wavelengths 365 (II) and 436 (III) nm. Scale bars in the images are of 500 μm. (b) Normalized steady-state PL spectra collected from T–$Mn_3O_4$ NPs with four different excitation wavelengths of 315, 370, 440, and 760 nm at pH~12. (c) Excitation spectra of T–$Mn_3O_4$ NPs at different PL maxima of 410, 470, 515, and 834 nm. (d) Picosecond-resolved PL transients of T–$Mn_3O_4$ NPs in water measured at emission wavelengths of 410, 470, and 515 nm upon excitation with excitation source of 293, 375, and 445 nm wavelengths. (e) Field-dependent magnetization (M vs. H) at room temperature (300 K) of T–$Mn_3O_4$. The distinct hysteresis loops observed confirms ferromagnetic activation of the NPs upon functionalization with carboxylate ligands. (f) Plots of relative concentration ($C_t/C_o$) versus time for the photodegradation of MB (monitored at 660 nm) alone and in presence of T–$Mn_3O_4$ NPs, $H_2O_2$, T–$Mn_3O_4$@$H_2O_2$ and T–$Mn_3O_4$@$H_2O_2$@EtOH, are shown. (Reproduced with permission from Giri et al.[80])

#### *2.6.2.2 Unprecedented Catalytic Activity Owing to Interfacial Charge Transfer Dynamics of $Mn_3O_4$ Nanoparticles: Potential Lead of a Sustainable Therapeutic Agent for Hyperbilirubinemia*

To further utilize the enhanced surface reactivity of functionalized $Mn_3O_4$ NPs, carboxylate-rich biocompatible ligand sodium citrate was employed.[85,86] The citrate–$Mn_3O_4$4 NPs (C–$Mn_3O_4$) are found to be nearly spherical with an average diameter of 3 nm. The UV–vis absorption spectrum of C–$Mn_3O_4$ reveals a peak at around 290 nm (possible high energy LMCT involving citrate–$Mn^{4+}$ interaction) and the other bands at 430, 565, and 752 nm are attributed to the J–T distorted d–d transitions centered over $Mn^{3+}$ ions in C–$Mn_3O_4$ NPs.[85] The C–$Mn_3O_4$ NPs are found to be a highly efficient catalyst for the decomposition of aqueous BR solution in absence of any photoactivation due to its mixed valence states of Mn (+2, +3 and +4), along with the functional groups on the surface coordinating ligands. BR, an orange-yellow pigment produced from the breakdown of the heme group in the red blood cells leads to jaundice due to its increased concentration in blood. The catalytic decomposition of BR follows a first-order exponential rate equation and 92% BR is degraded in the presence of C–$Mn_3O_4$ without any photoactivation. The unprecedented catalytic activity could be recycled without any significant loss of activity (tested for 20 consecutive cycles). The mechanistic insight of the catalytic process hypothesized to be initialized by the conversion of $Mn^{3+}$ to $Mn^{4+}$ states at the NPs surface and the subsequent formation of ROSs (such as OH. radicals) ultimately leads to the decomposition of the analyte BR. The exceptional catalytic activity of C–$Mn_3O_4$ was tested as a symptomatic therapeutic agent for alternative rapid treatment of hyperbilirubinemia through direct removal of BR from blood in mice.64 The functionalized NPs were found to generate ROS for eight consecutive cycles and the ROS is responsible for the catalytic performance. The *in vitro* mimic of the blood system consists of HSA protein in the presence of NPs.[86] The BR decomposition activity of C–$Mn_3O_4$ was found to be decreased in the presence of HSA. The intraperitoneally injected and orally administered C–$Mn_3O_4$ NPs could degrade BR very fast and in a specific nontoxic way compared with conventional drugs in a mouse model, ensured by various biochemical tests and histopathological studies. Moreover, ultrahigh efficacy of orally administrated C–$Mn_3O_4$ NPs in the treatment of chronic liver diseases such as hepatic fibrosis and cirrhosis in mice[87] compared with conventional medicine silymarin was recently discovered, without any toxicological implications.

## 2.7 Conclusion

The interface between nanomaterials (organic or inorganic) and biological systems (essentially organic) comprises a dynamic series of molecular level interactions between nanomaterial surfaces and biological entities. These interactions are shaped by a large number of forces that could determine nanomaterial's colloidal stability, physicochemical behavior, photophysical properties, etc. The most relevant parameter which dictates the functionalities of nanohybrids is the interfacial dynamical process at the inorganic–organic junction. The works covered in this chapter demonstrate the impressive array of experimental techniques emphasizing the beauty of ultrafast spectroscopy to optimize the synthesis methodologies of various nanohybrids and their further surface modulation to impart novel optical and/or magnetic properties. Moreover, employing a multitude of spectroscopic tools, we have investigated the characteristic nature of the interfacial interaction present in nanohybrids. Finally, we have also discussed promising applications, particularly in the biological aspects of these developed nanohybrids. The applications include photodynamic therapy in various diseases and the eradication of jaundice. The correlation between the interfacial dynamics and improved functionalities in nanohybrids as indicated in this chapter are expected to be helpful in designing improved functional nanohybrids in the near future.

## Acknowledgments

D.B. thanks the Department of Science and Technology (DST, India) for the INSPIRE fellowship. We would like to thank the Department of Science & Technology (DST) (DST-SERB EMR/2016/004698), the Department of Biotechnology (DBT, India) (BT/PR11534/NNT/28/766/2014) and ICMR (5/3/8/247/2014ITR) for financial grants. We thank DST, India for financial grants DST-TM-SERI-FR-117. We would like to thank our colleagues and collaborators whose contributions over the years, acknowledged in the references, have been priceless in the successful evolution of the work in this area. In particular, we thank Dr. Anupam Giri, Dr. Nirmal Goswami, Dr. Samim Sardar, Dr. Siddhi Chaudhuri, Dr. Nabarun Polley, Dr. Susobhan Choudhury, Dr. Abhinandan Makhal, Dr. Soumik

Sarkar, Mr. Ramesh Nandi, Ms. Priya Singh. We thank Dr. Saleh A Ahmed, Dr. Indranil Banerjee, Dr. Partha Saha, Dr. Omar F. Mohammed, Dr. Krishna Das Saha, Prof. Peter Lemmens, and Prof. Joydeep Dutta for collaborative work.

## REFERENCES

1. C. R. Kagan, D. B. Mitzi, C. D. Dimitrakopoulos, *Science*, 1999, **286**, 945–947.
2. N. Aich, J. Plazas-Tuttle, J. R. Lead, N. B. Saleh, *Environmental Chemistry*, 2014, **11**, 609–623.
3. P. Gómez-Romero, C. Sanchez, *Functional Hybrid Materials*, John Wiley & Sons, Hoboken, NJ, 2006.
4. A. E. Nel, L. Mädler, D. Velegol, T. Xia, E. M. Hoek, P. Somasundaran, F. Klaessig, V. Castranova, M. Thompson, *Nature Materials*, 2009, **8**, 543–557.
5. J. E. Gagner, S. Shrivastava, X. Qian, J. S. Dordick, R. W. Siegel, *The Journal of Physical Chemistry Letters*, 2012, **3**, 3149–3158.
6. P. J. Hagrman, D. Hagrman, J. Zubieta, *Angewandte Chemie*, 1999, **38**, 2638–2684.
7. D. Horn, J. Rieger, *Angewandte Chemie*, 2001, **40**, 4330–4361.
8. S. Kango, S. Kalia, A. Celli, J. Njuguna, Y. Habibi, R. Kumar, *Progress in Polymer Science*, 2013, **38**, 1232–1261.
9. H. Ishii, K. Sugiyama, E. Ito, K. Seki, *Advanced Materials*, 1999, **11**, 605–625.
10. K. G. Sharp, *Advanced Materials*, 1998, **10**, 1243–1248.
11. P. J. Hagrman, D. Hagrman, J. Zubieta, *Angewandte Chemie – International Edition*, 1999, **38**, 2638–2684.
12. T. E. Mallouk, P. Yang, *Journal of the American Chemical Society*, 2009.
13. D. Bagchi, T. K. Maji, S. Sardar, P. Lemmens, C. Bhattacharya, D. Karmakar, S. K. Pal, *Physical Chemistry Chemical Physics: PCCP*, 2017, **19**, 2503–2513.
14. J. K. Suh, H. W. Matthew, *Biomaterials*, 2000, **21**, 2589–2598.
15. T. K. Maji, D. Bagchi, P. Kar, D. Karmakar, S. K. Pal, *Journal of Photochemistry and Photobiology A: Chemistry*, 2017, **332**, 391–398.
16. M. C. Shih, S. S. Li, C. H. Hsieh, Y. C. Wang, H. D. Yang, Y. P. Chiu, C. S. Chang, C. W. Chen, *Nano Letters*, 2017, **17**, 1154–1160.
17. F. Teale, *Macmillan Magazines Ltd.*, 1984.
18. S. K. Pal, A. H. Zewail, *Chemical Reviews*, 2004, **104**, 2099–2123.
19. S. K. Pal, J. Peon, B. Bagchi, A. H. Zewail, *Journal of Physical Chemistry B*, 2002.
20. N. Goswami, A. Giri, M. S. Bootharaju, P. L. Xavier, T. Pradeep, S. K. Pal, *Analytical Chemistry*, 2011, **83**, 9676–9680.
21. N. Polley, P. K. Sarkar, S. Chakrabarti, P. Lemmens, S. K. Pal, *Chemistry Select*, 2016, **1**, 2916–2922.
22. S. Sardar, S. Sarkar, M. T. Z. Myint, S. Al-Harthi, J. Dutta, S. K. Pal, *Physical Chemistry Chemical Physics: PCCP*, 2013, **15**, 18562–18570.
23. P. Kar, S. Sardar, E. Alarousu, J. Sun, Z. S. Seddigi, S. A. Ahmed, E. Y. Danish, O. F. Mohammed, S. K. Pal, *Chemistry - A European Journal*, 2014, **20**, 10475–10483.
24. D. Bagchi, A. Ghosh, P. Singh, S. Dutta, N. Polley, I. I. Althagafi, R. S. Jassas, S. A. Ahmed, S. K. Pal, *Scientific Reports*, 2016, **6**, 34399.
25. W. I. Goldburg, *American Journal of Physics*, 1999, **67**, 1152–1160.
26. Z. S. Seddigi, S. A. Ahmed, S. Sardar, S. K. Pal, *Solar Energy Materials and Solar Cells*, 2015, **143**, 63–71.
27. A. Giri, N. Goswami, M. Bootharaju, P. L. Xavier, R. John, N. T. Thanh, T. Pradeep, B. Ghosh, A. Raychaudhuri, S. K. Pal, *The Journal of Physical Chemistry C*, 2012, **119**, 25623–25629.
28. N. Goswami, A. Giri, S. K. Pal, *Langmuir: the ACS Journal of Surfaces and Colloids*, 2013, **29**, 11471–11478.
29. A. Giri, N. Goswami, P. Lemmens, S. K. Pal, *Materials Research Bulletin*, 2012, **47**, 1912–1918.
30. P. Singh, D. Bagchi, S. K. Pal, *Journal of Biosciences*, 2018, 43, 485–498.
31. P. Kar, S. Sardar, S. Ghosh, M. R. Parida, B. Liu, O. F. Mohammed, P. Lemmens, S. K. Pal, *Journal of Materials Chemistry C*, 2015, **3**, 8200–8211.
32. S. Sardar, P. Kar, H. Remita, B. Liu, P. Lemmens, S. K. Kumar Pal, S. Ghosh, *Scientific Reports*, 2015, **5**, 17313.

33. S. Ghosh, H. Remita, P. Kar, S. Choudhury, S. Sardar, P. Beaunier, P. S. Roy, S. K. Bhattacharya, S. K. Pal, *Journal of Materials Chemistry A*, 2015, **3**, 9517–9527.
34. R. Saha, P. K. Verma, S. Rakshit, S. Saha, S. Mayor, S. K. Pal, *Scientific Reports*, 2013, **3**, 1580.
35. S. Rakshit, R. Saha, A. Chakraborty, S. K. Pal, *Langmuir: the ACS Journal of Surfaces and Colloids*, 2013, **29**, 1808–1817.
36. S. Banerjee, M. Tachiya, S. K. Pal, *The Journal of Physical Chemistry. B*, 2012, **116**, 7841–7848.
37. S. Banerjee, N. Goswami, S. K. Pal, *ChemPhysChem: A European Journal of Chemical Physics and Physical Chemistry*, 2013, **14**, 3581–3593.
38. D. Banerjee, S. K. Pal, *The Journal of Physical Chemistry. B*, 2007, **111**, 5047–5052.
39. F. Debuigne, L. Jeunieau, M. Wiame, J. B. Nagy, *Langmuir*, 2000, **16**, 7605–7611.
40. H. Otsuka, Y. Nagasaki, K. Kataoka, *Advanced Drug Delivery Reviews*, 2012, **64**, 246–255.
41. D. Bagchi, S. Dutta, P. Singh, S. Chaudhuri, S. K. Pal, *ACS Omega*, 2017, **2**, 1850–1857.
42. R. S. Stern, R. B. Armstrong, T. F. Anderson, D. R. Bickers, N. J. Lowe, L. Harber, J. Voorhees, J. A. Parrish, *Journal of the American Academy of Dermatology*, 1986, **15**, 546–552.
43. D. Kanne, K. Straub, H. Rapoport, J. E. Hearst, *Biochemistry*, 1982, **21**, 861–871.
44. N. J. Lowe, D. Weingarten, T. Bourget, L. S. Moy, *Journal of the American Academy of Dermatology*, 1986, **14**, 754–760.
45. Y. T. Zhang, L. N. Shen, Z. H. Wu, J. H. Zhao, N. P. Feng, *International Journal of Pharmaceutics*, 2014, **471**, 449–452.
46. S. Doppalapudi, A. Jain, D. K. Chopra, W. Khan, *European Journal of Pharmaceutical Sciences: Official Journal of the European Federation for Pharmaceutical Sciences*, 2017, **96**, 515–529.
47. W. W. Mantulin, P. S. Song, *Journal of the American Chemical Society*, 1973, **95**, 5122–5129.
48. G. D. Cimino, H. B. Gamper, S. T. Isaacs, J. E. Hearst, *Annual Review of Biochemistry*, 1985, **54**, 1151–1193.
49. S. K. Pal, D. Sukul, D. Mandal, K. Bhattacharyya, *The Journal of Physical Chemistry B*, 2000, **104**, 4529–4531.
50. P. Sen, S. Mukherjee, A. Patra, K. Bhattacharyya, *The Journal of Physical Chemistry. B*, 2005, **109**, 3319–3323.
51. D. Bose, D. Ghosh, P. Das, A. Girigoswami, D. Sarkar, N. Chattopadhyay, *Chemistry and Physics of Lipids*, 2010, **163**, 94–101.
52. N. A. Khatune, M. E. Islam, M. E. Haque, P. Khondkar, M. M. Rahman, *Fitoterapia*, 2004, **75**, 228–230.
53. P. Singh, S. Choudhury, S. Kulanthaivel, D. Bagchi, I. Banerjee, S. A. Ahmed, S. K. Pal, *Colloids and Surfaces B, Biointerfaces*, 2018, **162**, 202–211.
54. C. Weber, F. Rustemeyer, H. Dürr, *Advanced Materials*, 1998, **10**, 1348–1351.
55. N. Goswami, R. Saha, S. K. Pal, *Journal of Nanoparticle Research*, 2011, **13**, 5485–5495.
56. T. Sen, S. Mandal, S. Haldar, K. Chattopadhyay, A. Patra, *The Journal of Physical Chemistry C*, 2011, **115**, 24037–24044.
57. S. Chaudhuri, S. Sardar, D. Bagchi, S. S. Singha, P. Lemmens, S. K. Pal, *The Journal of Physical Chemistry A*, 2015, **119**, 4162–4169.
58. W. H. Walker, E. B. Kearney, R. L. Seng, T. P. Singer, *European Journal of Biochemistry*, 1971, **24**, 328–331.
59. T. L. Jennings, M. P. Singh, G. F. Strouse, *Journal of the American Chemical Society*, 2006, **128**, 5462–5467.
60. A. Makhal, S. Sarkar, S. K. Pal, *Inorganic Chemistry*, 2012, **51**, 10203–10210.
61. H. Yang, P. H. Holloway, B. B. Ratna, *Journal of Applied Physics*, 2003, **93**, 586–592.
62. T. Bora, K. K. Lakshman, S. Sarkar, A. Makhal, S. Sardar, S. K. Pal, J. Dutta, *Beilstein Journal of Nanotechnology*, 2013, **4**, 714–725.
63. S. Sarkar, A. Makhal, S. Baruah, M. A. Mahmood, J. Dutta, S. K. Pal, *The Journal of Physical Chemistry C*, 2012, **116**, 9608–9615.
64. L. Guo, Y. L. Ji, H. Xu, P. Simon, Z. Wu, *Journal of the American Chemical Society*, 2002, **124**, 14864–14865.
65. H. Zeng, G. Duan, Y. Li, S. Yang, X. Xu, W. Cai, *Advanced Functional Materials*, 2010, **20**, 561–572.
66. A. B. Djurišić, Y. H. Leung, K. H. Tam, Y. F. Hsu, L. Ding, W. K. Ge, Y. C. Zhong, K. S. Wong, W. K. Chan, H. L. Tam, K. W. Cheah, W. M. Kwok, D. L. Phillips, *Nanotechnology*, 2007, **18**, 1–8.

67. S. Sardar, S. Chaudhuri, P. Kar, S. Sarkar, P. Lemmens, S. K. Pal, *Physical Chemistry Chemical Physics: PCCP*, 2015, **17**, 166–177.
68. J. W. Rasmussen, E. Martinez, P. Louka, D. G. Wingett, *Expert Opinion on Drug Delivery*, 2010, **7**, 1063–1077.
69. D. A. Bass, J. W. Parce, L. R. Dechatelet, P. Szejda, M. C. Seeds, M. Thomas, *Journal of Immunology*, 1983, **130**, 1910–1917.
70. S. Chaudhuri, S. Sardar, D. Bagchi, S. Dutta, S. Debnath, P. Saha, P. Lemmens, S. K. Pal, *ChemPhysChem: A European Journal of Chemical Physics and Physical Chemistry*, 2016, **17**, 270–277.
71. A. C. B. P. Costa, V. M. C. Rasteiro, C. A. Pereira, R. D. Rossoni, J. C. Junqueira, A. O. C. Jorge, *Mycoses*, 2012, **55**, 56–63.
72. C. J. Gomer, M. Luna, A. Ferrario, N. Rucker, *Photochemistry and Photobiology*, 1991, **53**, 275–279.
73. R. R. Allison, G. H. Downie, R. Cuenca, X. H. Hu, C. J. Childs, C. H. Sibata, *Photodiagnosis and Photodynamic Therapy*, 2004, **1**, 27–42.
74. E. Zenkevich, E. Sagun, V. Knyukshto, A. Shulga, A. Mironov, O. Efremova, R. Bonnett, S. P. Songca, M. Kassem, *Journal of Photochemistry and Photobiology B: Biology*, 1996, **33**, 171–180.
75. Q. Yuan, S. Hein, R. D. Misra, *Acta Biomaterialia*, 2010, **6**, 2732–2739.
76. R. Nandi, S. Mishra, T. K. Maji, K. Manna, P. Kar, S. Banerjee, S. Dutta, S. K. Sharma, P. Lemmens, K. D. Saha, S. K. Pal, *Journal of Materials Chemistry B*, 2017, **5**, 3927–3939.
77. D. Peer, J. M. Karp, S. Hong, O. C. Farokhzad, R. Margalit, R. Langer, *Nature Nanotechnology*, 2007, **2**, 751–760.
78. W. Yan, H. Fan, C. Yang, *Materials Letters*, 2011, **65**, 1595–1597.
79. M. H. Han, G. Y. Kim, Y. H. Yoo, Y. H. Choi, *Toxicology Letters*, 2013, **220**, 157–166.
80. A. Giri, N. Goswami, M. Pal, M. T. Zar Myint, S. Al-Harthi, A. Singha, B. Ghosh, J. Dutta, S. K. Pal, *Journal of Materials Chemistry C*, 2013, **1**, 1885–1895.
81. W. Wu, Q. He, C. Jiang, *Nanoscale Research Letters*, 2008, **3**, 397–415.
82. A. K. Gupta, M. Gupta, *Biomaterials*, 2005, **26**, 3995–4021.
83. W. S. Seo, H. H. Jo, K. Lee, B. Kim, S. J. Oh, J. T. Park, *Angewandte Chemie*, 2004, **43**, 1115–1117.
84. Y. Li, H. Tan, X. Y. Yang, B. Goris, J. Verbeeck, S. Bals, P. Colson, R. Cloots, G. Van Tendeloo, B. L. Su, *Small*, 2011, **7**, 475–483.
85. A. Giri, N. Goswami, C. Sasmal, N. Polley, D. Majumdar, S. Sarkar, S. N. Bandyopadhyay, A. Singha, S. K. Pal, *RSC Advances*, 2014, **4**, 5075–5079.
86. N. Polley, S. Saha, A. Adhikari, S. Banerjee, S. Darbar, S. Das, S. K. Pal, *Nanomedicine*, 2015, **10**, 2349–2363.
87. A. Adhikari, N. Polley, S. Darbar, D. Bagchi, S. K. Pal, *Future Science OA*, 2016, **2**, FSO146.

# 3

# *Multi-Functional Nanomaterials for Biomedical Applications*

**Balaprasad Ankamwar, Saee Gharpure, and Aman Akash**

**CONTENTS**



## 3.1 Introduction

Recent advances have resulted in synthesis as well as characterization of a large number of nanomaterials with novel properties. The applicability of these nanomaterials is tremendous due to the fact that they have great potential because of the unique properties they feature unlike their corresponding bulk materials [1, 2]. However, individual nanomaterials have their associated drawbacks as well. It is because of these reasons that the direct use of these nanomaterials without further functionalization or modification is difficult [3]. A lot of advancement in the field of nanotechnology has encouraged synthesizing as well as using nanohybrids (NHs) by combining two or more types of nanomaterials [4]. The hybrids are substances or composites made up of two or more materials which consist of at least one of the materials on the nanometer scale. Various kinds of NHs can be synthesized by using four major kinds of materials, that is, metals, polymers, organics, and non-metal inorganics by various permutations and combinations. This includes NHs with two different metals or alloys, NHs with metals and polymers, NHs with non-metal inorganics and metals, NHs with both polymers, etc. In addition to these categories, metallo-organic frameworks (MOFs) act as exceptions as they do not fit into any of these categories [5].

Various kinds of manipulations can be done on the properties of NHs such as size, surface, microstructure, interface, or quantum confinement, depending upon the associated applications. Morphology, as well as composition of components, are varied individually during synthesizing NHs [6]. Thus, it is crucial to minutely monitor conditions required for synthesis in order to get desired outcomes. Methods used to synthesize NHs include the displacement process, the deposition process, epitaxial growth process, etc. [7–9] by manipulating thermodynamic properties like electro-negativities, activation energy, and reduction potential, as well as kinetic properties such as concentration, temperature, diffusivity, time, nature of the solvent, etc. [10–12]. Problems such as mismatch present at the lattice structure among individual components as well as interfacial stress as a result of wide curvature present at nano level are very difficult to deal with. Maintenance of uniformity of thickness of the shell is also a challenging task in the case of the direct deposition process. Solvo-thermal and hydrothermal processes, in combination with *in situ* reduction methods, have been employed for synthesizing complex NHs [10, 12].

Newer methods of synthesis like anion reduction as well as co-ordination methods [13], liquid precipitation along with its reduction in multiple steps [14], non-epitaxial growth methods [15], deposition as well as coating methods for atomic layer [16], reduction and oxidation methods taking place at high temperature [17], etc. are being developed currently so as to solve problems associated with the currently available methods. Detailed analysis of the mentioned methods makes it clear that critical monitoring of thermodynamic parameters during each and every step of NH synthesis is crucial so as to get morphology as well as structure of desired specifications after NH synthesis is completed. Thus, NHs synthesized under highly controlled conditions are currently used in a wide range of applications including electronics, catalysis, energy, biomedicine, environmental remediation, optical imaging, etc. [18].

## 3.2 Biomedical Applications of Nanohybrids

### 3.2.1 Cancer-Related Applications

In the present scenario, the incidence of life-threatening diseases worldwide has increased dramatically due to the alarming increase in pollution as well as complications due to aging. Of these diseases, cancer has been found to be among the topmost members [19, 20]. Treatment of cancer has been tricky even though various types of chemotherapy measures have been used in cancer therapeutics [21]. This is because of the multi-drug resistance which the tumor develops against various available anticancer agents, thus limiting their biological activity [22–24]. The use of nanotechnology for cancer detection as well as its treatment serves as a very efficient alternative compared to the traditionally available chemotherapeutic agents. Nanotechnology-based systems work by developing drug carriers which help in the administration of the concerned drug. This approach has an upper hand over the traditional approach as it shows high drug solubilization, efficient control on drug release, enhanced drug stability, and specific targeting abilities for the tumor tissue [25–27].

NHs with unique magnetic as well as optical properties have been used for cancer diagnostics as well as treatment for a long time. NHs showing a potential size in the range 50–200 nm, which have been used as carriers in drug delivery systems, include polymerosomes, liposomes, metal nanoparticles, inorganic materials like mesoporous silica, microgels, carbon-containing materials, etc. The nanoparticles which are a part of the NHs function as a framework for imaging as well as signal transduction [28, 29] (see Figure 3.1).

These nanoparticles show optical as well as magnetic properties which are increased after the formation of NHs. Photothermal, surface-enhanced Raman scattering (SERS) as well as the photoacoustic properties of Au nanorods, iron oxide nanoparticles showing magnetic properties, fluorescent characteristics of Au nanoclusters are some examples [30, 31]. It is by virtue of these properties that these NHs can be used for cancer diagnostics with increased sensitivity as well as cancer therapy with high efficiency [32]. Biomacromolecules like proteins, nucleic acids, polysaccharides, etc. which constitute the other part, act as binding scaffolds. These biomaterials act as an envelope for nanoparticles as well as drugs to be administered. It is because of these biomaterials that the life span of the NH being retained in blood circulation increases. Moreover, it also enhances the tumor-targeting capabilities of the NHs [33]. Once these NHs get anchored upon the tumor tissue, they can be disintegrated into individual nanoparticles in response to specific stimuli. These stimuli can be internal as well as external cues. Some stimuli which are actively used include a change in pH, enzyme activity, induction by a magnetic field, near-infrared (NIR) laser, etc. These individual nanoparticles now show increased effectivity due to the depth of its penetration ability as well as even distribution of nanoparticles throughout the tumor tissue. It is because

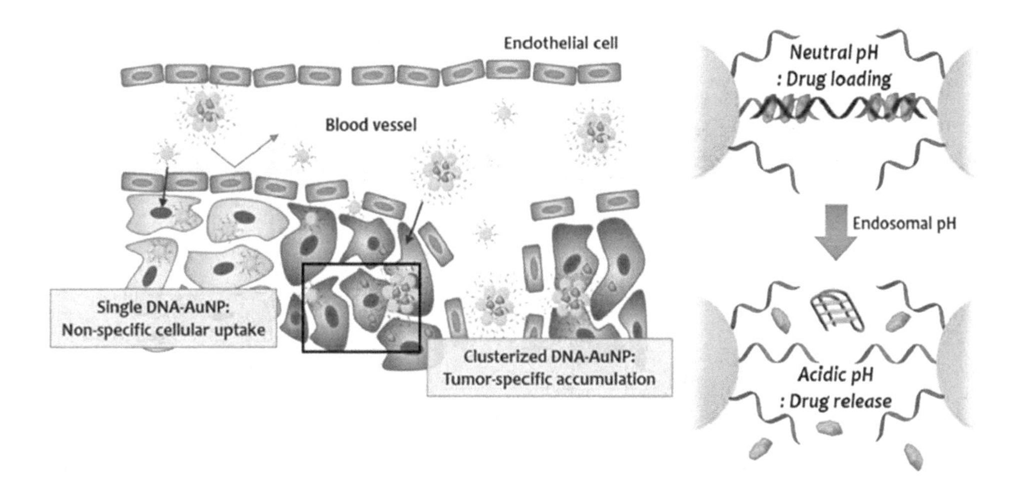


**FIGURE 3.1** Schematic representation of accumulation of cAuNPs in tumor tissue followed by cellular uptake for drug delivery with increased efficiency. (Reproduced with permission from Kim et al. [29]. American Chemical Society.)

of these properties that these NHs can serve as potential candidates in the treatment as well as the diagnosis of cancer [33, 34].

These NHs possess some characteristic advantages like enhanced penetration and retention (EPR) resulting in increased accumulation within the tumor, increase in bioavailability, prolonged circulation in blood, and a decrease in clearance via decreased reticulo-endothelial system (RES) [35, 36]. The smaller size of these NHs enhances their penetration capacity within the tumor tissue as the smaller the nanomaterials, the higher its tissue penetration capability [36, 37]. One more advantage of smaller particle size is that the particle gets flushed out of the system very easily, thus avoiding further complications due to long-term retention in the body which would cause toxicity to the healthy tissues as well. However, there are several disadvantages associated with the use of these NHs as well. These disadvantages include less efficient drug loading, lack of specificity during drug accumulation, constraints in stability, and burst release, etc. These factors have a great effect during the development of a specific nano-carrier system for drug delivery. It is therefore essential to develop a system with increased capacities for loading of the drug, controlled targeting in an *in vitro* as well as *in vivo* environment, and increased retention time in addition to efficient drug disposal, etc. [38–40].

### 3.2.1.1 Tumor-Targeted Delivery

All the attractive properties of NHs can only be exploited if the nanoparticles released upon disintegration of these NHs show efficient accumulation at the site of tumor formation. It is only then that these NHs can be utilized for effective diagnosis as well as the treatment of cancer [41]. However, this task is not as easy as it seems because of the tremendous complexity associated with the biological systems. Enhancement in NH accumulation has been achieved through the aid of intrinsic as well as extrinsic factors such as the size of the particle, externally applied magnetic fields, associated ligands, etc. [28].

#### *3.2.1.1.1 Size-Mediated Cancer Targeting*

Accumulation of nanoparticles at the site of tumor tissue can only be achieved if the nanoparticles have an appropriate size by virtue of the EPR effect [42]. Even though nanoparticles with smaller size are seen to achieve intra-tumor penetration, they have a smaller retention time as they are wiped from blood circulation in 24 hours. However, in the case of larger nanoparticles, in spite of the great accumulation of nanoparticles at the tumor site, penetration into the tumor tissue is a problem [43]. Thus, a balance needs to be attained in order to achieve both properties. A 16 nm zwitterionic gold nanoparticle of pH-sensitive nature has been engineered so that nanoparticles of small size can be retained at the site of tumor tissue. These nanoparticles have shown more effective dispersion under the influence of pH

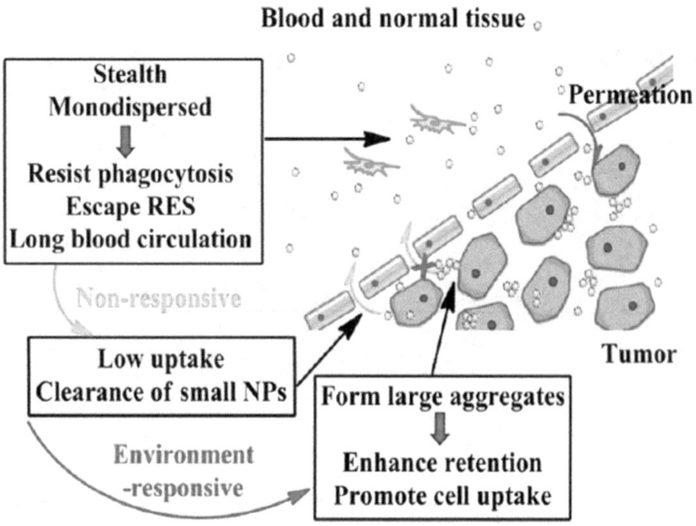

FIGURE 3.2 Schematic representation showing enhancement in retention as well as cellular uptake of small nanoparticles in tumor tissue upon induction of nano-aggregation within the tumor microenvironment. (Reproduced with permission from Liu et al. [44]. American Chemical Society.)

acting as "stealth" nanoparticles by the virtue of non-fouling properties of the zwitterionic nanoparticles [44] (see Figure 3.2).

Rapid aggregation of these nanoparticles up to the size of about 300 nm is observed at acidic pH. As is known, tumor tissue is an acidic environment. Thus, these nanoparticles in the form of nano-aggregates show enhanced confinement by entrapment into the extracellular matrix as well as internalization into the tumor cells. Thus, these hybrid nano-aggregates induced by a change in pH can show significant enhancement in accumulation, retention, and cellular uptake of the Au nanoparticles with small size ones present in the tumor microenvironment compared to solitary Au nanoparticles [14]. When the uptake of NHs was compared with that of individual nanoparticles, NH uptake was seen to have significantly increased as opposed to single nanoparticles. *In vitro* uptake of Au NHs into the endocytic compartment of the human monocytes was seen to be far greater compared to those with free Au nanoparticles. Smaller sizes of NHs increase the chance of the drug reaching its destined tumor site. On reaching the tumor site, the smaller size will ensure the efficient delivery of the drug at the site of action, ultimately killing the cell. As in the case of nanoparticle/nanorod (NP/NR) nano-assemblies, their small size ensures targeted delivery into the nucleus even though nuclear localization signal is not present within the assembly. The use of photocontrollable as well as size-transformed drug delivery systems has tremendously helped in the effective transport of nanodrugs through the cell membrane into the nucleus in a well-regulated manner. This has facilitated targeted delivery as well as the accumulation of various drugs into the nucleus which has caused a tremendous increase in the therapeutic effectivity of the drug [45].

#### *3.2.1.1.2 Ligand-Mediated Cancer Targeting*

In addition to the use of the passive approach, NHs have also been used for actively targeting tumors of a specific type. This can be accomplished by the functionalization of NHs using ligands specific for a particular tumor type. Various kinds of small-sized ligands have been attached to the surface of the NHs by means of chemical reaction. This has been done so as to ensure the targeting effect of the tumor tissue of a specific type in the active form. There are many such examples of active targeting of NHs. Functionalization of iron oxide–protein NHs with folic acid (FA) was carried out by a two stepEDC/Sulfo NH covalent coupling method [46]. When human nasopharyngeal carcinoma (KB cells) were tested for inhibition under *in vitro* as well as *in vivo* conditions, FA–iron oxide–proteins NHs exhibited higher inhibition compared to the non-targeted group. Also, nanovesicles functionalized using FA had the ability to specifically recognize an FA receptor which had been overexpressed in breast cancer cells

(MDA-MB-435) [47]. Bortezomib (BTZ) which is a novel anticancer drug has been labeled by biotin connected via catechol polymers acting as linkers. The presence of biotin facilitates targeted entry of the NH into the cell, thus increasing the efficiency of drug delivery. This ultimately increases the effectivity of the drug [48].

Active targeting has also been done by functionalization of the NHs using antibodies. Gelatin–iron oxide NHs, which are amphiphilic in nature, have been functionalized by attachment of Herceptin (HER) (trastuzumab) monoclonal antibodies against HER2-positive metastatic breast cancer cells [49]. Efficient internalization of these NHs with HER has been observed in HER2 overexpression breast cancer cell lines (SKBr3 cells) compared to non-targeted NHs. This shows the increase in efficiency of the functionalized NHs to recognize HER2 receptors so as to differentiate between cells showing differential expression of HER2 receptors in HER2 overexpression breast cancer cell lines (SKBr3 cells). Active targeting has also been achieved by using synthetic peptides for the functionalization of NHs. Simultaneous synthesis and capping of gold nanoparticles with cell penetrating peptides (CPP) have been achieved ultimately resulting in the formation of Au–CPP NHs. These NHs have shown increased penetration ability as well as considerable cytotoxicity when tested on HeLa cells, unlike the unlabelled gold nanoparticles, thus serving as a promising candidate for targeted drug delivery [50] (see Figure 3.3).

Cyclic peptides such as Arg–Gly–Asp which are abbreviated as cRGDyK have been attached to dendrimer–iron oxide NHs. These cyclic peptides attach to avb3–integrin molecules which are seen to be upregulated in various tumor cells. Thus, functionalization with these cyclic peptides enhanced the capacity of the NHs to specifically recognize and bind to avb3 receptors. The affinity of NHs toward these receptors has enabled its use in efficient targeting *in vitro* as well as *in vivo*. This has been shown in C6 cells which overexpress avb3 receptors. Functionalized NHs have been shown to recognize these receptors with increased efficiency which has been proven by monitoring cellular intake and localization studies under *in vitro* conditions as well as by magnetic resonance imaging (MRI) under *in vivo* conditions [51].

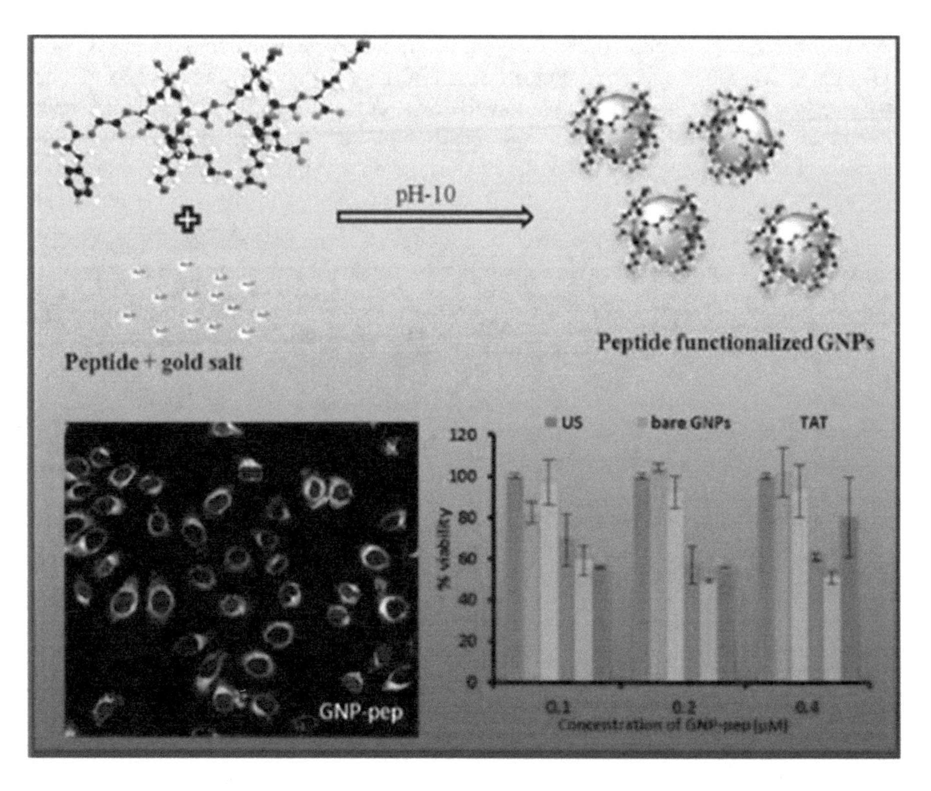


**FIGURE 3.3** **(See color insert.)** Schematic representation of efficient penetration ability and cytotoxicity of peptide-conjugated NHs. (Reproduced with permission from Bansal et al. [50] American Chemical Society.)

#### *3.2.1.1.3 Magnetic Field-Mediated Cancer Targeting*

The magnetic properties of nanoparticles have been exploited for the diagnosis as well as the treatment of cancer. Iron oxide nanoparticles are one of the most widely used nanoparticles due to their attributed magnetic properties. These nanoparticles can undergo magnetization under the influence of an external magnetic field. These magnetized nanoparticles can then be driven to the site of the tumor within the body which can be externally controlled [52, 53]. There are various merits associated with magnetic targeting of these nanoparticles. There are no side effects associated with their use. Also, these nanoparticles are required at very low doses. If the associated NHs have been positioned at a local site, it avoids direct contact with the RES, which is one of the major advantages. NHs consisting of magnetite with colloidal properties have been synthesized. NHs based upon reticulocyte derived exosomes have been used for the targeted treatment of cancer by exploiting the magnetic properties of these NHs.

Exosomes which have been derived from reticulocytes have been embedded by the iron oxide nanoparticles, which have magnetic properties, by the interaction between transferrin and transferrin receptor. An increase in the sensitivity of the response to an external magnetic field has been observed in the case of these NHs compared to individual iron oxide nanoparticles (see Figure 3.4). This property of these NHs has been used for the targeting of cancer cells under the influence of an external magnetic field [54].

NHs consisting of superparamagnetic iron oxide (SIO) nanoparticles integrated into nano-assemblies made up of polymeric core–shell particles have been used for the magnetic targeting of tumor cells in mice under the influence of an external magnetic field [55]. The synergistic effect of SIO nanoparticle-loading as well as the application of an external magnetic field was observed under *in vivo* conditions in this study. Thus, magnetic targeting has shown enhanced efficacy in treating tumor cells, as well as improved survival rates compared to passive targeting approach. This approach has also allowed filling the gap between medical research and its translation into therapy.

### *3.2.1.2 Enhanced Cancer Imaging Performances*

It has been observed that different types of imaging techniques have been exploited in case of detection in clinical research as well as disease diagnostics. All of these techniques can be classified depending upon their energy type, like ultrasound, photons, X-rays, etc., in order to get visible attributes, additional information in terms of molecular, anatomical or physiological details, spatial resolution, etc. [56]. The imaging of cancerous tissue has always been an indispensable component for cancer diagnosis as well as therapy. NHs have been widely used in cancer imaging by the virtue of associated properties

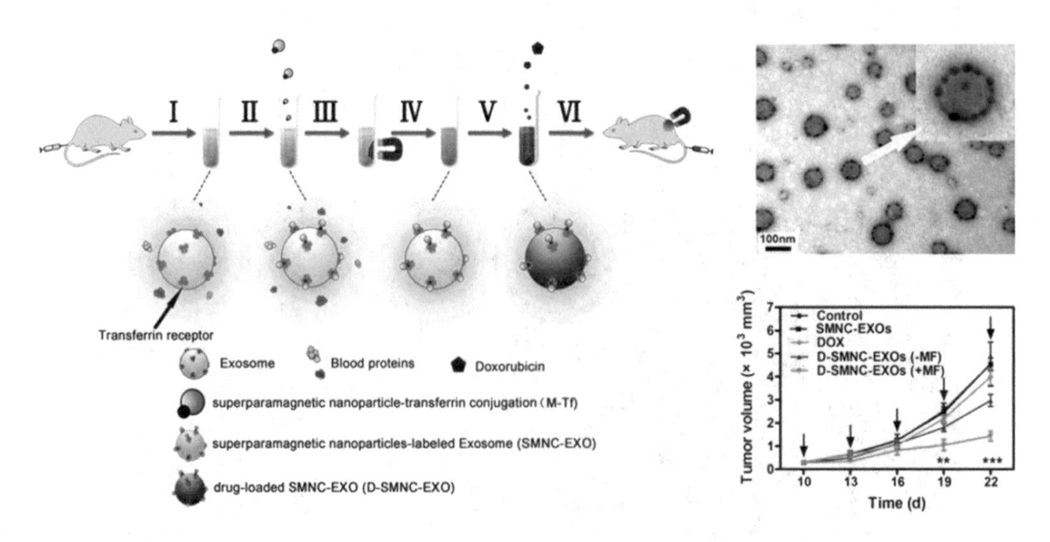


**FIGURE 3.4** **(See color insert.)** Schematic representation of synthesis of dual-functional exosome-based superparamagnetic nanoparticles clusters and its use as a targeted drug delivery carrier for cancer therapy. (Reproduced with permission from Qi et al. [54] American Chemical Society.)

of metal nanoparticles which constitute a part of these NHs [57]. The NHs so formed are a combination of multiple constituents which modulate both the physical and chemical properties with distinguishing characteristics. These NHs, upon modification of their novel chemical, electronic, and optical characteristics can serve as probable candidates in various applications, of which biomedical imaging is among the major ones [58]. It has been observed that signals obtained upon imaging of cancerous cells or tissues have been amplified upon the use of these NHs in place of associated metal nanoparticles. These amplified signals have been associated with imaging techniques such as fluorescence imaging [59], photoacoustic imaging [60], and SERS [61, 62], resulting in an increase in the sensitivity of these techniques. In addition to this, applications in clinical studies have resulted in the improvement of techniques such as MRI [63], positron emission tomography (PET) [64], and computed tomography (CT) [65].

#### *3.2.1.2.1 Photoacoustic Imaging of Cancer*

Photoacoustic imaging (PAI), also known as optoacoustic imaging, has emerged as one of the latest imaging techniques used in biomedical applications. PAI makes use of ultrasound generated using a laser source. It is a combination of optical imaging showing increased contrast as well as enhanced specificity in terms of spectroscopic sensitivity in comparison with ultrasound imaging showing enhanced dimensional resolution [66]. This technique involves the advantages of both ultrasound as well as optical imaging. This hybrid PA image yields enhanced contrast, efficient infiltration into the tissue/cells along with increased resolution upon conversion of the excitation signal coming from the laser source into ultrasound emission [32].

The synthesis of multi-faceted Cy5.5–HANP/CuS NHs obtained upon associating copper sulfide (CuS) with hyaluronic acid nanoparticles (HANPs) conjugated with Cy5.5 has been reported (Figure 3.5A). The fluorescent signal of Cy5.5 is kept quenched by the CuS as long as the complete NH is intact. When these NHs come into contact with the hyaluronidase enzyme present in the tumor cells, they are disintegrated, resulting in Cy5.5 giving a strong fluorescent signal. Moreover, CuS has strong absorbance in the NIR region which acts as an efficient contrast background for PAI, as well as succeeding photothermal therapy guided with the help of PAI (Figure 3.5B) [67].

Gold nanorods (AuNR) were constituted into nanovesicles by the use of poly (ethylene glycol) (PEG)-functionalized AuNRs along with poly (lactic-co-glycolic acid) (PLGA). As the AuNRs present in the nanovesicles show enhanced plasmonic coupling, the associated NHs can be used for drug delivery in

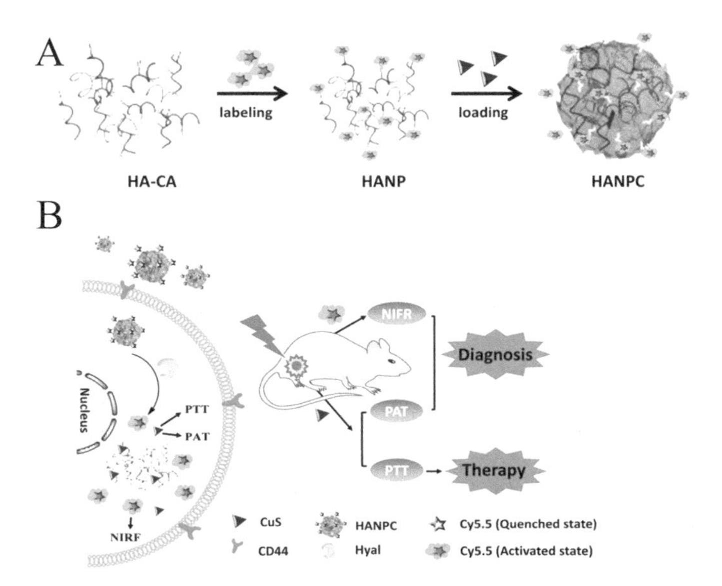


**FIGURE 3.5** (A) Schematic representation of the synthesis of the multi-faceted Cy5.5-HANP/CuS (HANPC) NHs (B) Applications of HANPC NHs in NIR fluorescence as well as PAI-targetted photothermal therapy under *in vivo*. (Reproduced with permission from Zhang et al. [67] American Chemical Society.)

cancer treatment via photoacoustic imaging. These nanovesicles have been observed to show remarkable enhancement in photoacoustic signals compared to the corresponding AuNRs when tested at equal optical density under the laser source at 808 nm. Uninterrupted monitoring of two-dimensional as well as three-dimensional PA images showed that these nanovesicles accumulate to a larger level in the tumor regions at the same time point as against AuNRs, which showed basal levels of accumulation. This suggests that uptake of nanovesicles in tumor sites is significantly higher compared to individual AuNRs [68].

#### *3.2.1.2.2 Fluorescence Imaging of Cancer*

Fluorescence imaging has been considered among the most robust fundamental methods used for clinical as well as biomedical analyses. The most important characteristic of this technique is its non-invasive nature which makes its use favorable even in case of rare samples. Fluorescence imaging enables the display of only the entity to be focused on against a black background [69]. A perfect fluorescent probe should be highly photo-stable as well as dispersible under various biological conditions, have high resistance against degradation in these biological environments, and be biologically compatible, apart from providing enhanced yields in terms of fluorescence intensities. The use of organic dyes and biocompatible fluorescent proteins for fluorescence imaging has been well documented for a long time. However, these molecules show problems like photo-bleaching which restrict their use in extended as well as instantaneous imaging of biological samples. Nanoparticles serve as a promising alternative to fluorescent probes due to their unique physical, chemical, and optical properties. However, the use of nanoparticles is limited due to their associated toxicity as well as bio-accumulation issues [70]. It is because of these reasons that the development and use of NH systems with high fluorescence yields as well as reduced toxicity issues is one of the most lucrative areas in research.

Recently, gold nanoclusters (AuNCs) have been considered one of the probable candidates to be used as fluorescent probes for the imaging of cancer cells under *in vivo* as well as *in vitro* conditions. This is due to their unique characteristics such as ultrasmall size, easy functionalization, high biological compatibility in living systems, increased life span, enhanced fluorescence intensities, and increased stability compared to the corresponding quantum dots [71, 72]. According to the recent reports, AuNC nano-assemblies have been observed to show an increase in fluorescence intensities upon the aggregation of these NHs. However, the exact mechanism behind this still remains unknown [73]. AuNC nano-assemblies have been used in different ways for detecting fluorescence in cancer cells. For example, NHs with AuNCs coated upon by PAA/m$SiO_2$ so as to form core–shell particles have been used for cancer cell imaging using fluorescence. These NHs with aggregation-enhanced fluorescence (AEF) properties have been used in cancer diagnosis and therapy by fluorescence imaging-mediated drug delivery under *in vitro* as well as *in vivo* conditions [74].

Tumor sites are known to have reduced oxygen levels in comparison with normal cells. Such reducing of the microenvironment causes hypoxic conditions in the cancerous cells. This results in increased levels of reactive oxygen species (ROS), reactive nitrogen species (RNS), free radicals, etc. Thus, in order to maintain the homeostatic environment of the cell, cancer cells are seen to have high levels of glutathione peroxidase, reducing equivalents like NADH, NADPH, etc., and cysteine-rich proteins in comparison with those present in the normal cells [75]. It is as a result of these properties that metals like platinum which are oxidizing in nature show maximum activity in such a microenvironment. Conventional fluorescent probes made up of platinum, however, show various problems such as quick photo-bleaching, reduced photostability, and increased aggregation of intrinsic proteins. This limits their use in fluorescence imaging of biological systems for a prolonged time span. Therefore, bioimaging of fluorescent NHs which have been synthesized within the cancer cells themselves serves as an efficient alternative. Biosynthesis of platinum NCs within cancer cells/tumor cells has been reported. These fluorescent NCs facilitate the bioimaging of cancer cells themselves and thereby helps in the development of treatment of cancer through photothermal therapy aided by fluorescence imaging by a synergistic approach by making use of water-soluble porphyrin tetrakis(sulfonatophenyl)porphyrin (TSPP). This results in the enhancement of therapeutic efficacy, thus serving as a promising strategy [59] (see Figure 3.6).

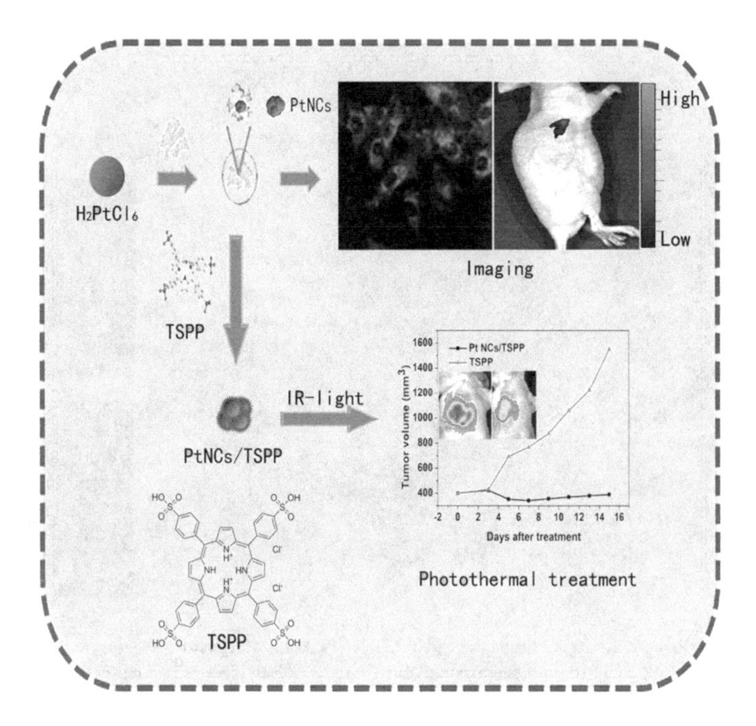


**FIGURE 3.6** **(See color insert.)** Schematic representation of the synthesis of platinum NCs within cancer cells themselves and its application in fluorescence imaging guided photothermal treatment. (Reproduced with permission from Chen et al. [59] American Chemical Society.)

### *3.2.1.2.3 Magnetic Resonance Imaging of Cancer*

The principle behind the functioning of MRI is the detection of a difference in relaxation times of proton moieties present in water upon the application of an external magnetic field. There exists a remarkable difference in these relaxation values because of the diversified distribution of water in various tissues present throughout the body. MRI scans are made more sensitive by the use of compounds which decrease the time of relaxation at longitudinal (T1) as well as transverse (T2) levels [76, 77]. The normal tissues are differentiated from the abnormal tissues through differences in contrast which are amplified by the use of MRI contrast agents (CAs). Discovery and investigation of novel CAs with increased accuracy as well as sensitivity are being done constantly, resulting in the addition to the CAs which are currently used for imaging applications [78]. Nanoparticle-aided CAs serve as the best possible platforms so as to achieve modalities at different levels. This is because respective nanoparticles can be associated internally as well as externally with considerably high concentrations of CAs. Different imaging multimodalities thus provide propitious platforms as they cater information to its maximum extent leading to morphological as well as functional relevance [79].

Synthesis and characterization of the multi-faceted NHs abbreviated as LA-LAPNHs have been reported for targeted use in magnetic resonance imaging/computed X-ray tomography (MRI/CT) as well as in the treatment of hepatocarcinoma by photothermaltherapy (see Figure 3.7) [80].

LA-LAPNHs have been proposed as an effective therapeutic agent in the diagnosis as well as the treatment of cancer through their dual use in photothermal therapy of hepatocarcinoma as well as targeted imaging by means of MRI/CT. LA-LAPNHs are NHs consisting of core–shell particles with AuNPs functionalized by polydopamine (PDA) at the core, which are coated by indocyanine green (ICG). These core–shell structures act as an effective agent for photothermal therapy. Further modification by gadolinium–1, 4, 7, 10-tetraacetic acid as well as lactobionicacid (LA) upon the outer shell makes them suitable to be used as CAs in MRI/CT. These LA-LAPNHs have been observed to show remarkable cytotoxicity against hepatocarcinomas by photothermal therapy as well as an enhancement in MRI/CT as they show a decrease in T1 relaxation time [80].

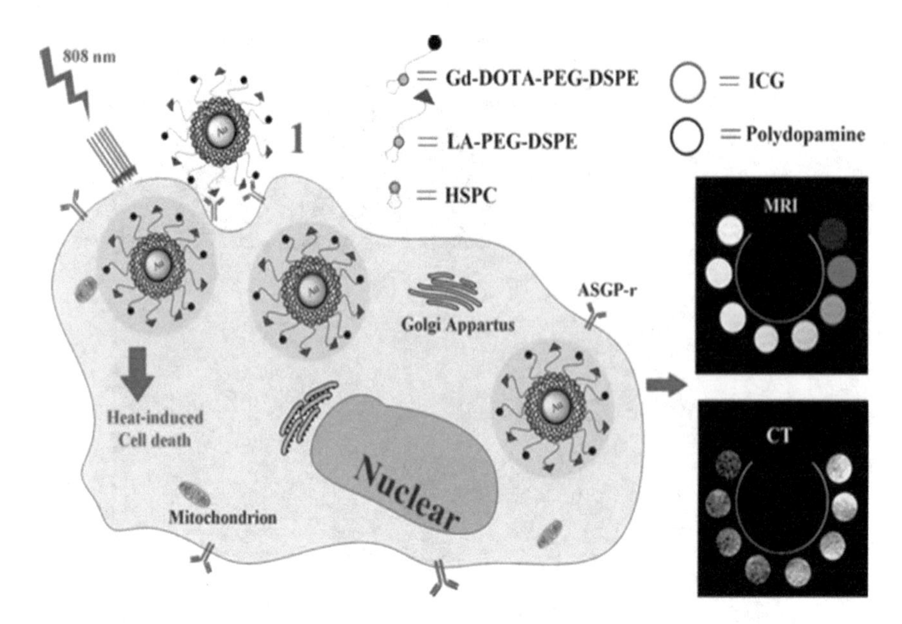


**FIGURE 3.7** Schematic representation of the dual use of LA-LAPNHs as MRI contrasts agent as well as their application in photothermal therapy. (Reproduced with permission from Zeng et al. [80] American Chemical Society.)

#### *3.2.1.2.4 PET and CT Imaging of Cancer*

PET is an imaging technique which is widely used in clinical as well as biomedical studies due to its non-invasive and highly sensitive nature. PET imaging monitors the concentration and the distribution of the positrons emitted by radionuclides after the injection of radiolabelled imaging probes into the body [81]. The use of radiolabeled nanoparticles as PET imaging probes has been very important in recent times. PET imaging using radiolabeled nanoparticles also provides an insight into the distribution of associated nanoparticles, thus checking their effectiveness in targeted drug delivery [82]. Chen et al. [83] have reported the synthesis of UiO-66 nanoscale metal–organic frameworks (nMOF) containing isotope zirconium-89 ($^{89}Zr$), which is frequently used in PET imaging as it emits positrons. These nMOFs have been attached with pyrene-derived polyethylene glycol (Py–PGA-PEG) followed by further functionalization using a peptide ligand (F3) so as to target nucleolin to be used in the case of triple-negative breast cancer cells (see Figure 3.8A1). Targeted delivery of these nMOFs loaded with doxorubicin to the specific tumor site has been monitored by PET imaging. These NHs have been found to be efficient in tumor-specific targeted drug delivery, which can be monitored *in vivo* via PET imaging (see Figure 3.8A2).

X-ray CT is among the most widely used imaging techniques in biomedical as well as clinical studies. This is because CT provides information about the tissues under investigation in three dimensions depending upon the extent of absorption of X-rays emitted by the source. Conventional CT CAs, however, suffer drawbacks such as short retention times and the renal toxicity caused as these CAs are flushed out rapidly through the kidney [84, 85]. It is for of these reasons that nanoparticles are being considered as a viable alternative to be used as CT CAs. Shen et al. [65] have reported the design and fabrication of multi-functional upconversion nanoparticles (MUNCPs) and their use as CT CAs in *in vivo* bioimaging. During the synthesis of MUCNPs, $NaYF_4$:Yb/Tm UCNPs act as the core structure and are then coated using $NaLuF_4$ giving CT active properties followed by a $NaYF_4$ layer which imparts MR functional characteristics. The outermost coating has been done by $NaGdF_4$ to bring out increased proton relaxation times of the water present in the tissues of interest [86] (see Figure 3.8B). Figure 3.8C depicts the enhanced brightness when PAA-MUNCPs have been used as CT CAs against increasing mass concentrations. *In vivo* bioimaging has been done after the injection of PAA-MUNCPs (Figure 3.8D1) as well as two hours post-injection (see Figure 3.8D2) and shows notable enhancement in CT signals of vital organs with excellent contrast effect.

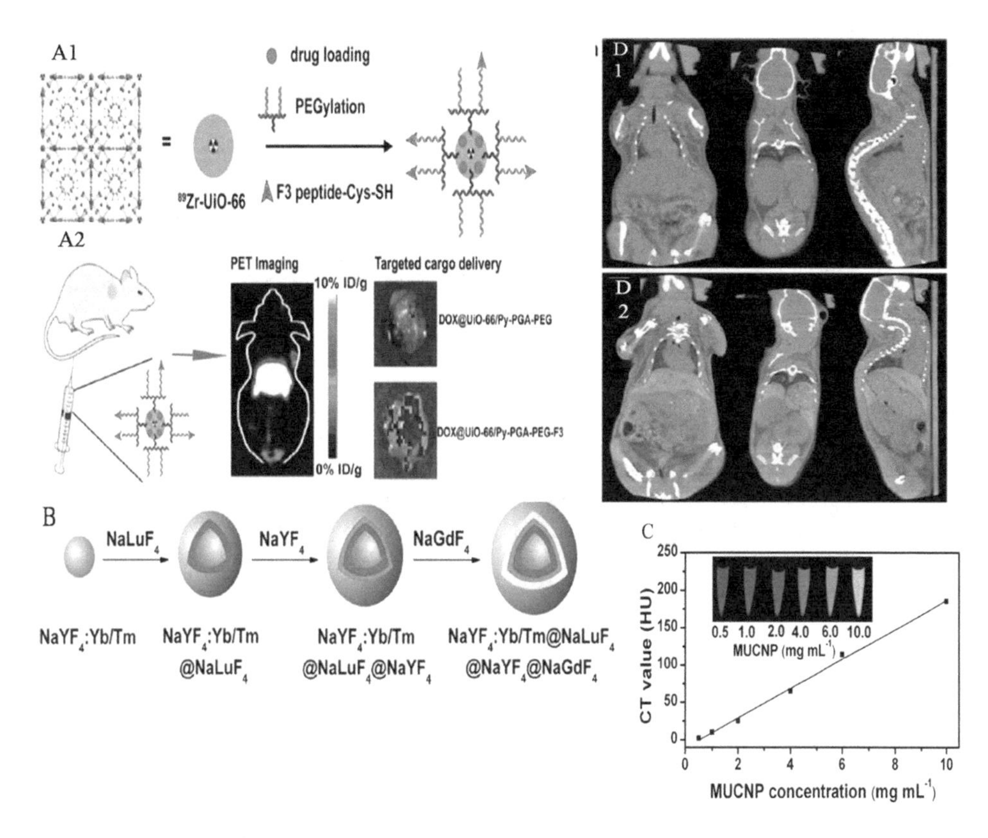


**FIGURE 3.8** **(See color insert.)** (A) Schematic representation of the synthesis of UiO-66 nMOFs attached with pyrene-derived polyethylene glycol (Py–PGA-PEG) followed by functionalization using peptide ligand (F3) (A1) and its application in PET imaging assisted drug delivery (A2). (Reproduced with permission from Chen et al. [83]). American Chemical Society. (B) Design for synthesis and functionalization of multi-functional upconversion nanoparticles (MUNCPs) (C) Efficiency of MUNCPs at different concentrations at CT contrast agent (D) CT imaging of Kunming mouse using PAA-functionalized MUNCPs before injection (D1) as well as 2 hrs after injection (D2). (Reproduced with permission from Shen et al. [65] American Chemical Society.)

### *3.2.1.2.5 Surface-Enhanced Raman Scattering Imaging of Cancer*

Recently, Raman spectroscopy is being exploited on a large scale for sensing applications as well as bioimaging because of its increased sensitivity and high spatial resolution [87, 88]. The use of Raman scattering has been found to be more beneficial than fluorescence as it shows efficient excitation and detection in the NIR range [89, 90] as well as showing comparatively reduced photo-bleaching [91]. Nanoparticles have been exploited so as to enhance the efficiency of diagnostics by employing Raman spectroscopy under *in vitro* as well as *in vivo* conditions. Raman active probes functionalized by NHs enhance the SERS signal over that of conventional probes, thereby making them a suitable alternative for use in cancer diagnosis and therapy [61].

Enhancement in SERS signals has been observed in the use of NHs over that of individual nanoparticles. This is because properties of NHs can be modulated in order to increase their sensitivity as Raman probes, which is not possible in the case of individual nanoparticles [31]. Song et al. [92] have reported the synthesis of a CNTR@AuNP NH which is carried out by coating gold nanoparticles (AuNPs) on the carbon nanotube ring (CNTR) which serves as the template. Determination of the plasmonic coupling in between the inner as well as the outer surface of the gold nanoparticles would help in further modulation of the structural properties of these NHs. Tunable surface plasmon wavelength within the NIR range enables the use of CNTR@AuNP NHs in cancer cell imaging. The SERS signals of these NHs have been observed to be enhanced by 110 times compared to CNTRs and AuNPs individually, thus increasing their efficient use in cancer imaging as well as photothermal therapy for cancer treatment.

### 3.2.1.3 High-Efficacy Cancer Therapy

Treatment of cancer through the use of NHs can be achieved via various modes of action such as drug-mediated therapy involving chemotherapeutic agents, nucleic acids as well as proteins, photothermal therapy, photodynamic therapy, etc. [93, 94]. Drug-mediated therapy supplemented by photothermal therapy resulting in drug release into the cell, death of cancerous cells caused by the formation of cytotoxic free radicals induced by the externally applied laser or cytotoxic heat generation within the cancer cells causes an increase in efficacy of the cancer treatment [95]. The combination of two or more modes of treatment has been reported to cause a reduction in respective drawbacks of individual therapies, thus increasing the efficacy of the cancer treatment.

#### *3.2.1.3.1 Chemotherapy*

The sole use of drug-mediated chemotherapy has decreased in recent times because of some major disadvantages. One of the major shortcomings involves the lack of specificity of these drugs which ultimately results in the killing of both cancerous cells and healthy cells.[89] Some of these disadvantages can be overcome by conjugating these chemotherapeutic drugs with nanoparticles and then using them for drug delivery. Drugs administered in this way are taken up by the cell via endocytosis which prevents their repeated uptake by the same cells, thus reducing associated side effects [96].

The use of hydrophilic drugs is still limited due to bioavailability-associated problems, because of which lipophilic carriers are still preferred. Chemotherapeutic drugs like sorafenib which are insoluble in water have been loaded upon diamond NHs functionalized using lipids so as to improve their bioavailability and efficacy for use in cancer therapy. The use of sorafenib coated nanodiamonds (SND) has caused an evident increase in the drug concentrations present within the major body organs, in comparison with the drug present in a suspension. Tumor cells treated with SNDs have caused effective inhibition in cancer cell growth over that of drug suspensions, even under *in vivo* conditions. Thus, SND-mediated treatment can be considered an effective platform for the targeted delivery of chemotherapeutic drugs showing good bioavailability in the case of lipophilic drugs to be used in cancer therapy [97].

In the case of hydrophilic drugs, different approaches have been applied so as to ensure efficient drug delivery in a targeted manner into the cancer cell in order to use them for cancer treatment. For example, Khandelia et al. [98] have used nano-agglomerates made up of gold nanoparticles (AuNPs) and lysozyme which have been stabilized by using albumin for functionalization. These nano-agglomerates were encapsulated by using hydrophilic doxorubicin (DOX) along with the hydrophobic pyrene derivatives (PYR) in order to ensure increased efficiency of drug delivery at the tumor site. It has been observed that the nano-agglomerates loaded with DOX and PYR show initial internalization into the cancer cells followed by drug release. This leads to an increase in the efficacy of cancer cell death compared to drug suspension. The non-toxic nature of the NHs proves an added advantage as it does not show any inherent toxicity. Thus, these NHs have proven to be a promising alternative due to their effective action in targeting the release of chemotherapeutic drugs to the cancer cells.

#### *3.2.1.3.2 Gene Therapy*

Gene therapy makes use of DNA and/or siRNA which is transported to the cytoplasm or nucleus of the cancer cells, which brings about its action. For gene therapy against cancer, a large number of nanoparticles have been developed to deliver DNA and/or siRNA into the cancer cells. This includes the use of different kinds of NHs as well [99]. A great amount of effort is being made to overcome the associated shortcomings such as reduced specificity, a decrease in transfection rate, and cytotoxicity problems associated with the gene delivery systems. Mahor et al. [99] have developed NHs made up of natural as well as synthetic polymers in combination, which are functionalized using compounds containing mannose residues. This has been done to increase the efficiency of transfection, the specificity toward macrophages, and cell viability. Carbodiimide chemistry has been used for the synthesis of these NHs by employing hyaluronic acid (HA) as well as branched polyethyleneimine (bPEI). The functionalization of the polyethyleneiminehyaluronic acid (bPEI-HA) copolymer so formed has been done by using mannopyranosylphenylisothiocyanate at the terminal end, ultimately forming a mannosylated-bPEI-HA (Man-bPEI-HA) copolymer. These NHs have been observed to form polyplexes on association with DNA,

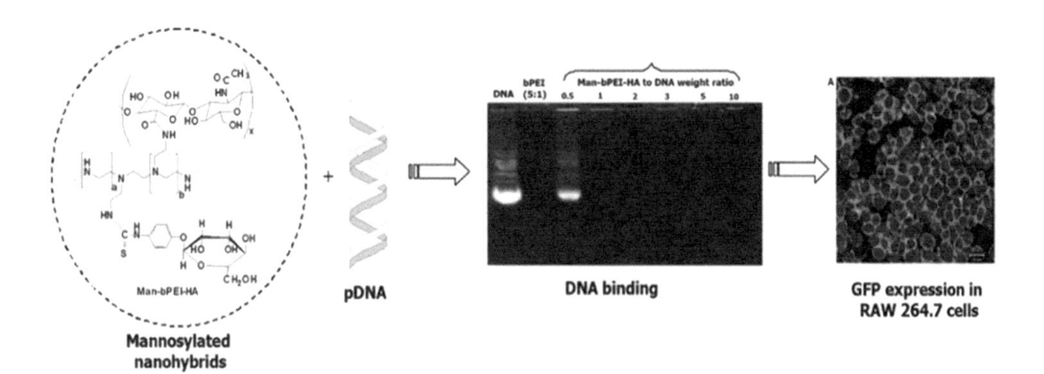


**FIGURE 3.9** Schematic representation of specific targeting of mannosylated NHs functionalized with DNA to be used in gene therapy for cancer in the case of macrophages. (Reproduced with permission from Mahor et al. [99] American Chemical Society.)

which have shown a significant reduction in cytotoxicity compared to individual bPEI. Localization studies using reporter plasmids containing gaussia luciferase (GLuc) and green fluorescent protein (GFP) have shown an observable increase in specificity in the case of mannose-associated NHs when tested on murine as well as human macrophage-like cells (RAW 264.7) and human acute monocytic leukemia cells (THP1), respectively (see Figure 3.9). Thus, these NHs can serve as effective vehicles for gene therapy against cancer.

#### *3.2.1.3.3 Protein Drug Therapy*

Protein drug therapy possesses several advantages over that of other modes of treatment by virtue of its complicated function and high biocompatibility. Protein therapy is considered to be safer compared to gene therapy because it does not involve irreversible or random changes at the genetic level [100, 101]. However, protein delivery still remains a challenging task due to a number of shortcomings such as large size, smaller retention time, permeability issues from the cell membrane, degradation due to the action of cellular enzymes, etc. [102, 103].

Omar et al. [104] have reported the synthesis of silica@iron oxide nanovectors with large mesopores. These mesopores, ranging in size from 20 to 60 nm show increased loading as well as delivery capacities for bulky proteins like mTFPFerritin (about 534 kDa in size) targeted toward the cancer cells. These NHs, consisting of iron oxide nanoparticles which are incorporated into silica walls, show biodegradable characteristics with a retention time of three days. The void space present within the NH can accommodate about 23% by weight of proteins making it a unique delivery system for larger proteins. As these proteins are present within the NHs, they remain protected from enzyme degradation, which is one of the major advantages associated with the protein delivery system.

Yahia-Ammar et al. [31] have reported the synthesis and characterization of AuNCs which self-assemble to form AuNPs by the cationic polymer-mediated method. The monodisperse nanoparticles so formed have shown swelling characteristics dependent upon pH change, enhanced fluorescence intensities, and a significant increase in photostability. The evident enhancement in cellular intake of these NHs functionalized using proteins was observed in human monocytes under *in vitro* conditions compared to individual AuNCs (see Figure 3.10). The incidence of low cytotoxicity during delivery makes its use more favorable for protein delivery.

#### *3.2.1.3.4 Photothermal Therapy*

In recent times, photothermal therapy (PTT) has been developed as a novel strategy for treating cancer. This involves the use of optical radiations present within the NIR range from 700 nm to 2000 nm. After the laser source is focused upon a particular area within the affected tissue, absorption of photon energy occurs within the cell as well as its intracellular compartments. The absorbed photon energy is then converted to generate heat energy. This causes an increase in the corresponding temperature which

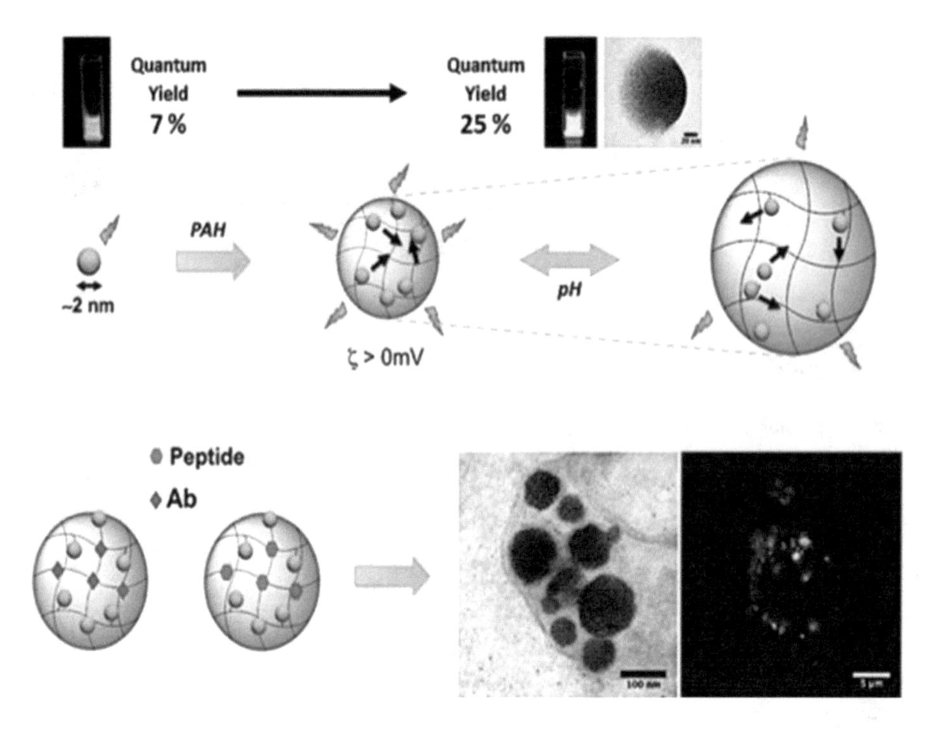


**FIGURE 3.10** Schematic representation of the synthesis of AuNCs using a cationic polymer (PAH) and their application to protein delivery in cells. (Reproduced with permission from Yahia-Ammar et al. [31] American Chemical Society.)

ultimately causes cell death [105, 106]. The use of PTT in the treatment of cancer has a lot of benefits, such as an increase in specificity, negligibly invasive nature, accurate selective features in terms of spatio-temporal properties, etc. [107–111]. PTT can effectively kill cancerous cell existing at the tumor site or control local metastasis in order to treat cancer in the initial stages. PTT has been combined with other modes of treatment to make it more effective in treating cancer at the metastatic sites [112–115]. However, conversion of light into appropriate levels of heat energy using PTT agents governs the effectivity of PTT in cancer treatment. Some nanotherapeutics which are currently used in PTT are metal nanoparticles, nanoparticles consisting of transition metal sulfides and oxides, carbon-containing nanomaterials, organic nanoparticles, etc. [107, 116]. PTT can be used individually or in combination with other modes of therapy in order to make it more effective in treating cancer [117–120].

The synthesis of AuNRs coated with mesoporous silica (AuNR@$SiO_2$) has been reported, which would function as a nanoplatform to be used for cancer treatment. These NHs function in imaging-guided targeted drug release and PTT at the tumor site when induced using a NIR laser [121]. These NHs were further functionalized using poly (N isopropylacrylamide-co-acrylic acid) which is a polymer with thermal and pH-sensitive properties (see Figure 3.11A). These NHs were used for imaging as well as the treatment of cancer cells. They were observed to show enhanced accumulation of the drug at the tumor site when injected intravenously, after which tumor growth was inhibited completely. This makes these NHs a promising alternative in the use of cancer treatment as an effective PTT agent as well as a drug delivery system (see Figure 3.11B) [122].

#### *3.2.1.3.5 Combinatorial Therapy*

With an increase in the incidence of cancer cases, it has become crucial to develop an efficient system for drug delivery which is competent enough to enhance therapeutic efficiency as well as increase satisfaction of the patient during therapy, along with long-lasting recovery. These targets can only be achieved using a synergistic approach in cancer therapy by combining two or more modes of treatment so as to reduce the associated disadvantages of individual modalities as well as related complications. The use of nanoparticles proves of great assistance in the development of such methodologies because of their

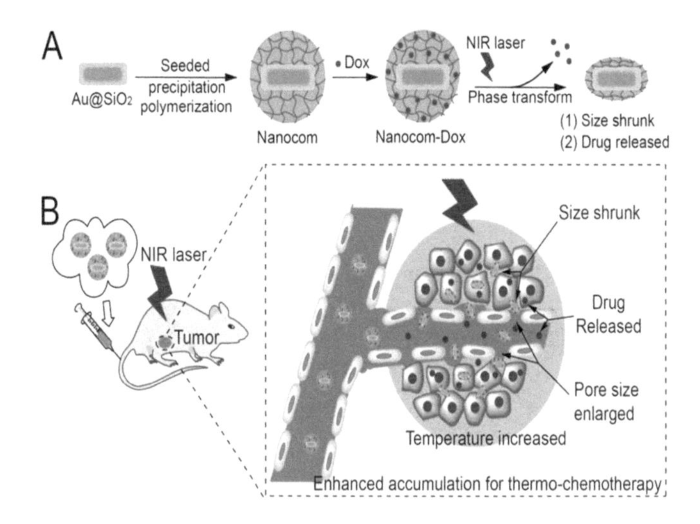


**FIGURE 3.11 (See color insert.)** (A) Schematic representation of the synthesis and functionalization of AuNR@$SiO_2$ (B) Application of these NHs in chemotherapy as well as photothermal therapy for treating cancer cells. (Reproduced with permission from Zhang et al. [122] American Chemical Society.)

multi-faceted nature. The therapeutic potential of these nanoplatforms has been observed to increase in coupling with a diagnostic approach. This further helps in the error-free targeting of tumor cells along with simultaneous tracking of the treatment progress [118, 123, 124]. Silica nanoparticles with mesopores have long been known as hybrid nanomaterials because of their unique properties, which include a porous framework, a stable nature, and congeniality with biological systems [125, 126]. After the loading of the therapeutic drug upon mesoporous silica nanoparticles (mSi), these NHs are functionalized by a variety of chemical agents which can be used as stimulus-triggered systems resulting in opening or closing of systems in response to internal or external signals [126–129].

Baek et al. [130] have developed a novel system by combining cancer diagnosis with therapeutics. Enhancement in the therapeutic efficiency of this system has been brought about by combining drug treatment as well as photothermal therapy with nanoparticles. Figure 3.12 shows the synthesis and

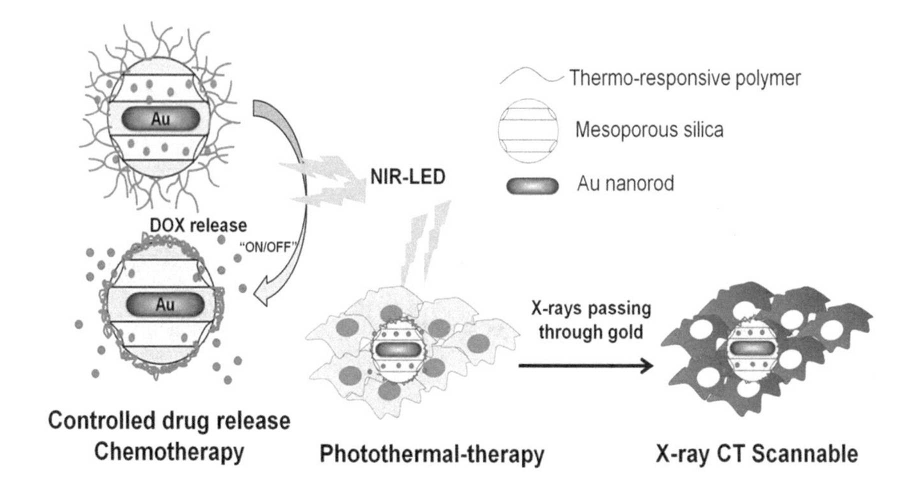


**FIGURE 3.12** Schematic representation showing the synthesis and fabrication of silica-coated gold nanorods (AuNRs) with thermo-responsive as well as pH-sensitive polymer and its application in cancer treatment by targeted drug delivery, photothermal therapy along with simultaneous monitoring via CT imaging with response to stimuli. (Reproduced with permission from Baek et al. [130] American Chemical Society.)

fabrication of silica-coated AuNRs with thermo-responsive as well as pH-sensitive polymers and their application in cancer treatment by targeted drug delivery and photothermal therapy, along with simultaneous monitoring by CT imaging with response to stimuli.

This system consists of AuNRs responsive to NIR radiation which have been coated in mSi nanomaterials. Functionalization of these NHs has been done using a thermo-responsive poly (N-isopropylacrylamide)-based N-butyl imidazolium (poly (NIPAAm-co-BVIm)) copolymer which allows continuous monitoring by CT imaging. Controlled release of doxirubicin (DOX) upon exposure to NIR radiation acts by an ON/OFF system when the loading of the drug onto the NHs is accomplished. When the system is exposed to NIR radiation for about five minutes, the opening of the mesopores present in the silica nanomaterials is triggered. This causes an increase in the temperature of the microenvironment to about 43°C, making these NHs suitable for photothermal therapy of cancer cells. The hydrophilic nature of the polymer so formed causes an enhancement in the uptake rate of the NHs. When these NHs are taken up at the tumor site, the acidic environment of the cancer cells causes the release of DOX. Thus, synergism between targeted chemotherapy as well as photothermal therapy with guided monitoring by CT imaging increases the treatment efficacy in comparison to those with individual treatments.

### 3.2.2 Biosensors

Biosensors are analytical devices mainly composed of a biological sensing part and a transducer. Biosensors are mainly used to produce a measurable electrical signal which is directly proportional to the amount or concentration of any analyte having biological importance. Simply put, biosensors are devices which convert biological or chemical signals into electrical signals. The produced electrical signal is then used to interpret the response. Biosensors are the product of multidisciplinary fields of biology, chemistry, and engineering. The most important aspect of a biosensor is that it should be specific and should yield a response to a particular analyte. It should also be not disturbed and not produce inaccurate signals under varying factors such as pH and temperature. Biosensors are of various types in which the major types are enzyme-based biosensors, cell-based biosensors, immunosensors, thermal biosensors, piezoelectric biosensors, and DNA biosensors [131, 132]. Because of their high specificity, biosensors have applications in a number of fields including:

(i) Food processing
(ii) Fermentation
(iii) Medical fields
(iv) Biochemical engineering
(v) Plant biology
(vi) Biodefense among other applications

The use of biosensors in the biomedical field is growing rather rapidly. Their use for detecting blood sugar levels has gained worldwide recognition and has gained household popularity. They are also used to detect several types of infectious diseases. With the substantial increase of toxigenic, mutagenic, and carcinogenic substances in our immediate environment, the need for detection of these substances has consequently increased. Conventional methods of detecting these harmful substances lack in detecting the overall impact on human health which can be covered using biosensors. Some medical uses of biosensors are:

(i) Detection of endocrine disrupting compounds which in turn can regulate hormone levels
(ii) Detection of genotoxicity
(iii) Detection of pathogens such as pathogenic bacteria and viruses
(iv) Detection of bacteriophages
(v) Detection of biofilms using specific markers

Additionally, these biosensors can be made smaller in size for incorporation in minute devices and cost-effectiveness [133]. Among various types of biosensors, electrochemical biosensors have been widely used for the longest period of time, starting with the first biosensors in the 1960s. Enzyme-based biosensors are mainly electrochemical in type and constitute most of the biosensors commonly used today. In this type of biosensor, an enzyme is mainly immobilized and acts as a catalyst for the analyte in the sample and produces electrical signals corresponding to the concentration of the analyte. Various types of biosensors are present within these criteria. In the biomedical field, enzyme-based biosensors have several advantages over other conventional methods such as rapid tests, real-time diagnosis, and a high degree of sensitivity and specificity, while being cost-effective at the same time. Because of these reasons, enzyme-based sensors are the most used biosensors. The most common example of the enzyme-based biosensors are the "glucose biosensors" which are available as strip tests nowadays. Other commonly used biosensors include sensors for cholesterol, lactate, and alcohol present in blood and urine [134, 135]. Another class of popular biosensors in the biomedical field are the cell-based biosensors. In cell-based biosensors, biochemical changes are detected directly by the incorporated living cells. With the help of transducers, the deteced biochemical signals are then converted into electrical signals. Cell-based biosensors offer advantages such as rapid and *in situ* monitoring of biochemical analyte. Cell-based biosensors also provide better results for bioactive compounds compared to molecule-based biosensors. They provide non-invasive, label-free, and quick-response tests, because of which they have been widely employed in pharmaceutical monitoring, biomedical diagnosis, physiological analysis, and other uses. Cell-based systems can also be used for drug screening and toxin detection. *In situ* monitoring is a very important aspect and increases the potential of cell-based biosensors in the biomedical diagnosis field [136]. Figure 3.13 represents a cell coupled sensor with modified surface morphologies of various kinds including proteins, extracellular matrix, self-assembled monolayer, etc.

Immunosensors incorporate the use of antibody/antigen interaction to produce electrical signals, i.e., the events of the formation of an antibody-antigen complex are converted into measurable electrical signals. Antibodies are proteins which have an affinity for specific molecules known as antigens. This makes immunosensors highly specific. The receptor is immobilized on the biosensor surface for the analyte to interact with the receptor. This immobilized receptor has shown reusability. Apart from being highly specific, immunosensors also allow real-time monitoring and are cost-effective due to their property of reusability. Overall, they offer speed, cost-effectiveness, non-invasiveness and multiple assays, which make them ideal for diagnostic purposes. Immunosensors are mainly used for the detection of antibodies, biomarkers, toxins, and microorganisms, among other things. They have also shown potential in the detection of tumor biomarkers, thus leading to the early detection of cancer [137–139].

The previously mentioned classes of biosensors are extensively used for biomedical applications. However, these types of biosensors also suffer from drawbacks and other limitations. The incorporation of nanotechnology into the field of biosensors has shown great potential and can be used to overcome several of these limitations. Due to the nanoscale size of nanomaterials, they have revolutionized the

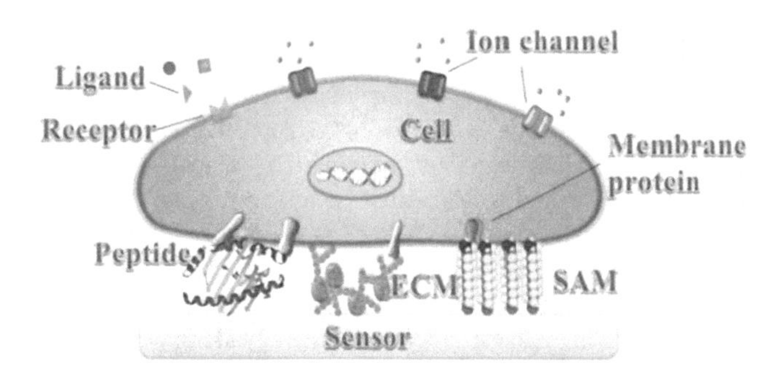


**FIGURE 3.13** Schematic illustration of cell-coupled sensor with modified surface morphologies of various kinds including proteins, extracellular matrix, self-assembled monolayer, etc. (Reproduced with permission from Liu et al. [136] American Chemical Society.)

field of biosensors. The use of nanotechnology in the making of biosensors can result in much smaller biosensors, which in turn can be used for non-invasive measurements. Also, the use of nanomaterials can increase the sensitivity and specificity of the biosensors. The lower detection limit of the biosensor is also reduced with the incorporation of nanomaterials [140]. Different types of nanomaterials are incorporated in various biosensors according to their properties. Nanomaterials such as nanoparticles, nanotubes, and quantum dots are usually used among other nanostructured materials. Nanoparticle-based biosensors incorporate the use of nanoparticles of different elements which confer different properties. Metallic nanoparticles are rather famous due to their electronic and electrocatalytic properties. Nanoparticle-based biosensors are comparatively easy to synthesize and can be produced in bulk. Nanoparticles with magnetic properties have shown potential in the rapid detection of biological targets. The use of carbon nanotubes has gained popularity in the last two decades with the rise in nanoscience. Studies have shown that carbon nanotubes can enhance the electrocatalytic activity of biosensors and promote electron transfer for improved results. Nanoparticles synthesized from biological compounds or incorporating biological compounds show greater sensitivity. The presence of biological compounds makes the biosensors highly specific. Natural as well as synthetic biological compounds are used for this purpose [141]. The incorporation of nanomaterials can benefit the biological sensor element, transducer or both. This has made the overall mechanism a lot quicker, more cost-effective, and more user-friendly. Plus, the advancement of nanotechnology has led to subsequent miniaturization and has improved their application in sensing key events of specific regulatory pathways. The stated abilities of nanomaterials have led to significant contributions in sensitive detection of biomarkers, nucleic acid, biological analyte, etc. [142].

As already discussed, NHs are nothing but multicomponent assemblies of two or more nanostructures. The conjugation of different nanomaterials imparts the properties of individual nanomaterials. Thus, these assemblies where two or more nanomaterials are conjugated to possess multi-functionality are known as NHs. These NHs show alteration in properties of the incorporated nanomaterials. This alteration of the initial properties results in the emergence of novel characteristics. These novel characteristics make NHs potent in a lot of areas including biosensing. These novel characteristics lead to easy functionalization of NHs and have novel electrochemical configuration [18].

A lot of research has already been done integrating novel NHs in the field of biosensing. Biosensors made from various different combinations of NHs have been made and tested. However, the integration of NHs in biosensors is still relatively new and is expanding with every passing day. One such hybrid film is made using NHs of reduced graphene oxide to form a hybrid assembly in which graphene oxide is dispersed throughout "Nafion," which is a functional polymer. Figure 3.14 shows the design of reduced graphene oxide/nafion (RGON) hybrids followed by its subsequent application, as electrochemical biosensors. This NH film assembly showed enhanced electron transfer and low interfacial resistance. It showed great mechanical and electrochemical properties making it suitable for use in biosensors. It also showed a reduction in response time and increase in sensitivity. The hybrid film is also flexible, deeming it useful in strip tests as well [143].

Focusing on the most popular form of biosensors, i.e., glucose biosensors, the use of NHs for the development of non-enzymatic biosensors has been of keen interest in the field of research. As known, the usual glucose biosensors incorporate the use of enzymes such as glucose oxidase for the detection of glucose in the blood. However, the enzyme-based biosensors show thermal and chemical instability, limiting their use. This can be overcome by using non-enzymatic biosensors. In one such piece of research, palladium nanoparticles were conjugated with graphene to form palladium NP–graphene NHs dispersed in Nafion. Palladium is used because of its excellent catalytic properties and its NPs are already used for glucose sensing [144]. Graphene, due to its previously stated properties of high conductivity and ease of functionalization, was used with palladium nanoparticles embedded on a Nafion polymer. This NH-based glucose biosensor showed high electrocatalytic activity along with great glucose sensitivity. A linear dependency was also observed with glucose concentration change. These features made palladium NP–graphene NH-based sensors highly viable for use in glucose biosensors [145]. Like glucose biosensors, urea biosensors are also used in a similar fashion for the determination of urea in a patient's body for the diagnosis of diabetes and other pathological problems, such as dysfunction of the kidneys and other related organs, as these diseases result in a variation of urea levels in the urine as well as the blood.

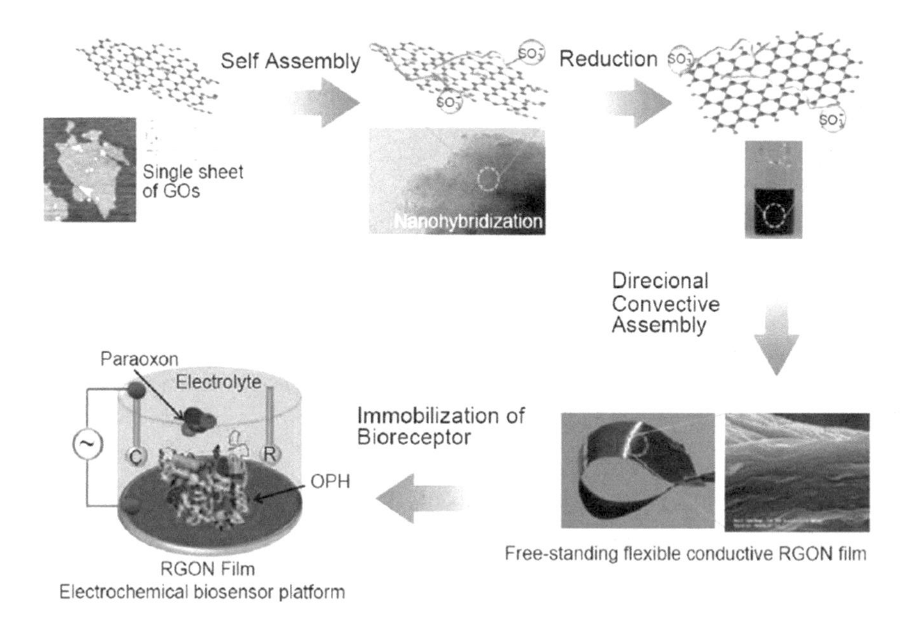


**FIGURE 3.14** Schematic representation showing process employing design of RGON hybrids followed by their subsequent application as electrochemical biosensors. (Reproduced with permission from Choi et al. [143] American Chemical Society.)

Layered double hydroxides (LDH) have great value when used with biomolecules due to their anionic exchange properties, as biomolecules are usually negatively charged. In one such preparation, urease is immobilized in zinc–aluminum LDH i.e., (Zn–AL)LDH which is synthesized using Zn–Al NHs creating a bio-NH material. This preparation allows a substantial amount of urease to be immobilized. The structure and activity of the enzyme were also preserved. Thus, it showed potential use of the synthesized (Zn–Al)LDH NH in the preparation of urea biosensors [146].

Nucleic acids make up the genetic material of our body and contain vital information and thus is required for essential life processes such as growth, hereditary, and overall morphology. Therefore, detection of specific DNA is very important to find any kind of mutation which can lead to severe genetic disorders. The early diagnosis of such mutations can help in immediate treatment and lead to a healthy life. AuNCs can be synthesized in the presence of graphene to synthesize AuNC–graphene NHs. Graphene is used as the base for the distribution of AuNCs because of its excellent conductivity. Together, graphene and AuNCs have great potential in biosensing applications due to the great conductivity and non-toxicity of AuNCs. This novel NH is coupled with Exo III-aided cascade target recycling for DNA recognition. In exo-aided target cycling, cleavage takes place in the presence of the target and target binding initiates hydrolyzes after which the target is released. The final product of this cleavage is then picked by the electrochemical biosensor with a AuNP–graphene NH as the interface for enzyme-catalysis. Exo III specifically targets the 3' end of a DNA to produce target products. This novel composition for DNA biosensor has advantages like ultra sensitivity and selective determination of target sequence [147].

### 3.2.3 Tissue Engineering

The tissues and bones present in our body are made up of complex structural features and specific architecture which perform respective biological functions in our body. This architecture also allows the growth and proliferation of cells of the tissue and the formation of extra cellular matrices (ECMs). Damage to these tissues due to an injury or disease results in the loss of their function. This is usually overcome by replacing the tissue or the organ itself. However, the major drawbacks are the immunological problems faced during transplantation and the significant lack in the number of organ donors.

This problem can be solved by implementing tissue engineering. Tissue engineering incorporates both engineering sciences and biosciences for the treatment of damaged tissues. It incorporates the study of biochemical and physical factors for providing tissue-like structures or neotissue formation. It has basically three types of approaches:

(i) Replacement of damaged tissue by isolating cells or other cell substitutes
(ii) Delivery of substances which induce tissue growth and differentiation
(iii) Growth of cells in three-dimensional scaffolds

The major goal of tissue engineering is not only replacing the damaged tissue but also to improve and maintain tissue function [148, 149].

Scaffolds are three-dimensional constructs made possible by tissue engineering. Scaffolds provide a framework for growing cells by inducing cell growth and synthesis of ECM and other biological compounds [150]. Certain properties are required for constructing a successful scaffold. A scaffold:

(i) Must be biocompatible, so it doesn't induce an immune response
(ii) Must be biodegradable to be later diffused and let the natural tissue take over
(iii) Should have intended mechanical strength
(iv) Should have a porous structure for the synthesis and flow of ECM for improved tissue regeneration

In the case of bone and cartilage generation, scaffolds also provide the framework for a specific shape, maintenance of cell proliferation, and regaining of biological function. Hence, scaffolds have become a promising technique in the field of orthopedics. Research has already been done to find biodegradable scaffolds for bone tissue engineering in which natural compounds such as chitin, collagen, and hydroxyapatite (HAP) have gained popularity because of their biocompatibility and biodegradability [151]. Using polymers for designing scaffolds has major advantages over non-polymer scaffolds as several characteristics such as biodegradability, strength, and porosity and its structure can be easily controlled using various polymers. Using biomaterials for designing scaffolds makes them biocompatible, which is crucial for the acceptance of a scaffold within a body. Biological substances such as HAP, chitosan (CS), calcium phosphate (CaP), and types of silicates have several advantages but have some disadvantages as well, such as their brittle nature and reduced biodegradability. These can be overcome by conjugating these biological materials with synthetic material such as polyglycolide, polylactic acid etc. to form conjugates [152].

As discussed, naturally occurring compounds such as HAP, collagen, CaP, CS, etc. are readily biocompatible when compared to synthetic polymers. Hence, the use of these natural bioactive compounds has gained a large focus in the manufacturing of scaffolds to be used in tissue engineering.

#### 3.2.3.1 Hydroxyapatite

HAP is the mineral form of calcium apatite which occurs naturally and is found to be similar to many bone phosphates, which makes it very significant among other promising scaffold materials. It is also the most stable at pH 4–14 at room temperature compared to other CaP derivatives. These and more features of HAP make it viable for many applications. Biomedical applications associated with HAP are mainly due to its bioactive and biocompatible characteristics. Furthermore, it has been found to increase osteogenesis and thus helps in bone tissue regeneration [153].

#### 3.2.3.2 Collagen

Collagen is a naturally occurring protein which is present in abundance in ECMs. It possesses a fibrillar structure which contributes to its extracellular scaffolding. Collagen provides physical support to the cells as well as playing an essential role in maintaining the integrity of the ECM. It is also the major structural protein present in soft as well as hard tissue. Due to these features, collagen is present in the ECM of important tissues, such as blood vessels, bones, cartilage, skin, tendons, ligaments, etc. Collagen has several advantageous properties such as good permeability, biocompatibility, and biodegradability and it boasts a porous structure. Based on these properties, collagen can support and regulate adhesion, differentiation, and migration of cells. These functions of collagen make it viable to be used in the manufacturing of scaffolds [154].

### 3.2.3.3 Calcium Phosphate

CaP plays a major role in our daily lives as it is the main constituent of bone and teeth. Several derivatives of CaP are present in our body with different structures and functions [155]. The main function of CaP is providing mechanical strength to these tissues. CaP derivatives, such as tricalcium phosphate and HAP, are widely known because of their potential use in the field of tissue engineering. CaP-based materials are gaining popularity because of their potential to enhance osteoblast differential. These materials can degrade at required rates and gets completely replaced by the original tissue. Another type of CaP, biphasic calcium phosphate (BCP) can be synthesized such that its degradation rates can be controlled. CaP material possesses the features required for making viable scaffolds such as osteoinductivity, porosity, osteoconductibility, biodegradability, and biocompatibility. The above factors make CaP materials significant in the formation of scaffolds [156].

### 3.2.3.4 Chitosan

CS is a linear polysaccharide made of mainly two units, namely, n-acetyl D-glucosamine and D-glucosamine. CS is a naturally occurring polymer but is not extensively present in our surroundings. However, CS can be derived from chitin which is present in abundance in nature [157]. CS is known to have biocompatible and biodegradable properties and it is soluble in aqueous as well as organic solvents. It is a bioactive polymer, having high flexibility, low toxicity, and some antimicrobial properties. The physiochemical properties of CS can be controlled at the time of de-acetylation of chitin to form CS. It can also be synthesized in many forms such as nanoparticles, gels, nano- and microfibers, membranes, and beads. Because of its stated features, CS has several applications in tissue engineering and the manufacturing of scaffolds with different properties [158].

As discussed, a scaffold must have certain features which should promote cell attachment, differentiation, and proliferation. Thus, the materials chosen for the making of these scaffolds are of great importance as scaffolds made of different materials exhibit different features. Also, scaffolds are mainly made by mixing two or more different materials to reduce the drawbacks and improve the overall performance. Synthetic polymers can be conjugated with natural polymers to exhibit properties of the synthetic polymers and have great biocompatibility. This opens up a wide range of combinations featuring different properties, and thus advantages [159].

Nanostructures have shown great importance in the field of tissue engineering. It has been also found that introducing nanostructures in scaffolds mimics the deposition of collagen and apatite fibers in ECMs, and thus affects the cell response. Thus, nanofibrous scaffolds overall mimic natural ECMs and promote differentiation and proliferation. The presence of nanoparticles in the scaffolds also regulates the cell regeneration [160]. Thus, it can be said that incorporating nanoscale material can enhance the overall functionality of the scaffolds.

The repair of injured or lost bone tissue due to an injury is still a big problem in subsequent treatment and recovery. This can be overcome by the use of a suitable scaffold for repair and regeneration of damaged bone tissue. Bone present in our bodies is a biomineralized structure in which HAP nanocrystals are deposited on collagen fibers. A similar scaffold mimicking natural bone formation can be used in the treatment of injured bone tissue. In one such piece of research, CS was used as the basis of the scaffold due to its bioactivity and its biocompatible and biodegradable properties. The pH-dependent solubility and the precipitation of HAP can be exploited for the synthesis of NH scaffolds using a freeze-drying method. It was found that nanoHAP of size 90 to 200 nm was scattered homogenously in the scaffold mimicking natural bone. The resultant NH scaffold was found to be non-toxic and bone marrow cells showed migration and proliferation and further infiltration in the synthesized scaffold. This confirmed the potential use of these NHs as a biocomposite scaffold in the field of tissue engineering [161]. Further incorporation of other bioactive compounds in the scaffolds can enhance its functions, and thus overall applicability. As stated previously, HAP and CS are already known for their properties and are used in the manufacturing of scaffolds. Chondroitin sulfate (CSA) is a compound which is a component of ECM and may improve secretion of proteoglycan and collagen. A NH scaffold of hydroxyapatite, CS, and CSA was prepared by a fabrication and freeze-drying technique which was found to be suitable for cell attachment, differentiation, and proliferation. Using an MTT assay, it was also seen that the introduction of nano-crystalline HAP in CS/CSA framework enhances the proliferation and differentiation of osteoblasts [162].

Damaged bone treatment is often done by replacing the tissue by either autologous or allogenous tissue, which is limited in supply and which may transmit diseases, respectively. As we know, using scaffolds for repair and regeneration of tissue causes no ailments and is readily available. Another piece of research, in which a polyelectrolytic complex (PEC) of HYA and CS combined with nanoHAP to form a composite scaffold for bone tissue engineering was performed. PECs are formed when polymers of opposite charges are reacted with each other. These PECs have several features, because of which they are used in various fields such as tissue engineering, drug delivery, prosthetics, etc. HA is mainly present in the ECM and readily absorbs and retains water and enhances other cellular function such as adhesion, migration, and proliferation. NanoHAP was found in the porous PEC skeleton and ranged in size from 54 to 147 nm. MTT assay showed more proliferation of bone marrow cells in the HA/CS/nanoHAP scaffold compared to a HA/CS scaffold confirming that nanostructured HAP plays an important role in the proliferation of cells. The distribution assay confirmed the migration and differentiation of the bone marrow cells in the PEC and nanoHAP complex, which shows good pore structure and high porosity of the designed scaffold. These results confirm the viability of these scaffolds in bone tissue formation [163].

Composite scaffolds have many advantages such as stability, osteoconductivity, biocompatibility, etc. but many lack the property of osteoinductivity. This can be solved by loading cytokines on the composite scaffolds to induce osteogenetic differentiation. However, cytokines are replaced by bioactive trace elements, such as $Sr^{2+}$, $Zn^{2+}$, $Mg^{2+}$, $Cu^{2+}$, and $Si^{4+}$ due to uncontrolled release of cytokines. Among these, strontium (Sr) has been known to enhance neotissue formation and due to its ionic radius, similar to that of calcium, thus can be used to substitute calcium in HAP. Nanocrystals of strontium–HAP (SrHAP) were first synthesized using a co-precipitation method and NH scaffold of SrHAP nanocrystals with CS using a freeze-drying method. The crystal size of SrHAP nanocrystals was found to range from 70.4 nm to 46.7 nm when the number of substituted Sr ions were increased from 0 to 1, then 5, and lastly to 10. These nanocrystals were found to be uniformly distributed along the microporous scaffold. The synthesized NH scaffold showed great cytocompatibilty and further aided the adhesion, migration, and proliferation of bone marrow cells. The release of Sr ions also helped in cell proliferation and osteogenic differentiation [164].

Scaffolds are not only useful in case of bone repairs but also in the treatment of dental-related injuries. Dentin is one such component present beneath the enamel which provides support to the enamel, in the absence of which enamel would fracture upon mastication. Dentin is mainly composed of collagen, carbonated apatite, and dentinal fluid. The synthetic material commonly used for filling/replacing dental tissue doesn't show similarity to actual bone tissue, because of which complete restoration can subsequently fail. Therefore, a nanocomposite of ethyl methacrylate (EMA) and hydroxyethyl acrylate (HEA) was chosen because of their properties of mechanical strength and hydrophilicity, respectively. This nanocomposite was used to prepare copolymer scaffold, i.e., P(EMA-co-HEA). Additionally, 15% w/w of silica was added because of its property to confer bioactivity to form a P(EMA-co-HEA)$SiO_2$ nanocomposite scaffold. The scaffold with silica showed a covering of HAP after being submerged in simulated body fluid (SBF). The prepared scaffold showed physical and histological similarity to natural dentin tissue. It was found that cell colonization and viability was enhanced in the presence of HAP, and thus the HAP-incorporated P(EMA-co-HEA)$SiO_2$ scaffold showed better results compared to P(EMA-co-HEA)$SiO_2$ without HAP coating and P(EMA-co-HEA) scaffolds [165].

In the case of diabetes, wounds caused by injury can result in death. Therefore, diabetic wounds are still a major cause of death among diabetic patients. Diabetic wound healing is slow and incorporates many factors; however, inflammation due to infection can cause weakening and delay wound healing. Collagen scaffolds have gained recognition because of its already stated features; however, due to its instability, it is not an ideal material. This can be improved by crosslinking with other polymers. Thus, the incorporation of alginate in the collagen matrix has been done to improve its stability. Furthermore, curcumin was loaded in a nanoparticle solution of CS to exploit its anti-inflammatory and antioxidant properties. DOX is known for its broad range of antimicrobial activity and can decrease the risk of infection. Thus, curcumin-loaded CS nanoparticles (CUR–CSNPs) were incorporated into collagen alginate (COL–ALG) scaffold along with DOX. This scaffold showed retention of drugs up to 15 days in the CUR–CSNPs incorporated into the COL–ALG scaffold. The combination of DOX and CUR resulted in a significant reduction in inflammation and also protected the collagen from degrading. Thus, the synthesized NH scaffold fulfill the requirements of an ideal scaffold to be used in diabetic wound healing featuring needed physical properties, such as mechanical strength, porosity, biocompatibility, and biodegradability. The scaffold also showed controlled release of drug, cell adhesion, proliferation, and anti-inflammatory and anti-bacterial properties which can be deemed crucial in diabetic wound healing [166].

#### *3.2.3.5 Vaccine Development*

Nanomaterials have been used for various biomedical applications for quite some time due to their unique and adjustable properties. However, nanomaterials can be toxic and non-specific, and thus have limited functionality. This can be overcome by functionalization of the nanomaterials using biological compounds and thus forming NHs [167]. As discussed in the previous section, NHs can be used for drug delivery because of their characteristic of combining different materials for a targeted and constant release of drugs *in vivo*. This property of NHs is also applicable in the field of vaccines. Vaccines for different kinds of diseases can be conjugated or loaded inside these hybrids for better delivery. Vaccines coated on solid surfaces can result in conformational changes in conjugates and decrease the activity of the vaccine, for example, coating on alum (aluminum hydroxide), one of the most used adjuvants for creating vaccines [168]. Thus, the need for new carriers is increased, which can improve the overall interfacial interaction. Lipid-based nanocarriers have become a focus of attention due to their high biocompatibilty and low toxicity to cells and tissues [169].

Sepiolite is one such organic derivative of silicate which can yield hybrid materials as it interacts with various organic compounds either on the outer surface or internal surface, as shown in Figure 3.15.

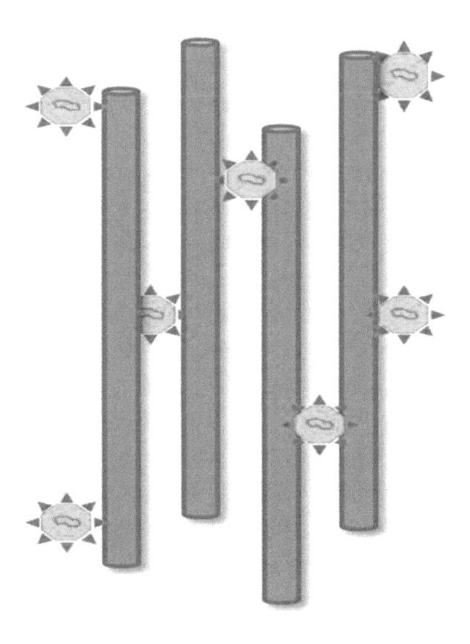

**FIGURE 3.15** Schematic representation of a vaccine conjugated on sepiolite, a derivative of silicate, for vaccine delivery.

Due to its microfibrous nature, it can be essentially used as the building block of these hybrids [170]. Influenza is a very common disease that can be deadly, having three subtypes, A, B, and C. Influenza A emergse from avian sources with its own subtypes. It has been known to cause many pandemics in the past with a recent case being due to the newly found H1N1 strain. Therefore, vaccination against influenza A is necessary for decreasing the possible risks [171]. Sepiolite in conjugation with a lipid, i.e., phosphatidylcholine, was used to form SEP–PC NHs as an adjuvant for the influenza A vaccine. LDHs of Mg/Al were also used in place of sepiolite with phosphatidylcholine to form Mg/Al DLH-PC for further studies while working with influenza A vaccine. Both of these bionanohybrids showed better enzyme activity of the attached virus as compared to alum which is as said, is generally used. Thermal stability tests confirmed SEP–PC hybrids showed better immobilization conditions compared to the alum adjuvant, which could be due to the presence of lipid membrane. A strong immunogenic response was also recorded in mice with an increase in specific antibodies (of TH1 profile) [172].

Mostly, the vaccines available today are delivered intravenously with the use of conventional injections, which have several drawbacks, such as needle injuries and the chance of spreading infections. Additionally, administration of these vaccines requires an expert practitioner. Thus, the need for needle-free vaccination techniques is growing and is becoming incorporated into the basic description of an ideal vaccine. Mucosal administration is one such needle-free vaccine delivery with several benefits such as safety, low cost, and easy administration. Until now, only a few suitable mucosal vaccine adjuvants have been approved for human use. This is primarily because of the absence of effective and safe adjuvant [173]. Although, studies have shown that intranasal vaccination can initiate increased immune response in the patient's body as approximately 75% of the immune cells are present in the lining of the nasal cavity and furthermore, it is known that a particulate delivery system with a size of less than 250 nm is ideal for intranasal vaccination [174]. Thus, the use of NHs can increase the efficiency of such vaccinations due to its small size and larger surface area, resulting in more interaction with the target tissues. CaP is one such formulation which has shown increased biocompatibility and non-toxicity. It was also found to be almost as potent as alum adjuvant which is usually used in humans when tested on guinea pigs [175]. Previous studies also reported that amorphous CaP formulations have good adhesion to cells as well as tissues. On the basis of such reports, a study based on virus (DENV2) plaque formulation assay, ELISA etc., intranasal administrations and adsorption tests were also performed in four to six week old female BALB/c mice using the same NH formulation. Results of these tests confirmed the release of DENV2 from the CaP shell and showed similar antigenicity as the native vaccine. Nasal adhesion tests showed a 2.5-fold increase in the adhesion as compared to a free virus, and thus suggested that this NH formulation increased its interaction with the nasal tissues. An increase in the levels of nasal antibody IgA against DENV2 was also detected as compared to the free vaccine confirming an efficient viral release in the nasal tissue [176].

As discussed earlier, CaP is one such biomineral used as an adjuvant because of its non-toxicity, biocompatibility, and biodegradability. Furthermore, its inability to produce immunogenic response makes it more viable for use as an adjuvant for various vaccines [177]. As tested on animals, recombinant adenovirus serotype 5 (rAd5) vector-based vaccines have shown a highly increased response. rAd5-based vaccines have the potential to be used against various infectious diseases. However, a major limitation in using these particular vaccines is the already acquired immunity in most of the human population against adenovirus serotype vectors due to natural exposure. This acquired immunity decreases the overall immunogenicity and efficiency of the vaccine [178]. This can be overcome by using a biomineralizing technique, i.e., forming a degradable core–shell structure incorporating a rAd5 envelope protein (rAd5-env) core and CaP shell. The CaP shell surrounds the viral surface and bypasses the immune response which can be elicited by a naked virus and decreases the efficiency of the vaccine. Thus, the CaP shell retains its viral activity. The hybrid structure was confirmed using TEM (transmission electron microscopy) and the size of the core–shell structure was found to be 70–80 nm. Dot blot assay was also performed to support the core–shell structure as the hybrid was not detected by its specific antibody, whereas rAd5 was detected by the same. Under denatured conditions, the hybrid was also detected affirming the pH sensitivity of the CaP shell, which was found to be readily degraded at $pH < 6$ making it more efficient as an adjuvant. ELISA and ELISPOT assays were also performed which further confirmed that the hybrid preserves the immunogenicity and activity of the vaccine. Cellular uptake tests were also performed using RAW264.7 cell lines with both rAd5 and rAd5–CaP, which proved that the

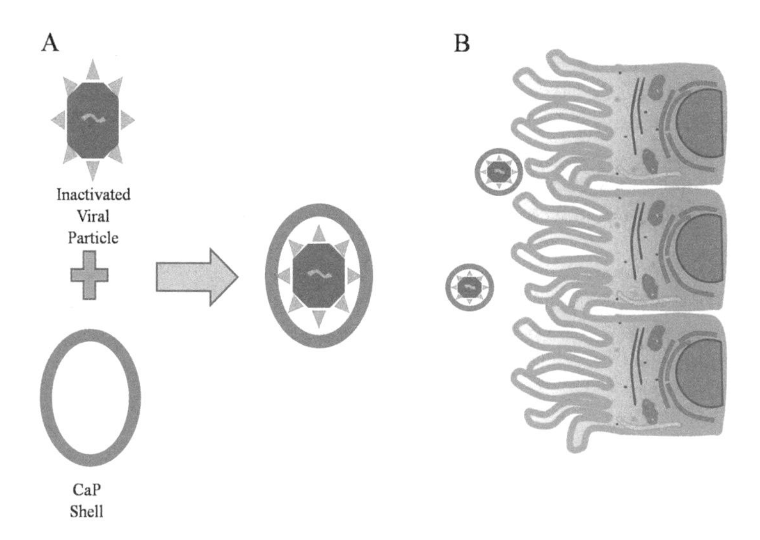


**FIGURE 3.16** (A) Encapsulation of vaccine inside a degradable CaP shell to form a functional bionanohybrid. (B) Attachment of CaP shells containing the vaccine on the cell surface for better vaccine delivery.

intake was greater with rAd5–CaP compared to native rAd5, therefore confirming the importance of the hybrid structure in subsequent vaccine delivery [179]. To illustrate this in a more effective manner, the formation of a CaP NH consisting of the vaccine to be targeted (see Figure 3.16A) along with its effective delivery to the cell has been shown (see Figure 3.16B).

## 3.3 Conclusions

NHs have been observed to show various novel properties in comparison with their individual components, thus increasing their overall effectivity in a wide range of applications. NHs serve as an excellent framework to be used for the functionalization of biomacromolecules. These functionalized NHs have shown remarkable applications in the field of biomedicine which includes the detection and treatment of cancer, vaccine development, biosensing, and tissue regeneration. These functionalized NHs then show enhanced properties which are exploited for applications in various fields.

However, there are some shortcomings associated with the use of these NHs which need to be addressed so as to make their applicability more versatile, which have been listed as follows:

(i) NHs which are available currently such as nanovesicles consisting of AuNPs range in particle size from 100 to 200 nm. A remarkable reduction in particle size of these NHs is needed so as to bring them within the range of 50 to 100 nm. NHs of this size will be feasible to be used as delivery systems as well as tissue scaffolds.

(ii) Detailed monitoring of the action of NHs on a quantitative basis needs to be done apart from applications involving "proof of concept" experiments. The effectivity of these NHs along with their extent of aggregation is still critical to understanding the exact functioning of these NHs.

(iii) Issues such as biocompatibility, retention time, and biodistribution need to be addressed with utmost priority as most of the reports on NHs document extracellular properties, cellular responses under *in vitro* conditions. *In vivo* studies in animal models need to be done to make them suitable for clinical trials.

## REFERENCES

1. Sur UK, Ankamwar B, Karmakar S, Halder A, Pulak Das P (2018) Green synthesis of silver nanoparticles using the plant extract of Shikakai and Reetha. *Mater. Today* 5: 2321–2329.

2. Kirtiwar S, Gharpure S, Ankamwar B (2019) Effect of nutrient media on antibacterial activity of silver nanoparticles synthesized using Neolamarckia Cadamba. *J. Nanosci. Nanotechnol.* 19(4):1923–1933.
3. Jin Y, Li A, Hazelton SG, Liang S, John CL, Selid PD, Pierce DT, Zhao JX (2009) Amorphous silica nanohybrids: Synthesis, properties and applications. *Coord. Chem. Rev.* 253: 2998–3014.
4. Aich N, Masud A, Sabo-Attwood T, Plazas-Tuttle J, Saleh NB (2017) Dimensional variations in nanohybrids: property alterations, applications and considerations for toxicological implications. In: Lockwood DJ (ed) *Anisotropic and Shape-Selective Nanomaterials: Structure-Property Relationships*, Springer International Publishing AG, Switzerland.
5. Wang J, Song Y (2017) Microfluidic synthesis of nanohybrids. *Small* 13(18): 1604084.
6. Sur UK, Ankamwar B (2016) Optical, dielectric, electronic and morphological study of biologically synthesized zinc sulphide nanoparticles using *Moringa oleifera* leaf extracts and quantitative analysis of chemical components present in the leaf extract. *RSC Adv.* 6: 95611–95619.
7. Kim H, Achermann M, Balet LP, Hollingsworth JA, Klimov VI (2005) Synthesis and characterization of Co/CdSe core/shell nanocomposites: Bifunctional magnetic-optical nanocrystals. *J. Am. Chem. Soc.* 127: 544–546.
8. Gaur S, Wu H, Stanley GG, More K, Kumar CSSR, Spivey JJ (2013) CO oxidation studies over cluster-derived Au/$TiO_2$ and AUROlite™ Au/$TiO_2$ catalysts using DRIFTS. *Catal. Today* 208: 72–81.
9. Carroll KJ, Hudgins DM, Spurgeon S, Kemner KM, Mishra B, Boyanov MI, Brown LW, Taheri ML, Carpenter EE (2010) One-pot aqueous synthesis of Fe and Ag core/shell nanoparticles. *Chem. Mater.* 22: 6291–6296.
10. Zhang ZC, Xu B, Wang X (2014) Engineering nanointerfaces for nanocatalysis. *Chem. Soc. Rev.* 43: 7870–7886.
11. Song Y, Ji S, Song YJ, Li R, Ding J, Shen X, Wang R, Xu R, Gu X (2013) In situ redox microfluidic synthesis of core–shell nanoparticles and their long-term stability. *J. Phys. Chem. C* 117: 17274–17284.
12. Jiang R, Li B, Fang C, Wang J (2014) Metal/semiconductor hybrid nanostructures for plasmon-enhanced applications. *Adv. Mater.* 26: 5274–5309.
13. Zhang J, Tang Y, Lee K, Ouyang M (2010) Nonepitaxial growth of hybrid core-shell nanostructures with large lattice mismatches. *Science* 327: 1634–1638.
14. Serpell CJ, Cookson J, Ozkaya D, Beer PD (2011) Core@shell bimetallic nanoparticle synthesis via anion coordination. *Nat. Chem.* 3: 478–483.
15. Xu Z, Hou Y, Sun S (2007) Magnetic core/shell $Fe_3O_4$/Au and $Fe_3O_4$/Au/Ag nanoparticles with tunable plasmonic properties. *J. Am. Chem. Soc.* 129: 8698–8699.
16. Li JF, Huang YF, Ding Y, Yang ZL, Li SB, Zhou XS, Fan FR, Zhang W, Zhou ZY, Wu DY, Ren B, Wang ZL, Tian ZQ (2010) Shell-isolated nanoparticle-enhanced Raman spectroscopy. *Nature* 464: 392–395.
17. Chen G, Zhao Y, Fu G, Duchesne PN, Gu L, Zheng Y, Weng X, Chen M, Zhang P, Pao C-W, Lee J-F, Zheng N (2014) Interfacial effects in iron-nickel hydroxide–platinum nanoparticles enhance catalytic oxidation. *Science* 344: 495–499.
18. Aich N, Plazas-Tuttle J, Lead JR, Saleh NB (2014) A critical review of nanohybrids: Synthesis, applications and environmental implications. *Environ. Chem.* 11: 609–623.
19. Zhou B, Wu B, Wang J, Qian Q, Wang J, Xu H, Yang S, Feng P, Chen W, Li Y, Jiang J, Han B (2018) Drug-mediation formation of nanohybrids for sequential therapeutic delivery in cancer cells. *Colloids Surf. B Biointerfaces* 163: 284–290.
20. Wang X, Yang L, Chen Z, Shin DM (2008) Application of nanotechnology in cancer therapy and imaging. *CA Cancer J. Clin.* 58: 97–110.
21. Duggan ST, Keating GM (2011) Pegylated liposomal doxorubicin: a review of its use in metastatic breast cancer, ovarian cancer, multiple myeloma and AIDS-related Kaposi's sarcoma. *Drugs* 71: 2531–2558.
22. Yin Q, Shen J, Zhang Z, Yu H, Li Y (2013) Reversal of multidrug resistance by stimuli-responsive drug delivery systems for therapy of tumor. *Adv. Drug Deliv. Rev.* 65: 1699–1715.
23. Gonçalves M, Figueira P, Maciel D, Rodrigues J, Shi X, Tomás H, Li Y (2014) Antitumor efficacy of doxorubicin-loaded laponite/alginate hybrid hydrogels. *Macromol. Biosci.* 14: 110–120.
24. Gonçalves M, Figueira P, Maciel D, Rodrigues J, Qu X, Liu C, Tomás H, Li Y (2014) pH-sensitive laponite (R)/doxorubicin/alginate nanohybrids with improved anticancer efficacy. *Acta Biomater.* 10: 300–307.
25. Ankamwar B, Lai TC, Huang JH, Liu RS, Hsiao M, Chen CH, Hwu YK (2010) Biocompatibility of $Fe_3O_4$ nanoparticles evaluated by in vitro cytotoxicity assays using normal, glia and breast, cancer cells. *Nanotechnology* 21: 075102.

26. Schroeder A, Heller DA, Winslow MM, Dahlman JE, Pratt GW, Langer R, Jacks T, Anderson DG (2012) Treating metastatic cancer with nanotechnology. *Nat. Rev. Cancer* 12: 39–50.
27. Rostami M, Aghajanzadeh M, Zamani M, Manjili HK, Danafar H (2018) Sono-chemical synthesis and characterization of $Fe_3O_4$@m$TiO_2$-GO nanocarriers for dual-targeted colon drug delivery. *Res. Chem. Intermed.* 44: 1889–1904.
28. Elzoghby AO, Hemasa AL, Freag MS (2016) Hybrid protein-inorganic nanoparticles: From tumor-targeted drug delivery to cancer imaging. *J. Control. Release* 243: 303–322.
29. Kim J, Lee YM, Kang Y, Kim WJ (2014) Tumor-homing, size-tunable clustered nanoparticles for anticancer therapeutics. *ACS Nano* 8: 9358–9367.
30. Ankamwar B, Kamble V, Sur UK, Santra C (2016) Spectrophotometric evaluation of surface morphology dependent catalytic activity of biosynthesized silver and gold nanoparticles using UV-visible spectra: A comparative kinetic study. *Appl. Surf. Sci.* 366: 275–283.
31. Yahia-Ammar A, Sierra D, Mérola F, Hildebrandt N, Le Guével XL (2016) Self-assembled gold nanoclusters for bright fluorescence imaging and enhanced drug delivery. *ACS Nano* 10: 2591–2599.
32. Wang S, Lin J, Wang T, Chen X, Huang P (2016) Recent advances in photoacoustic imaging for deep-tissue biomedical applications. *Theranostics* 6: 2394–2413.
33. Song J, Huang P, Duan H, Chen X (2015) Plasmonic vesicles of amphiphilic nanocrystals: Optically active multifunctional platform for cancer diagnosis and therapy. *Acc. Chem. Res.* 48: 2506–2515.
34. Huang P, Lin J, Li W, Rong P, Wang Z, Wang S, Wang X, Sun X, Aronova M, Niu G, Leapman RD, Nie Z, Chen X (2013) Biodegradable gold nano vesicles with an ultrastrong plasmonic coupling effect for photoacoustic imaging and photothermal therapy. *Angew. Chem. Int. Ed.* 52: 13958–13964.
35. Wang Y, Wang F, Shen Y, He Q, Guo S (2018) Tumor-specific disintegratable nanohybrids containing ultrasmall inorganic nanoparticles: from design and improved properties to cancer applications. *Mater. Horiz.* 5: 184–205.
36. Liang H, Ren X, Qian J, Zhang X, Meng L, Wang X, Li L, Fang X, Sha X (2016) Size-shifting micelle nanoclusters based on a cross-linked and pH sensitive framework for enhanced tumor targeting and deep penetration features. *ACS Appl. Mater. Interfaces* 8: 10136–10146.
37. Li H, Du J, Du X, Xu C, Sun C, Wang H, Cao Z, Yang X, Zhu Y, Nie S, Wang J (2016) Stimuli-responsive clustered nanoparticles for improved tumor penetration and therapeutic efficacy. *Proc. Natl Acad. Sci. U.S.A.* 113: 4164–4169.
38. Kolosnjaj-Tabi J, Javed Y, Lartigue L, Volatron J, Elgrabli D, Marangon I, Pugliese G, Caron B, Figuerola A, Luciani N, Pellegrino T, Alloyeau D, Gazeau F (2015) The one year fate of iron oxide coated gold nanoparticles in mice. *ACS Nano* 9: 7925–7939.
39. Sunoqrot S, Bugno J, Lantvit D, Burdette JE, Hong S (2014) Prolonged blood circulation and enhanced tumor accumulation of folate-targeted dendrimer-polymer hybrid nanoparticles. *J. Control. Release* 191: 115–122.
40. Shankar SS, Rai A, Ankamwar B, Singh A, Ahmad A, Sastry M (2004) Biological synthesis of triangular gold nanoprisms. *Nat. Mater.* 3: 482–488.
41. Duncan R, Gaspar R (2011) Nanomedicine(s) under the microscope. *Mol. Pharmaceutics* 8: 2101–2141.
42. Duan X, Li Y (2013) Physico-chemical characteristics of nanoparticles affects circulation, biodistribution, cellular internalization, and trafficking. *Small* 9: 1521–1532.
43. Perrault SD,Walkey C, Jennings T, Fischer HC, Chan WCW (2009) Mediating tumor targeting efficiency of nanoparticles through design. *Nano Lett.* 9: 1909–1915.
44. Liu X, Chen Y, Li H, Huang N, Jin Q, Ren K, Ji J (2013) Enhanced retention and cellular uptake of nanoparticles in tumors by controlling their aggregation behavior. *ACS Nano* 7: 6244–6257.
45. Qiu L, Chen T, Öcsoy I, Yasun E, Wu C, Zhu G, You M, Han D, Jiang J, Yu R, Tan W (2015) A cell-targeted, size-photocontrollable, nuclear-uptake nanodrug delivery system for drug-resistant cancer therapy. *Nano Lett.* 15: 457–463.
46. Bartczak D, Kanaras AG (2011) Preparation of peptide-functionalized gold nanoparticles using one pot EDC/sulfo-NHS coupling. *Langmuir* 27: 10119–10123.
47. Song J, Fang Z, Wang C, Zhou J, Duan B, Pu L, Duan H (2013) Photo labile plasmonic vesicles assembled from amphiphilic gold nanoparticles for remote-controlled traceable drug delivery. *Nanoscale* 5: 5816–5824.
48. Su J, Chen F, Cryns VL, Messersmith PB (2011). *J. Am. Chem. Soc.* 133: 11850–11853.
49. Li W, Chiang C, Huang W, Su C, Chiang M, Chen J, Chen S (2015) Amifostine-conjugated pH-sensitive calcium phosphate-covered magnetic-amphiphilic gelatin nanoparticles for controlled intracellular dual drug release for dual-targeting in HER-2-overexpressing breast cancer. *J. Control. Release* 220: 107–118.

50. Bansal K, Aqdas M, Kumar M, Bala R, Singh S, Agrewala JN, Katare OP, Sharma RK, Wangoo N (2018) A facile approach for synthesis and intracellular delivery of size tunable cationic peptide functionalized gold nanohybrids in cancer cells. *Bioconjug. Chem.* 29: 1102–1110.
51. Yang J, Luo Y, Xu Y, Li J, Zhang Z, Wang H, Shen M, Shi X, Zhang G (2015) Conjugation of iron oxide nanoparticles with RGD-modified dendrimers for targeted tumor MR imaging. *ACS Appl. Mater. Interfaces* 7: 5420–5428.
52. Tian Y, Jiang X, Chen X, Shao Z, Yang W (2014) Doxorubicin-loaded magnetic silk fibroin nanoparticles for targeted therapy of multidrug-resistant cancer. *Adv. Mater.* 26: 7393–7398.
53. Yu J, Chen F, Gao W, Ju Y, Chu X, Che S, Sheng F, Hou Y (2017) Iron carbide nanoparticles: An innovative nanoplatform for biomedical applications. *Nanoscale Horiz.* 2: 81–88.
54. Qi H, Liu C, Long L, Ren Y, Zhang S, Chang X, Qian X, Jia H, Zhao J, Sun J, Hou X, Yuan X, Kang C (2016) Blood exosomes endowed with magnetic and targeting properties for cancer therapy. *ACSnano* 10: 3323–3333.
55. Al-Jamal KT, Bai J, Wang JT, Protti A, Southern P, Bogart L, Heidari H, Li X, Cakebread A, Asker D, Al-Jamal WT, Shah A,Bals S, Sosabowski J, Pankhurst QA (2016) Magnetic drug targeting: Preclinical in vivo studies, mathematical modeling and extrapolation to humans. *Nano Lett.* 16: 5652–5660.
56. Weissleder R, Pittet MJ (2008) Imaging era of molecular oncology. *Nature* 452: 580–589.
57. Chien YH, Chou YL, Wang SW, Hung ST, Liau MC, Chao YJ, Su CH, Yeh CS (2013) Near-Infrared light photo-controlled targeting, bioimaging, and chemotherapy with caged upconversion nanoparticles in vitro and in vivo. *ACS Nano* 7: 8516–8528.
58. Sreejith S, Huong TTM, Borah P, Zhao Y (2015) Organic–inorganic nanohybrids for fluorescence, photoacoustic and Raman bioimaging. *Sci. Bull.* 60: 665–678.
59. Chen D, Zhao C, Ye J, Li Q, Liu X, Su M, Jiang H, Amatore C, Selke M, Wang X (2015) In situ biosynthesis of fluorescent platinum nanoclusters: Toward self-bioimaging-guided cancer theranostics. *ACS Appl. Mater. Interfaces* 7: 18163–18169.
60. Song J, Yang X, Jacobson O, Lin L, Huang P, Niu G, Ma Q, Chen X (2015) Sequential drug release and enhanced photothermal and photoacoustic effect of hybrid reduced graphene oxide-loaded ultrasmall gold nanorod vesicles for cancer therapy. *ACS Nano* 9: 9199–9209.
61. Henry A, Sharma B, Cardinal MF, Kurouski D,Van Duyne RP (2016) Surface-enhanced Raman spectroscopy biosensing: In vivo diagnostics and multimodal imaging. *Anal. Chem.* 88: 6638–6647.
62. Ankamwar B, Sur UK, Das P (2016) SERS study of bacteria using biosynthesized silver nanoparticles as SERS substrate. *Anal. Methods* 8: 2335–2340.
63. Narayanan S, Sathy BN, Mony U, Koyakutty M, Nair SV, Menon D (2012) Biocompatible magnetite/gold nanohybrid contrast agents via green chemistry for MRI and CT bioimaging. *ACS Appl. Mater. Interfaces* 4: 251–260.
64. Sun X, Cai W, Chen X (2015) Positron emission tomography imaging using radiolabeled inorganic nanomaterials. *Acc. Chem. Res.* 48: 286–294.
65. Shen JW, Yang CX, Dong LX, Sun HR, Gao K, Yan XP (2013) Incorporation of computed tomography and magnetic resonance imaging function into NaYF4: Yb/Tm upconversion nanoparticles for in vivo trimodal bioimaging. *Anal. Chem.* 85: 12166–12172.
66. Beard P (2011) Bio-medical photoacoustic imaging. *Interface Focus* 1: 602–631.
67. Zhang L, Gao S, Zhang F, Yang K, Ma Q, Zhu L (2014) Activatable hyaluronic acid nanoparticle as a theranostic agent for optical/photoacoustic image-guided photothermal therapy. *ACS Nano* 8: 12250–12258.
68. Song J, Yang X, Jacobson O, Huang P, Sun X, Lin L, Yan X, Niu G, Ma Q, Chen X (2015) Ultrasmall gold nanorod vesicles with enhanced tumor accumulation and fast excretion from the body for cancer therapy. *Adv. Mater.* 27: 4910–4917.
69. Lichtman JW, Conchello J (2005) Fluorescence microscopy. *Nat. Methods* 2: 910–919.
70. Shen L (2011) Bio-compatible polymer/quantum dots hybrid materials: Current status and future developments. *J. Funct. Biomater.* 2: 355–372.
71. Chen L, Wang C, Yuan Z, Chang H (2015) Fluorescent gold nanoclusters: Recent advances in sensing and imaging. *Anal. Chem.* 87: 216–229.
72. Ankamwar B, Damle C, Ahmad A, Sastry M (2005) Biosynthesis of gold and silver nanoparticles using *Emblica officinalis* fruit extract, their phase transfer and transmetallation in an organic solution. *J. Nanosci. Nanotechnol.* 5: 1665–1671.

73. Luo Z, Yuan X, Yu Y, Zhang Q, Leong DT, Lee JY, Xie J (2012) From aggregation-induced emission of Au(I)–thiolate complexes to ultrabright Au(0)@Au(I)–Thiolate core–shell nanoclusters. *J. Am. Chem. Soc.* 134: 16662–16670.
74. Wu X, Li L, Zhang L, Wang T, Wang C, Su Z (2015) Multifunctional spherical gold nanocluster aggregate@polyacrylicacid@mesoporous silica nanoparticles for combined cancer dual-modal imaging and chemo-therapy. *J. Mater. Chem.* B3: 2421–2425.
75. Reisner E, Arion VB, Keppler BK, Pombeiro AJL (2008) Electron-transfer activated metal-Based anticancer drugs. *Inorg. Chim. Acta* 361: 1569–1583.
76. Beija M, Li Y, Duong HT, Laurent S, Elst LV, Muller RN, Lowe AB, Davis TP, Boyer C (2012) Polymer–gold nanohybrids with potential use in bimodal MRI/CT: Enhancing the relaxometric properties of Gd(III) complexes. *J. Mater. Chem.* 22: 21382–21386.
77. Geraldes CF, Laurent S (2009) Classification and basic properties of contrast agents for magnetic resonance imaging. *Contrast Media Mol. Imaging* 4: 1–23.
78. Xiao YD, Paudel R, Liu J, Ma C, Zhang ZS, Zhou SK (2016) MRI contrast agents: Classification and application. *Int. J. Mol. Med.* 38: 1319–1326.
79. Lee DE, Koo H, Sun IC, Ryu JH, Kim K, Kwon IC (2012) Multifunctional nanoparticles for multimodal imaging and theragnosis. *Chem. Soc. Rev.* 41: 2656–2672.
80. Zeng Y, Zhang D, Wu M, Liu Y, Zhang X, Li L, Li Z, Han X, Wei X, Liu X (2014) Lipid-AuNPs@PDA nanohybrid for MRI/CT imaging and photothermal therapy of hepatocellular carcinoma. *ACS Appl. Mater. Interfaces* 6: 14266–14277.
81. Gambhir SS (2002) Molecular imaging of cancer with positron emission tomography. *Nat. Rev. Cancer* 2: 683–693.
82. Li Z, Barnes JC, Bosoy A, Stoddart JF, Zink JI (2012) Mesoporoussilica nanoparticles in biomedical applications. *Chem. Soc. Rev.* 41: 2590–2605.
83. Chen D, Yang D, Dougherty CA, Lu W, Wu H, He X, Cai T, Van Dort MEV, Ross BD, Hong H (2017) In vivo targeting and positron emission tomography imaging of tumor with intrinsically radioactive metal–organic frameworks nanomaterials. *ACS Nano* 11: 4315–4327.
84. Kelly J, Raptopoulos V, Davidoff A, Waite R, Norton P (1989) The value of non-contrast-enhanced CT in blunt abdominal trauma. *Am. J. Roentgenol.* 152: 41–48.
85. Hallouard F, Anton N, Choquet P, Constantinesco A, Vandamme T (2010) Iodinated blood pool contrast media for preclinical X-ray imaging applications review. *Biomaterials* 31: 6249–6268.
86. Chen F, Bu WB, Zhang SJ, Liu XH, Liu JN, Xing HY, Xiao QF, Zhou LP, Peng WJ, Wang LZ, Shi JL (2011) Positive and negative lattice shielding effects co-existing in Gd(III) ion-doped bifunctional upconversion nanoprobes. *Adv. Funct. Mater.* 21: 4285–4294.
87. Xu W, Ling X, Xiao J, Dresselhaus MS, Kong J, Xu H, Liu Z, Zhang J (2012) Surface enhanced Raman spectroscopy on a flat graphene surface. *Proc Natl Acad Sci U S A* 109: 9281–9286.
88. Ling X, Huang S, Deng S, Mao N, Kong J, Dresselhaus MS, Zhang J (2015) Lighting up the Raman signal of molecules in the vicinity of graphene related materials. *Acc. Chem. Res.* 48: 1862–1870.
89. Weissleder R (2001) A clearer vision for in vivo imaging. *Nat. Biotechnol.* 19: 316–317.
90. von Maltzahn G, Centrone A, Park JH, Ramanathan R, Sailor MJ, Hatton TA, Bhatia SN (2009) SERS-coded gold nanorods as a multifunctional platform for densely multiplexed near-infrared imaging and photothermal heating. *Adv. Mater.* 21: 3175–3180.
91. Porter MD, Lipert RJ, Siperko LM, Wang G, Narayanan R (2008) SERS as a bioassay platform: Fundamentals, design, and applications. *Chem. Soc. Rev.* 37: 1001–1011.
92. Song J, Wang F, Yang X, Ning B, Harp MG, Culp SH, Hu S, Huang P, Nie L, Chen J, Chen X (2016) Gold nanoparticle coated carbon nanotube ring with enhanced Raman scattering and photothermal conversion property for theranostic applications. *J. Am. Chem. Soc.* 138: 7005–7015.
93. Yi C, Zhang S, Webb KT, Nie Z (2017) Anisotropic self-assembly of hairy inorganic nanoparticles. *Acc. Chem. Res.* 50: 12–21.
94. Cheng L, Wang C, Feng L, Yang K, Liu Z (2014) Functional nanomaterials for phototherapies of cancer. *Chem. Rev.* 114: 10869–10939.
95. Li F, Wang Y, Zhang Z, Shen Y, Guo S (2017) A chemo/photo-co-therapeutic system for enhanced multidrug resistant cancer treatment using multifunctional mesoporous carbon nanoparticles coated with poly (curcumin-dithiodipropionic acid). *Carbon* 122: 524–537.

96. Wang Y, Zhang Z, Xu S, Wang F, Shen Y, Huang S, Guo S (2017) pH, redox and photothermal tri-responsive DNA/polyethylenimine conjugated gold nanorods as nanocarriers for specific intracellular co-release of doxorubicin and chemosensitizer pyronaridine to combat multidrug resistant cancer. *Nanomedicine* 13: 1785–1795.
97. Zhang Z, Niu B, Chen J, He X, Bao X, Zhu J, Yu H, Li Y (2014) The use of lipid-coated nanodiamond to improve bioavailability and efficacy of sorafenib in resisting metastasis of gastric cancer. *Biomaterials* 35: 4565–4572.
98. Khandelia R, Jaiswal A, Ghosh SS, Chattopadhyay A (2013) Gold nanoparticle-protein agglomerates as versatile nanocarriers for drug delivery. *Small* 9: 3494–3505.
99. Mahor S, Dash BC, Connor SO, Pandit A (2012) Mannosylated polyethylene imine–hyaluronan nanohybrids for targeted gene delivery to macrophage-like cell lines. *Bioconjug. Chem.* 23: 1138–1148.
100. Gonçalves C, Pereira P, Gama M (2010) Self-assembled hydrogel nanoparticles for drug delivery applications. *Materials* 3: 1420–1460.
101. Liu G, An Z (2014) Frontiers in the design and synthesis of advanced nanogels for nanomedicine. *Polym. Chem.* 5: 1559–1565.
102. Lu Y, Sun W, Gu Z (2014) Stimuli-responsive nanomaterials for therapeutic protein delivery. *J. Control. Release* 194: 1–19.
103. Herrera Estrada LP, Champion JA (2015) Protein nanoparticles for therapeutic protein delivery. *Biomater. Sci.* 3: 787–799.
104. Omar H, Croissant JG, Alamoudi K, Alsaiari S, Alradwan I, Majrashi MA, Anjum DH, Martins P, Moosa B, Almalik A, Khashab NM (2016) Biodegradable magnetic silica@iron oxide nanovectors with ultra-large mesopores for high protein loading, magnetothermal release, and delivery. *J. Control. Release* 259: 187–194.
105. Zou L, Wang H, He B, Zeng L, Tan T, Cao H, He X, Zhang Z, Guo S, Li Y (2016) Current approaches of photothermal therapy in treating cancer metastasis with nanotherapeutics. *Theranostics* 6: 762–772.
106. Cortezon-Tamarit F, Gel H, Mirabello V, Theobald MBM, Calatayud DG, Pascu SI (2017) Carbon nanotubes and related Nanohybrids incorporating inorganic transition metal compounds and radioactive species as synthetic scaffolds for nanomedicine design. In: Lo KK (ed) *Inorganic and Organometallic Transition Metal Complexes with Biological Molecules and Living Cells*, Academic Press.
107. Shanmugam V, Selvakumar S, Yeh CS (2014) Near-infrared light-responsive nanomaterials in cancer therapeutics. *Chem. Soc. Rev.* 43: 6254–6287.
108. Alkilany AM, Thompson LB, Boulos SP, Sisco PN, Murphy CJ (2012) Gold nanorods: Their potential for photothermal therapeutics and drug delivery, tempered by the complexity of their biological interactions. *Adv. Drug Deliver. Rev.* 64: 190–199.
109. Zhang Z, Wang J, Chen C (2013) Near-infrared light-mediated nano platforms for cancer thermo-chemotherapy and optical imaging. *Adv. Mater.* 25: 3869–3880.
110. Hu SH, Fang RH, Chen YW, Liao BJ, Chen IW, Chen SY (2014) Photoresponsive protein–graphene–protein hybrid capsules with dual targeted heat-triggered drug delivery approach for enhanced tumor therapy. *Adv. Funct. Mater.* 24: 4144–4155.
111. Ma Y, Liang X, Tong S, Bao G, Ren Q, Dai Z (2013) Gold nanoshell nanomicelles for potential magnetic resonance imaging, light-triggered drug release and photothermal therapy. *Adv. Funct. Mater.* 23: 815–822.
112. Wang DG, Xu ZA, Yu HJ, Chen XZ, Feng B, Cui ZR, Lin B, Yin Q, Zhang Z, Chen C, Wang J, Zhang W, Li Y (2014) Treatment of metastatic breast cancer by combination of chemotherapy and photothermal ablation using doxorubicin-loaded DNA wrapped gold nanorods. *Biomaterials* 35: 8374–8384.
113. He X, Bao X, Cao H, Zhang Z, Yin Q, Gu W, Chen L, Yu H, Li Y (2015) Tumor-penetrating nanotherapeutics loading a near-infrared probe inhibit growth and metastasis of breast cancer. *Adv. Funct. Mater.* 25: 2831–2839.
114. Zhou FF, Li XS, Naylor MF, Hode T, Nordquist RE, Alleruzzo L, Raker J, Lam SSK, Du N, Shi L, Wang X, Chen WR (2015) In CVAX- A novel strategy for treatment of late-stage, metastatic cancers through photoimmunotherapy induced tumor-specific immunity. *Cancer Lett.* 359: 169–177.
115. Chen Q, Liang C, Wang C, Liu Z (2015) An imagable and photothermal "Abraxane-like" nanodrug for combination cancer therapy to treat subcutaneous and metastatic breast tumors. *Adv. Mater.* 27: 903–910.

116. Orecchioni M, Cabizza R, Bianco A, Delogu LG (2015) Graphene as cancer theranostic tool: Progress and future challenges. *Theranostics* 5: 710–723.
117. Cai XJ, Jia XQ, Gao W, Zhang K, Ma M, Wang SG, Zheng Y, Shi JL, Chen HR (2015) A versatile nanotheranostic agent for efficient dual-mode imaging guided synergistic chemo-thermal tumor therapy. *Adv. Funct. Mater.* 25: 2520–2529.
118. Wang Y, Yang T, Ke H, Zhu A, Wang Y, Wang J, Shen J, Liu G, Chen C, Zhao Y, Chen H (2015) Smart albumin-biomineralized nanocomposites for multimodal imaging and photothermal tumor ablation. *Adv. Mater.* 27: 3874–3882.
119. Su S, Ding Y, Li Y, Wu Y, Nie G (2016) Integration of photothermal therapy and synergistic chemotherapy by a porphyrin self-assembled micelle confers chemo sensitivity in triple-negative breast cancer. *Biomaterials* 80: 169–178.
120. Chen Q, Wang C, Zhan Z, He W, Cheng Z, Li Y, Liu Z (2014) Near-infrared dye bound albumin with separated imaging and therapy wavelength channels for imaging-guided photothermal therapy. *Biomaterials* 35: 8206–8214.
121. Veiseh O, Gunn JW, Zhang M (2010) Design and fabrication of magnetic nanoparticles for targeted drug delivery and imaging. *Adv. Drug Deliv. Rev.* 62: 284–304.
122. Zhang Z, Wang J, Nie X, Wen T, Ji Y, Wu X, Zhao Y, Chen C (2014) Near infrared laser-induced targeted cancer therapy using thermo responsive polymer encapsulated gold nanorods. *J. Am. Chem. Soc.* 136: 7317–7326.
123. Zheng M, Liu S, Li J, Qu D, Zhao H, Guan X, Hu X, Xie Z, Jing X, Sun Z (2014) Integrating oxaliplatin with highly luminescent carbon dots: An unprecedented theranostic agent for personalized medicine. *Adv. Mater.* 26: 3554–3560.
124. Ankamwar B (2012) Size and shape effect on biomedical applications of nanomaterials. In: Hudak R (ed) *Biomedical Engineering*, Intech Open, Europe.
125. Vivero-Escoto JL, Slowing II, Trewyn BG, Lin VS-Y (2010) Mesoporous silica nanoparticles for intracellular controlled drug delivery. *Small* 6: 1952–1967.
126. Baek S, Singh RK, Khanal D, Patel KD, Lee EJ, Leong KW, Chrzanowski W, Kim HW (2015) Smart multifunctional drug delivery towards anticancer therapy harmonized in mesoporous nanoparticles. *Nanoscale* 7: 14191–14216.
127. Slowing II, Viveroescoto JL, Wu CW, Lin VSY (2008) Mesoporous silica nanoparticles as controlled release drug delivery and gene transfection carriers. *Adv. Drug Deliv. Rev.* 60: 1278–1288.
128. Manzano M, Vallet-Regí M (2010) New developments in ordered mesoporous materials for drug delivery. *J. Mater. Chem.* 20: 5593–5604.
129. Chan A, Orme RP, Fricker RA, Roach P (2013) Remote and local control of stimuli responsive materials for therapeutic applications. *Adv. Drug Deliv. Rev.* 65: 497–514.
130. Baek S, Singh RK, Kim TH, Seo JW, Shin US, Chrzanowski W, Kim HW (2016) Triple hit with drug carriers: pH- and temperature-responsive theranostics for multimodal chemo- and photothermal therapy and diagnostic applications. *ACS Appl. Mater. Interfaces* 8: 8967–8979.
131. Mehrotra P (2016) Biosensors and their applications–A review. *J. Oral Biol. Craniofac. Res.* 6: 153–159.
132. Turner A, Karube I, Wilson GS (1987), *Biosensors: Fundamentals and applications*, Oxford University Press, New York.
133. Alhadrami HA (2018) Biosensors: Classifications, medical applications and future perspective. *Biotechnol. Appl. Biochem.* 65: 497–508.
134. Rocchitta G, Spanu A, Babudieri S, Latte G, Madeddu G, Galleri G, Nuvoli S, Bagella P, Demartis MI, Fiore V, Manetti R, Serra P (2016) Enzyme biosensors for biomedical applications: Strategies for safeguarding analytical performances in biological fluids. *Sensors* 16: 780–801.
135. Ispas CR, Crivat G, Andreescu S (2012) Review: Recent developments in enzyme-based biosensors for biomedical analysis. *Anal. Lett.* 45: 168–186.
136. Liu Q, Wu C, Cai H, Hu N, Zhou J, Wang P (2014) Cell-based biosensors and their application in biomedicine. *Chem. Rev.* 114: 6423–6461.
137. Moina C, Ybarra G (2012) Fundamentals and applications of immunosensors. In: Chiu NHL (ed) *Advances in Immunoassay Technology*. InTech, Europe.
138. Wu J, Fu Z, Yan F, Ju H (2007) Biomedical and clinical applications of immunoassays and immunosensors for tumor markers. *Trends Analyt. Chem.* 26: 679–688.

139. Bojorge Ramírez N, Salgado AM, Valdman B (2009) The evolution and developments of immunosensors for health and environmental monitoring: Problems and perspectives. *Braz. J. Chem. Eng.* 26: 227–249.
140. Holzinger M, Le Goff A, Cosnier S (2014) Nanomaterials for biosensing applications: A review. *Front. Chem.* 2: 63.
141. Sagadevan S, Periasamy M (2014) Recent trends in nanobiosensors and their applications—A review. *Rev. Adv. Mater. Sci.* 36: 62–69.
142. Pandit S, Dasgupta D, Dewan N, Prince A (2016) Nanotechnology based biosensors and its application. *Pharma. Innov.* 5: 18–25.
143. Choi BG, Park H, Park TJ, Yang MH, Kim JS, Jang SY, Heo NS, Lee SY, Kong J, Hong WH (2010) Solution chemistry of self-assembled graphene nanohybrids for high-performance flexible biosensors. *ACS Nano* 4: 2910–2918.
144. Miao F, Tao B, Sun L, Liu T, You J, Wang L, Chu PK (2009) Amperometric glucose sensor based on 3D ordered nickel–palladium nanomaterial supported by silicon MCP array. *Sens. Actuators* 141: 338–342.
145. Lu LM, Li HB, Qu F, Zhang XB, Shen GL, Yu RQ (2011) In situ synthesis of palladium nanoparticle–graphene nanohybrids and their application in nonenzymatic glucose biosensors. *Biosens. Bioelectron.* 26: 3500–3504.
146. Vial S, Forano C, Shan D, Mousty C, Barhoumi H, Martelet C, Jaffrezic N (2006) Nanohybrid-layered double hydroxides/urease materials: Synthesis and application to urea biosensors. *Mater. Sci. Eng. C* 26: 387–393.
147. Wang W, Bao T, Zeng X, Xiong H, Wen W, Zhang X, Wang S (2017) Ultrasensitive electrochemical DNA biosensor based on functionalized gold clusters/graphene nanohybrids coupling with exonuclease III-aided cascade target recycling. *Biosens. Bioelectron.* 91: 183–189.
148. Berthiaume F, Maguire TJ, Yarmush ML (2011) Tissue engineering and regenerative medicine: History, progress, and challenges. *Annu. Rev. Chem. Biomol. Eng.* 2: 403–430.
149. Murugan R, Ramakrishna S (2006) Nano-featured scaffolds for tissue engineering: A review of spinning methodologies. *Tissue Engineering* 12: 435–447.
150. Ma PX (2004) Scaffolds for tissue fabrication. *Mater. Today* 7: 30–40.
151. Hutmacher DW (2000) Scaffolds in tissue engineering bone and cartilage. *Biomaterials* 21: 2529–2543.
152. Dhandayuthapani B, Yoshida Y, Maekawa T, Kumar DS (2011) Polymeric scaffolds in tissue engineering application: A review. *Int. J. Polym. Sci.* 2011: Article ID 290602.
153. Haider A, Haider S, Han SS, Kang IK (2017) Recent advances in the synthesis, functionalization and biomedical applications of hydroxyapatite: A review. *RSC Adv.* 7: 7442–7458.
154. Dong C, Lv Y (2016) Application of collagen scaffold in tissue engineering: Recent advances and new perspectives. *Polymers* 8: 4.
155. Habraken W, Habibovic P, Epple M, Bohner M (2016) Calcium phosphates in biomedical applications: Materials for the future. *Mater. Today* 19: 69–87.
156. Lu J, Yu H, Chen C (2018) Biological properties of calcium phosphate biomaterials for bone repair: A review. *RSC Adv.* 8: 2015–2033.
157. Croisier F, Jérôme C (2013) Chitosan-based biomaterials for tissue engineering. *Eur. Polym. J.* 49: 780–792.
158. Ahmed S, Annu, Ali A, Sheikh J (2018) A review on chitosan centred scaffolds and their applications in tissue engineering. *Int. J. Biol. Macromol.* 116: 849–862.
159. O'brien FJ (2011) Biomaterials and scaffolds for tissue engineering. *Mater. Today* 14: 88–95.
160. Smith IO, Liu XH, Smith LA, Ma PX (2009) Nanostructured polymer scaffolds for tissue engineering and regenerative medicine. *Wiley Interdiscip. Rev. Nanomed. Nanobiotechnol.* 1: 226–236.
161. Chen J, Nan K, Yin S, Wang Y, Wu T, Zhang Q (2010) Characterization and biocompatibility of nanohybrid scaffold prepared via in situ crystallization of hydroxyapatite in chitosan matrix. *Colloids Surf. B Biointerfaces* 81: 640–647.
162. Fan T, Chen J, Pan P, Zhang Y, Hu Y, Liu X, Shi X, Zhang Q (2016) Bioinspired double polysaccharides-based nanohybrid scaffold for bone tissue engineering. *Colloids Surf. B Biointerfaces* 147: 217–223.
163. Chen J, Yu Q, Zhang G, Yang S, Wu J, Zhang Q (2012) Preparation and biocompatibility of nanohybrid scaffolds by in situ homogeneous formation of nano hydroxyapatite from biopolymer polyelectrolyte complex for bone repair applications. *Colloids Surf. B Biointerfaces* 93: 100–107.

164. Lei Y, Xu Z, Ke Q, Yin W, Chen Y, Zhang C, Guo Y (2017) Strontium hydroxyapatite/chitosan nanohybrid scaffolds with enhanced osteoinductivity for bone tissue engineering. *Mater. Sci. Eng. C* 72: 134–142.
165. Vallés-Lluch A, Novella-Maestre E, Sancho-Tello M, Pradas MM, Ferrer GG, Batalla CC (2010) Mimicking natural dentin using bioactive nanohybrid scaffolds for dentinal tissue engineering. *Tissue Eng. A* 16: 2783–2793.
166. Parani M, Lokhande G, Singh A, Gaharwar AK (2016) Engineered nanomaterials for infection control and healing acute and chronic wounds. *ACS Appl. Mater. Interfaces* 8: 10049–10069.
167. Cai Z, Zhang H, Wei Y, Cong F (2017) Hyaluronan-inorganic nanohybrid materials for biomedical applications. *Biomacromolecules* 18: 1677–1696.
168. Clapp T, Siebert P, Chen D, Braun LJ (2011) Vaccines with aluminum-containing adjuvants: Optimizing vaccine efficacy and thermal stability. *J. Pharm. Sci.* 100: 388–401.
169. Carbone C, Leonardi A, Cupri S, Puglisi G, Pignatello R (2014) Pharmaceutical and biomedical applications of lipid-based nanocarriers. *Pharm. Pat. Anal.* 3: 199–215.
170. Ruiz-Hitzky E, Aranda P, Darder M, Rytwo G (2010) Hybrid materials based on clays for environmental and biomedical applications. *J. Mater. Chem.* 20: 9306–9321.
171. Taubenberger JK, Morens DM (2010) Influenza: The once and future pandemic. *Public Health Rep.* 125: 15–26.
172. Wicklein B, delBurgo M, Yuste M, Darder M, Llavata CE, Aranda P, Ortin J, del Real G, Ruiz-Hitzky E (2012) Lipid-based bio-nanohybrids for functional stabilisation of influenza vaccines. *Eur. J. Inorg. Chem.* 2012: 5186–5191.
173. Giudice EL, Campbell JD (2006) Needle-free vaccine delivery. *Adv. Drug Deliv. Rev.* 58: 68–89.
174. Lycke N (2012) Recent progress in mucosal vaccine development: Potential and limitations. *Nat. Rev. Immunol.* 12: 592–605.
175. Maughan CN, Preston SG, Williams GR (2015) Particulate inorganic adjuvants: Recent developments and future outlook. *J. Pharm. Pharmacol.* 67: 426–449.
176. Wang X, Yang D, Li S, Xu X, Qin CF, Tang R (2016) Biomineralized vaccine nanohybrid for needle-free intranasal immunization. *Biomaterials* 106: 286–294.
177. Wang X, Deng Y, Li S, Wang G, Qin E, Xu X, Tang R, Qin C (2012) Biomineralization-based virus shell-engineering: Towards neutralization escape and tropism expansion. *Adv. Healthc. Mater.* 1: 443–449.
178. Barouch DH, Pau MG, Custers JH, Koudstaal W, Kostense S, Havenga MJ, Truitt DM, Sumida SM, Kishko MG, Arthur JC, Korioth-Schmitz B, Newberg MH, Gorgone DA, Lifton MA, Panicali DL, Nabel GJ, Letvin NL, Goudsmit J (2004) Immunogenicity of recombinant adenovirus serotype 35 vaccine in the presence of pre-existing anti-Ad5 immunity. *J. Immunol.* 172: 6290–6297.
179. Wang X, Sun C, Li P, Wu T, Zhou H, Yang D, Liu Y, Ma X, Song Z, Nian Q, Feng L, Qin C, Chen L, Tang R (2016) Vaccine engineering with dual-functional mineral shell: A promising strategy to overcome pre-existing immunity. *Adv. Mater. Weinheim* 28: 694–700.

# 4

# *Diagnostics and Therapy Based on Photo-Activated Nanoparticles*

**Erving Clayton Ximendes, Uéslen Rocha Silva, and Carlos Jacinto da Silva**

**CONTENTS**



## 4.1 Introduction

Many of the natural processes taking place in the biological world are strongly affected by temperature. In this sense, it is easy to understand why temperature measurement is something demanded on a daily basis. When reviewing the literature on methods for thermal sensing, we can confidently say that optical temperature sensors offer a variety of advantages over conventional electric ones. Among these advantages, one can mention freedom from electromagnetic interference, electrical passiveness, greater sensitivity and remote detection. Over the last decade, there has been a particular interest in providing thermal sensing capabilities to the nanomaterials that so often find biological applications. This, in turns, led to the design of the so-called luminescent nanothermometers (LNThs). Their use has actualized applications not only in biology but also in fields such as microfluidics and microelectronics.

Furthermore, the discovery of unique physicochemical properties of nanomaterials has been a leading factor in the advance of nanoparticle-based theragnostics, i.e., the integration of therapeutics with diagnostics by means of suitable nanoparticles (NPs) in order to develop effective and safe medical treatments. Among the therapeutic applications of nanomaterials, hyperthermia is probably the one receiving the most attention from the scientific community, mainly due to its promising achievements in the battle against cancer. On the other hand, NP-based diagnostic methods rely, in most part, on either luminescence imaging or luminescence thermometry. In this chapter, we will briefly describe how photo-activated NPs provide the basis for reliable diagnostics and therapies by means of their optical properties.

## 4.2 Methods of Diagnostics Based on Photo-Activated NPs

With the increasing incidence of fatal diseases, the growing need for accurate methods of diagnosis has made researchers all over the world turn their eyes to nanomedicine. As a natural consequence, photo-activated nanomedicine has ended up gaining prominence in the battle against cancer and many other diseases (Rai et al. 2010).

The various modalities of diagnostics based on photo-activated NPs are extensively used *in vivo* imaging and targeted detection of tumors. The most common techniques would include (i) imaging and (ii) luminescence thermometry. Since thermometry usually provides more information on the characteristics properties of the tissue (such as density, specific heat, blood perfusion, etc.) when compared to the imaging techniques, a more thorough discussion on luminescent thermometry will be provided in subsection 4.2.6.

### 4.2.1 Optical Imaging

Optical imaging involves the detection of light photons transmitted through tissues in order to non-invasively monitor the progression of diseases (Jiang et al. 2010). Even though conventional fluorophores were used in the primordial stages of optical imaging, the recent interest in fluorescent NPs has made them gain prominence over their predecessors. The operating principle is the following: after administering the drug-loaded NP inside the body, one tracks the delivery vehicle in order to determine whether the drug is effectively delivered to the desired organ or tissue (Licha and Olbrich 2005).

### 4.2.2 Molecular Imaging

Since virtually all types of diseases originate at the molecular and cellular level, imaging systems capable of operating at these scales would be most beneficial to medicine. As such, optics has played a fundamental role, providing not only sufficient resolution but also wavelength sensitivity, enabling both the discrimination of morphological features at the cellular level and the localization of the optical signal from spectroscopically specific transfected genetic sequences, endogenous molecules, exogenous molecules or probes (Boppart et al. 2005). The informational content, on the other hand, is greatly limited if the presence of spectroscopically identifiable probes cannot be linked to some physiologically or functionally meaningful parameter. Thus, in order to add molecular specificity to the imaging techniques, one needs to functionalize and target probes to specific cellular and molecular sites. In this sense, the field of molecular imaging has traditionally been defined as the detection of a molecularly specific signal, either from an endogenous molecule or some targeted exogenous agent, in a living system. Several different NPs designs using light-sensitive novel imaging agents have been developed aiming the identification and characterization of various fundamental processes at the organ, tissue, cellular and molecular levels.

### 4.2.3 Optical Coherence Tomography

Optical coherent tomography (OCT) is an imaging technique sometimes described as the optical analog to ultrasound. It produces high resolution (typically 10–15 μm), real-time, cross-sectional images through biological tissues. OCT detects the reflections of a low coherence light source directed into a tissue and determines at what depth the reflections occurred (Loo et al. 2004). Thus, it is used for examining complex structures to dimensionally characterize and optimize cell growth (Tomlins and Wang 2005). Polymers and NPs made of gold or iron oxide have been frequently used as OCT contrast agents (Aaron et al. 2006, Au et al. 2011).

### 4.2.4 Photoacoustic Imaging

The working principle behind this modality is that after being illuminated by a short-pulsed laser, the biological sample in consideration absorbs the light and a subsequent thermoelastic expansion of the absorbent generates an ultrasonic acoustic signal. This signal, in turn, is detected by wideband transducers surrounding the object and used to determine its geometry. This technique has been used to image blood vessels, tumors, hemoglobin oxygenation and tumor angiogenesis. For photoacoustic imaging, gold nanoshells with a silica core and gold nanorods are frequently used (Jiang et al. 2010).

### 4.2.5 Multimodal Imaging

Multimodal imaging has been developed to overcome the many issues faced by biomedical technology, which would include the heterogeneity of a disease and the patient. The efficacy of personalized therapies can be improved as the progress of the disease is different for each patient, depending on the genetic and environmental factors or the characteristics of the disease (Lee et al. 2012). The combination of imaging modalities, such as positron emission tomography/computed tomography (PET/CT) or positron emission tomography/magnetic resonance imaging (PET/MRI), has gained attention to improve currently used imaging instruments for diagnosis. Thus, multimodal imaging has several advantages over single imaging modalities, like high sensitivity, multicolor imaging, reduced tissue penetrating limits and so on. Multimodal NPs containing upconversion NPs as the core, a layer of iron-oxide particles as the intermediate layer and an outer-most layer of gold have been developed and used for upconversion luminescence (UCL)/magnetic resonance (MR) multimodal imaging and photothermal ablation of tumors (Cheng et al. 2012).

### 4.2.6 Luminescence Thermometry

Briefly, luminescence thermometry is based on the use of Luminescent Nanoparticles (LNPs) whose spectroscopic properties show appreciable temperature dependence in a certain range of interest. Historically, three main types of LNThs were used to determine temperature: (i) LNThs based on emission intensity, using the integrated intensity of a single transition or of a pair of transitions, (ii) LNThs based on the spectral shift of a given transition and (iii) lifetime based LNThs, using the decay profiles of emitting excited states (Figure 4.1).

With some exceptions, most LNThs have to be referred to a well-known temperature for their calibration (Brites et al. 2012). The usual calibration procedure requires an independent measurement of the temperature (using, for instance, a heating plate or a thermographic infrared camera) in order to convert

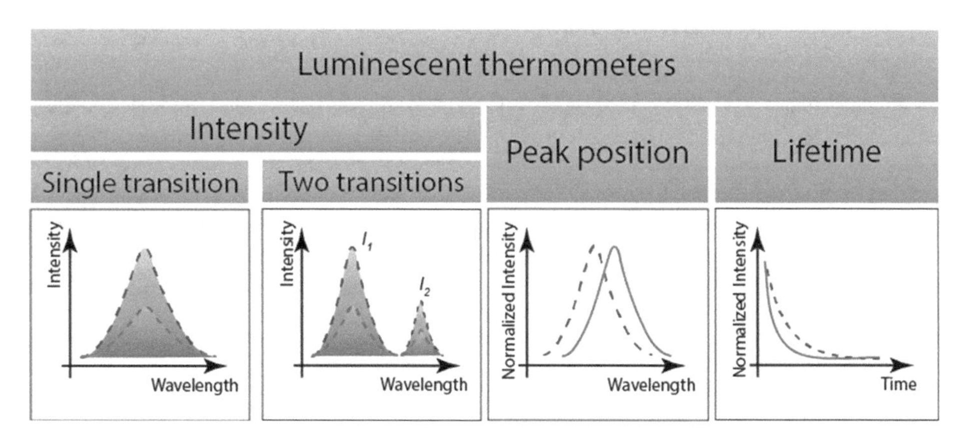


**FIGURE 4.1** Classification of luminescent thermometers.

it to its thermometric parameter (i.e., intensity, peak position, lifetime, intensity ratio, etc.). The necessity of an independent measurement of temperature, in turns, implies in a new calibration procedure for every new medium the thermometer will operate in. However, since this is not always possible to achieve, a single calibration is often assumed to be valid independently of the medium.

#### 4.2.6.1 Performance of LNThs

The comparison between distinct LNThs is made using various parameters. The most frequently used is the so-called relative thermal sensitivity, $S_R$. It stands for the relative change of $R$, the thermometric parameter, per degree of temperature change and is defined as:

$$S_R = \frac{1}{R}\left|\frac{dR}{dT}\right| \tag{4.1}$$

This parameter is usually expressed in units of % change per Kelvin (or Celsius), % $K^{-1}$. Compared to the absolute thermal sensitivity, $S_a = dR/dT$, $S_R$ presents the advantage of being independent of the thermometer's nature.

A second parameter is what we call temperature uncertainty (or temperature resolution), $\Delta T$. It is defined as the smallest temperature change detectable in a certain measurement or the limit of detection of the sensor. Assuming that the temperature uncertainty of a thermometer results only from changes in $R$, $\Delta T$ is given by the Taylor series expansion of the temperature as a function of $R$:

$$\delta T = \frac{\partial T}{\partial R}\delta R + \frac{1}{2!}\frac{\partial^2 T}{\partial R^2}(\delta R)^2 + \ldots \tag{4.2}$$

where $\delta R$ is the uncertainty in the determination of $R$, usually determined by the experimental conditions. Considering only the first term of the expansion, $\delta T$ becomes (Baker et al. 2005):

$$\delta T = \frac{\partial T}{\partial R}\delta R \tag{4.3}$$

The $\delta T$ values are experimentally determined from the distribution of temperature readouts of a luminescence thermometer when at a certain reference temperature. $\delta T$ is then defined as the standard deviation of the resulting temperature histogram. For typical portable detection systems, $\delta R/R$ can reach, in the best case scenario, the value of 0.1%, meaning that typical sensitivities of 1–10% $K^{-1}$ (Brites et al. 2012) correspond to temperature uncertainties of 0.01–0.1 K, respectively.

#### 4.2.6.2 Accessing the Basic Properties of a Living Tissue via Luminescence Thermometry

The reason for the success of luminescence thermometry in recent years relies on the fact that heat transfer in biological systems is characterized by the effects of blood flowing through the vascular circulatory system. As the blood moves through the microcapillary system, a stabilizing effect takes place, i.e., the perfuse tissue returns to the natural equilibrium reference state (when there are no externally applied fluxes). This is the so-called perfusion. In an exchange of thermal energy between the microcapillary network and the tissue, perfusion contributes in bringing the local tissue temperature closer to the body's core temperature. For instance, when a tissue's thermal energy is increased by an externally applied heat flux, perfusion aims to cool down the tissue by removing blood at the elevated temperature and by replacing it with blood the temperature of which is closer to the one of the core.

If one considers the distribution of microcapillaries to be uniform and the relative size scale of the capillary diameter to be small when compared to the length scale characterizing heat transport, then perfusion can be viewed as a volumetric phenomenon. In other words, any local fluctuations in blood temperature resulting from the thermal exchange between an individual capillary and the tissue cannot be distinguished. The idea is that the network's blood flow, when viewed from a scale much larger than the diameter of an individual capillary, has no particular direction and permeates the tissue in a generally homogenous manner. In mathematical terms, the conduction of heat into perfuse is commonly represented by Pennes' bioheat equation (Pennes 1948):

$$\rho c_p \frac{\partial T_p}{\partial t} = \nabla \cdot \left(k_p \nabla T_p\right) - \rho_b c_b \omega_p \left(T_p - T_a\right) + g \tag{4.4}$$

where ρ, $c_p$ and $k_p$ stand for the density, specific heat and thermal conductivity of the tissue, respectively. The *g* term stands for volumetric heat generation, which can be a function of both time and position. Heat generation can be a result of (i) low levels of metabolic sources or (ii) some higher intensity externally induced source. The variable $\omega_p$ stands for the blood perfusion rate and is representative of the volume rate at which blood passes through the tissue per unit volume of tissue ($m^3$ blood $s^{-1}$)/($m^3$ tissue). The terms $c_b$ and $\rho_b$ refer to the specific heat and the density of blood, respectively. The symbol $T_a$ stands for the temperature of the blood at the body's core; hence, in simple words, the perfusion term aims to bring the local tissue temperature closer to the body's core temperature. The reason why Pennes' bioheat equation is widely used by the scientific community lies in its simplicity and the accuracy provided in numerical analysis, despite many alternative improved models in the literature.

A successful example of LNThs used for getting access to the basic properties of a living tissue via Pennes' bioheat equation was provided by Nd@Yb $LaF_3$ NPs. These NPs are constituted by a $Nd^{3+}$-doped core and a $Yb^{3+}$-doped shell. Due to the spatial separation between $Nd^{3+}$ and $Yb^{3+}$ ions, the spectral distribution of the overall emission spectrum, when the system is optically excited at 790 nm, results in a strong temperature dependence. The near-infrared (NIR) thermal sensitivity provided by the ratiometric analysis of the luminescence generated by $LaF_3$:Nd@$LaF_3$:Yb NPs was close to 0.4% $K^{-1}$, which is good enough to achieve subdegree thermal readings. In fact, their ability to measure *in vivo* subcutaneous thermal transients (see the scheme of the experiment in Figure 4.2a) as a potential diagnostic tool was successfully

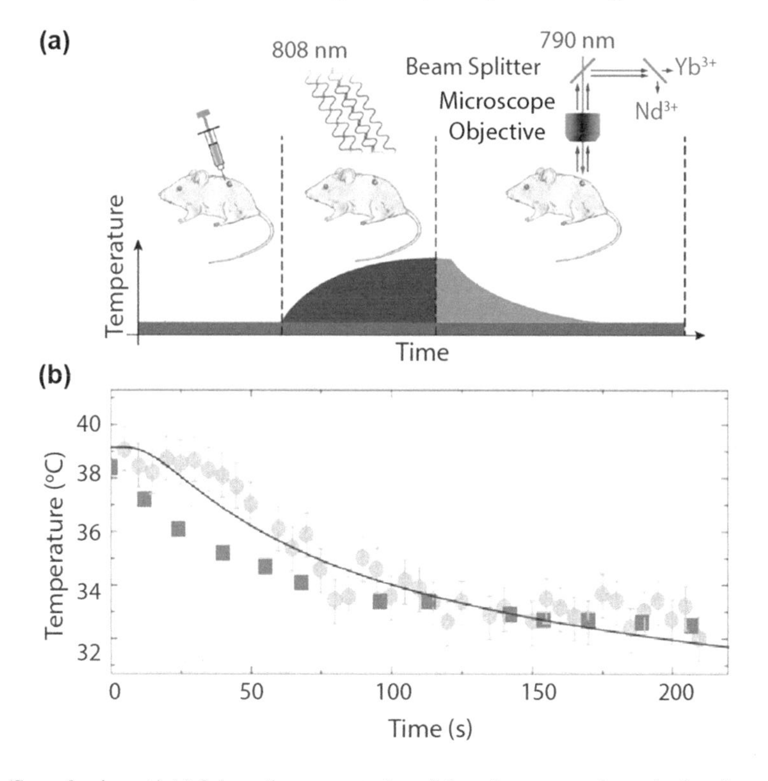


**FIGURE 4.2 (See color insert.)** (a) Schematic representation of the subcutaneous thermal relaxation experiments. (b) Time evolution of the temperatures measured by the subcutaneous luminescent thermometer (grey) and by the IR thermal camera (yellow). Dots are experimental subcutaneous (circles) and skin (squares) temperatures, whereas the solid line is the best fit. (Adapted with permission from Ximendes et al. 2016c. © 2016, American Chemical Society.)

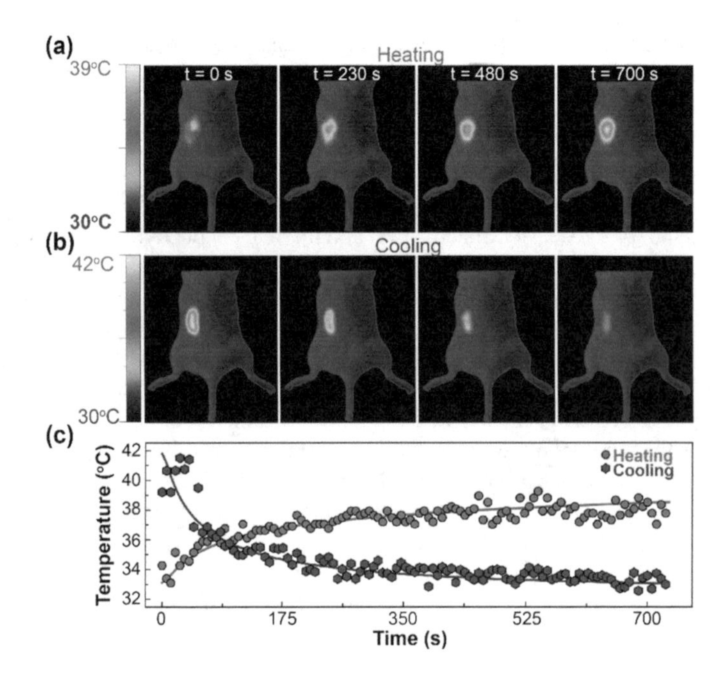


**FIGURE 4.3 (See color insert.)** (a) and (b) Schematic representations of the *in vivo* 2D-SDTI experiment. An optical figure of the anesthetized mouse was superimposed. (c) Time evolution of the average temperature of the injection area as measured by the subcutaneous LNThs during heating and thermal relaxation processes. (Reproduced with permission from Ximendes et al. 2017. © 2017, Wiley.)

explored by Ximendes et al. (2016c). The physical principle behind the experiment relied on the fact that the thermal relaxation of a tissue (temperature decrease in the absence of any heating source, see Figure 4.2a,b) strongly depends on the intrinsic properties of the tissue. Thus, its accurate measurement is capable of providing information on the tissue status and could be used to detect anomalies caused by incipient diseases such as dehydration, inflammation, tumor growth or even ischemia, as recently demonstrated (Ximendes et al. 2017a). Indeed, analysis of the thermal relaxation curve of a tissue (included in Figure 4.2b) provides, simultaneously, accurate and consistent values for the tissue thermal conductivity and absorption.

Despite the success of obtaining *in vivo* thermal transients, the measurements were still being performed on a single point inside the tissue. The acquisition of two-dimensional subcutaneous dynamical thermal imaging (2D-SDTI), on the other hand, has been a constant requirement for thermal-based diagnosis and, in principle, demands LNThs with higher relative thermal sensitivity. In view of that, $LaF_3$:Er,Yb@$LaF_3$:Tm,Yb core/shell NPs were recently proposed to overcome this limitation (Ximendes et al. 2017b). Ratiometric thermal sensitivities as large as 4% $K^{-1}$ were achieved when exciting the system at 690 nm and when monitoring the intensity ratio between the emitted intensity of $Yb^{3+}$ ions at 1000 nm and that of $Tm^{3+}$ ions at around 1230 nm. Figure 4.3a,b includes the ratiometric thermal *in vivo* images obtained at different times during heating and cooling processes. The time evolution of the subtissue temperature (Figure 4.3c) allowed the estimation of the thermal diffusivity of the living tissue.

For all purposes, we assumed the validity of Equation 4.4 for the *in vivo* applications of the LNThs. This book chapter will not address the relative strengths and shortcomings of Pennes' bioheat equation. Notwithstanding, from the promising results that were here presented by the NPs, it is very much likely that, in the long-run, when studying biological systems with a highly complex heat distribution, LNThs will shed some light on the extent of the validity of Equation 4.4 and maybe find some innovative applications.

## 4.3 Therapies based on Photo-Activated NPs

In addition to imaging and diagnostics, the design of nanomaterials has also aimed several light-based therapeutic uses. Most of them can be classified into two categories: photodynamic therapy (PDT) or photothermal therapy (PTT). In this section, we will briefly describe the working principles of each category and provide examples of NPs corresponding to that use.

### 4.3.1 Photodynamic Therapy

PDT is an important approach that makes use of light-activated drugs for the treatment of several ailments (Wilson and Patterson 2008). This method involves three essential components: light, photosensitizers and oxygen (Jeong et al. 2011). A photosensitizer absorbs a photon of light and is excited from ground state to a short-lived singlet state (Figure 4.4). The excited photosensitizer may either decay back to the ground state by emitting fluorescence, a property that is usually exploited for imaging and photodetection; or it can undergo intersystem crossing whereby the spin of its excited electron inverts to form a relatively long-lived triplet state (3 PS*) (Castano et al. 2004). The photosensitizer can then directly interact with a substrate and form a radical anion or cation, which then reacts with oxygen to produce hydroxyl radicals, superoxide anion radicals and hydrogen peroxides (type I reaction). Alternatively, the energy of the excited photosensitizer can be directly transferred to molecular oxygen, to form $^1O_2$ (type II reaction). The byproducts formed as a result of the type I and type II reactions are responsible for the cell-killing and therapeutic effect in PDT. Even though both type I and type II reactions can occur simultaneously, most of the studies indicate type II reactions, hence $^3O_2$ play a dominant role in PDT (Pineiro et al. 2002, Ding et al. 2011).

NPs represent an emerging technology in the field of PDT that can overcome most of the limitations of classic photosensitizers (Prasad 2004). In recent years, there has been an emphasis for utilizing NPs for PDT. These NPs were generally categorized into passive and active depending on the absence or presence of any targeting moieties on their surface, or based on their role in the excitation of photosensitizers (Konan et al. 2002, Bechet et al. 2008, Chatterjee et al. 2008, Paszko et al. 2011). In this section, since we are dealing with photo-activated NPs, we have narrowed our discussion to include only PDTs based on NPs, which are themselves photosensitizers.

#### *4.3.1.1 Titanium Dioxide NPs*

Since the discovery of photo-induced decomposition of water on $TiO_2$ electrodes under UV light, $TiO_2$ has been established as a foundation for many important energy-conversion processes (Fujishima and Honda 1972). In recent years, $TiO_2$ has been considered a potential photosensitizing agent for PDT due to its biocompatibility, physiological inertness, low toxicity, and unique photocatalytic activity (Cai et al. 1992, Wamer et al. 1997, Zhang 2004, Rozhkova et al. 2009). Under UV irradiation with shorter than 385 and 400 nm wavelength, the electrons in the valence band of $TiO_2$ are excited to the conduction band, resulting in photo-induced hole–electron pairs (Figure 4.5). These photo-induced electrons and holes possess strong reduction and oxidation properties that can interact with surrounding $O_2$ and $H_2O$ molecules to generate superoxide anion radicals ($OH^{\bullet -}$), hydroxyl radicals ($OH^{\bullet}$) or hydrogen peroxide ($H_2O_2$) (Nosaka et al. 2002, Kubo and Tatsuma 2004, Murakami et al. 2006).

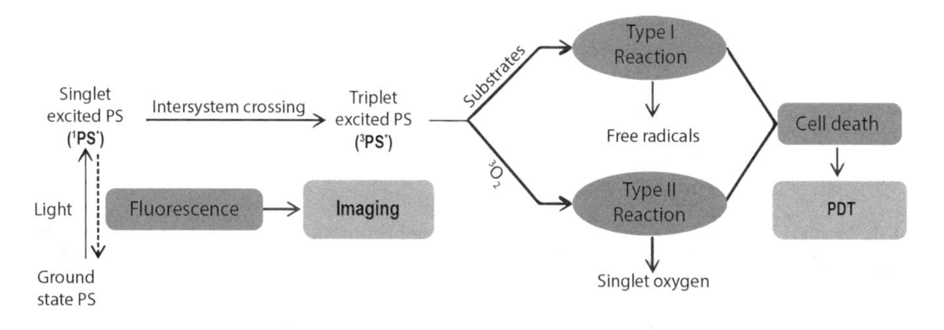


**FIGURE 4.4** Schematic illustration of a typical photodynamic reaction. (Adapted from Lucky et al. 2015. © 2015, American Chemical Society.)

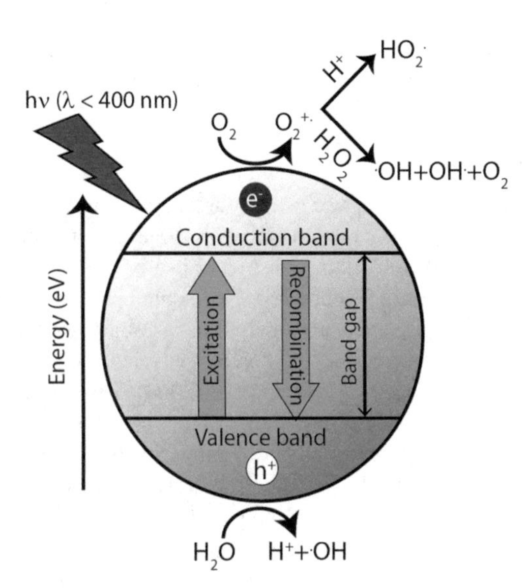


**FIGURE 4.5** Schematic illustration of photodynamic reaction upon light irradiation on $TiO_2$ NP.

The first report of the photokilling of malignant HeLa cells grown on a polarized illuminated $TiO_2$ film electrode was reported by A. Fujishima (Fujishima 1986). Later on, as research continued, the detailed mechanism of UV induced phototoxic effect of $TiO_2$ NPs was reported on a series of human cancer cells such as bladder cancer cells (T24), adenocarcinoma cells (SPC-A1), colon carcinoma cells (Ls174-t), monocytic leukemia cells (U937), breast epithelial cancer cells (MCF-7, MDA-MB468) and glioma cells (U87) (Kubota et al. 1994, Zhang 2004, Lagopati et al. 2010, Wang et al. 2011).

#### *4.3.1.2 Zinc Oxide NPs*

Due to the fact that nanosized zinc oxide (ZnO), a well-known photocatalyst, is comparable to $TiO_2$ in terms of band gap energy (3.2 eV) and photocatalytic activity, it has also been proposed as a potential photosensitizing agent for PDT. The phototoxic effect of ZnO NPs of different sizes following UV irradiation was compared by Li et al., who studied the synergetic cytotoxicity of the anticancer agent daunorubicin (DNR) with ZnO NPs against hepatocellular carcinoma cells (SMMC-7721) *in vitro* (Li et al. 2010). The size was found to have an important role in the phototoxic effect. At the same time, a greater cell-kill efficacy was achieved when SMMC-7721 cancer cells were treated with a combination of DNR and ZnO NPs. In addition to its photosensitizing capabilities, ZnO was also used as a delivery agent of anticancer drugs for combination therapy. PEGylated ZnO (ZnO/PEG) nanospheres loaded with the anticancer drug DOX were fabricated by Hariharan et al. They studied both its antibacterial and antitumor activity (Hariharan et al. 2012). DOX-ZnO/PEG nanocomposites proved to be capable of photodynamically inactivating of Gram-positive microorganisms under visible light. Furthermore, the nanocomposites exhibited dose-dependent toxicity toward HeLa cell lines, as well as demonstrated improved therapeutic efficacy upon UV irradiation, thus minimizing the side-effects.

### 4.3.2 Photothermal Therapy

Thermal therapy consists of treating malignant tumors by producing irreversible damage to cancer cells while minimizing the effect on the surrounding and healthy tissues. This is done by increasing the temperature in a well-defined location of the tissue. Its effects depend on both the magnitude and the duration of the temperature increment, as shown in Figure 4.6, which summarizes the main effects of treatments at different temperatures. For values lower than 41°C, no cellular damage occurs. Irreversible damage requires long exposures (longer than 1 hour) at moderate temperatures (41–48°C) or shorter exposures

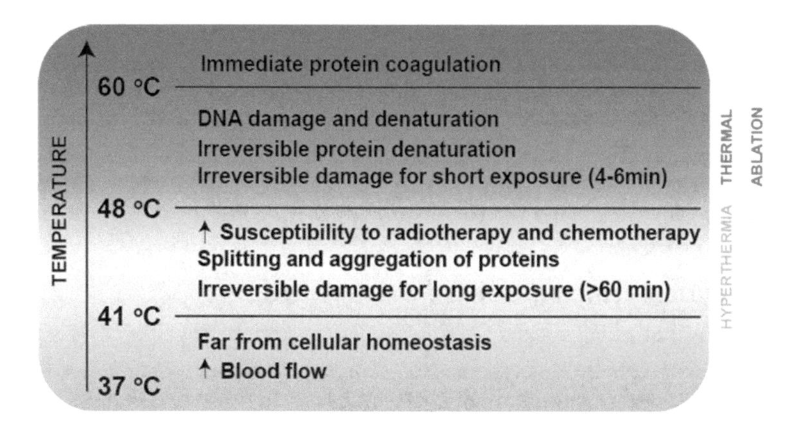


**FIGURE 4.6 (See color insert.)** Biological effects of heating procedures. Summary of the main effects of thermal therapies on treated tissues. Commonly used ranges for thermal therapy alone (thermal ablation, 48–60°C) and in combination with conventional treatments (hyperthermia, 41–48°C) are indicated.

(4–6 minutes) at higher temperatures (48–60°C). This latter temperature range, known as the thermal ablation range, is the one that is generally used in tumor therapy treatments. Therapy can also be used as an adjunct to conventional treatments (radiation therapy, chemotherapy). In this case, one usually works at lower temperatures (41–48°C, corresponding to the range of hyperthermia indicated in Figure 4.6) (Takahashi et al. 2002, Chicheł et al. 2007).

Although the possibility of treating tumors using thermal therapy has been known for decades, these types of treatments have regained interest with nanotechnology development. By locating nanoheaters (NHs) inside a tumor, it is possible to perform a minimally invasive treatment, something impossible for traditional heating systems (radiofrequency, microwave and ultrasound). Although magnetic nanomaterials' ability to treat tumors in a non-invasive way has been widely demonstrated in numerous *in vivo* experiments, the rapid development of synthesis methods has led to the appearance of a large number of materials capable of generating heat under optical excitation. As a result, PTT has attracted a great deal of attention from the scientific community in the last decade.

The main advantage of PTT is its simplicity in operation and its low cost, while its main limitation lies in the low penetration of light into the tissues. This problem, as we have stated before, can be overcome by operating with photothermal agents that can be excited within the biological windows (BWs). Additionally, in order to be a good photothermal agent, the nanomaterial must also have high photothermal conversion efficiency ($\eta$). This is defined as the fraction of absorbed energy (under a certain excitation wavelength) that is re-emitted as heat. Currently, there is a great variety of nanomaterials with reported high photothermal conversion efficiencies, including carbon nanostructures, semiconductors, organic NPs and hybrid systems in which various types of materials are combined. In terms of their applications in animal models, different NPs, among them gold nanostructures, carbon nanotubes (CNTs) and graphene NPs have demonstrated their potential to treat tumors effectively in animal models (Jaque et al. 2014). They have also been successfully used in combined therapies (photothermal therapy + chemotherapy or radiotherapy) in animal models (Hainfeld et al. 2004, Shen et al. 2013).

### *4.3.2.1 Heating NPs*

The battle against cancer is continuously bringing about new challenges that can only be overcome if faced from a multidisciplinary point of view. Among the different ongoing research areas focusing on cancer, we can safely say that nanotechnology is standing out by the development of new materials and techniques that would improve its detection and therapy (Ferrari 2005, Nie et al. 2007). If, at a certain moment in the past few years, only NPs with large fluorescence quantum yields (FQYs—which basically state the connection between the number of emitted photons and the number of absorbed photons [Jacinto et al. 2006, Rocha et al. 2013]) were desired, as they would operate as good luminescent probes, today we can also find applications for LNPs with low FQYs. In fact, when a LNP has a low FQY, a great fraction of the excitation radiation is transformed into heat and the NP behaves as an NH (Henderson and Imbusch 1989). This is, for example, the case of gold NPs (GNPs) and CNTs that have been extensively used as photothermal agents for both *in vitro* and *in vivo* experiments (Elsayed et al. 2006, Moon et al. 2009). The fact that many authors had already reported on successful laser-induced tumor treatment by using NHs led the scientific community to seriously consider NP-based PTTs as an actual alternative therapy against cancer.

#### 4.3.2.2 Example: Heat Generation in $Nd^{3+}$-Doped NPs

A recent but nonetheless important demonstration of heating by LNPs is found in the $Nd^{3+}$-doped $LaF_3$ NPs. The physical principle behind this system is found in the role of the concentration of optically active ions on the properties of rare-earth (RE)-doped NPs. In fact, it has been experimentally shown that for NPs of identical size, both fluorescence intensity and lifetime present a meaningful dependence with the concentration of dopant ions (and $Nd^{3+}$-doped $LaF_3$ NPs are no exceptions). In general, it is observed that the fluorescence intensity of nanocrystals doped with $Nd^{3+}$ ions increases with $Nd^{3+}$ concentration but, only up to a certain value, from which additional increases result in a decrease of intensity. Additionally, it was observed that the average fluorescence lifetimes monotonously decreased with the concentration of dopant ions within the NPs. This is because, by increasing the amount of $Nd^{3+}$ ions in the matrix, the distance between them is reduced and the energy transfer (ET) processes become more relevant (Henderson and Imbusch 1989, Jacinto et al. 2006).

Figure 4.7 contains the schematic representation of (i) the possible de-excitation processes of a free $Nd^{3+}$ ion when excited from the ground state up to the $^4F_{5/2}$ level by an 808 nm wavelength radiation and (ii) the ET processes between same-species of ions, which can be of two types: crossed relaxation (Figure 4.7b) and energy migration (Figure 4.7c). We see in this figure that, for the case of an isolated ion, non-radiative de-excitation processes occur through the emission of phonons, which implies in heat generation in the system. Both the transition between the excited state and the metastable state from which the radiative emissions occur ($^4F_{5/2} \rightarrow {}^4F_{3/2}$) and the transitions between the final states of the radiative processes $^4I_J$ (J = 15/2, 13/2, 11/2) and the ground state $^4I_{9/2}$ take place in a non-radiative manner. This indicates that, even in the case of NPs doped with a low concentration of $Nd^{3+}$ ions, a certain generation of heat will occur under optical excitation. The ET processes also provide additional mechanisms of non-radiative de-excitation:

- In the case of cross-relaxation processes between two $Nd^{3+}$ ions, which are schematically represented in Figure 4.7b, the $^4F_{3/2} \rightarrow {}^4I_{15/2}$ de-excitation of one of them is achieved through the excitation of the other from the ground state ($^4I_{9/2}$) to the state $^4I_{15/2}$. Both ions are then de-excited through multiphonon processes to the ground state.

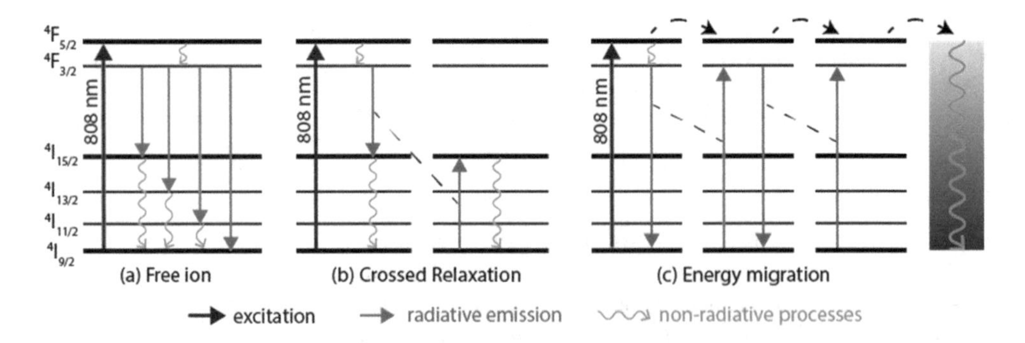


**FIGURE 4.7 (See color insert.)** Schematic representation of energy transfer processes between $Nd^{3+}$ ions. (a) Radiative and non-radiative de-excitation processes of free $Nd^{3+}$ ions (b) Cross-relaxation process between two adjacent $Nd^{3+}$ ions. (c) Energy migration process between $Nd^{3+}$ ions, which ends in a non-radiative de-excitation from a non-emitting center.

- The energy migration processes consist, as shown in Figure 4.7c, in the relaxation of one of the ions by transferring its energy to an adjacent ion. This ET between nearby ions continues until a non-radiative center is reached (usually associated with host defects), which is de-energized by heat generation. Since these defects are more frequent on the surface of materials, these processes are more relevant for nanometric systems (Quintanilla et al. 2013). Additionally, in the case of NPs dispersed in aqueous media, which is the case of many systems when thinking in biological applications, it is necessary to take into account the interactions between the optically active ions located on the surface and the solvent molecules. In this aspect, the hydroxyl groups (–OH) play a significant role due to the vibrational modes of high energies, so that the $Nd^{3+}$ ions can be de-excited by a non-radiative interaction with them (Kumar et al. 2007, Orlovskii et al. 2013). The NPs doped with a high concentration of $Nd^{3+}$ ions will therefore present a high probability of ET processes due to the small distance between ions. These processes will frequently result in non-radiative de-excitations both within the NP and in its surface and in the subsequent production of heat.

The relevance of ET processes in the case of NPs heavily doped with $Nd^{3+}$ ions is found in the decrease of the FQY of their emitting level (Jacinto et al. 2006, Rocha et al. 2013). This indicates that the higher the proportion of doping ions, the greater the ability of heat generation by $Nd^{3+}$ doped NPs. Indeed, this was experimentally shown in $LaF_3$:$Nd^{3+}$ and $NaYF_4$:$Nd^{3+}$ NPs (doped with molar percentages in the 1–25% range) (Bednarkiewicz et al. 2011, Rocha et al. 2014a).

## 4.4 Multifunctional NPs

The net effects caused on cancer tumors during PTTs strongly depend on both the magnitude of the heating as well as the treatment duration. In this regard, in order to achieve an effective treatment and keep the collateral damage at a minimum it is extremely necessary to have a temperature reading during NP-based PTTs. In order to do this, infrared thermography is commonly used in most *in vivo* experiments. This particular method has several advantages, such as simplicity, low cost, and relative precision (around 1°C) but its major drawback is that it only provides information about the surface thermal status. There is no guarantee, however, that the surface temperature corresponds to the one inside the tumor, especially if one considers the presence of heating NPs, blood perfusion, and various heat dissipation mechanisms on the surface and in the treated areas. Aiming to overcome this limitation, an increasing interest in the design of multifunctional LNPs capable of simultaneous heating and thermal sensing under single beam excitation has emerged, as they would constitute significant building blocks towards the achievement of real controlled PTTs (Ximendes et al. 2016).

First, the heating and sensing particle should operate under single beam excitation. Second, the optical excitation wavelength should avoid nonselective cellular damage. The first requirement relates to the complexity of the system of measurement, i.e., it must be as simple as possible. The second one has to do with undesired effects that may be produced by the excitation laser, such as non-localized heating. If one looks into the literature, one will find 808 nm is the optimum excitation wavelength as it minimizes both the laser-induced thermal loading of the tissue and the intracellular photochemical damage (Nyk et al. 2008, Hong et al. 2012, Zhang et al. 2012, Li et al. 2013). Additionally, the existence of high power, cost-effective laser diodes operating at 808 nm makes this specific wavelength interesting from the technical point of view (Zhan et al. 2011).

In the following sections, two single NPs capable of remote heating and thermal sensing under single 808 nm optical excitation are presented. The first one was specifically designed to introduce the concept of a single multifunctional NP. Its importance, therefore, was not so much on whether its emission lay in one of the BWs but rather on the fact that it was indeed possible to construct a simultaneous nanoheater and nanothermometer in a single NP by means of the core/shell engineering. The second one, on the other hand, was meant to present a broader applicability. Its focus was not only placed on the light-to-heat conversion efficiency but also on the spectral position of its thermal-sensitive emission (specifically in the second BW, BW-II).

### 4.4.1 Intratumoral Thermal Reading during PTT

As mentioned before, LNPs have proven to be fundamental building-blocks for advanced bio-imaging and sensing and consequently diagnosis and treatment through light-activated therapies in both *in vitro* and *in vivo* levels. All these applications have become possible due to the constant refinement and optimization of preparation techniques that have led to the ability to tailor and pre-design the physical and optical properties of LNPs with size ranging from 1 to 100 nm as effective agents for fluorescence imaging, sensing and therapies of small animals, as well as individual cells. These possibilities have led many groups around the world to focus their researches on the synthesis of multifunctional LNPs aiming to combine imaging, sensing and treatment capabilities into a single nanostructure. Thus, it is probably one of the most promising and challenging battlefronts.

A good example of the potential application of inorganic multifunctional NPs was presented by Carrasco et al. (2015). In their research, the authors demonstrated how single core relatively highly doped (5.6 at.%) $LaF_3:Nd^{3+}$ NPs under 808 nm light excitation are capable (at the same time) of deep tissue imaging, subtissue thermal sensing and acting as an efficient photothermal agent for *in vivo* hyperthermia treatment of human tumors (breast cancer, MDA-MB-231 line) using a xenograft murine animal model (Rocha et al. 2013, 2014, Carrasco et al. 2015). The authors selected 5.6 at.% of Nd ions as an ideal multifunctional agent after a systematic study of the light-to-heat efficiency conversion and emitted intensity of $LaF_3:Nd^{3+}$ NPs by varying the doping neodymium concentration levels under 808 nm excitation. As stated in section 4.3.2.2, the mechanism of heat delivery to the environment under NIR excitation is produced as a result of multiphonon relaxation from the $Nd^{3+}$ excited states or by emission quenching mediated by closely located non-radiative centers (killers). The authors also studied the temperature dependence behavior of the ${}^4F_{3/2} \rightarrow {}^4I_{9/2}$ emission band of $Nd^{3+}$ ions (corresponding to the 850–930 nm range) in the physiological temperature range (30–40°C) and a linear dependence of the emission peaks ratio at 865 and 885 nm ($R = I_{865} / I_{885}$) was found, with a thermal sensitivity of 0.25% $K^{-1}$ and a temperature accuracy/resolution close to $\Delta T = 2$°C. By means of this emission band, real control and temperature sensing was achieved during the *in vivo* photothermal treatment (Henderson and Imbusch 1989, Bednarkiewicz et al. 2011, Graham et al. 2013, Jaque et al. 2014, Rocha et al. 2014, Carrasco et al. 2015, Abadeer and Murphy 2016).

Figure 4.8a shows a mouse bearing two tumors (one tumor per flank) being submitted to light-activated thermal therapy. The tumor in the left side was treated with an injection of 5.6 at.% $LaF_3:Nd^{3+}$ NPs dispersed

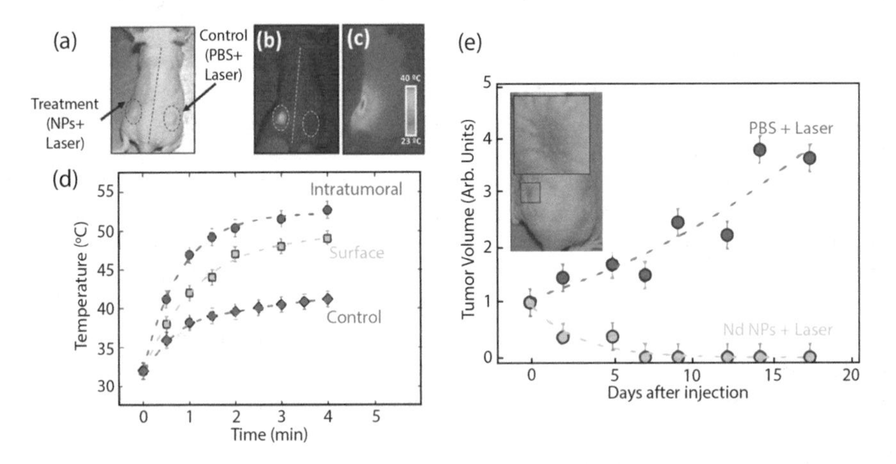


**FIGURE 4.8 (See color insert.)** (a) Optical image of a mouse bearing two tumors. Tumor at left side was treated with $LaF_3:Nd^{3+}$ NPs, while tumor at right side was used as a control (without NPs). (b) and (c) Correspond to the fluorescence image and thermal image (after 4 minutes of 808 nm laser irradiation with a power density of 4 W $cm^{-2}$) of the same mouse shown in (a). (d) The time evolution with irradiation laser of intratumoral temperature as obtained from the analysis of sub-tissue fluorescence with irradiation of the left side and tumor surface temperature (as measured by infrared thermal imaging). The evolution of the surface temperature of the control tumor is included for comparison. (e) The evolution time of the sizes post treatment of both treated and control tumors. The inset shows the optical image of reduced tumor post treatment. (Adapted with permission from Carrasco et al. 2015. © 2015, Wiley-VCH.)

in PBS at a concentration of 10% in mass (direct intratumoral injection), while the other one (right side) was used as a control (without NPs). In these experiments, the injected volume was half of the estimated total tumor volume and the number of NPs inside the injected volume was estimated at $7.65 \times 10^{13}$ NPs. As can be observed in Figure 4.8b, only the tumor injected with NPs exhibited NIR fluorescence; this clearly shows the ability of these NPs for subcutaneous fluorescence imaging. The heating function is shown in Figure 4.8c, as only the tumor containing the NPs is substantially heated under 808 nm (4 W/cm$^2$) laser irradiation.

As a matter of fact, Figure 4.8d reveals the importance of precise tumor temperature control during photothermal treatment. As can be seen, the intratumoral temperature and the surface temperature of the treated tumor were plotted as a function of the irradiated time during a 4-minute-long photothermal treatment (the time evolution of the surface temperature of a laser only control tumor, which was injected with an equivalent amount of pure PBS, has also been included for the sake of comparison). On the one hand, the intratumoral temperature was monitored by a proper spectral analysis of the $^4F_{3/2} \rightarrow ^4I_{9/2}$ emission band and on the other, the surface tumor temperature of the treated tumor was obtained by thermographic imaging. The results show a large difference between intratumoral and surface temperatures, which clearly demonstrates the need to control the temperature at the tumor site (injection region) instead of that at the skin surface, in order to avoid undesired damage due to excessive heating. The successful light activated therapy based on multifunctional using 5.6 at.% doped $LaF_3$:$Nd^{3+}$ NPs is shown in Figure 4.8e, where it can be seen that the tumor size reduces to zero 6 days post-treatment. Meanwhile, the control (non-treated) tumor continues to increase with time.

Despite the good results obtained, it is important to mention, that the work carried out by Carrasco et al. constitutes the initial "proof-of-concept" that 5.6 at.% doped $LaF_3$:$Nd^{3+}$ NPs can play a dual role as an efficient photothermal agent for the treatment of tumors and as a LNThs at the same time. However, it is reasonable that the authors performing direct intratumoral injection and real tumor treatments in preclinical stages require that Nd:$LaF_3$ NPs reach the tumor by delivery active accumulation after intravenous injection.

### 4.4.2 Self-Monitored Photothermal Core/Shell NPs (Er-Yb@Nd $LaF_3$ NPs)

Even though $LaF_3$ Nd: NPs had shown some ability for remote thermal sensing during PTT, achieving temperature resolutions below 1°C was extremely hard due to their relatively low thermal sensitivity ($10^{-3}$% $K^{-1}$) (Weber 1971, Weissleder 2001, Smith et al. 2009, Li et al. 2013, Rocha et al. 2013, 2014, Wang et al. 2013b, Zhao et al. 2015). In order to overcome this problem, different routes have been proposed. For instance, combination with other emitting units such as quantum dots (QDs), co-doping with other RE ions and structural improvement by post-synthesis processes are some of the routes that have been tried (Sidiroglou et al. 2003, Cerón et al. 2015, Marciniak et al. 2015). Although some improvements have in fact been achieved, their application in sub-degree measurements has still been unattainable.

Thus, by realizing the limitations imposed on Nd:NPs, Ximendes et al. (2016) explored the alternative of improving their thermal sensitivity by means of core/shell engineering. The proposal was ignited by the $Nd^{3+}$ ions' ability of acting as sensitizers of rare earth ions commonly used in highly sensitive luminescence thermometry (whose thermal sensitivity lies in the $10^{-2}$% $K^{-1}$ level)—specifically, $Yb^{3+}$ and $Er^{3+}$ (Rotman 1990, Barbosa-García et al. 1997, Wang et al. 2013a). It was well known in the literature that the combination of $Nd^{3+}$, $Yb^{3+}$ and $Er^{3+}$ into a single NP provided an efficient green emission generated by $Er^{3+}$ ions under single 808 nm optical excitation of $Nd^{3+}$ ions thanks to both $Nd^{3+} \rightarrow Yb^{3+}$ and $Yb^{3+} \rightarrow Er^{3+}$ ET processes, with efficiencies superior to 50% (Huang and Lin 2015). Nevertheless, some requirements had to be satisfied. Indeed, the efficient excitation of the thermosensitive $Er^{3+}$ ions could only be achieved if they were spatially separated from the $Nd^{3+}$ sensitizing ions (Chen et al. 2012, Wang et al. 2013b, Rocha et al. 2014). The proposed NPs are schematically drawn in Figure 4.9a where one sees a $Nd^{3+}$-doped shell surrounding an $Er^{3+}$,$Yb^{3+}$-co-doped core. Under this approach, the shell acts as a donor unit absorbing the 808 nm radiation through the $^4I_{9/2} \rightarrow ^4F_{5/2}$ $Nd^{3+}$ absorption (see Figure 4.9b). Then, non-radiative ET between $Nd^{3+}$ and $Yb^{3+}$ ions at the core/shell interface results in the $^2F_{7/2} \rightarrow ^2F_{5/2}$ excitation of $Yb^{3+}$ ions inside the core (Ostroumov et al. 1998, Tu et al. 2011). Finally, ET between $Yb^{3+}$ and $Er^{3+}$ ions allocated at core leads to the excitation of erbium ions up to their $^2H_{11/2}$, $^4S_{3/2}$ thermally coupled states (Figure 4.9b), from which the thermal sensitive luminescence is generated.

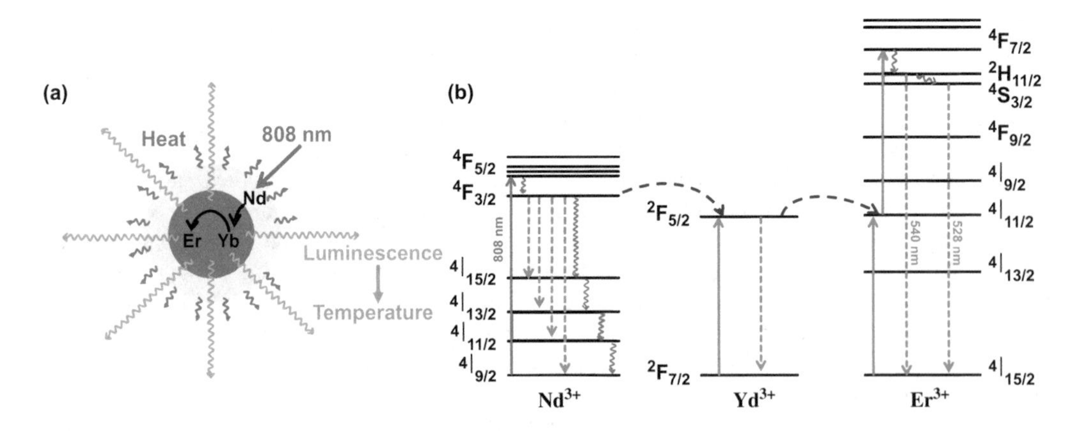


**FIGURE 4.9** **(See color insert.)** (a) Schematic diagram of the active-core/active-shell nanoparticles specifically designed for simultaneous heating and thermal sensing activated by 808 nm light. In this design, the shell acts as a heating center whereas the core provides thermal sensitivity to the structure. (b) Detail of the energy level diagrams of $Nd^{3+}$, $Yb^{3+}$ and $Er^{3+}$ ions. The heat produced by $Nd^{3+}$ ions in the shell is represented by red wave arrows whereas the green emission, resultant of the subsequent energy transfer between (i) $Nd^{3+}$ and $Yb^{3+}$ ions and (ii) between $Yb^{3+}$ and $Er^{3+}$, is represented by green dashed arrows. (Adapted from Ximendes et al. 2016. © 2016, Royal Society of Chemistry.)

Under this scheme, if the concentration of $Nd^{3+}$ ions is high enough, the non-radiative delivered energy (the waving arrows in Figure 4.9a) is sufficiently high to make the shell behaves as a heating unit (see Section 4.3.2.1.1) (Kumar et al. 2007, Pavel et al. 2008).

In order to evaluate the potential of Er:Yb@Nd $LaF_3$ core/shell NPs in *in vivo* controlled photothermal processes, the authors designed a simple experiment aiming to constitute an *ex vivo* proof of concept. An amount of 0.2 ml of an aqueous solution containing Er-Yb@Nd $LaF_3$ NPs was injected into a chicken breast. The injection depth was estimated to be around 1.0 mm. Optical excitation and subsequent visible luminescence collection were both performed by using a single long working distance objective. The spectral analysis of the subcutaneous NPs fluorescence was used to estimate the injection's temperature. An infrared thermographic camera was coupled to the optical setup with the aim of recording the superficial temperature of the tissue and to evidence its difference to the subcutaneous one (the former being expected to be lower due to heat diffusion processes) (Rocha et al. 2014, Carrasco et al. 2015). The steady state subcutaneous and surface temperatures were recorded for different 808 nm laser powers, and in both cases, a linear relation was experimentally found (Figure 4.10b). As expected, there were

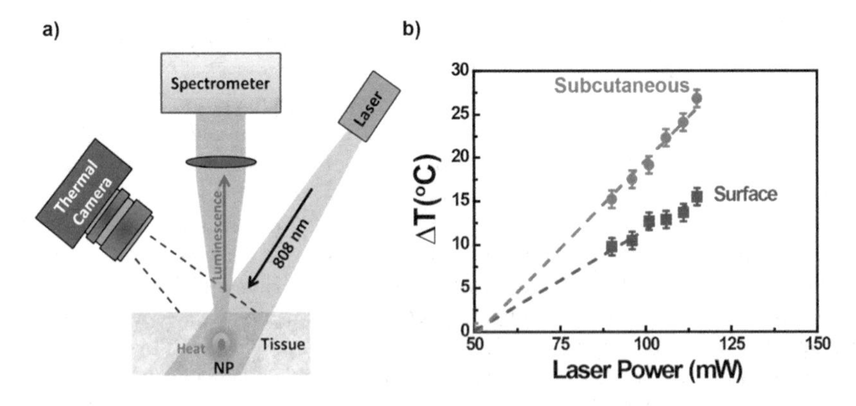


**FIGURE 4.10** (a) Schematic diagram of the *ex vivo* single beam self-referenced thermal loading experiments. The spectral analysis of the injected core/shell nanoparticles provided the subcutaneous temperature whereas the surface temperature was recorded with an infrared thermographic camera. (b) Surface and subcutaneous temperatures as obtained for different 808 nm laser powers. Dots are experimental data and dashed lines are linear fits. Notice that for laser power lower than 90 mW, the NPs fluorescence was too weak to be considered. (Reproduced with permission from Ximendes et al. 2016. © 2016, Royal Society of Chemistry.)

remarkable differences between the subcutaneous and the superficial temperatures. In fact, this was found to be as large as 38%. Although experimental data included in Figure 4.10b reveals the potential of Er-Yb@Nd $LaF_3$ for thermally controlled subcutaneous PTTs, it should be mentioned that their application would be limited by the optical penetration of the green $Er^{3+}$ fluorescence (1–2 mm depth) into tissues. For the performance of controlled PTTs at larger depths, the heating/sensing core/shell structure should work (emit) in the NIR. In fact, that was the problem faced in the design of the NPs described in the following subsection.

### 4.4.3 Core/Shell NPs for Subcutaneous Heating and Thermal Sensing in BW-II

Even though the possibility of constructing single NPs capable of both heating and thermal sensing by means of the core/shell nanoengineering was demonstrated, their potential to work subcutaneously was still disputed. The next multifunctional agent to be proposed aimed to join the heating capacity of heavily $Nd^{3+}$-doped $LaF_3$ NPs and the infrared luminescence thermal sensing of $Nd^{3+}$,$Yb^{3+}$-co-doped core/shell $LaF_3$ NPs in a single nanostructure. The proposed NPs were based on a core/shell structure with an $Yb^{3+}$-doped core (concentration of 10 mol. %) and a highly $Nd^{3+}$-doped shell (concentration of 25 mol. %).

The attentive reader will notice that, under this proposal, the heating capacity of the NPs is conferred by the same principles described in Section 4.4.2 for the heating shell. The thermal sensing capability, on the other hand, has very different foundations. In simple terms, the thermal sensing ability comes from the fact that the relative contribution of $Nd^{3+}$ and $Yb^{3+}$ emissions to the overall emission of the core/shell NPs depends on the $Nd^{3+} \rightarrow Yb^{3+}$ ET efficiency as well as on the back ET, which are both expected to be temperature dependent (de Sousa et al. 2001, González-Pérez et al. 2007). A value of $S_r = 0.74 \pm 0.02\%$ $K^{-1}$ at 20°C was reported for the relative thermal sensitivity.

In order to evaluate the potential application of Yb@Nd* $LaF_3$, two simple experiments, aiming to constitute clear proofs of concepts, were designed (see Figure 4.11a). In both experiments, an amount of 0.2 mL of an aqueous solution of Yb@Nd* $LaF_3$ NPs (10% in mass) was injected into a chicken breast sample (injection depth estimated to be close to 2 mm). An 808 nm laser beam was focused into the injection by using a single long-working-distance microscope objective (40X, N.A. of 0.25). The Yb@Nd* $LaF_3$ NPs would partially convert the 808 nm laser radiation into heat, producing a well localized temperature increment inside the tissue. The amount of this heating at the injection volume was estimated by means of the intensity ratio $\Delta$ (as defined above), from the infrared spectra generated at the injection volume.

For the first experiment, an infrared thermal camera was coupled to the optical set-up in order to record the superficial temperature of the tissue and to contrast its value with the one obtained subcutaneously (the latter is expected to be lower due to heat diffusion processes) (Rocha et al. 2014, Carrasco et al. 2015, Ximendes et al. 2016). The steady state subcutaneous and surface temperatures were recorded for different 808 nm power densities, and in both cases, a linear relationship was experimentally found (Figure 4.11b). As expected, remarkable differences between subcutaneous and surface temperatures were observed, revealing, therefore, that the infrared thermal camera fails to provide subcutaneous thermal reading. Thus, to achieve an accurate and remote control over PTTs, the use of LNThs was extremely necessary.

Additionally, the same experimental setup was used to demonstrate the capability of Yb@Nd* NPs for real-time subcutaneous thermal measurements. The second experiment consisted in taking the luminescence spectra generated by the core/shell NPs in intervals of 400 ms for a time period of approximately 3 min (when thermal stabilization was achieved) at a fixed laser power density, starting on the moment the 808 nm laser was turned on. Figure 4.11c shows the time dependence of the subcutaneous temperature variation under two different excitation laser power densities. This temperature was obtained by computing the time evolution of the intensity ratio ($\Delta$) from the subcutaneous emission spectra. Fitting the experimental data, obtained with low (1.2 $W.cm^{-2}$) and high laser power densities (3.5 $W.cm^{-2}$), to the corresponding solution of Pennes' bioheat equation provided the parameters $R = 2.0 \pm 0.1$ mm for 1.2 $W.cm^{-2}$ and $R = 1.9 \pm 0.1$ mm for 3.5 $W.cm^{-2}$ (Huang and Liu 2009). These parameters were, indeed, close to the estimated depth of injection, i.e., the distance between the heating source (NP injection) and the heat sink (tissue–air interface). Thus, R defined the region where the set of NPs were simultaneously

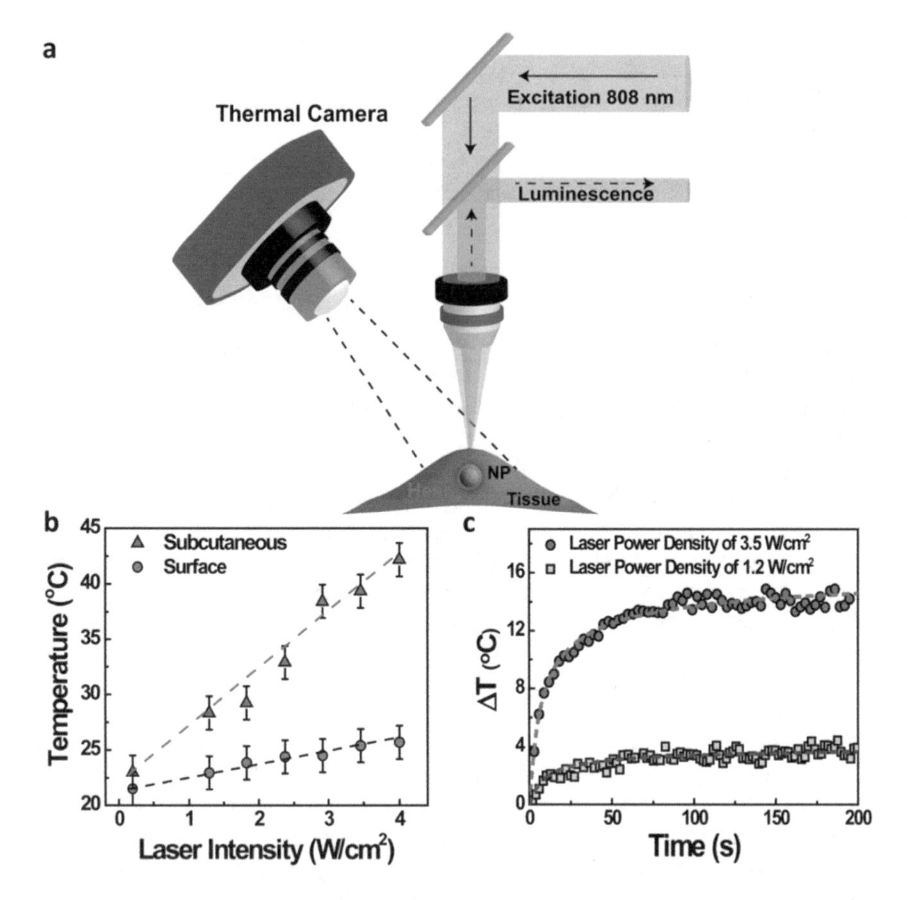


**FIGURE 4.11** (a) Schematic diagram of the *ex vivo* single beam experiment. The laser beam size is not drawn to scale. The spectral analysis of the injected core/shell nanoparticles provided the subcutaneous temperature whereas the surface temperature was recorded with an infrared thermographic camera. (b) Surface and subcutaneous temperatures as obtained for different 808 nm laser power densities. Dots are experimental data and dashed lines are linear fits. (c) Temperature stabilization of a tissue when exposed to an incident 808 nm laser beam at constant power density (1.2 $W.cm^{-2}$ and 3.5 $W.cm^{-2}$ for blue squares and red circles, respectively). (Reproduced with permission from Ximendes et al. 2016. © 2016, American Institute of Physics [AIP].)

sensing temperature and meaningfully heating the medium. Data included in Figure 4.11 demonstrates that our core/shell NPs are capable of providing a steady state and time-resolved thermal reading over a subcutaneous heating process activated by themselves, constituting, therefore, a proof for the potential of Yb@Nd* $LaF_3$ NPs for thermally controlled subcutaneous PTTs.

## 4.5 Conclusions and Remarks

In summary, we have revisited some examples of diagnosis and therapy based on photo-activated nanoparticles. Lanthanides-based luminescence thermometry is a versatile technique operating in a wide range of the optical spectrum from the UV to the IR and with a large number of parameters to be exploited, such as intensity ratios, peak shift, lifetime, polarization, etc. In fact, the temperature reading in intra-cell or in sub-micro systems was only possible using luminescence nanothermometer (Vetrone et al. 2010). Furthermore, other fields like microelectronics, microfluidics, microoptics, etc. can and do use this approach. Conclusions can be drawn that lanthanide-doped nanoparticles present great promise in theranostics as they offer a large number of advantages over conventional biomarkers and therapeutic agents due to the feature of the 4f orbital of lanthanide ions and the biocompatibility brought by the intrinsic merits of nanoscaled materials.

However, despite the promising progress, the research on luminescence thermometry can be considered as in its early stages, and more basic knowledge is still needed before prototypes become a commercial reality. The main challenges currently found by the scientists of the field are:

- To design luminescence nanomaterials for bioapplications with superior relative thermal sensitivity, around or larger than 1% $K^{-1}$, and whose emission and absorption bands lie within the biological windows (650–950, 1000–1350, 1600–1750 nm), where tissue scattering and absorption are minimized. These finds would enable bi- and tri-dimensional thermal images.
- To understand the temperature dependence of the main mechanisms/processes that determine the thermal sensitivity of co-doping or dual-center systems. For example, ion–ion, host–ion, center–ion, ligand–ion energy transfer versus temperature.
- To develop a multifunctional NP with the features set out above and still be an efficient NH operating under a single excitation source.
- The nanothermometers pointed out above still need to be of easy calibration (preferably self-calibrated) and with well-established equations.
- Maybe the most difficult, to develop such bio-NPs with targeting effects for specific problems, for instance, tumoral cells, thus avoiding secondary effects such as the destruction of healthy tissues during treatment.
- The current bioimaging and therapeutic models are mainly focused on cells (*in vitro*) and mice (*in vivo*). Other biological models including large animals should also be considered to enrich the potentiality of the applications.

Although the field of optical nanothermometry for theranostics is still in its infancy, mainly for commercial applications, many excellent reviews, book chapters and complete books have been published and deserve attention. For example, L. D. Carlos's group (Brites et al. 2012, 2016), D. Jaque's group (Jaque and Vetrone 2012, Jaque et al. 2014, 2016, Labrador-Páez et al. 2018) and others (Dong et al. 2015, Wang et al. 2015) have given great contributions.

## Acknowledgments

We would like to thank the Brazilian funding agencies for the partial support of these studies: CNPq (Conselho Nacional de Desenvolvimento Científico e Tecnológico) by means of the Grant nr. 304479/2014-4; FINEP (Financiadora de Estudos e Projetos) by means of CT-INFRA projects (INFRAPESQ-11 and INFRAPESQ-12); CAPES (Coordenação de Aperfeiçoamento de Pessoal de Ensino Superior) Grant PNPD-CAPES and Project PVE A077/2013; and FAPEAL (Fundação de Amparo à Pesquisa do Estado de Alagoas), Grant 60030-000384/2017. E. C. Ximendes was supported by a PhD scholarship from CNPq and by the PVE A077/2013 project by means of a PhD sandwich program developed at the Universidad Autonoma de Madrid, Spain.

# REFERENCES

Aaron, J.S., Oh, J., Larson, T.A., Kumar, S., Milner, T.E., and Sokolov, K.V., 2006. Increased optical contrast in imaging of epidermal growth factor receptor using magnetically actuated hybrid gold/iron oxide nanoparticles. *Optics Express*, 14 (26), 12930.

Abadeer, N.S., and Murphy, C.J., 2016. Recent progress in cancer thermal therapy using gold nanoparticles. *Journal of Physical Chemistry C*, 120 (9), 4691–4716.

Au, K.M., Lu, Z., Matcher, S.J., and Armes, S.P., 2011. Polypyrrole nanoparticles: A potential optical coherence tomography contrast agent for cancer imaging. *Advanced Materials*, 23 (48), 5792–5795.

Baker, S.N., McCleskey, T.M., and Baker, G.A., 2005. An ionic liquid-based optical thermometer. In: R.D. Rogers and K.R. Seddon, eds. *Ionic Liquids IIIB: Fundamentals, Progress, Challenges, and Opportunities*. Washington, DC: American Chemical Society, 171–181.

Barbosa-García, O., McFarlane, R.A., Birnbaum, M., and Díaz-Torres, L.A., 1997. Neodymium-to-erbium nonradiative energy transfer and fast initial fluorescence decay of the $^4F_{3/2}$ state of neodymium in garnet crystals. *Journal of the Optical Society of America B*, 14 (10), 2731.

Bechet, D., Couleaud, P., Frochot, C., Viriot, M.-L., Guillemin, F., and Barberi-Heyob, M., 2008. Nanoparticles as vehicles for delivery of photodynamic therapy agents. *Trends in Biotechnology*, 26 (11), 612–621.

Bednarkiewicz, A., Wawrzynczyk, D., Nyk, M., and Strek, W., 2011. Optically stimulated heating using Nd3+ doped NaYF4 colloidal near infrared nanophosphors. *Applied Physics B*, 103 (4), 847–852.

Boppart, S.A., Oldenburg, A.L., Xu, C., and Marks, D.L., 2005. Optical probes and techniques for molecular contrast enhancement in coherence imaging. *Journal of Biomedical Optics*, 10 (4), 041208.

Brites, C.D.S., Lima, P.P., Silva, N.J.O., Millán, A., Amaral, V.S., Palacio, F., and Carlos, L.D., 2012. Thermometry at the nanoscale. *Nanoscale*, 4 (16), 4799.

Brites, C.D.S., Millán, A., and Carlos, L.D., 2016. Lanthanides in luminescent thermometry. In: J.-C. G. Bünzli and V. Pecharsky, eds. *Handbook on the Physics and Chemistry of Rare Earths*. Elsevier, 339–427.

Cai, R., Kubota, Y., Shuin, T., Sakai, H., Hashimoto, K., and Fujishima, A., 1992. Induction of cytotoxicity by photoexcited TiO2 particles. *Cancer Research*, 52 (8), 2346–2348.

Carrasco, E., del Rosal, B., Sanz-Rodríguez, F., de la Fuente, Á.J., Gonzalez, P.H., Rocha, U., Kumar, K.U., Jacinto, C., Solé, J.G., and Jaque, D., 2015. Intratumoral thermal reading During photo-thermal therapy by multifunctional fluorescent nanoparticles. *Advanced Functional Materials*, 25 (4), 615–626.

Castano, A.P., Demidova, T.N., and Hamblin, M.R., 2004. Mechanisms in photodynamic therapy: Part one—Photosensitizers, photochemistry and cellular localization. *Photodiagnosis and Photodynamic Therapy*, 1 (4), 279–293.

Cerón, E.N., Ortgies, D.H., del Rosal, B., Ren, F., Benayas, A., Vetrone, F., Ma, D., Sanz-Rodríguez, F., Solé, J.G., Jaque, D., and Rodríguez, E.M., 2015. Hybrid nanostructures for high-sensitivity luminescence nanothermometry in the second biological window. *Advanced Materials*, 27 (32), 4781–4787.

Chatterjee, D.K., Fong, L.S., and Zhang, Y., 2008. Nanoparticles in photodynamic therapy: An emerging paradigm. *Advanced Drug Delivery Reviews*, 60 (15), 1627–1637.

Chen, G., Shen, J., Ohulchanskyy, T.Y., Patel, N.J., Kutikov, A., Li, Z., Song, J., Pandey, R.K., Ågren, H., Prasad, P.N., and Han, G., 2012. (α-NaYbF4 :Tm3+)/CaF2 core/shell nanoparticles with efficient near-infrared to near-infrared upconversion for high-contrast deep tissue bioimaging. *ACS Nano*, 6 (9), 8280–8287.

Cheng, L., Yang, K., Chen, Q., and Liu, Z., 2012. Organic stealth nanoparticles for highly effective *in vivo* near-infrared photothermal therapy of cancer. *ACS Nano*, 6 (6), 5605–5613.

Chicheł, A., Skowronek, J., Kubaszewska, M., and Kanikowski, M., 2007. Hyperthermia – Description of a method and a review of clinical applications. *Reports of Practical Oncology and Radiotherapy*, 12 (5), 267–275.

de Sousa, D.F., Batalioto, F., Bell, M.J.V., Oliveira, S.L., and Nunes, L.A.O., 2001. Spectroscopy of Nd3+ and Yb3+ codoped fluoroindogallate glasses. *Journal of Applied Physics*, 90 (7), 3308–3313.

Ding, H., Yu, H., Dong, Y., Tian, R., Huang, G., Boothman, D.A., Sumer, B.D., and Gao, J., 2011. Photoactivation switch from type II to type I reactions by electron-rich micelles for improved photodynamic therapy of cancer cells under hypoxia. *Journal of Controlled Release*, 156 (3), 276–280.

Dong, H., Du, S.-R., Zheng, X.-Y., Lyu, G.-M., Sun, L.-D., Li, L.-D., Zhang, P.-Z., Zhang, C., and Yan, C.-H., 2015. Lanthanide nanoparticles: From design toward bioimaging and therapy. *Chemical Reviews*, 115 (19), 10725–10815.

Elsayed, I., Huang, X., and Elsayed, M., 2006. Selective laser photo-thermal therapy of epithelial carcinoma using anti-EGFR antibody conjugated gold nanoparticles. *Cancer Letters*, 239 (1), 129–135.

Ferrari, M., 2005. Cancer nanotechnology: Opportunities and challenges. *Nature Reviews Cancer*, 5 (3), 161–171.

Fujishima, A., 1986. Behavior of tumor cells on photoexcited semiconductor surface. *Photomedicine Photobiological*, 8, 45–46.

Fujishima, A., and Honda, K., 1972. Electrochemical photolysis of water at a semiconductor electrode. *Nature*, 238 (5358), 37–38.

González-Pérez, S., Martín, I.R., Rivera-López, F., and Lahoz, F., 2007. Temperature dependence of Nd3+→Yb3+ energy transfer processes in co-doped oxyfluoride glass ceramics. *Journal of Non-Crystalline Solids*, 353 (18–21), 1951–1955.

Graham, E.G., Macneill, C.M., and Levi-Polyachenko, N.H., 2013. Review of metal, carbon and polymer nanoparticles for infrared photothermal therapy. *NANO Life*, 3 (3), 1330002.

Hainfeld, J.F., Slatkin, D.N., and Smilowitz, H.M., 2004. The use of gold nanoparticles to enhance radiotherapy in mice. *Physics in Medicine and Biology*, 49 (18), N309–N315.

Hariharan, R., Senthilkumar, S., Suganthi, A., and Rajarajan, M., 2012. Synthesis and characterization of doxorubicin modified ZnO/PEG nanomaterials and its photodynamic action. *Journal of Photochemistry and Photobiology B: Biology*, 116, 56–65.

Henderson, B., and Imbusch, G.F., 1989, *Optical Spectroscopy of Inorganic Solids*. Oxford [Oxfordshire]; New York: Clarendon Press; Oxford University Press.

Hong, G., Robinson, J.T., Zhang, Y., Diao, S., Antaris, A.L., Wang, Q., and Dai, H., 2012. In vivo fluorescence imaging with Ag2 S quantum dots in the second near-infrared region. *Angewandte Chemie International Edition*, 51 (39), 9818–9821.

Huang, L., and Liu, L.-S., 2009. Simultaneous determination of thermal conductivity and thermal diffusivity of food and agricultural materials using a transient plane-source method. *Journal of Food Engineering*, 95 (1), 179–185.

Huang, X., and Lin, J., 2015. Active-core/active-shell nanostructured design: An effective strategy to enhance Nd3+/Yb3+ cascade sensitized upconversion luminescence in lanthanide-doped nanoparticles. *Journal of Materials Chemistry C*, 3 (29), 7652–7657.

Jacinto, C., Oliveira, S.L., Nunes, L.A.O., Myers, J.D., Myers, M.J., and Catunda, T., 2006. Normalized-lifetime thermal-lens method for the determination of luminescence quantum efficiency and thermo-optical coefficients: Application to Nd3+-doped glasses. *Physical Review B*, 73 (12).

Jaque, D., Martínez Maestro, L., del Rosal, B., Haro-Gonzalez, P., Benayas, A., Plaza, J.L., Martín Rodríguez, E., and García Solé, J., 2014. Nanoparticles for photothermal therapies. *Nanoscale*, 6 (16), 9494–9530.

Jaque, D., Richard, C., Viana, B., Soga, K., Liu, X., and García Solé, J., 2016. Inorganic nanoparticles for optical bioimaging. *Advances in Optics and Photonics*, 8 (1), 1.

Jaque, D., and Vetrone, F., 2012. Luminescence nanothermometry. *Nanoscale*, 4 (15), 4301.

Jeong, H., Huh, M., Lee, S.J., Koo, H., Kwon, I.C., Jeong, S.Y., and Kim, K., 2011. Photosensitizer-conjugated human serum albumin nanoparticles for effective photodynamic therapy. *Theranostics*, 1, 230–239.

Jiang, S., Gnanasammandhan, M.K., and Zhang, Y., 2010. Optical imaging-guided cancer therapy with fluorescent nanoparticles. *Journal of the Royal Society Interface*, 7 (42), 3–18.

Konan, Y.N., Gurny, R., and Allémann, E., 2002. State of the art in the delivery of photosensitizers for photodynamic therapy. *Journal of Photochemistry and Photobiology B: Biology*, 66 (2), 89–106.

Kubo, W., and Tatsuma, T., 2004. Detection of H2O2 released from TiO2 photocatalyst to air. *Analytical Sciences*, 20 (4), 591–593.

Kubota, Y., Shuin, T., Kawasaki, C., Hosaka, M., Kitamura, H., Cai, R., Sakai, H., Hashimoto, K., and Fujishima, A., 1994. Photokilling of T-24 human bladder cancer cells with titanium dioxide. *British Journal of Cancer*, 70 (6), 1107–1111.

Kumar, G.A., Chen, C.W., Ballato, J., and Riman, R.E., 2007. Optical characterization of infrared emitting rare-earth-doped fluoride nanocrystals and their transparent nanocomposites. *Chemistry of Materials*, 19 (6), 1523–1528.

Labrador-Páez, L., Ximendes, E.C., Rodríguez-Sevilla, P., Ortgies, D.H., Rocha, U., Jacinto, C., Martín Rodríguez, E., Haro-González, P., and Jaque, D., 2018. Core–shell rare-earth-doped nanostructures in biomedicine. *Nanoscale*, 10 (27), 12935–12956.

Lagopati, N., Kitsiou, P.V., Kontos, A.I., Venieratos, P., Kotsopoulou, E., Kontos, A.G., Dionysiou, D.D., Pispas, S., Tsilibary, E.C., and Falaras, P., 2010. Photo-induced treatment of breast epithelial cancer cells using nanostructured titanium dioxide solution. *Journal of Photochemistry and Photobiology A: Chemistry*, 214 (2–3), 215–223.

Lee, D.-E., Koo, H., Sun, I.-C., Ryu, J.H., Kim, K., and Kwon, I.C., 2012. Multifunctional nanoparticles for multimodal imaging and theragnosis. *Chemical Society Reviews*, 41 (7), 2656–2672.

Li, J., Guo, D., Wang, X., Wang, H., Jiang, H., and Chen, B., 2010. The photodynamic effect of different size ZnO nanoparticles on cancer cell proliferation in vitro. *Nanoscale Research Letters*, 5 (6), 1063–1071.

Li, X., Wang, R., Zhang, F., Zhou, L., Shen, D., Yao, C., and Zhao, D., 2013. Nd3+ sensitized up/down converting dual-mode nanomaterials for efficient in-vitro and in-vivo bioimaging excited at 800 nm. *Scientific Reports*, 3 (1).

Licha, K., and Olbrich, C., 2005. Optical imaging in drug discovery and diagnostic applications. *Advanced Drug Delivery Reviews*, 57 (8), 1087–1108.

Loo, C., Lin, A., Hirsch, L., Lee, M.-H., Barton, J., Halas, N., West, J., and Drezek, R., 2004. Nanoshell-enabled photonics-based imaging and therapy of cancer. *Technology in Cancer Research and Treatment*, 3 (1), 33–40.

Lucky, S.S., Soo, K.C., and Zhang, Y., 2015. Nanoparticles in photodynamic therapy. *Chemical Reviews*, 115 (4), 1990–2042.

Marciniak, ł., Bednarkiewicz, A., Stefanski, M., Tomala, R., Hreniak, D., and Strek, W., 2015. Near infrared absorbing near infrared emitting highly-sensitive luminescent nanothermometer based on Nd3+ to Yb3+ energy transfer. *Physical Chemistry Chemical Physics*, 17 (37), 24315–24321.

Moon, H.K., Lee, S.H., and Choi, H.C., 2009. *In vivo* near-infrared mediated tumor destruction by photothermal effect of carbon nanotubes. *ACS Nano*, 3 (11), 3707–3713.

Murakami, Y., Kenji, E., Nosaka, A.Y., and Nosaka, Y., 2006. Direct detection of OH radicals diffused to the gas phase from the UV-irradiated photocatalytic TiO2 surfaces by means of laser-induced fluorescence spectroscopy. *Journal of Physical Chemistry B*, 110 (34), 16808–16811.

Nie, S., Xing, Y., Kim, G.J., and Simons, J.W., 2007. Nanotechnology applications in cancer. *Annual Review of Biomedical Engineering*, 9 (1), 257–288.

Nosaka, Y., Nakamura, M., and Hirakawa, T., 2002. Behavior of superoxide radicals formed on TiO2 powder photocatalysts studied by a chemiluminescent probe method. *Physical Chemistry Chemical Physics*, 4 (6), 1088–1092.

Nyk, M., Kumar, R., Ohulchanskyy, T.Y., Bergey, E.J., and Prasad, P.N., 2008. High contrast in vitro and in vivo photoluminescence bioimaging using Near infrared to Near infrared up-conversion in Tm 3+ and Yb 3+ doped fluoride nanophosphors. *Nano Letters*, 8 (11), 3834–3838.

Orlovskii, Y.V., Popov, A.V., Platonov, V.V., Fedorenko, S.G., Sildos, I., and Osipov, V.V., 2013. Fluctuation kinetics of fluorescence hopping quenching in the Nd3+:Y2O3 spherical nanoparticles. *Journal of Luminescence*, 139, 91–97.

Ostroumov, V., Jensen, T., Meyn, J.-P., Huber, G., and Noginov, M.A., 1998. Study of luminescence concentration quenching and energy transfer upconversion in Nd-doped LaSc3(BO3)4 and GdVO4 laser crystals. *Journal of the Optical Society of America B*, 15 (3), 1052.

Paszko, E., Ehrhardt, C., Senge, M.O., Kelleher, D.P., and Reynolds, J.V., 2011. Nanodrug applications in photodynamic therapy. *Photodiagnosis and Photodynamic Therapy*, 8 (1), 14–29.

Pavel, N., Kränkel, C., Peters, R., Petermann, K., and Huber, G., 2008. In-band pumping of Nd-vanadate thin-disk lasers. *Applied Physics B*, 91 (3–4), 415–419.

Pennes, H.H., 1948. Analysis of tissue and arterial blood temperatures in the resting human forearm. *Journal of Applied Physiology*, 1 (2), 93–122.

Pineiro, M., Pereira, M.M., Formosinho, S.J., and Arnaut, L.G., 2002. New halogenated Phenylbacteriochlorins and their efficiency in singlet-oxygen sensitization. *Journal of Physical Chemistry A*, 106 (15), 3787–3795.

Prasad, P.N., 2004. Polymer science and technology for new generation photonics and biophotonics. *Current Opinion in Solid State and Materials Science*, 8 (1), 11–19.

Quintanilla, M., Núñez, N.O., Cantelar, E., Ocaña, M., and Cussó, F., 2013. Energy transfer efficiency in YF3 nanocrystals: Quantifying the Yb3+ to Tm3+ infrared dynamics. *Journal of Applied Physics*, 113 (17), 174308.

Rai, P., Mallidi, S., Zheng, X., Rahmanzadeh, R., Mir, Y., Elrington, S., Khurshid, A., and Hasan, T., 2010. Development and applications of photo-triggered theranostic agents. *Advanced Drug Delivery Reviews*, 62 (11), 1094–1124.

Rocha, U., Jacinto da Silva, C., Ferreira Silva, W., Guedes, I., Benayas, A., Martínez Maestro, L., Acosta Elias, M., Bovero, E., van Veggel, F.C.J.M., García Solé, J.A., and Jaque, D., 2013. Subtissue thermal sensing based on neodymium-doped $LaF_3$ nanoparticles. *ACS Nano*, 7 (2), 1188–1199.

Rocha, U., Kumar, K.U., Jacinto, C., Villa, I., Sanz-Rodríguez, F., del Carmen Iglesias de la Cruz, M., Juarranz, A., Carrasco, E., van Veggel, F.C.J.M., Bovero, E., Solé, J.G., and Jaque, D., 2014a. Neodymium-doped LaF 3 nanoparticles for fluorescence bioimaging in the second biological window. *Small*, 10 (6), 1141–1154.

Rocha, U., Upendra Kumar, K., Jacinto, C., Ramiro, J., Caamaño, A.J., García Solé, J., and Jaque, D., 2014b. Nd3+ doped LaF3 nanoparticles as self-monitored photo-thermal agents. *Applied Physics Letters*, 104 (5), 053703.

Rotman, S.R., 1990. Analysis of neodymium-to-erbium energy transfer in yttrium aluminum garnet with a nonuniform-distribution model. *Optics Letters*, 15 (4), 230.

Rozhkova, E.A., Ulasov, I., Lai, B., Dimitrijevic, N.M., Lesniak, M.S., and Rajh, T., 2009. A high-performance NanoBio photocatalyst for targeted brain cancer therapy. *Nano Letters*, 9 (9), 3337–3342.

Shen, S., Tang, H., Zhang, X., Ren, J., Pang, Z., Wang, D., Gao, H., Qian, Y., Jiang, X., and Yang, W., 2013. Targeting mesoporous silica-encapsulated gold nanorods for chemo-photothermal therapy with near-infrared radiation. *Biomaterials*, 34 (12), 3150–3158.

Sidiroglou, F., Wade, S.A., Dragomir, N.M., Baxter, G.W., and Collins, S.F., 2003. Effects of high-temperature heat treatment on Nd3+-doped optical fibers for use in fluorescence intensity ratio based temperature sensing. *Review of Scientific Instruments*, 74 (7), 3524–3530.

Smith, A.M., Mancini, M.C., and Nie, S., 2009. Second window for in vivo imaging: Bioimaging. *Nature Nanotechnology*, 4 (11), 710–711.

Takahashi, I., Emi, Y., Hasuda, S., Kakeji, Y., Maehara, Y., and Sugimachi, K., 2002. Clinical application of hyperthermia combined with anticancer drugs for the treatment of solid tumors. *Surgery*, 131 (1 Suppl), S78–S84.

Tomlins, P.H., and Wang, R.K., 2005. Theory, developments and applications of optical coherence tomography. *Journal of Physics D: Applied Physics*, 38 (15), 2519–2535.

Tu, D., Liu, L., Ju, Q., Liu, Y., Zhu, H., Li, R., and Chen, X., 2011. Time-resolved FRET biosensor based on amine-functionalized lanthanide-doped NaYF4 nanocrystals. *Angewandte Chemie International Edition*, 50 (28), 6306–6310.

Vetrone, F., Naccache, R., Zamarrón, A., Juarranz de la Fuente, A., Sanz-Rodríguez, F., Martinez Maestro, L., Martín Rodriguez, E., Jaque, D., García Solé, J., and Capobianco, J.A., 2010. Temperature sensing using fluorescent nanothermometers. *ACS Nano*, 4 (6), 3254–3258.

Wamer, W.G., Yin, J.J., and Wei, R.R., 1997. Oxidative damage to nucleic acids photosensitized by titanium dioxide. *Free Radical Biology and Medicine*, 23 (6), 851–858.

Wang, C., Cao, S., Tie, X., Qiu, B., Wu, A., and Zheng, Z., 2011. Induction of cytotoxicity by photoexcitation of TiO2 can prolong survival in glioma-bearing mice. *Molecular Biology Reports*, 38 (1), 523–530.

Wang, F., Liu, X., Liu, L., Yuan, Y., and Cai, Y., 2013a. Experimental study of the scintillation index of a radially polarized beam with controllable spatial coherence. *Applied Physics Letters*, 103 (9), 091102.

Wang, X., Liu, Q., Bu, Y., Liu, C.-S., Liu, T., and Yan, X., 2015. Optical temperature sensing of rare-earth ion doped phosphors. *RSC Advances*, 5 (105), 86219–86236.

Wang, Y.-F., Liu, G.-Y., Sun, L.-D., Xiao, J.-W., Zhou, J.-C., and Yan, C.-H., 2013b. Nd3+ -sensitized upconversion nanophosphors: Efficient *in vivo* bioimaging probes with minimized heating effect. *ACS Nano*, 7 (8), 7200–7206.

Weber, M.J., 1971. Optical properties of Yb3+ and $Nd^{3+}$ - $Yb^{3+}$ energy transfer in YAlO3. *Physical Review B*, 4 (9), 3153–3159.

Weissleder, R., 2001. A clearer vision for in vivo imaging. *Nature Biotechnology*, 19 (4), 316–317.

Wilson, B.C., and Patterson, M.S., 2008. The physics, biophysics and technology of photodynamic therapy. *Physics in Medicine and Biology*, 53 (9), R61–R109.

Ximendes, E.C., Rocha, U., del Rosal, B., Vaquero, A., Sanz-Rodríguez, F., Monge, L., Ren, F., Vetrone, F., Ma, D., García-Solé, J., Jacinto, C., Jaque, D., and Fernández, N., 2017a. In vivo ischemia detection by luminescent nanothermometers. *Advanced Healthcare Materials*, 6 (4), 1601195.

Ximendes, E.C., Rocha, U., Jacinto, C., Kumar, K.U., Bravo, D., López, F.J., Rodríguez, E.M., García-Solé, J., and Jaque, D., 2016a. Self-monitored photothermal nanoparticles based on core–shell engineering. *Nanoscale*, 8 (5), 3057–3066.

Ximendes, E.C., Rocha, U., Kumar, K.U., Jacinto, C., and Jaque, D., 2016b. LaF 3 core/shell nanoparticles for subcutaneous heating and thermal sensing in the second biological-window. *Applied Physics Letters*, 108 (25), 253103.

Ximendes, E.C., Rocha, U., Sales, T.O., Fernández, N., Sanz-Rodríguez, F., Martín, I.R., Jacinto, C., and Jaque, D., 2017b. In vivo subcutaneous thermal video recording by supersensitive infrared nanothermometers. *Advanced Functional Materials*, 27 (38), 1702249.

Ximendes, E.C., Santos, W.Q., Rocha, U., Kagola, U.K., Sanz-Rodríguez, F., Fernández, N., Gouveia-Neto, AdS., Bravo, D., Domingo, A.M., del Rosal, B., Brites, C.D.S., Carlos, L.D., Jaque, D., and Jacinto, C., 2016c. Unveiling in vivo subcutaneous thermal dynamics by infrared luminescent nanothermometers. *Nano Letters*, 16 (3), 1695–1703.

Zhan, Q., Qian, J., Liang, H., Somesfalean, G., Wang, D., He, S., Zhang, Z., and Andersson-Engels, S., 2011. Using 915 nm laser excited Tm3+/Er3+/Ho3+-doped NaYbF4 upconversion nanoparticles for *in vitro* and deeper *in vivo* bioimaging without overheating irradiation. *ACS Nano*, 5 (5), 3744–3757.

Zhang, A.-P., 2004. Photocatalytic killing effect of TiO2 nanoparticles on Ls-174-t human colon carcinoma cells. *World Journal of Gastroenterology*, 10 (21), 3191.

Zhang, Y., Hong, G., Zhang, Y., Chen, G., Li, F., Dai, H., and Wang, Q., 2012. Ag2S quantum dot: A bright and biocompatible fluorescent nanoprobe in the second near-infrared window. *ACS Nano*, 6 (5), 3695–3702.

Zhao, Y., Zhan, Q., Liu, J., and He, S., 2015. Optically investigating Nd3+-Yb3+ cascade sensitized upconversion nanoparticles for high resolution, rapid scanning, deep and damage-free bio-imaging. *Biomedical Optics Express*, 6 (3), 838.

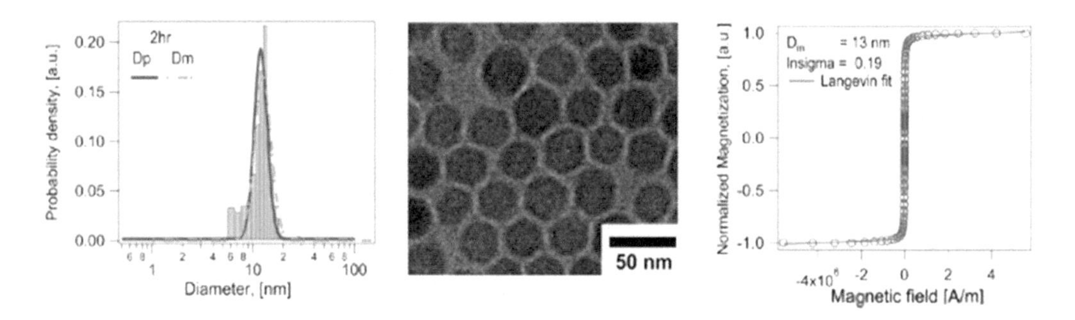


FIGURE 1.3 Controlled growth of both physical and magnetic diameter for iron oxide nanoparticles obtained using the Extended LaMer mechanism based synthesis thermal decomposition synthesis in the presence of molecular oxygen for 2 h. Reused with permission from Unni et al., 2017, American Chemical Society.

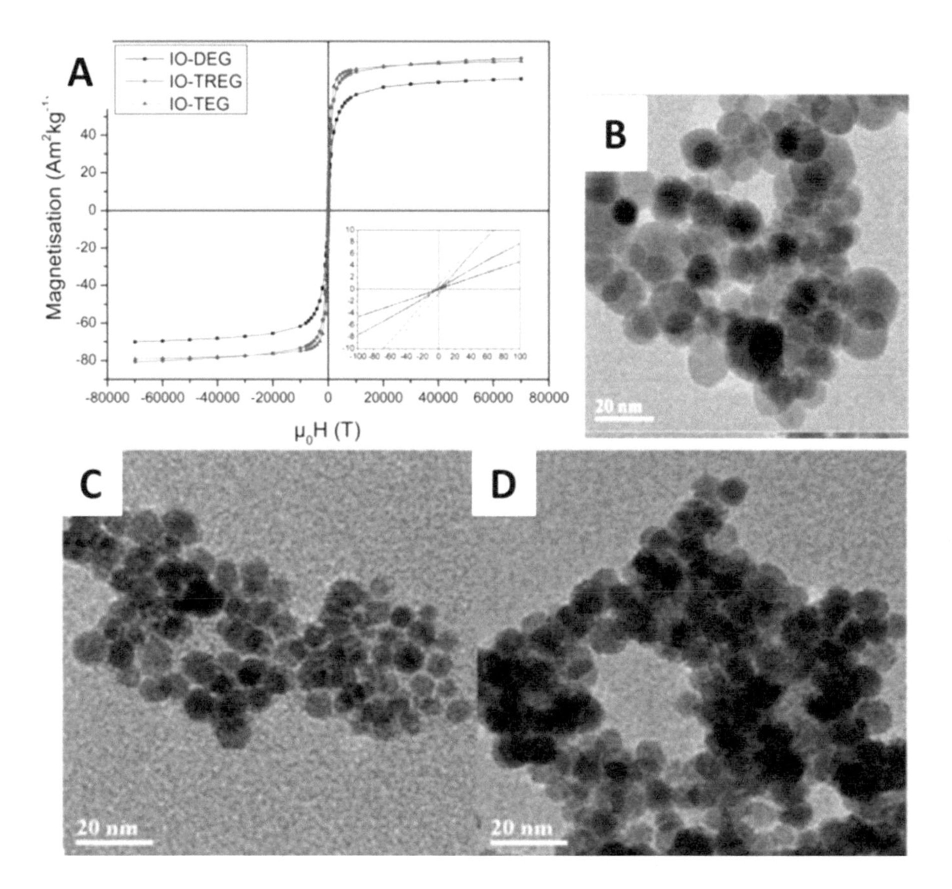


FIGURE 1.4 (A) Magnetization curves of iron oxide NPs obtained with different polyols (DEG, TREG and TEG). (B) TEM images of iron oxide nanoparticles synthesized using TEG. (C) and (D) TEM images of iron oxide nanoparticles synthesized using different reaction times in tri (EG): (C) 1 h, (D) 2 h. (Reproduced from Hachani et al., 2015. With permission.)

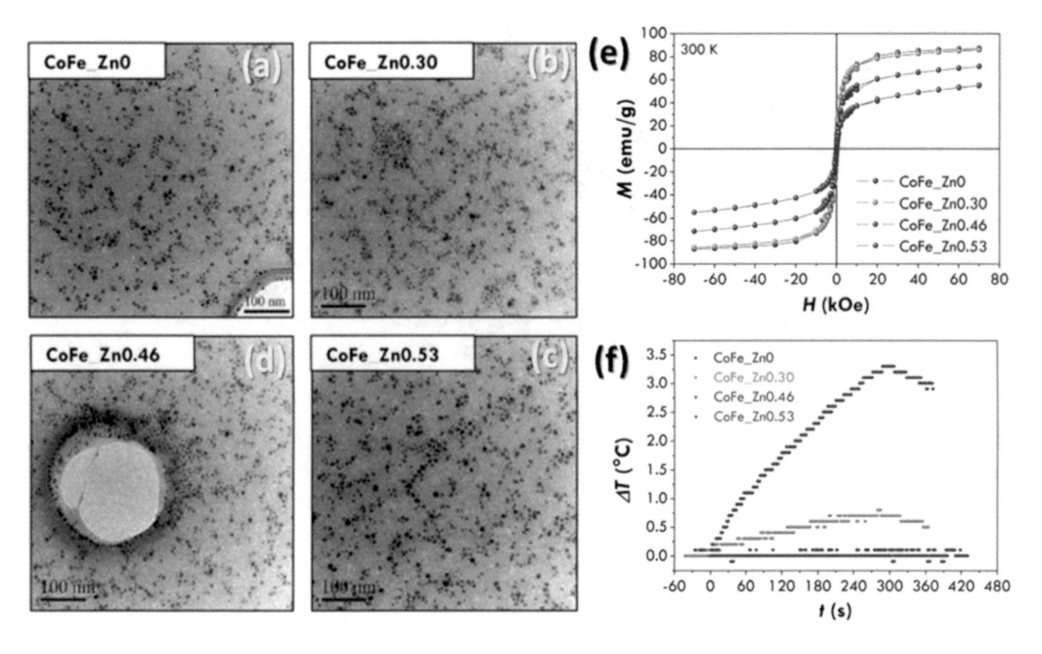


**FIGURE 1.5** (a-d) Morphological properties/TEM images of $Zn_xCo_{1-x}Fe_2O_4$ (with x = 0, 0.30, 0.46, 0.53) samples using the JEM (2010) UHR, (e) Magnetization versus magnetic field curves of $Zn_xCo_{1-x}Fe_2O_4$ (with x = 0, 0.30, 0.46, 0.53) samples measured at 300 K. (f) Heating curves of $Zn_xCo_{1-x}Fe_2O_4$ (with x = 0, 0.30, 0.46, 0.53) samples at 25°C, obtained under a magnetic field of 183 kHz and 17 kA $m^{-1}$. Reused with permission from open access reference Mameli et al., 2016.

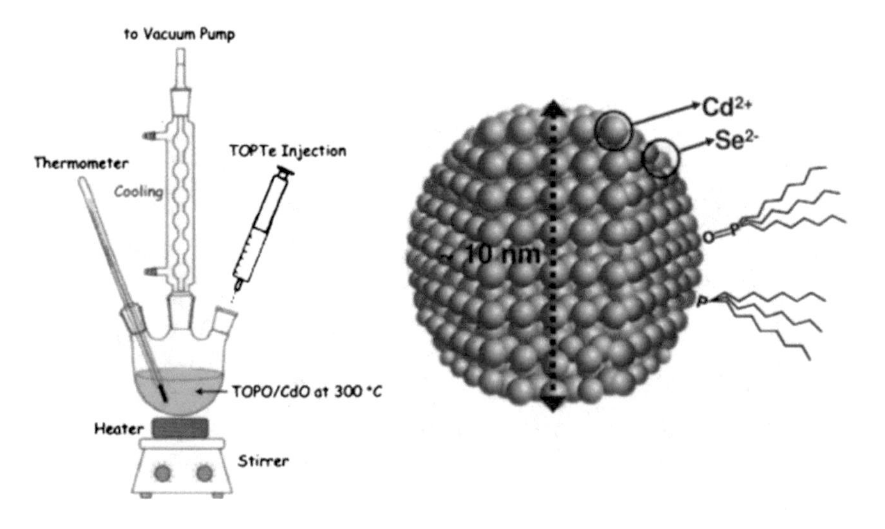


**FIGURE 1.6** Figure presents an experimental set-up for QD synthesis by high temperature organometallic method, along with a schematic illustration of CdSe QD capped with TOPO (bound to $Cd^{2+}$) and TOP (bound to $Se^{2+}$). (Adapted from Kuno, 2008.)

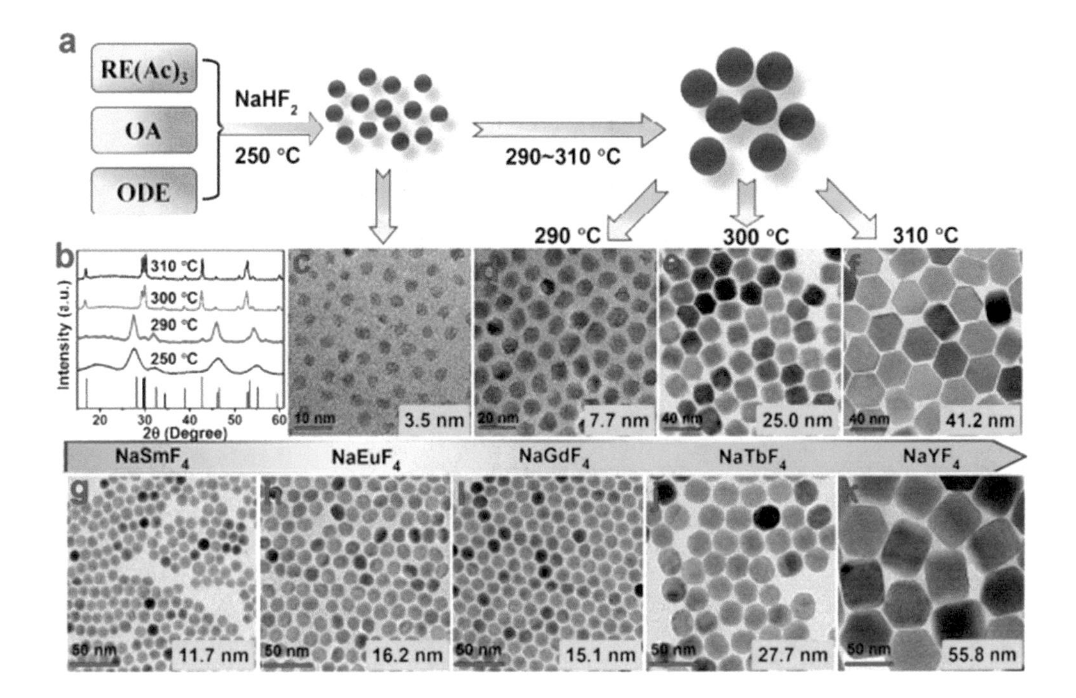


**FIGURE 1.8** (a) Schematic procedure for the synthesis of $Ln^{3+}$-doped $NaREF_4$ NPs. (b) XRD of $NaGdF_4$:Yb,Er NPs synthesized at different temperature. The black and red vertical lines in (b) represent the standard pattern of β-$NaGdF_4$ (JCPDS No. 027-0699) and α-$NaGdF_4$ (JCPDS No. 027-0697), respectively. (c) TEM image of α-$NaGdF_4$:Yb,Er NPs synthesized at 250°C. (d–f) TEM images of $NaGdF_4$:Yb,Er NCs synthesized at 290, 300, and 310°C, respectively. (g–k) TEM images of the as-prepared β-$NaSmF_4$, β-$NaEuF_4$, β-$NaGdF_4$, β-$NaTbF_4$ and β-$NaYF_4$ NOs, respectively. (Reproduced with permission from You et al., 2018.)

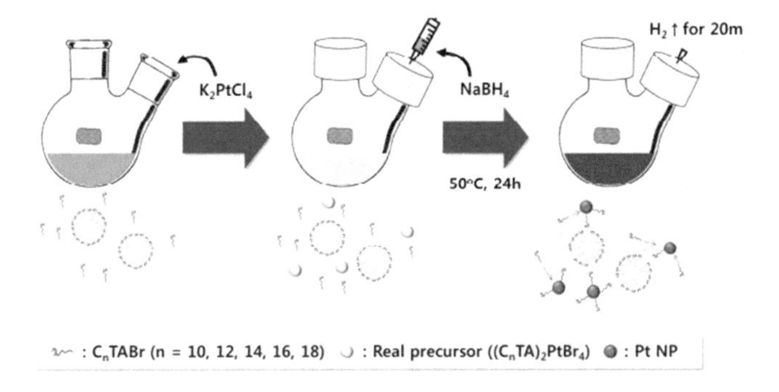


**FIGURE 1.12** Schematic flowchart of the synthetic procedure for the Pt nanoparticles (NPs). (From Seo et al., 2018. With permission.)

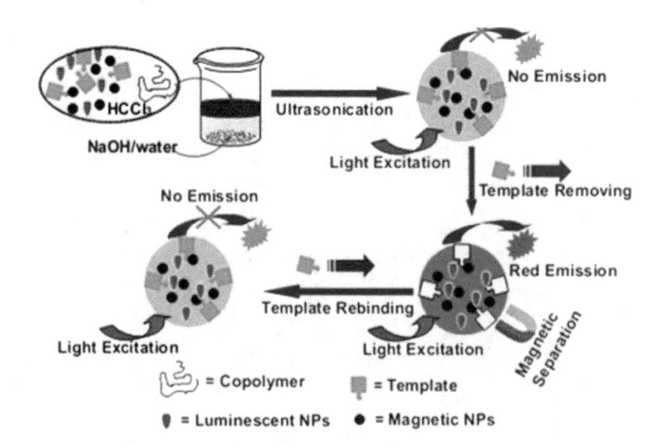


**FIGURE 1.13** Scheme for the fabrication of multifunctional MIP composite nanospheres and the luminescence response via the removal and rebinding of template analytes. (From Ma et al., 2012. With permission.)

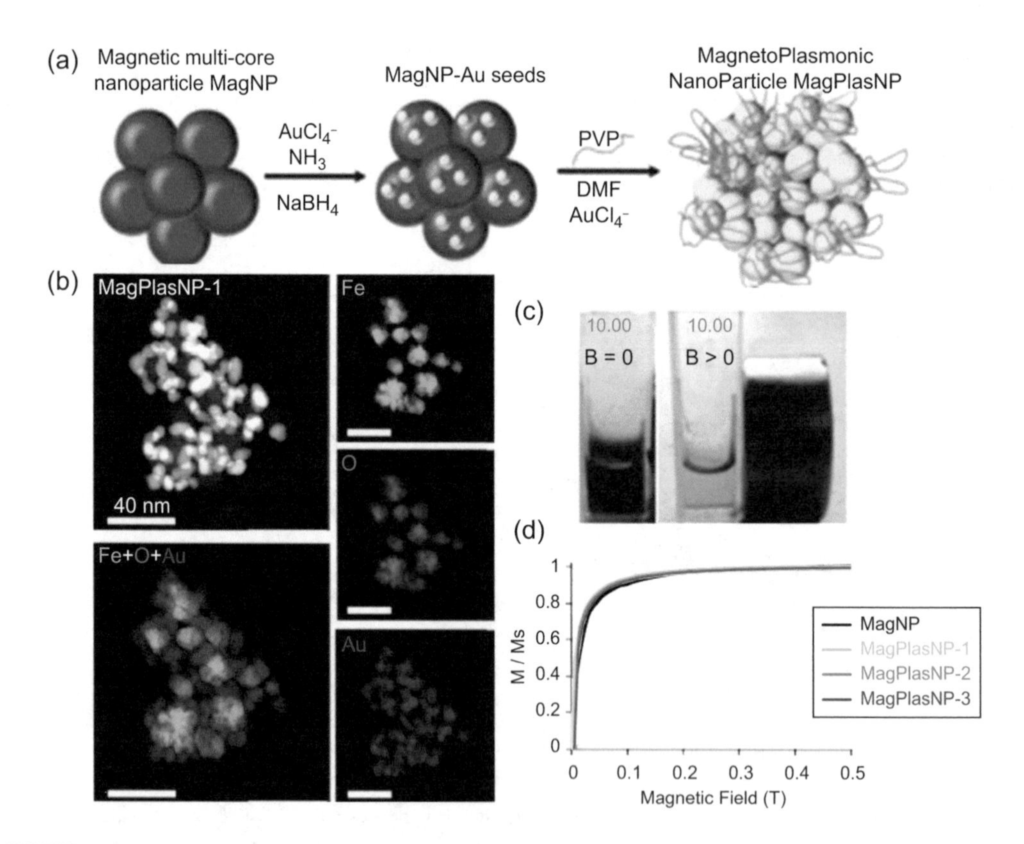


**FIGURE 1.15** (a) The seeding-growth process used for the synthesis of the magneto-plasmonic NPs (MagPlasNP) of different ratio Au/Fe. (b) Elemental analysis of different MagPlasNP-1. HAADF image, EELS Fe map (green), EELS O map (blue), EELS Au map (red) and overlaid maps. Superparamagnetic behavior of the nanohybrid: (c)Attraction of MagPlasNP-2 by a permanent magnet and (d) Magnetization curves of MagNP and MagPlasNP-1,2, and 3. (From Espinosa et al., 2015. With permission.)

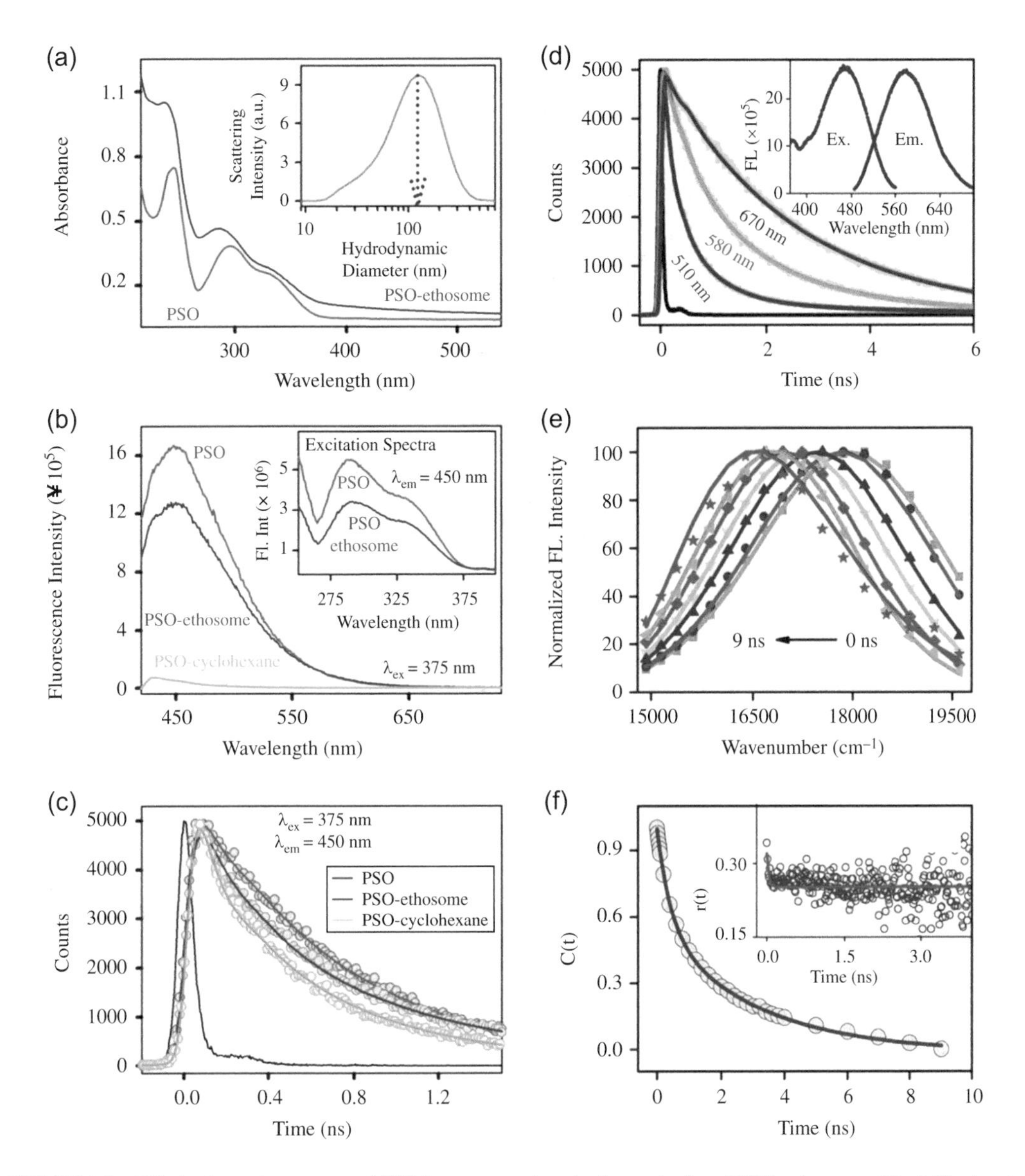


**FIGURE 2.1** UV-vis absorption spectra of PSO in a water–ethanol mixture (red) and PSO–ethosomes (blue). The inset shows the hydrodynamic diameter of the ethosomes measured by DLS. (b) Room-temperature emission spectra of PSO (red), PSO–ethosomes (blue), and PSO in cyclohexane (green). The excitation wavelength is 375 nm. The inset shows the excitation spectra of PSO (red) and PSO–ethosome (blue). (c) Fluorescence transients of PSO in a water–ethanol mixture (red), PSO–ethosomes (blue), and PSO in cyclohexane (green). (d) Fluorescence transients at different wavelengths for DCM–ethosomes. The inset shows excitation and emission spectra of DCM. (e) TRES. (f) Decay of the solvation correlation function, C (t), with time. The inset shows temporal decay of fluorescence anisotropy, r(t), of DCM–ethosomes. (Reproduced from Bagchi et al.[41] With permission.)

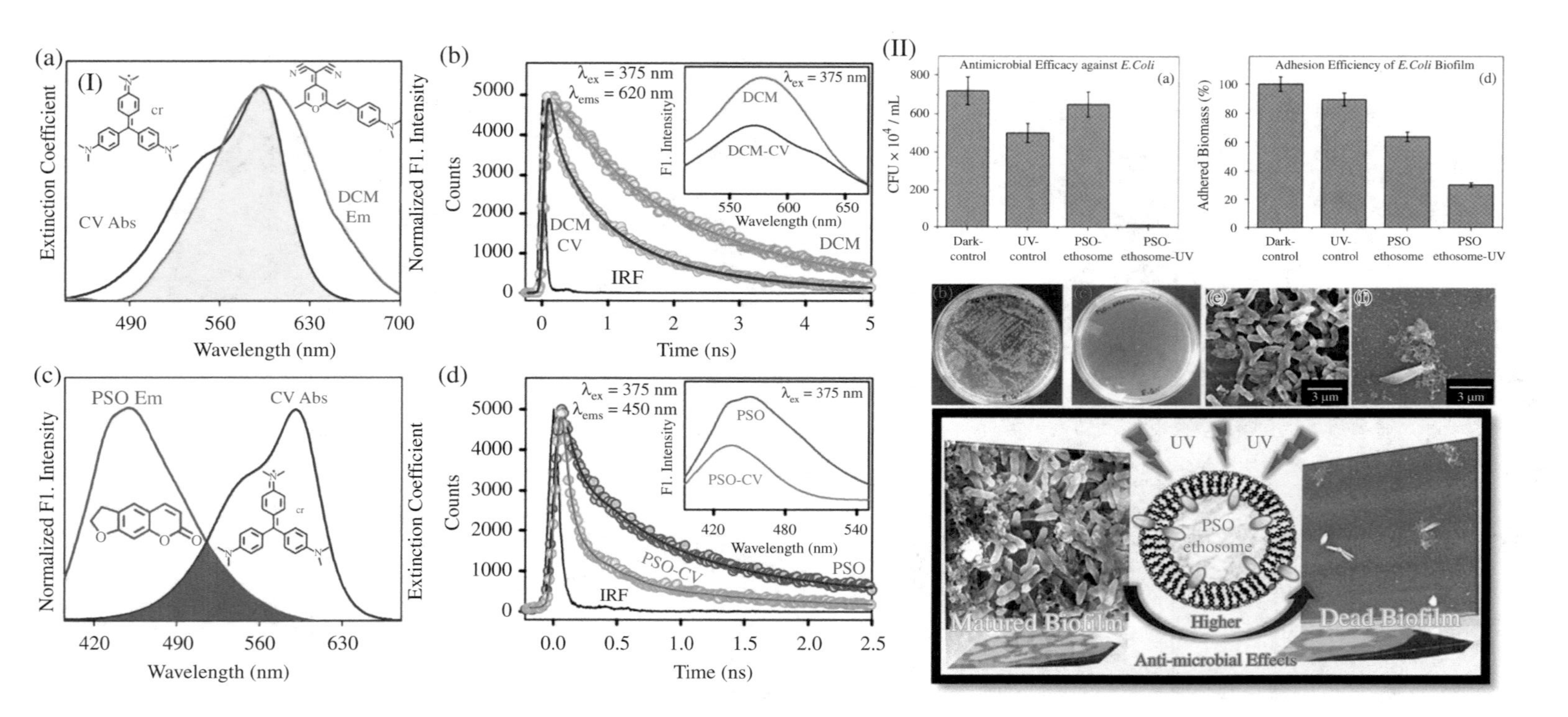


**FIGURE 2.2** (I) (a) Overlap of DCM–ethosome emission and CV absorption in water. (b) Picosecond-resolved fluorescence transients of DCM–ethosomes (excited at 375 nm) in the absence (green) and presence (blue) of CV collected at 620 nm. The inset shows steady-state emission spectra of DCM–ethosomes in the absence (green) and presence (blue) of CV. (c) Overlap of PSO–ethosome emission and CV absorption in water. (d) Picosecond-resolved fluorescence transients of PSO–ethosome (excited at 375 nm) in the absence (red) and presence (cyan) of CV collected at 450 nm. The inset shows steady-state emission spectra of PSO–ethosomes in the absence (red) and presence (cyan) of CV. (II) (a) Antibacterial activity of PSO–ethosomes against *E. coli* in the absence and presence of UVA. (b, c) Images of PSO–ethosome-treated E. coli culture plates before and after UVA irradiation, respectively. (d) Adhesion efficiency of PSO–ethosome-treated *E. coli* biofilms in the absence and presence UVA irradiation. SEM images of an *E. coli* biofilm (e) without treatment and (f) treated with PSO–ethosomes followed by UVA illumination for 30 min. The lower panel depicts a schematic representation of the processes. (Reproduced from Bagchi et al.[41] With permission.)

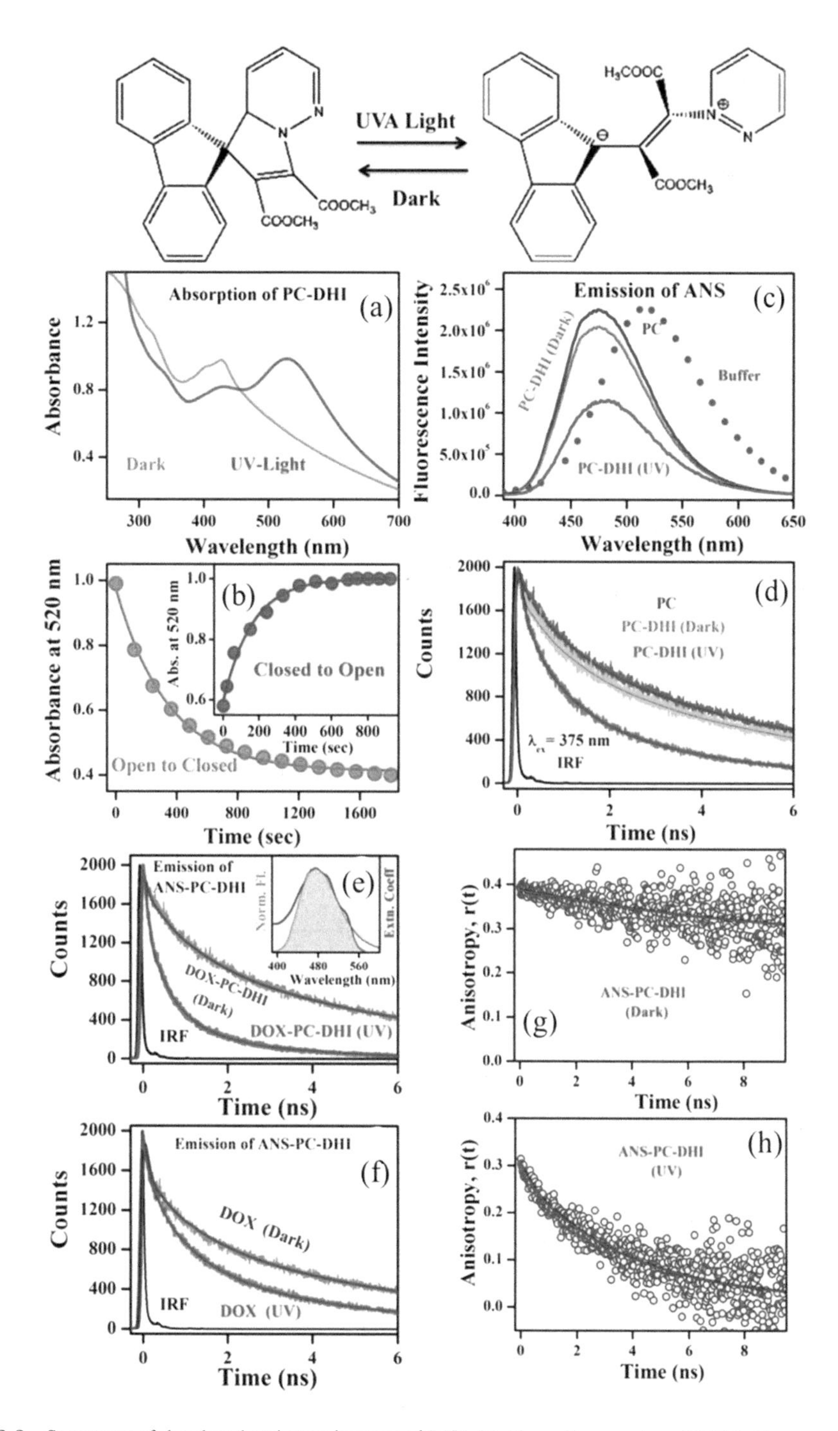


**FIGURE 2.3** Structures of the closed and open isomers of DHI. (a) Absorption spectra of DHI in liposome: closed and open isomers. (b) Kinetics of the open to closed transition of DHI in liposome (SD = ±0.008, $n$ = 3). Inset shows the corresponding closed to open conversion rate (SD = ±0.003, $n$ = 3). (c) Steady-state emission spectra of ANS in buffer, ANS bound to PC and PC-DHI in presence and absence of UVA light. (d) Time-resolved transients of ANS, ANS bound to PC and ANS bound to PC-DHI in the presence and absence of UVA light. Picosecond-resolved transients of the donor-acceptor in the absence and presence of UVA light, (e) donor is (ANS-PC-DHI) and acceptor is (DOX-PC-DHI) and (f) donor is (ANS-PC-DHI) and acceptor is free DOX. Insets depict the corresponding spectral overlap between donor (ANS-PC-DHI complex) emission and acceptor (DOX-PC-DHI) absorbance. Time-resolved anisotropy of ANS bound to PC-DHI in presence of (g) dark and (h) UVA light. (Reproduced with permission from P. Singh et al.[53], © 2017 Elsevier, B.V. All rights reserved.)

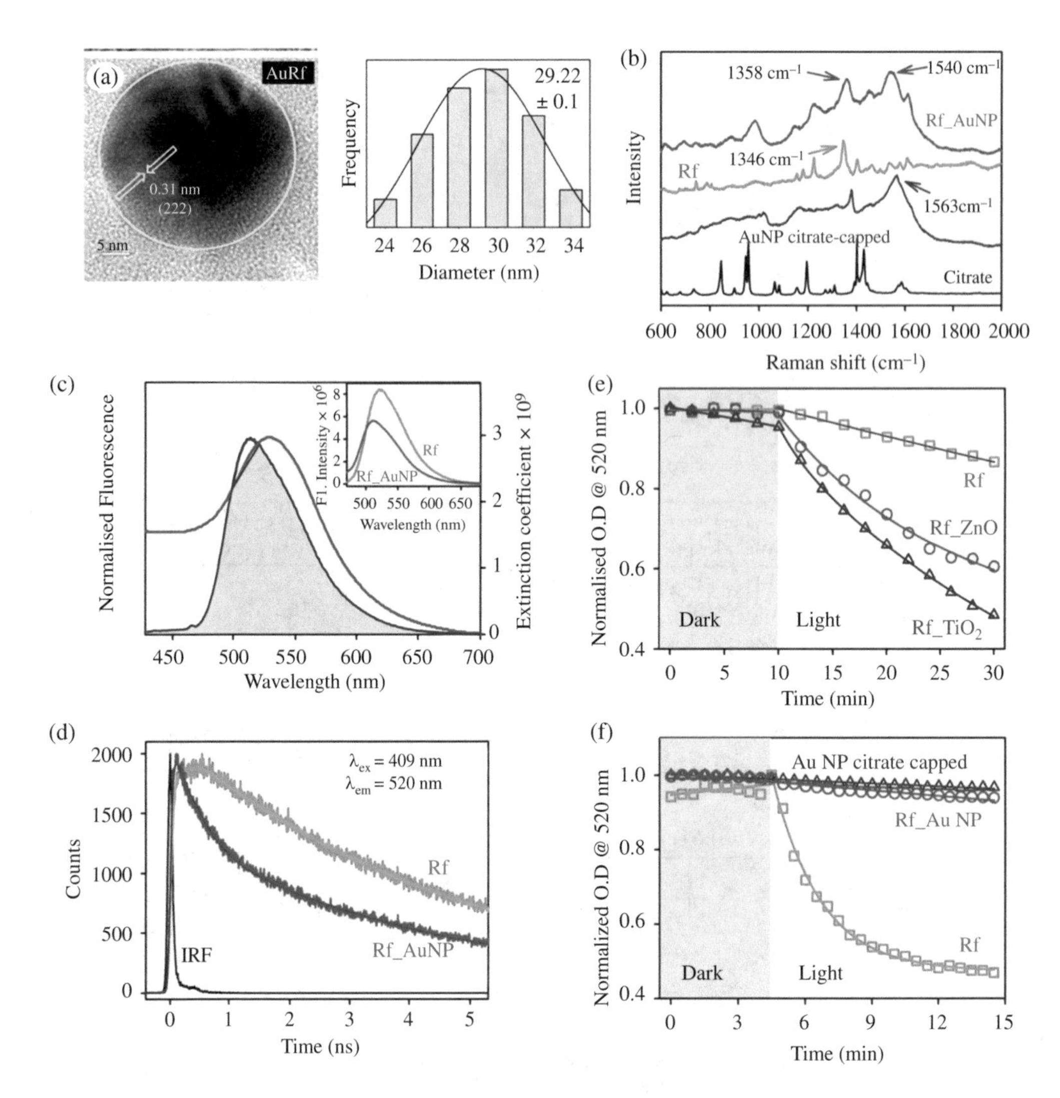


**FIGURE 2.6** (a) HRTEM images of Au nanoparticles. The right panel shows the corresponding particle size distribution of the NPs. (b) Raman spectra of Rf and Rf–Au NPs in the solid phase. (c) Spectral overlap of donor Rf (dark green) and acceptor Au NPs (pink). The inset shows the steady-state emission quenching of Rf after complexation with Au NP. (d) The picosecond-resolved fluorescence transients of Rf and Rf–Au NPs (excitation at 409 nm). All the spectra were taken in aqueous medium. Absorption kinetics (at 520 nm) of DPPH in the presence of (e) Rf–ZnO, Rf–$TiO_2$ in ethanol and (f) Rf–AuNPs in ethanol:water (1:1) mixture under blue light irradiation. Part (f) shows the degradation kinetics under dark only up to 5 min for the representation as antioxidant activity under dark for a longer time window of 10 min shows no significant activity. (Reproduced with permission from Chaudhuri et al.[57], © 2015, American Chemical Society.)

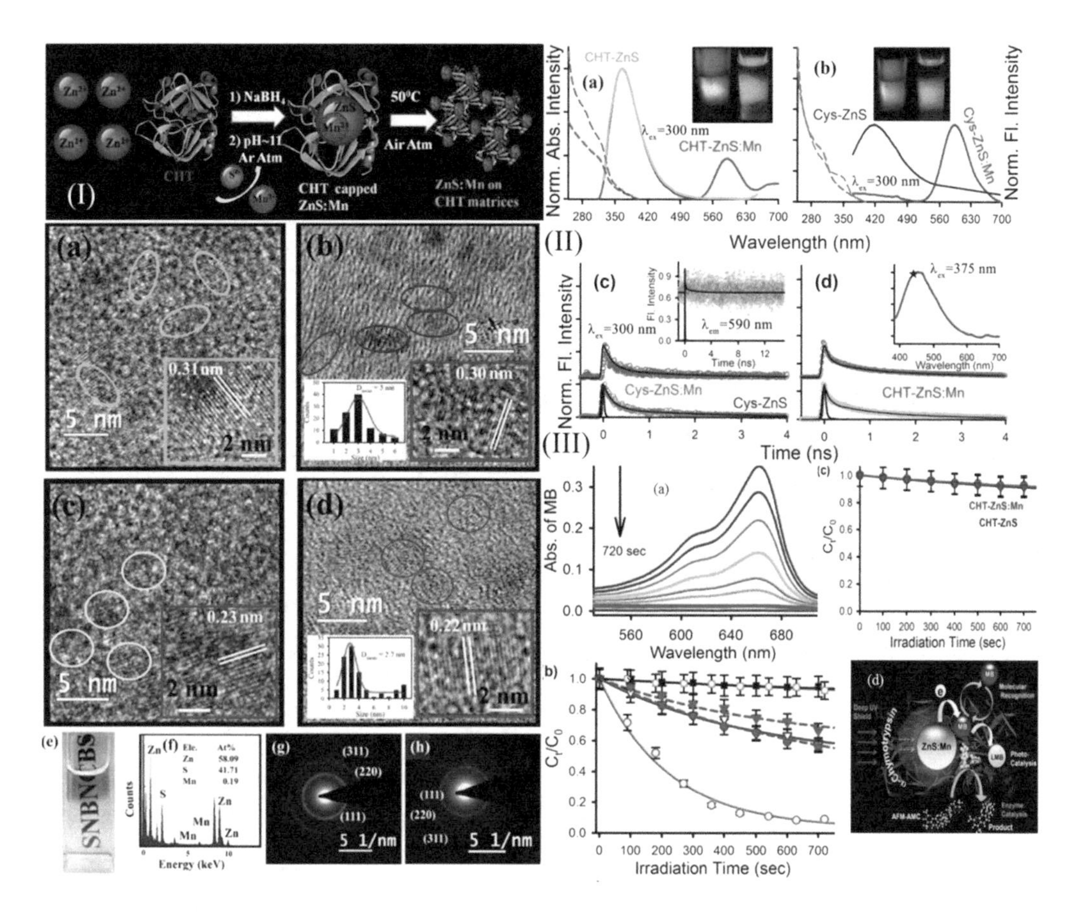


FIGURE 2.7 (I) Synthetic strategy of enzyme-mediated Mn-doped ZnS BNCs. TEM and HRTEM images (inset) of (a) CHT-ZnS, (b) CHT-ZnS:Mn, (c) Cys-ZnS, (d) Cys-ZnS:Mn NPs. Inset left of panels (b) and (d) represent the size distribution analysis of CHT-ZnS NPs and Cys-ZnS:Mn NPs, respectively. (e) Optically transparent solution of CHT-ZnS:Mn BNCs under daylight. (f) EDAX analysis and atomic percentages elements, (g) and (h) SAED analysis of CHT-ZnS and CHT-ZnS:Mn BNCs, respectively. (II) Optical absorption and steady-state emission spectra of (a) CHT-ZnS and CHT-ZnS:Mn BNCs and (b) Cys-ZnS, Cys-ZnS:Mn NPs, respectively. Inset of (a) and (b) shows PL photos of the corresponding solutions upon 300 nm excitation. (c) The picosecond-resolved fluorescence transients of Cys-ZnS and Cys-ZnS:Mn NPs (excitation at 300 nm) collected at 420 nm and inset shows fluorescence transient of Cys-ZnS:Mn NPs collected at 590 nm. (d) The picosecond-resolved fluorescence transients of CHT-ZnS:Mn NPs (excitation at 375 nm) collected at 460 nm (green) and 590 nm (red). The inset shows PL spectra upon 375 nm excitation. (III) (a) Time-dependent UV–vis spectral changes of MB in the presence of CHT-ZnS BNCs under UV-light irradiation. (b) Plot of relative concentration ($C_t/C_0$) versus irradiation time for the degradation of MB (monitored at 655 nm) is shown. The degradation is performed in the presence of BNCs: CHT-ZnS (empty circle), CHT-ZnS:Mn (filled circle), Cys-ZnS (empty triangle), Cys-ZnS:Mn (filled triangle), only CHT (empty square), no catalysts (crossed). (c) Plot of $C_t/C_0$ versus irradiation time in the presence of CHT-ZnS (filled triangle) and CHT-ZnS:Mn (filled circle) upon selective excitation with a 350 nm high-pass filter. (Reproduced with permission from Makhal et al.[60], © 2012, American Chemical Society.)

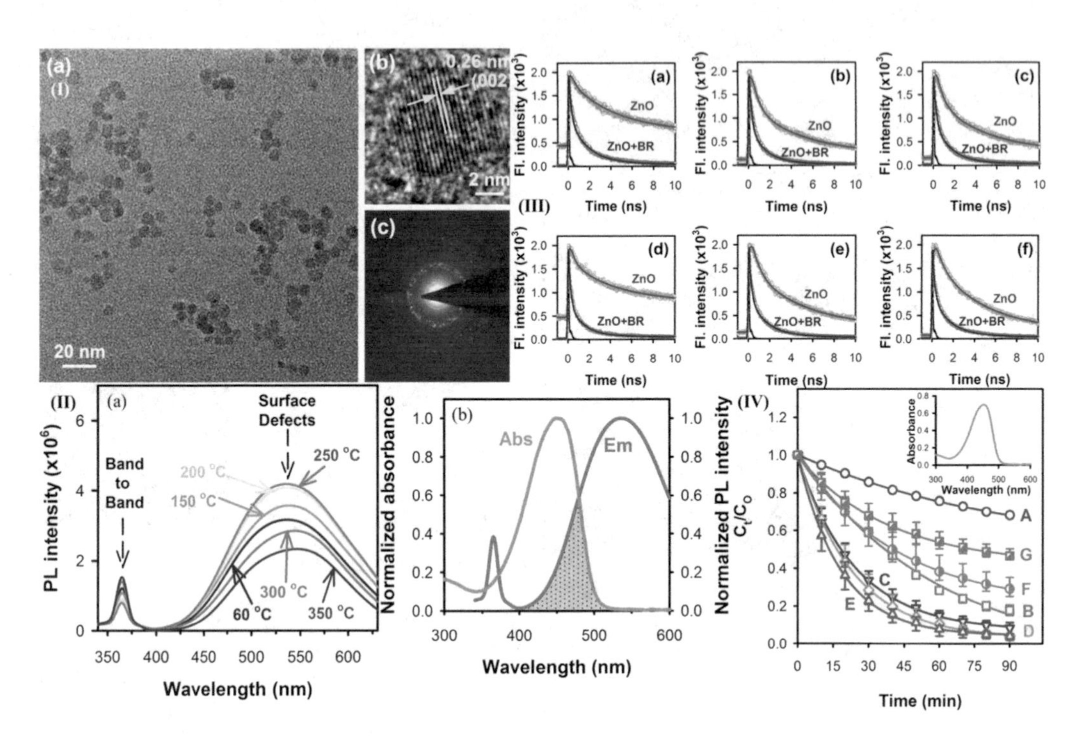


**FIGURE 2.8** (I) (a) Transmission electron micrograph, (b) HRTEM image of a single ZnO nanoparticle and (c) SAED pattern of the ZnO nanoparticles annealed at 250°C. (II) (a) Room temperature photoluminescence (PL) spectra of the ZnO nanoparticles annealed at various temperatures in air for 1 h (excitation wavelength: 320 nm) (b) Spectral overlap (shaded area) between the defect-mediated ZnO nanoparticle emission and the BR absorption. (III) The picosecond-resolved fluorescence transients of (a) as-synthesized (hydrolyzed at 60°C), (b) 150°C, (c) 200°C, (d) 250°C, (e) 300°C and (f) 350°C annealed ZnO nanoparticles in the presence (blue) and in the absence (red) of BR. The fluorescence decay was monitored at 540 nm with an excitation wavelength of 375 nm. (IV) Relative concentration ($C_t/C_0$) of BR with varying UV irradiation time during photocatalytic degradation (monitored for peak absorbance at 450 nm) (A) in the absence and in the presence of (B) as-synthesized particles, (C) annealed at 150°C, (D) annealed at 200°C, (E) annealed at 250°C, (F) annealed at 300°C and (G) annealed at 350°C particles. The inset shows the UV/Vis optical absorption spectrum of BR solution. (Reproduced with permission from Bora et al.[62])

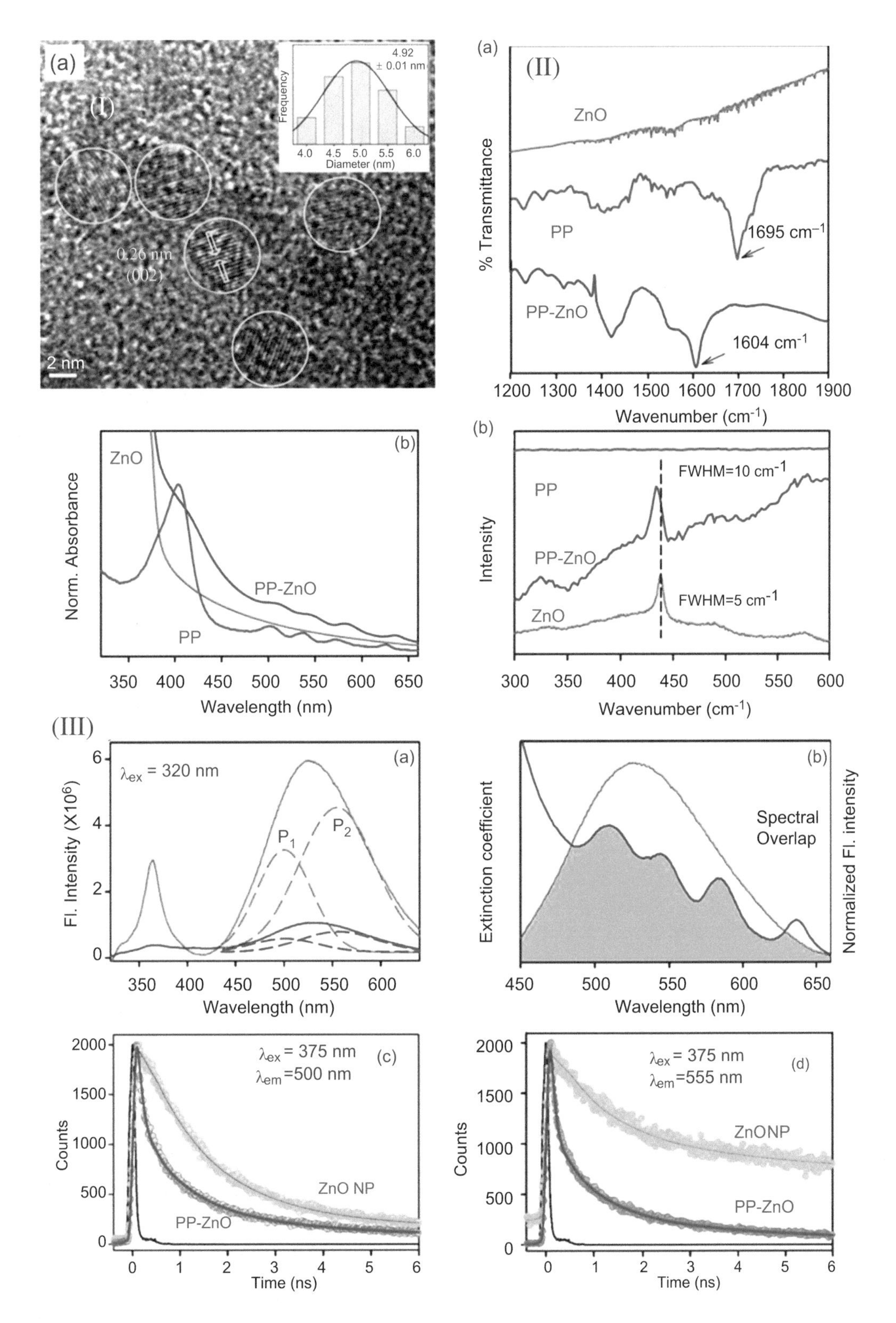


**FIGURE 2.9** (I) (a) HRTEM images of ZnO NPs. The inset shows the size distribution of the ZnO NPs. (b) UV-Vis absorption of ZnO NPs (green), PP (red) and PP–ZnO (blue) in the DMSO-ethanol mixture. (II) (a) FTIR and (b) Raman spectra of PP (red), ZnO NPs (green) and PP–ZnO composites (blue). (III) (a) Room temperature PL spectra of ZnO NPs (green) and PP–ZnO composites (blue) are shown. The excitation wavelength was at 300 nm. The broad emission band is composed of two components, P1 (500 nm) and P2 (555 nm). (b) Shows the overlap of the ZnO NP emission and PP absorption. The picosecond-resolved fluorescence transients of ZnO NPs (excitation at 375 nm) in the absence (green) and in the presence of PP (blue) collected at (c) 500 nm and (d) 555 nm are shown. (Reproduced with permission from Sardar et al.[67])

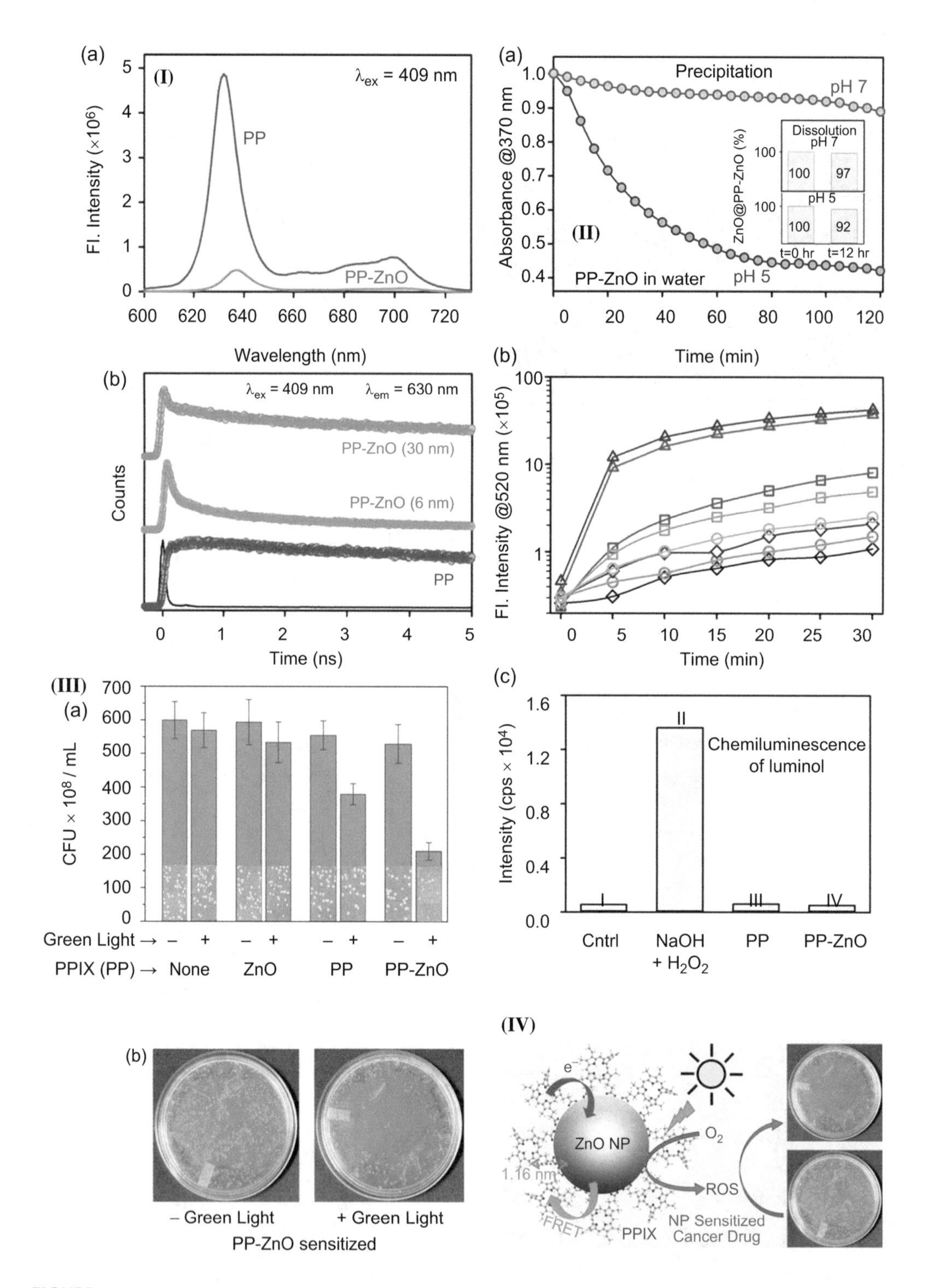


**FIGURE 2.10** (I) (a) Room-temperature PL spectra (excitation at 409 nm) of free PP (red) and PP–ZnO (green). (b) Fluorescence decay profiles of PP (red), PP–ZnO (5 nm) (blue) and PP–ZnO (30 nm) (cyan) at 630 nm (excitation at 409 nm). (II) (a) Stability of the nanohybrid from absorbance at 370 nm of PP– ZnO dispersed in water at pH 7 and pH 5. Inset shows the dissolution of PP–ZnO nanohybrids in water at pH 5 and pH 7 after 12 hours. (b) The DCFH oxidation with time in the presence of PP–ZnO (blue at pH 5 and pink at pH 7), PP (red at pH 5 and dark yellow at pH 7), ZnO (green at pH 5 and orange at pH 7) and DCFH only (black at pH 5 and dark red at pH 7) under green light irradiation. The excitation was at 488 nm. (c) Chemiluminescence of luminol after green light irradiation for 15 minutes in the presence of (I) control, (II) NaOH+H2O2, (III) PP and (IV) PP–ZnO. (III) (a) Antibacterial activity of PP–ZnO, PP and ZnO in the presence and absence of green light. (b) Images of E. coli. in PP–ZnO sensitized plates before and after green light irradiation. (IV) Schematic representation of the entire process. (Reproduced with permission from Sardar et al.[67])

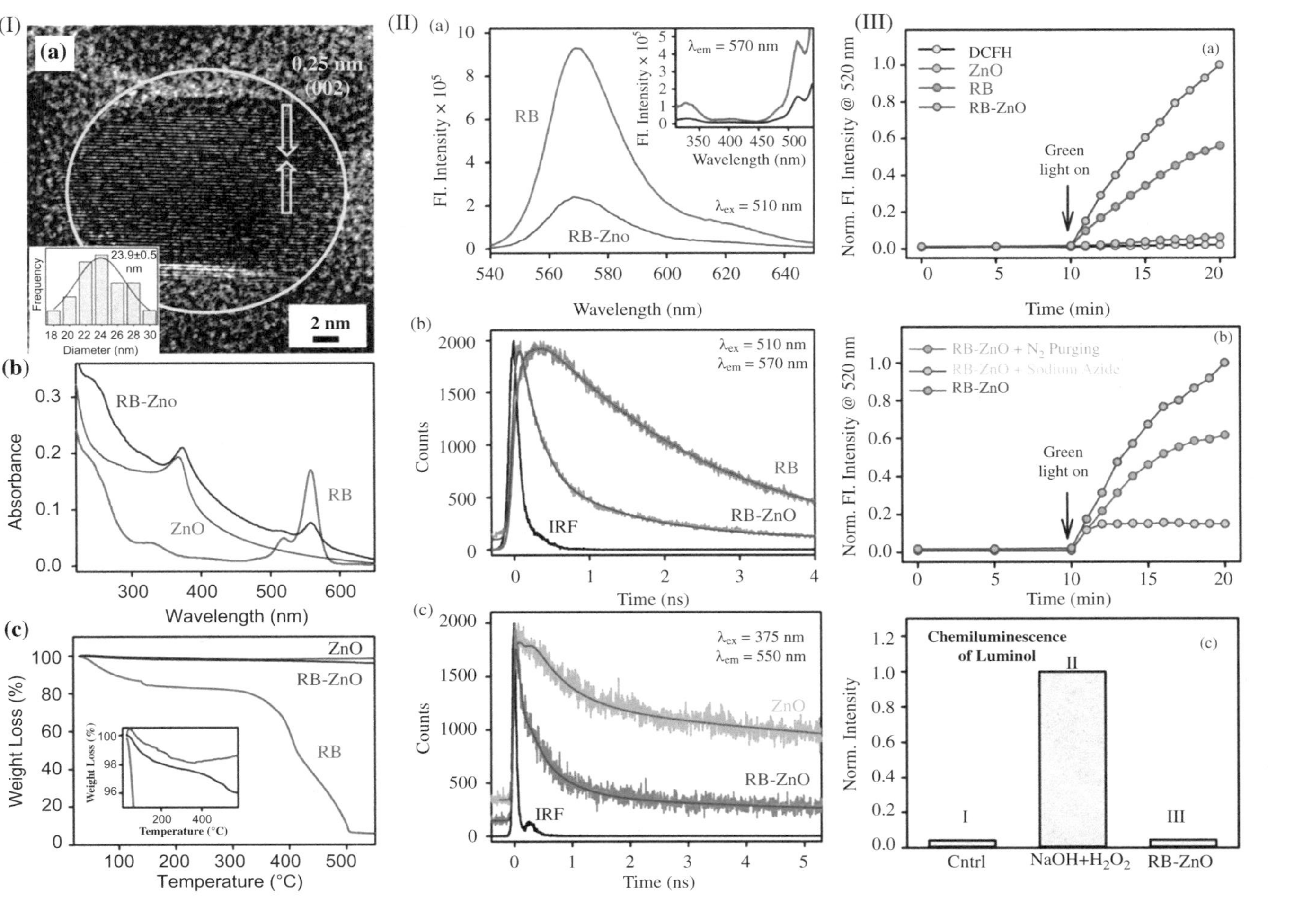


**FIGURE 2.11** I (a) HRTEM image of ZnO NPs. Inset: The size distribution of ZnO NPs. (b) Absorption spectra of RB–ZnO (blue), RB (red), and ZnO (green). (c) TGA profile of RB–ZnO (blue), RB (red), and ZnO (green). II (a) Room temperature PL spectra of RB (red) and RB–ZnO (30 nm; blue) upon excitation at 510 nm. Inset: The excitation spectra of RB (red) and RB–ZnO (30 nm; blue) at detection wavelength 570 nm. (b) Fluorescence decay profiles of RB (red) and RB–ZnO (30 nm; blue) upon excitation at 510 nm and detection wavelength at 570 nm. (c) Fluorescence decay profiles of ZnO (5 nm; green) and RB–ZnO (5 nm; blue) upon excitation at 375 nm and at a detection wavelength at 550 nm. III a) DCFH oxidation with respect to time with the addition of RB–ZnO (blue), RB (red), ZnO (green), and control DCFH (black) under dark with subsequent green-light irradiation. b) DCFH oxidation with respect to time with RB–ZnO addition in an atmosphere of purged nitrogen (pink), sodium azide (green), and a control (blue) under dark with subsequent green-light irradiation c) Chemiluminescence of luminol prior to green-light illumination for 15 min for the control (I), NaOH+H2O2 (II) and RB–ZnO (III). (Reproduced with permission from Chaudhuri et al.[70] © 2015, John Wiley and Sons.)

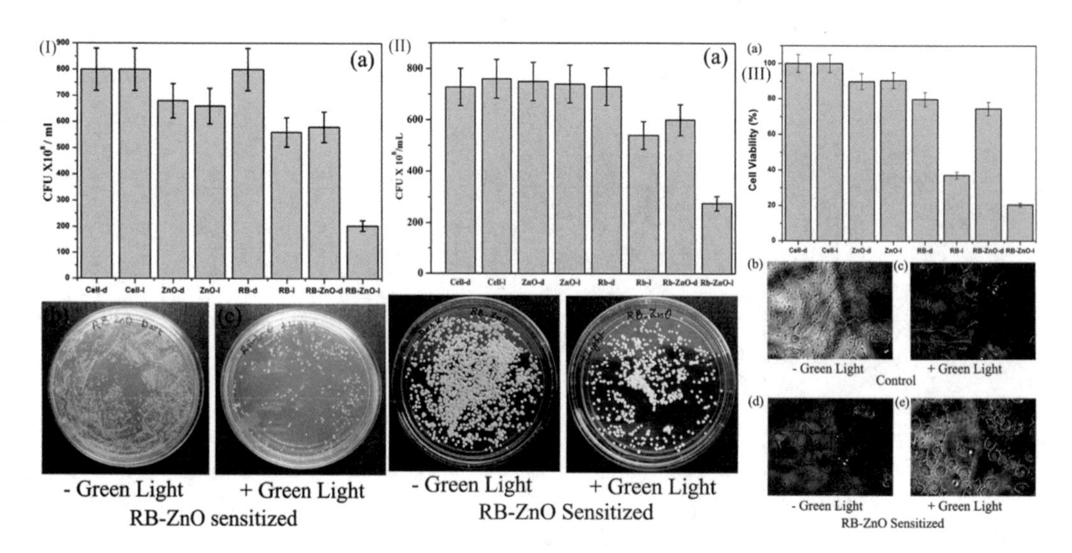


FIGURE 2.12 (I) (a) Antibacterial activity of RB–ZnO, RB, and ZnO in the presence (with suffix -l) and absence (with suffix -d) of green light. Images of RB–ZnO-treated *E. coli* plates in the absence (b) and presence (c) of green light. (II) a) Antifungal activity of RB–ZnO, RB, and ZnO in the presence (with suffix -l) and absence (with suffix -d) of green light. Images of RB–ZnO-treated *C. albicans* plates in the absence (b) and presence (c) of green light. (III) (a) *In vitro* cytotoxicity assay in HeLa cells with RB–ZnO, RB, and ZnO with MTT as an indicator dye in the presence (with suffix -l) and absence (with suffix -d) of green light. Images of control and RB–ZnO-treated HeLa cells in the absence (b, d) and the presence (c, e) of green light. (Reproduced with permission from Chaudhuri et al.[70] © 2015, John Wiley and Sons.)

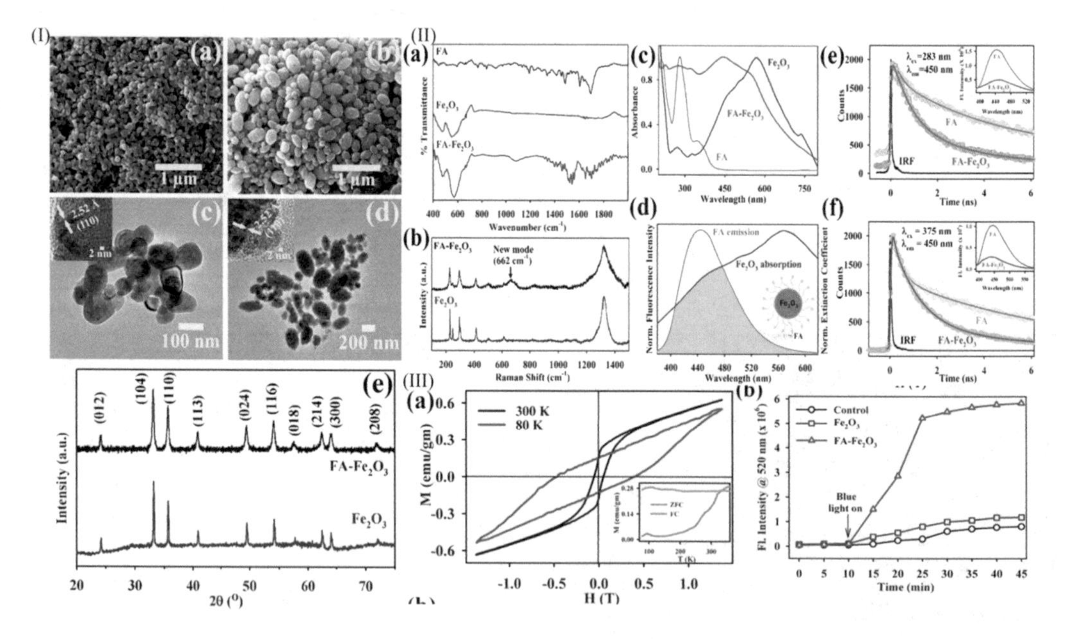


FIGURE 2.13 (I) FESEM image of (a) $Fe_2O_3$ and (b) FA– $Fe_2O_3$ NPs. HRTEM image of (c) $Fe_2O_3$ and (d) FA–$Fe_2O_3$ NPs. Inset of (c) and (d) shows the corresponding HRTEM images at higher magnification. (g) XRD pattern of $Fe_2O_3$ and FA–$Fe_2O_3$ NPs. (II) (a) FTIR spectra of FA, $Fe_2O_3$, and FA–$Fe_2O_3$. (b) Raman spectra of FA–$Fe_2O_3$ and $Fe_2O_3$. (c) Absorption spectra of FA, $Fe_2O_3$, and FA–$Fe_2O_3$. (d) Spectral overlap between the emission of FA and absorption of $Fe_2O_3$. (e) and (f) are the picosecond-resolved fluorescence transient spectra of FA in the absence and presence of $Fe_2O_3$, recorded at 450 nm upon excitation at 283 nm and 375 nm, respectively. Insets show the corresponding steady-state emission spectra. (III) (a) Field-dependent magnetization (M vs. H) of FA– $Fe_2O_3$ at 300 K and at 80 K. Bottom right inset shows temperature dependence of $M_{ZFC}$ and $M_{FC}$ curves measured at H = 1.6 T. (b) DCFH oxidation with respect to time in addition of $Fe_2O_3$, FA–$Fe_2O_3$ and control under dark with subsequent blue light irradiation. (Reproduced with permission from Nandi et al.[76])

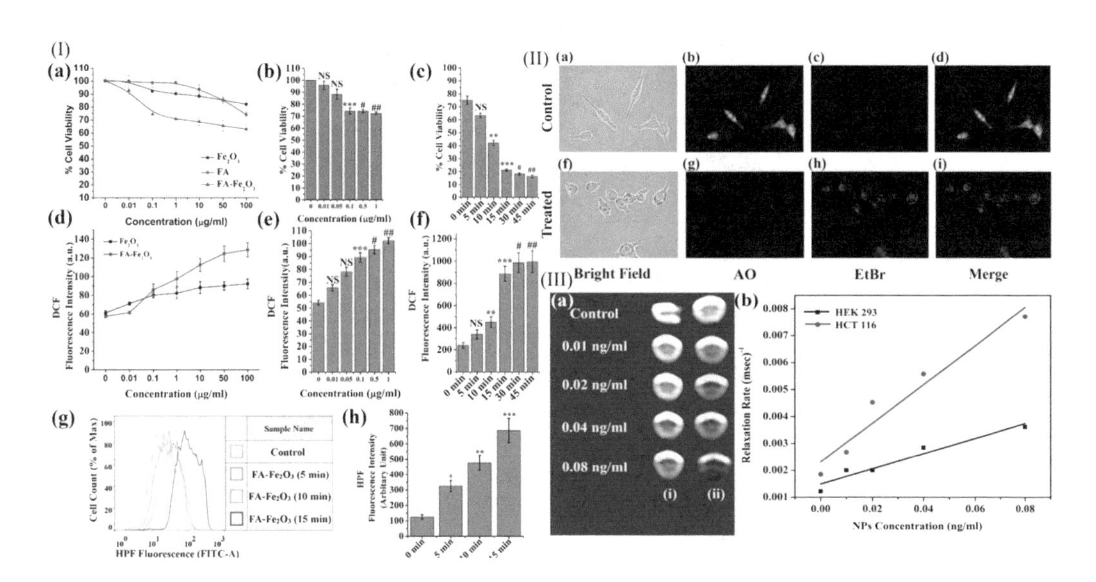


FIGURE 2.14 (I) HCT 116 cells were treated with the requisite amount of drugs for 24 h prior to various experiments. (a) MTT assay quantified cell viability with different concentrations of FA, $Fe_2O_3$, and FA–$Fe_2O_3$ in the absence of blue light. (b) The same at different concentrations of only FA– $Fe_2O_3$. (c) Light-induced cytotoxicity after treatment with 0.1 mg $ml^{-1}$ of FA– $Fe_2O_3$ followed by blue light irradiation for different time durations (0–45 min). (d) $Fe_2O_3$ and FA– $Fe_2O_3$ treated dose-dependent (0–100 mg $ml^{-1}$) intracellular ROS level in terms of DCF fluorescence intensity without blue light treatment. (e) FA– $Fe_2O_3$ treated dose-dependent (0–1 mg $ml^{-1}$) intracellular ROS level in terms of DCF fluorescence intensity without blue light treatment. (f) Intracellular ROS measurement after treatment with 0.1 mg $ml^{-1}$ FA– $Fe_2O_3$ followed by time-dependent (0–45 min) blue light exposure. (g) Intracellular hydroxyl radical (OH) determination by HPF staining after treatment with FA– $Fe_2O_3$ followed by time-dependent (0–15 min) blue light exposure, using flow cytometer. (h) Flow Jo analyzed data of hydroxyl radical (OH) determination. (II) Morphological changes of FA–$Fe_2O_3$ (0.1 mg ml1 for 24 h)-treated HCT 116 cells followed by blue light exposure (15 min). (a) and (f) control and treated cells under bright field. (b) and (g) AO stained control and treated cells. (c) and (h) EtBr staining of control and treated cells. (d) and (i) Merged image of AO and EtBr stained cells. (III) (a) T2 weighted MRI phantom images of (i) HEK 293 and (ii) HCT 116 cells treated with different concentrations of FA– $Fe_2O_3$. (b) Relaxivity study of FA–Fe2O3 incubated in HEK 293 and HCT 116 cell lines. (Reproduced with permission from Nandi et al.[76])

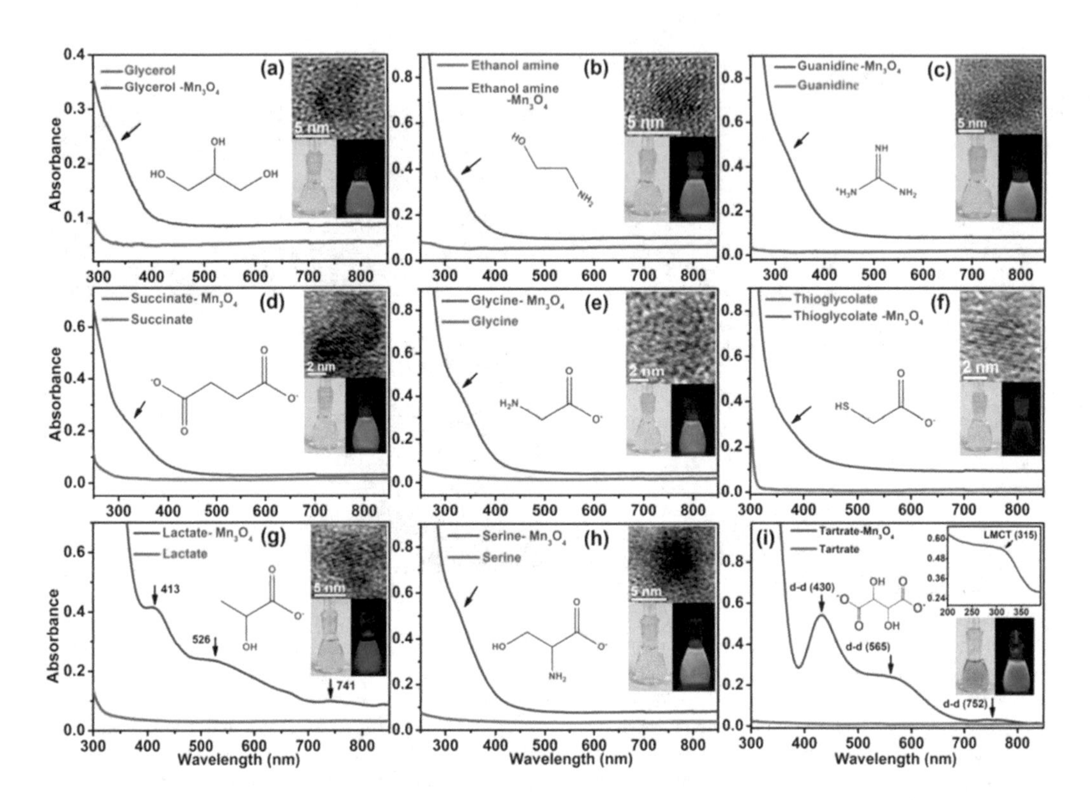


FIGURE 2.15 (a)–(i) represents the UV-vis absorption spectra of ligand-functionalized-$Mn_3O_4$ NPs in aqueous solution at pH~7. Different combinations of ligand functional groups have been employed in order to activate the J–T splitting of $Mn^{3+}$ ions in the NPs surface and to bring out optimal optical responses from the functionalized NPs. (a) –OH (hydroxyl) group of glycerol (b) –OH & –$NH_2$ (hydroxyl and amine) groups of ethanol amine, (c) –$NH_2$ group of guanidine, (d) –COO– (carboxylate) group of succinate, (e) –COO– and –$NH_2$ groups of glycine, (f) –COO– and –SH (carboxylate and thiol) groups of thioglycolate, (g) –COO– and –OH (at α position) groups of lactate, (h) –COO– and –OH (at β position) groups of serine and (i) –COO– and –OH (two α hydroxyl groups) groups of tartrate have been used respectively, to functionalize the as-prepared $Mn_3O_4$ NPs. Upper inset of Figure (a)–(h) show the corresponding HRTEM image of various ligand functionalized $Mn_3O_4$NPs. Photographs under visible (left) and UV light (right) of various ligand functionalized $Mn_3O_4$ NPs have been shown in the lower inset. (Reproduced with permission from Giri et al.[80])

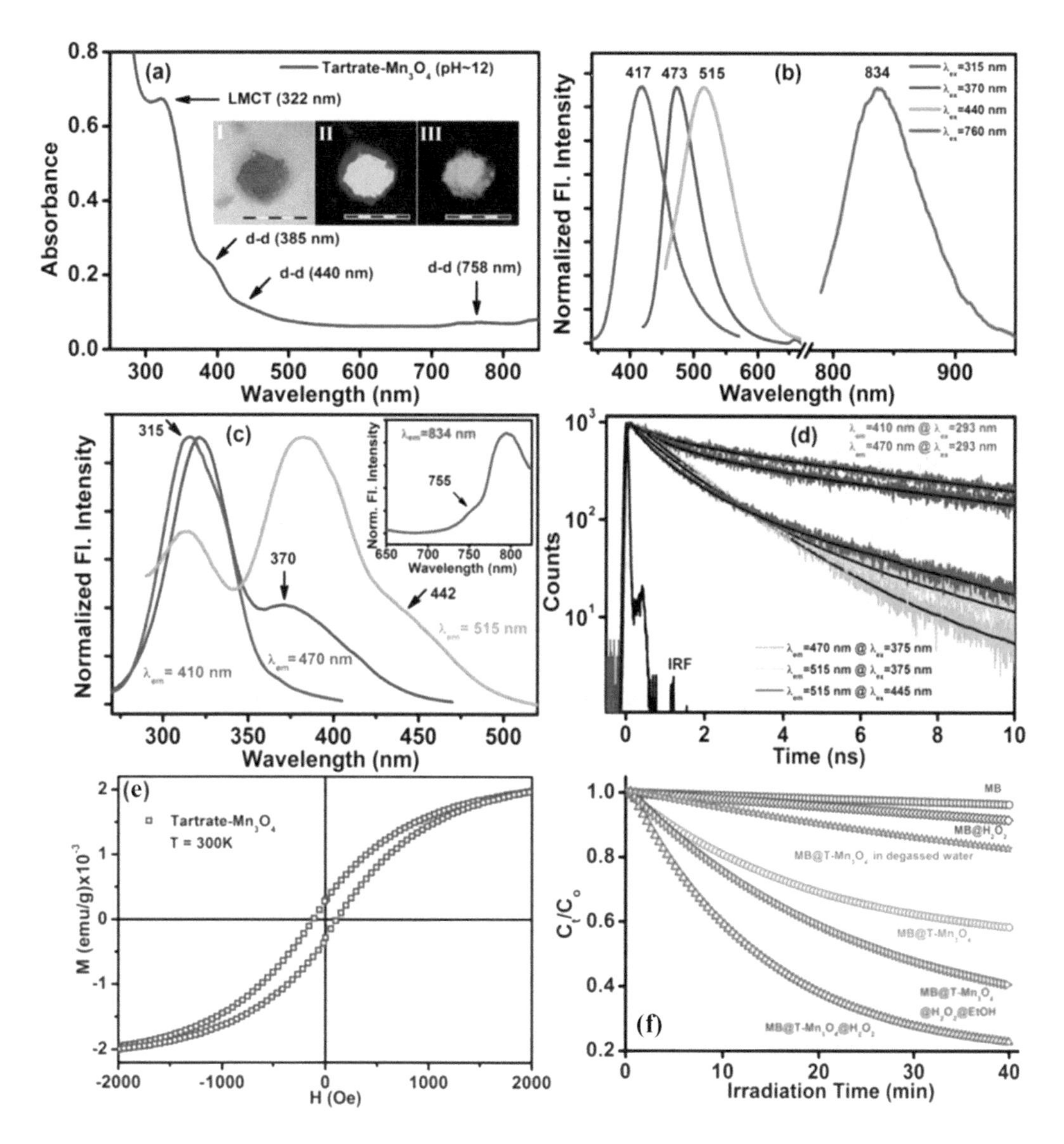


**FIGURE 2.16** (a) UV–vis absorption spectrum of T–$Mn_3O_4$ NPs after treatment (at pH ~12 and 70°C for 12 hrs). Inset shows the fluorescence microscopic images of the same under irradiation of white light (bright field, I) and light of two different wavelengths 365 (II) and 436 (III) nm. Scale bars in the images are of 500 μm. (b) Normalized steady-state PL spectra collected from T–$Mn_3O_4$ NPs with four different excitation wavelengths of 315, 370, 440, and 760 nm at pH~12. (c) Excitation spectra of T–$Mn_3O_4$ NPs at different PL maxima of 410, 470, 515, and 834 nm. (d) Picosecond-resolved PL transients of T–$Mn_3O_4$ NPs in water measured at emission wavelengths of 410, 470, and 515 nm upon excitation with excitation source of 293, 375, and 445 nm wavelengths. (e) Field-dependent magnetization (M vs. H) at room temperature (300 K) of T–$Mn_3O_4$. The distinct hysteresis loops observed confirms ferromagnetic activation of the NPs upon functionalization with carboxylate ligands. (f) Plots of relative concentration (Ct/C0) versus time for the photodegradation of MB (monitored at 660 nm) alone and in presence of T–$Mn_3O_4$ NPs, $H_2O_2$, T–$Mn_3O_4$@$H_2O_2$ and T–$Mn_3O_4$@$H_2O_2$@EtOH, are shown. (Reproduced with permission from Giri et al.[80])

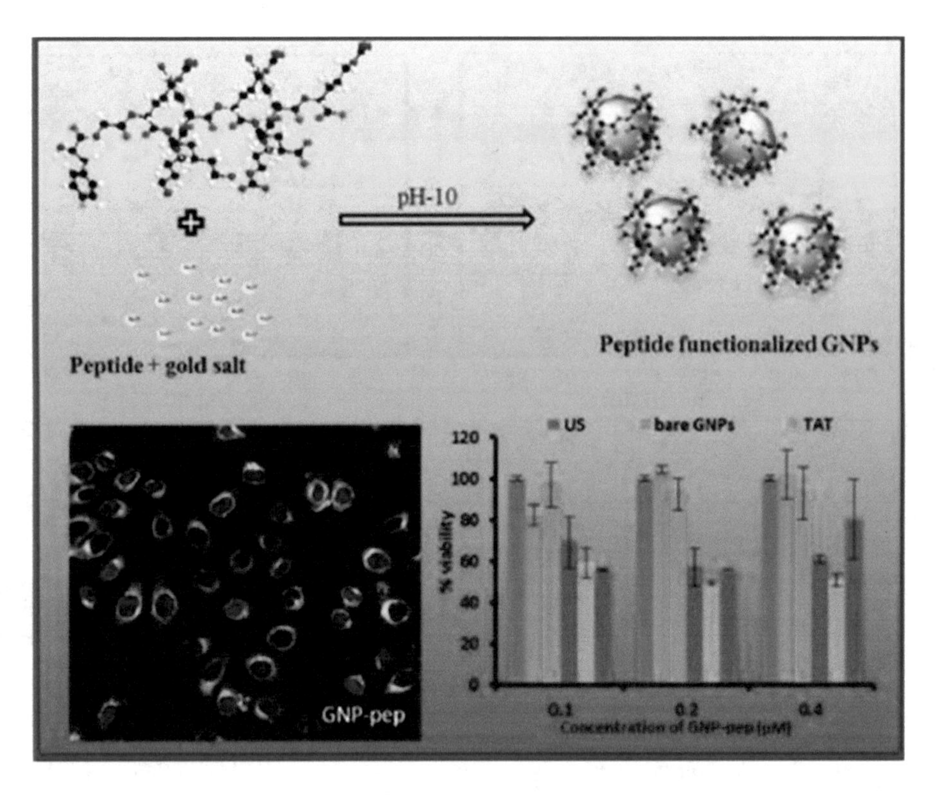


FIGURE 3.3 Schematic representation of efficient penetration ability and cytotoxicity of peptide-conjugated NHs. (Reproduced with permission from Bansal et al. [50]. American Chemical Society.)

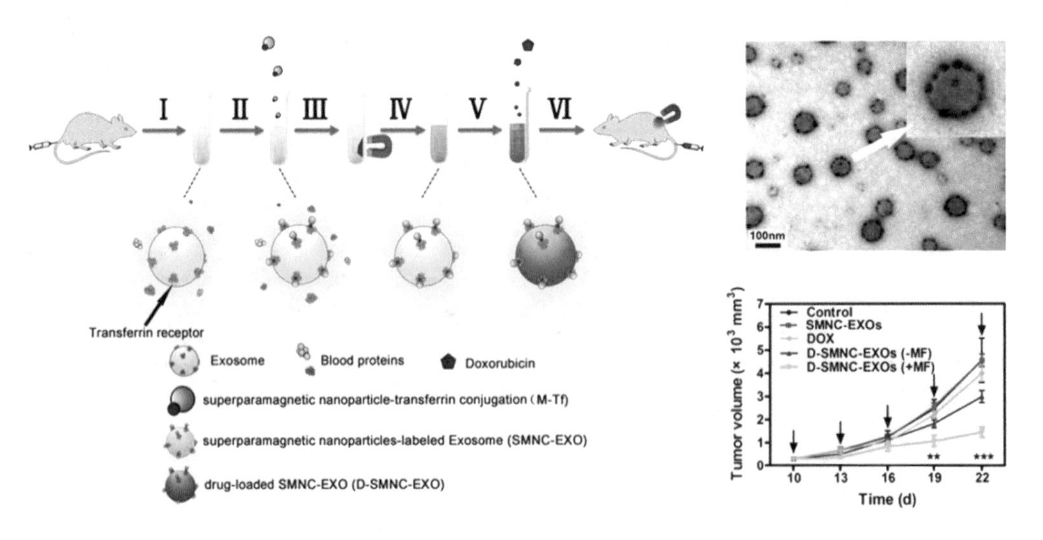


FIGURE 3.4 Schematic representation of synthesis of dual-functional exosome-based superparamagnetic nanoparticles clusters and its use as a targeted drug delivery carrier for cancer therapy. (Reproduced with permission from Qi et al. [54]. American Chemical Society.)

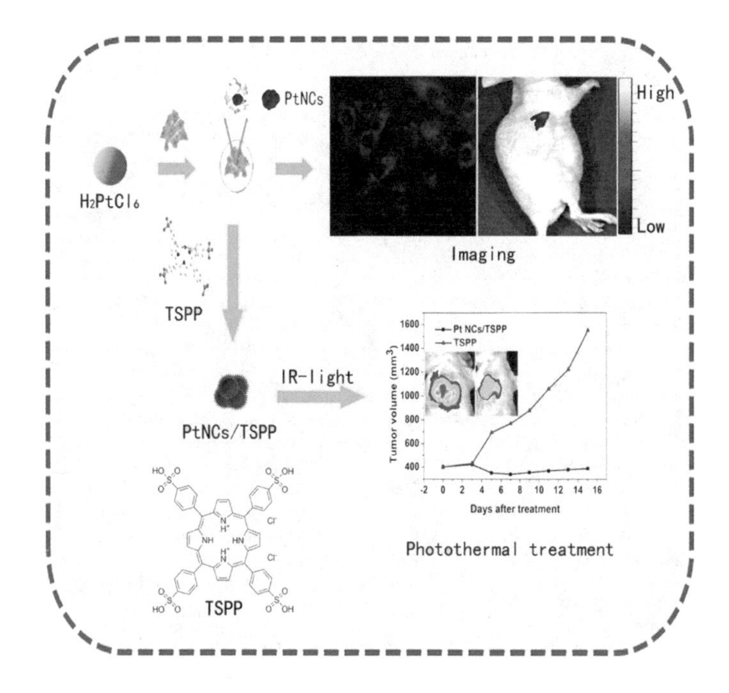


FIGURE 3.6 Schematic representation of the synthesis of platinum NCs within cancer cells themselves and its application in fluorescence imaging guided photothermal treatment. (Reproduced with permission from Chen et al. [59]. American Chemical Society.)

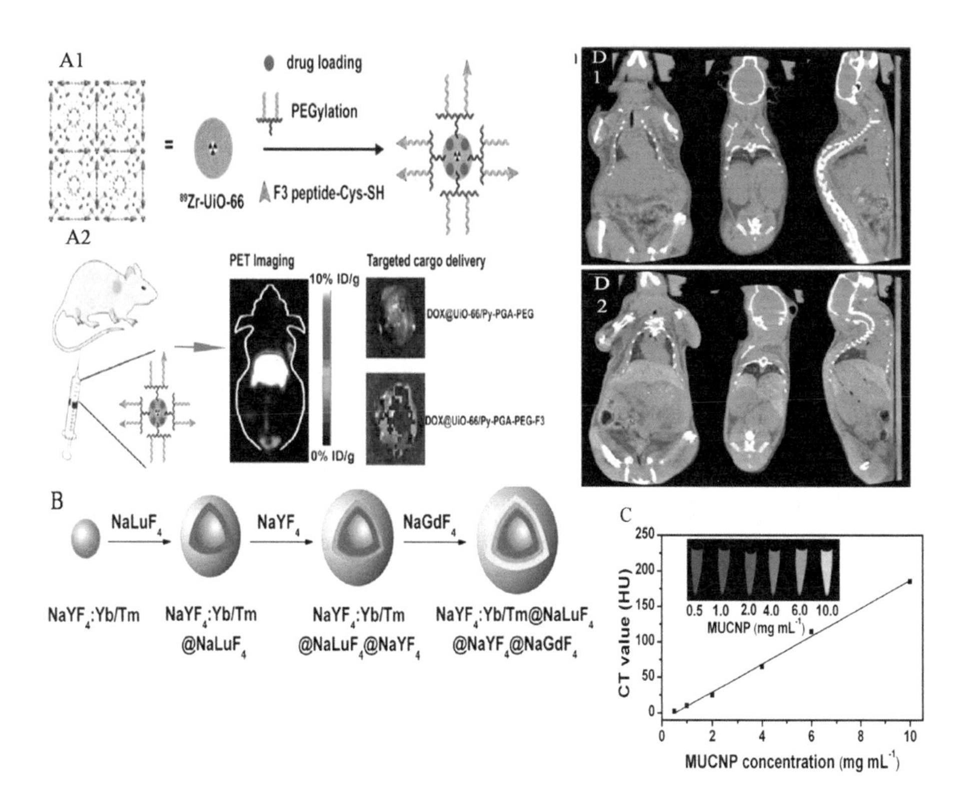


FIGURE 3.8 (A) Schematic representation of the synthesis of UiO-66 nMOFs attached with pyrene-derived polyethylene glycol (Py–PGA-PEG) followed by functionalization using peptide ligand (F3) (A1) and its application in PET imaging assisted drug delivery (A2). (Reproduced with permission from Chen et al. [83]). American Chemical Society. (B) Design for synthesis and functionalization of multi-functional upconversion nanoparticles (MUNCPs) (C) Efficiency of MUNCPs at different concentrations at CT contrast agent (D) CT imaging of Kunming mouse using PAA-functionalized MUNCPs before injection (D1) as well as 2 hrs after injection (D2). (Reproduced with permission from Shen et al. [65] American Chemical Society.)

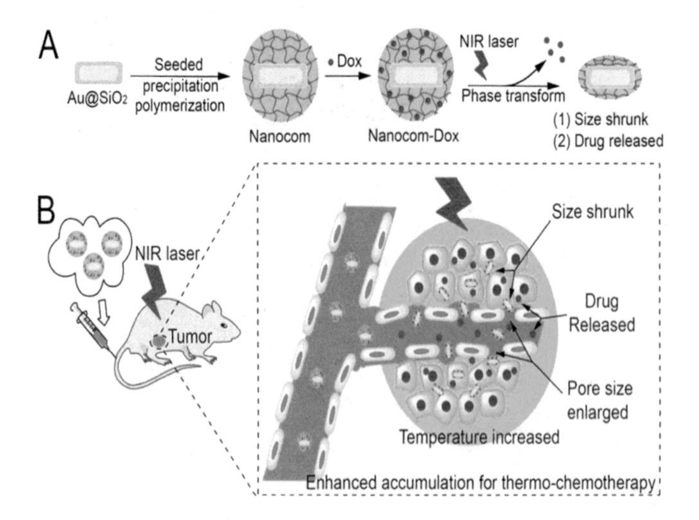


**FIGURE 3.11** (A) Schematic representation of the synthesis and functionalization of AuNR@$SiO_2$ (B) Application of these NHs in chemotherapy as well as photothermal therapy for treating cancer cells. (Reproduced with permission from Zhang et al. American Chemical Society.)

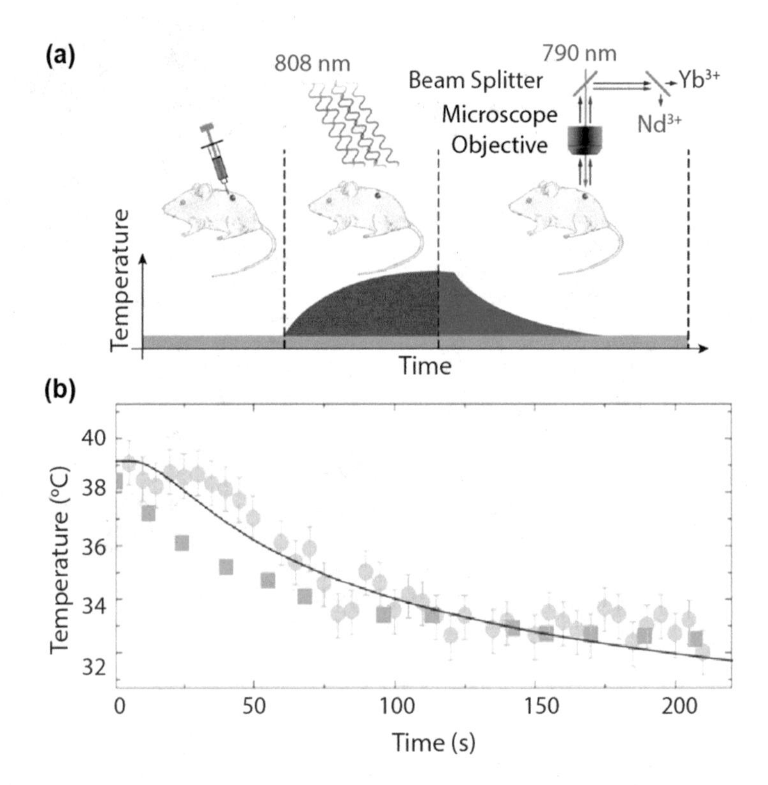


**FIGURE 4.2** (a) Schematic representation of the subcutaneous thermal relaxation experiments. (b) Time evolution of the temperatures measured by the subcutaneous luminescent thermometer (grey) and by the IR thermal camera (yellow). Dots are experimental subcutaneous (circles) and skin (squares) temperatures, whereas the solid line is the best fit. Adapted with permission from Ximendes et al. 2016. © 2016, American Chemical Society.

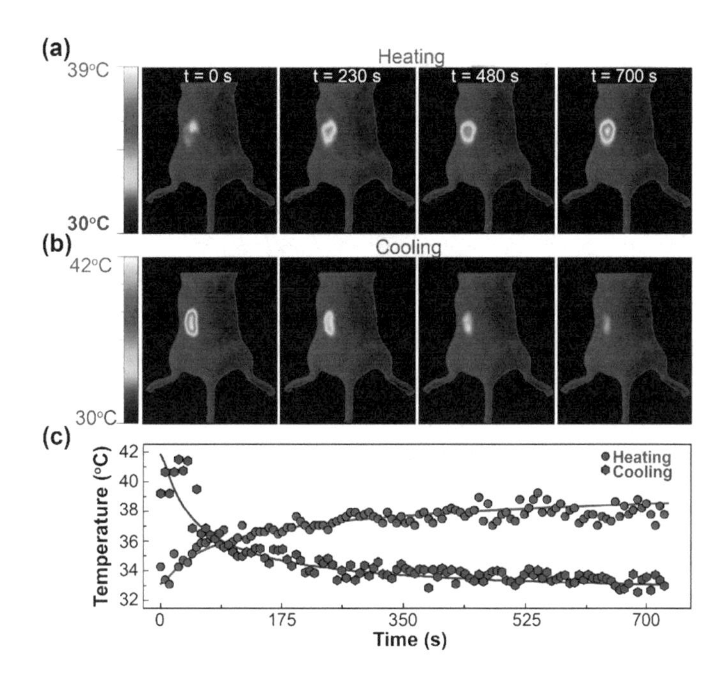


FIGURE 4.3 (a) and (b) Schematic representations of the *in vivo* 2D-SDTI experiment. An optical figure of the anesthetized mouse was superimposed. (c) Time evolution of the average temperature of the injection area as measured by the subcutaneous LNThs during heating and thermal relaxation processes. (Reproduced with permission from Ximendes et al. 2017. © 2017, Wiley.)

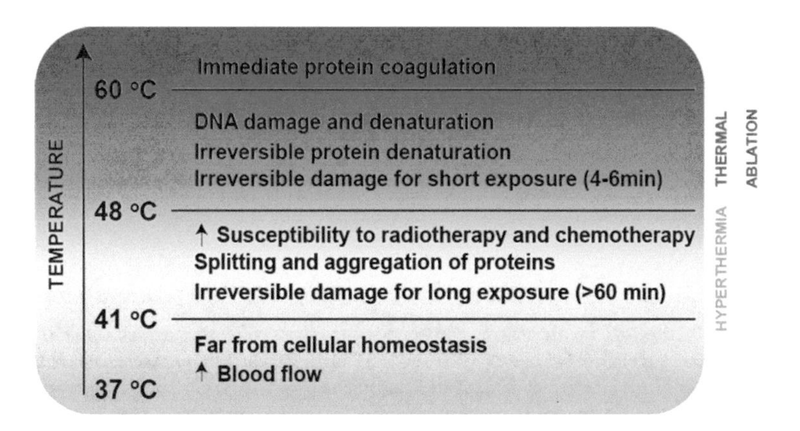


FIGURE 4.6 Biological effects of heating procedures. Summary of the main effects of thermal therapies on treated tissues. Commonly used ranges for thermal therapy alone (thermal ablation, 48–60°C) and in combination with conventional treatments (hyperthermia, 41–48°C) are indicated.

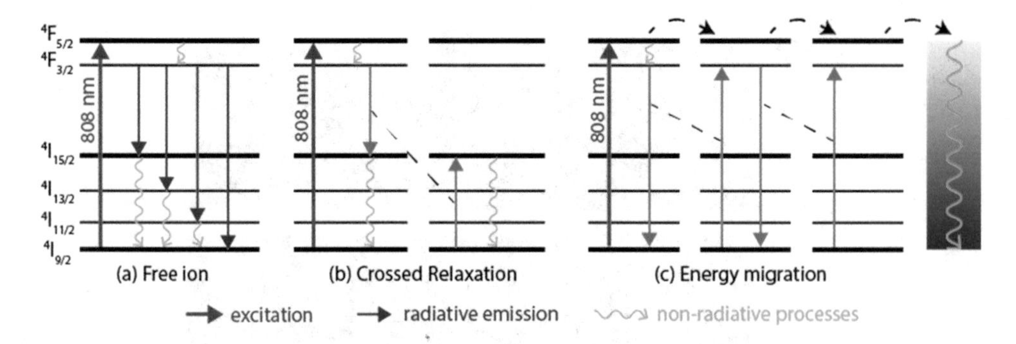


**FIGURE 4.7** Schematic representation of energy transfer processes between $Nd^{3+}$ ions. (a) Radiative and non-radiative de-excitation processes of free $Nd^{3+}$ ions (b) Cross-relaxation process between two adjacent $Nd^{3+}$ ions. (c) Energy migration process between $Nd^{3+}$ ions, which ends in a non-radiative de-excitation from a non-emitting center.

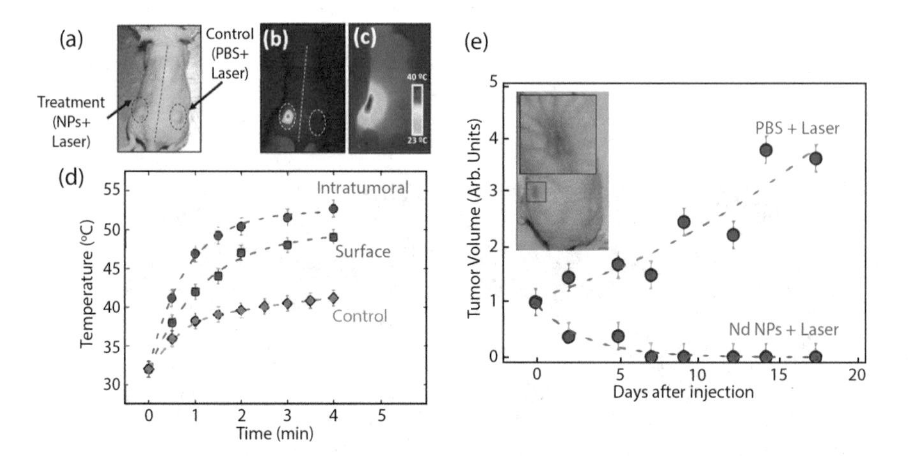


**FIGURE 4.8** (a) Optical image of a mouse bearing two tumors. Tumor at left side was treated with $LaF_3$:$Nd^{3+}$ NPs, while tumor at right side was used as a control (without NPs). (b) and (c) Correspond to the fluorescence image and thermal image (after 808 nm 4 W $cm^{-2}$, 4 minutes laser irradiation) of the same mouse shown in (a). (d) The time evolution with irradiation laser of intratumoral temperature as obtained from the analysis of sub-tissue fluorescence with irradiation of the left side and tumor surface temperature (as measured by infrared thermal imaging). The evolution of the surface temperature of the control tumor is included for comparison. (e) The evolution time of the sizes post treatment of both treated and control tumors. The inset shows the optical image of reduced tumor post treatment. (Adapted with permission from Carrasco et al. 2015. © 2015, Wiley-VCH.)

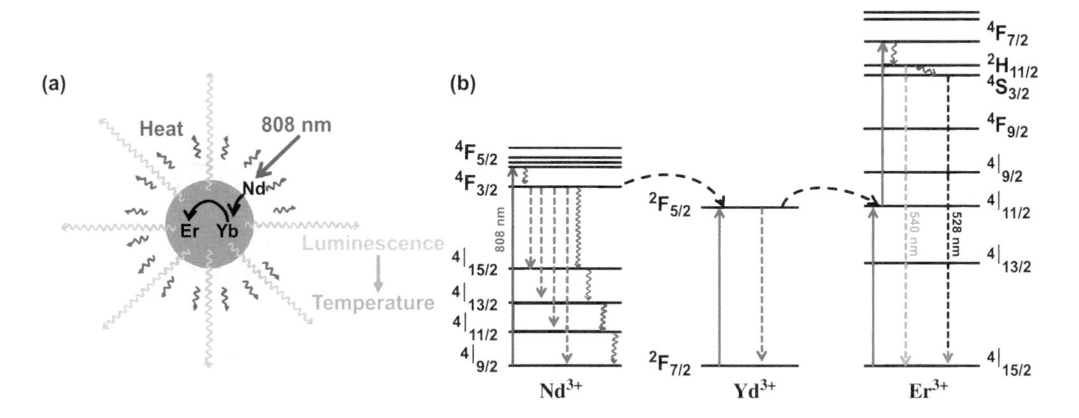


**FIGURE 4.9** (a) Schematic diagram of the active-core/active-shell nanoparticles specifically designed for simultaneous heating and thermal sensing activated by 808 nm light. In this design, the shell acts as a heating center whereas the core provides thermal sensitivity to the structure. (b) Detail of the energy level diagrams of $Nd^{3+}$, $Yb^{3+}$ and $Er^{3+}$ ions. The heat produced by $Nd^{3+}$ ions in the shell is represented by red wave arrows whereas the green emission, resultant of the subsequent energy transfer between (i) $Nd^{3+}$ and $Yb^{3+}$ ions and (ii) between $Yb^{3+}$ and $Er^{3+}$, is represented by green dashed arrows. (Adapted from Ximendes et al. 2016. © 2016, Royal Society of Chemistry.)

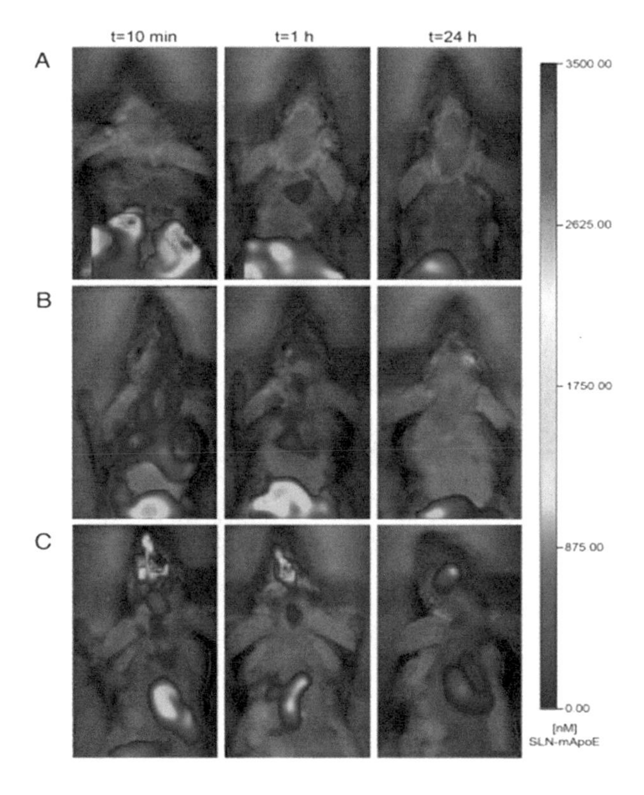


**FIGURE 5.10** Fluorescence molecular tomographic images to access biodistribution of SLN in the brain administered by different routes (A) intraperitoneally, (B) intravenously and (C) intratracheally. Images were acquired at time 10 mins, 1 hr, 24 hr. (Reproduced with permission from Dal Magro et al., 2017.)

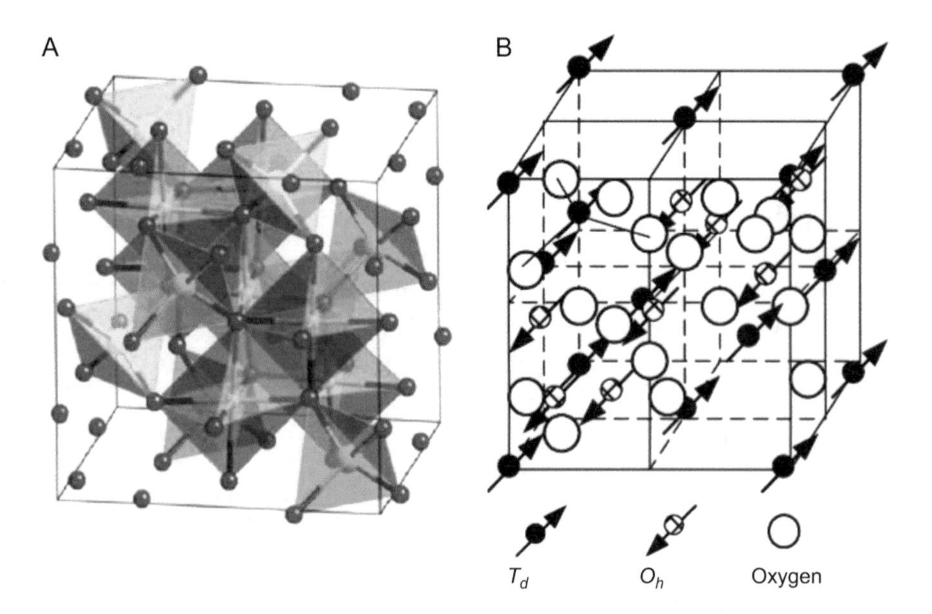


FIGURE 6.1 (**A**) Crystal structure of $MnFe_2O_4$ spinel ferrite with $T_d$ and $O_h$ sites in yellow and blue, respectively; oxygen atoms are in red (Adapted from Errandonea, 2014, with permission from Springer Nature). (**B**) Ferrimagnetic arrangement in ferrite spinel structure. (Adapted from Mathew and Juang, 2007, with permission from Elsevier.)

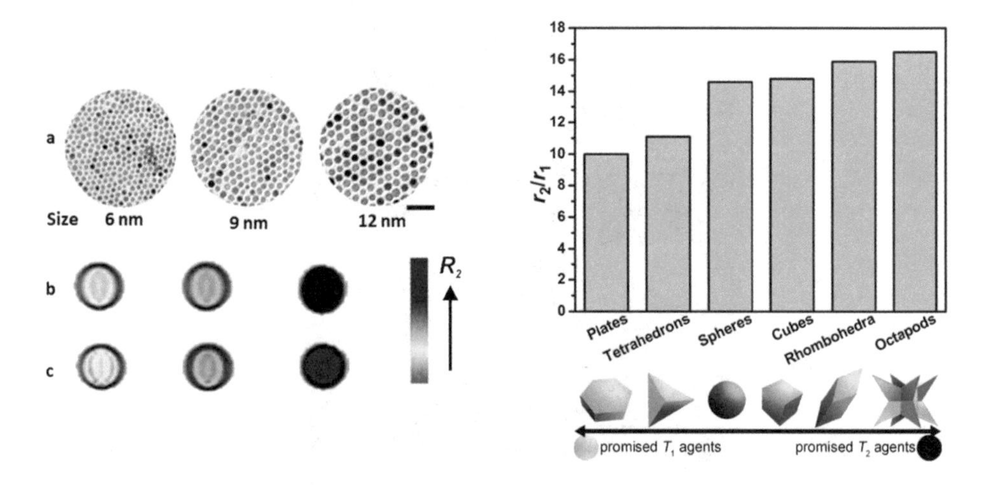


FIGURE 6.7 *Left:* Size-dependent MR contrast effect of $MnFe_2O_4$ NPs: (**a**) TEM images, (**b**) $T_2$-weighted MR images and (**c**) color maps of 6, 9 and 12 nm (scale bar: 50 nm). (Adapted from Lee et al., 2007, with permission from Springer Nature.) *Right:* The role of ferrite NP morphology on the $R_1/R_2$ ratio. (Reproduced from Yang et al., 2018, with permission from American Chemical Society.)

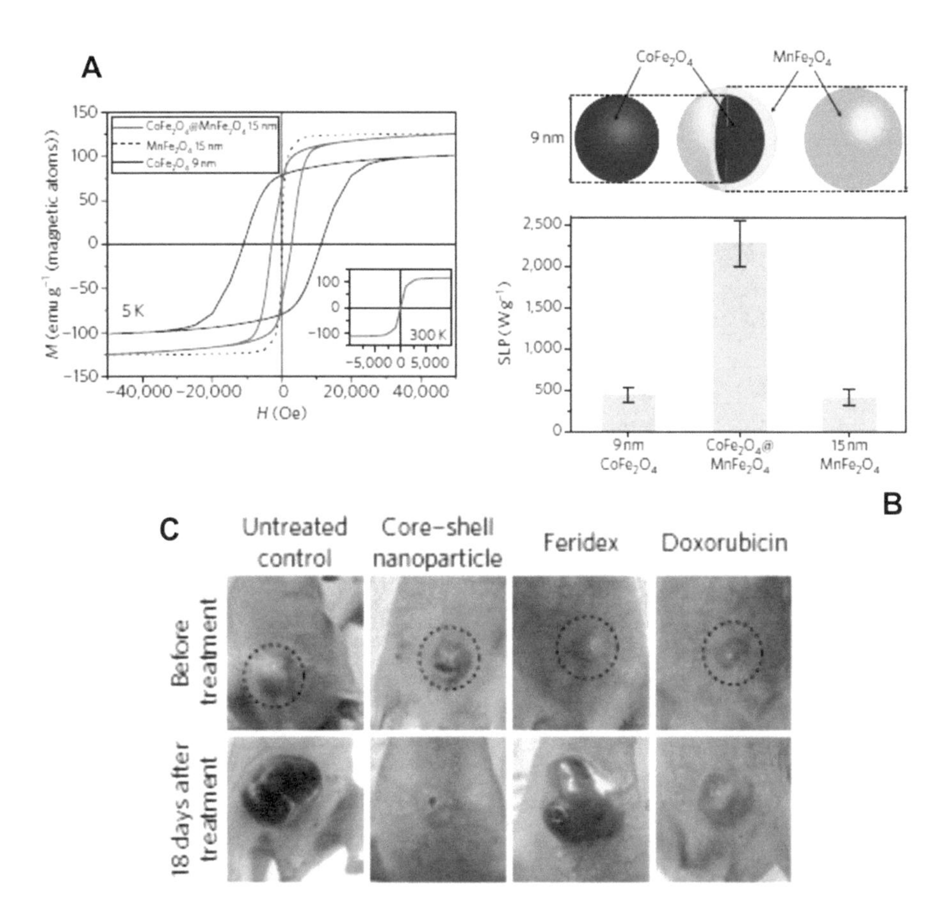


FIGURE 6.8 Interfacial exchange-coupled magnetic NPs as high-performance MH agents. (**A**) M-H curve of 15 nm $CoFe_2O_4@MnFe_2O_4$, 15 nm $MnFe_2O_4$ and 9 nm $CoFe_2O_4$ NPs measured at 5 K (Inset: M-H curve of $CoFe_2O_4@MnFe_2O_4$ at 300 K). (**B**) Schematic representation $CoFe_2O_4@MnFe_2O_4$ NPs and its SLP value compared with the values for the single components. (**C**) Nude mice xenografted with cancer cells (U87MG) before treatment (upper row, dotted circle) and 18 days after treatment (lower row) with $CoFe_2O_4@MnFe_2O_4$ hyperthermia, Feridex hyperthermia and DOX, respectively. (Adapted from Lee et al., 2011, with permission from Springer Nature.)

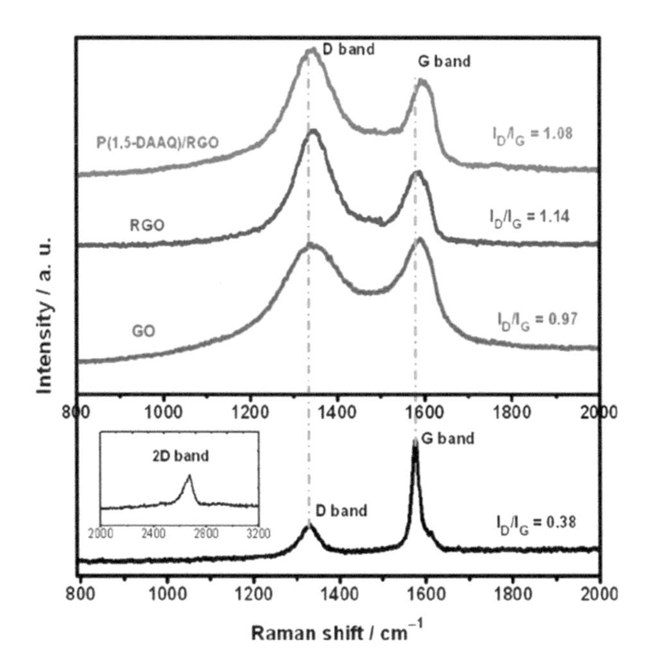


FIGURE 7.4 Raman spectra of graphite, GO, rGO and the P(1,5-DAAQ)/RGO nanohybrid. The inset shows the 2D band of graphite. (Figure reprinted/adapted with permission from Ref. [43].)

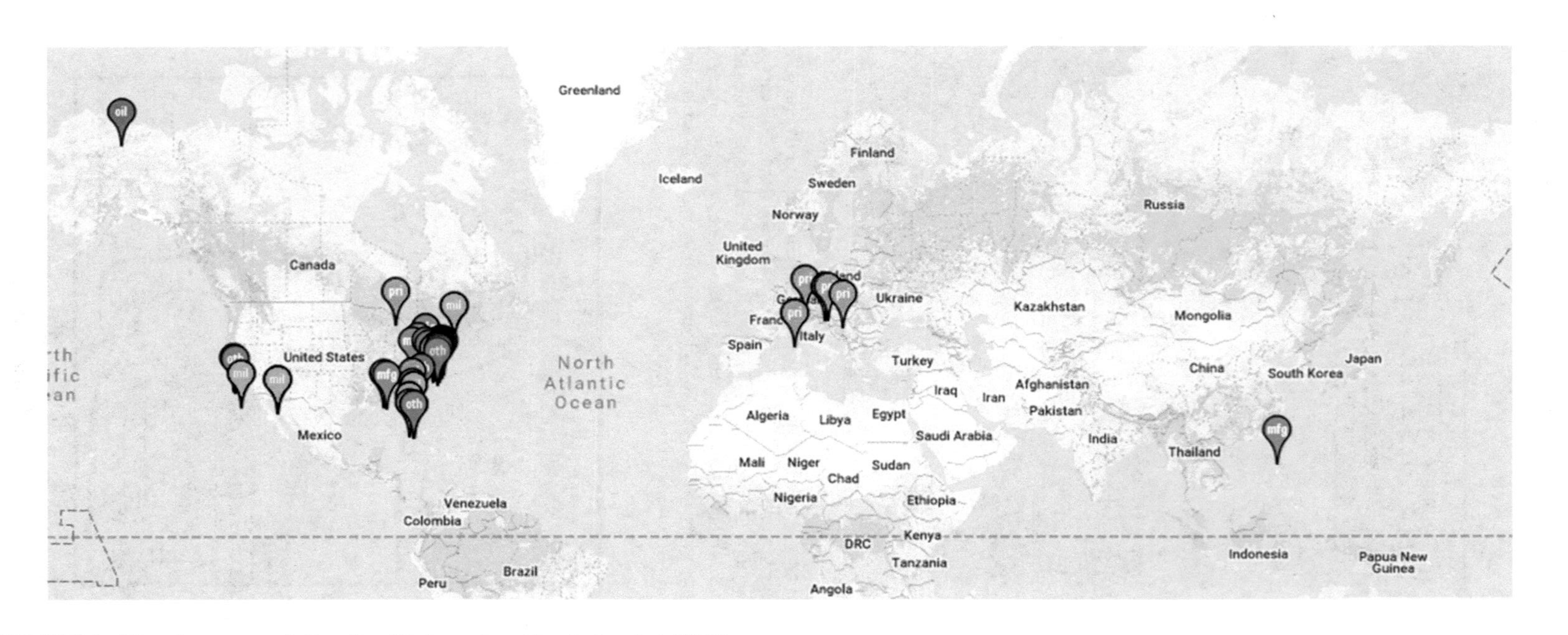


FIGURE 8.1 Map of nano-remediation sites. (Redrawn from Nanotechproject, 2009.)

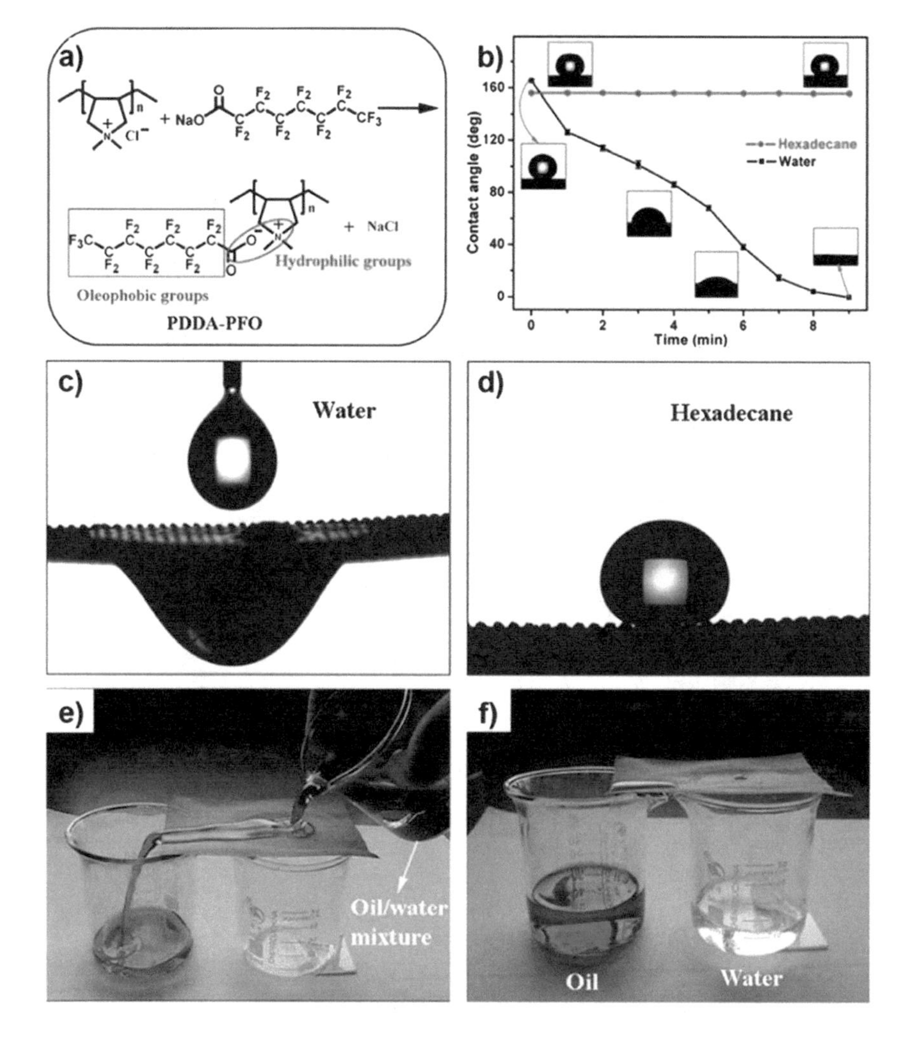


FIGURE 9.2 The PDDA–PFO/$SiO_2$-coated mesh film shows special wettability, with both superhydrophilic and superoleophobic properties. (a) Chemical reaction scheme used for the synthesis of PDDA–PFO. (b) Time dependence of contact angles for water and hexadecane on the PDDA–PFO/$SiO_2$ coating. (c) Water droplet spreading on and permeating through the mesh. (d) Shape of a hexadecane droplet on the mesh with a contact angle of 157°. (e and f) Oil–water separation experiment was performed on the PDDA–PFO/$SiO_2$-coated mesh. (Reproduced from Yang et al., *J. Mater. Chem.*, 2012b, 22, 2834–2837. Reprinted with permission from The Royal Society of Chemistry.)

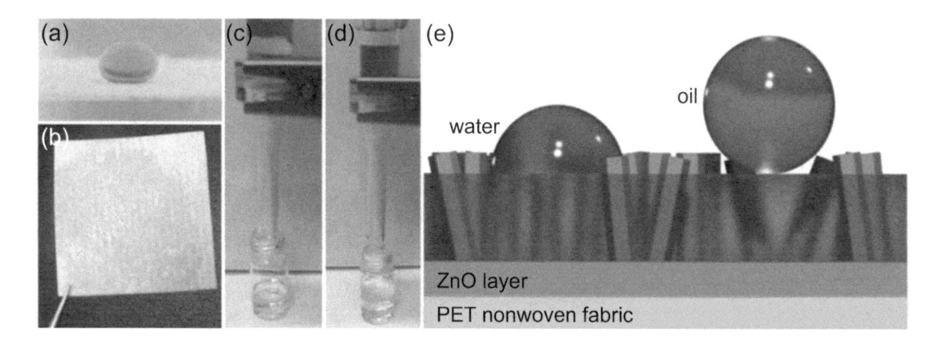


FIGURE 9.4 (a) Photograph of $CCl_4$ droplet, colored blue is placed in water on top of the functionalized PET fabric, (b) no residual $CCl_4$ left on the PET fabric, (c–d) diesel oil (dyed blue) is blocked in the filtration cell by water pre-wetted PET fabric, water is filtering through the fabric, and (e) the diagram showing the mechanism of water-permeable process: the blue rods/layer and the green layer represent ZnO nanostructure and pre-wetted liquid layer of water, respectively; oil droplet (brown) is staying on top of the water layer. (Reproduced from Xiong et al., *Journal of Membrane Science* 493 (2015) 478–485. Reprinted with permission from Elsevier B.V.)

FIGURE 9.5 A series of photos for the process of absorption and collection of lubricating oil (dyed with oil red) from the water. (Reproduced from Kong et al., *Journal of Industrial and Engineering Chemistry*, 58 (2018) 369–375. Reprinted with permission from the Elsevier B.V. on behalf of The Korean Society of Industrial and Engineering Chemistry.)

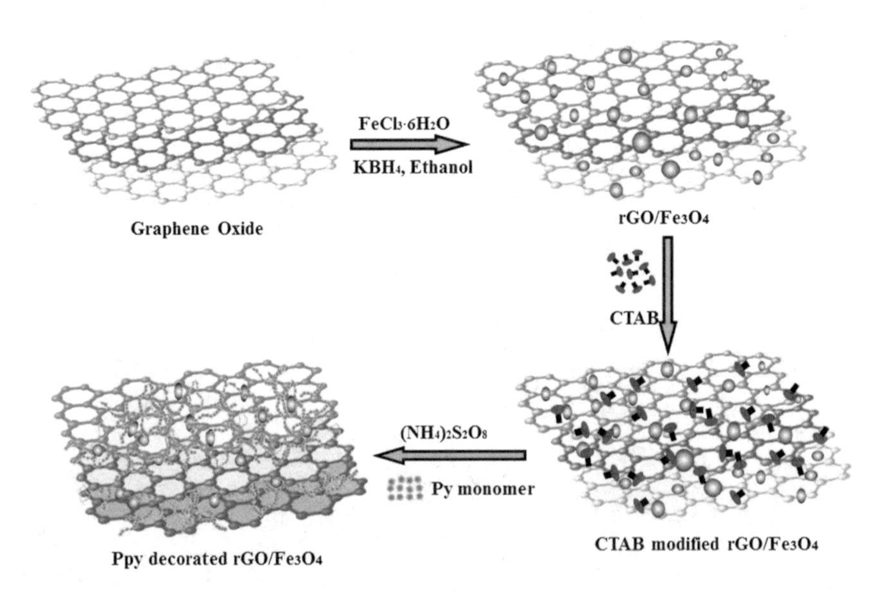


FIGURE 10.1 Schematic illustration of the ternary composites preparation. (Reprinted with permission from Hou Wang et al. [24]. © (2015) Elsevier.)

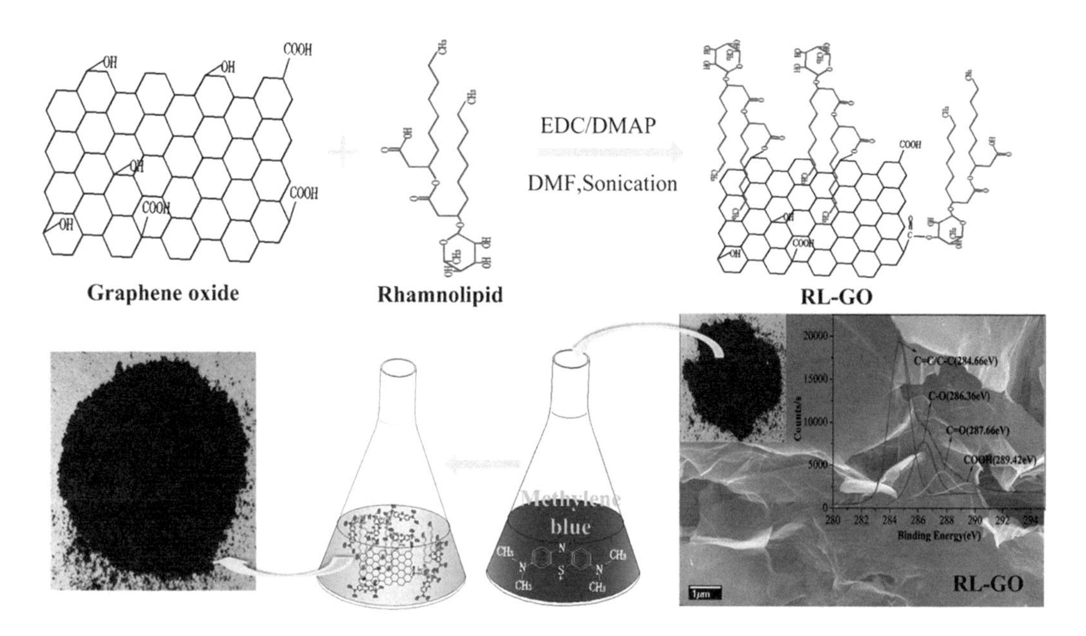


FIGURE 10.2 Schematic depiction of the formation of RL–GO and application for removal of MB. (Reprinted with permission from Zhibin Wu et al. [25]. © (2014) Elsevier.)

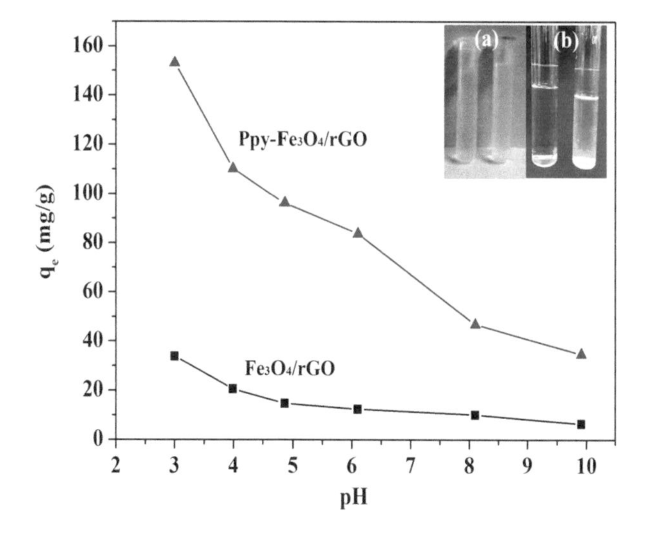


FIGURE 10.10 Cr (VI) adsorption and the pH effect for $Fe_3O_4$/rGO and Ppy-$Fe_3O_4$/rGO. C (Cr(VI)) initial = 48.4 mg/L, $m/V$ = 0.25 g/L, T = 303 K. In the inserted photograph presented: (a) the Cr(VI) removal ability of Ppy-$Fe_3O_4$/rGO and (b) the chemical experiment for $SO_4^{-2}$ and after $Ba^{2+}$ injection. (Reprinted with permission from Hou Wang et al. [24]. © (2015) Elsevier.)

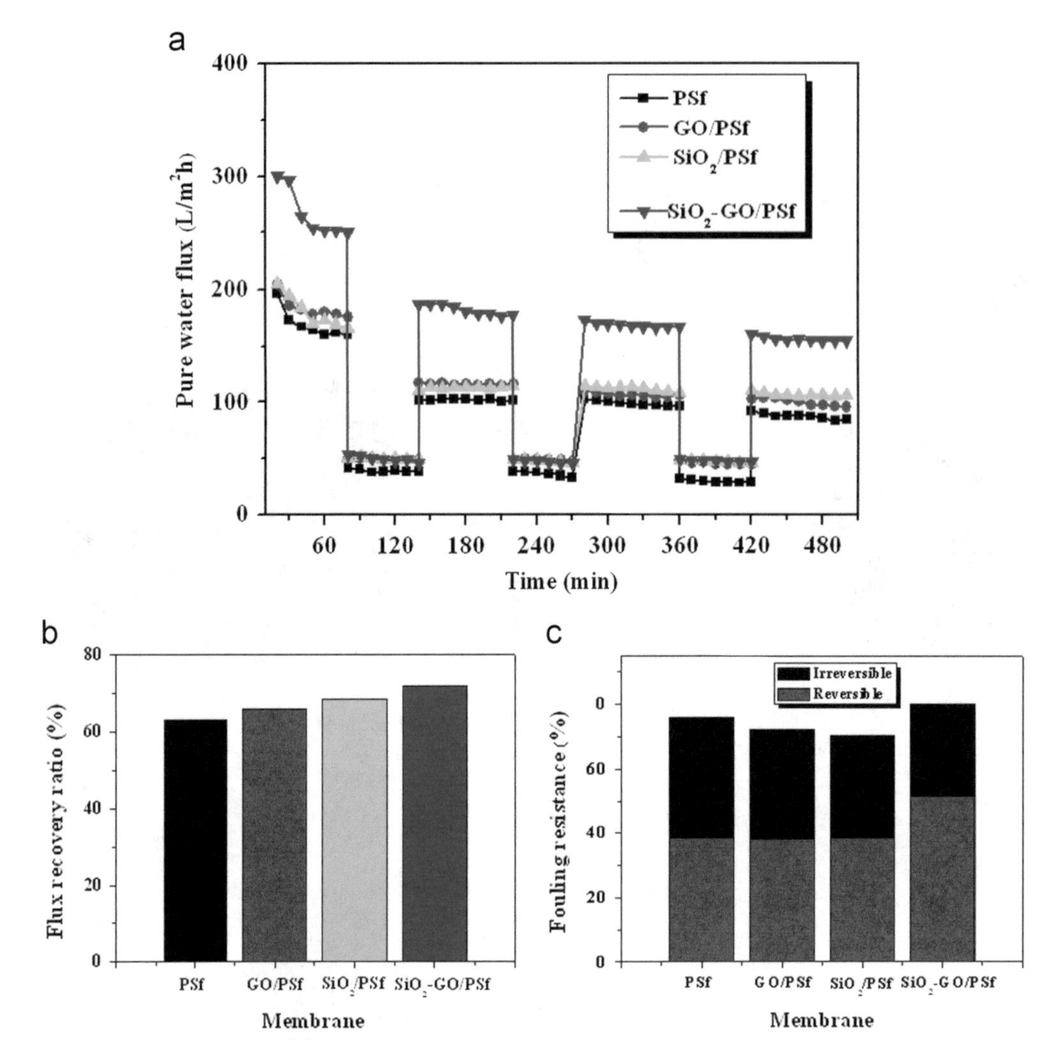


**FIGURE 10.13** Membranes with different inorganic materials and the time-dependent fluxes during an anti-fouling experiment with BSA filtration (1 mg/mL, pH = 7.4) at 0.2 MPa, (b) FRR of the prepared membranes and (c) Fouling resistance of the prepared membranes. (Reprinted with permission from Huiqing Wu et al. [41]. © (2014) Elsevier.)

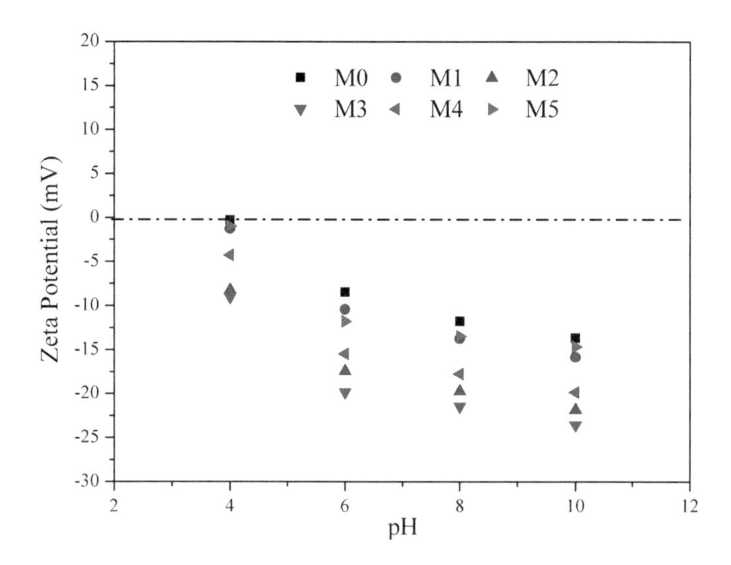


**FIGURE 10.14** Zeta potential of PVDF/GO/LiCl nanohybrid membranes. (Reprinted with permission from Zhenya Zhu et al. [42]. © (2017) Elsevier.)

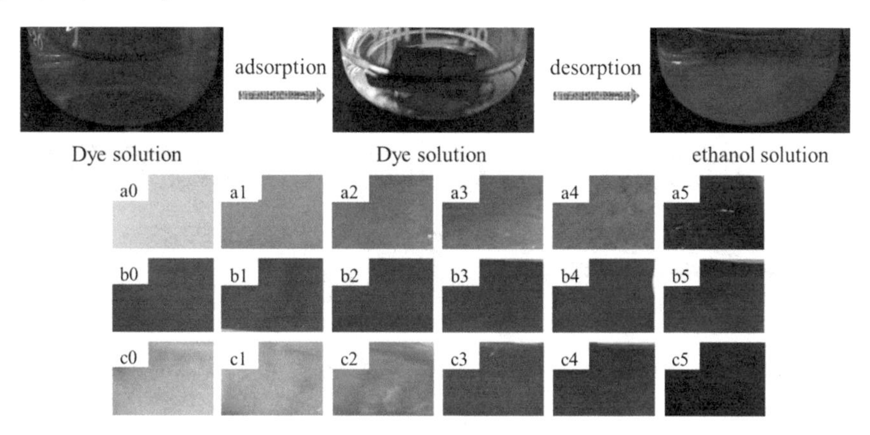


**FIGURE 10.15** The images of PVDF/GO/LiCl nanohybrid membranes (a0–a5: pristine membranes, b0–b5: after immersing the Rhodamine B solution for 36 h, c0–c5: after decoloration by ethanol alcohol). (Reprinted with permission from Zhenya Zhu et al. [42]. © (2017) Elsevier.)

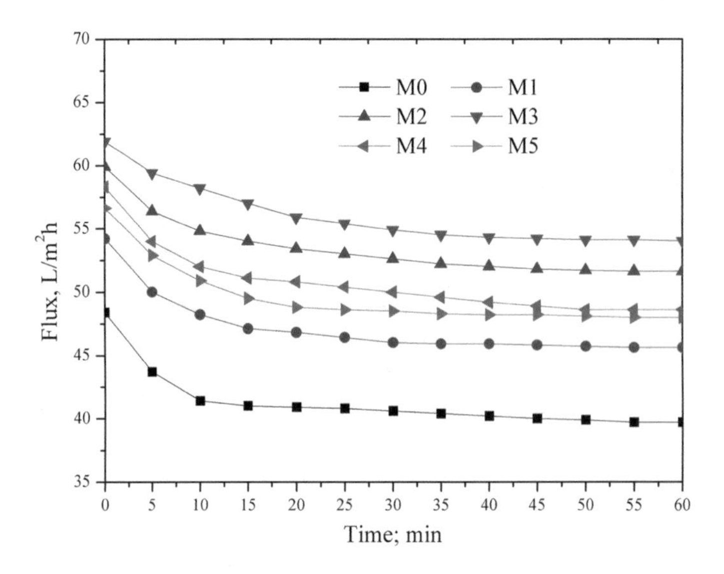


FIGURE 10.16 M0–M5 PVDF/GO/LiCl nanohybrid membranes and the flux decline curves at 100 kPa. (Reprinted with permission from Zhenya Zhu et al. [42]. © (2017) Elsevier.)

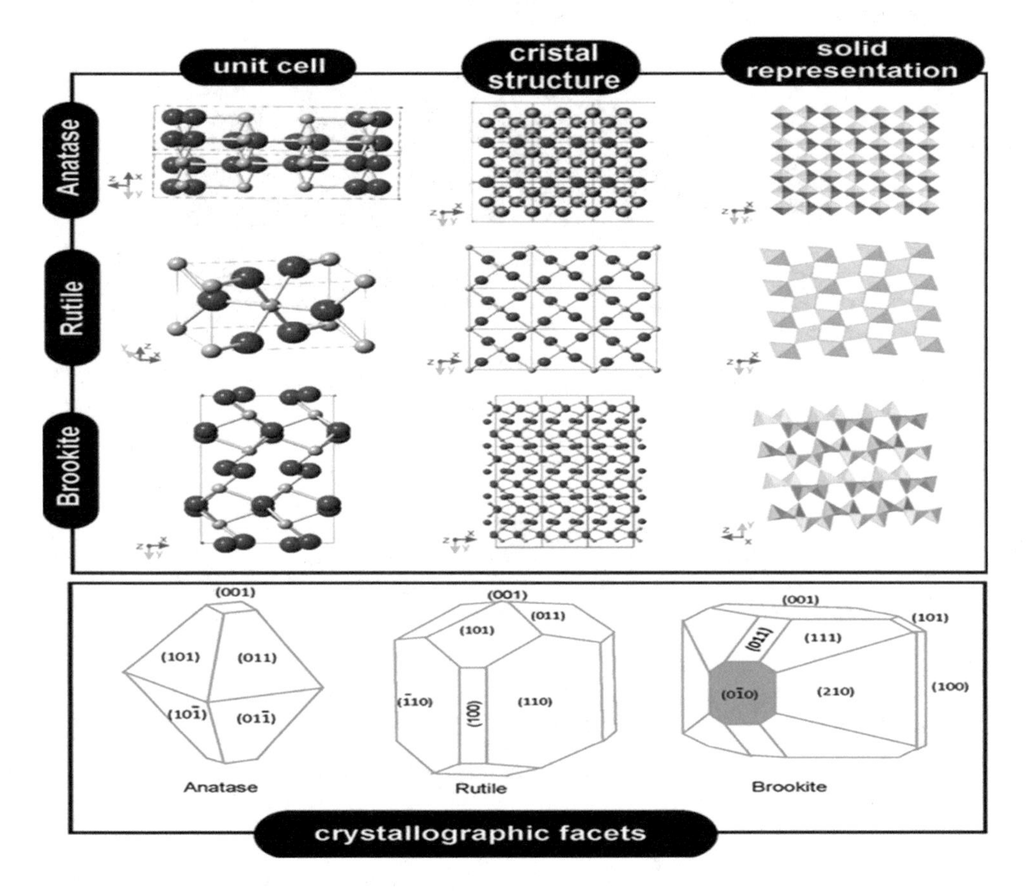


FIGURE 11.3 Structures of $TiO_2$and crystal images. (Adapted from Pimentel et al., 2016.)

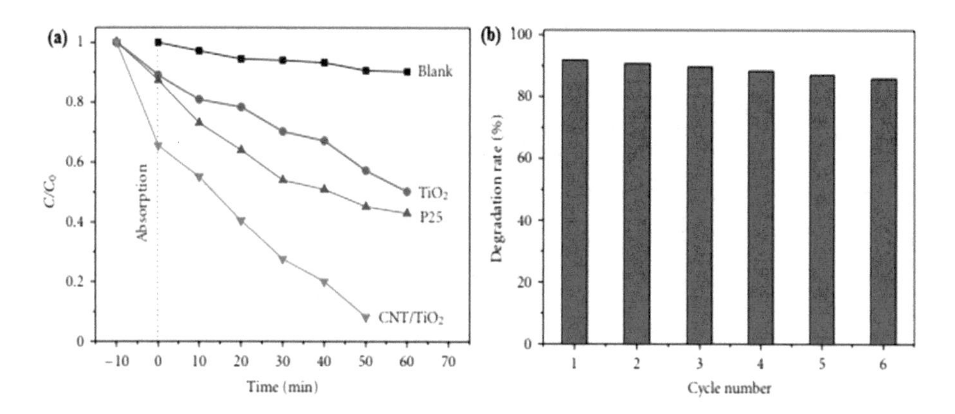


**FIGURE 11.6** (a) The photocatalytic degradation of RhB in the absence of any photocatalysts (the blank test) and in the presence of different photocatalysts. (b) Six cycles of degradation of RhB using CNT/$TiO_2$ nanohybrids as the photocatalyst. (Adapted from Xie et al., 2012.)

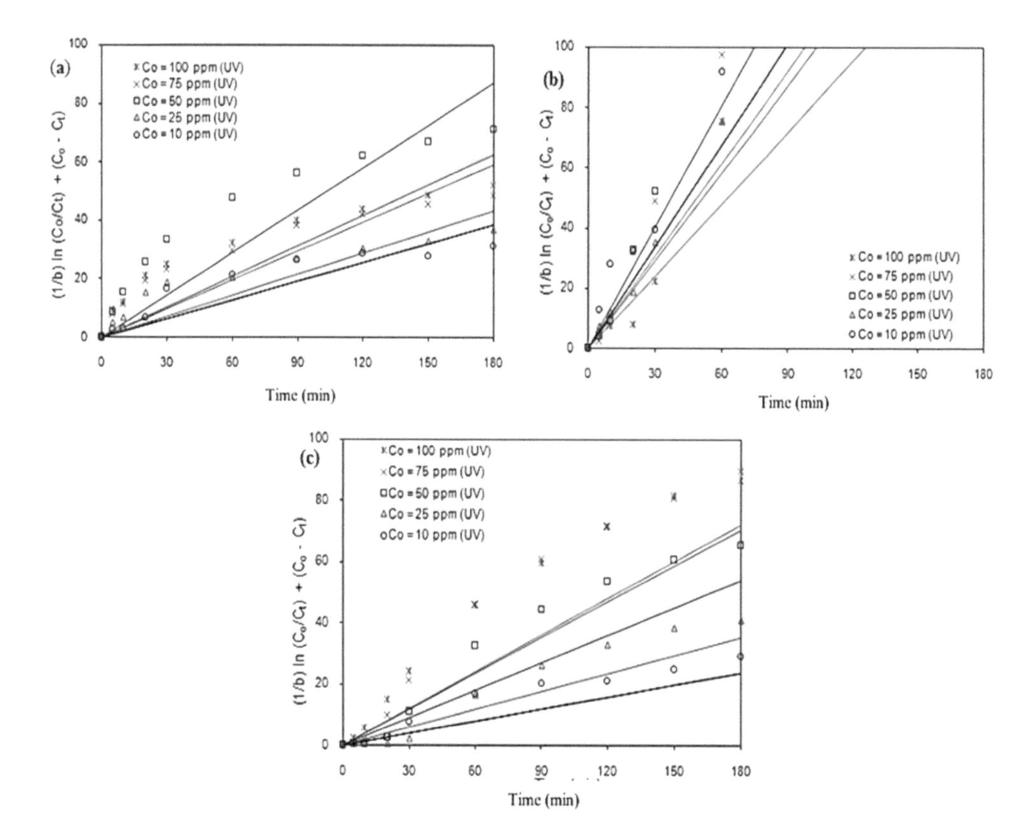


**FIGURE 11.11** Effect of $TiO_2$/CS films on dye removal for various types and concentration of dye using Langmuir–Hinshelwood model, (a) RR 120, (b) RY 17, and (c) RB 220. (Adapted from Norranattrakul et al., 2013.)

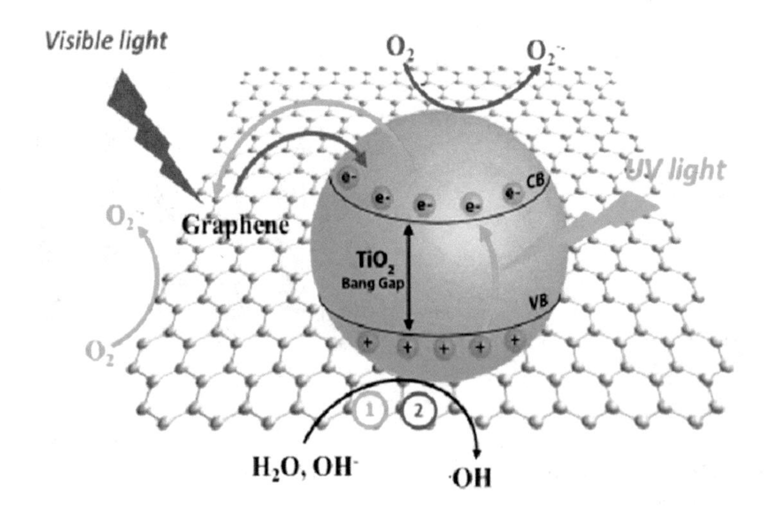


FIGURE 11.14 Mechanisms of UV and visible light activation of $TiO_2$ with graphene. (Adapted from Giovannetti et al., 2017.)

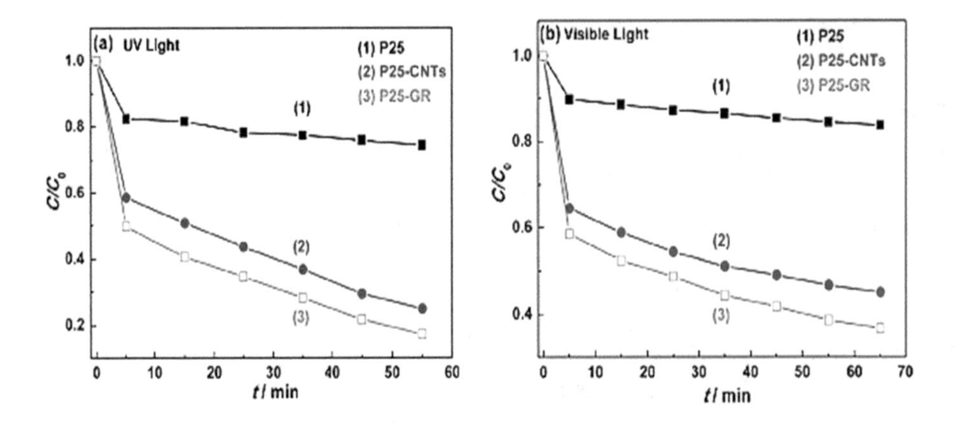


FIGURE 11.15 Photodegradation of MB under (a) UV light and (b) visible light (400 nm) over (1) P25, (2) P25/CNTs, and (3) P25/GR photocatalysts, respectively. (Reproduced with the permission from Zang et al., 2010.)

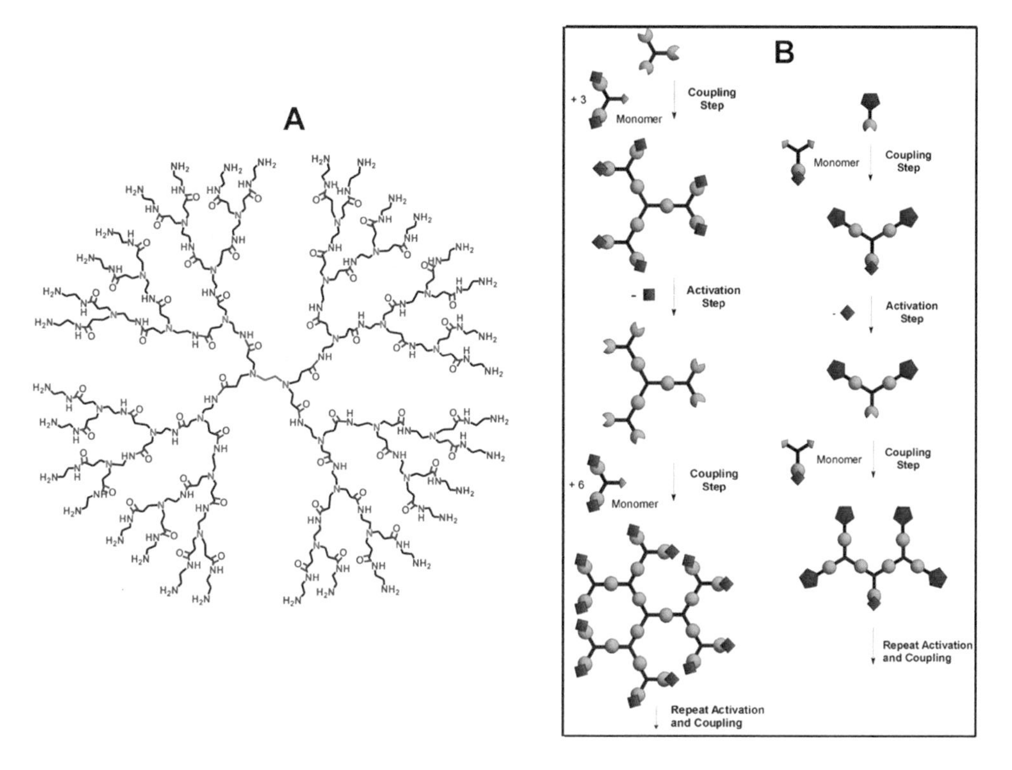


FIGURE 12.1 (**A**) Structure of PAMAM-G4 dendrimer. (**B**) Schematic representation of divergent (*left*) and convergent (*right*) strategies for dendrimer synthesis. (Adapted with permission from Grayson and Frechet, 2001, American Chemical Society.)

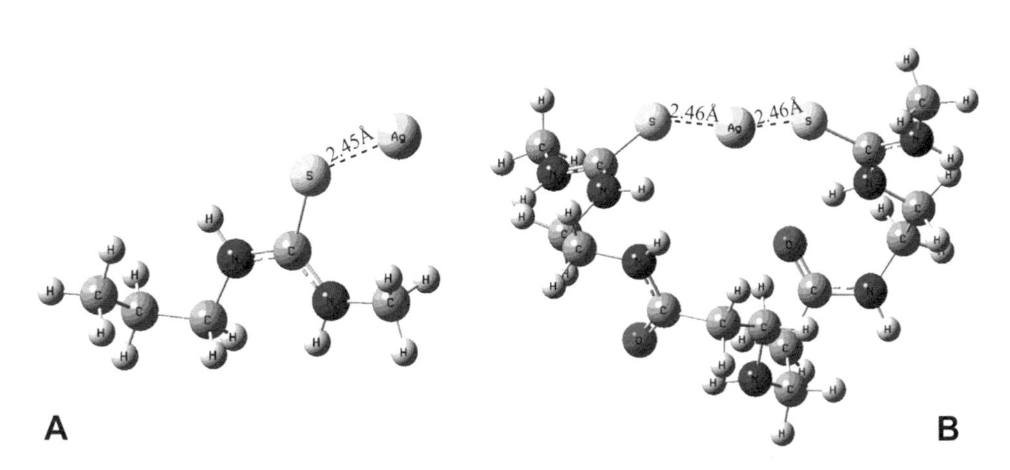


FIGURE 12.9 DFT optimized geometries of the complexes formed by the interaction of Ag(I) with PAMAM-G0 (**A**) and G1 (**B**) generations. (Adapted from Zhang et al., 2018, with permission from Elsevier.)

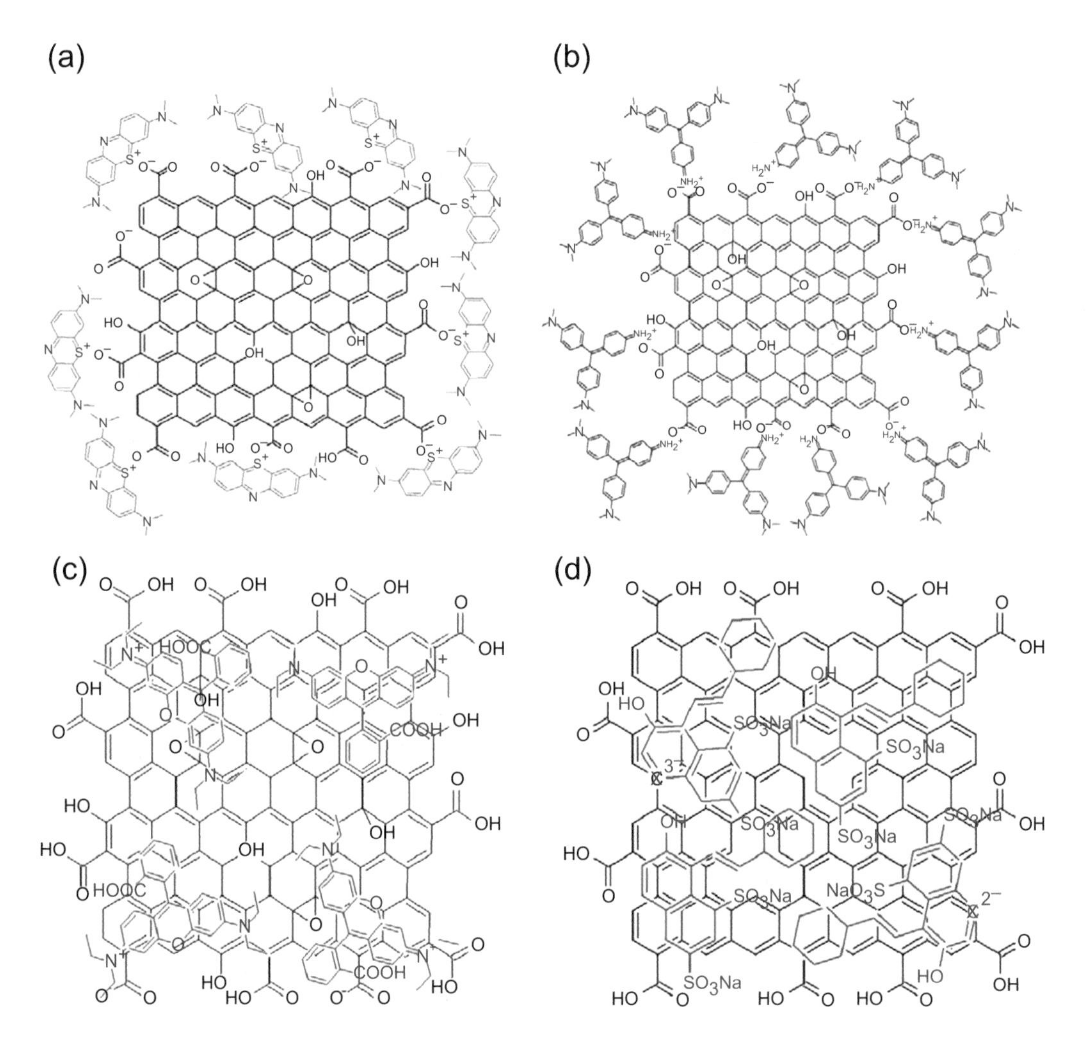


**FIGURE 13.5** Schematic representation of exfoliated graphene oxide interaction with organic dyes (a) methylene blue, (b) methyl violet, (c) rhodamine B, and (d) orange G. (Reproduced from Ramesha et al.[52] With permission from Elsevier.)

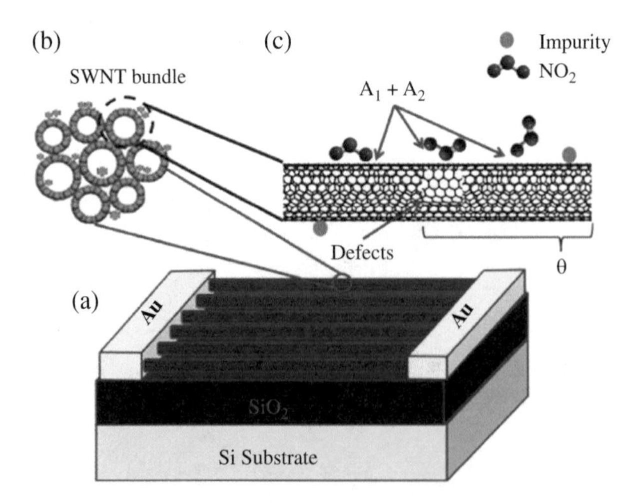


**FIGURE 13.6** Schematic representation of SWCNT-gas sensor for $NO_2$ monitoring. (Reproduced from Kumar et al.[59] With permission from Elsevier.)

# 5

# *Solid-Lipid Hybrid Nanostructures and Their Biomedical Applications*

**Kanwal Akhtar, Yasir Javed, Hafeez Anwar, Khuram Ali, and Naveed A. Shad**

**CONTENTS**



## 5.1 Introduction

A cell membrane primarily consists of different types of lipids in the form of functional proteins and two asymmetric leaflets. Membranes provide a cellular boundary along with basic regulatory platform for many biological processes such as signal transduction, materials transport, pathogenic pathways and intracellular organizations. These processes involve the primary arrangement of various membrane constituents with conformational changes, which induce spatial and temporal heterogeneity in the cell membranes. Such complex and dynamical nature of cell membranes provide experimental uncertainty. Chemical and physical specifications of cell membranes are accessed to reduce this issue (Castellana and Cremer, 2006, Chan and Boxer, 2007).

In 1964, phospholipid vesicles were observed for the first time by Bangham. These are super molecular assemblies which can be effectively synthesized with different sizes, compositions and shapes.

Physical properties of the native cell membrane are similar to lipid vesicles in which the external environment is separated from the cellular interior through protein-rich and lipid-dynamic boundaries (Tanaka and Sackmann, 2005). Self-assembled and supported bilayer lipid membranes were developed by McConnell and Tamm (1985). Supported lipid bilayer (SLB) membrane is a platform which provides artificial cell membranes to probe the possible structural changes in the surface membrane through various optical characterization techniques (Liu et al., 2011a). SLBs have been manipulated using coated solid support, membrane associated molecules, coupled microfluidic and many microarray techniques. These advances lead towards membrane mimicking systems based on SLB for sensing membrane reactions, separating membrane species and intracellular signaling (Nam et al., 2006, Lee and Nam, 2012). To stimulate cell membrane-based processes, molecular interactions and subcellular lateral heterogeneity should be monitored with high sensitivity and reproducibility. Nanomaterials have garnered great interest because of their dimension, shape and compositional and size-dependent properties compared with biological structures and molecules, which allow the complete investigation of many biological phenomena at atomic and subcellular level. Many recent advances have been made in synthesis protocols for targeted nanostructures (Henzie et al., 2012).

Current progress in the synthesis methods for targeted nanostructures leads towards the development of new methodologies and platforms that are designed for extracting the useful information from many complex biological processes. The variation in different factors such as shape, composition, size and architecture leads toward enhancement of many physiochemical properties including luminescence, superparamagnetism, higher signal to noise ratio, enhanced surface plasmon resonance and higher surface to volume ratio (Nel et al., 2009). With the use of plain lipid structures as solid support, they can be transformed into various hybrid structures. Solid-lipid hybrid nanostructures exhibit new properties and structures which cannot be achieved with a single component. Preparation of solid-lipid hybrid nanostructures permits the formation of an artificial cell membrane environment at the nanomaterials surface. This creates nano-surfaces on the cells, having enhanced chemical and physical properties, and results in developing a possible interface in living/non-living platforms (Lee et al., 2012).

Hybrid systems have wide applications in biosensors for the detection of many molecular reactions that are occurring at surface membrane level. Development of new processes based on unique properties such as lateral fluidity and electrical resistances is still challenging. Lipid bilayer hybrid nanomaterials provide many exciting opportunities such as sensitivity, controllability and spatial resolution. However, many synthesis methods for formation of hybrid structures have not been studied yet. The settlement of this critical issue for practical and reliable use of nanostructure-based solid-lipid hybrids in biological and materials science is still under discussion (Jesorka and Orwar, 2008, Bally et al., 2010).

In this chapter, we review the preparation techniques, design and biomedical applications of solid-lipid hybrid nanostructures. We will mainly focus on the biomedical applications of the solid-lipid hybrid nanostructure for detection of the interaction of biomolecules with cognate membrane receptors for investigation of all the possible intracellular processes occurring at the membrane junctions.

## 5.2 Need of Solid-Lipid-Based Nano-Systems

Solid lipid nanoparticles (SLNs) were presented for the first time in 1991 and they introduced an alternative nanocarrier system for traditional available colloidal carriers including liposomes, emulsion, polymeric micron and nanoparticles. Solid lipid hybrid nanoparticles attracted interest as novel drug carriers for biomedical applications in many alternative carrier systems. SLNs are colloidal vesicles in the micron range varies from 50–1,000 nm and composed of many physiological lipids immersed in aqueous or water-based surfactant solutions (Ugazio et al., 2002, Mehnert and Mäder, 2012). SLNs are preferred to use for improvement of pharmaceutical performance due to their small size, large surface area, phase interactions at interface and high drug loading capacities. To overcome limitations related with the liquid phase of the oil droplets, solid-lipid was replaced with the liquid-lipid phase that transformed gradually into SLNs (Figure 5.1) (Müller et al., 2000).

Reasons for the need of lipid-based system include following:

- Lipids reduce the plasma profile variability
- Through oral route of administration, poorly water-soluble drugs show increased bioavailability

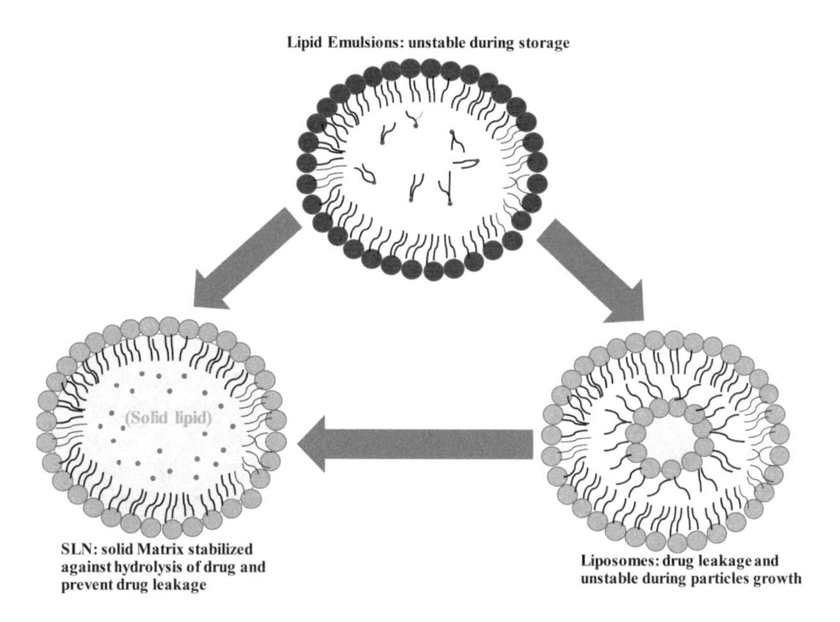


**FIGURE 5.1** A schematic representation and advantages of SLN over liposomes and lipid emulsions. (Modified with permission from Wang et al., 2014.)

- Better lipid excipients characterization
- Improved ability to resolve key issues of synthesis and technological transfer along with their characterization
- Act as an active colloidal carrier system to a variety of materials and polymers that is like oil in water emulsion process for parenteral nutrition
- Solid lipids provide low toxicity, good biocompatibility, physical stability and better lipophilic drug delivery
- At room temperature, solid state provides high physiological stability and tolerated lipid components
- Controlled targeted drug delivery
- Enhanced and high drug contents
- Easy to sterilize and scale up
- Better control over kinetics of released encapsulated compounds
- Chemical protection of incorporated labile components
- Easy-to-synthesize biopolymeric nanoparticles
- No requirement for solvents
- Long-term high stability

Potential disadvantages of solid-lipid hybrid nanomaterials include high water contents (70.0–99.9%) in solid-lipid dispersions, drug repulsion after polymeric transitions, unpredictable gelation tendency, anticipated dynamic of polymeric transitions and poor drug loading capacities (i.e., poorly soluble drug in lipid melt, polymeric state and lipid matrix structure) (Mukherjee et al., 2009). If the lipid matrix is composed of similar molecules (tripalmitin or tristearin), this results in the formation of crystals with some defects. To resolve these issues, a drug is usually incorporated between lipid layers or fatty acid chains or in imperfect crystals (Figure 5.2). A huge amount of drugs are not accommodated in the highly ordered crystal lattice. Hence, complex lipids are more sensible for drug loadings (Jenning et al., 2000).

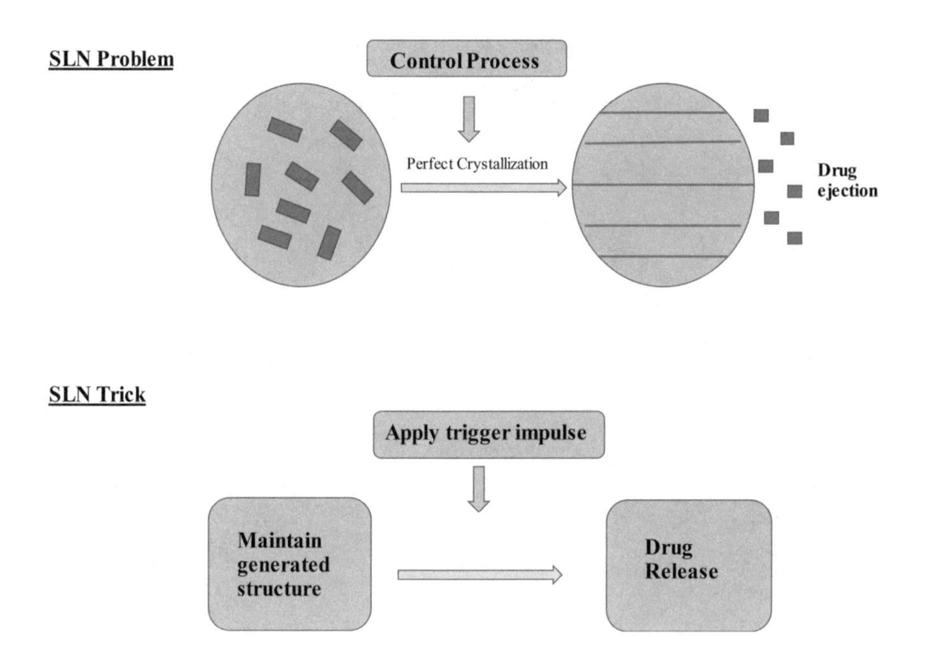


**FIGURE 5.2** Schematic of triggered release of drug from the SLN. (Reproduced with permission from Müller et al., 2002.)

## 5.3 Lipids: Structure and Classification

Study of lipids revolutionized the research field because of their use in multiple biological applications including pathology, physiology and cell biology, and now nanotechnology is becoming more important. Lipid pathways and active metabolites facilitated the promising approach for quantification, pathway and detection mechanisms for the reconstruction of lipids, genes and proteins at a biological level (Fahy et al., 2011). A lipid can be defined as group of organic compounds which are insoluble in water but soluble in organic compounds. The presence of chemical features is not restricted to a few molecules; they are present in many molecules such as sterols, sphingo lipids, fatty acids, terpenes and phospholipids. Lipids are based on the collection of extremely heterogeneous molecules from a functional and structural stand point; surprisingly, they are not significantly the same with regard to organization, scope and current classification schemes (Fathi et al., 2012). Along with sugar, protein and nucleic acid, lipids are considered a fourth major fundamental group of biomolecules. For different cells/cell organelles, lipid compositions can vary considerably. Lipids have been classified into eight major classes: glycerolipids, fatty acyls, polyketides, saccharolipids, sterol lipids, sphingolipids, glycerophospholipids and prenol lipids. Each class is divided further into subclasses. Cells consist of various types of lipids. However, three major groups that can be distinguished are sphingolipids, glycerophospholipids and cholesterol (Figure 5.3) (Müller et al., 2000, Fahy et al., 2005).

*Glycerophospholipids:* Most of the membranes are composed of glycerol backbone, which is an important group of lipids. This group is linked to one phosphate and two fatty acid groups. Phosphate groups are esterified to any of the ethanolamine, alcohol choline, inositol and serine. A unipolar acyl chain is based upon hydrophobic interior of the bilayer lipid; while the head group is charged, therefore, it is hydrophilic (Cajka and Fiehn, 2014).

*Sphingolipids:* These are similar in structural to the glycerophospholipids but are composed of a sphingosine molecule that is linked with only one fatty acid. Hence this type of lipid is not used widely (Köberlin et al., 2015).

*Cholesterol:* Structure of cholesterol is different from the above mentioned two groups. Its hydrophobic portion is a rigid ring-like system, i.e., a planar and small hydrocarbon chain, while the hydrophilic

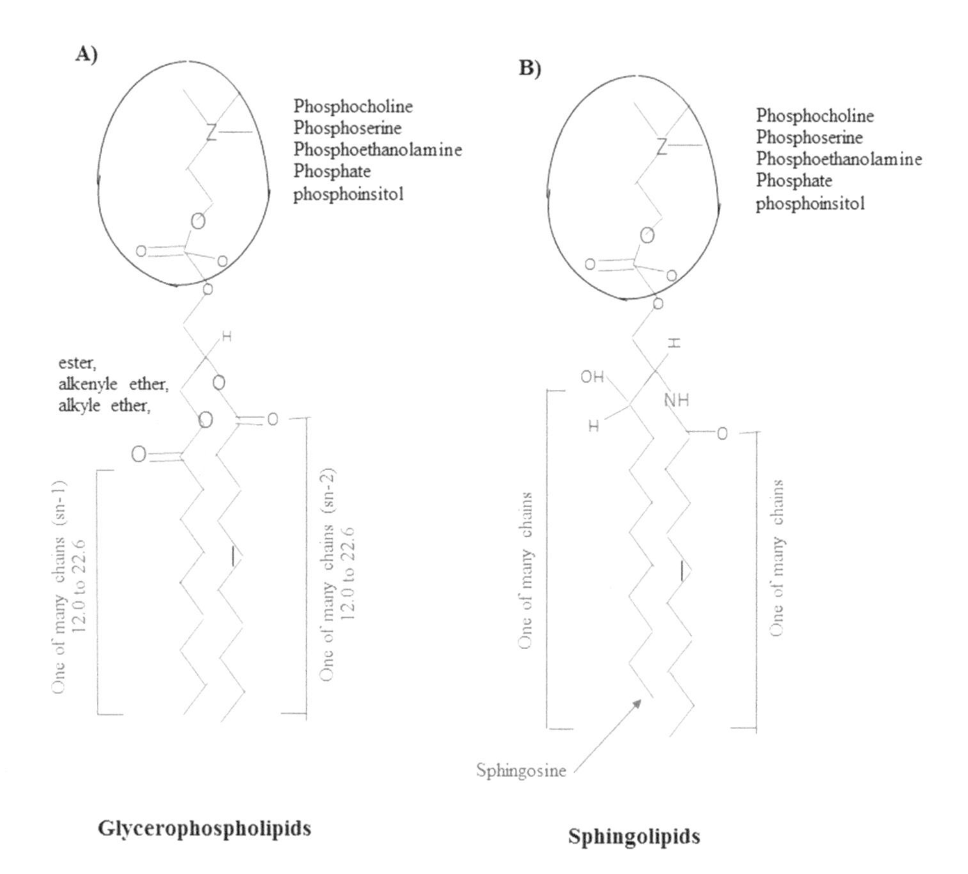


**FIGURE 5.3** Sketch of typical types of lipids, (A) Glycerophospholipids, (B) Sphingolipids. (Reproduced with permission from Coskun and Simons, 2011.)

portion is based on single hydroxyl group. Cholesterol doesn't form a bilayer and is insoluble in water. It can show a flip flop movement when it is incorporated in a lipid bilayer. Incorporation in a fluid bilayer progressively increases the chain ordering which leads toward well ordered liquid phase with increased thickness (Köberlin et al., 2015).

## 5.4 Nanomaterials Used in SLNs

Metal-based nanomaterials have been extensively investigated due to their novel properties which make them significantly contrasting to the bulk materials. Small clusters (usually consisting of <50 metal atoms) behave like large molecules but the larger ones (consisting of >300 atoms) possess the properties of bulk samples. Between these two extremes, materials are usually referred as nanomaterials with enhanced physical, chemical, electronic, magnetic and optical properties. In the recent past, lipid-based inorganic nanomaterials (i.e., gold, iron, silica nanoparticles, quantum dots and carbon nanotubes) gained attention due to high reactivity and stability for biomedical applications (Gupta and Gupta, 2005, Na et al., 2009), (Connor et al., 2005, Chen et al., 2006a). Han et al. (2002) prepared lipid protected gold nanoparticles with a diameter of 6.5 nm. Synthesized didodecyldimethylammonium bromide-based gold nanoparticles facilitated the direct electron transfer reactions in hemoglobin. Reported value for formal potential was −169 mV vs. Ag/AgCl with the stability duration of 8 months (Han et al., 2002). Li et al. prepared gold nanoparticles coated with the lipid bilayer of di-methyl-di-octa-decyl-ammonium bromide for effective delivery of DNA plasmid in embryonic cells of human kidney. They reported five times higher transfection efficiency of gold nanoparticles with the diameter of 14 (±4) nm (Li et al., 2008).

Jiang reported lipidoids coated with the iron oxide nanoparticles (50–100 nm) for optimal delivery of siRNA and DNA to culture cells in presence of external magnetic field (Jiang et al., 2013). Kong et al. (2013) reported the polymer–lipid hybrid nanoparticles-based system is composed of magnetic beads of iron and checked its effective drug responsive by applying radio frequency remotely. Reported hybrid nanoparticles showed enhanced stability in terms of the polydispersity index and crystallite size in phosphate-buffered saline. Controlled loading of iron and camptothecin in hybrid nanoparticles along with controlled drug release from these particles were demonstrated (Kong et al., 2013). Ravetti-Duran et al.(2012) prepared hierarchical macro-mesoporous silica by a co-templated method combined with micelles through a co-operative templating mechanism, stabilized with block copolymer/nonionic polysorbate surfactants. They reported macropores with size 3, 5 or 9 nm, which depend upon SLN size (250 ± 150 nm) (Ravetti-Duran et al., 2012). Pasc et al. (2011) reported the SLN templating silica with enhanced biocompatibility that was achieved by Tween 20 and cetylpalmitate through solvent injection method. They reported the microporous material contained silica beads with size range of 0.5–1.5 μm in SLN (5–400 nm) (Pasc et al., 2011). In general, inorganic nanomaterials are providing an excellent platform for SCL as a constituent.

## 5.5 Preparation Methods for Solid-Lipid Hybrid Nanostructures

For the synthesis of a solid-lipid hybrid, many preparation methods such as evaporation, low/elevated temperature-based high pressure homogenization (Silva et al., 2011), supercritical critical fluid extraction (Chattopadhyay et al., 2007), high-speed homogenization (Liu et al., 2007), spray drying (Chaubal and Popescu, 2008), solvent emulsification (Blasi et al., 2007) and ultrasonication are commonly used (Muchow et al., 2008).

*High pressure homogenization (HPH):* High-pressure homogenization is a potent and well-established method commonly used for the synthesis of solid-lipid hybrid nanoparticles due to improved product stability, absence of organic solvents and improved drug-loading capacity. High-pressure homogenization involves two processes, i.e., hot homogenization and cold homogenization, for large-scale production of SLNs. In high-pressure homogenization, high pressure (100–200 bar) is applied to raise the fluid in the small gap homogenizer. Sub-micron sized particles can prepare through HPH but the high temperature and pressure conditions are challenging for different applications (Liu et al., 2007, Wang et al., 2008).

*Hot homogenization:* Hot homogenization of lipids usually occurs at high temperature ranges compared to their melting points. In hot aqueous surfactants, melted drug-loaded lipids are dispersed through a mixing device to form pre-emulsions as shown in Figure 5.4. At a higher temperature, particle size becomes lesser due to reduced viscosity. The hot homogenization process has three major problems, i.e., drug penetration into aqueous phase, complicated crystallization steps toward super cooled melts of the nanoemulsions and temperature-dependent transformations of the drugs (Dingler and Gohla, 2002, Souto and Müller, 2006).

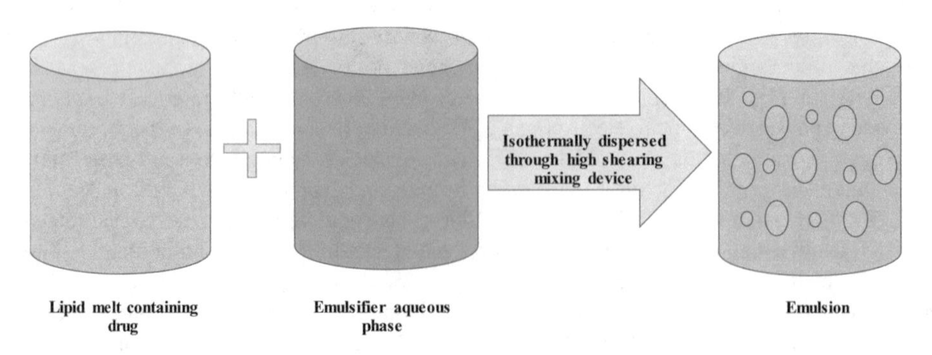


**FIGURE 5.4** A diagrammatic representation of hot homogenization process. (Modified after permission from Ganesan and Narayanasamy, 2017.)

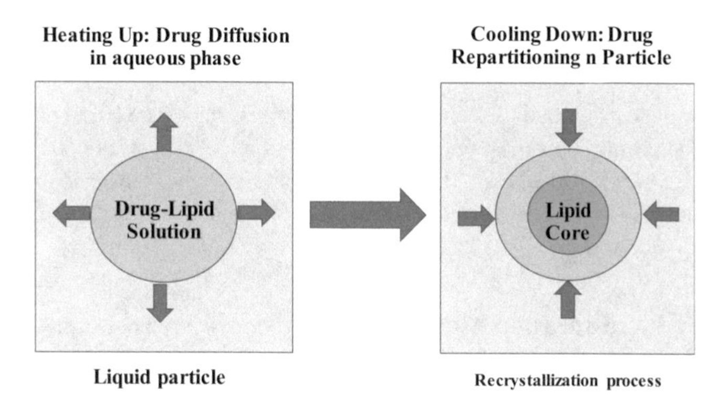


**FIGURE 5.5** Illustrative redistribution of drug during the heating and cooling phase of SLN formation in hot homogenization method. (Modified after permission from Müller et al., 2002.)

*Cold homogenization:* In cold homogenization, a drug is usually dispersed in melted lipid and then cooled down through dry ice and liquid nitrogen rapidly. Nanoparticles formed through milling have a size ranging from 50–100 nm in the dispersible cold surfactant phase for the preparation of pre-suspensions. To breakdown the nanoparticles into solid-lipid nanoparticles, ambient pressure is used (Pardeike et al., 2009). A schematic of drug redistribution during cooling and heating phases is shown in Figure 5.5.

*Supercritical fluid extraction of emulsions:* Supercritical fluid extraction is a novel approach for the synthesis of solid-lipid hybrids because of commonly used carbon dioxide as a supercritical fluid that is non-dissolvable for numerous drugs, that's why as alternative method named as a supercritical antisolvent precipitation is commonly used. For solvent extraction from the emulsions, supercritical fluid (for example, carbon dioxide0 is commonly used (Chattopadhyay et al., 2007). This technique is purely based on the emulsion diffusion method to modify the processing steps of supercritical fluid to remove the need of organic solvents from the obtained nanosuspensions. In this way, flexibility of the synthesized nanoparticles through emulsion systems can be combined with high scalability and efficiency of supercritical fluid extraction. A short processing duration is required to facilitate more optimized, consistent particles with narrow size distribution. During extraction, the interaction between the lipid and supercritical fluid leads towards the suppression of melting points and plasticization with different physical properties and structures (Chattopadhyay et al., 2007, Campardelli et al., 2013).

*Spray drying:* Spray drying acts as a substitute process to lyophilization method which results in the production of many pharmaceutical products from aqueous SLN dispersions. The spray drying process is considered a cost-effective process, but it is not used for lipid production. Particles can agglomerate in the spray drying process because of the involvement of shear forces and high temperature ranges. Suitable reported melting point for lipids in previous studies is 70°C for spray drying process (Freitas and Müller, 1998b, Okuyama et al., 2006). In the spray drying process, aqueous dispersion is converted into dry powder which can be stored for a longer period. Granules formed after redispersion can be acceptable according to their distribution and toxicity for the intravenous administration. Hence, before drying, physiologically acceptable excipients including methanol, ethanol and carbohydrates were added in SLN dispersions. Particle size depends upon the redispersion medium, type of spraying, carbohydrates and chemical nature of lipid phase (Lee et al., 2011).

*Evaporation/solvent emulsification:* In an organic solvent, the lipid is dissolved during the solvent emulsification process. An emulsion contained surfactant in an aqueous phase is formed.

In solvent emulsification process, the lipids are dissolved in a water immiscible organic solvent. From the emulsion, removal of the solvent is done under reduced pressure. Dispersion of nanoparticles through evaporation is done in the aqueous phase. Particle size varies according to the use of different surfactants. Solvent emulsification is widely used because of non-thermal stress in prepared particles; however, use of an organic solvent restricts the process (Liu et al., 2011b, Battaglia et al., 2014).

## 5.6 Lipid Nanocarriers: Lipid Nanoparticles with Solid Matrix

In 1990, solid–liquid nanostructures (SLNs) were presented as a substitute carrier system for polymeric nanoparticles, emulsions and liposomes. On average, SLNs range in size from 40 to 1,000 nm (Pardeike et al., 2009). SLNs are composed of a solid phase with (0.1–30% w/w) that is dispersed in an aqueous phase. To enhance stability, different types of surfactants with a concentration of about 0.5–5% are usually used (Pardeshi et al., 2012).

### 5.6.1 Active Drug Incorporation Methods in Solid-Lipid Hybrid Nanostructures

Drug loading, particle size, long-term stability and behavior of release are the factors that can be affected by the type of surfactants and lipids. For synthesis of solid-lipid hybrid nanostructures, lipids such as waxes, triglacerides, monoglycerides, steroids and diglycerides are used. Use of prepared solid–liquid hybrid nanostructures in hydrophobic and hydrophilic drugs depend on the synthesis methods (Blasi et al., 2007). SLNs are considered promising candidates for many biomedical applications because of their high-scale production, good release profile, low cost, excellent physical stability, reliability and biodegradability. A major disadvantage of SLNs include dynamics of polymorphic transitions, lipid particle growth, tendency to gelation and low incorporation rate of solid-lipids due to their crystalline structure (Kim et al., 2005, Wang et al., 2012). Based on SLNs production, there are three promising drug incorporation models, i.e., core-shell model (enriched drug core), core-shell model (enriched drug shell) and solid solution model (Figure 5.6). Drug loading capacity is affected by the miscibility of lipid/drug melt, drug solubility in lipid melt, physical and chemical properties of solid-lipid matrix and polymorphic state of the lipid-based materials (Lima et al., 2013, Wang et al., 2014) (Table 5.1).

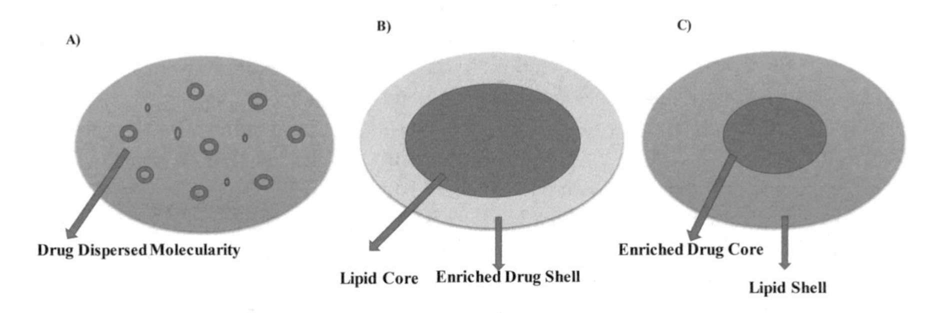


**FIGURE 5.6** Schematic of drug incorporation models in SLN. (Modified from Naseri et al., 2015.)

**TABLE 5.1**

Drug Incorporation Models

| Core–Shell Model (Enriched Drug Shell) | Solid Solution Model | Core–Shell Model (Enriched Drug Core) |
|---|---|---|
| Lipid core formation at specified range of temperature | No use of drug solubilizing surfactants | Precipitate formation of drug in the melted lipid |
| Concentration of drugs in the surrounding membrane | Drug lipid strong interaction | Formation of enriched drug core |
| Formation of enriched drug shell model through hot homogenization technique | Formation of solid solution model through cold homogenization technique | Supersaturation of drug through dispersion cooling in dissolved lipid |
| Repartitioning of drug lipid phase through cooling process | Drug dispersion in lipid matrix | Recrystallization of lipid because of further cooling |

*Source:* Das and Chaudhury (2011) and Beloqui et al. (2014).

### 5.6.2 Structural Classification of Solid-Lipid Hybrid Nanostructures

Lipid-based nanoparticles are broadly categorized into nanostructured lipid carrier (NLCs) and SLNs. At ambient temperature, NLCs are considered as modified SLNs for next-generation lipid nanoparticles because the lipidic phase consists of both liquid (oil) and solid (fat) (Doktorovova et al., 2014). A formless matrix is produced by mixing liquid and solid phases to improve loading capacity and stability. NLCs are classified into three major classes (Figure 5.7) (Jaiswal et al., 2016). Class I is usually referred to as the imperfect type in which liquid and solid fats are heterogeneously mixed in different lipid structures (Araujo et al., 2010). The imperfect lipid matrix presents a structural gap between chains of triglyceride fatty acids and thus increases drug's ability to enter into the matrix. Class II is non-crystalline matrix which possesses no crystalline structure and prevents all loaded drug expulsions. Crystals of this type are formed due to cooling processes in the presence of specific lipid mixtures. The third class is referred to as multiple type, which has greater liquid-lipid solubility (Selvamuthukumar and Velmurugan, 2012, Cipolla et al., 2014).

### 5.6.3 Effect of Different Parameters on Quality Product of Solid-Lipid Hybrid Nanostructures

There are different parameters that can affect the physical stability of solid-lipid hybrid nanostructures. Few of these are discussed below:

*Effect of particle size:* Physical stability, drug release rate and bio-fate of solid-lipid nanostructures are greatly affected by the particle size. Therefore, size control of solid-lipid nanoparticles over a reasonable range is of great importance. According to colloidal particles' definition, formulated systems such as nanoparticles, liposomes and nano-spheres should possess a size distribution in the submicron range. Various parameters such as compositional formulations (drug incorporation, surfactant/surfactant mixtures and lipid properties) and conditions during synthesis protocols (cycle number, lyophilization, sterilization, pressure, time and temperature) can affect the size of prepared solid-lipid nanoparticles (Mehnert and Mäder, 2012). Different synthesis methods are used to obtain different particle size i.e., a particle size smaller than 500 nm can be prepared through hot homogenization and narrow size distribution of particles can be compared with the particles obtained through cold homogenization. Reduced polydispersity index values and low mean particle size are reported at hot homogenization process until 1500 nm except in three to five cycles (Del Pozo-Rodríguez et al., 2007).

*Effect of lipids:* Lipids with high melting points possess increased size of SLNs through the hot homogenization process. However, many other parameters such as lipid hydrophilicity (effect of shapes on lipid crystals and self-emulsifying properties), lipid crystallization and surface area are critically different for other lipids. With an increase in lipid contents from 5–10%, broad size distribution of the large particle is usually reported (Chen et al., 2006b, Mehnert and Mäder, 2012).

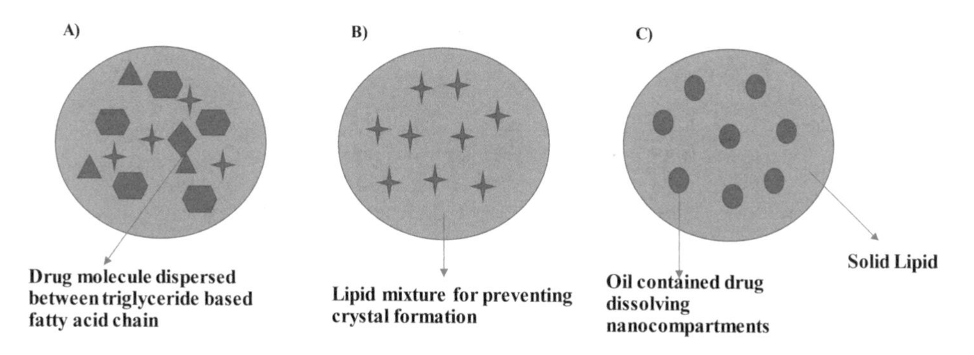


**FIGURE 5.7** Different structures of NLC: (A) Imperfect (B) formless (C) multiple type. Adequate gap is required in NLC to overcome the limitation of SLN. (Modified from Selvamuthukumar and Velmurugan, 2012.)

*Effect of emulsifiers:* Generally, small particle size was observed at a higher concentration of surfactant/lipid ratio. During storage, increased particle size is observed at decreased concentration of surfactant. Surface tension between particles is reduced due to the use of surfactants (Mehnert and Mäder, 2012).

### 5.6.4 High-Concentrated Solid-Lipid Nanoparticle Dispersions with Enhanced Physical Stability

Solid-lipid hybrid systems are composed of oily lipids, solid-lipid, solvent/water and emulsifier. Lipids may be partial glycerides, triglycerides, fatty acids, waxes and steroids. For stabilization of lipid dispersions, various emulsifiers/co-emulsifiers in combination with Lecithins, F 127, Pluronic F 68 and Tweens have been used (Müller et al., 1996). Emulsifiers can be used effectively to minimize the agglomeration of particles. Various surfactant and lipid-based materials used in SLN formulations have been listen in the Table 5.2.

Similar to other nanoparticle suspensions, the suboptimal choice of concentration and stabilizer results in aggregation of SLN dispersions during their long-term storage. By exchanging different types of surfactant, physical stability of SLN dispersions can be enhanced. Higher concentration of SLN dispersions showed enhanced physical stability for storage purposes while the lower concentration of dispersion revealed the aggregation problems. Physical stability of the SLN's dispersions have been analyzed by measuring the particle size (laser diffraction and photon correlation spectroscopy), charge (zeta potential) and thermal analysis (differential scanning calorimeter). Westesen et al. reported that optimized aqueous SL dispersions possessed the physical stability of more than one year (Westesen et al., 1997, Westesen, 2000). But zur Mhlen reported the stability of SLN dispersions prepared from trobehenate and glyceryl palmitostearate of up to three years (zur Mühlen et al., 1998). Average diameter of the solid-lipid hybrid nanostructures was in the range of 160–220 nm for the specified investigated duration. Müller and Freitas investigated the effect of temperature and light on the stability of SLN dispersions composed of 1.2% poloxamer 188 and 10% tribehenate. They observed that growth of particles might be induced through kinetic energy (temperature or light) to the system. Gelation of the system under artificial light can be achieved within seven days of storage, within three months during daylight and after four months in darkness. Gelation was done with the decrease in zeta potential from 24.7 to 18 mV. Effect of storage temperature on particles size was analyzed (Freitas and Müller, 1998a). These studies confirmed that physical stability of unstable SLN formulations can be improved through optimal storage conditions. Reported data confirmed the time for SLN-based calixarenes around one year. In another study, Shahgaldian and co-workers reported the effect of ionic strength on physical stability. They observed the strong destabilization due to sulphate ions in the SLN dispersions. To achieve more physical stability, SLN dispersions can be lyophilized or dried apart from optimal storage conditions. Melting points reported in case of spray drying for lipid matrix was >70°C (Shahgaldian et al., 2003).

**TABLE 5.2**
Material Used for SLN Formulations

| Lipids | Products |
|---|---|
| Oils | Miglyol 812 (capric caprylic triglyceride) and oleic acid |
| Triglycerides | Tristearin,, trimyristin, tricaprin, softisan 142 (hydrogenated coco glycerides) |
| Emulsifiers/coemulsifiers | Sodium cholate, taurocholic acid with dioctyl sodium sulfosuccinate, sodium salt, bytyric acid, butyric,<br>Egg lecithin (lipoid E80), Soybean lecithin (lipoid S100, Lipoid S75), phosphatidylcholine (epikuron 200, epikuron 170) |
| Hard fat type | Cetyl palmitate, decanoic acid, stearic acid, behenic acid, palmitic acid<br>witepsol E 85, witepsol H 35, Witepsol W 35,<br>Glyceryl behenate (compritol ATO 888), glyceryl monostearate (lmwitor 900) |

*Source:* Müller et al., 1996.

### 5.6.5 Biodistribution of Solid-Lipid Hybrid Nanostructures Through Different Routes of Administrations

*In vivo* distribution of SLN depends on route of administration. Interaction of the biological environment with SLN depends on enzymatic and distribution processes (biological materials adsorption on surface of particles and desorption of components of SLNs into its components).

*Oral administration:* Controlled release of solid-lipid nanostructures are enabled to bypass the intestinal and gastric degradation of encapsulated drug along with their transport and uptake through intestinal mucosa. In gastrointestinal fluids, stability of colloidal carriers for prediction of suitability of oral administration is important. Oral administration of SLNs may involve traditional dosage forms (i.e., pellets, tablets and capsules) or aqueous dispersions. To enhance bioavailability, SLNs in the form of powders/granulates can be employed into soft or hard gelatin capsules, incorporated into pellets or compressed into tablets. For tablets, aqueous dispersions of SLNs can be employed in granulation processes in the form of granulation fluids. Spray dried SLNs can alternatively be added to powder mixtures and tablets (Vrignaud et al., 2011). In extrusion processes, aqueous dispersions of SLNs can be used as wetting agents. SLNs for peptide drugs through oral drug delivery gained much attention in the pharmaceutical industries. Therapeutically active peptides (i.e., insulin, somatostatin, cyclosporine A, calcitonin and insulin), protein drugs (i.e., lysozyme and bovine serum albumin) and antigen proteins (i.e., malaria antigens and hepatitis B) have been incorporated widely in SLNs (Yang et al., 1999). Prolonged plasma levels and increased bioavailability have been reported after oral administration of lipid-based nano-dispersions comprised of cyclosporine in animals. Müller et al.(2006) formulated lipid nanoparticles of 157 nm size with 2% cyclosporine. Improved bioavailability due to fast absorption rate (>1000 ng/ml with 96.1% encapsulation rate in pigs) as compared to commercially available cyclosporin A formulations were reported (Müller et al., 2006). Zhuang et al. (2010) developed optimized lipid nanostructure carriers through high-pressure homogenization for the investigation of bioavailability of poorly soluble drug named as vinpocetine. Reported encapsulation efficiency of vinpocetine loaded NLCs was $94.9 \pm 0.4\%$. Improved bioavailability up to 32.2% of vinpocetin NLCs through oral administration in wistar rats was reported, as compared to commercially available vinposetin suspensions (Zhuang et al., 2010).

*Rectal administration:* In some cases, when rapid pharmacological action is required, rectal or parenteral administration is usually preferred. For pediatric patients, the conventional rectal route of administration is mostly followed. At the same doses, higher therapeutic efficacy and plasma level were reported in rectally administrated drugs comparison to the intramuscularly or orally administrated drugs (Joshi and Müller, 2009). In literature, few reports of rectal administration of drug are reported. Sznitowska et al. (2001) reported the mechanism of incorporation of diazepam on SLN through rectal administration for rapid pharmacological action. They investigated the prepared SLN dispersions on a group of rabbits to evaluate the bioavailability. They observed that the solid-lipid matrix was not advantageous at body temperature for diazepam rectal delivery. They recommended the use of a lipid that melted at body temperature (Sznitowska et al., 2001). This area needs more investigation to understand the benefits of rectal administration. PEG coatings appear as a more revolutionizing approach towards rectal delivery for increased bioavailability.

*Parenteral administration:* In market, proteins and peptides are usually available for parenteral use. In the gastrointestinal track, the conventional oral route of administration is not suitable due to enzymatic degradation. Increased bioavailability of an incorporated drug through parenteral applications reduces many side effects. Use as a cure for central nervous system disorders including AIDS, brain tumors and psychiatric and neurological disorders are constrained because of the inability of active drugs to pass through the blood brain barrier (BBB). Hydrophilic colloidal coatings can improve tissue distribution and transport mechanism through BBB (Yang et al., 1999). Fundaro et al. (2000) synthesized the doxorubicin-loaded non-stealth- and stealth-based solid-lipid nanocarriers. They preferred PEG 2000 for stabilization and modification of stearic acid SLN. After 24 hours of intravenous administration, they observed a higher concentration of non-stealth in comparison to stealth nanoparticles. After the administration of a detectable concentration of stealth SLN, doxorubicin was determined. In such cases, the drugs were trapped easily through BBB. Stealth SLN demonstrated a high distribution volume, low clearance and high area under the curve (AUC) as compared to non-stealth SLN. They studied the higher concentration of doxorubicin in brain and lung tissues with stealth SLN whereas low concentration was reported in the spleen, kidney, liver and heart with non-stealth SLN. Increased drug availability was reported in tissues of non-reticuloendothelial (RES) system because of decreased uptake by tissues of RES system (Fundarò et al., 2000). Research has suggested that a coated SLN carrying anticancer drugs with hydrophilic molecules can effectively reach in cancerous cells and solid tumors as compared to healthy tissues (Zara et al., 2002).

*Topical administration:* Tremendous use of SLN due to the attractive characteristics of colloidal systems is suitable because of low toxic and non-irritant lipids for many skin applications. Topically applied lipid hybrid nanoparticles showed enhanced efficacy of UV molecular blockers with minimum side effects, increased hydration, drug release modifications, wrinkle smoothing, reduced pigment effect and protection from various chemical compounds. In many cases, incorporation of SLN dispersions in gel/ointment can be necessary to for topical administration. Jain et al. (2010) reported that miconazole nitrate formulation possessed enhanced retention in hydrogel of SLNs as compared to miconazole nitrate hydrogel and miconazole nitrate suspensions (Jain et al., 2010). They performed *in vivo* studies on infected rats with candida species. They observed that for treatment of candidiasis, miconazole nitrate-loaded hydrogel of SLNs were more efficient. SLN formulations with active compounds such as clotrimazole (Souto et al., 2004), retinol (Jenning et al., 2000), triptolide (Mei et al., 2003), ascorbyl palmitate, tocopherol acetate (Wissing and Müller, 2001), nonsteroidal antiandrogen RU 58841 and phodphyllotoxin (Chen et al., 2006b) for topical applications have been studied from the last few years.

## 5.7 Formation and Structural Properties of Solid-Lipid Hybrid Nanostructures

By the mechanical support of lipid films with nanostructured substrates, the synthesized hybrid nanoparticles can be utilized for hybrid structure determination because of duplicate lipid layers-maintain morphologies until they are in proximity of the close surface. Lipid molecules are subsequently rearranged after adsorption through self-assembly on nanostructures surfaces. Persistence of biological functions after the integration of proteins and lipid membranes with the nanostructured surfaces induce promising effects on the nanotopographic features. Selection of a suitable method can affect the rational design, formation kinetics, co-existence of gel-fluid region, dynamics and stability of lipid molecules (Lee et al., 2013).

### 5.7.1 Modified Lipid-Bilayer Nano-Topographic Substrates

Using a conventional vesicle fusion method, planar SLBs can be formed. The incubation of unilamellar vesicles having the diameter of 100 nm with the hydrophilic solid support shows deformed interaction. Induced support stress may result in enhancing the adsorption of these vesicles (Richter et al., 2006, Jonsson et al., 2007). Formation of a continuous lipid bilayer of the vesicles at a specific range of surface density will result in critical vesicular coverage (Figure 5.8). For the formation of lipid nanostructure hybrids, the vesicle fusion method is significantly used. However, the kinetics and physics behind the solid-lipid hybrid nanostructures are quite different form the planer SLBs. Formation of lipid membranes and loss of structural integrity is due to an induced degree of surface curvature (Pfeiffer et al., 2008, Roiter et al., 2008).

Atomic force microscopy confirmed that the supported silica substrate nanoparticles with diameter 1.2–22 nm cut the bilayer of L-a-dimyristoyl phosphatidylcholine, however formation of an L-admyristoyl phophatidylcholine membrane supported particles with narrow pore size was reported (Jonsson et al., 2007). Unfavorable bending energies of the membranes are responsible for the formation of incomplete bilayers (Roiter et al., 2009). Cryo-electron microscopy presented detailed rapture mechanism of unilamellar vesicles, which form bilayer structures on curved surfaces of silica nanoparticles. Liposomes with low negative, neutral and positive charge were used for SLB formation on the surface of silica nanoparticles, whereas highly negative net charged lysosomes remained un-raptured. Small sized unilamellar

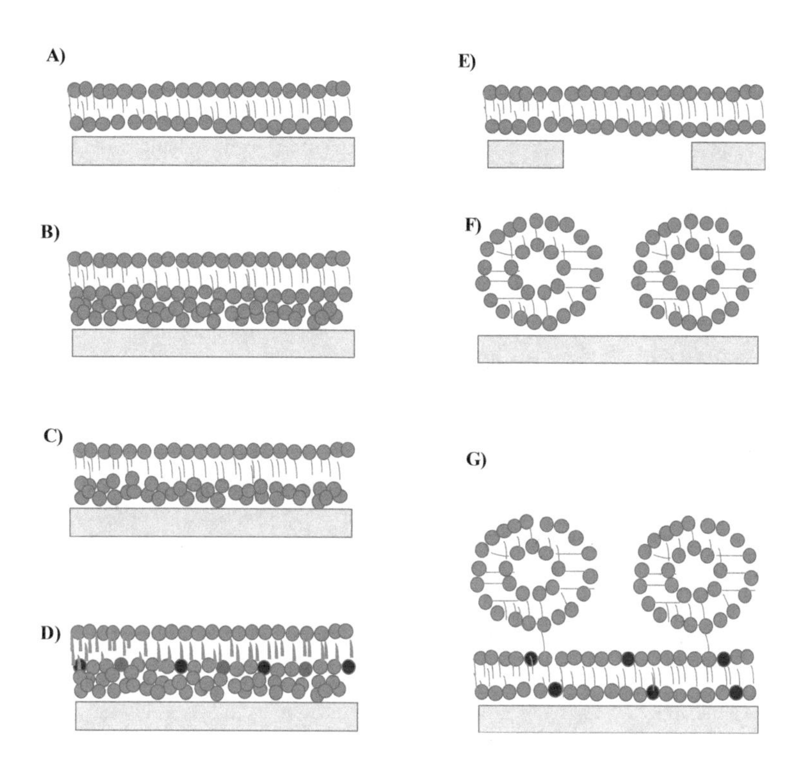


**FIGURE 5.8** Surface-confined membrane models of lipid bilayers: (A) Solid supported, (B) polymer aided, (C) hybrid (monolayer + Au or Si), (D) tethered, (E) freely suspended, (F) vesicular supported. (Modified from Richter et al., 2006.)

vesicles were found to have promising effects on the possible kinetics during the transformations of bilayer vesicles (Mornet et al., 2005, Baciu et al., 2008).

## 5.7.2 Lipid Bilayer Formation on Nanopatterns

Lipid bilayers are assembled as fluids on specified materials, typically, at the surface of silica. Formation of sliding lipid bilayer molecules can be used for diffusion barriers. Boxer group extensively studied the fabrication methodologies for lipid bilayer patterns through different patterning techniques (Hovis and Boxer, 2000). To confine the molecular motions, incorporation of small-size unilamellar vesicles on patterned substrate leads toward the lipid membrane formations, which is used for the confinement of molecular motion. Recent advances in patterning techniques with different materials provide favorable chances for the fabrication of nano-patterns that greatly enhance the motion of nanometric membrane species (Groves and Boxer, 2002).

To fabricate the nanopattern of chromium, electron beam lithography is commonly used which disrupts the lipid bilayer completely. Therefore, it serves as a passive barrier which promotes the lateral diffusion of the lipids. A chromium barrier that is formed through electron beam lithography is a few nanometers high and hundreds of nanometers wide (Nair et al., 2011). For interfacing the cell membrane receptors with the living cells, spatial organization of the barriers plays an important role. These methods are used for fabrication of patterns in small areas of micrometers with the substrate exposure to a high vacuum environment. However, limits the multi components and direct patterning of biomolecules (Kam and Boxer, 2000).

Din pen nanolithography (DPN) is a versatile scanning-based probe lithography which delivers the ink molecules from the sharp tip directly by the atomic force microscopy to the substrate. For fabrication of multiplexed patterns of molecular prints and molecules, DPN techniques have been widely used. For patterning of lipid-based molecules, poly(diallydimethylammonium chloride) as inking molecules is introduced in DPN which acts as bio-functionalization site and diffusion barrier in supported lipid bilayers (SLB). The resulting pattern of poly(diallydimethlammonium chloride) provides a template for inhibition of lipid diffusion and protein adsorption (Lenhert et al., 2007, Lohmüller et al., 2011). In some cases, disjointed lines of poly(diallydimethlammonium chloride) were formed and these structures can restrict the cell diffusion study to understand the behavior of membrane bounded molecules during non-Brownian diffusion (Salaita et al., 2007). A major part of the cell membrane such as phospholipid can be transferred directly to form the pattern of fluid lipid films through the DPN method. With appropriate scanning speed and humidity conditions, thickness of deposited layers of lipid can be controlled to reduce the lateral resolution up to 100 nm with single bilayer thickness (Jang et al., 2010).

Direct and massive delivery of multi-component lipophilic and lipids through DPN to a solid surface (on a scale of multiple lengths) allows the heterogeneous mimicking of cell membranes at the micro- and nano-levels. The tip of the AFM is usually used to remove the pre-existing thin films, named as nanoshaving lithography. The shaved region is filled again with other lipid layers which result in forming the narrow lipid pattern with a thickness of 50 nm (Shi et al., 2008). In some cases, the site of protein nanoarrays is modified through copolymer block micelle nanolithography. Self-assembly of gold containing copolymer diblock micelles with a hexagonal array arrangement have been produced through air plasma treatment with a size of 5–7 nm (Lenhert et al., 2007, Sekula et al., 2008). This approach allows the controlled spacing of nanoparticles through the tuning of arrays with polymer molecular weight with complex patterning procedures. SLB-embedded gold nanoparticles were used for the modification of many biological molecules through thiol linkers. This approach is used as functionalization of ligand and proteins to lipid fluid components over large areas to fix nanoparticles, lipid-based nanoparticles or both (Narui and Salaita, 2012).

### 5.7.3 1D Lipid Hybrid Nanostructures

One-dimensional (1D) nanostructures are efficiently used in integrated functional devices because of their unique geometric features, high density array organization and highly sensitive surface bindings at nanoscale. These features make them a promising candidate for many biomedical applications. Introduction of lipid bilayer at the surface of 1D nanostructures allow the integration of many structures and involve the associated membrane functioning of biomolecules on the 1D array nanostructures (Richard et al., 2003, Zheng et al., 2005). Mechanical and chemical properties of cell membrane critically limit the synthesis of 1D nanostructures. Beyond the critical concentration of micellar, the transmission electron microscope can be used to observe supermolecular self-assembled lipids on carbon nanotubes. Adsorption at the surface of carbon nanotubes (CNTs) greatly depends on the diameter and shape of helix, ring and double helix symmetries. Non-covalent surface formation with amphiphilies is considered a promising approach for maintaining properties of CNTs and rendering them in a soluble aqueous environment (Hurtig et al., 2004, Artyukhin et al., 2005). Static lipid CNTs hybrid structures offer many advantages for fluid lipid bilayer formation on CNTs but functioning of membrane associated biomolecules is still under discussion. Self-assembled lipid molecules in fluidic 1D nanostructures on CNTs template was reported in the literature (Huang et al., 2007). The lateral lipid fluidity was tested by tuning the diameter of CNTs. For the support of hydrophilic groups on lipid bilayer, CNTs were covered in a layer-by-layer arrangement with polyelectrolytes. In a 1D configuration, electrostatically assembled polyelectrolytes were stabilized in the lipid bilayer with control over the diameter of the structures (Stamou et al., 2003). Template 1D CNTs with lipid bilayer showed structural continuity with sustained lateral mobility for the curved structure of the lipid molecules. However, their diffusion coefficients in comparison to other supported polymers, i.e., planar lipid bilayers, were two orders smaller in magnitude. Amorphous silica coated CNTs as solid hydrophilic support was suspended on micro fabricated channels at the substrate of $Si/SiO_2$. These structures provided smooth walls, ultra-narrow distribution and presented a platform for the probing of lipid mobility on curved substrates (Bolinger et al., 2004, Jesorka et al., 2011). The diffusion coefficient on tubular nanostructures of lipids was determined to be in the range of 2–10 $\mu m^2 s^{-1}$. Trends of these values can be predicted with the free space diffusion model, according to which additional space induces between lipid head groups and highly curved surfaces (Pick et al., 2005). Total significant volume of lipid vesicles that can be achieved with conventional microfluidic devices make them important for many applications. Stamou et al. (2003) synthesized the assembled molecular vessel in the specified ordered array of glass substrate through specific ligand receptor interactions and microcontact printing techniques. Immobilized lipid vesicles at the solid surface have been employed as nanoreactors for the isolation of very small-sized solution volume with products and reactants. Self-enclosed immobilized volume of vesicles through controlled triggering and mixing is considered highly beneficial but still challenging (Cisse et al., 2007, Bolinger et al., 2008).

## 5.8 Biomedical Applications of Solid-Lipid Hybrid Nanostructures

### 5.8.1 Role of SLN for Cancer Chemotherapy

Many chemotherapeutic agents have been incorporated in suitable solid-lipid nanoparticles and their *in vivo* and *in vitro* efficacy has been critically evaluated. Many studies have shown reduced side effects and improved efficiency of chemotherapy drugs. Reduced *in vitro* toxicity, enhanced drug efficacy, diversified physiochemical properties, improved stability of drugs and improved pharmokinetics are the promising features of solid-lipid nanoparticles which make them promising candidates for chemotherapeutic drug delivery carrier (Luo et al., 2006). Many limitations of anticancer compounds, including poor specificity, high incidence of active drug resistant tumor cells and normal tissue toxicity, are partially resolved through delivering them with SLNs. Removal of these colloidal particles with macrophages of reticuloendothelial system is main issue of tissue targeting drug delivery such as solid tumors and bone marrow. Lu et al. (2006) synthesized local injections of Mitoxantrone SLN with mean diameter of 62 nm and encapsulation yield of $87.23 \pm 2.16\%$ through ultrasonication method and evaluated the bioavailability of the optimized composite design on mice. An *in vitro* study revealed the cumulative release rate of $Q_{24h} = 25.86 \pm 0.82\%$, $t_{90} = (28.38 \pm 4.50)d$ and $t_{50} = (5.25 \pm 1.10)d$ while no serious toxicity of lungs and liver was reported (Lu et al., 2006). Tamoxifen is an anticancer drug used for prolonged release by incorporating it in SLN after IV administration. Huang et al. (2008) evaluated the feasibility of SLNs for the active camptothecin delivery. They obtained the controlled release of the camptothecin modified by the lipid matrix and showed high cytotoxicity as compared to the free drug for melanoma treatment (Huang et al., 2008). Tran et al.(2014) developed hyaluronic acid-coated vorinostat loaded solid-lipid nanostructures (HA-VRS-SLNs) with the narrow size distribution of (~100 nm) and negative charge (−9 mV) for the controlled targeted delivery. *In vitro* release of HA-VRS-SLNs revealed a bi-phasic pattern. Longer blood circulation time of HA shell and reduced clearance VRS rate in rats with higher bioavailability and plasma concentration was reported. They clearly revealed the potential of hyaluronic acid-coated nanoparticles as a nano-sized drug carrier for chemotherapy (Tran et al., 2014). Linalool-loaded SLNs that consisted of different lipids were used to obtain controlled release and higher encapsulation efficiency. Cell viability was determined through MTT assay. Cell viability for varying concentration of linalool is in Figure 5.9 (Rodenak-Kladniew et al., 2017).

### 5.8.2 Role of SLN for Targeted Drug Delivery

Solid-lipid nanoparticles with a size less than 50 nm can be utilized for the drug targeting. Small sized carriers favorably reduce the reticuloendothelial system uptake. Drug targeting might be enhanced due to surface modifications of SLN. Drug delivery research employed through nanoparticles and micelles have wide applications in gene delivery and ultrasonic drug (Valo et al., 2010). Nano vehicles actively deliver cytotoxic drugs with high concentration to the treatment of selective tissues with reduced side effects. Traditionally, ultrasounds are used in diagnostic fields in drug delivery through nanoparticles. The non-invasive nature of SLN along with active pharmacological agents make them a promising candidate to focus on targeted tissues. Solid-lipid-based nano-systems such as lipid micro-particles, solid-lipid nanoparticles and lipospheres have been employed widely as secondary carriers for many therapeutic proteins, antigens and peptides. In the recent past, research confirmed that under controlled and optimized condition's incorporation of hydrophilic and hydrophobic proteins could be utilized in many particulate carrier systems (Valo et al., 2010). Antigens and proteins can be adsorbed or incorporated on SLN for many therapeutic purposes and further administration through different routes.

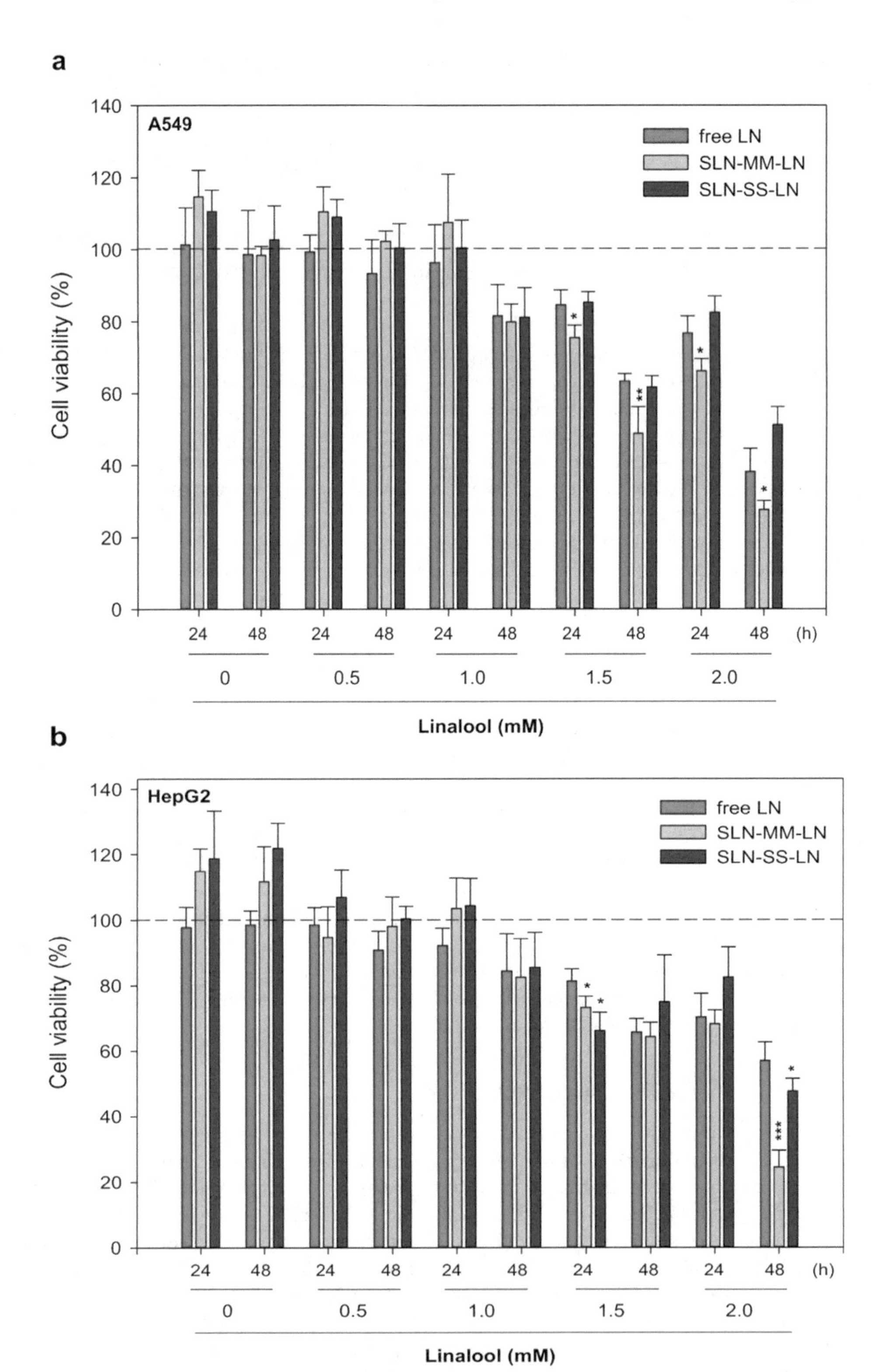


**FIGURE 5.9** Cytotoxicity studies of Linalool-loaded SLN in A549 (a) and HepG2 (b) cancer cells. Cell viability was studied for different concentrations of Linalool. (Reproduced with permission from Rodenak-Kladniew et al., 2017.)

SLN formulation avoids proteolytic degradation, great protein stability and sustained release of molecules. Incorporation of peptides such as insulin, cyclosporine, somatostatin and calcitonin into SLNs are still under consideration. Several systemic and local therapeutic applications might be anticipated such as immunization with infectious disease treatment, protein antigens, cancer therapy and chronic diseases (Cavalli et al., 1999, Müller et al., 2000). SLNs can improve drug penetration ability through the BBB and as a result can be useful for treatment purposes of central nervous system disorders. To resolve the issue of limited supply, 5-fluoro-2'-deoxyuridine drug to the brain and 3',5'-dioctanoyl-5-fluoro-2'-deoxyuridine were prepared and then incorporated in SLNs (Wang et al., 2002). For targeting the brain, surfactant-coated poly alkylcyanoacrylate-based nanoparticles were designed specifically by controlling the transport of solid-lipid matrices. Physiological characteristics of SLNs are used to address many critical issues for developing suitable formulations for brain targeting (Reis et al., 2006). Lv et al. (2009) prepared penciclovir loaded SLN with a size of 254.9 nm by double emulsion method. The reported values of drug loading, zeta potential and entrapment efficiency of prepared SLNs were 4.62%, 92.40% and –25 mV respectively. *In vivo* studies were carried out on a group of rats. A deposited amount of penciclovir in epidermic after the administration of SLNs for the durations of 2, 6 and 12 hours was noted. Increased uptake of penciclovir in dermis indicated good targeting skin effects (Lv et al., 2009). Doktorovova et al. (2014) investigated the camptothecin-loaded SLN which confirmed the mean size of 200 nm, negative surface charge (–20 mV), low polydispersity index and high association capability of camptothecin (>94%). Cell death was reported in the cytotoxicity studies against macrophage and glioma human cells, which revealed the higher anti-tumor activity of the camptothecin-loaded SLN. However, biodistribution studies were performed on rats that confirmed their positive role on brain targeting. Higher accumulation rate of camptothecin as compared to non-encapsulated drugs were observed (Doktorovova et al., 2014). Magro et al. evaluated the ligand surface modifications through the warm microemulsion process

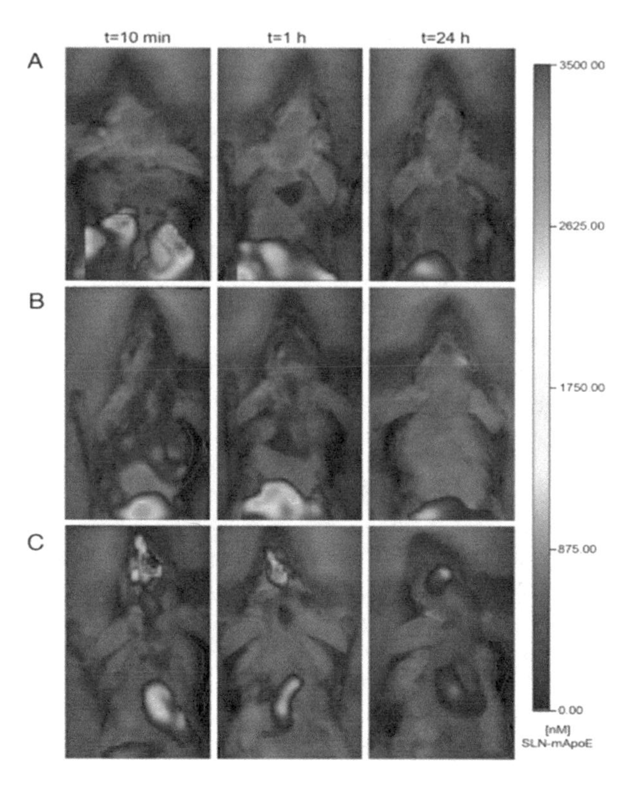


**FIGURE 5.10 (See color insert.)** Fluorescence molecular tomographic images to access biodistribution of SLN in the brain administered by different routes (A) intraperitoneally, (B) intravenously and (C) intratracheally. Images were acquired at time 10 mins, 1 hr, 24 hr. (Reproduced with permission from Dal Magro et al., 2017.)

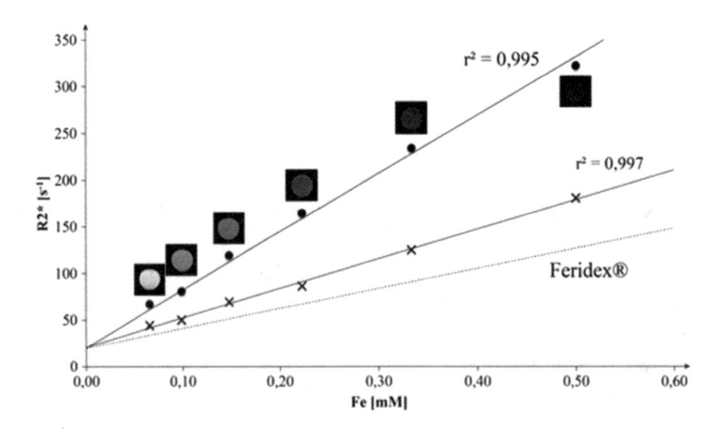


**FIGURE 5.11** Contrast enhancement studies of SLN for magnetic resonance imaging: Graph between transverse relaxation and iron concentration for negatively charged SLN (dot line), positively charged SLN (cross line) and Feridex® showed high relaxation time for both negatively and positively charged SLN than clinically used Feridex®. (Reproduced with permission from Oumzil et al., 2016. © 2016 American Chemical Society.)

of ApoE-SLN for effective BBB targeting. They investigated the bioavailability of ApoE-modified SLN through a different route of administration (Figure 5.10) (Dal Magro et al., 2017).

### 5.8.3 Role of SLN in Diagnostic Applications

In the emerging field of molecular and cellular magnetic resonance imaging, many new strategies are required for the synthesis of contrast agents which can be altered in terms of composition and size. For molecular imaging applications, relaxivity values of the contrast agents should be very high to resolve issues of low sensitivity. Lipid-based nanoparticles, including micelles or liposomes, have been extensively employed as drug carrier vehicles. Targeting ligands were incorporated in lipids with functional moiety, which allowed specific interactions with molecular makers to achieve the agglomeration of those particles at the tumor sites. Liposomal-based MRI contrast agents include liposomes entrapped paramagnetic agents (i.e., Gadolinium- dipalmitoylphosphatidylcholine (Tilcock et al., 1989), Gadolinium-dipalmitoylphosphatidylcholine-omniscan (Fossheim et al., 2000), manganese chloride (Magin et al., 1986, Koenig et al., 1988), manganese-diethylene triamine penta acetic acid (Caride et al., 1984) and gadolinium-[10-(2-hydroxypropyl)-1,4,7,10-tetraazacyclododecane-1,4,7-triacetic acid (Fossheim et al., 1999) in aqueous lumen (Unger et al., 1989). Recently, doxorubicin- and gadodiamide-based liposomes were investigated for liposomal distribution after enhanced-convection delivery to brain tumors were carried out on groups of rats (Saito et al., 2004) and mice (Mamot et al., 2004). Furthermore, liposomal carriers based on gadobutrol linked with specified ligands were investigated in v3-integrin-expressing cells followed by *in vitro* studies. Although paramagnetic-based liposomal formulations have been utilized successfully (Heverhagen et al., 2004), low relaxivity values were reported due to restricted exchange of contrast agents with bulk water. This exchange might depend on the lipid concentration, permeability of liposomal membranes, length of lipid chain and saturation level of lipid chains. Better relaxivity across the bilayer indicates the more permeability of the membrane (Unger et al., 1991). Enhanced relaxivity values of iron-loaded SLN as compared to other contrast agents used clinically such as Feridex (Figure 5.11) (Oumzil et al., 2016).

## 5.9 Conclusion

Solid-lipid hybrid nanostructures are considered as complex systems with many advantages and disadvantages in comparison to other colloidal nanocarriers. Due to drug expulsion and stability issues, new synthesis techniques for controlled designing of SLNs still present challenges. Based on organization and compositional effects of drugs and their incorporation in nanoparticles, wide collection in structural forms has been investigated. The extremely disordered lipid-based matrix structures of NLC due to stability, enhanced drug encapsulation and good profile releasing capacities made them promising candidates in many biomedical application and nano-pharmaceutical industries. Preparation of these lipid-based nanoparticles in laboratories is practicable and possible on a large scale. Additional efforts are required for the dynamic *in vitro* and *in vivo* applications of SLN on molecular phases.

## REFERENCES


Araújo, J., Gonzalez-Mira, E., Egea, M.A., Garcia, M.L. & Souto, E.B., 2010. Optimization and physicochemical characterization of a triamcinolone acetonide-loaded NLC for ocular antiangiogenic applications. *International Journal of Pharmaceutics*, 393, 168–176.

Artyukhin, A.B., Shestakov, A., Harper, J., Bakajin, O., Stroeve, P. & Noy, A., 2005. Functional one-dimensional lipid bilayers on carbon nanotube templates. *Journal of the American Chemical Society*, 127, 7538–7542.

Baciu, C.L., Becker, J., Janshoff, A. & SöNnichsen, C., 2008. Protein–membrane interaction probed by single plasmonic nanoparticles. *Nano Letters*, 8, 1724–1728.

Bally, M., Bailey, K., Sugihara, K., Grieshaber, D., Vörös, J. & Städler, B., 2010. Liposome and lipid bilayer arrays towards biosensing applications. *Small*, 6, 2481–2497.

Bangham, A.D. & Horne, R.W., 1964. Negative staining of phospholipids and their structural modification by surface-active agents as observed in the electron microscope. *Journal of Molecular Biology*, 8(5), 660–668, IN2–IN10.

Battaglia, L., Gallarate, M., Panciani, P.P., Ugazio, E., Sapino, S., Peira, E. & Chirio, D., 2014. Techniques for the preparation of solid-lipid nano and microparticles. In: A.D. Sezer, ed. *Application of Nanotechnology in Drug Delivery*. Intech.

Beloqui, A., Solinís, M.Á., Des Rieux, Ad, Préat, V. & Rodríguez-Gascón, A., 2014. Dextran–protamine coated nanostructured lipid carriers as mucus-penetrating nanoparticles for lipophilic drugs. *International Journal of Pharmaceutics*, 468, 105–111.

Blasi, P., Giovagnoli, S., Schoubben, A., Ricci, M. & Rossi, C., 2007. Solid lipid nanoparticles for targeted brain drug delivery. *Advanced Drug Delivery Reviews*, 59, 454–477.

Bolinger, P.Y., Stamou, D. & Vogel, H., 2004. Integrated nanoreactor systems: Triggering the release and mixing of compounds inside single vesicles. *Journal of the American Chemical Society*, 126, 8594–8595.

Bolinger, P.Y., Stamou, D. & Vogel, H., 2008. An integrated self-assembled nanofluidic system for controlled biological chemistries. *Angewandte Chemie*, 47, 5544–5549.

Cajka, T. & Fiehn, O., 2014. Comprehensive analysis of lipids in biological systems by liquid chromatography-mass spectrometry. *Trends in Analytical Chemistry*, 61, 192–206.

Campardelli, R., Cherain, M., Perfetti, C., Iorio, C., Scognamiglio, M., Reverchon, E. & Della Porta, G., 2013. Lipid nanoparticles production by supercritical fluid assisted emulsion–diffusion. *The Journal of Supercritical Fluids*, 82, 34–40.

Caride, V.J., Sostman, H.D., Winchell, R.J. & Gore, J.C., 1984. Relaxation enhancement using liposomes carrying paramagnetic species. *Magnetic Resonance Imaging*, 2, 107–112.

Castellana, E.T. & Cremer, P.S., 2006. Solid supported lipid bilayers: From biophysical studies to sensor design. *Surface Science Reports*, 61, 429–444.

Cavalli, R., Peira, E., Caputo, O. & Gasco, M.R., 1999. Solid lipid nanoparticles as carriers of hydrocortisone and progesterone complexes with β-cyclodextrins. *International Journal of Pharmaceutics*, 182, 59–69.

Chan, Y.-H.M. & Boxer, S.G., 2007. Model membrane systems and their applications. *Current Opinion in Chemical Biology*, 11, 581–587.

Chattopadhyay, P., Shekunov, B.Y., Yim, D., Cipolla, D., Boyd, B. & Farr, S., 2007. Production of solid-lipid nanoparticle suspensions using supercritical fluid extraction of emulsions (SFEE) for pulmonary delivery using the AERx system. *Advanced Drug Delivery Reviews*, 59, 444–453.

Chaubal, M.V. & Popescu, C., 2008. Conversion of nanosuspensions into dry powders by spray drying: A case study. *Pharmaceutical Research*, 25, 2302–2308.

Chen, C.-C., Lin, Y.-P., Wang, C.-W., Tzeng, H.-C., Wu, C.-H., Chen, Y.-C., Chen, C.-P., Chen, L.-C. & Wu, Y.-C., 2006a. DNA–gold nanorod conjugates for remote control of localized gene expression by near infrared irradiation. *Journal of the American Chemical Society*, 128, 3709–3715.

Chen, H., Chang, X., Du, D., Liu, W., Liu, J., Weng, T., Yang, Y., Xu, H. & Yang, X., 2006b. Podophyllotoxin-loaded solid-lipid nanoparticles for epidermal targeting. *Journal of Controlled Release*, 110, 296–306.

Cipolla, D., Shekunov, B., Blanchard, J. & Hickey, A., 2014. Lipid-based carriers for pulmonary products: Preclinical development and case studies in humans. *Advanced Drug Delivery Reviews*, 75, 53–80.

Cisse, I., Okumus, B., Joo, C. & Ha, T., 2007. Fueling protein–DNA interactions inside porous nanocontainers. *Proceedings of the National Academy of Sciences*, 104, 12646–12650.

Connor, E.E., Mwamuka, J., Gole, A., Murphy, C.J. & Wyatt, M.D., 2005. Gold nanoparticles are taken up by human cells but do not cause acute cytotoxicity. *Small*, 1, 325–327.

Coskun, Ü. & Simons, K., 2011. Cell membranes: The lipid perspective. *Structure*, 19, 1543–1548.

Dal Magro, R., Ornaghi, F., Cambianica, I., Beretta, S., Re, F., Musicanti, C., Rigolio, R., Donzelli, E., Canta, A., Ballarini, E., Cavaletti, G., Gasco, P. & Sancini, G., 2017. ApoE-modified solid-lipid nanoparticles: A feasible strategy to cross the blood-brain barrier. *Journal of Controlled Release: Official Journal of the Controlled Release Society*, 249, 103–110.

Das, S. & Chaudhury, A., 2011. Recent advances in lipid nanoparticle formulations with solid matrix for oral drug delivery. *AAPS PharmSciTech*, 12, 62–76.

Del Pozo-Rodríguez, A., Delgado, D., Solinís, M.A., Gascón, A.R. & Pedraz, J.L., 2007. Solid lipid nanoparticles: Formulation factors affecting cell transfection capacity. *International Journal of Pharmaceutics*, 339, 261–268.

Dingler, A. & Gohla, S., 2002. Production of solid-lipid nanoparticles (SLN): Scaling up feasibilities. *Journal of Microencapsulation*, 19, 11–16.

Doktorovova, S., Souto, E.B. & Silva, A.M., 2014. Nanotoxicology applied to solid-lipid nanoparticles and nanostructured lipid carriers–a systematic review of in vitro data. *European Journal of Pharmaceutics and Biopharmaceutics*, 87, 1–18.

Fahy, E., Cotter, D., Sud, M. & Subramaniam, S., 2011. Lipid classification, structures and tools. *Biochimica et Biophysica Acta*, 1811, 637–647.

Fahy, E., Subramaniam, S., Brown, H.A., Glass, C.K., Merrill, A.H., Murphy, R.C., Raetz, C.R.H., Russell, D.W., Seyama, Y., Shaw, W., Shimizu, T., Spener, F., van Meer, G., VanNieuwenhze, M.S., White, S.H., Witztum, J.L. & Dennis, E.A., 2005. A comprehensive classification system for lipids. *Journal of Lipid Research*, 46, 839–862.

Fathi, M., Mozafari, M.R. & Mohebbi, M., 2012. Nanoencapsulation of food ingredients using lipid based delivery systems. *Trends in Food Science and Technology*, 23, 13–27.

Fossheim, S.L., Fahlvik, A.K., Klaveness, J. & Muller, R.N., 1999. Paramagnetic liposomes as MRI contrast agents: Influence of liposomal physicochemical properties on the in vitro relaxivity. *Magnetic Resonance Imaging*, 17, 83–89.

Fossheim, S.L., Il'yasov, K.A., Hennig, J. & Bjørnerud, A., 2000. Thermosensitive paramagnetic liposomes for temperature control during MR imaging-guided hyperthermia: In vitro feasibility studies. *Academic Radiology*, 7, 1107–1115.

Freitas, C. & Müller, R.H., 1998a. Effect of light and temperature on zeta potential and physical stability in solid-lipid nanoparticle (SLN™) dispersions. *International Journal of Pharmaceutics*, 168, 221–229.

Freitas, C. & Müller, R.H., 1998b. Spray-drying of solid-lipid nanoparticles (SLNTM). *European Journal of Pharmaceutics and Biopharmaceutics*, 46, 145–151.

Fundarò, A., Cavalli, R., Bargoni, A., Vighetto, D., Zara, G.P. & Gasco, M.R., 2000. Non-stealth and stealth solid-lipid nanoparticles (SLN) carrying doxorubicin: Pharmacokinetics and tissue distribution after iv administration to rats. *Pharmacological Research*, 42, 337–343.

Ganesan, P. & Narayanasamy, D., 2017. Lipid nanoparticles: Different preparation techniques, characterization, hurdles, and strategies for the production of solid-lipid nanoparticles and nanostructured lipid carriers for oral drug delivery. *Sustainable Chemistry and Pharmacy*, 6, 37–56.

Groves, J.T. & Boxer, S.G., 2002. Micropattern formation in supported lipid membranes. *Accounts of Chemical Research*, 35, 149–157.

Gupta, A.K. & Gupta, M., 2005. Synthesis and surface engineering of iron oxide nanoparticles for biomedical applications. *Biomaterials*, 26, 3995–4021.

Han, X., Cheng, W., Zhang, Z., Dong, S. & Wang, E., 2002. Direct electron transfer between hemoglobin and a glassy carbon electrode facilitated by lipid-protected gold nanoparticles. *Biochimica et Biophysica Acta*, 1556, 273–277.

Henzie, J., Grünwald, M., Widmer-Cooper, A., Geissler, P.L. & Yang, P., 2012. Self-assembly of uniform polyhedral silver nanocrystals into densest packings and exotic superlattices. *Nature Materials*, 11, 131–137.

Heverhagen, J.T., Graser, A., Fahr, A., Müller, R. & Alfke, H., 2004. Encapsulation of gadobutrol in AVE-based liposomal carriers for MR detectability. *Magnetic Resonance Imaging*, 22, 483–487.

Hovis, J.S. & Boxer, S.G., 2000. Patterning barriers to lateral diffusion in supported lipid bilayer membranes by blotting and stamping. *Langmuir*, 16, 894–897.

Huang, S.C., Artyukhin, A.B., Martinez, J.A., Sirbuly, D.J., Wang, Y., Ju, J.W., Stroeve, P. & Noy, A., 2007. Formation, stability, and mobility of one-dimensional lipid bilayers on polysilicon nanowires. *Nano Letters*, 7, 3355–3359.

Huang, Z.R., Hua, S.C., Yang, Y.L. & Fang, J.Y., 2008. Development and evaluation of lipid nanoparticles for camptothecin delivery: A comparison of solid-lipid nanoparticles, nanostructured lipid carriers, and lipid emulsion. *Acta Pharmacologica Sinica*, 29, 1094–1102.

Hurtig, J., Karlsson, M. & Orwar, O., 2004. Topographic SU-8 substrates for immobilization of three-dimensional nanotube–vesicle networks. *Langmuir: the ACS Journal of Surfaces and Colloids*, 20, 5637–5641.

Jain, S., Jain, S., Khare, P., Gulbake, A., Bansal, D. & Jain, S.K., 2010. Design and development of solid-lipid nanoparticles for topical delivery of an anti-fungal agent. *Drug Delivery*, 17, 443–451.

Jaiswal, P., Gidwani, B. & Vyas, A., 2016. Nanostructured lipid carriers and their current application in targeted drug delivery. *Artificial Cells, Nanomedicine, and Biotechnology*, 44, 27–40.

Jang, J.W., Zheng, Z., Lee, O.S., Shim, W., Zheng, G., Schatz, G.C. & Mirkin, C.A., 2010. Arrays of nanoscale lenses for subwavelength optical lithography. *Nano Letters*, 10, 4399–4404.

Jenning, V., Gysler, A., Schäfer-Korting, M. & Gohla, S.H., 2000. Vitamin A loaded solid-lipid nanoparticles for topical use: Occlusive properties and drug targeting to the upper skin. *European Journal of Pharmaceutics and Biopharmaceutics*, 49, 211–218.

Jesorka, A. & Orwar, O., 2008. Liposomes: Technologies and analytical applications. *Annual Review of Analytical Chemistry*, 1, 801–832.

Jesorka, A., Stepanyants, N., Zhang, H., Ortmen, B., Hakonen, B. & Orwar, O., 2011. Generation of phospholipid vesicle-nanotube networks and transport of molecules therein. *Nature Protocols*, 6, 791–805.

Jiang, S., Eltoukhy, A.A., Love, K.T., Langer, R. & Anderson, D.G., 2013. Lipidoid-coated iron oxide nanoparticles for efficient DNA and siRNA delivery. *Nano Letters*, 13, 1059–1064.

Jonsson, M.P., Jönsson, P., Dahlin, A.B. & Höök, F., 2007. Supported lipid bilayer formation and lipid-membrane-mediated biorecognition reactions studied with a new nanoplasmonic sensor template. *Nano Letters*, 7, 3462–3468.

Joshi, M.D. & Müller, R.H., 2009. Lipid nanoparticles for parenteral delivery of actives. *European Journal of Pharmaceutics and Biopharmaceutics*, 71, 161–172.

Kam, L. & Boxer, S.G., 2000. Formation of supported lipid bilayer composition arrays by controlled mixing and surface capture. *Journal of the American Chemical Society*, 122, 12901–12902.

Kim, B.D., Na, K. & Choi, H.K., 2005. Preparation and characterization of solid-lipid nanoparticles (SLN) made of cacao butter and curdlan. *European Journal of Pharmaceutical Sciences*, 24, 199–205.

Köberlin, M.S., Snijder, B., Heinz, L.X., Baumann, C.L., Fauster, A., Vladimer, G.I., Gavin, A.C. & Superti-Furga, , 2015. A conserved circular network of coregulated lipids modulates innate immune responses. *Cell*, 162, 170–183.

Koenig, S.H., Brown, R.D., Kurland, R. & Ohki, S., 1988. Relaxivity and binding of Mn2+ ions in solutions of phosphatidylserine vesicles. *Magnetic Resonance in Medicine*, 7, 133–142.

Kong, S.D., Sartor, M., Hu, C.M., Zhang, W., Zhang, L. & Jin, S., 2013. Magnetic field activated lipid–polymer hybrid nanoparticles for stimuli-responsive drug release. *Acta Biomaterialia*, 9, 5447–5452.

Lee, J.H., Kim, G.H. & Nam, J.M., 2012. Directional synthesis and assembly of bimetallic nanosnowmen with DNA. *Journal of the American Chemical Society*, 134, 5456–5459.

Lee, S.H., Heng, D., Ng, W.K., Chan, H.K. & Tan, R.B., 2011. Nano spray drying: A novel method for preparing protein nanoparticles for protein therapy. *International Journal of Pharmaceutics*, 403, 192–200.

Lee, Y.K., Lee, H. & Nam, J.-M., 2013. Lipid-nanostructure hybrids and their applications in nanobiotechnology. *NPG Asia Materials*, 5, e48.

Lee, Y.K. & Nam, J.M., 2012. Electrofluidic lipid membrane biosensor. *Small*, 8, 832–837.

Lenhert, S., Sun, P., Wang, Y., Fuchs, H. & Mirkin, C.A., 2007. Massively parallel dip-pen nanolithography of heterogeneous supported phospholipid multilayer patterns. *Small*, 3, 71–75.

Li, P., Li, D., Zhang, L., Li, G. & Wang, E., 2008. Cationic lipid bilayer coated gold nanoparticles-mediated transfection of mammalian cells. *Biomaterials*, 29, 3617–3624.

Lima, A.M., Dal Pizzol, C.D., Monteiro, F.B., Creczynski-Pasa, T.B., Andrade, G.P., Ribeiro, A.O. & Perussi, J.R., 2013. Hypericin encapsulated in solid-lipid nanoparticles: Phototoxicity and photodynamic efficiency. *Journal of Photochemistry and Photobiology B, Biology*, 125, 146–154.

Liu, C., Monson, C.F., Yang, T., Pace, H. & Cremer, P.S., 2011a. Protein separation by electrophoretic–electroosmotic focusing on supported lipid bilayers. *Analytical Chemistry*, 83, 7876–7880.

Liu, D., Jiang, S., Shen, H., Qin, S., Liu, J., Zhang, Q., Li, R. & Xu, Q., 2011b. Diclofenac sodium-loaded solid-lipid nanoparticles prepared by emulsion/solvent evaporation method. *Journal of Nanoparticle Research*, 13, 2375–2386.

Liu, J., Hu, W., Chen, H., Ni, Q., Xu, H. & Yang, X., 2007. Isotretinoin-loaded solid-lipid nanoparticles with skin targeting for topical delivery. *International Journal of Pharmaceutics*, 328, 191–195.

Lohmüller, T., Triffo, S., O'donoghue, G.P., Xu, Q., Coyle, M.P. & Groves, J.T., 2011. Supported membranes embedded with fixed arrays of gold nanoparticles. *Nano Letters*, 11, 4912–4918.

Lu, B., Xiong, S.B., Yang, H., Yin, X.D. & Chao, R.B., 2006. Solid lipid nanoparticles of mitoxantrone for local injection against breast cancer and its lymph node metastases. *European Journal of Pharmaceutical Sciences*, 28, 86–95.

Luo, Y., Chen, D., Ren, L., Zhao, X. & Qin, J., 2006. Solid lipid nanoparticles for enhancing vinpocetine's oral bioavailability. *Journal of Controlled Release*, 114, 53–59.

Lv, Q., Yu, A., Xi, Y., Li, H., Song, Z., Cui, J., Cao, F. & Zhai, G., 2009. Development and evaluation of penciclovir-loaded solid-lipid nanoparticles for topical delivery. *International Journal of Pharmaceutics*, 372, 191–198.

Magin, R.L., Wright, S.M., Niesman, M.R., Chan, H.C. & Swartz, H.M., 1986. Liposome delivery of NMR contrast agents for improved tissue imaging. *Magnetic Resonance in Medicine*, 3, 440–447.

Mamot, C., Nguyen, J.B., Pourdehnad, M., Hadaczek, P., Saito, R., Bringas, J.R., Drummond, D.C., Hong, K., Kirpotin, D.B., Mcknight, T., Berger, M.S., Park, J.W. & Bankiewicz, K.S., 2004. Extensive distribution of liposomes in rodent brains and brain tumors following convection-enhanced delivery. *Journal of Neuro-Oncology*, 68, 1–9.

McConnell, H.M. & Tamm, L.K., 1985. Supported phospholipid bilayers. *Biophysical Journal*, 47(1), 105–113.

Mehnert, W. & Mäder, K., 2012. Solid lipid nanoparticles: Production, characterization and applications. *Advanced Drug Delivery Reviews*, 64, 83–101.

Mei, Z., Chen, H., Weng, T., Yang, Y. & Yang, X., 2003. Solid lipid nanoparticle and microemulsion for topical delivery of triptolide. *European Journal of Pharmaceutics and Biopharmaceutics*, 56, 189–196.

Mornet, S., Lambert, O., Duguet, E. & Brisson, A., 2005. The formation of supported lipid bilayers on silica nanoparticles revealed by cryoelectron microscopy. *Nano Letters*, 5, 281–285.

Muchow, M., Maincent, P. & Müller, R.H., 2008. Lipid nanoparticles with a solid matrix (SLN®, NLC®, LDC®) for oral drug delivery. *Drug Development and Industrial Pharmacy*, 34, 1394–1405.

Mukherjee, S., Ray, S. & Thakur, R.S., 2009. Solid lipid nanoparticles: A modern formulation approach in drug delivery system. *Indian Journal of Pharmaceutical Sciences*, 71, 349–358.

Müller, B.G., Leuenberger, H. & Kissel, T., 1996. Albumin nanospheres as carriers for passive drug targeting: An optimized manufacturing technique. *Pharmaceutical Research*, 13, 32–37.

Müller, R.H., Mäder, K. & Gohla, S., 2000. Solid lipid nanoparticles (SLN) for controlled drug delivery—A review of the state of the art. *European Journal of Pharmaceutics and Biopharmaceutics*, 50, 161–177.

Müller, R.H., Radtke, M. & Wissing, S.A., 2002. Solid lipid nanoparticles (SLN) and nanostructured lipid carriers (NLC) in cosmetic and dermatological preparations. *Advanced Drug Delivery Reviews*, 54, S131–S155.

Müller, R.H., Runge, S., Ravelli, V., Mehnert, W., Thünemann, A.F. & Souto, E.B., 2006. Oral bioavailability of cyclosporine: Solid lipid nanoparticles (SLN®) versus drug nanocrystals. *International Journal of Pharmaceutics*, 317, 82–89.

Na, H.B., Song, I.C. & Hyeon, T., 2009. Inorganic nanoparticles for MRI contrast agents. *Advanced Materials*, 21, 2133–2148.

Nair, P.M., Salaita, K., Petit, R.S. & Groves, J.T., 2011. Using patterned supported lipid membranes to investigate the role of receptor organization in intercellular signaling. *Nature Protocols*, 6, 523–539.

Nam, J.M., Nair, P.M., Neve, R.M., Gray, J.W. & Groves, J.T., 2006. A fluid membrane-based soluble ligand-display system for live-cell assays. *ChemBioChem*, 7, 436–440.

Narui, Y. & Salaita, K.S., 2012. Dip-pen nanolithography of optically transparent cationic polymers to manipulate spatial organization of proteolipid membranes. *Chemical Science*, 3, 794–799.

Naseri, N., Valizadeh, H. & Zakeri-Milani, P., 2015. Solid lipid nanoparticles and nanostructured lipid carriers: Structure, preparation and application. *Advanced Pharmaceutical Bulletin*, 5, 305–313.

Nel, A.E., Mädler, L., Velegol, D., Xia, T., Hoek, E.M., Somasundaran, P., Klaessig, F., Castranova, V. & Thompson, M., 2009. Understanding biophysicochemical interactions at the nano–bio interface. *Nature Materials*, 8, 543–557.

Okuyama, K., Abdullah, M., Lenggoro, I.W. & Iskandar, F., 2006. Preparation of functional nanostructured particles by spray drying. *Advanced Powder Technology*, 17, 587–611.

Oumzil, K., Ramin, M.A., Lorenzato, C., Hemadou, A., Laroche, J., Jacobin-Valat, M.J.E., Mornet, S., Roy, C.-E., Kauss, T. & Gaudin, K., 2016. Solid-lipid nanoparticles for image-guided therapy of atherosclerosis. *Bioconjugate Chemistry*, 27, 569–575.

Pardeike, J., Hommoss, A. & Müller, R.H., 2009. Lipid nanoparticles (SLN, NLC) in cosmetic and pharmaceutical dermal products. *International Journal of Pharmaceutics*, 366, 170–184.

Pardeshi, C., Rajput, P., Belgamwar, V., Tekade, A., Patil, G., Chaudhary, K. & Sonje, A., 2012. Solid lipid based nanocarriers: An overview. *Acta Pharmaceutica*, 62, 433–472.

Pasc, A., Blin, J.-L., Stébé, M.-J. & Ghanbaja, J., 2011. Solid lipid nanoparticles (SLN) templating of macroporous silica beads. *RSC Advances*, 1, 1204–1206.

Pfeiffer, I., Seantier, B., Petronis, S., Sutherland, D., Kasemo, B. & Zäch, M., 2008. Influence of nanotopography on phospholipid bilayer formation on silicon dioxide. *Journal of Physical Chemistry. B*, 112, 5175–5181.

Pick, H., Schmid, E.L., Tairi, A.P., Ilegems, E., Hovius, R. & Vogel, H., 2005. Investigating cellular signaling reactions in single attoliter vesicles. *Journal of the American Chemical Society*, 127, 2908–2912.

Ravetti-Duran, R., Blin, J.-L., Stébé, M.-J., Castel, C. & Pasc, A., 2012. Tuning the morphology and the structure of hierarchical meso–macroporous silica by dual templating with micelles and solid-lipid nanoparticles (SLN). *Journal of Materials Chemistry*, 22, 21540–21548.

Reis, C.P., Neufeld, R.J., Ribeiro, A.J. & Veiga, F., 2006. Nanoencapsulation I. Methods for preparation of drug-loaded polymeric nanoparticles. *Nanomedicine : Nanotechnology, Biology, and Medicine*, 2, 8–21.

Richard, C., Balavoine, F., Schultz, P., Ebbesen, T.W. & Mioskowski, C., 2003. Supramolecular self-assembly of lipid derivatives on carbon nanotubes. *Science*, 300, 775–778.

Richter, R.P., Bérat, R. & Brisson, A.R., 2006. Formation of solid-supported lipid bilayers: An integrated view. *Langmuir*, 22, 3497–3505.

Rodenak-Kladniew, B., Islan, G.A., De Bravo, M.G., Durán, N. & Castro, G.R., 2017. Design, characterization and in vitro evaluation of linalool-loaded solid-lipid nanoparticles as potent tool in cancer therapy. *Colloids and Surfaces. B, Biointerfaces*, 154, 123–132.

Roiter, Y., Ornatska, M., Rammohan, A.R., Balakrishnan, J., Heine, D.R. & Minko, S., 2008. Interaction of nanoparticles with lipid membrane. *Nano Letters*, 8, 941–944.

Roiter, Y., Ornatska, M., Rammohan, A.R., Balakrishnan, J., Heine, D.R. & Minko, S., 2009. Interaction of lipid membrane with nanostructured surfaces. *Langmuir*, 25, 6287–6299.

Saito, R., Bringas, J.R., Mcknight, T.R., Wendland, M.F., Mamot, C., Drummond, D.C., Kirpotin, D.B., Park, J.W., Berger, M.S. & Bankiewicz, K.S., 2004. Distribution of liposomes into brain and rat brain tumor models by convection-enhanced delivery monitored with magnetic resonance imaging. *Cancer Research*, 64, 2572–2579.

Salaita, K., Wang, Y. & Mirkin, C.A., 2007. Applications of dip-pen nanolithography. *Nature Nanotechnology*, 2, 145–155.

Sekula, S., Fuchs, J., Weg-Remers, S., Nagel, P., Schuppler, S., Fragala, J., Theilacker, N., Franzreb, M., Wingren, C., Ellmark, P., Borrebaeck, C.A., Mirkin, C.A., Fuchs, H. & Lenhert, S., 2008. Multiplexed lipid dip-pen nanolithography on subcellular scales for the templating of functional proteins and cell culture. *Small*, 4, 1785–1793.

Selvamuthukumar, S. & Velmurugan, R., 2012. Nanostructured lipid carriers: A potential drug carrier for cancer chemotherapy. *Lipids in Health and Disease*, 11, 159.

Shahgaldian, P., Da Silva, E., Coleman, A.W., Rather, B. & Zaworotko, M.J., 2003. Para-acyl-calix-arene based solid-lipid nanoparticles (SLNs): A detailed study of preparation and stability parameters. *International Journal of Pharmaceutics*, 253, 23–38.

Shi, J., Chen, J. & Cremer, P.S., 2008. Sub-100 nm patterning of supported bilayers by nanoshaving lithography. *Journal of the American Chemical Society*, 130, 2718–2719.

Silva, A.C., González-Mira, E., García, M.L., Egea, M.A., Fonseca, J., Silva, R., Santos, D., Souto, E.B. & Ferreira, D., 2011. Preparation, characterization and biocompatibility studies on risperidone-loaded solid-lipid nanoparticles (SLN): High pressure homogenization versus ultrasound. *Colloids and Surfaces B, Biointerfaces*, 86, 158–165.

Souto, E.B. & Müller, R.H., 2006. Investigation of the factors influencing the incorporation of clotrimazole in SLN and NLC prepared by hot high-pressure homogenization. *Journal of Microencapsulation*, 23, 377–388.

Souto, E.B., Wissing, S.A., Barbosa, C.M. & Müller, R.H., 2004. Development of a controlled release formulation based on SLN and NLC for topical clotrimazole delivery. *International Journal of Pharmaceutics*, 278, 71–77.

Stamou, D., Duschl, C., Delamarche, E. & Vogel, H., 2003. Self-assembled microarrays of attoliter molecular vessels. *Angewandte Chemie*, 115, 5738–5741.

Sznitowska, M., Gajewska, M., Janicki, S., Radwanska, A. & Lukowski, G., 2001. Bioavailability of diazepam from aqueous-organic solution, submicron emulsion and solid-lipid nanoparticles after rectal administration in rabbits. *European Journal of Pharmaceutics and Biopharmaceutics*, 52, 159–163.

Tanaka, M. & Sackmann, E., 2005. Polymer-supported membranes as models of the cell surface. *Nature*, 437, 656–663.

Tilcock, C., Unger, E., Cullis, P. & Macdougall, P., 1989. Liposomal Gd-DTPA: Preparation and characterization of relaxivity. *Radiology*, 171, 77–80.

Tran, T.H., Choi, J.Y., Ramasamy, T., Truong, D.H., Nguyen, C.N., Choi, H.G., Yong, C.S. & Kim, J.O., 2014. Hyaluronic acid-coated solid-lipid nanoparticles for targeted delivery of vorinostat to CD44 overexpressing cancer cells. *Carbohydrate Polymers*, 114, 407–415.

Ugazio, E., Cavalli, R. & Gasco, M.R., 2002. Incorporation of cyclosporin A in solid-lipid nanoparticles (SLN). *International Journal of Pharmaceutics*, 241, 341–344.

Unger, E., Shen, D.K., Wu, G.L. & Fritz, T., 1991. Liposomes as MR contrast agents: Pros and cons. *Magnetic Resonance in Medicine*, 22, 304–308; discussion 313.

Unger, E.C., Winokur, T., Macdougall, P., Rosenblum, J., Clair, M., Gatenby, R. & Tilcock, C., 1989. Hepatic metastases: Liposomal Gd-DTPA-enhanced MR imaging. *Radiology*, 171, 81–85.

Valo, H.K., Laaksonen, P.H., Peltonen, L.J., Linder, M.B., Hirvonen, J.T. & Laaksonen, T.J., 2010. Multifunctional hydrophobin: Toward functional coatings for drug nanoparticles. *ACS Nano*, 4, 1750–1758.

Vrignaud, S., Benoit, J.P. & Saulnier, P., 2011. Strategies for the nanoencapsulation of hydrophilic molecules in polymer-based nanoparticles. *Biomaterials*, 32, 8593–8604.

Wang, J.X., Sun, X. & Zhang, Z.R., 2002. Enhanced brain targeting by synthesis of 3′, 5′-dioctanoyl-5-fluoro-2′-deoxyuridine and incorporation into solid-lipid nanoparticles. *European Journal of Pharmaceutics and Biopharmaceutics*, 54, 285–290.

Wang, S., Chen, T., Chen, R., Hu, Y., Chen, M. & Wang, Y., 2012. Emodin loaded solid-lipid nanoparticles: Preparation, characterization and antitumor activity studies. *International Journal of Pharmaceutics*, 430, 238–246.

Wang, S., Su, R., Nie, S., Sun, M., Zhang, J., Wu, D. & Moustaid-Moussa, N., 2014. Application of nanotechnology in improving bioavailability and bioactivity of diet-derived phytochemicals. *Journal of Nutritional Biochemistry*, 25, 363–376.

Wang, X., Jiang, Y., Wang, Y.W., Huang, M.T., Ho, C.T. & Huang, Q., 2008. Enhancing anti-inflammation activity of curcumin through O/W nanoemulsions. *Food Chemistry*, 108, 419–424.

Westesen, K., 2000. Novel lipid-based colloidal dispersions as potential drug administration systems–expectations and reality. *Colloid and Polymer Science*, 278, 608–618.

Westesen, K., Bunjes, H. & Koch, M.H.J., 1997. Physicochemical characterization of lipid nanoparticles and evaluation of their drug loading capacity and sustained release potential. *Journal of Controlled Release*, 48, 223–236.

Wissing, S.A. & Müller, R.H., 2001. A novel sunscreen system based on tocopherol acetate incorporated into solid-lipid nanoparticles. *International Journal of Cosmetic Science*, 23, 233–243.

Yang, S., Zhu, J., Lu, Y., Liang, B. & Yang, C., 1999. Body distribution of camptothecin solid-lipid nanoparticles after oral administration. *Pharmaceutical Research*, 16, 751–757.

Zara, G.P., Cavalli, R., Bargoni, A., Fundarò, A., Vighetto, D. & Gasco, M.R., 2002. Intravenous administration to rabbits of non-stealth and stealth doxorubicin-loaded solid-lipid nanoparticles at increasing concentrations of stealth agent: Pharmacokinetics and distribution of doxorubicin in brain and other tissues. *Journal of Drug Targeting*, 10, 327–335.

Zheng, G., Patolsky, F., Cui, Y., Wang, W.U. & Lieber, C.M., 2005. Multiplexed electrical detection of cancer markers with nanowire sensor arrays. *Nature Biotechnology*, 23, 1294–1301.

Zhuang, C.Y., Li, N., Wang, M., Zhang, X.N., Pan, W.S., Peng, J.J., Pan, Y.S. & Tang, X., 2010. Preparation and characterization of vinpocetine loaded nanostructured lipid carriers (NLC) for improved oral bioavailability. *International Journal of Pharmaceutics*, 394, 179–185.

Zur Mühlen, A., Schwarz, C. & Mehnert, W., 1998. Solid lipid nanoparticles (SLN) for controlled drug delivery–drug release and release mechanism. *European Journal of Pharmaceutics and Biopharmaceutics*, 45, 149–155.

# 6

# *Hybrid Manganese Spinel Ferrite Nanostructures: Synthesis, Functionalization and Biomedical Applications*

**S. Del Sol Fernández, Oscar F. Odio, E. Ramón-Gallegos, and Edilso Reguera**

**CONTENTS**



## 6.1 General Introduction

Manganese ferrite ($MnFe_2O_4$) nanoparticles (NPs) are a current hot topic in materials science for biomedical applications (Ahmad et al., 2018, Enoch et al., 2018, He et al., 2018, Monaco et al., 2018, Neto et al., 2018, Shi and Shen, 2018, Vetr et al., 2018, Wang et al., 2018, Yang et al., 2018, Zhou et al., 2018). Among many other ferrites, nanostructured $MnFe_2O_4$ occupy a special place due their unique and tunable magnetic properties and an excellent chemical stability. Besides, due to their high reactivity toward several organic groups, their ferrite surface offers a great versatility for ligand functionalization, which in many cases defines the ultimate application. Therefore, there has been a growing interest toward developing a method to coat $MnFe_2O_4$ for their effective use in several applications involving drug- (Rodrigues et al., 2016, Enoch et al., 2018, Wang et al., 2018) or gene delivery

(Lee et al., 2009), protein immobilization (Figueroa-Espi et al., 2011, Rashid et al., 2017), magnetic resonance imaging (MRI) (Lee et al., 2007, Yang et al., 2017, Zhang et al., 2017a, Enoch et al., 2018, Yang et al., 2018), magnetic hyperthermia (MH) (Lee et al., 2011, Le et al., 2016, Oh et al., 2016, He et al., 2018) and theranostics (Jing et al., 2018, Sheng et al., 2018, Zhou et al., 2018). Currently, there are available in the literature several extensive reviews covering these issues in detail for magnetic nanomaterials (Blanco-Andujar et al., 2016, Wu et al., 2016, Zhang et al., 2018). However, to the best of our knowledge, this is the first review completely dedicated to $MnFe_2O_4$ hybrid nanosystems for biomedical applications.

The rest of this section briefly covers the structure and magnetic properties of $MnFe_2O_4$ NPs. In the next sections we discuss different synthetic methods and surface engineering strategies for $MnFe_2O_4$ nanohybrids. Also, attention is paid to the *in vitro* and *in vivo* biodistribution, pharmakokinetics and toxicity behavior of $MnFe_2O_4$ nanohybrids, and how physicochemical properties directly affect these parameters. Hereafter, an outline covering the biomedical applications of surface-engineered $MnFe_2O_4$ NPs is presented. Finally, we discuss the major challenge in realizing this technology as well as future research directions.

### 6.1.1 Structure and Magnetic Properties of Manganese Spinel Ferrites

Spinel metal ferrites ($MFe_2O_4$ where M=Fe(II), Ni(II), Mg(II), Mn(II), Zn(II), Co(II), etc.) can be considered as oxygen-packed, face-centered cubic lattices with tetrahedral ($T_d$) and octahedral ($O_h$) sites occupied by the metal cations (Figure 6.1A). The structure of $MnFe_2O_4$ is a mixed spinel ($T_d$ sites are occupied by $Mn^{2+}{}_{1-x}Fe^{3+}{}_x$ and $O_h$ sites are occupied by $Mn^{2+}{}_x$ $Fe^{3+}{}_{2-x}$), in which the inversion parameter ($x$), which describes the relative concentration of the cations in both sites, has a value of 0.2. Thus, its structural formula is $Mn_{0.8}{}^{2+}Fe_{0.2}{}^{3+}[Mn_{0.2}{}^{2+}Fe_{1.8}{}^{3+}]O_4{}^{2-}$. (Grimes et al., 1989, Lee et al., 2007, Mathew and Juang, 2007, Odio and Reguera, 2017).

Spinel ferrites display ferrimagnetic ordering. Under an external magnetic field, spins in $O_h$ sites align in parallel with the direction of the field, while those in $T_d$ sites align antiparallel (Figure 6.1B). Since spins in both lattices are generally uncompensated, the resulting net magnetic moment causes the

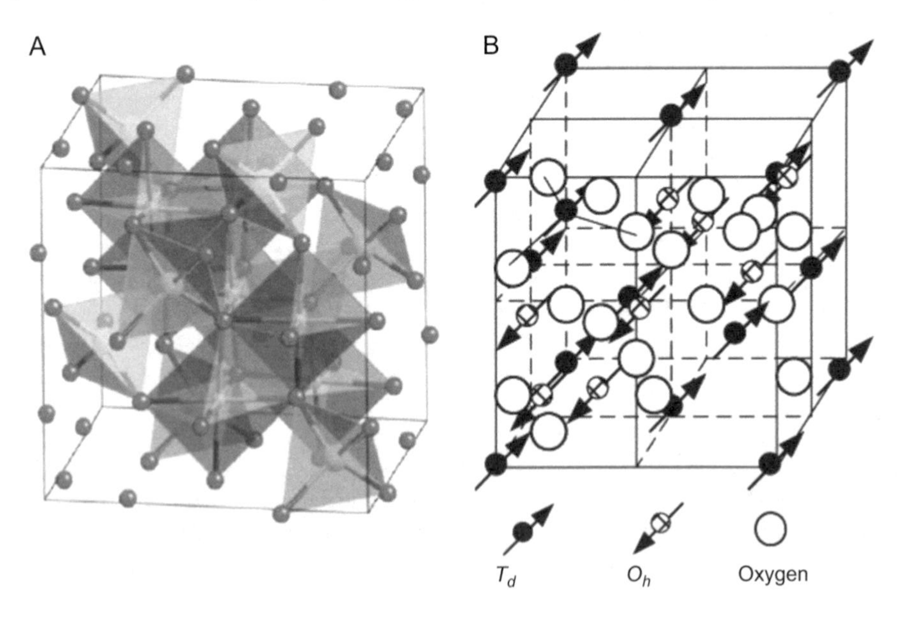


**FIGURE 6.1 (See color insert.)** (A) Crystal structure of $MnFe_2O_4$ spinel ferrite with $T_d$ and $O_h$ sites in yellow and blue, respectively; oxygen atoms are in red (Adapted from Errandonea, 2014, with permission from Springer Nature). (B) Ferrimagnetic arrangement in ferrite spinel structure. (Adapted from Mathew and Juang, 2007, with permission from Elsevier.)

material to display ferrimagnetic behavior (Mathew and Juang, 2007). By replacing $Fe^{2+}$ ($d^6$) ions with $Mn^{2+}$ ($d^5$), $Co^{2+}$ ($d^7$) or $Ni^{2+}$ ($d^8$), the net magnetic moment of the $MFe_2O_4$ unit can be tuned from 4 μB to 5, 3 or 2 μB, respectively (Lee et al., 2007, Yang et al., 2017). Therefore, $MnFe_2O_4$ had the highest magnetic susceptibility with a value of ca. 5 μB (Lee et al., 2007). As we can see, the choice of metal cation and the distribution of ions between the $T_d$ and $O_h$-sites therefore offer a tunable magnetic system that opens a big research field to exploration.

As magnetic materials' size is reduced to the nanometer scale, their magnetic properties usually display some unique features, such as superparamagnetism. This phenomenon occurs below a certain critical size at which magnetic walls are no longer energy affordable and the magneto-crystalline energy is overcome by the thermal motion. Under these conditions, a given particle behaves as a magnetic monodomain with all spins pointing in the same direction, which leads to a huge magnetic moment that can fluctuate coherently like a typical paramagnet in the time frame of the experiment. For a random assembly of superparamagnetic NPs, paramagnetic behavior occurs above a certain temperature known as the blocking temperature ($T_B$) with no coercivity and remanence.

Superparamagnetic $MnFe_2O_4$ NPs have great potential for technological applications due to their high saturation magnetization ($M_S$) and low $T_B$; in addition, this material presents a low coercive field ($H_C$), high permeability, low magnetorestriction, low losses and low anisotropy (Mathew and Juang, 2007). In conjunction, all these properties make $MnFe_2O_4$ NPs very attractive for biomedical applications, as will be shown in Section 6.4.

## 6.2 Synthetic Methods

Synthetic methodologies play an important role in controlling the key parameters to make NPs suitable for potential applications. To be successful in biomedical applications, magnetic NPs should be monodisperse. Also, they should display high magnetization and large susceptibility. They should also exhibit hydrodynamic diameters below 50 nm for extravasation ability and to be stable against uptake by the reticuloendothelial system (RES). We briefly introduce five customary and representative methods for the scalable synthesis of $MnFe_2O_4$ NPs devoted to biomedical use. For other synthetic strategies, readers are encouraged to consult several references covering electrochemical (Mazarío et al., 2015, Mosivand and Kazeminezhad, 2015), sol-gel (Yue et al., 2000, Chen and He, 2001), combustion (Silambarasu et al., 2018) and biomimetic (Maeda et al., 2015, Maeda et al., 2016) routes.

### 6.2.1 Coprecipitation

The general strategy consists in the mixing of ferric and divalent metal precursor ions in a basic solution under an inert atmosphere at room temperature or at elevated temperature (Kafshgari et al., 2018, Akhtar and Younas, 2012, AlTurki, 2018). Based on this method, Zipare et al. (2015) reported an economical and efficient large-scale synthesis of superparamagnetic $MnFe_2O_4$ NPs with controlled size and high saturation magnetization. Coprecipitation is highly dependent on the reaction parameters (e.g. iron precursor, molar ratio of $Fe^{3+}$ and $M^{2+}$, reaction temperature, solution pH and reaction medium), which enables the tuning of the size, shape, composition and magnetic properties of the products. For example, Pereira et al. (2012) used alkanolamines acting as both alkaline agents and complexing ligands that controlled the particle size (4–12 nm) and improved the spin rearrangement at the surface, while Iranmanesh et al. (2017) found that increasing the solution pH results in smaller particles.

Coprecipitation is an inexpensive, rapid and energy-saving method for large scale preparation of NPs. However, they tend to be rather polydisperse without an adequate shape control, which makes these materials not suitable for biomedical applications. In this sense, the addition of surfactant molecules like sodium dodecyl sulfate (Vadivel et al., 2015) and poly(acrylic) acid (PAA) chains (Surendra et al., 2014) allows for the control of the nucleation and growth of NPs. Different approaches entailed the use of graphene oxide (GO) (Le et al., 2016) or carbon nanotubes (CNT) in $MnFe_2O_4$ nanohybrids (Pereira et al., 2018) through one-pot *in situ* coprecipitation process. The resulting materials possess high surface area since GO and CNT avoid ferrite NP agglomeration; however, the improvement in the control over the size and distribution of the CNT/GO grafted NPs continues to be a challenging milestone.

### 6.2.2 Thermal Decomposition

Thermal decomposition has been the most used strategy for obtaining monodisperse $MnFe_2O_4$ NPs with high crystallinity and varied sizes and shapes (Sun et al., 2004, Zeng et al., 2004, Mohapatra et al., 2011, Mohapatra et al., 2013, Yang et al., 2018, Li et al., 2014, Peng et al., 2014). This method uses thermal decomposition of organic metal complexes in a high boiling point solvent and in presence of a surfactant, which also acts as NP surface capping ligand. Oleic acid (OA) and oleylamine (OAm) are typical surfactants in the synthesis of magnetic nanostructures from metal acetylacetonate (acac) (Sun et al., 2004, Xie et al., 2006). This approach allows for the fine-tuning of NP size and morphology by controlling several parameters such as the surfactant composition, heating rate, precursor concentration, decomposition temperature and reaction time. For example, Li et al. (2014) obtained different morphologies of monodisperse $Mn_xFe_{1-x}O$ nanocrystals by changing the Mn/Fe ratio in the precursor solutions. Similar results were obtained by strictly controlling the amount of sodium oleate, OA and the heating temperature (Yang et al., 2018). Though there are numerous reports on surfactant-assisted synthesis of magnetic nanostructures with varied shapes, studies on shape formation mechanisms are rather scarce (Xie et al., 2013). A critical parameter for size-controlled synthesis is the thermal stability of metal precursors. For instance, Song et al. (2007) controlled the size of $MnFe_2O_4$ from 3 to 12 nm through the substitution of metal acetylacetonate by benzoylacetonate (bzac) ligands. Following other strategies, monodispersed $MnFe_2O_4$ NPs with average size of 6, 9 and 12 nm were obtained by the non-hydrolytic reaction between $MnCl_2$ and iron tris-2,4-pentadionate (Lee et al., 2007), while ultrasmall NPs (less than 4 nm) were obtained by dynamic simultaneous decomposition of iron erucicate and manganese oleate complexes (Zhang et al., 2017a).

Monodisperse $MnFe_2O_4$ NPs synthesized by thermal decomposition have recently become a conventional scaffold for biomedical applications. For instance, Liu et al. (2014) prepared highly monodispersed NPs with super high $r_2$ relaxivity for Ultrasensitive Magnetic Resonance Imaging (UMRI) and Oh et al. (2016) obtained cube-like 18 nm NPs for localized magnetic hyperthermia. Other interesting approaches can be found elsewhere (Huh et al., 2007, Casula et al., 2016).

In summary, thermal decomposition methods offer many advantages for preparing highly monodisperse NPs with a narrow size distribution; however, the resultant products are dispersed only in organic media, which requires extra procedures before their use in biomedical applications.

### 6.2.3 Hydrothermal and Solvothermal Synthesis

Hydrothermal synthesis is based on the high-pressure reaction of metal precursors in water or water–alcoholic solutions, which is often performed in reactors or autoclaves. Particle size and shape can be effectively tuned by varying metal concentration, solvent composition, temperature and reaction time (Wu et al., 2016, Odio and Reguera, 2017). Besides, the addition of biocompatible polymers as surfactants like polyethylene glycol (PEG), dextran (D) and chitosan (CH) can change the shape of NPs, aid in controlling growth and avoid particle agglomeration and the onset of magnetic interactions (Zahraei et al., 2016). Another paper describes the synthesis of $MnFe_2O_4$ using a gelatinous medium at 175°C; the resulting NPs display one of the best reported heating efficiency for manganese ferrite with values comparable to those of commercial magnetic NPs (Cruz et al., 2017).

Solvothermal synthesis refers to the same hydrothermal method accomplished in an organic solvent. For instance, heptane in combination with cetyltrimethylammonium bromide (CTAB), sodium dodecylbenzene sulfonate (SDBS) and sodium dodecyl sulfate (SDS) has been employed for the preparation of mixed ferrite NPs with several compositions and varying sizes (5–7 nm), which were tuned as a function of the reaction time, temperature and surfactant dosage (Bateer et al., 2014). Bastami et al. (2014), introduced PEG and PVP as polymeric surfactants, which bind preferentially at the surface of near-normal

$MnFe_2O_4$ ferrite. Other approaches reported the use of diol molecules as solvents (Sahoo et al., 2014, Bastami et al., 2014, Aslibeiki et al., 2016).

### 6.2.4 Microemulsion

Microemulsion (ME) methods have been used to prepare a wide range of spinel NPs with narrow size distributions (Vestal and Zhang, 2004, Mathew and Juang, 2007). Microemulsions are clear and thermodynamically stable colloidal dispersions in which two liquids (typically, water and oil) that are initially immiscible co-exist in one phase because of the presence of surfactant molecules (Wu et al., 2016). Three types of MEs can be prepared: direct (oil dispersed in water, o/w), reversed (water dispersed in oil, w/o) and a bi-continuous type (Boutonnet et al., 2008). In this methodology, the morphology and size of the products may be affected by several factors including the molar ratio of water to oil or surfactant, temperature, iron concentration and aging time. For example, $MnFe_2O_4$ NPs with diameters in the range of 4 to 15 nm were prepared by reversed micelles using SDBS as surfactant; NP size can be controlled by adjusting the water/toluene volume ratio (Liu et al., 2000, Rondinone et al., 2001). Several $MnFe_2O_4$-based nanostructures for biomedical applications have been prepared by this method (Vestal and Zhang, 2003, Figueroa-Espi et al., 2011); however, its limited use relies on tedious procedures prior to their further application.

## 6.3 Surface Modification for Biomedical Applications

Surface coatings are important to prevent ferrite NPs from agglomeration in a physiological environment. Additionally, they act as barriers effectively shielding the magnetic core against the attack of chemical species and can bear functional groups (e.g. amine, carboxyl) that could serve as anchor points for further attachment of functional moieties such as targeting ligands and fluorescent molecules. In addition, coatings can also be tailored to improve NP intracellular behaviors, therapeutic loading and releasing properties. A large variety of biocompatible materials have been used to modify $MnFe_2O_4$ NPs for biomedical applications; some of them are summarized in this section

### 6.3.1 Inorganic Coating

#### *6.3.1.1 Normal and Reverse Core–Shell Structures*

A critical strategy for expanding the application field of spinel ferrite NPs is their overgrowth with a shell of extra functional materials. Core–shell structures are a conventional array in manganese doped iron oxide-based nanocomposites, for which *silica coating* is the most popular approach in biomedical applications. $SiO_2$ coating offers two main benefits: (1) it improves the biocompatibility by reducing toxicity and enhancing hydrophilicity as well as colloidal stability; (2) it enables surface functionalization with targeting moieties (e.g. antibodies, peptides). The Stöber and reverse ME methods are frequently used for fabricating iron oxide@$SiO_2$ NPs (Ahmad et al., 2018, He et al., 2018). The $SiO_2$ shell thickness can be controlled by varying the reaction time, the TEOS/NPs ratio and NP concentration. For instance, Choi et al. (2010) obtained $MnFe_2O_4$@$SiO_2$@$Gd_2O(CO_3)_2$ structures; the $SiO_2$ shell thickness were adjusted between 4–20 nm by varying the amount of TEOS (Figure 6.2). Following similar procedures, Sheng et al. (2018) tuned the $SiO_2$ shell thickness on $MnFe_2O_4$ surface from 30 to 5 nm; further reduction of the shell thickness provokes NPs aggregation.

In addition, mesoporous silica ($mSiO_2$) has become a research focus because it can load drugs directly. The most popular approach to obtain $MnFe_2O_4$@$mSiO_2$ is based on the base-catalyzed hydrolysis–condensation of TEOS in presence of CTAB-functionalized cores; then, mesoporous $SiO_2$ can be obtained by extracting CTAB from the $SiO_2$ matrix with acidic/ethanol solution or calcination (Sahoo et al., 2012, Sahoo et al., 2014, Sheng et al., 2018) (Figure 6.3). Furthermore, $SiO_2$ coating can be functionalized with several organosilanes containing suitable groups for future bioconjugation (Monaco et al., 2018). $MnFe_2O_4$ can also be used as a shell layer when coated on the surface of other materials, thus forming a

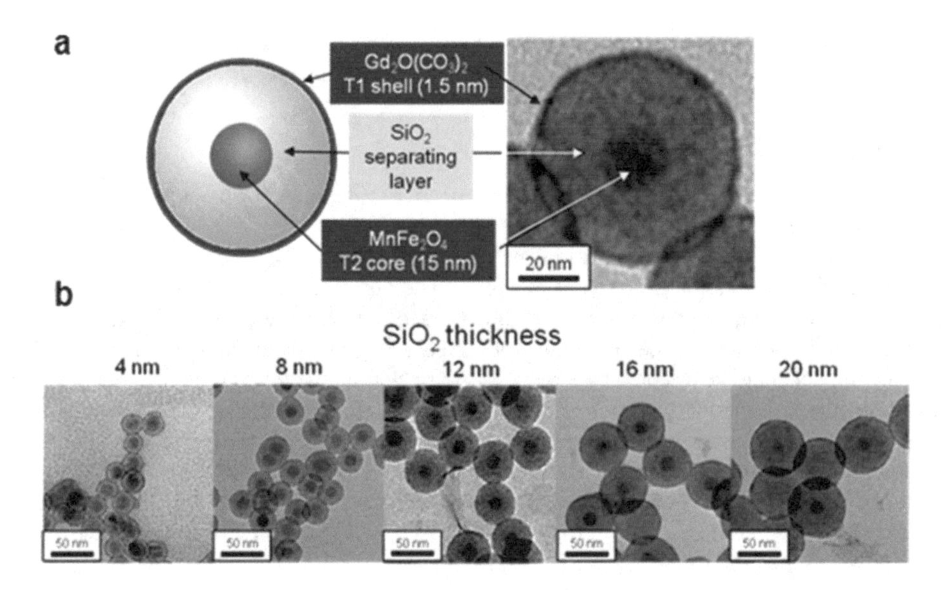


**FIGURE 6.2** TEM images of core–shell–shell $MnFe_2O_4@SiO_2@Gd_2O(CO_3)_2$ NPs with variable $SiO_2$ shell thickness. (Reproduced from Choi et al., 2010, with permission from American Chemical Society.)

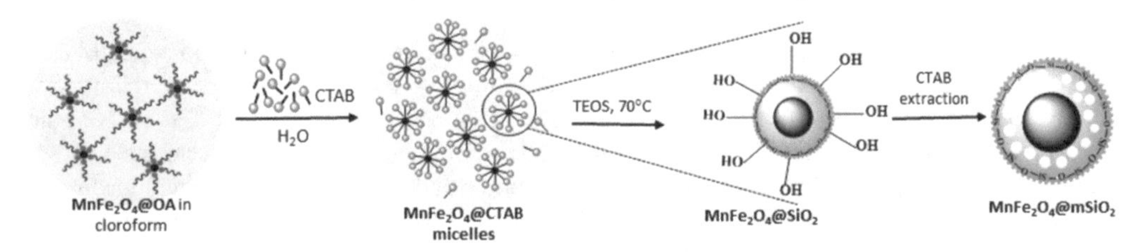


**FIGURE 6.3** Schematic illustration of the steps for the fabrication of mesoporous $MnFe_2O_4@SiO_2$ NPs.

reverse core-shell structure. For example, monodispersed $CoFe_2O_4@MnFe_2O_4$ NPs has been synthesized as suitable materials for MRI; the $CoFe_2O_4$ core enhances the magnetic anisotropy while $MnFe_2O_4$ raises the $M_S$ value (Zhang et al., 2017b).

## 6.3.2 Organic Coating

### *6.3.2.1 Small Molecules and Surfactants*

The presence of numerous $OH^-$ groups on the surface of the NPs enables the direct attachment of numerous small molecules and surfactants. Such anchoring conserves a small NP hydrodynamic radius and retains the original magnetic properties. Various conventional surfactants have been employed for stabilizing $MnFe_2O_4$ NPs against aggregation during the preparation or post-synthetic process such as CTAB (Vamvakidis et al., 2014), SDBS and SDS (Bateer et al., 2014). In presence of OA coated $MnFe_2O_4$ NPs, these surfactants form a bilayer array of interdigitated molecules, which displays excellent water stability. In addition, NPs prepared by thermal decomposition are typically stabilized by monofunctional fatty acid molecules (e.g. OA, lauric and stearic acid [Pradhan et al., 2007]); these molecules hold a long tail with a cis-double bond in the middle that prevents NP aggregation. Since they are only soluble in apolar organic solvents, transfer to aqueous medium is mandatory in order to be suitable for biological applications. This task is usually accomplished through ligand exchange reactions whith small biocompatible molecules like citric acid (Zahraei et al., 2016, Bellusci et al., 2014), folic acid (Safdar et al., 2017), tartrate (Pal et al., 2014, Huang et al., 2014), DMSA (Huh et al., 2007, Oh et al., 2016, Lee et al., 2007, Mazarío et al., 2015, Jang et al., 2009), dopamine (Hu et al., 2011), etc. Drawbacks might include incomplete ligand replacement and reduced NP stability (Xu et al., 2011).

### *6.3.2.2 Polymers*

Polymer coating increases repulsive forces that decrease magnetic interaction and van der Waals attractive forces between naked ferrite NPs. Moreover, polymer-engineered NPs exhibit improved blood circulation time, stability and biocompatibility; thus, polymer coating is often preferred over functionalization with small organic compounds and surfactants for biomedicine uses.

Many polymers are employed for surface functionalization of naked $MnFe_2O_4$ NPs, including natural (chitosan [Oh et al., 2016, Zahraei et al., 2016, Haghiri and Izanloo, 2018], dextran [Zahraei et al., 2016] and gelatin [Cruz et al., 2017]) and synthetic polymers (poly[ethylene glycol] [PEG] [Zahraei et al., 2016, Pernia Leal et al., 2015, Zhang et al., 2017a], poly[methyl acrylate] [PMA], poly[acrylic acid] [PAA] [Casula et al., 2016], polyvinylpyrrolidone [PVP] [Wang et al., 2018], polyamidoamine [PAMAM] dendrimers [Haribabu et al., 2016, Liu et al., 2014, Ertürk and Elmacı, 2018] and several copolymers [Lu et al., 2009, Tromsdorf et al., 2007, Shah et al., 2012, Materia et al., 2017]). Some methodologies have been developed to combine NPs with polymers, such as physical adsorption, direct grafting after atom transfer radical polymerization (ATRP), surface-initiated controlled polymerization (SICP), self-assembly and self-association (Lu et al., 2009, Quarta et al., 2012, Wu et al., 2016). In particular, self-assembly is a simple approach that enhances NP biocompatibility; it involves the use of amphiphilic block copolymers that self-assembly in water media by forming stable core/shell micellar NPs where terminal hydrophobic groups serve as carriers for the inorganic cores and terminal hydrophilic groups promote NP stabilization in the water phase (Figure 6.4).

### *6.3.2.3 Liposomes*

Magnetoliposomes (MLs) result from the encapsulation of magnetic NPs into liposomes. They are promising multifunctional nanosystems that can be used in dual cancer therapy (hyperthermia and drug delivery) assisted by magnetic guidance to the therapeutic site of interest (Béalle et al., 2012, Hervault and Thanh, 2014, Rodrigues et al., 2018). There are several strategies to fabricate distinct structures of magnetoliposomes. Magnetic NPs can be encapsulated directly into the aqueous lumen of the liposomes (Rodrigues et al., 2014, Rodrigues et al., 2016), embedded in between the lipid bilayer (Amstad et al., 2011, Bonnaud et al., 2014), or be surface-conjugated with the liposomes (Floris et al., 2011).

Size and magnetic properties are key features of MLs. Some recent results have shown that the presence of small magnetic NPs have a negligible influence on the size of AMLs (Rodrigues et al., 2016). Regarding magnetic properties, AMLs based on $MnFe_2O_4$ NPs have shown to be superparamagnetic at room temperature, but with low Ms (Rodrigues et al., 2017); in this sense, MLs with NP forming a cluster inside the cavity are preferable (Rodrigues et al., 2016, Rodrigues et al., 2018) since cluster structure retains the magnetic properties of neat NPs (Zhang et al., 2012). MLs based on $MnFe_2O_4$ have successfully applied for hyperthermia treatment (Pradhan et al., 2007), antitumor drug nanocarriers (Rodrigues et al., 2016) and MRI (Tromsdorf et al., 2007). Liposomes have been described as ideal nanoencapsulation system that protect and transport loaded compounds to the sites of interest. However, they still

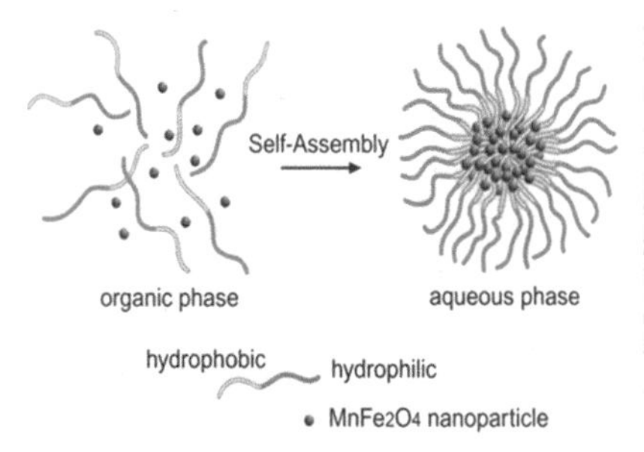


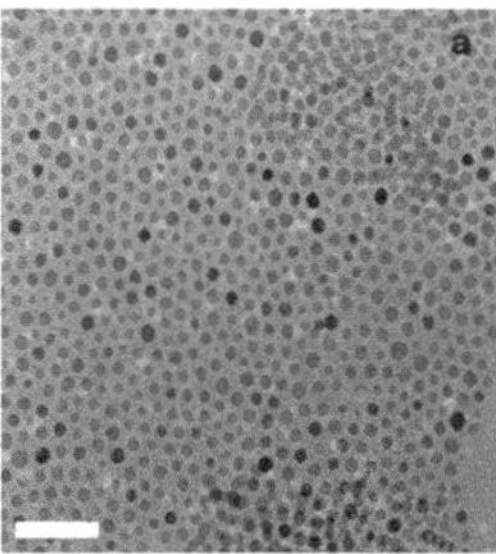


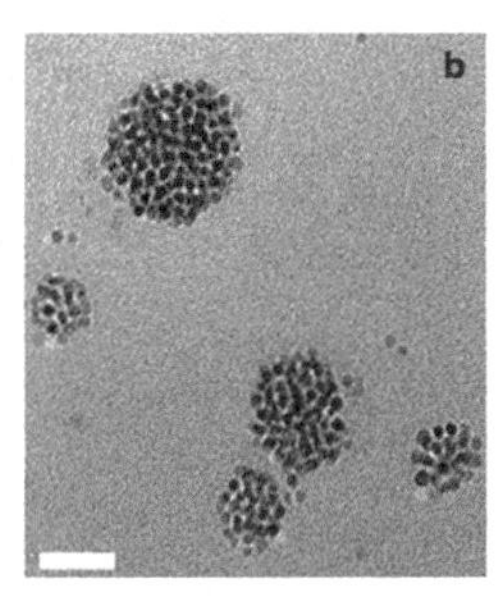


**FIGURE 6.4** *Left:* Schematic illustration of core–shell micellar NPs formed by self-assembled amphiphilic block copolymers. *Right:* TEM images of $MnFe_2O_4$ NPs before (a) and after (b) mPEG-*b*-PCL self-assembly (scale bar: 50 nm) (Adapted from Lu et al., 2009, with permission from Elsevier).

present some issues for *in vivo* application, namely its recognition and capture by the immune system and the difficulties to be located in therapeutic sites for drug release (Seow and Wood, 2009, Allen and Cullis, 2013).

### 6.3.3 Bioconjugation

Surface functionalization can render NPs with water-solubility, functionality and stability in physiological environment. Subsequently, NPs are required to conjugate with biomolecules of interest (e.g. proteins, antibodies, DNA, polypeptides, biotin, avidin, enzymes, etc.). Many conjugation strategies have been developed based on covalent coupling, physical or direct adsorption and biological interactions.

*Covalent coupling* provides specific and stable conjugation of biomolecules with NPs. As a rule, functional groups on the NPs' surface, including carboxylic, amine and thiol groups, are bonded with biomolecules through various coupling strategies. The most common approach involves amine functional groups which are coupled to carboxylic acids by means of carbodiimide agents like 1-ethyl-3-(dimethylaminopropyl) carbodiimide (EDC) aided by N-hydroxysuccinimide (NHS), thus forming "zero length" amide bonds. This strategy has been applied for the coupling of various proteins (e.g. enzyme, peptides and antibodies). For example, Lee et al. (2013) attached hyaluronic acid (HA) on $MnFe_2O_4$ NP functionalized with aminated polysorbate80 (P80) through EDC/NHS coupling for efficient CD44-targeted MRI of breast cancer cells. A similar procedure was employed for the immobilization of monoclonal antibodies (Rashid et al., 2017) and anti-α-fetoprotein (AFP) (Hu et al., 2018). Carboxylic groups can also be activated by NHS chemistry. Gong et al. (2016) conjugated CL1555 peptide to the terminal carboxylic groups of amphiphilic block copolymer $MnFe_2O_4$-PCL-b-PEG-COOH.

Thiol groups can be selectively conjugated with primary amine functions; the reaction starts rapidly under the mediation of maleimides and iodoacetamides. Reaction with sulfhydryl groups generates a stable 3-thiosuccinimidyl ester linkage. A typical coupling reagent is sulfosuccinimidyl-4-(maleimidomethyl cyclohexane-1-carboxylate (sulfo-SMCC). Following this idea, Huh et al. (2007) designed adenovirus-$MnFe_2O_4$ hybrid NPs with dual-functional capabilities as target-specific MR imaging and gene delivery. The strategy comprised first the conversion of adenoviruses lysine residues to maleimide groups through sulfo-SMCC coupling; then, these groups were allowed to react with DMSA-modified $MnFe_2O_4$ NPs in order to get the desired adenovirus/$MnFe_2O_4$ hybrid NPs (Figure 6.5). One of the main limitations of this procedure is that the maleimide ring may hydrolyze in aqueous buffer to a non-reactive cis-maleamic acid derivative over long reaction times or at pH > 8.0 (Conde et al., 2014). Nevertheless, this type of conjugation shows great potential for a large number of biomolecules bearing reactive thiol or amino groups.

Another common approach for bioconjugation is the so called "click chemistry." Copper(I)-catalyzed azide-alkyne cycloaddition (CuAAC) click reaction has been recognized as a facile and versatile

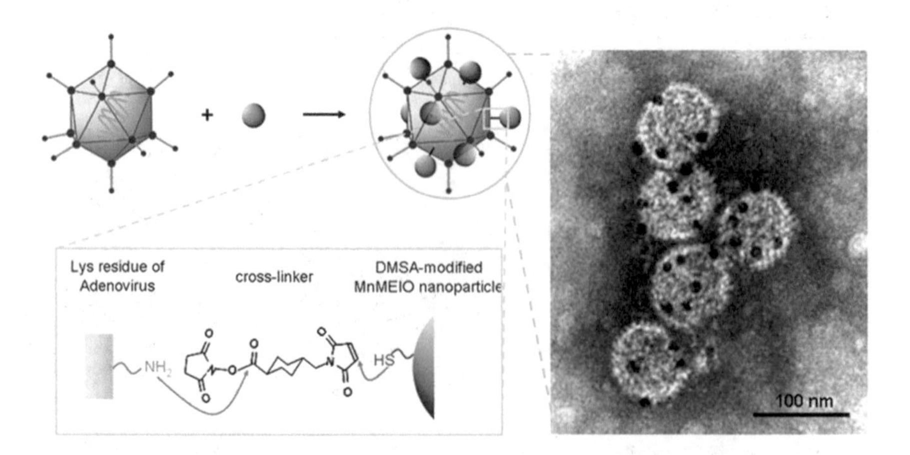


**FIGURE 6.5** *Left:* Scheme of adenovirus/$MnFe_2O_4$ hybrid NPs formation. *Right:* TEM image of resulting hybrid NPs. (Reproduced from Huh et al., 2007, with permission from John Wiley and Sons.)

chemistry for bioconjugation. The reaction occurs at room temperature showing a high degree of solvent and pH insensitivity and a high chemoselectivity. CuAAC occurs between an azide and a terminal acetylene group. The resulting cyclic product is a triazole. Cu(II) ions are reduced *in situ* to produce Cu(I) at the beginning of the coupling (Conde et al., 2014). Click chemistry sometimes refers to a group of reactions that are fast, simple to use, easy to purify, versatile, region–specific and give high product yields; however, they have several limitations. First, if the azide group is too electron deficient, the reaction does not proceed; second, a more common problem is alkyne homocoupling; moreover, the most obvious drawback is the requirement for a Cu-based catalyst due to copper toxicity for human health (Wang and Guo, 2006, Hein et al., 2008). In a very recently work, click chemistry reaction was performed in order to bind gold nanorods (GNRs) on fluorescent core–shell $MnFe_2O_4$@$SiO_2$ NPs (Monaco et al., 2018). Authors first modified the $SiO_2$ shell with an amino-terminated silane (APTMS); resulting terminal $-NH_2$ groups are further converted to azido groups. In parallel, GNRs are funtionalized with alkyne-pendant groups. Finally, both materials are coupled by CuAAC under mild conditions yielding $MnFe_2O_4$@$SiO_2$@GNRs NPs.

*Biological interactions* include electrostatic, hydrophobic and affinity interactions. These interactions have several advantages, such as the ease of functionalization, speed of binding and that neither the biomolecules nor the NPs must be modified in case of electrostatic and hydrophobic interactions. However, conjugation is less stable and reproducible when compared to covalent methods. Moreover, it is difficult to control the amount and orientation of bound molecules (Sperling and Parak, 2010). In the context of this chapter, *direct absorption* refers to the specific attachment of certain biological molecules to NPs with non-covalent interactions (Yu et al., 2012). Such interactions are usually weak and nonspecific and have limited uses.

## 6.4 Biomedical Applications

Among the wide variety of nanomaterials being investigated for biomedical applications, $MnFe_2O_4$ has drawn considerable attention because of its intrinsic magnetic properties. Moreover, it has a nontoxicity profile and a reactive surface that can be readily modified with other biocompatible materials as well as targeting, imaging and therapeutic molecules for improving and extending *in vivo* application fields (e.g. for targeted drug delivery [TDD], MRI and hyperthermia treatment [HT]). There are some other important application for $MnFe_2O_4$ nanohybrids such as photodynamic therapy (PDT) (Wang et al., 2004, Kim et al., 2017), near-infrared (NIR) photothermal therapy (PTT) (Yang et al., 2016, Jing et al., 2018, Zhou et al., 2018) and the less-known immunotherapy and cell therapy (Neto et al., 2018).

$MnFe_2O_4$ nanohybrids are also very well suited for *in vitro* application such as extraction and sorting (Qi et al., 2016). The dual properties of magnetic fluorescent $MnFe_2O_4$ allow combining local heating, temperature sensing or cell guidance with subsequent fluorescent analysis (Zhang et al., 2014, Zhang et al., 2017b). Several biomolecules, e.g. antibodies or DNAs, can also be analyzed by this way (Figueroa-Espi et al., 2011, Rashid et al., 2017, Hu et al., 2018). The immobilization of specific antibodies to the NPs surface takes place by conjugation via several reactive groups mentioned in Section 3.3. In addition, a calorimetric method was developed that was based on $MnFe_2O_4$ NPs and other ferrites as optical sensors in the catalytic oxidation of o-phenylenediamine (OPD) to 2,3–diaminophenazine (DAP) using $H_2O_2$ as oxidant at room temperature (Vetr et al., 2018). The oxidation of OPD to DAP is of great importance in immunoassay determination of enzyme-catalyzed reactions such as oxygenase/oxidases and peroxidases. From this study, $MnFe_2O_4$ NPs provide the highest catalytic activity in the oxidation of OPD to DAP.

Many magnetic nanoplatforms are reported for the applications mentioned before, but here we only focus on those systems based on $MnFe_2O_4$ nanohybrids for TDD, MRI and hyperthermia treatment since these applications are the most common and well documented so far.

### 6.4.1 Target Drug Delivery

The development of suitable processes to formulate drug-loaded magnetic nanoplatforms is of paramount importance. Stimuli used as release triggers can be found in the interior of biological systems (pH, temperature, redox potential and biomolecules) or can be applied externally to biological systems (light or magnetic field) (Colilla et al., 2013, Shah et al., 2013).

The use of pH-responsive gatekeepers as release triggers is possible because certain altered body tissues present a more acidic pH than blood or healthy tissues. To achieve successful pH-responsive delivery systems, the carrier must be stable at physiological pH (c.a. 7.4) but release its loading charge in acidic environments. Some nanoplatforms based on $MnFe_2O_4$ for drug delivery have been designed with only pH stimuli (Wang et al., 2013, ,Sahoo et al., 2014, Wang et al., 2018). On the other hand, temperature-responsive gatekeepers are based in the fact that many tumors have a slightly higher temperature than the rest of the body. Hence, temperature-responsive delivery systems must be able to release the load at temperatures higher than 37°C, while preserving the drugs during the circulation time. For these systems, it is customary to employ poly(N-isopropylacrylamide) (PNIPAM) and derived copolymers as smart thermo-sensitive macromolecules, since they present a lower critical solution temperature (LCST) phase transition near the physiological temperature. When the temperature remains below the LCST value, the chains are hydrated and adopt an extended conformation that prevents drug release, but when the temperature exceeds the LCST value, the chains become hydrophobic and shrink abruptly, enabling a fast drug release. A fine tune of the LCST value can be achieved by varying the comonomer and the copolymer composition. Within this strategy, temperatures higher than LCST can be achieved through the heat induced when magnetic NPs are subjected to an alternate magnetic field (AMF) (see Figure 6.6). For instance, Shah et al. (2012) reported a fast release of the anticancer drug doxorubicin (DOX) above 39°C by using $MnFe_2O_4$ NPs coated with a PNIPAM-co-AM copolymer; DOX release is not significant at 37°C. In another approach, Enoch et al. (2018) developed a dual pH and temperature stimuli-responsive gatekeeper based on a new combination of $Mn_{0.98}Fe_2.0_{2O4}$-β-cyclodextrin(CD)-PEG nanohybrid for the delivery of the anticancer drug camptothecin (CPT). CD-PEG strands allow CPT loading through host -guest interactions. These interactions become less intense either when the solution pH decreases or when the temperature increases, leading to enhanced CPT release.

Redox potential is another important internal stimuli -responsive gatekeeper. Since inside tumor cells the glutathione (GSH) concentration is typically 100–1000 times higher than in the extracellular space, a natural redox potential is developed between both sides of the cellular membrane. Relying on this knowledge, a redox potential-responsive release system based on $MnFe_2O_4$ NPs carrying therapeutic small interfering RNA (siRNA) has been developed by Lee and co-workers (Lee et al., 2009). siRNAs is conjugated to the

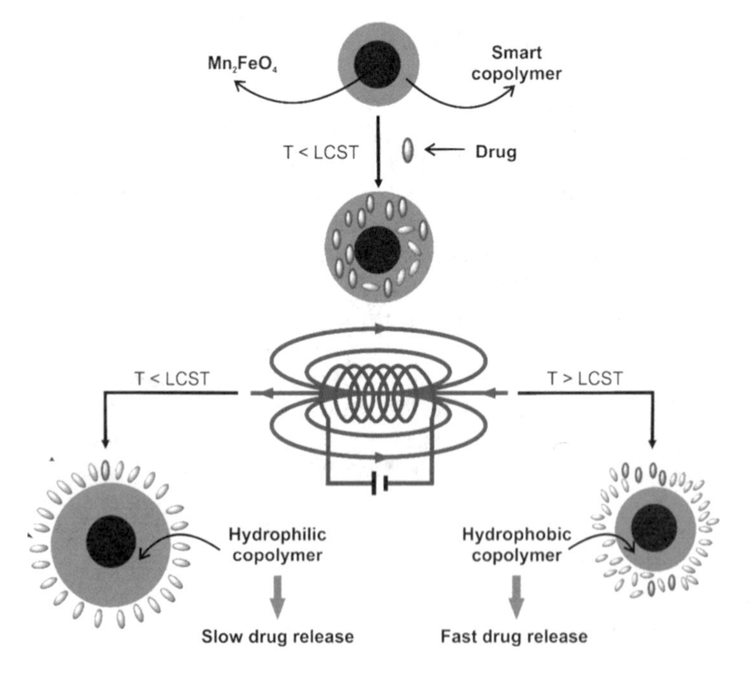


**FIGURE 6.6** Mechanism of drug release in a temperature-responsible copolymer-based $MnFe_2O_4$ NPs triggered by a HFMF; release is faster when temperature is above LCST.

ferrite by disulfide bonds, which are preferentially cleaved inside the tumor cells due to the reducing environment at the cellular cytoplasm, leading to siRNA discharge only inside the targeted cells.

Because external processes involving light activation are rapid and directional, light-responsive controlled release systems are receiving increasing interest. For example, Yang et al. (2016) reported GO/$MnFe_2O_4$ nanohybrids for DOX delivery. DOX is efficiently loaded on GO through strong $\pi$ -$\pi$ stacking and hydrophobic interactions (Zhang et al., 2010). Since the GO/$MnFe_2O_4$ system displays a remarkable photothermal effect, exposure to a NIR laser triggers a rapid release of DOX due to a local temperature increase that weakens GO-DOX interactions. This process is also enhanced in acidic environments.

## 6.4.2 MRI Contrast Agents

MRI is widely employed in clinical diagnoses because of its excellent spatial resolution and real-time monitoring features. This technique is based on the physical phenomenon of nuclear magnetic resonance of hydrogen atoms. However, in most tissues, the intrinsic contrast is not sufficient to provide relevant structural information and external agents are used to enhance it. These contrast agents (CAs) are used to improve the visibility of tissues by changing the proton relaxation time of the water molecules around them. Ferromagnetic or superparamagnetic NPs ($T_2$ CAs) have attracted increasing attention because they can alter their spin-spin relaxation energy due to the induced local field inhomogeneities, resulting in shorter $T_1$ (spin-lattice) and $T_2$ (spin-spin) relaxation times (Blanco-Andujar et al., 2016, Wu et al., 2016, Estelrich et al., 2015, Stanicki et al., 2017). $T_2$ CAs decrease the MR signal intensity in the region nearby; as a consequence, they produce hypointense signals in $T_2$- and $T_2$*-weighted images, and thus the affected regions appear darker.

There are important parameters controlling contrast enhancement such as the magnetic moment ($\mu_c$) or $M_S$, the effective or aggregate NP size, and the water diffusion constant around the magnetic core, which is mainly related to the nature of the organic coating. In turn, $M_S$ itself depends on many factors such as size, composition and particle shape. For example, Zhang and co-workers tested ultrasmall $MnFe_2O_4$ NPs coated with phosphorylated mPEG. At this size range (less than 5 nm), $T_1$ relaxation dominates due to low values of $\mu_c$. Measured relaxitivities ($R = 1/T$) $R_1$ gives higher values than other commercial CAs (Zhang et al., 2017a). On the other hand, for larger NPs with higher values of $\mu_c$, $T_2$ mechanism prevails; under this condition, $R_2$ is proportional to the square of $\mu_c$ and inversely proportional to the NP radius; at the same time, $\mu_c$ is proportional to NP volume, hence, $R_2$ increases with the NP size (Figure 6.7 *left*) (Lee et al., 2007, Huang et al., 2014, Pernia Leal et al., 2015). In other work, Yang et al. (2017) revealed the possible role of Mn atoms in contrast abilities. They demonstrated that $M_S$ and $R_2$ relaxivities of $Mn_xFe_{3-x}O_4$ NPs increase as the Mn doping level rises, reaches a maximum when $x = 0.43$, and decrease

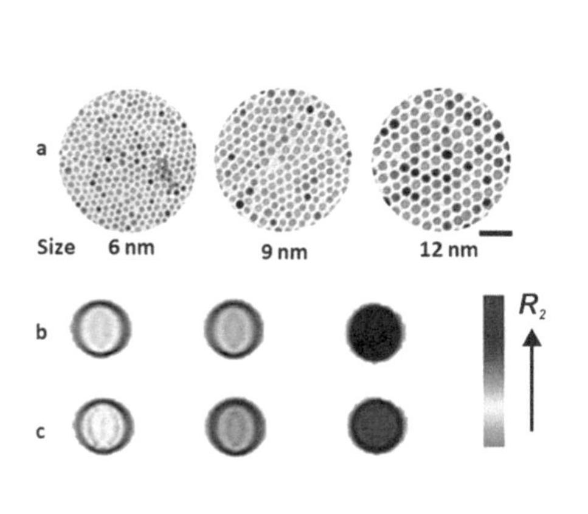


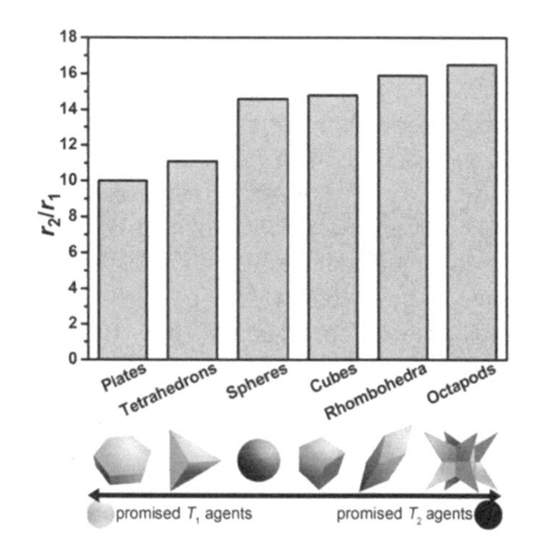


**FIGURE 6.7 (See color insert.)** *Left:* Size-dependent MR contrast effect of $MnFe_2O_4$ NPs: **(a)** TEM images, **(b)** $T_2$-weighted MR images and **(c)** color maps of 6, 9 and 12 nm (scale bar: 50 nm). (Adapted from Lee et al., 2007, with permission from Springer Nature.) *Right:* The role of ferrite NP morphology on the $R_1/R_2$ ratio. (Reproduced from Yang et al., 2018, with permission from American Chemical Society.)

as the Mn doping level continues to augment. Finally, $R_2$ can be optimized by varying NP shape since $M_S$ is directly influenced by the effective particle anisotropy, which increases by the addition of shape anisotropy. In a recent report, Yang et al. (2018) analyzed six uniform morphologies of $MnFe_2O_4$ NPs and their role on $R_2$ (Figure 6.7 right). Also, shape and composition can be simultaneously tuned to achieve an enhanced effect on $R_2$ (Yoon et al., 2011). In an interesting report, Venkatesha et al. (2015) introduced $MnFe_2O_4$-$Fe_3O_4$ core–shell NPs as CAs based on the fact that the differences in the magneto-crystalline anisotropies between the two magnetic phases induces additional magnetic inhomogeneties in the NP; such field inhomogeneties do significantly reduce $T_2$ relaxation of surrounding water protons (Ahmad et al., 2018). Other studies concerning the variation of contrast performance with nanoferrite properties are found elsewhere (Pöselt et al., 2012).

The development of a new generation of dual mode contrast agents (DMCA) has been exploited during recent years. A good DMCA is defined by a high $R_1$ value but a low $R_2/R_1$ ratio (close to 1). Currently, three approaches are assessed to obtain dual CAs: doping the magnetic NPs with paramagnetic $T_1$ ions, grafting $T_1$ ions onto the NP surface, and core–shell structures (Blanco-Andujar et al., 2016). So far, only doped NPs have shown an $R_2/R_1$ ratio close to 1, but at the expense of low $R_2$ (Xiao et al., 2014). Higher values of $R_2$ were achieved with core–shell structures, yet the magnetic coupling between the $T_1$ and $T_2$ layers had a detrimental effect on $R_1$ (Yang et al., 2011, Cui et al., 2014). To overcome this problem, Choi et al. (2010) added a diamagnetic layer of $SiO_2$ with variable thicknesses between the $MnFe_2O_4$ ($T_2$ CAs) core and the $Gd_2O(CO_3)_2$ shell ($T_1$ CAs) in order to the attenuate the interlayer magnetic coupling (cf. Figure 6.2). The resulting enhancement of $T_1$ contrast with the $SiO_2$ thickness layer is significant, while at the same time $T_2$ contrast decreases more gradually. In another work, a targeted peptide bound to PEG–$MnFe_2O_4$ NPs achieved the conjugation of tumor vascular endothelial cells and at the same time induced dark and bright contrast in MR imaging; this system acts as a novel molecular probe for $T_1$ and $T_2$ enhanced MR imaging of tumor angiogenesis (Gong et al., 2016).

### 6.4.3 Magnetic Fluid for Hyperthermia

Magnetic fluid hyperthermia (MFH) is a way to achieve localized heating (42–46°C) where magnetic NPs are used to kill or damage tumor cells. Cancer cells are much more sensitive to heat shock than normal healthy cells, and when exposed to temperatures above 43–45°C, their proliferation and metabolic activity is inhibited, which can lead to either apoptosis or necrosis (Muckle and Dickson, 1971). The capacity of a material to generate heat under the influence of an AC magnetic field is characterized by its specific absorption rate (SAR) or specific loss power (SLP). Two main mechanisms are responsible for heat dissipation by magnetic NPs under the effect of an AC magnetic field: hysteresis and relaxation losses. Since superparamagnetic NPs have zero coercivity, heating losses occurs through relaxation mechanisms. There are two types of relaxation processes: Néel and Brownian relaxation. Brown relaxation dominates for large NPs in settings with low viscosity, while Néel relaxation dominates for smaller NP systems in viscous solutions (Soukup et al., 2015). However, the predominance of either relaxation phenomenon is also highly dependent on the magnetic anisotropy (Blanco-Andujar et al., 2016). Lee et al. (2011) employed exchange-coupled 15 nm $CoFe_2O_4@MnFe_2O_4$ NPs to test the efficacy of anti-tumor hyperthermia therapy. The interfacial coupling between the hard Co ferrite and the Mn soft ferrite gives rise to intermediate coercive fields (Figure 6.8A); as a result, the SLP value of the core-shell structure (2,280 W/g) was one order of magnitude higher compared with single-component NPs (Figure 6.8B). *In vivo* studies indicate a high anti-cancer hyperthermia efficiency of the exchange-coupled NPs, since the tumor subjected to 10 min hyperthermia treatment with these particles was eliminated on day 18 (Figure 6.8C) (Lee et al., 2011). A recent study also employed an exchange-couple magnet for tuning the effective anisotropy and the barrier of magnetization reversal (He et al., 2018).

Many polymers have been employed for coating $MnFe_2O_4$ NPs and for use in MH agents (Casula et al., 2016, Zahraei et al., 2016, Shah et al., 2012). An enhanced SAR value was found with a smaller thickness of the polymer layer because of improvements to the Brownian loss and thermal conductivity. In very viscous media, Brown relaxation time is longer, thus the Néel mechanism is expected to be the dominant source of heat, as was corroborated by Le et al. (2016). Other *in vitro* studies using $MnFe_2O_4$-based NPs can be found elsewhere (Peng et al., 2014, Zahraei et al., 2016).

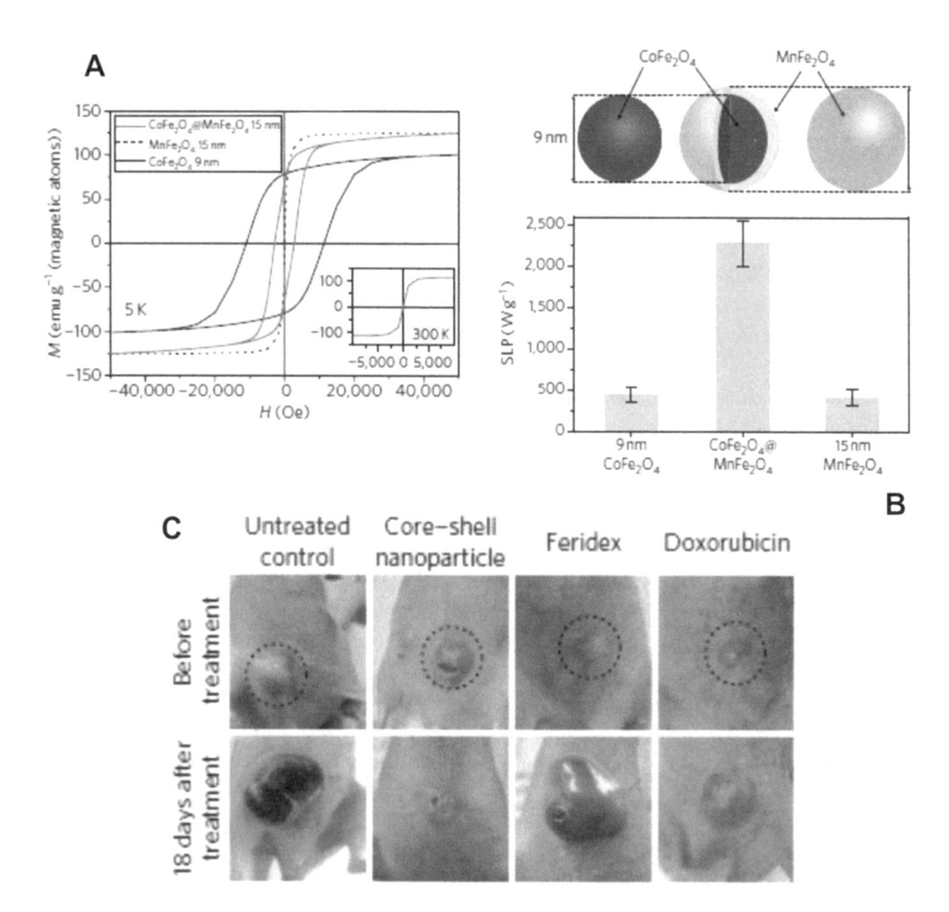


**FIGURE 6.8 (See color insert.)** Interfacial exchange-coupled magnetic NPs as high-performance MH agents. (**A**) M-H curve of 15 nm $CoFe_2O_4$@$MnFe_2O_4$, 15 nm $MnFe_2O_4$ and 9 nm $CoFe_2O_4$ NPs measured at 5 K (Inset: M–H curve of $CoFe_2O_4$@$MnFe_2O_4$ at 300 K). (**B**) Schematic representation $CoFe_2O_4$@$MnFe_2O_4$ NPs and its SLP value compared with the values for the single components. (**C**) Nude mice xenografted with cancer cells (U87MG) before treatment (upper row, dotted circle) and 18 days after treatment (lower row) with $CoFe_2O_4$@$MnFe_2O_4$ hyperthermia, Feridex hyperthermia and DOX, respectively. (Adapted from Lee et al., 2011, with permission from Springer Nature.)

As we can see the design and synthesis of magnetic NPs for MH should take into consideration the following constraints: first of all, they should have the highest possible SLP value within the field and frequency range deemed safe for human body; second, they should be close to superparamagnetic behavior with low magnetostatic interactions to avoid agglomeration; and third, they should be biocompatible with low cytotoxicity.

## 6.4.4 Multifunctional Platforms

Advances in nanotechnology have opened new possibilities for theranostics, which is defined as the combination of therapy and imaging within a single platform. In addition, multifunctional $MnFe_2O_4$ NPs enable multimodal imaging with the combination of two or more imaging modalities (Zhang et al., 2017b, Monaco et al., 2018, Shi and Shen, 2018, Yang et al., 2016, Zhou et al., 2018, Peng et al., 2012, Huh et al., 2007, Wang et al., 2013, Sahoo et al., 2014, Casula et al., 2016, Yang et al., 2016, Zahraei et al., 2016).

A minimally invasive theranostic platform integrating therapeutic and imaging modalities has emerged as a significant technological development due to the following benefits: (1) synergy between the strengths from individual function, (2) reduced diagnosing and treating time due to the co-registration of multiple functions, (3) reduced exposure risks and inconveniences to the patients (Lim et al., 2018).

Zhou et al. (2018) reported a multifunctional porous core–shell-$MnFe_2O_4$/PB nanocomposites for application in combined MRI-guided PTT and MHT. Prussian blue (PB) belongs to the latest generation of agents for both MR imaging and photothermal tumor ablation (Cheng et al., 2014). Water molecules trapped in the porous structure greatly improved MR contrast signals. Besides, the porous shell helps the NIR absorbance of the core, which leads to an enhanced thermal effect due to the synergic effect of PTT and MHT. Jing et al. (2018) designed a multifunctional nanomaterial through the loading of chitosan and metformin on $MnFe_2O_4$@$MoS_2$ nanoflowers for simultaneous multimodal imaging and combination of high-sensitive Chemo-PTT treatment of liver cancer. $MnFe_2O_4$ cores are used as simultaneous $T_1$-$T_2$ agents while $MoS_2$ nanosheets are used as effective NIR photothermal conversion agents for PTT therapy. Furthermore, metformin dose led to the suppression of photothermally sensitized hepatoma cells without damage to normal liver cells. In other recently work, Sheng et al. (2018) designed a multifunctional theranostic nanoplatform with MHT and dual-modal MRI/fluorescence imaging by integrating $MnFe_2O_4$/$SiO_2$-mPEG NPs with fluorescein, making this nanohybrids capable to emit bright green fluorescence under UV excitation.

$MnFe_2O_4$-based nanoplatforms could also include drug transportation facilities. The obtained theranostic nanoparticles could allow combined drug delivery/MRI/PTT (Yang et al., 2016), MRI/drug delivery (Wang et al., 2013) and MH/drug delivery (Patra et al., 2015). For instance, Sahoo et al. (2014), designed nanosystems based on superparamagnetic $MnFe_2O_4$@m$SiO_2$@FA NPs with fluorescent moieties attached. These nanostructures can be used in cancer cell imaging and can deliver doxorubicin (DOX) in the presence of the cancer cells. In addition, it was proven that this nanosystem was specifically retained by HeLa cancer cells in contrast to normal cells. *In vitro* biological experiments revealed that NPs with attached folate and loaded with DOX presented high efficiency for both targeting and killing cancer cells. *In vivo* studies using mice confirmed the usefulness and the biocompatibility of the described nanosystems. A similar all-in-one strategy is used for siRNA delivery and simultaneous *in vivo* transportation monitoring. Bovine serum albumin (BSA)-coated $MnFe_2O_4$ NPs were modified with Cy5 dye for subcellular imaging and cyclic RGD peptides as cancer cell-targeted delivery. The multifunctional theranostic nanovector was designed to enable highly accurate imaging to be carried out simultaneously with the delivery of therapeutic siRNA, which can minimize invasiveness and deleterious side effects (Lee et al., 2009). Other examples of theranostic applications comprising $MnFe_2O_4$ NPs can be found elsewhere (Kim et al., 2017).

## 6.5 *In Vitro/In Vivo* Toxicity Evaluation of Manganese Ferrite Nanohybrids

The current section aims to explain the correlation between the mechanism of toxicity of $MnFe_2O_4$ NPs and the major physicochemical factors responsible for *in vitro/in vivo* toxicity (Patil et al., 2015). Properties like size, shape and surface chemistry contribute to generate reactive oxygen species (ROS) in cells, which are recognized as the main cause of cellular death.

*In vitro* nanotoxicity assessments can produce reliable and reproducible results without the use of animals; also, they comprise simple, rapid and inexpensive procedures. The MTT (3-[4,5-dimethylthiazol-2-yl]-2,5-diphenyltetrazolium bromide) or MTS (3-[4,5-dimethylthiazol-2-yl]-5-[3-carboxymethoxyphenyl]-2-[4-sulfophenyl]-2*H*-tetrazolium) assays measure the mitochondrial function, while a lactate dehydrogenase (LDH) assay measures the cell membrane integrity; in conjunction, these protocols are widely used in studying the toxicity of iron oxide NPs.

In Table 6.1 we present for the first time a summary of the physicochemical parameters of $MnFe_2O_4$ nanohybrids and their influence on the viability (*in vitro*) of different kind of cells (i.e. human and nonhuman) by means of several cytotoxicity assays. The viability was mostly measured between 24 and 72 h. It can be seen that some data are omitted in many works, mainly related to colloidal stability parameters such as hydrodynamic diameter or surface charge. Considering the significant impact of these physicochemical properties on the absorption, distribution, metabolism and excretion (ADME) pattern of iron oxide NPs, they should be clarified. From Table 6.1 it is apparent that reproducibility is highly affected by certain parameters such as the NP coating, kind of studied cells (Materia et al., 2017), cell culture conditions and assay protocols (Bellusci et al., 2014). Increased toxicity was observed

**TABLE 6.1**

Brief Overview of Recent *In Vitro* Cytotoxic Studies of $MnFe_2O_4$ NPs, Organized with Emphasis on Physicochemical Parameters

| Core Shape | Coating Agent | Diameter (nm) | | Charge (mV) | Cell Type | Assay | Dose(μg/ml) | Incubation Time | Brief Result | Ref. |
|---|---|---|---|---|---|---|---|---|---|---|
| | | $D_{TEM}$ | $D_h$ | | | | | | | |
| Cube-like | DMSA-Chitosan | 17.63 ± 1.23 | 91.37 (water) 105.78 (DMEM) | +10.02 | MDA-MB 231 | WST-1, | 0.1–1.5 | 12, 24 and 48 h | No significant cytotoxic effect | Oh et al. (2016) |
| | Dopamine | 11 | | | HeLa | MTT | 0–200 | 12 h | No cytotoxic effect even with the highest dose | Hu et al. (2011) |
| | Aminated P80 -HA | | 233.8 ± 5.2 | −32.8 ± 0.5 | MDA-MB-231 | WST-1 | 0.02–1.25 | 24 h | Marked non-cytotoxic effect | Lee et al. (2013) |
| | CTAB | 21 | | | L929 | MTT | 7.8–500 | 24 h | No cytotoxic effect | Yang et al. (2015) |
| Spherical | Sodium tartrate | 5–12 | 10.49–22.28 | | HeLa | MTT | 0.39–100 | 24 h | No cytotoxic effect even with the highest dose | Huang et al. (2014) |
| | DMSA-Herceptin | 6–12 | | | HeLa and HepG2 | MTT | 0–200 | | Not cytotoxic effect | Lee et al. (2007) |
| | CA | | 95 | −25 | HeLa | MTT | 30–200 | 24 h | No cytotoxic effect | Mazarío et al. (2015) |
| | CA | | 59 ± 5 | −37 | Balb/3T3 | MTT and CFA | 5–100 | 24 and 48 h | Cell viability and proliferation significantly decreased up to 50 and 20 μg/mL | Bellusci et al. (2014) |
| | Sodium citrate | 18.5 | >50 | | SMMC-7721 | MTT | 0–120 | 24 h | No cytotoxic effect | Yang et al. (2017) |
| | mPEG-*g*-PEI, phosphorylated mPEG | 6 6 | 30.6–32.6 11.2 | +20.3, +42.7 -8.4 | MDA-MB-231 | MTS | 1.5–30 | 8 h | $MnFe_2O_4$-mPEG-*g*-PEI showed increasing cytotoxicity with Fe and PEI content $MnFe_2O_4$-mPEG showed no cytotoxic | Liu et al. (2014) |
| | Gallol-PEG | 6–14 | 20–32 | −9.7 to −19.5 | PC-3 | MTT and LDH | 0.1–100 | | No cytotoxic effect by MTT. Mild decrease in cell number by LDH activity at the highest concentration | Pernia Leal et al. (2015) |

*(Continued)*

**TABLE 6.1 (CONTINUED)**

Brief Overview of Recent *In Vitro* Cytotoxic Studies of $MnFe_2O_4$ NPs, Organized with Emphasis on Physicochemical Parameters

| Core Shape | Coating Agent | Diameter (nm) | | Charge (mV) | Cell Type | Assay | Dose(µg/ml) | Incubation Time | Brief Result | Ref. |
|---|---|---|---|---|---|---|---|---|---|---|
| | | $D_{TEM}$ | $D_h$ | | | | | | | |
| | PVP | 7.5 | | | HeLa | MTT | 0–150 | 12 and 24 h | Not cytotoxic effect | Wang et al. (2018) |
| | mPEG | 2, 3 and 3.9 | 7.81 ± 1.72, 8.79 ± 1.06 12.50 ± 1.24 | | Chang liver cells, HepG2 and endothelium cells | MTS | 6.25–100 | 24 h | No cytotoxic effect except for Chang liver and endothelium cells. Slight decrease in cell viability for HepG2 with smaller NPs and high Fe concentration (>50 µg/mL) | Zhang et al. (2017a) |
| Clustering micelles | mPEG-b-PCL | 8.5 ± 0.9 | 79.6 ± 29.4 | | Raw 264.7 and HepG2 | MTT | 0.1–140 | 4 h | No cytotoxic effect | Lu et al. (2009) |
| | PMAO | | 80–120 | | HeLa-WT and A431 | LDH | 6–24 | 12 and 24 h | No cytotoxic effect for HeLa cells. Slight toxicity (25%) after 24 h | Materia et al. (2017) |
| | PMA-PEG-FA | | 25.7 ± 1.1 | −28.5 ± 2.0 | MGC-803 | Resazurin | 42–1,000 | 24 h | Cell viability decreases to 55% up to 444 µg/mL | Zhang et al. (2017b) |
| | PEG-CL (1555) | 146.7 ± 25.9 | 183.4 ± 26.5 | | TVECs | CCK-8 | 0–4.5 mM | 24 h | No cytotoxic effect | Gong et al. (2016) |
| Core–shell | Fe3O4-chitosan | 12.8 ± 1.1 | | | MCF-7 | MTT | 6.25–100 | 24 h | No evident toxicity | Venkatesha et al. (2015) |
| | Au | 6.3 ± 0.7 | | | H9c2, MCF-7 | MTS | 25–500 | 24 h | No cytotoxic effect | Mohammad et al. (2010) |
| | $mSiO_2$-FA | 200–300 | | | L929, HeLa | MTT | 1–20 | 24 h | No cytotoxic effect | Sahoo et al. (2014) |
| | PB-CA | | 168 | | 4T1, Hela | MTT | 0–400 | 24 h | No cytotoxic effect | Zhou et al. (2018) |
| | $SiO_2$-mPEG-NHS | 9.6 ± 0.8 and 13.9 ± 1.0 | 92 | | NIH/3T3 | CCK-8 | 0–200 | 24 h | The viability was kept above 80% for all doses | Sheng et al. (2018) |

(Continued)

**TABLE 6.1 (CONTINUED)**

Brief Overview of Recent *In Vitro* Cytotoxic Studies of $MnFe_2O_4$ NPs, Organized with Emphasis on Physicochemical Parameters

| Core Shape | Coating Agent | Diameter (nm) | | Charge (mV) | Cell Type | Assay | Dose(μg/ml) | Incubation Time | Brief Result | Ref. |
|---|---|---|---|---|---|---|---|---|---|---|
| | | $D_{TEM}$ | $D_h$ | | | | | | | |
| Nearly spherical attached on thin GO sheets | GO | 4–7 (NPs) 70–310 (GO sheets) | | | HeLa and L929 | MTT | 0–200 | 12 and 24 h | No cytotoxic effect | Yang et al. (2016) |
| Octahedrally | GO sheets | 18.5 ± 2.9 | 50.6 ± 0.3 | 44.9 ± 1.2 | NIH/3T3 | CCK-8 | | 24 h | No cytotoxic effect | Peng et al. (2014) |

DMSA: Dimercaptosuccinic acid, HA: hialuronic acid, CTAB: Hexadecyltrimethylammonium bromide, CA: citric acid, PVP: polyvinylpyrrolidone, PMAO: poly(maleic anhydride-alt-1-octadecene), MTT: 3-(4,5-dimethylthiazol-2-yl)-2,5-diphenyltetrazolium bromide, CCK-8: 2-(2-methoxy-4-nitrophenyl)-3-(4-nitrophenyl)-5-(2,4-disulfophenyl)-2H-tetrazolium, PB: prussian blue, FA: folic acid, GO: graphene oxide, PEG-b-PCL: polyethylene glycol-block-poly(ε-caprolactone), CL 1555: name of a peptide, TVECs: tumor vascular endothelial cells.

with polyethyleneimine (PEI) as a coating polymer, whereas inclusion of PEGylation and acetylation eliminated cytotoxicity in MDA-MB-231 cells according to MTS assay (Liu et al., 2014). In general, the surface coatings of $MnFe_2O_4$ have shown cytotoxicity reduction, which are related to changes in NP size. For example, Zhang, et al. (2017a) studied 2, 3 and 3.9 nm $MnFe_2O_4$ @mPEG nanohybrids and their effect on the viability of Chang liver, HepG2 and endothelium cells. They found insignificant toxicity with a cell viability of more than 80% for Chang liver and endothelium cells, but for HepG2 cells a decrease in cell viability was observed for smaller NPs (>50 µg/mL), which may be associated with their nuclear internalization. Furthermore, the type of functional groups adds another complication to the *in vitro* test results. Amine-modified $MnFe_2O_4$ have been found to be more lethal in *in vitro* tests; such behavior owes to the presence of positive surface charges that can induce disruption and solubilization of cell membranes by electrostatic interactions (Lee et al., 2013). Even if the *in vitro* toxicity measurements of NPs on various cell lines may reveal preliminary information concerning the safety of the NPs, these assays have limited relevance, as they do not represent the real *in vivo* conditions. This is partly due to the inability of *in vitro* studies to mimic the complex environment and homeostasis mechanisms maintained by clearance organs such as the kidney and liver. Noteworthy is the fact that cytotoxicity may, however, be more pronounced *in vitro* than *in vivo* because in cell culture conditions the NPs and/or their degradation products (which can also affect cell viability) remain in close contact with the cells without disposal routes. On the contrary, in *in vivo* tests NPs are continuously eliminated from the body providing they are biodegradable (Reddy et al., 2012).

Similar to the *in vitro* studies, many physicochemical factors such as surface chemistry, size, shape and NP surface charge play an important role in animal studies. *In vivo* analyses are considered a critical step to study pharmacokinetic parameters such as absorption, distribution, metabolism, and excretion of NPs. There are some *in vivo* studies about biodistribution, pharmacokinetic and toxicity of $MnFe_2O_4$ nanohybrids in the literature (Yang et al., 2010, Bellusci et al., 2014, Yang et al., 2016). Different types of animal models have displayed variable toxicity and biodistribution profiles of iron oxide NPs. Organs enriched with reticuloendothelial systems (liver, spleen and lungs) take up the majority of iron oxide NPs introduced by most of the administration routes. Iron oxide NPs administered by inhalation route and intravenous route accumulated in the liver, spleen, brain, testis and lung, whereas iron oxide NPs administered by an intravenous route in mice were accumulated in the kidney, spleen and brain (Bellusci et al., 2014) or in liver, spleen and kidney in rats (Zhang et al., 2017a). Changes in the levels of aspartate transaminase (AST), alanine transaminase (ALT), which is indicative of liver function, total bilirubin (TBIL) and two indicators for kidney functions (creatinine and blood urea nitrogen [BUN]) were reported in mice upon intravenous administration of $MnFe_2O_4$/GO (Yang et al., 2016). Their results showed no significant differences in the indicators for renal and hepatic functions as compared with the control after 2 h and 5 days post-injection. In physiological conditions Mn is a cofactor for several enzymes required in neuronal and glial cell function, but an overexposure to this metal may cause neurotoxicity. In this sense, data on $MnFe_2O_4$ NPs are insufficient. Recently, 50 nm $MnFe_2O_4$ NPs administrated via inhalatory were evaluated on it permeation properties of the meningeal membrane. They found that $MnFe_2O_4$ meningeal permeation is negligible, since concentrations were extremely low (Mauro et al., 2018). NPs can potentially interact more with biological fluids, with subsequent toxicological effects still poorly evaluated. Detailed pharmacokinetic and toxicity studies of the NPs at extended time periods are necessary for successful transition to clinical setting.

## 6.6 Final Remarks and Perspectives

Surface-engineered $MnFe_2O_4$ nanohybrids are excellent candidates for biomedical applications because of their potential as multifunctional platforms, owing to a unique combination of suitable magnetic properties and versatile surface decoration. Several synthesis methodologies afford monodispersed ferrite NPs and allow for the tailoring of their size, shape and magnetism, which opens a wide range of potential uses. However, the development of simpler, cheaper and greener strategies that satisfy the biomedical requirements in terms of water stability, yield and control over the size or shape is still a pendant task.

$MnFe_2O_4$ NP surface allows for a great number of conjugation possibilities with either organic/inorganic compounds or biomolecules that lead to fascinating advanced nanomaterials which can be used in MR imaging, targeted drug delivery systems, magnetic hyperthermia and other diagnostic and therapeutic techniques. Precisely, the design of nanohybrid structures for theranostics is a promising research field that entails great challenges due to difficulties in achievement high performance in both modalities. Besides, although many efforts have been devoted to cytotoxic studies of these materials, inconsistencies in the results seem to keep growing; in this sense, more *in vitro* and *in vivo* reproducible experiments are mandatory in combination with a detailed physicochemical characterization of the nanoplatforms to stablish confident relationships between NP surface chemistry and induced toxicity.

## REFERENCES

Ahmad, A., Bae, H. & Rhee, I. 2018. Highly stable silica-coated manganese ferrite nanoparticles as high-efficacy T2 contrast agents for magnetic resonance imaging. *AIP Advances*, 8, 055019.

Akhtar, M. J. & Younas, M. 2012. Structural and transport properties of nanocrystalline MnFe2O4 synthesized by co-precipitation method. *Solid State Sciences*, 14, 1536–1542.

Allen, T. M. & Cullis, P. R. 2013. Liposomal drug delivery systems: From concept to clinical applications. *Advanced Drug Delivery Reviews*, 65, 36–48.

Alturki, A.-M. 2018. Superparamagnetic MnFe2O4 and MnFe2O4 NPs/ABS nanocomposite: Preparation, thermal stability and exchange bias effect. *Indian Journal of Science and Technology*, 11, 37–45.

Amstad, E., Kohlbrecher, J., Müller, E., Schweizer, T., Textor, M. & Reimhult, E. 2011. Triggered release from liposomes through magnetic actuation of iron oxide nanoparticle containing membranes. *Nano Letters*, 11, 1664–1670.

Aslibeiki, B., Kameli, P., Ehsani, M. H., Salamati, H., Muscas, G., Agostinelli, E., Foglietti, V., Casciardi, S. & Peddis, D. 2016. Solvothermal synthesis of MnFe2O4 nanoparticles: The role of polymer coating on morphology and magnetic properties. *Journal of Magnetism and Magnetic Materials*, 399, 236–244.

Bastami, T. R., Entezari, M. H., Kwong, C. & Qiao, S. 2014. Influences of spinel type and polymeric surfactants on the size evolution of colloidal magnetic nanocrystals (MFe2O4, M=Fe, Mn). *Frontiers of Chemical Science and Engineering*, 8, 378–385.

Bateer, B., Tian, C., Qu, Y., Du, S., Yang, Y., Ren, Z., Pan, K. & Fu, H. 2014. Synthesis, size and magnetic properties of controllable MnFe2O4 nanoparticles with versatile surface functionalities. *Dalton Transactions*, 43, 9885–9891.

Béalle, G. L., Di Corato, R., Kolosnjaj-Tabi, J., Dupuis, V., Clément, O., Gazeau, F., Wilhelm, C. & Ménager, C. 2012. Ultra magnetic liposomes for MR imaging, targeting and hyperthermia. *Langmuir*, 28, 11834–11842.

Bellusci, M., La Barbera, A., Padella, F., Mancuso, M., Pasquo, A., Grollino, M. G., Leter, G., Nardi, E., Cremisini, C., Giardullo, P. & Pacchierotti, F. 2014. Biodistribution and acute toxicity of a nanofluid containing manganese iron oxide nanoparticles produced by a mechanochemical process. *International Journal of Nanomedicine*, 9, 1919–1929.

Blanco-Andujar, C., Walter, A., Cotin, G., Bordeianu, C., Mertz, D., Felder-Flesch, D. & Begin-Colin, S. 2016. Design of iron oxide-based nanoparticles for MRI and magnetic hyperthermia. *Nanomedicine*, 11, 1889–1910.

Bonnaud, C., Monnier, C. A., Demurtas, D., Jud, C., Vanhecke, D., Montet, X., Hovius, R., Lattuada, M., Rothen-Rutishauser, B. & Petri-Fink, A. 2014. Insertion of nanoparticle clusters into vesicle bilayers. *ACS Nano*, 8, 3451–3460.

Boutonnet, M., Lögdberg, S. & Svensson, E. E. 2008. Recent developments in the application of nanoparticles prepared from w/o microemulsions in heterogeneous catalysis. *Current Opinion in Colloid and Interface Science*, 13, 270–286.

Casula, M. F., Conca, E., Bakaimi, I., Sathya, A., Materia, M. E., Casu, A., Falqui, A., Sogne, E., Pellegrino, T. & Kanaras, A. G. 2016. Manganese doped-iron oxide nanoparticle clusters and their potential as agents for magnetic resonance imaging and hyperthermia. *Physical Chemistry Chemical Physics: PCCP*, 18, 16848–16855.

Chen, D.-H. & He, X.-R. 2001. Synthesis of nickel ferrite nanoparticles by sol-gel method. *Materials Research Bulletin*, 36, 1369–1377.

Cheng, L., Gong, H., Zhu, W., Liu, J., Wang, X., Liu, G. & Liu, Z. 2014. Pegylated Prussian Blue nanocubes as a theranostic agent for simultaneous cancer imaging and photothermal therapy. *Biomaterials*, 35, 9844–9852.

Choi, J. S., Lee, J. H., Shin, T. H., Song, H. T., Kim, E. Y. & Cheon, J. 2010. Self-confirming "AND" logic nanoparticles for fault-free MRI. *Journal of the American Chemical Society*, 132, 11015–11017.

Colilla, M., González, B. & Vallet-Regí, M. 2013. Mesoporous silica nanoparticles for the design of smart delivery nanodevices. *Biomaterials Science*, 1, 114–134.

Conde, J., Dias, J. T., Grazú, V., Moros, M., Baptista, P. V. & De La Fuente, J. M. 2014. Revisiting 30 years of biofunctionalization and surface chemistry of inorganic nanoparticles for nanomedicine. *Frontiers in Chemistry*, 2, 48.

Cruz, M. M., Ferreira, L. P., Ramos, J., Mendo, S. G., Alves, A. F., Godinho, M. & Carvalho, M. D. 2017. Enhanced magnetic hyperthermia of CoFe2O4 and MnFe2O4 nanoparticles. *Journal of Alloys and Compounds*, 703, 370–380.

Cui, X., Belo, S., Krüger, D., Yan, Y., De Rosales, R. T., Jauregui-Osoro, M., Ye, H., Su, S., Mathe, D., Kovács, N., Horváth, I., Semjeni, M., Sunassee, K., Szigeti, K., Green, M. A. & Blower, P. J. 2014. Aluminium hydroxide stabilised MnFe2O4 and Fe3O4 nanoparticles as dual-modality contrasts agent for MRI and PET imaging. *Biomaterials*, 35, 5840–5846.

Enoch, I. V. M. V., Ramasamy, S., Mohiyuddin, S., Gopinath, P. & Manoharan, R. 2018. Cyclodextrin–PEG conjugate-wrapped magnetic ferrite nanoparticles for enhanced drug loading and release. *Applied Nanoscience*, 8, 273–284.

Errandonea, D. 2014. AB2O4 compounds at high pressures. In: F. J. Manjon, I. Tiginyanu & V. Ursaki, eds., *Pressure-Induced Phase Transitions in AB2X4 Chalcogenide Compounds*, pp. 53–73. Springer.

Ertürk, A. S. & Elmacı, G. 2018. PAMAM dendrimer functionalized manganese ferrite magnetic nanoparticles: Microwave-assisted synthesis and characterization. *Journal of Inorganic and Organometallic Polymers and Materials*, 28, 2100–2107.

Estelrich, J., Sánchez-Martín, M. J. & Busquets, M. A. 2015. Nanoparticles in magnetic resonance imaging: from simple to dual contrast agents. *International Journal of Nanomedicine*, 10, 1727–1741.

Figueroa-Espí, V., Alvarez-Paneque, A., Torrens, M., Otero-González, A. J. & Reguera, E. 2011. Conjugation of manganese ferrite nanoparticles to an anti Sticholysin monoclonal antibody and conjugate applications. *Colloids and Surfaces A: Physicochemical and Engineering Aspects*, 387, 118–124.

Floris, A., Ardu, A., Musinu, A., Piccaluga, G., Fadda, A. M., Sinico, C. & Cannas, C. 2011. SPION@ liposomes hybrid nanoarchitectures with high density SPION association. *Soft Matter*, 7, 6239–6247.

Gong, M., Yang, H., Zhang, S., Yang, Y., Zhang, D., Li, Z. & Zou, L. 2016. Targeting T1 and T2 dual modality enhanced magnetic resonance imaging of tumor vascular endothelial cells based on peptides-conjugated manganese ferrite nanomicelles. *International Journal of Nanomedicine*, 11, 4051–4063.

Grimes, R. W., Anderson, A. B. & Heuer, A. H. 1989. Predictions of cation distributions in AB2O4 spinels from normalized ion energies. *Journal of the American Chemical Society*, 111, 1–7.

Haghiri, M. E. & Izanloo, A. 2018. Design and Characterization of colloidal solution of manganese ferrite nanostructure coated with carboxymethyl chitosan. *Materials Chemistry and Physics*, 216, 265–271.

Haribabu, V., Farook, A. S., Goswami, N., Murugesan, R. & Girigoswami, A. 2016. Optimized Mn-doped iron oxide nanoparticles entrapped in dendrimer for dual contrasting role in MRI. *Journal of Biomedical Materials Research Part B, Applied Biomaterials*, 104, 817–824.

He, S., Zhang, H., Liu, Y., Sun, F., Yu, X., Li, X., Zhang, L., Wang, L., Mao, K., Wang, G., Lin, Y., Han, Z., Sabirianov, R. & Zeng, H. 2018. Maximizing specific loss power for magnetic hyperthermia by hard–soft mixed ferrites. *Small*, 14, e1800135.

Hein, C. D., Liu, X. M. & Wang, D. 2008. Click chemistry, a powerful tool for pharmaceutical sciences. *Pharmaceutical Research*, 25, 2216–2230.

Hervault, A. & Thanh, N. T. K. 2014. Magnetic nanoparticle-based therapeutic agents for thermo-chemotherapy treatment of cancer. *Nanoscale*, 6, 11553–11573.

Hu, H., Tian, Z. Q., Liang, J., Yang, H., Dai, A. T., An, L., Wu, H. X. & Yang, S. P. 2011. Surfactant-controlled morphology and magnetic property of manganese ferrite nanocrystal contrast agent. *Nanotechnology*, 22, 085707.

Hu, T., Wang, Z., Lv, P., Chen, K. & Ni, Z. 2018. One-pot synthesis of a highly selective carboxyl-functionalized superparamagnetic probes for detection of alpha-fetoprotein. *Sensors and Actuators B: Chemical*, 266, 270–275.

Huang, G., Li, H., Chen, J., Zhao, Z., Yang, L., Chi, X., Chen, Z., Wang, X. & Gao, J. 2014. Tunable T 1 and T 2 contrast abilities of manganese-engineered iron oxide nanoparticles through size control. *Nanoscale*, 6, 10404–10412.

Huh, Y. -M., Lee, E. -S., Lee, J. -H., Jun, Y. -w, Kim, P. -H., Yun, C. -O., Kim, J. -H., Suh, J. -S. & Cheon, J. 2007. Hybrid nanoparticles for magnetic resonance imaging of target-specific viral gene delivery. *Advanced Materials*, 19, 3109–3112.

Iranmanesh, P., Saeednia, S., Mehran, M. & Dafeh, S. R. 2017. Modified structural and magnetic properties of nanocrystalline MnFe2O4 by pH in capping agent free co-precipitation method. *Journal of Magnetism and Magnetic Materials*, 425, 31–36.

Jang, J. T., Nah, H., Lee, J. H., Moon, S. H., Kim, M. G. & Cheon, J. 2009. Critical enhancements of MRI contrast and hyperthermic effects by dopant-controlled magnetic nanoparticles. *Angewandte Chemie*, 121, 1260–1264.

Jing, X., Zhi, Z., Wang, D., Liu, J., Shao, Y. & Meng, L. 2018. Multifunctional nanoflowers for simultaneous multimodal imaging and high-sensitivity chemo-photothermal treatment. *Bioconjugate Chemistry*, 29, 559–570.

Kafshgari, L. A., Ghorbani, M. & Azizi, A. 2018. Synthesis and characterization of manganese ferrite nanostructure by co-precipitation, sol-gel, and hydrothermal methods. *Particulate Science and Technology*, 1–7.

Kim, J., Cho, H. R., Jeon, H., Kim, D., Song, C., Lee, N., Choi, S. H. & Hyeon, T. 2017. Continuous O2–evolving MnFe2O4 nanoparticle-anchored mesoporous silica nanoparticles for efficient photodynamic therapy in hypoxic cancer. *Journal of the American Chemical Society*, 139, 10992–10995.

Surendra, M. K., Annapoorani, S., Ansar, E. B., Harikrishna Varma, P. R. & Ramachandra Rao, M. S. 2014. Magnetic hyperthermia studies on water-soluble polyacrylic acid-coated cobalt ferrite nanoparticles. *Journal of Nanoparticle Research*, 16, 2773.

Le, A. T., Giang, C. D., Tam, le T., Tuan, T. Q., Phan, V. N., Alonso, J., Devkota, J., Garaio, E., García, J. Á, Martín-Rodríguez, R., Fdez-Gubieda, M. L., Srikanth, H. & Phan, M. H. 2016. Enhanced magnetic anisotropy and heating efficiency in multi-functional manganese ferrite/graphene oxide nanostructures. *Nanotechnology*, 27, 155707.

Lee, J. H., Huh, Y. M., Jun, Y. W., Seo, J. W., Jang, J. T., Song, H. T., Kim, S., Cho, E. J., Yoon, H. G., Suh, J. S. & Cheon, J. 2007. Artificially engineered magnetic nanoparticles for ultra-sensitive molecular imaging. *Nature Medicine*, 13, 95–99.

Lee, J. H., Jang, J. T., Choi, J. S., Moon, S. H., Noh, S. H., Kim, J. W., Kim, J. G., Kim, I. S., Park, K. I. & Cheon, J. 2011. Exchange-coupled magnetic nanoparticles for efficient heat induction. *Nature Nanotechnology*, 6, 418–422.

Lee, J. H., Lee, K., Moon, S. H., Lee, Y., Park, T. G. & Cheon, J. 2009. All-in-one target-cell-specific magnetic nanoparticles for simultaneous molecular imaging and siRNA delivery. *Angewandte Chemie*, 121, 4238–4243.

Lee, T., Lim, E.-K., Lee, J., Kang, B., Choi, J., Park, H. S., Suh, J.-S., Huh, Y.-M. & Haam, S. 2013. Efficient CD44-targeted magnetic resonance imaging (MRI) of breast cancer cells using hyaluronic acid (HA)-modified MnFe 2 O 4 nanocrystals. *Nanoscale Research Letters*, 8, 149.

Li, Z., Ma, Y. & Qi, L. 2014. Controlled synthesis of $Mn_xFe_{1-x}O$ concave nanocubes and highly branched cubic mesocrystals. *CrystEngComm*, 16, 600–608.

Lim, J.-W., Son, S. U. & Lim, E.-K. 2018. Recent advances in bioimaging for cancer research. In: M. S. Ghamsari, ed., *State of the Art in Nano-Bioimaging*, pp. 11–33. IntechOpen.

Liu, C., Zou, B., Rondinone, A. J. & Zhang, Z. J. 2000. Reverse micelle synthesis and characterization of superparamagnetic MnFe2O4 spinel ferrite nanocrystallites. *The Journal of Physical Chemistry B*, 104, 1141–1145.

Liu, X. L., Wang, Y. T., Ng, C. T., Wang, R., Jing, G. Y., Yi, J. B., Yang, J., Bay, B. H., Yung, L. Y. L. & Fan, D. D. 2014. Coating engineering of MnFe2O4 nanoparticles with superhigh T2 relaxivity and efficient cellular uptake for highly sensitive magnetic resonance imaging. *Advanced Materials Interfaces*, 1, 1300069.

Lu, J., Ma, S., Sun, J., Xia, C., Liu, C., Wang, Z., Zhao, X., Gao, F., Gong, Q., Song, B., Shuai, X., Ai, H. & Gu, Z. 2009. Manganese ferrite nanoparticle micellar nanocomposites as MRI contrast agent for liver imaging. *Biomaterials*, 30, 2919–2928.

Maeda, Y., Wei, Z., Ikezoe, Y. & Matsui, H. 2015. Enzyme-mimicking peptides to catalytically grow ZnO nanocrystals in non-aqueous environments. *ChemNanoMat*, 1, 319–323.

Maeda, Y., Wei, Z., Ikezoe, Y., Tam, E. & Matsui, H. 2016. Biomimetic crystallization of MnFe2O4 mediated by peptide-catalyzed esterification at low temperature. *ChemNanoMat*, 2, 419–422.

Materia, M. E., Pernia Leal, M., Scotto, M., Balakrishnan, P. B., Kumar Avugadda, S., García-Martín, M. L., Cohen, B. E., Chan, E. M. & Pellegrino, T. 2017. Multifunctional magnetic and upconverting nanobeads as dual modal imaging tools. *Bioconjugate Chemistry*, 28, 2707–2714.

Mathew, D. S. & Juang, R.-S. 2007. An overview of the structure and magnetism of spinel ferrite nanoparticles and their synthesis in microemulsions. *Chemical Engineering Journal*, 129, 51–65.

Mauro, M., Crosera, M., Bovenzi, M., Adami, G., Baracchini, E., Maina, G. & Filon, F. L. 2018. In vitro meningeal permeation of MnFe2O4 nanoparticles. *Chemico-Biological Interactions*, 293, 48–54.

Mazarío, E., Sánchez-Marcos, J., Menéndez, N., Cañete, M., Mayoral, A., Rivera-Fernández, S., De La Fuente, J. M. & Herrasti, P. 2015. High specific absorption rate and transverse relaxivity effects in manganese ferrite nanoparticles obtained by an electrochemical route. *The Journal of Physical Chemistry C*, 119, 6828–6834.

Mohammad, F., Balaji, G., Weber, A., Uppu, R. M. & Kumar, C. S. S. R. 2010. Influence of gold nanoshell on hyperthermia of superparamagnetic iron oxide nanoparticles. *The Journal of Physical Chemistry C*, 114, 19194–19201.

Mohapatra, J., Mitra, A., Bahadur, D. & Aslam, M. 2013. Surface controlled synthesis of MFe 2 O 4 (M= Mn, Fe, Co, Ni and Zn) nanoparticles and their magnetic characteristics. *CrystEngComm*, 15, 524–532.

Mohapatra, S., Rout, S. R. & Panda, A. B. 2011. One-pot synthesis of uniform and spherically assembled functionalized MFe2O4 (M=Co, Mn, Ni) nanoparticles. *Colloids and Surfaces A: Physicochemical and Engineering Aspects*, 384, 453–460.

Monaco, I., Armanetti, P., Locatelli, E., Flori, A., Maturi, M., Del Turco, S., Menichetti, L. & Comes Franchini, M. C. 2018. Smart assembly of Mn-ferrites/silica core–shell with fluorescein and gold nanorods: Robust and stable nanomicelles for in vivo triple modality imaging. *Journal of Materials Chemistry B*, 6, 2993–2999.

Mosivand, S. & Kazeminezhad, I. 2015. A novel synthesis method for manganese ferrite nanopowders: The effect of manganese salt as inorganic additive in electrosynthesis cell. *Ceramics International*, 41, 8637–8642.

Muckle, D. S. & Dickson, J. A. 1971. The selective inhibitory effect of hyperthermia on the metabolism and growth of malignant cells. *British Journal of Cancer*, 25, 771–778.

Neto, L. M. M., Zufelato, N., De Sousa-Júnior, A. A., Trentini, M. M., Da Costa, A. C., Bakuzis, A. F., Kipnis, A. & Junqueira-Kipnis, A. P. 2018. Specific T cell induction using iron oxide based nanoparticles as subunit vaccine adjuvant. *Human Vaccines and Immunotherapeutics*, 14, 1–16.

Odio, O. F. & Reguera, E. 2017. Nanostructured spinel ferrites: Synthesis, functionalization, nanomagnetism and environmental applications. In: M. S. Seehra, ed., *Magnetic Spinels-Synthesis, Properties and Applications*, pp. 185–216. InTech.

Oh, Y., Lee, N., Kang, H. W. & Oh, J. 2016. In vitro study on apoptotic cell death by effective magnetic hyperthermia with chitosan-coated MnFe2O4. *Nanotechnology*, 27, 115101.

Pal, M., Rakshit, R. & Mandal, K. 2014. Surface modification of MnFe2O4 nanoparticles to impart intrinsic multiple fluorescence and novel photocatalytic properties. *ACS Applied Materials and Interfaces*, 6, 4903–4910.

Patil, U. S., Adireddy, S., Jaiswal, A., Mandava, S., Lee, B. R. & Chrisey, D. B. 2015. In vitro/in vivo toxicity evaluation and quantification of iron oxide nanoparticles. *International Journal of Molecular Sciences*, 16, 24417–24450.

Peng, E., Choo, E. S. G., Chandrasekharan, P., Yang, C. T., Ding, J., Chuang, K. H. & Xue, J. M. 2012. Synthesis of manganese ferrite/graphene oxide nanocomposites for biomedical applications. *Small*, 8, 3620–3630.

Peng, E., Ding, J. & Xue, J. M. 2014. Concentration-dependent magnetic hyperthermic response of manganese ferrite-loaded ultrasmall graphene oxide nanocomposites. *New Journal of Chemistry*, 38, 2312–2319.

Pereira, C., Costa, R. S., Lopes, L., Bachiller-Baeza, B., Rodríguez-Ramos, I., Guerrero-Ruiz, A., Tavares, P. B., Freire, C. & Pereira, A. M. 2018. Multifunctional mixed valence N-doped CNT@ MFe2O4 hybrid nanomaterials: from engineered one-pot coprecipitation to application in energy storage paper supercapacitors. *Nanoscale*, 10, 12820–12840.

Pereira, C., Pereira, A. M., Fernandes, C., Rocha, M., Mendes, R., Fernández-García, M. P., Guedes, A., Tavares, P. B., Grenèche, J.-M., Araújo, J. P. & Freire, C. 2012. Superparamagnetic MFe2O4 (M= Fe, Co, Mn) nanoparticles: Tuning the particle size and magnetic properties through a novel one-step coprecipitation route. *Chemistry of Materials*, 24, 1496–1504.

Pernia Leal, M., Rivera-Fernández, S., Franco, J. M., Pozo, D., de la Fuente, J. M. & García-Martín, M. L. 2015. Long-circulating pegylated manganese ferrite nanoparticles for MRI-based molecular imaging. *Nanoscale*, 7, 2050–2059.

Pöselt, E., Kloust, H., Tromsdorf, U., Janschel, M., Hahn, C., Maßlo, C. & Weller, H. 2012. Relaxivity optimization of a PEGylated iron-oxide-based negative magnetic resonance contrast agent for T 2-weighted spin–echo imaging. *ACS Nano*, 6, 1619–1624.

Pradhan, P., Giri, J., Banerjee, R., Bellare, J. & Bahadur, D. 2007. Preparation and characterization of manganese ferrite-based magnetic liposomes for hyperthermia treatment of cancer. *Journal of Magnetism and Magnetic Materials*, 311, 208–215.

Qi, H., Liu, C., Long, L., Ren, Y., Zhang, S., Chang, X., Qian, X., Jia, H., Zhao, J., Sun, J., Hou, X., Yuan, X. & Kang, C. 2016. Blood exosomes endowed with magnetic and targeting properties for cancer therapy. *ACS Nano*, 10, 3323–3333.

Quarta, A., Curcio, A., Kakwere, H. & Pellegrino, T. 2012. Polymer coated inorganic nanoparticles: Tailoring the nanocrystal surface for designing Nanoprobes with biological implications. *Nanoscale*, 4, 3319–3334.

Rashid, Z., Soleimani, M., Ghahremanzadeh, R., Vossoughi, M. & Esmaeili, E. 2017. Effective surface modification of MnFe2O4@ SiO2@ PMIDA magnetic nanoparticles for rapid and high-density antibody immobilization. *Applied Surface Science*, 426, 1023–1029.

Reddy, L. H., Arias, J. L., Nicolas, J. & Couvreur, P. 2012. Magnetic nanoparticles: Design and characterization, toxicity and biocompatibility, pharmaceutical and biomedical applications. *Chemical Reviews*, 112, 5818–5878.

Rodrigues, A. R. O., Almeida, B. G., Araújo, J. P., Queiroz, M.-J. R., Coutinho, P. J. & Castanheira, E. M. 2018. Magnetoliposomes for dual cancer therapy. In: A. M. Grumezescu, ed., *Inorganic Frameworks as Smart Nanomedicines*, pp. 489–527. Elsevier.

Rodrigues, A. R. O., Almeida, B., Rodrigues, J. M., Queiroz, M. J. R., Calhelha, R., Ferreira, I. C., Pires, A., Pereira, A., Araújo, J. P. & Coutinho, P. J. 2017. Magnetoliposomes as carriers for promising antitumor thieno [3, 2-b] pyridin-7-arylamines: Photophysical and biological studies. *RSC Advances*, 7, 15352–15361.

Rodrigues, A. R. O., Gomes, I. T., Almeida, B. G., Araújo, J. P., Castanheira, E. M. S. & Coutinho, P. J. G. 2014. Magnetoliposomes based on nickel/silica core/shell nanoparticles: Synthesis and characterization. *Materials Chemistry and Physics*, 148, 978–987.

Rodrigues, A. R. O., Ramos, J. M. F., Gomes, I. T., Almeida, B. G., Araújo, J. P., Queiroz, M. J. R. P., Coutinho, P. J. G. & Castanheira, E. M. S. 2016. Magnetoliposomes based on manganese ferrite nanoparticles as nanocarriers for antitumor drugs. *RSC Advances*, 6, 17302–17313.

Rondinone, A. J., Liu, C. & Zhang, Z. J. 2001. Determination of magnetic anisotropy distribution and anisotropy constant of manganese spinel ferrite nanoparticles. *The Journal of Physical Chemistry B*, 105, 7967–7971.

Safdar, M. H., Hasan, H., Anees, M. & Hussain, Z. 2017. Folic acid-conjugated doxorubicin-loaded photosensitizing manganese ferrite nanoparticles: Synthesis, characterization and anticancer activity against human cervical carcinoma cell line (HELA). *International Journal of Pharmacy and Pharmaceutical Sciences*, 9, 60–67.

Sahoo, B., Devi, K. S. P., Dutta, S., Maiti, T. K., Pramanik, P. & Dhara, D. 2014. Biocompatible mesoporous silica-coated superparamagnetic manganese ferrite nanoparticles for targeted drug delivery and MR imaging applications. *Journal of Colloid and Interface Science*, 431, 31–41.

Sahoo, B., Sahu, S. K., Nayak, S., Dhara, D. & Pramanik, P. 2012. Fabrication of magnetic mesoporous manganese ferrite nanocomposites as efficient catalyst for degradation of dye pollutants. *Catalysis Science and Technology*, 2, 1367–1374.

Seow, Y. & Wood, M. J. 2009. Biological gene delivery vehicles: Beyond viral vectors. *Molecular Therapy*, 17, 767–777.

Shah, S. A., Asdi, M. H., Hashmi, M. U., Umar, M. F. & Awan, S.-U. 2012. Thermo-responsive copolymer coated MnFe2O4 magnetic nanoparticles for hyperthermia therapy and controlled drug delivery. *Materials Chemistry and Physics*, 137, 365–371.

Shah, S. A., Majeed, A., Rashid, K. & Awan, S.-U. 2013. PEG-coated folic acid-modified superparamagnetic MnFe2O4 nanoparticles for hyperthermia therapy and drug delivery. *Materials Chemistry and Physics*, 138, 703–708.

Sheng, Y., Li, S., Duan, Z., Zhang, R. & Xue, J. 2018. Fluorescent magnetic nanoparticles as minimally-invasive multi-functional theranostic platform for fluorescence imaging, MRI and magnetic hyperthermia. *Materials Chemistry and Physics*, 204, 388–396.

Shi, X. & Shen, L. 2018. Integrin αvβ3 receptor targeting PET/MRI dual-modal imaging probe based on the 64Cu labeled manganese ferrite nanoparticles. *Journal of Inorganic Biochemistry*, 186, 257–263.

Silambarasu, A., Manikandan, A., Balakrishnan, K., Jaganathan, S. K., Manikandan, E., Aanand, J. S. 2018. Comparative study of structural, morphological, magneto-optical and photo-catalytic properties of magnetically reusable spinel MnFe2O4 nano-catalysts. *Journal of Nanoscience and Nanotechnology*, 18, 3523–3531.

Song, Q., Ding, Y., Wang, Z. L. & Zhang, Z. J. 2007. Tuning the thermal stability of molecular precursors for the nonhydrolytic synthesis of magnetic MnFe2O4 spinel nanocrystals. *Chemistry of Materials*, 19, 4633–4638.

Soukup, D., Moise, S., Céspedes, E., Dobson, J. & Telling, N. D. 2015. In situ measurement of magnetization relaxation of internalized nanoparticles in live cells. *ACS Nano*, 9, 231–240.

Sperling, R. A. & Parak, W. J. 2010. Surface modification, functionalization and bioconjugation of colloidal inorganic nanoparticles. *Philosophical Transactions of the Royal Society of London A: Mathematical, Physical and Engineering Sciences*, 368, 1333–1383.

Stanicki, D., Vander Elst, L., Muller, R. N., Laurent, S., Felder-Flesch, D., Mertz, D., Parat, A., Begin-Colin, S., Cotin, G. & Greneche, J.-M. 2017. Iron-oxide nanoparticle-based contrast agents for MRI. *Molecular Pharmaceutics*, 14, 1352–1364.

Sun, S., Zeng, H., Robinson, D. B., Raoux, S., Rice, P. M., Wang, S. X. & Li, G. 2004. Monodisperse MFe2O4 (M= Fe, Co, Mn) nanoparticles. *Journal of the American Chemical Society*, 126, 273–279.

Tromsdorf, U. I., Bigall, N. C., Kaul, M. G., Bruns, O. T., Nikolic, M. S., Mollwitz, B., Sperling, R. A., Reimer, R., Hohenberg, H., Parak, W. J., Förster, S., Beisiegel, U., Adam, G. & Weller, H. 2007. Size and surface effects on the MRI relaxivity of manganese ferrite nanoparticle contrast agents. *Nano Letters*, 7, 2422–2427.

Vadivel, M., Babu, R. R., Arivanandhan, M., Ramamurthi, K. & Hayakawa, Y. 2015. Role of SDS surfactant concentrations on the structural, morphological, dielectric and magnetic properties of CoFe 2 O 4 nanoparticles. *RSC Advances*, 5, 27060–27068.

Vamvakidis, K., Katsikini, M., Sakellari, D., Paloura, E., Kalogirou, O. & Dendrinou-Samara, C. 2014. Reducing the inversion degree of MnFe2O4 nanoparticles through synthesis to enhance magnetization: evaluation of their 1 H NMR relaxation and heating efficiency. *Dalton Transactions*, 43, 12754–12765.

Venkatesha, N., Pudakalakatti, S. M., Qurishi, Y., Atreya, H. S. & Srivastava, C. 2015. MnFe 2 O 4–Fe 3 O 4 core–shell nanoparticles as a potential contrast agent for magnetic resonance imaging. *RSC Advances*, 5, 97807–97815.

Vestal, C. R. & Zhang, Z. J. 2003. Synthesis and magnetic characterization of Mn and Co spinel ferrite-silica nanoparticles with tunable magnetic core. *Nano Letters*, 3, 1739–1743.

Vestal, C. R. & Zhang, Z. J. 2004. Magnetic spinel ferrite nanoparticles from microemulsions. *International Journal of Nanotechnology*, 1, 240–263.

Vetr, F., Moradi-Shoeili, Z. & Özkar, S. 2018. Oxidation of o-phenylenediamine to 2, 3-diaminophenazine in the presence of cubic ferrites MFe2O4 (M= Mn, Co, Ni, Zn) and the application in colorimetric detection of H2O2. *Applied Organometallic Chemistry*, 32, e4465.

Wang, A., Qi, W., Wang, N., Zhao, J., Muhammad, F., Cai, K., Ren, H., Sun, F., Chen, L., Guo, Y., Guo, M. & Zhu, G. 2013. A smart nanoporous theranostic platform for simultaneous enhanced MRI and drug delivery. *Microporous and Mesoporous Materials*, 180, 1–7.

Wang, G., Zhao, D., Ma, Y., Zhang, Z., Che, H., Mu, J., Zhang, X. & Zhang, Z. 2018. Synthesis and characterization of polymer-coated manganese ferrite nanoparticles as controlled drug delivery. *Applied Surface Science*, 428, 258–263.

Wang, S., Gao, R., Zhou, F. & Selke, M. 2004. Nanomaterials and singlet oxygen photosensitizers: Potential applications in photodynamic therapy. *Journal of Materials Chemistry*, 14, 487–493.

Wang, T. & Guo, Z. 2006. Copper in medicine: Homeostasis, chelation therapy and antitumor drug design. *Current Medicinal Chemistry*, 13, 525–537.

Wu, W., Jiang, C. Z. & Roy, V. A. 2016. Designed synthesis and surface engineering strategies of magnetic iron oxide nanoparticles for biomedical applications. *Nanoscale*, 8, 19421–19474.

Xiao, N., Gu, W., Wang, H., Deng, Y., Shi, X. & Ye, L. 2014. T1–T2 dual-modal MRI of brain gliomas using pegylated Gd-doped iron oxide nanoparticles. *Journal of Colloid and Interface Science*, 417, 159–165.

Xie, J., Peng, S., Brower, N., Pourmand, N., Wang, S. X. & Sun, S. 2006. One-pot synthesis of monodisperse iron oxide nanoparticles for potential biomedical applications. *Pure and Applied Chemistry*, 78, 1003–1014.

Xie, J., Yan, C., Zhang, Y. & Gu, N. 2013. Shape evolution of "multibranched" Mn–Zn ferrite nanostructures with high performance: A transformation of nanocrystals into nanoclusters. *Chemistry of Materials*, 25, 3702–3709.

Xu, Y., Qin, Y., Palchoudhury, S. & Bao, Y. 2011. Water-soluble iron oxide nanoparticles with high stability and selective surface functionality. *Langmuir*, 27, 8990–8997.

Yang, G., He, F., Lv, R., Gai, S., Cheng, Z., Dai, Y. & Yang, P. 2015. A cheap and facile route to synthesize monodisperse magnetic nanocrystals and their application as MRI agents. *Dalton Transactions*, 44, 247–253.

Yang, H., Zhuang, Y., Sun, Y., Dai, A., Shi, X., Wu, D., Li, F., Hu, H. & Yang, S. 2011. Targeted dual-contrast T1-and T2-weighted magnetic resonance imaging of tumors using multifunctional gadolinium-labeled superparamagnetic iron oxide nanoparticles. *Biomaterials*, 32, 4584–4593.

Yang, H., Zhang, C., Shi, X., Hu, H., Du, X., Fang, Y., Ma, Y., Wu, H. & Yang, S. 2010. Water-soluble superparamagnetic manganese ferrite nanoparticles for magnetic resonance imaging. *Biomaterials*, 31, 3667–3673.

Yang, L., Xin, L., Li, J., Sun, A., Wei, C., Ren, R., Chen, B. W., Lin, Z. H. & Gao, J. 2017. Composition tunable manganese ferrite nanoparticles for optimized T 2 contrast ability. *Chemistry of Materials*, 29, 3038–3047.

Yang, L., Wang, Z., Ma, L., Li, A., Xin, J., Wei, R., Lin, H., Wang, R., Chen, Z. & Gao, J. 2018. The roles of morphology on the relaxation rates of magnetic nanoparticles. *ACS Nano*, 12, 4605–4614.

Yang, Y., Shi, H., Wang, Y., Shi, B., Guo, L., Wu, D., Yang, S. & Wu, H. 2016. Graphene oxide/manganese ferrite nanohybrids for magnetic resonance imaging, photothermal therapy and drug delivery. *Journal of Biomaterials Applications*, 30, 810–822.

Yoon, T. J., Lee, H., Shao, H. & Weissleder, R. 2011. Highly magnetic core–shell nanoparticles with a unique magnetization mechanism. *Angewandte Chemie*, 50, 4663–4666.

Yu, M. K., Park, J. & Jon, S. 2012. Targeting strategies for multifunctional nanoparticles in cancer imaging and therapy. *Theranostics*, 2, 3–44.

Yue, Z., Zhou, J., Li, L., Zhang, H. & Gui, Z. 2000. Synthesis of nanocrystalline NiCuZn ferrite powders by sol–gel auto-combustion method. *Journal of Magnetism and Magnetic Materials*, 208, 55–60.

Zahraei, M., Marciello, M., Lazaro-Carrillo, A., Villanueva, A., Herranz, F., Talelli, M., Costo, R., Monshi, A., Shahbazi-Gahrouei, D., Amirnasr, M., Behdadfar, B. & Morales, M. P. 2016. Versatile theranostics agents designed by coating ferrite nanoparticles with biocompatible polymers. *Nanotechnology*, 27, 255702.

Zeng, H., Rice, P. M., Wang, S. X. & Sun, S. 2004. Shape-controlled synthesis and shape-induced texture of MnFe2O4 nanoparticles. *Journal of the American Chemical Society*, 126, 11458–11459.

Zhang, H., Huang, H., He, S., Zeng, H. & Pralle, A. 2014. Monodisperse magnetofluorescent nanoplatforms for local heating and temperature sensing. *Nanoscale*, 6, 13463–13469.

Zhang, H., Li, L., Liu, X. L., Jiao, J., Ng, C.-T., Yi, J. B., Luo, Y. E., Bay, B.-H., Zhao, L. Y., Peng, M. L., Gu, N. & Fan, H. M. 2017a. Ultrasmall ferrite nanoparticles synthesized via dynamic simultaneous thermal decomposition for high-performance and multifunctional T 1 magnetic resonance imaging contrast agent. *ACS Nano*, 11, 3614–3631.

Zhang, H., Liu, X. L., Zhang, Y. F., Gao, F., Li, G. L., He, Y., Peng, M. L. & Fan, H. M. 2018. Magnetic nanoparticles based cancer therapy: Current status and applications. *Science in China (Life Sciences)*, 61, 400–414.

Zhang, L., Xia, J., Zhao, Q., Liu, L. & Zhang, Z. 2010. Functional graphene oxide as a nanocarrier for controlled loading and targeted delivery of mixed anticancer drugs. *Small*, 6, 537–544.

Zhang, Q., Yin, T., Gao, G., Shapter, J. G., Lai, W., Huang, P., Qi, W., Song, J. & Cui, D. 2017b. Multifunctional core@ shell magnetic Nanoprobes for enhancing targeted magnetic resonance imaging and fluorescent labeling in vitro and in vivo. *ACS Applied Materials and Interfaces*, 9, 17777–17785.

Zhang, S., Niu, H., Zhang, Y., Liu, J., Shi, Y., Zhang, X. & Cai, Y. 2012. Biocompatible phosphatidylcholine bilayer coated on magnetic nanoparticles and their application in the extraction of several polycyclic aromatic hydrocarbons from environmental water and milk samples. *Journal of Chromatography A*, 1238, 38–45.

Zhou, X., Lv, X., Zhao, W., Zhou, T., Zhang, S., Shi, Z., Ye, S., Ren, L. & Chen, Z. 2018. Porous MnFe 2 O 4-decorated PB nanocomposites: A new theranostic agent for boosted T 1/T 2 MRI-guided synergistic photothermal/magnetic hyperthermia. *RSC Advances*, 8, 18647–18655.

Zipare, K., Dhumal, J., Bandgar, S., Mathe, V. & Shahane, G. 2015. Superparamagnetic manganese ferrite nanoparticles: Synthesis and magnetic properties. *Journal of Nanoscience and Nanoengineering*, 1, 178–182.

# Part II

# Environmental Applications

# 7

# $TiO_2$-Based Nanohybrids for Energy and Environmental Applications

**Khuram Ali, Syedda Shaher Bano, and Yasir Javed**

**CONTENTS**



## 7.1 Introduction

Nanohybrid materials are among the most important elements that grabbed the attention of scientists over the past few years owing to their unique properties in the field of nanoparticles [1]. Nanohybrid materials are basically mixtures of two or more materials possessing different properties, resulting in a new material with modified properties and functions at atomic or subatomic level. The process of formation of nanohybrid materials takes place at a nanometric range and is formed by covalent bonds between polymer and organic or inorganic materials, optimally serving a specific engineering purpose. Addition of inorganic and organic materials on a nanometric scale makes this area a point of concern for many researchers, physicists and scientists [2]. This review will briefly describe titanium as a nanohybrid material and its vast applications at the atomic and subatomic levels. Titanium dioxide is a naturally occurring nanomaterial which is extracted from the core of the Earth and utilized at an industrial level and in many useful products. Titanium dioxide has been used as a nanohybrid along with zinc oxide in order to enhance the photolytic activity at nanoscale [3]. By studying the combined properties of zinc oxide and titanium dioxide, higher catalytic activities are achieved. This combination of the properties of both materials has opened new ways for the formation of materials that can solve many technical problems in future. This will help in finding how nanohybrid structures will help in finding the technological solutions to design new materials for future implantations [4, 5]. According to a scientist named Hagiwara, nanohybrid materials represent an intellectual combination of materials co-operating with each other to have high functional properties which the individual components did not bear. According to his point of view, the difference between nanohybrid and composite materials lies in their properties and functions and the way they act as a unit [6]. When compared with standard nanocomposite, nanohybrids materials bear many useful properties. Nanohybrid materials can be distinguished by three main classes relying upon their structural and bonding properties:

- Structurally nanohybridized materials
- Functionally hybridized nanomaterials
- Materials with hybridized structure having chemical bonding.

Recently a product is launched which is titanium dioxide (RUTILE) TR996. This product has a nanohybrid structure manufactured by inorganic and organic surface treatment technology followed by sulphate process. This nanohybrid structure has very efficient dispersion, very good opacity and remarkable anti chalking performance [7].

## 7.2 Nanohybrid Composites Materials

Many scientists are curious to find the difference between nanocomposites and nanohybrid materials. In the meantime, they have discovered a new class of hybridized nanocomposites as a combination of both composites and hybrid materials at the nanometric range, having more than two fibers or a nanocomposite consisting of thin foils metals and fiber reinforced metals (FRM) [8]. In nanohybrid composite materials, the meaning of hybrid represents the macroscopic geometry on a large scale. When discussing functional materials, hybrid nanomaterials are preferable compared to composites at a nanometric scale. In the preset discussion, nanohybrid composite materials account for the hybridization of the mixture at a very small scale on an order of a few nanometers [9]. The functional properties of these materials can be easily explained by the combination of all the properties of its maternal material, and these materials should be called structurally hybridized nanomaterials [10]. On the other hand, the composites on the macroscopic range have quite different properties than those of composites in the nanoscale range. The fine microstructure and the grain boundaries of nanohybrids make them best suited to be used in many products. Some nanohybrid materials have remarkable properties, owing to the strong bonding between them present at the interfaces between different component materials [11]. On the basis of geometrical differences in the hybridization of organic and inorganic materials, nanohybrid materials can be divided into three classes. The first example of organic and inorganic nanohybrids includes the modified silicate deposited by the sol–gel technique. Owing to the existence of strong bonding between organic molecules and silica atoms at the molecular/micro level, this nanohybrid has remarkable mechanical properties [12]. Nanohybrid materials are deposited in such a way that not only are organic clusters dispersed into polymer materials but they are also distinguishable by the presence of chemical bonding between both molecules when compared with traditional nanohybrids. There are many different examples regarding organic and inorganic nanohybrid materials. A clay polymer hybrid is one example of nanohybrid organic or inorganic material. Owing to the strong bonding present between silicate layers and polymers, these nanohybrids have much improved mechanical properties than those of other materials [13].

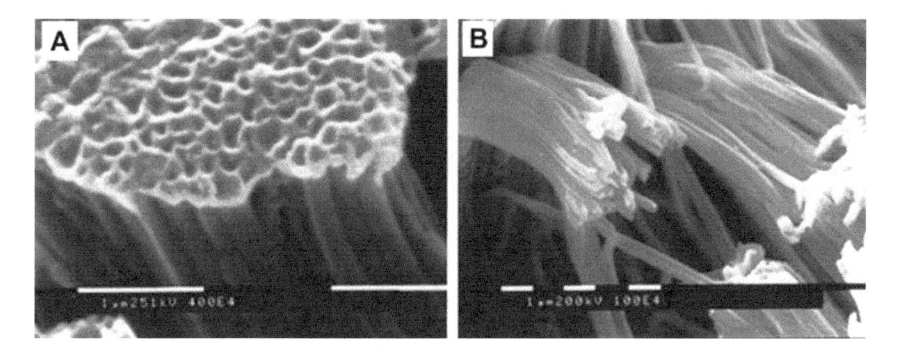


**FIGURE 7.1** SEM images of $TiO_2$ nanotubules prepared by sol–gel methods, before (A) and after (B) filling with the polypyrrole nanowires. Outer diameter of tubular composite is 200 nm. (Figure reprinted/adapted with permission from Ref. [17].)

In recent years, the study of one-dimensional (1D) nanomaterials such as nanorods, nanowires, etc., have entered in the era of fast developmental approach. Owing to the unique properties of 1D nanomaterials, their way of formation, geometry and interaction among other materials can be explained briefly by their complete geometrical analysis. These unique structures play a vital role in understanding the basic building blocks and functional units in the deposition of various optical and electrical devices at nanoscale [14]. The combination of active organic and inorganic material components represents the new and vast area of materials science which will produce remarkable effects on the development and helps in understanding the design of various materials with well-define function in nanoscale range. This combination of organic and inorganic nanohybrids made a class of nanohybrid composites which helps in understanding the formation of nanohybrid composite materials [15]. A number of factors are responsible for understanding the one-dimensional inorganic and organic hybrid materials. When compared with inorganic materials many one-dimensional nanohybrid materials have high stability, thus reducing the surface tension and even improving the rate of solubility in the solution. Stability of the nanohybrid materials is the primary requirement for understanding the future applications involving the utilization of organic and inorganic nanohybrid materials [16]. $TiO_2$ nanotubules prepared by sol–gel methods, before and after filling with the polypyrrole nanowires are shown in Figure 7.1.

## 7.3 Titanium Dioxide as a Nanohybrid

Semiconductor materials prove to be very useful materials in various fields including the degradation of pollutants, emission of hydrogen gas from dispersion of water, fabrication processes for different substrates and photo induced gas sensors [18]. Titanium dioxide offers a very high binding energy for excitation and has various crystal geometries, different phases and a huge band gap (3 to 3.26 eV) [19, 20]. Titanium dioxide is unable to react with other elements at normal temperature, stable to light, contains no toxicity and is cheap and easily available [21]. Titanium dioxide may also be employed for many useful applications and have distinguishable characteristics when compared with single crystals geometry and thin films. Titanium dioxide acts as an efficient photocatalyst for the removal of air pollution. Titanium dioxide shows immediate response to the harmful ultraviolet radiations which contains only a few percent of solar radiations [22]. While in the presence of visible light, this photocatalytic activity of titanium dioxide is controlled by the fact that there are more chances of high combination rates of electrons and holes transformation. Titanium dioxide is currently prepared as nanoparticles, core shells and nanorods using low-cost methods. To avoid poor photocatalytic performance under solar radiations, titanium dioxide acts as a functional source to enhance that photocatalytic activity [23]. During the manufacture of dyes, wastewater becomes polluted by more than 20% dye content on an industrial scale and makes the water harmful for consumers. Thus, there is a need for a low-cost and efficient technique for the removal of dyes from polluted water to clean water on a large scale. Thus, there is a need for photocatalysts like titanium dioxide which can be utilized to clean the water using a nontoxic and low-cost technique [24]. The main objective of titanium dioxide nanohybrid materials is to enhance the environmental stability and quantum efficiency under different pH and utilize visible light for the removal of organic pollutants. In this chapter, we have reviewed recent advances of titanium dioxide nanohybrids along with the polymer and carbon-based nanohybrids, their methods of preparation and photocatalytic applications. The catalytic activity of titanium dioxide is enhanced by varying its morphology and compositions. Decomposition of organic pollutants, charge carrier separation and charge carrier combinations are the major factors responsible for the high photocatalytic activity of titanium dioxide.

## 7.4 Titanium Dioxide and Graphene as a Nanohybrid

The reduction of nonrenewable assets due to climatic changes has forced the scientists to explore alternate sources of energy. Investigations are being carried out for the replacement of many costly metals with cheap and highly stable non-noble metals to make microbial fuel cells more practical and useful. Nowadays many nanohybrids materials such as reduced graphene oxide (rGO) and titanium dioxide oxide ($TIO_2$) have gathered a remarkable attention as catalysts in many chemical reactions [25]. rGO and titanium dioxide ($TiO_2$) for catalyst applications are shown in Figure 7.2. The properties of titanium dioxide along with the properties of graphene have resulted in its particular applications in depositing composite electrodes for microbial fuel cells. However, because of the deposition of electrodes, binding or cross-linking materials make the electrodes very expensive and the process much more complicated [26]. This work is then more proficient with the aim of entering efficient cathodes compared to graphite. The overall electrochemical performance can be improved by the conductivity of graphene and titanium dioxide nanohybrids.

### 7.4.1 Titanium Dioxide as a Photocatalyst

The process of photocatalysis has been mostly used for the purification of water and air. Titanium dioxide oxide proves to be the most consistent material for the degradation of many organic compounds owing to its non-toxicity, high reactivity and chemical stability. However the recovery of titanium dioxide nanoparticles from treated water is still a great challenge. Immobilization of titanium dioxide has been noticed on glass, ceramics, cellulose and activated carbon [28]. This property greatly enhances the stability of titanium dioxide and helps in the separation of catalyst from aqueous solution [29].

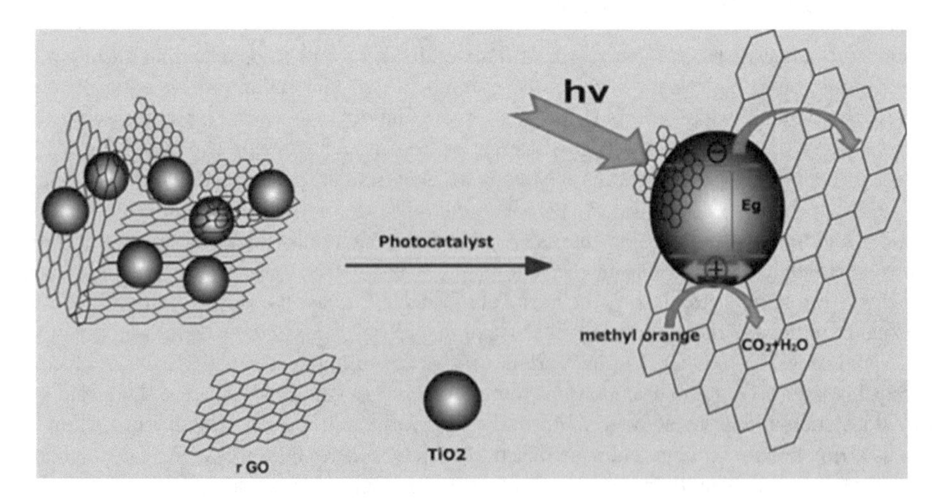


**FIGURE 7.2** Illustration of the rGO supported $TiO_2$ for the photocatalyst. (Figure reprinted/adapted with permission from Ref. [27].)

The process of immobilizating titanium dioxide leads to a considerable decrease in the surface area which will ensure unique properties that help in the separation and recovery of these materials. However, the synthesis of titanium dioxide requires a high temperature and the assembling of organic templates. There is still a need to develop an entirely new generation of photocatalysts that endure high efficiency and recovery at fairly low temperatures. Titanium dioxide is an excellent photocatalyst that helps in the breakdown of many organic compounds [30]. Titanium dioxide has self-sterilizing surfaces that can be used for killing bacteria. Photocatalysis of titanium dioxide involves the medical diagnosis of many diseases such as cancer. Treatment of cancer is one of the major possibilities linked with photo catalysis. Many radiological, physiotherapeutic, chemotherapeutic and surgical problems have been cured and are contributing much to the treatment of patients; however, cancer has remained the main cause of death over the past 20 years [28]. Illuminated titanium dioxide has been used for killing many tumor cells, as it possesses high oxidizing power. In some experiments, polymerized illuminated titanium dioxide has been used in the form of a film electrode, which helps in the effective killing of cancerous cells. Using a polymerized microelectrode of titanium dioxide, a single cancerous cell may also be killed. Titanium dioxide has unique photocatalytic properties that are utilized in many applications [31].

### 7.4.2 Graphene as a Photocatalyst

Graphene has attracted many scientists to study its unique photocatalytic property because of its high mobility of charge carriers, good electricity and thermal conduction and large surface area [32]. In order to construct composite materials and improve the photocatalytic properties, many materials are deposited on graphene sheets by utilizing the unique properties of graphene [33]. Most of the metal oxides and metals have been deposited on graphene sheets and as a result of this deposition the resultant composites endure high photocatalytic activity. Owing to strong forces of attraction, graphene is insoluble in water and polar organic solvents easily aggregate due to this property. The deposition of metal oxide on graphene is difficult owing to the hydrophilic incompatibility between inorganic compounds and graphene. Thus, a well-defined method is used for solving this incompatibility issue [34]. Moreover, there is difficulty in depositing graphene on a large scale, which becomes a barrier in its applications. Graphene oxide (GO) sheets contain a large amount of reactive oxygen functional groups on their surface area that makes them suitable for metal oxide and metals. To develop graphene sheet deposition on a large scale and in order to solve the incompatibility issues, the hydrothermal method has been used. Thus, by depositing the metals and metal oxide particles on the surface of grapheme, the photocatalytic properties of graphene can be enhanced [35].

Recently, GO has attracted many scientists to study its photocatalytic properties. The GO band gap can be adjusted by changing the oxidation level. Partially oxidized GO is used as a semi conductor while fully oxidized GO is used as an insulator. Normally, the synthesis of GO is carried out by the oxidation of graphite powder using Hummer's methods. After that brown colloidal suspension is formed. Further, graphite powder is mixed in sulfuric acid for specified amount of time.

After some time, potassium permanganate ($KMnO_4$) is added drop by drop into this solution at a temperature below 20°C and then hydrogen per oxide is added in this suspension of distilled water. At the end of this process graphite oxide suspension is washed by centrifuging it several times. Centrifugation is carried out by 4% of HCL then washed with distilled water. After getting a neutral pH level, nanostructures of GO are formed by adding an excess amount of water to the precipitates. At the end of process, uniform suspension is achieved by sonication method [36]. GO has a unique carbon structure, which can bare high electronic properties. GO is used as a mediator for photocatalytic splitting of water by tuning the electronic structure of GO alone or with other materials.

### 7.4.3 Titanium Dioxide and Graphene Oxide Nanohybrids as Photocatalysts

A number of well defined methods have been used in order to synthesize titanium dioxide and graphite oxide as a highly well-organized photocatalyst. These are prepared by depositing titanium dioxide on GO nanosheets by liquid phase deposition [37]. This process is carried out by calcinations treatment at a temperature above 200°C. GO and titanium dioxide bear high photocatalytic activity but depend upon

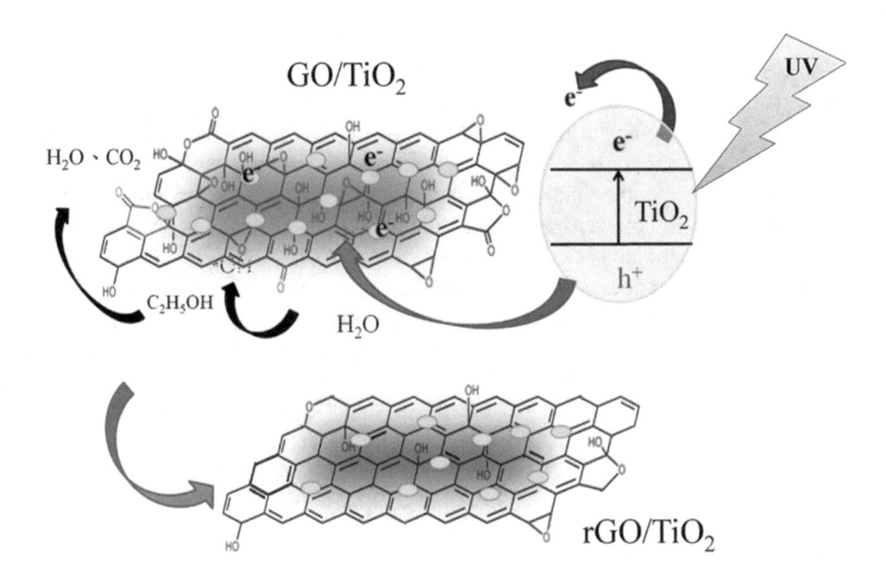


FIGURE 7.3 rGO–$TiO_2$ composites via the UV-assisted photocatalytic reduction of GO. (Figure reprinted/adapted with permission from Ref. [41].)

post-calcination temperatures, pH solution and GO contents. Remarkable enhancement of the photocatalytic properties of titanium dioxide has been achieved by using two-dimensional GO nanosheets [38, 39]. The sheet contains a large surface area with enhanced absorption capacity and the ability to transfer the electrons of thermally rGO. The combination of titanium dioxide and GO makes it more suitable to be utilized in a number of applications such as gas sensors, solar cells, optical and electronic devices and diagnosis of many diseases [40]. rGO–$TiO_2$ composites used in UV assisted photocatalytic reduction process of GO are shown in Figure 7.3.

### 7.4.4 Temperature Affecting the Photocatalytic Activity of Nanohybrids

The effect of temperature on the photocatalytic activity of nanohybrids has been studied using Raman spectroscopy. Raman spectroscopy is used to study the misalignment of the crystal structure of many materials [42]. Raman spectra of graphite (G), GO, rGO and the rGO nanohybrid is shown in Figure 7.4. Band G shows the recorded values for the Raman spectra of GO and titanium dioxide before and after the heat treatments. In the Raman spectra, before the heat treatments, the intensity increases to a very large value, seen at 1597 and 1350 cm. The peak intensity of the G and D bands is observed at 1607 and 1345 respectively. There is a decrease in the size of the domain after calcinations and the peak intensity ratios are measured from one to eight. This could be done by removing the oxygen containing groups after heat treatments. It has been noted that the resulting peaks of titanium dioxide and graphene composites are certified with no peak found on 460.01 and 465.8.

## 7.5 Carbon Dots and Titanium Dioxide as a Nanohybrids

Titanium dioxide have been used as one of the most important material having remarkable photocatalytic activity owing to its unique properties like high stability, non-toxicity and low cost. This photocatalytic activity originates from the electrons and holes which are photo generated charge carriers and are strong oxidizing agents [44]. However, due to a wide band gap of titanium dioxide, it gives a very efficient response to ultraviolet light, thus making the energy of solar spectrum unstable. To enhance the photocatalytic performance of titanium dioxide under visible light, many efforts have been made, such as doping to coupling with many other metal ions and semiconductors. Among

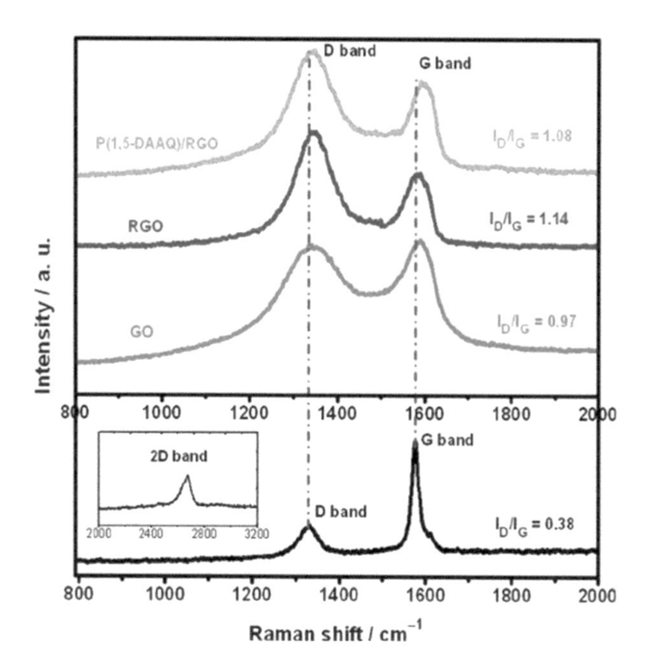


**FIGURE 7.4 (See color insert.)** Raman spectra of graphite, GO, rGO and the P(1,5-DAAQ)/rGO nanohybrid. The inset shows the 2D band of graphite. (Figure reprinted/adapted with permission from Ref. [43].)

these approaches, the basic idea is generation of energy levels in the band gap. Scientists have performed many experiments with sulfur, carbon and boron as doping material with titanium dioxide [45]. The recombination of different photo-induced charge carriers is responsible for the introduction of different defects and impurities [18]. Carbon nanotubes and graphene have been used to hinder the recombination of many charge carriers. But due to the limited contact between titanium dioxide and graphene nanoparticles, adverse effects are produced during the transfer of electrons from titanium dioxide to carbon materials. Many scientists have designed a core–shell of rGO and titanium dioxide which suppresses the decomposition of many organic pollutants [46]. By direct carbonization, the C/$TiO_2$ photocatalysts with high photocatalytic activity can be achieved. Titanium dioxide together with composites of amorphous carbon possessed the improved photocatalytic activities. When the process of annealing is performed, the best performance of photocatalysts could be obtained. Due to good bio-compatibility and excellent photoluminescence properties, carbon dots have attained considerable attentions of many scientists to produce titanium dioxide and carbon dots nanohybrids [23]. Recently, many microstructures like carbon nanorods, carbon nanosheets and titanium dioxide are combined with the carbon dots and showed a remarkable increase in photo-catalytic activity. It has been cleared that graphite derived carbon dots can be utilized for the generation of hydrogen. In fact, nickel-doped carbon dots have performed very well in photocatalytic activity for the degradation of methyl orange under visible light [47].

## 7.6 Carbon Nanotubes and Titanium Dioxide as Nanohybrids

A number of investigations have been carried out on hybrid materials such as titanium dioxide and carbon nanotubes for many electronics and optical applications. The formation of new hybrid materials requires the removal of many toxic metal ions from the solution. For this purpose, titanium dioxide and multi-walled carbon nanotubes are used for the removal of thorium from an aqueous solution. The process of fabrication of $TiO_2$ coated multi-walled carbon nanotubes is shown in Figure 7.5. Multi-walled carbon nanotubes also contain impurities like carbon contents, graphite sheets and other nanoparticles.

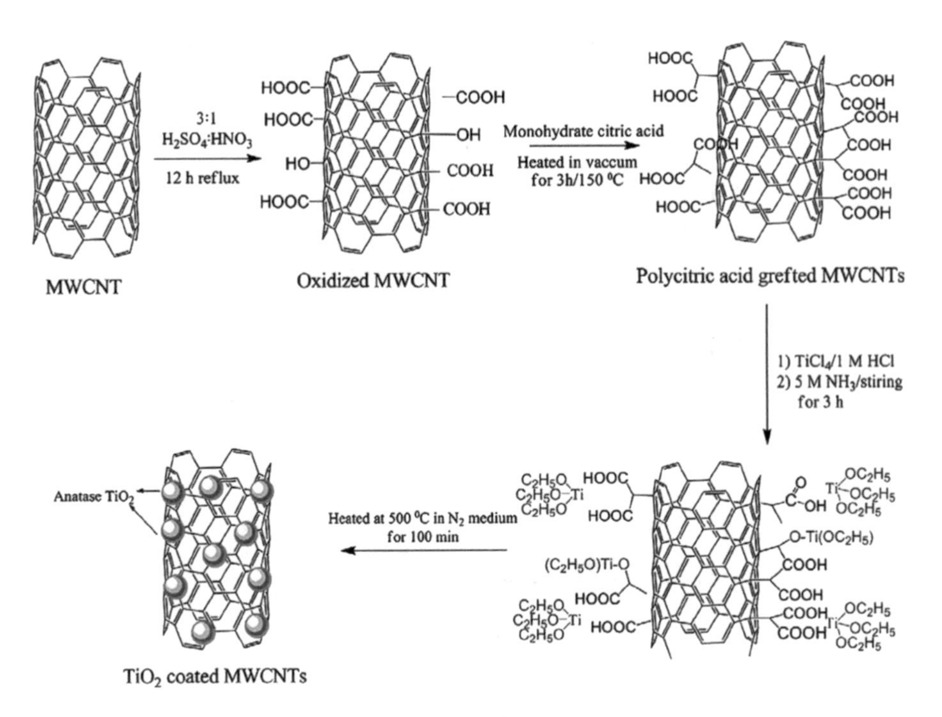


FIGURE 7.5 The schematic of preparation of $TiO_2$ coated MWCNTs. (Figure reprinted/adapted with permission from Ref. [49].)

By the process of oxidation, purified multi-walled carbon nanotubes can be obtained. For this purpose, oxygen is utilized to improve its hydrophilic properties and rate of adsorption on the inner side of carbon nanotubes. The pH of the solution also plays a very important role in the adsorption of many metal ions. By enhancing the pH, the amount of adsorption of thorium would first rise and then become constant. So as a result of increasing value of pH, the concentration of hydronium ions would start to decrease. This shows the presence of various functional groups on multi-walled nanotubes and on the hybrid materials and may contain a greater surface area as compared to those of titanium dioxide. Simultaneously the acidic functional groups present on the surface of carbon nanotubes ionizes with the increase in the pH value [48].

## 7.7 Silicon Dioxide and Titanium Dioxide as a Nanohybrid

All alkyl metals have very high hydrolysis rates except for silicone which results in the precipitation of many hydro-oxides before reacting with different alkyl oxides [50]. Therefore, the sol–gel method is better to prepare silicon dioxide and titanium oxide nanohybrids [51]. By carefully controlling the concentration of acetic acid the rate of hydrolysis could be easily tuned which results in the formation of titanium dioxide and silicone dioxide nanohybrids. In order to study the influence of size on the structure of nanohybrids, band gap energy would be measured by the resultant nanoparticles [52].

Band gap energy can also be calculated from the absorption edge. The values are obtained from the ultraviolet visible spectra for titanium dioxide, silicone dioxide and that of commercially available anatase nanocrystal of titanium dioxide. The strong chemical bonding between titanium, oxygen and silicone are responsible for the blue shift of band gap energy. As a result, nanohybrid materials are confirmed and the quantum size effect of nanocrystal anatase appears on the surface of silicone dioxide nanosphere for the fitting of the absorption peaks plots [53]. The band gap energy of commercially available titania crystals is greater than those of band gap of nanohybrids particles. Titanium dioxide–silicone dioxide nanohybrids are prepared by attaining the super critical drying of nanohybrids precursors on molecular level. As a result of drying process, resultant hybrids materials have a diameter of 140 nm. The measurement of the band gap energy of nanohybrids is in the range of 0.10 to 0.13 eV, when compared with the anatase crystal without quantum effects. From the above discussion, it is observed that nanohybrids have the ability to control the quantum size effects completely by controlling the drying process during the preparation of nanohybrids [54]. This process involved the preparation of hybrid sol by the process of the sol–gel method to get titanium dioxide on the top of a silicone dioxide nanosphere. As a result of drying the hard agglomeration of the nanohybrids sol is suppressed to obtain the titanium dioxide–silicone dioxide nanohybrids along with titanium alkyl oxide.

## 7.8 Titanium Dioxide/Polyaniline as a Nanohybrids

Polyaniline (PANI) is one of the most vital polymers because of its high processing and thermal and environmental stability and unique characteristics. PANI offers possible applications in the domain of solar cells, electrical and optical devices, motors and in many capacitors [55]. One of the major tasks of nanotechnology is to prepare the nanocomposites with metal composites that have multi-functional properties for different applications in the area of photo catalysis [56]. Encapsulation of metallic nanoparticles into conjugated polymers leads to the formation of functional nanocomposites that endure high thermal stability, catalytic activity and electrical properties [57]. For example, Pt–PANI endures remarkable catalytic activities because the polymer acts like a nucleus between the catalytic sites. Similarly, many composites along with PANI serve as a major factor to be used in many lithium batteries and act as a catalyst for the conversion of many alkenes into ketenes. $TiO_2$–PANI nanohybrids are prepared by oxidative polymerization of aniline in the presence of titanium dioxide nanoparticles [58]. The photocatalytic activity of these nanohybrid materials was calculated under the passage of ultraviolet light by the photocatalytic degradation of much organic waste such as phenol [58]. When compared with the pure titanium dioxide, titanium dioxide–PANI nanohybrids possess higher photo-catalytic activity which can be utilized for cleaning the contaminated water. A $TiO_2$–PANI nanohybrid catalyst is 20% titanium dioxide, which is responsible for highest photo-catalytic activity that can be used in various applications such as laser ablations and solar cells. The photo degradation of many pollutants takes place under the high photo-catalytic activity by the combined effect of nanohybrids as compared to pristine or titanium dioxide nanoparticles alone [59].

## 7.9 Role of Titanium Dioxide in Controlling Ultraviolet Radiations and Its Application

### 7.9.1 Role of $TiO_2$ against Ultraviolet Radiations

Protection of human skin from harmful radiations coming out from the sun such as ultraviolet rays is a major issue especially in European countries. Owing to the decrease in the thickness of ozone layer, most of the ultraviolet radiations pass through this layer and reach the ground level. This greatly enhances the chances of skin diseases. In order to stop the allergic reactions of ultraviolet radiation,s UV filters are used in different sunscreens [28, 60]. This could be done by using the physical components that have nanosized particles of zinc oxide and/or titanium dioxide. Titanium dioxide is used in many sunscreen creams in order to minimize the effect of ultraviolet radiations through absorption and scattering. The size of nanoparticles plays an important role in the effective interaction between these particles and ultraviolet light [61]. The main purpose to use titanium dioxide is to protect the human skin against these ultraviolet radiations. The UV radiations fall into three different ranges among which the most harmful ultraviolet radiations are completely absorbed by the ozone layer but the other two radiations reach the earth's atmosphere [62]. These two harmful ultraviolet radiations are responsible for sunburn and other skin diseases. By the use of titanium dioxide nanoparticles in different sun blocks greatly reduces the effect of these UV rays. But the effect of these nanoparticles may appear to be less because these particles may aggregate in sunscreens thus reducing their protective properties [63].

### 7.9.2 Formation of Titania and Its Vast Applications

Titanium dioxide is a naturally occurring semiconductor inorganic material which has many interesting properties and salient features. Titanium dioxide has been widely used in gas sensors and in wastewater treatments for degradation of organic waste. It is highly soluble in ethanol and chloroform. Titanium dioxide can be prepared in powder form by different methods such as micro-emulsion, CVD and sol–gel, etc. [64, 65]. To utilize titanium dioxide in powder form, the mechanical milling method is used, which helps in the generation of Titania powder on a large scale. The manufacturing time of nanopowder in this milling machine may vary from 0 to 60 hours. The mechanical milling process has proved to be more beneficial than other methods due to its low cost and high stability. The grinding process of titanium dioxide alone or with other alloys creates entirely new properties. Titanium dioxide doped with compounds such as $La_2O_3$ also shows unique industrial applications [66]. The nanocrystalline powder ofa titanium dioxide–$La_2O_3$ alloy is designed to improve the electrodes and energy storage applications. This nanoalloy shows high resistance against the corrosive effects produced by different reactive elements present in the atmosphere. Other alloys of titanium dioxide have also remarkable properties in the fields of ceramic and electronic industries [67].

### 7.9.3 Titanium Dioxide for Removal of Organic Wastes from Water

Titanium dioxide acts as a promising semiconductor photo-catalyst owing to its high structural, physical and optical properties under ultraviolet light. Titanium dioxide has been widely used for the purpose of the purification of air and treatment of water in order to avoid the harmful effects caused by polluted air and contaminated water [68]. Many researchers focus on VLA (visible light active) titanium dioxide photo catalysis and its vast industrial applications. The diverse group of substances that are responsible for the contamination of water can be removed by using nitrogen and carbon-doped titanium dioxide by the process of degradation [69]. The presence of water contaminants in excess amount can produce serious health issues for human beings and aquatic species. This contaminated water may cause endocrine disruption effects in living organisms that may result in developmental problems and reproduction disorders. So there is a need to purify this aquatic environment which can only be done by utilizing titanium dioxide along with suitable doping concentrations of carbon and nitrogen [70]. The photo-catalytic oxidation reactor is shown in Figure 7.6. This reactor is used for the purification of water by using titanium dioxide in combination with ultraviolet light [71]. The use of light-emitting diodes along with titanium dioxide offers many advantages including extended lifetime and flexibility. The CN–$TiO_2$ photocatalysts work for the removal of bis phenol with high effectiveness. Doped titanium dioxide materials show a higher extent for the removal of organic waste in water with the help of emitted white light from LED's. Neutral pH is preferred for the process of degradation of organic wastes from water [72]. VLA titanium dioxide has also been used for the photocatalytic degradation of cyano-toxins which are responsible for the contamination of water [73]. Water coating cyano-toxins are very dangerous for human health and also for plant life. Therefore, such VLA titanium dioxide appears to be more

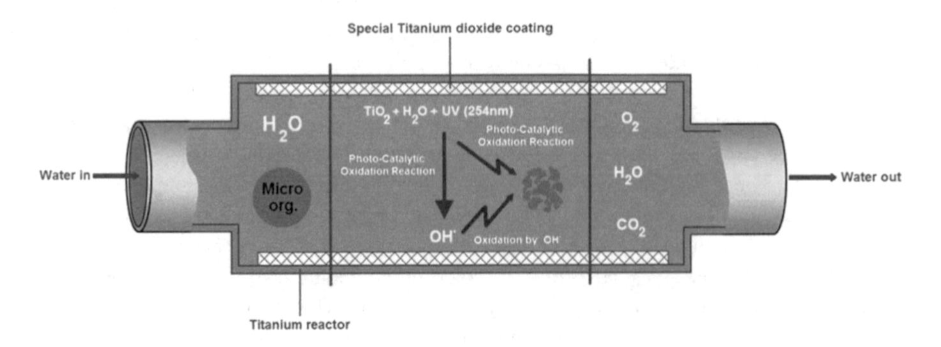


**FIGURE 7.6** The photo-catalytic oxidation reactor. (Figure reprinted/adapted with permission from Ref. [74].)

promising for the environmental sustainability and purification of contaminated water powered by solar light as an energy source.

### 7.9.4 Titanium Dioxide for the Purification of Aquatic Life

The aquatic life of several species has been greatly influenced by the presence of viruses, bacteria and algae that cause both direct and indirect issues to water quality. The presence of toxic bacteria has been proved to cause many adverse effects on human health through water consumption [75]. There are many factors responsible for the contamination of purified water, such as the intensity of stream flow, temperature of water and wildlife wastes. Most of the microbes could be easily sanitized by exposure to sun light. The traditional disinfection approach is carried out using chemical agents such as ozone, chlorine, etc., which are more effective than any other chemicals being used. Titanium dioxide-based photocatalysts hold best for the disinfection and degradation of contaminants [76]. Thus, they can be used for the treatment of water. Titanium dioxide photocatalysts have the ability to remove many organic pollutants and unwanted biological components through water treatments with high efficiency [77].

### 7.9.5 Titanium Dioxide as Photosensitive Polymer

Offset printing is one of the exceptional applications of titanium dioxide-based photocatalysts. This technique is commonly adopted for printing newspaper and magazines. Offset printing commonly uses a photosensitive polymer deposited on an aluminum plate enlightened by a photo mask, while the plate is designed in such a manner that can remove the protected area of the photosensitive polymer [78]. In this process, the plate is wet by water that covers particular areas and then this surface is completely covered by oil-based printing ink. The main purpose of using water on the surface of plate is to avoid the sticking of oil-based inks in these areas. At the end of this process, the specific design of ink is formed on the top of plate. As a result, this pattern is then transferred to the paper. But this process has to face a drawback that the printing plate has to be disposed of after the process of printing. Scientists have developed different kinds of offset printing plates that are based on super-hydrophilic and hydrophobic designs arranged by titanium dioxide-based photocatalysts [79]. These offset printing plates of titanium dioxide could be prepared by the fabrication of titanium dioxide on the surface of the plate and then the surface would be patterned by the self-assembled monolayer of ODS (octadodecyl phonic acid). Titanium dioxide-based offset printing is widely used as a renewable fabricated plate. Titanium dioxide offset printing is used to pattern the surface with higher resolutions using modified plates in its processing. The surface of the plate is moistened by sulfuric acid to achieve greater testability contrast. Owing to the high compatibility of titanium dioxide with modern technology, titanium dioxide-based photocatalysts are widely used nowadays. Many applications involving titanium dioxide have improved our lives from an environmental aspect [80].

### 7.9.6 Titanium Dioxide as a Ceramic

The initiation of any chemical reaction or the use of solar radiation for supplying energy is a well recognized idea. For the generation of electricity or to start chemical reactions, titanium dioxide semiconductors have a wide band gap that is irradiated by light, and the resultant excited electron–hole pairs are then employed in the solar cell, which in turn results in the generation of light [81]. According to the most recent research, the appearance of the photo reactivity of titanium dioxide can be used for the purification of air. Titanium dioxide-based tiles are also very effective tools to fight against bacterial action as well as against organic and inorganic materials. These titanium dioxide-based ceramic tiles provide better killing of bacterial at a very high speed in such a way that the number of killed bacteria should be larger then the growth of bacteria. The use of these ceramic coated tiles of titanium dioxide in hospitals and in other healthcare institutions can greatly reduce the spreading of many infections which may cause harmful effects to human health. Titanium dioxide-coated materials are used for the creation of self-cleaning building materials during construction [82].

## 7.10 Size Controlled Synthesis of Nanohybrid: Basic Concept

Although the formation of nanohybrid materials has been reported by many researchers, control of the distribution of size and shape is still challenging. Great effort is needed to offer flexible separation techniques to control the size of nanohybrids.

### 7.10.1 Size Controlled Synthesis through Non-Hydrolytic Route

A non-hydrolytic route is applied for the size controlled synthesis of oxide-based hybrid materials. Owing to the unique properties and high surface to volume ratio, iron oxide nanoparticles have been focused on by many researchers, and the non-hydrolytic tool is developed for its size controlled characteristics. Although many approaches have been used for the synthesis of iron oxide nanocrystals, which involves the co-precipitation method, hydrothermal method and micro emulsion process. By the thermal decomposition method, the controlled size of iron oxide nanohybrid is noticed below 50 nm. Here the content synthesis of iron oxide has been carried out with controlled morphology and size by adding FeO(OH) as a precursor in high boiling temperature solvent with additional co-surfactants. Iron oxide nanocrystals having different shapes such as cubic flower-like and potato-like geometry that is obtained by adding the measurable concentration of co-surfactant and specific additives by one pot synthesis route. Moreover, the reaction time is very important along with the amount of additives for controlling the size and morphology of iron oxide nanohybrids. Final materials must be capped with a stabilizing layer in order to synthesize non-polar solvents. The size, shape and magnetic properties of resulting iron oxide crystals are normally characterized using XRD and TEM [83].

### 7.10.2 Size and Shape Controlled Synthesis of Titania Nanohybrids by Wet Chemical Method

Size controlled synthesis of nanohybrids is needed to achieve high technological objectives. There are many factors that determine the properties of nanohybrids at the nanoscale. Semiconductor materials and metal nanohybrids are best suited for controlling the optical and electrical properties [84]. These nanohybrids can tune properties, which is possible by modifying shape and size in the controlled synthesis. For example, it has been observed that tips and sharp edges enhance the electric field properties in a particular direction [85].

The sol–gel method or solution-based techniques can be used to synthesis size controlled titania nanohybrids. Titania shows high chemical reactivity of pre-cursors during size controlled synthesis process. Many techniques have been carried out for controlling the reactivity of titania alk oxides and it can be achieved by isolating the poly-oxoalkoxide. During this chemical reaction, polymeric species are normally formed. These polymeric species may lead to the formation of various sizes and structures of amorphous solid materials [86]. The hydrolysis process is carried out in the presence of an alkaline solution. The hydrolysis process removes the precipitates of amorphous solids and remaining bulk dioxide leads to the formation of a gel-like structure. A peptization mechanism is needed to convert the polymers into small crystalline particles. However, these dissolution growth phenomena are difficult to achieve for the controlled size distributions of nanohybrids. For this purpose the whole process needs to be performed in an aqueous medium [87]. Thus at this stage poly condensation and the hydrolysis process of titanium alkoixde take place in the presence of tetra methylene ammonium hydroxide. The alkaline solution helps in stabilizing the anatase poly-anionic cores. These anatase clusters are self-assembled into nanocrystals and then turn into super lattice structures. This process is used to explain the formation of highly assembled titania films as shown in Figure 7.7. Larger anatase titania nanocrystals with controlled size and shape can be achieved by optimizing the relative concentration of base and titanium alkoxide. TEM and XRD techniques are normally used for the characterization of the size controlled titania anatase nanocrystals. Figure 7.7 shows the TEM image of triangular-shaped nanocrystals with a homogenous size distribution of titania nanohybrids.

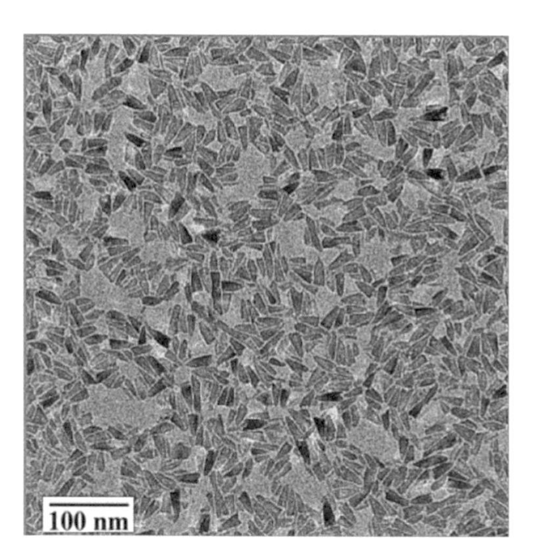


FIGURE 7.7 TEM image of triangular shaped titania nano hybrids with homogenous size distribution. (Figure reprinted/adapted with permission from Ref. [87].)

### 7.10.3 Seed Mediated Method for the Synthesis of Inorganic Nanohybrid

Inorganic hybrid nanomaterials are multi-component nanocrystals containing two or more nanocomponents that are embedded by strong chemical bonding. Over the past few years, inorganic hybrid materials have attracted many scientists, owing to their unique hybrid system, responsible for collective properties that are very difficult to attain from individual nanocomponents [88]. Inorganic nanohybrids materials not only have multi-tasking properties but also contain unique characteristics. The interactions of particles in such materials provide an easy electron transfer through the interface that could alter the electronic structure and develop synergetic properties of nanohybrids materials [89]. By controlling the geometries and composition of each component, the physiochemical characteristics of hybrid nanoparticles could be easily tuned. The surface of a nanocrystal with deposited metal clusters serves as a source for performing oxidation reactions. The distribution of charge between metal has played a vital role for improving the catalytic activity. The seed mediated method involves the micro emulsion process and thermal decomposition of nanocomponents in the solution. This method has been extended for the synthesis of multi components nanocrystals that contain the interaction of metal oxide species [90, 91]. By the reaction of the molecular precursor and growth under selected solvents, nanohybrids have been generated by the seed mediated method. Organic capping agents play a major role in stabilizing the solvent during nanohybrid configuration.

This type of stabilizer helps to regulate the super saturation level and to ensure the steady growth of nanohybrids. Nanohybrids with dumbell or core shapes can be formed by using above mentioned process. It has been noticed that the growth of different shapes of nanohybrids is possible despite their having the same lattice spacing [92]. There is also a possibility of forming core–shell structures that occur as a result of mismatches of lattices. However, the selection of suitable process for desireable shape and size of nanohybrids is still a great challenge [93]. Due to unique properties of nanohybrids, these are being used in medical diagnosis and treatments of disease, for biological tagging and in many electrical and optical devices [94].

## 7.11 Energy Conversion Using Graphene Oxide and Titanium Dioxide Nanohybrids

For the synthesis of titanium dioxide and graphene sheets, a green facile approach has been used for making nanocomposites using hydrothermal method. These nanocomposites are characterized using XRD and the reflectance spectra is observed using TEM. Two-dimensional nanohybrids can be used for transporting electrons in semiconductors [95, 96]. rGO has been used in energy conversion devices, such as pervoskite solar cells [97]. A schematic illustration of the fabrication of boron-doped rGO in a perovskite solar cell and current voltage characteristics of a boron-doped perovskite solar cell is shown in Figure 7.8. GO–$TiO_2$ nanohybrids are also used for the removal of pollutants and other harmful compounds in order to split the water into hydrogen and oxygen gas. Most of the applications involve the energy conversion using injection of electron. Many semiconductor nanostructures are combined with GO such as titanium dioxide, zinc oxide, copper oxide and cadmium sulphoid. A detailed study of graphene sheets on semiconductor oxides has been completed using XRD. According to previous studies, GO sheets can produce two-dimensional sources for the transportation and collection of electrons at the surface of electrode [98]. By studying the detailed analysis of graphene sheets in semi-conductor materials the catalytic performance of many oxides has been improved. Photodegradation of graphene-based semiconductor photocatalytics helps in the conversion of carbon dioxide to hydro-carbons fuels and photocatalytic conversion in water splitting reactions [99].

Output characteristics of boron-doped GO in a preovskite solar cell are illustrated in Figure 7.8. The results signify that the boron-doped rGO structure have greatest absorption efficiency as compared to $TiO_2$ and FTO. As compared to other nanohybrids, these boron-doped rGO papers can be used for fabricating large-surface area solar cells. The current voltage (*I–V*) characteristics of Figure 7.8 show the output parameters of pure rGO compared to the boron-doped rGO. The *I–V* curves of boron doped rGO and pure rGO were calculated at 60% humidity and under 100mW/cm$^2$ illumination conditions. It is evident from the results that boron doped rGO shows improved conversion efficiency of perviskite solar cells as compared to pure rGO-based solar cells.

## 7.12 Conclusion

A flexible, simple and versatile approach is demonstrated to design titanium dioxide as a nanohybrid with other materials. Titanium dioxide can act as a nanohybrid with different materials such as graphene, carbon dots, silicon dioxide and carbon nanotubes. Unique properties are observed by mixing two nanohybrids at controlled temperature and pressure. The optimization of size/shape control parameters of nanohybrids can help greatly in getting desired nanotechnological objectives. These parameters determine the properties of nanohybrids at the nanoscale. Nanohybrids appear to be promising materials for environmental sustainability. Hybrid materials can use solar energy for the purification of contaminated water. In addition, titanium dioxide and graphene nanohybrids are being used to increase the absorption efficiency of photovoltaic devices even at 60% humidity level. It is observed that nanohybrids are also

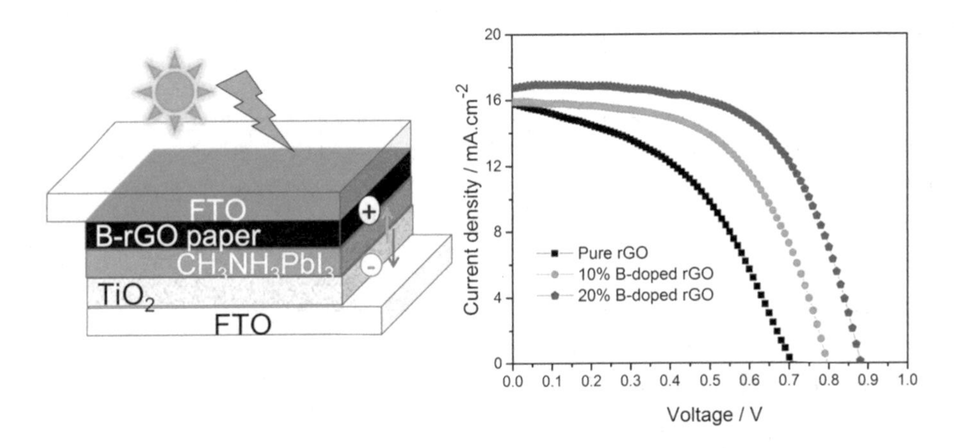


**FIGURE 7.8** Schematic illustration of the fabrication of boron doped rGO in perovskite solar cell (a) current voltage characteristics of boron-doped perovskite solar cell (b). (Figure reprinted/adapted with permission from Ref. [100].)

useful for enhancing the efficiency of perovskite solar cells. Titanium dioxide with graphene bears high electromagnetic properties that involve easy transportation of eddy currents. Current studies show that these nanohybrids can be used to enhance the efficiency of perovskite solar cells. Normally, nanohybrids are characterized by scanning electron microscopy (SEM) and transmission electron microscopy (TEM). Characterization techniques are used to study the results that help in calculating the particle size and surface morphology of nanohybrids. Based on these studies it is evident that nanohybrids may have a significant role in overcoming potential environmental uncertainties and can be used in future research and development of energy conversion studies.

## Acknowledgment

This work was supported by the Higher Education Commission (HEC) of Pakistan [Grant No: 21-811/SRGP/R&D/HEC/2016]. The authors also acknowledge the National Centre for Physics (NCP), Islamabad, Pakistan for providing assistance and Associate Membership to Dr. Khuram Ali.

## REFERENCES

1. M. Nanko, Definitions and categories of hybrid materials, *AZojomo*, 6 (2009) 1–8.
2. T.A. Elbokl, C. Detellier, Aluminosilicate nanohybrid materials. Intercalation of polystyrene in kaolinite, *Journal of Physics and Chemistry of Solids*, 67 (2006) 950–955.
3. C. Cheng, A. Amini, C. Zhu, Z. Xu, H. Song, N. Wang, Enhanced photocatalytic performance of TiO 2–ZnO hybrid nanostructures, *Scientific Reports*, 4 (2014) 4181.
4. S. Letaief, I.K. Tonle, T. Diaco, C. Detellier, Nanohybrid materials from interlayer functionalization of kaolinite. Application to the electrochemical preconcentration of cyanide, *Applied Clay Science*, 42 (2008) 95–101.
5. Y. Tao, Y. Lin, Z. Huang, J. Ren, X. Qu, DNA-templated silver nanoclusters–graphene oxide nanohybrid materials: A platform for label-free and sensitive fluorescence turn-on detection of multiple nucleic acid targets, *Analyst*, 137 (2012) 2588–2592.
6. C. Schulz-Drost, V. Sgobba, C. Gerhards, S. Leubner, R.M. Krick Calderon, A. Ruland, D.M. Guldi, Innovative inorganic–organic nanohybrid materials: Coupling quantum dots to carbon nanotubes, *Angewandte Chemie*, 49 (2010) 6425–6429.
7. F.-X. Qiu, Y.-M. Zhou, J.-Z. Liu, The synthesis and characteristic study of 6FDA–6FHP–NLO polyimide/SiO2 nanohybrid materials, *European Polymer Journal*, 40 (2004) 713–720.
8. G. Kickelbick, Hybrid Materials–Past, Present and Future, *Hybrid Materials*, 1 (2014) 39–51.
9. W. Ma, W.O. Yah, H. Otsuka, A. Takahara, Application of imogolite clay nanotubes in organic–inorganic nanohybrid materials, *Journal of Materials Chemistry*, 22 (2012) 11887–11892.
10. K. Kume, N. Kawasaki, H. Wang, T. Yamada, H. Yoshikawa, K. Awaga, Enhanced capacitor effects in polyoxometalate/graphene nanohybrid materials: A synergetic approach to high performance energy storage, *Journal of Materials Chemistry A*, 2 (2014) 3801–3807.
11. E. Ruiz-Hitzky, M. Darder, P. Aranda, An introduction to bio-nanohybrid materials. In: E. Ruiz-Hitzky, K. Ariga, and Y.M. Lvov, Eds. *Bio-Inorganic Hybrid Nanomaterials, Strategies, Syntheses, Characterization and Applications*. (2008) 1–32. Wiley-VCH, Weinheim.
12. V.-D. Dao, Y. Choi, K. Yong, L.L. Larina, H.-S. Choi, Graphene-based nanohybrid materials as the counter electrode for highly efficient quantum-dot-sensitized solar cells, *Carbon*, 84 (2015) 383–389.
13. A.V. Vadivel Murugan, Novel organic–inorganic poly (3, 4-ethylenedioxythiophene) based nanohybrid materials for rechargeable lithium batteries and supercapacitors, *Journal of Power Sources*, 159 (2006) 312–318.
14. J. Yuan, A.H.E. Müller, One-dimensional organic–inorganic hybrid nanomaterials, *Polymer*, 51 (2010) 4015–4036.
15. L.-H. Liu, R. Métivier, S. Wang, H. Wang, Advanced nanohybrid materials: Surface modification and applications, *Journal of Nanomaterials*, 2012 (2012) 1–2.
16. L.V. Kumar, S.A. Ntim, O. Sae-Khow, C. Janardhana, V. Lakshminarayanan, S. Mitra, Electro-catalytic activity of multiwall carbon nanotube-metal (Pt or Pd) nanohybrid materials synthesized using microwave-induced reactions and their possible use in fuel cells, *Electrochimica Acta*, 83 (2012) 40–46.

17. J.C. Hulteen, C.R. Martin, A general template-based method for the preparation of nanomaterials, *Journal of Materials Chemistry*, 7 (1997) 1075–1087.
18. K.R. Reddy, M. Hassan, V.G. Gomes, Hybrid nanostructures based on titanium dioxide for enhanced photocatalysis, *Applied Catalysis A: General*, 489 (2015) 1–16.
19. S.K. Yadav, S.R. Madeshwaran, J.W. Cho, Synthesis of a hybrid assembly composed of titanium dioxide nanoparticles and thin multi-walled carbon nanotubes using "click chemistry", *Journal of Colloid and Interface Science*, 358 (2011) 471–476.
20. K. Alireza, M.G. Ali, *Nanostructured Titanium Dioxide Materials: Properties, Preparation and Applications*, World Scientific (2011).
21. D.E. Giammar, C.J. Maus, L. Xie, Effects of particle size and crystalline phase on lead adsorption to titanium dioxide nanoparticles, *Environmental Engineering Science*, 24 (2007) 85–95.
22. P.-C. Chiang, W.-T. Whang, The synthesis and morphology characteristic study of BAO-ODPA polyimide/TiO2 nano hybrid films, *Polymer*, 44 (2003) 2249–2254.
23. D. Hazarika, N. Karak, Photocatalytic degradation of organic contaminants under solar light using carbon dot/titanium dioxide nanohybrid, obtained through a facile approach, *Applied Surface Science*, 376 (2016) 276–285.
24. S.H. Kim, S.-Y. Kwak, T. Suzuki, Photocatalytic degradation of flexible PVC/TiO2 nanohybrid as an eco-friendly alternative to the current waste landfill and dioxin-emitting incineration of post-use PVC, *Polymer*, 47 (2006) 3005–3016.
25. M. Mashkour, M. Rahimnejad, S.M. Pourali, H. Ezoji, A. ElMekawy, D. Pant, Catalytic performance of nano-hybrid graphene and titanium dioxide modified cathodes fabricated with facile and green technique in microbial fuel cell, *Progress in Natural Science: Materials International*, 27 (2017) 647–651.
26. M. Ding, D.C. Sorescu, A. Star, Photoinduced charge transfer and acetone sensitivity of single-walled carbon nanotube–titanium dioxide hybrids, *Journal of the American Chemical Society*, 135 (2013) 9015–9022.
27. Y.-C. Cao, Z. Fu, W. Wei, L. Zou, T. Mi, D. He, C. Yan, X. Liu, Y. Zhu, L. Chen, Y. Sun, Reduced graphene oxide supported titanium dioxide nanomaterials for the photocatalysis with long cycling life, *Applied Surface Science*, 355 (2015) 1289–1294.
28. Y. Haldorai, J.J. Shim, Novel chitosan-TiO2 nanohybrid: Preparation, characterization, antibacterial, and photocatalytic properties, *Polymer Composites*, 35 (2014) 327–333.
29. A. Fujishima, T.N. Rao, D.A. Tryk, Titanium dioxide photocatalysis, *Journal of Photochemistry and Photobiology C: Photochemistry Reviews*, 1 (2000) 1–21.
30. U.G. Akpan, B.H. Hameed, Parameters affecting the photocatalytic degradation of dyes using TiO2-based photocatalysts: A review, *Journal of Hazardous Materials*, 170 (2009) 520–529.
31. K. Nakata, A. Fujishima, TiO2 photocatalysis: Design and applications, *Journal of Photochemistry and Photobiology C: Photochemistry Reviews*, 13 (2012) 169–189.
32. Q. Xiang, J. Yu, M. Jaroniec, Graphene-based semiconductor photocatalysts, *Chemical Society Reviews*, 41 (2012) 782–796.
33. M.-Q. Yang, Y.-J. Xu, Basic principles for observing the photosensitizer role of graphene in the graphene–semiconductor composite photocatalyst from a case study on graphene–ZnO, *Journal of Physical Chemistry C*, 117 (2013) 21724–21734.
34. S.G. Babu, R. Vinoth, B. Neppolian, D.D. Dionysiou, M. Ashokkumar, Diffused sunlight driven highly synergistic pathway for complete mineralization of organic contaminants using reduced graphene oxide supported photocatalyst, *Journal of Hazardous Materials*, 291 (2015) 83–92.
35. X. Wu, L. Wen, K. Lv, K. Deng, D. Tang, H. Ye, D. Du, S. Liu, M. Li, Fabrication of ZnO/graphene flake-like photocatalyst with enhanced photoreactivity, *Applied Surface Science*, 358 (2015) 130–136.
36. X.-Y. Zhang, H.-P. Li, X.-L. Cui, Y. Lin, Graphene/TiO2 nanocomposites: Synthesis, characterization and application in hydrogen evolution from water photocatalytic splitting, *Journal of Materials Chemistry*, 20 (2010) 2801–2806.
37. M. Cao, P. Wang, Y. Ao, C. Wang, J. Hou, J. Qian, Photocatalytic degradation of tetrabromobisphenol A by a magnetically separable graphene–TiO2 composite photocatalyst: Mechanism and intermediates analysis, *Chemical Engineering Journal*, 264 (2015) 113–124.
38. G. Hu, B. Tang, Photocatalytic mechanism of graphene/titanate nanotubes photocatalyst under visible-light irradiation, *Materials Chemistry and Physics*, 138 (2013) 608–614.

39. H. Zhao, F. Su, X. Fan, H. Yu, D. Wu, X. Quan, Graphene-TiO2 composite photocatalyst with enhanced photocatalytic performance, *Chinese Journal of Catalysis*, 33 (2012) 777–782.
40. G. Jiang, Z. Lin, C. Chen, L. Zhu, Q. Chang, N. Wang, W. Wei, H. Tang, TiO2 nanoparticles assembled on graphene oxide nanosheets with high photocatalytic activity for removal of pollutants, *Carbon*, 49 (2011) 2693–2701.
41. W.-D. Yang, Y.-R. Li, Y.-C. Lee, Synthesis of r-GO/TiO2 composites via the UV-assisted photocatalytic reduction of graphene oxide, *Applied Surface Science*, 380 (2016) 249–256.
42. V.D. Dao, L.L. Larina, K.D. Jung, J.K. Lee, H.S. Choi, Graphene–NiO nanohybrid prepared by dry plasma reduction as a low-cost counter electrode material for dye-sensitized solar cells, *Nanoscale*, 6 (2014) 477–482.
43. H. Liu, G. Zhang, Y. Zhou, M. Gao, F. Yang, One-step potentiodynamic synthesis of poly (1, 5-diaminoanthraquinone)/reduced graphene oxide nanohybrid with improved electrocatalytic activity, *Journal of Materials Chemistry A*, 1 (2013) 13902–13913.
44. S. Sakthivel, H. Kisch, Daylight photocatalysis by carbon-modified titanium dioxide, *Angewandte Chemie*, 42 (2003) 4908–4911.
45. X. Li, J. Niu, J. Zhang, H. Li, Z. Liu, Labeling the defects of single-walled carbon nanotubes using titanium dioxide nanoparticles, *Journal of Physical Chemistry B*, 107 (2003) 2453–2458.
46. S. Hussain, S. Boland, A. Baeza-Squiban, R. Hamel, L.C. Thomassen, J.A. Martens, M.A. Billon-Galland, J. Fleury-Feith, F. Moisan, J.C. Pairon, F. Marano, Oxidative stress and proinflammatory effects of carbon black and titanium dioxide nanoparticles: Role of particle surface area and internalized amount, *Toxicology*, 260 (2009) 142–149.
47. F. Li, F. Tian, C. Liu, Z. Wang, Z. Du, R. Li, L. Zhang, One-step synthesis of nanohybrid carbon dots and TiO 2 composites with enhanced ultraviolet light active photocatalysis, *RSC Advances*, 5 (2015) 8389–8396.
48. R. Yavari, N. Asadollahi, M.A. Abbas Mohsen, Preparation, characterization and evaluation of a hybrid material based on multiwall carbon nanotubes and titanium dioxide for the removal of thorium from aqueous solution, *Progress in Nuclear Energy*, 100 (2017) 183–191.
49. S. Mallakpour, E. Khadem, Carbon nanotube–metal oxide nanocomposites: Fabrication, properties and applications, *Chemical Engineering Journal*, 302 (2016) 344–367.
50. J. Kischkat, S. Peters, B. Gruska, M. Semtsiv, M. Chashnikova, M. Klinkmüller, O. Fedosenko, S. Machulik, A. Aleksandrova, G. Monastyrskyi, Y. Flores, W.T. Masselink, Mid-infrared optical properties of thin films of aluminum oxide, titanium dioxide, silicon dioxide, aluminum nitride, and silicon nitride, *Applied Optics*, 51 (2012) 6789–6798.
51. T. Watanabe, A. Nakajima, R. Wang, M. Minabe, S. Koizumi, A. Fujishima, K. Hashimoto, Photocatalytic activity and photoinduced hydrophilicity of titanium dioxide coated glass, *Thin Solid Films*, 351 (1999) 260–263.
52. T. Ohno, S. Tagawa, H. Itoh, H. Suzuki, T. Matsuda, Size effect of TiO2–SiO2 nano-hybrid particle, *Materials Chemistry and Physics*, 113 (2009) 119–123.
53. M. Anpo, H. Nakaya, S. Kodama, Y. Kubokawa, K. Domen, T. Onishi, Photocatalysis over binary metal oxides. Enhancement of the photocatalytic activity of titanium dioxide in titanium-silicon oxides, *Journal of Physical Chemistry*, 90 (1986) 1633–1636.
54. J.-Y. Jeon, Y.-H. Hong, S.-G. Kim, Method for preparing copolyester resins using titanium dioxide/silicon dioxide coprecipitate catalyst in the form of suspension in glycol, Google Patents (2003).
55. S.-J. Su, N. Kuramoto, Processable polyaniline–titanium dioxide nanocomposites: Effect of titanium dioxide on the conductivity, *Synthetic Metals*, 114 (2000) 147–153.
56. Y. Qiao, S.J. Bao, C.M. Li, X.Q. Cui, Z.S. Lu, J. Guo, Nanostructured polyaniline/titanium dioxide composite anode for microbial fuel cells, *ACS Nano*, 2 (2008) 113–119.
57. J.-C. Xu, W.-M. Liu, H.-L. Li, Titanium dioxide doped polyaniline, *Materials Science and Engineering C*, 25 (2005) 444–447.
58. K.R. Reddy, K.V. Karthik, S.B.B. Prasad, S.K. Soni, H.M. Jeong, A.V. Raghu, Enhanced photocatalytic activity of nanostructured titanium dioxide/polyaniline hybrid photocatalysts, *Polyhedron*, 120 (2016) 169–174.
59. I.S. Lee, J.Y. Lee, J.H. Sung, H.J. Choi, Synthesis and electrorheological characteristics of polyaniline-titanium dioxide hybrid suspension, *Synthetic Metals*, 152 (2005) 173–176.

60. Y. Lan, Y. Lu, Z. Ren, Mini review on photocatalysis of titanium dioxide nanoparticles and their solar applications, *Nano Energy*, 2 (2013) 1031–1045.
61. H. Yang, S. Zhu, N. Pan, Studying the mechanisms of titanium dioxide as ultraviolet-blocking additive for films and fabrics by an improved scheme, *Journal of Applied Polymer Science*, 92 (2004) 3201–3210.
62. T.G. Smijs, S. Pavel, Titanium dioxide and zinc oxide nanoparticles in sunscreens: Focus on their safety and effectiveness, *Nanotechnology, Science and Applications*, 4 (2011) 95–112.
63. M. Thanihaichelvan, M.M.P.S. Kodikara, P. Ravirajan, D. Velauthapillai, Enhanced performance of nanoporous titanium dioxide solar cells using cadmium sulfide and poly (3-hexylthiophene) co-sensitizers, *Polymers*, 9 (2017) 467.
64. S. Yin, H. Hasegawa, D. Maeda, M. Ishitsuka, T. Sato, Synthesis of visible-light-active nanosize rutile titania photocatalyst by low temperature dissolution–reprecipitation process, *Journal of Photochemistry and Photobiology A: Chemistry*, 163 (2004) 1–8.
65. P. Cheng, M. Zheng, Y. Jin, Q. Huang, M. Gu, Preparation and characterization of silica-doped titania photocatalyst through sol–gel method, *Materials Letters*, 57 (2003) 2989–2994.
66. P. Cheng, W. Li, T. Zhou, Y. Jin, M. Gu, Physical and photocatalytic properties of zinc ferrite doped titania under visible light irradiation, *Journal of Photochemistry and Photobiology A: Chemistry*, 168 (2004) 97–101.
67. T. Theivasanthi, *Review on Titania Nanopowder-Processing and Applications*, arXiv Preprint ArXiv:1704.00981 (2017).
68. R.W. Matthews, An adsorption water purifier with in situ photocatalytic regeneration, *Journal of Catalysis*, 113 (1988) 549–555.
69. R.W. Matthews, Solar-electric water purification using photocatalytic oxidation with TiO2 as a stationary phase, *Solar Energy*, 38 (1987) 405–413.
70. G.K.-C. Low, R.W. Matthews, Flow-injection determination of organic contaminants in water using an ultraviolet-mediated titanium dioxide film reactor, *Analytica Chimica Acta*, 231 (1990) 13–20.
71. M.F.J. Dijkstra, H. Buwalda, A.W.F. De Jong, A. Michorius, J.G.M. Winkelman, A.A.C.M. Beenackers, Experimental comparison of three reactor designs for photocatalytic water purification, *Chemical Engineering Science*, 56 (2001) 547–555.
72. M. Pelaez, N.T. Nolan, S.C. Pillai, M.K. Seery, P. Falaras, A.G. Kontos, P.S.M. Dunlop, J.W.J. Hamilton, J.A. Byrne, K. O'shea, M.H. Entezari, D.D. Dionysiou, A review on the visible light active titanium dioxide photocatalysts for environmental applications, *Applied Catalysis B: Environmental*, 125 (2012) 331–349.
73. C.T. Nachtman, C.E. Chomka, J.R. Edwards, Water purifier, Google Patents (1999).
74. L. Yang, Z. Liu, J. Shi, H. Hu, W. Shangguan, Design consideration of photocatalytic oxidation reactors using TiO2-coated foam nickels for degrading indoor gaseous formaldehyde, *Catalysis Today*, 126 (2007) 359–368.
75. N. Guettaï, H.A. Ait Amar, Photocatalytic oxidation of methyl orange in presence of titanium dioxide in aqueous suspension. Part II: Kinetics study, *Desalination*, 185 (2005) 439–448.
76. F. Seitz, R.R. Rosenfeldt, S. Schneider, R. Schulz, M. Bundschuh, Size-, surface-and crystalline structure composition-related effects of titanium dioxide nanoparticles during their aquatic life cycle, *Science of the Total Environment*, 493 (2014) 891–897.
77. P.V. Laxma Reddy, B. Kavitha, P.A. Kumar Reddy, K.H. Kim, TiO2-based photocatalytic disinfection of microbes in aqueous media: A review, *Environmental Research*, 154 (2017) 296–303.
78. H. Tai, Y. Jiang, G. Xie, J. Yu, X. Chen, Fabrication and gas sensitivity of polyaniline–titanium dioxide nanocomposite thin film, *Sensors and Actuators. Part B: Chemical*, 125 (2007) 644–650.
79. R. Asahi, T. Morikawa, H. Irie, T. Ohwaki, Nitrogen-doped titanium dioxide as visible-light-sensitive photocatalyst: Designs, developments, and prospects, *Chemical Reviews*, 114 (2014) 9824–9852.
80. K. Nakata, T. Ochiai, T. Murakami, A. Fujishima, Photoenergy conversion with TiO2 photocatalysis: New materials and recent applications, *Electrochimica Acta*, 84 (2012) 103–111.
81. J. Määttä, M. Piispanen, H.-R. Kymäläinen, A. Uusi-Rauva, K.-R. Hurme, S. Areva, A.-M. Sjöberg, L. Hupa, Effects of UV-radiation on the cleanability of titanium dioxide-coated glazed ceramic tiles, *Journal of the European Ceramic Society*, 27 (2007) 4569–4574.
82. O. Carp, C.L. Huisman, A. Reller, Photoinduced reactivity of titanium dioxide, *Progress in Solid State Chemistry*, 32 (2004) 33–177.

83. W. Li, S.S. Lee, J. Wu, C.H. Hinton, J.D. Fortner, Shape and size controlled synthesis of uniform iron oxide nanocrystals through new non-hydrolytic routes, *Nanotechnology*, 27 (2016) 324002.
84. L.-s. Li, J. Hu, W. Yang, A.P. Alivisatos, Band gap variation of size-and shape-controlled colloidal CdSe quantum rods, *Nano Letters*, 1 (2001) 349–351.
85. M.P. Pileni, The role of soft colloidal templates in controlling the size and shape of inorganic nanocrystals, *Nature Materials*, 2 (2003) 145–150.
86. Y.W. Jun, M.F. Casula, J.H. Sim, S.Y. Kim, J. Cheon, A.P. Alivisatos, Surfactant-assisted elimination of a high energy facet as a means of controlling the shapes of TiO2 nanocrystals, *Journal of the American Chemical Society*, 125 (2003) 15981–15985.
87. C. Abdelkrim, M. Thomas, N. Titania, Control over nanocrystal structure, size, shape, and organization, *European Journal of Inorganic Chemistry*, 1999 (1999) 235–245.
88. M.Z.. Hussein, Z. Zainal, A.H. Yahaya, Synthesis of layered organic–inorganic nanohybrid material: An organic dye, naphthol blue black in magnesium–aluminum layered double hydroxide inorganic lamella, *Materials Science and Engineering B*, 88 (2002) 98–102.
89. T. Miyazaki, C. Ohtsuki, M. Tanihara, Synthesis of bioactive organic–inorganic nanohybrid for bone repair through sol–gel processing, *Journal of Nanoscience and Nanotechnology*, 3 (2003) 511–515.
90. E. Ruiz-Hitzky, K. Ariga, Y.M. Lvov, *Bio-Inorganic Hybrid Nanomaterials: Strategies, Synthesis, Characterization and Applications*, John Wiley & Sons (2008).
91. A.J. Patil, S. Mann, Self-assembly of bio–inorganic nanohybrids using organoclay building blocks, *Journal of Materials Chemistry*, 18 (2008) 4605–4615.
92. S. Förster, T. Plantenberg, From self-organizing polymers to nanohybrid and biomaterials, *Angewandte Chemie*, 41 (2002) 689–714.
93. S.-M. Paek, H. Jung, M. Park, J.-K. Lee, J.-H. Choy, An inorganic nanohybrid with high specific surface area: TiO2-pillared MoS2, *Chemistry of Materials*, 17 (2005) 3492–3498.
94. T.-D. Nguyen, T.-O. Do, Size- and shape-controlled hybrid inorganic nanomaterials and application for low-temperature CO oxidation, In: *Controlled Nanofabrication: Advances and Applications*, ed. Ru-Shi Liu, Pan Stanford Publishing (2013).
95. G. Williams, B. Seger, P.V. Kamat, TiO2-graphene nanocomposites. UV-assisted photocatalytic reduction of graphene oxide, *ACS Nano*, 2 (2008) 1487–1491.
96. A.C. Arango, L.R. Johnson, V.N. Bliznyuk, Z. Schlesinger, S.A. Carter, H.-H. Hörhold, Efficient titanium oxide/conjugated polymer photovoltaics for solar energy conversion, *Advanced Materials*, 12 (2000) 1689–1692.
97. C. Chen, W. Cai, M. Long, B. Zhou, Y. Wu, D. Wu, Y. Feng, Synthesis of visible-light responsive graphene oxide/TiO2 composites with p/n heterojunction, *ACS Nano*, 4 (2010) 6425–6432.
98. X. Wang, L. Zhi, K. Müllen, Transparent, conductive graphene electrodes for dye-sensitized solar cells, *Nano Letters*, 8 (2008) 323–327.
99. Y. Sun, Q. Wu, G. Shi, Graphene based new energy materials, *Energy and Environmental Science*, 4 (2011) 1113–1132.
100. D. Selvakumar, G. Murugadoss, A. Alsalme, A.M. Alkathiri, R. Jayavel, Heteroatom doped reduced graphene oxide paper for large area perovskite solar cells, *Solar Energy*, 163 (2018) 564–569.

# 8

# *Progress of Nanomaterials Application in Environmental Concerns*

**Lienda Handojo, Eduardus Budi Nursanto, and Antonius Indarto**

**CONTENTS**



## 8.1 Introduction

Rapid industrialization and urbanization around the world have led torecognition and understanding of the relationship between environmental contamination and public health. Among the different sources of environmental pollution, industrial waste discharged from different industries is considered as one of the major sources of environmental pollution (Ho et al., 2012). At high concentrations, toxic and hazardous wastes are able to damage the ecological function of the environment because they can kill microorganisms in the water and soil and change the balance of chemical species in our atmosphere. Therefore, for waste to be safe for the environment, proper handling is required. In most countries in the world, the emitted waste allowed to be released into the environment must meet the criteria set by the government regulator.

Currently, the decomposition process of waste is done through chemical, physical, and biological processes. Some conventional techniques used to eliminate the toxic contaminants are combustion (Jouhara et al., 2017), plasma (Keun Song et al., 2005; Indarto et al., 2008), ion exchange, selective absorption (Handy et al., 2014; Widodo et al., 2015), reverse osmosis, ultrafiltration, chemical scavenging, and others. The process usually has limitations and, in particular, it is expensive to remove waste down to very low concentration (Mahmoud et al., 2011). The above conventional techniques use a large number of chemicals and generate a large amount of sludge after treatment, both of which are classified as secondary pollutants in the environment. In some cases, for example, the elimination of waste by combustion or plasma, the reaction just changes the phase of the original waste into other forms of waste that possibly could have higher toxicity (Stanmore, 2004; Indarto, 2012a; Indarto, 2016). Further, to reduce the toxicity of a secondary effluent, the secondary process is still carried out in many complex waste treatment plants by adding chlorine and oxidizing with ozone. Although this method is efficient, the cost is significantly high. Moreover, its residue and excessive use of chemicals can threaten the environment as more microorganisms can be killed.

Following the above discussion, advanced techniques will be necessary to overcome the more complex problems in the future. One of the techniques that currently attracts many environmentalists is the introduction of nanotechnology for waste treatment. Nanomaterials are shown to be promising materials or catalysts for various applications in the environmental (Yunus et al., 2012) and energy sectors (Christian et al., 2013). By reducing the size of the solid particle down to the nano-size, the contact surface area of the adsorbent is increased and allows the nanoparticles to adsorb, react with, or detect high to low concentrations of chemical species. Moreover, by utilizing nanotechnology, the decomposition of waste becomes more effective, cheaper, and more environmentally friendly. Nowadays, thorough research and evaluation are conducted to study nanomaterials and ensure that they will not have an impact on the environment itself and not be the next contributor to pollution.

## 8.2 Current Applications of Nanotechnology for the Environment

Following the 2004 Environmental Protection Agency (EPA) report, it was estimated that the remediation process will take 30 to 35 years and cost up to $250 billion to clean up all hazardous waste areas around the United States (EPA, 2004). Later, the EPA suggested the development of technologies that could result in better, cheaper, and faster site cleanups. Nano-remediation has the potential not only to reduce the overall costs of cleaning up large-scale contaminated sites, but it also can reduce cleanup time, reduce some contaminant concentrations to near zero, and can be done *in situ*. Currently, more than 44 environmental remediation projects have been identified for remediation by the application of nanotechnology, as shown in Figure 8.1 (Nanotechproject, 2009). Those projects were divided into six categories: (1) oil and minefields, (2) manufacturing sites, (3) military, (4) private properties, (5) residence, and (6) others. Most of them were located in the United States and Europe and mostly dealt with wastewater and the soil remediation of chlorinated compounds, such as perchloroethylene (PCE), trichloroethene (TCE), or polychlorinated biphenyls (PCBs). Other pollutants include Cr(VI) and nitrate.

Most of the bench-scale research and field application of nanoparticles for remediation at full-scale have focused on nano-zero valence iron (nZVI) with the size of the particle ranging from 10 to 100 nanometers in diameter or slightly larger (Mu et al., 2017). Taking the example of Hamilton Township, New Jersey, where contaminated with TCE and its derivatives and an initial maximum volatile organic compound (VOC) concentration of 1,600 micrograms per liter (μg/L), the nZVI was injected in two phases over a total of 30 days and could decrease the concentration of chlorinated contaminants up to 90% (Fulekar, 2012). Currently, more and more researchers are developing a variety of nanomaterials for potential use in adsorbing or eliminating contaminants as part of either *in situ* or *ex situ* processes. The development of nanomaterials for environmental applications is one of the most active topics of research worldwide.

## 8.3 Potential of Nanotechnology for Solving Wastewater Problems

In general, two contaminant parameters that become targets of reduction are heavy metals and organic pollutants for wastewater. Heavy metals, such as cadmium, can have a great impact on the environment and health even in low concentration because of their toxic, carcinogenic, and undegradable natures. Cadmium is released through wastewater by the electroplating process, mining, and the production of battery, fertilizers, and alloys (Mahmoud et al., 2011). The World Health Organization (WHO) set a limit on the maximum concentration of cadmium in drinking water at 0.003 mg/L. Similar to metal contamination, some organics such as phenol, in wastewater are harmful to organisms even at low concentrations. The EPA issued a regulation for reducing the phenol content in wastewater to less than 1

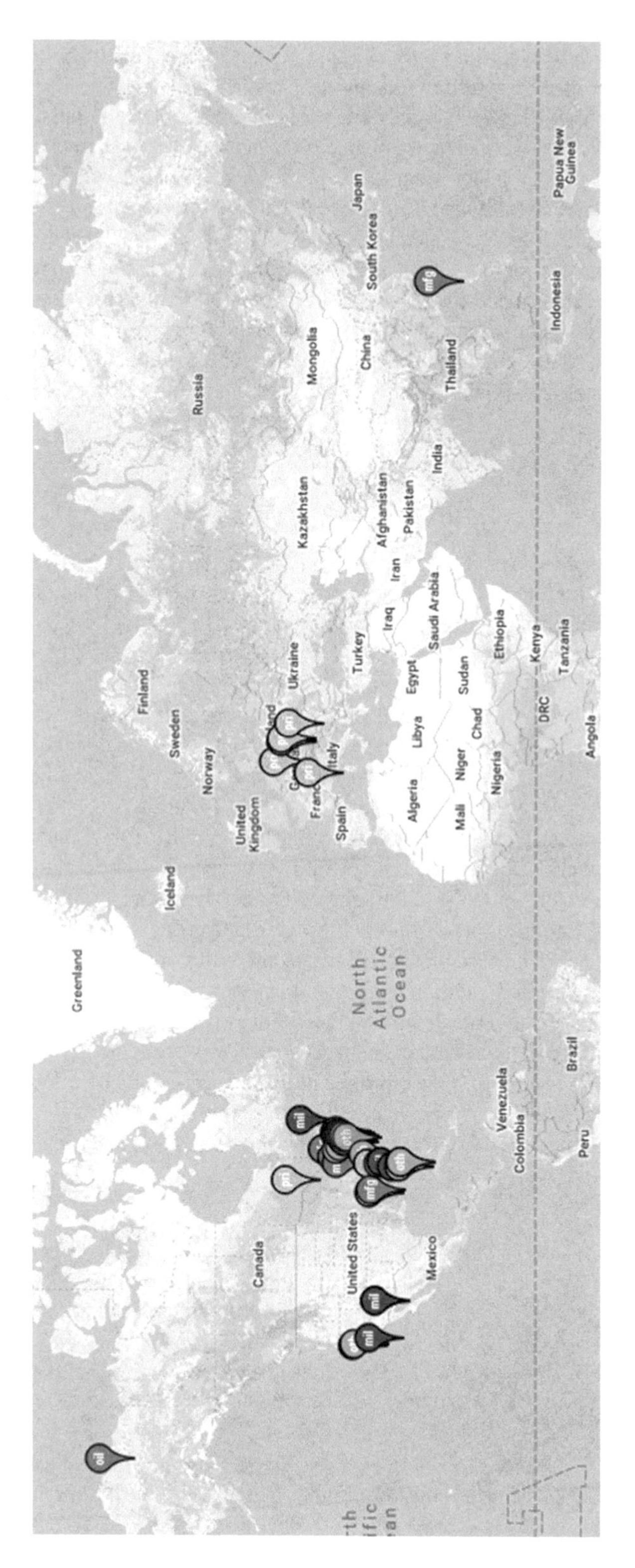


**FIGURE 8.1** **(See color insert.)** Map of nano-remediation sites. (Redrawn from Nanotechproject, 2009.)

ppm (Jin et al., 2011). The high concentration and poor biodegradability of organic pollutants and non-biodegradable nature of inorganic metal pollutants in industrial wastewaters pose a major challenge for environmental safety and human health protection and thus, adequate treatment of industrial wastewaters is required before their final disposal in the environment.

Currently, biological methods using microbes are becoming much more popular for the treatment of industrial wastewaters in wastewater treatment plants. Biological treatment methods employ microorganisms cultured in an aerated pond or a bioreactor as a catalyst for waste degradation. This method is considered to be an environmentally friendly and cost-effective method for wastewater treatment with a simple structural set-up, wider application, easy operation, and lower level of sludge production as compared to physico-chemical methods (Mendez-Paz et al., 2005; Pandey et al., 2007). The problem is that not all waste can be treated using this method, especially the waste coming from chemical processes because of its resistance or toxicity to the microorganism, such as pinene, carbol, etc. Decomposition of these wastes not only involves high costs but also results in other new forms of waste. One solution to deal with these problems is the utilization of nanotechnology, which is environmentally friendly and able to reduce the costs.

## 8.3.1 Absorption Process

### *8.3.1.1 Nano-Zero Valence Iron (nZVI)*

By making the size of the adsorbent particle very small, its contact area becomes huge. It makes the adsorption process becomes more efficient and effective for a low concentration of heavy metals. nZVI is a successful nanomaterial used to adsorb various contaminants and metal ions in water. nZVI particles are widely used to remediate groundwater and wastewater contaminated by heavy metals (Qiu et al., 2012), chlorinated organic compounds (Fu et al., 2014), nitroaromatic compounds (Yin et al., 2012), arsenic (Neumann et al., 2013), nitrate (Hwang et al., 2011), phenols (Shimizu et al., 2012), and dyes (Luo et al., 2013). In addition, nZVI was also applied to construct permeable reactive barriers (PRBs) of high reactivity to remove a large number of contaminants including chlorinated compounds, inorganic anions, and dissolved metals in groundwater (Mu et al., 2017). In almost all cases, the adsorption efficiency could reduce the contaminant by >90%. The iron acts as both a catalyst and an electron donor for the reduction reaction (Luo et al., 2013). The core metallic iron can act as the electron source and endow nZVI with a reductive character, while the ubiquitous iron oxide shell can facilitate the adsorption of solutes and the surface complexation, enhance/inhibit the electron transfer for direct reduction of pollutants and/or molecular oxygen activation, and thus affect the accessibility of pollutants and the degradation of organic contaminants. Adsorption of inorganic/organic species by nZVI was predominantly mediated by the oxide shell. As electrons play an important role, the interaction (sorption and reduction) of nZVI with the metal ion is depending and defined on the standard electrode potential ($E^0$), as shown in Figure 8.2.

In some cases, the adsorption efficiency needs to be improved, especially to reduce the removal period or reaction time. The process of adsorption is sensitive to the temperature as the adsorption rate will increase by raising the temperature. The thermodynamic process is generally endothermic and occurs spontaneously (Boparai et al., 2011).

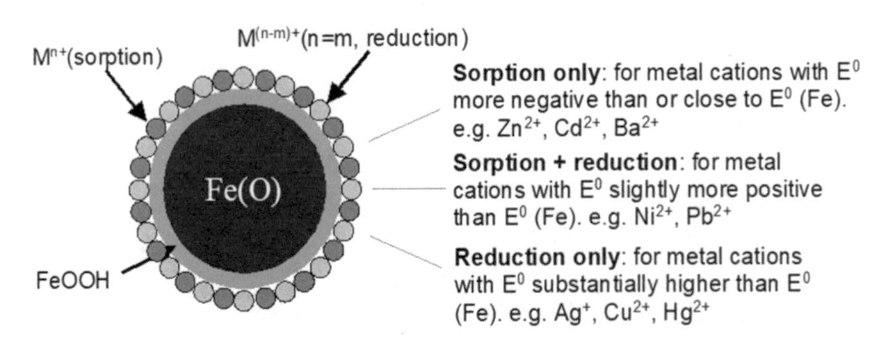


**FIGURE 8.2** Core-shell model of nano-zero valence iron for metal ion removal. (Redrawn from Boparai et al., 2011.)

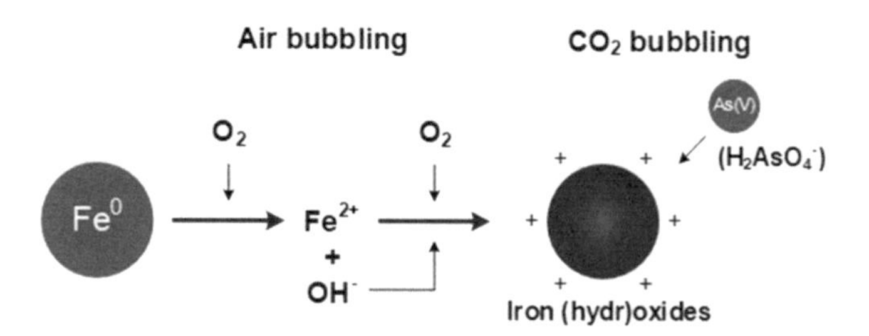


**FIGURE 8.3** Reaction mechanism of nZVI with As ion in the presence of $O_2$ and $CO_2$. (Redrawn from Tanboonchuy et al., 2011.)

Another technique for improving the adsorption process is by blowing $CO_2$ or oxygen gas into the contaminant water. Flowing oxygen will oxidize the $Fe^0$ into $Fe^{2+}$ and form a chelate to interact with the metal ion. With an oxygen injection, the removal rate of $As^{5+}$ increases five-fold (Tanboonchuy et al., 2011). The schematic mechanism is shown in Figure 8.3.

Modification of nZVI could also be conducted to match the specific removal characteristic such as nZVI for adsorbing nitrate ($NO_3^-$). In this case, the nanoparticles used are nZVI with a surface that was modified using copper (see the reaction mechanism in Figure 8.4). The removal efficiency could reach 95% at 2.5% of Cu addition on the surface of the nZVI. The mechanism, as shown in Figure 8.4, is rather too complex to be explained, but in general, the presence of doping metal was used to improve the potential parameter ($E^0$) of the nZVI surface.

### 8.3.1.2 Graphene-Based Nanomaterials

Currently, graphene-based nanomaterials (GBNs) attract many scientists and researchers for utilization in various potential applications. Although large-scale applications are still under development, e.g. for water purification, GBNs provide an excellent particle property due to their large surface area (2,630 $m^2/g$) and specific structure (aromatic structure). More specifically, nanographene oxide and reduced graphene oxide (variant types of GBNs) have received significant attention because of their low cost, facile fabrication, and outstanding heterogeneous structures, which contain abundant adsorption sites for different organic pollutants, such as dyes, polycyclic aromatic hydrocarbons (PAHs), and antibiotics. In general, adsorptions using nano-sized GO or rGO are related to the π-electron density of the adsorbate, indicating that π–π electron donor–acceptor interactions and surface capacity are an important mechanism controlling the adsorption of aromatic organics and metal ions on GBNs. Table 8.1 tabulates the performance of various GBNs to adsorb organic pollutants.

Carbon nanotubes (CNTs) are another graphene type of nanomaterial that have a curvature on the graphite sheet that gives the CNT surface a stronger and higher polarity surface than a plain graphite sheet. CNT materials have already replaced the activated carbon adsorbent in many daily uses and are

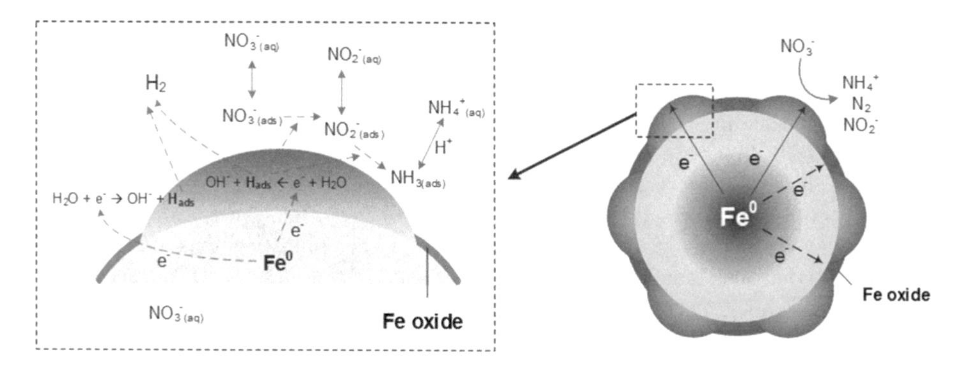


**FIGURE 8.4** Mechanism of nitrate absorption by using Cu/nZVI. (Redrawn from Mossa Hosseini et al., 2011.)

**TABLE 8.1**

Adsorption Capacity of GBNs to Various Organic Pollutants

| Adsorbents | Organic Pollutants | Adsorption Cap. (mg/g) | References |
|---|---|---|---|
| GO | methylene blue | 714 | Yang et al. (2011) |
| GO | basic yellow 28, basic red | 68.5–76.9 | Konicki et al. (2017) |
| GO | tetracycline, oxytetracycline | 212–398 | Gao et al. (2012) |
| GO | 17β-estradiol | 149 | Jiang et al. (2016) |
| Double oxidized GO | acetaminophen | 704 | Moussavi et al. (2016) |
| Sulfonated Graphene | naphthalene, 1-naphthol | 307–341 | Zhao et al. (2011) |
| Sulfonated Graphene | phenanthrene, methylene | 400–906 | Shen & Chen (2015) |
| Hydrazine reduced graphene | naphthalene, 1-naphthol | 145–269 | Wang & Chen (2015) |
| Annealing reduced graphene | naphthalene, 1-naphthol | 52.4–282 | Wang & Chen (2015) |
| Graphene | bisphenol A | 182 | Xu et al. (2012) |
| Activated graphene | p-nitrotoluene, naphthalene | 161–317 | Wang et al. (2016) |

able to adsorb various metals, such as Cu (Liang et al., 2005), Ni (Chen & Wang, 2006), Cd (Liang et al., 2004), and Pb (Li et al., 2006). CNTs are also very powerful adsorbents for a wide variety of organic compounds from water. Examples include dyes (Fugetsu et al., 2004), PAHs (Yang et al., 2006), dioxin (Long & Yang, 2001), chlorobenzenes and chlorophenols (Cai et al., 2005), trihalomethanes (Lu et al., 2006), polybrominated diphenyl ethers (PBDEs) (Wang et al., 2006), bisphenol A and nonylphenol (Cai et al., 2003), pesticides (thiamethoxam, imidacloprid, and acetamiprid) (Zhou et al., 2006a), and dichlorodiphenyltrichloroethane (DDT) (Zhou et al., 2006b). Cross-linked nanoporous polymers that have been copolymerized with functionalized CNTs have been demonstrated to have a very high sorption capacity for a variety of organic compounds such as p-nitrophenol and trichloroethylene (Salipira et al., 2007). Adsorption of organometallic compounds on pristine multi-walled CNTs was found to be stronger than for carbon black (Muñoz et al., 2005). The available adsorption space was found to be the cylindrical external surface; neither the inner cavity nor the inter-wall space of multi-walled CNT contributed to adsorption (Yang & Xing, 2007).

#### *8.3.1.3 Nanocrystalline Metal Oxide (NMO)*

The development of nanocrystalline metal oxides (NMOs) over the previous few years has focused on the adsorption capacity, and subsequently the capability of adsorption technologies for metal removal from wastewater.

Nanocrystalline aluminum(III) oxide has the ability to adsorb selenium oxy-anions, i.e. selenite $SeO_3^{2-}$ ($Se^{4+}$) and selenate $SeO_4^{2-}$ ($Se^{6+}$) following the mechanism proposed by Jordan et al. (see Figure 8.5). Yamani et al. (2014) successfully developed nanocrystalline aluminum oxide- and titanium dioxide-impregnated chitosan beads as an adsorbent for both ions. However, the presence of competing ions (such as sulfate and phosphate) hinders the adsorption capacity, due to the non-selective adsorption of anions by the aluminum oxide. In the case of selenite and selenite adsorption, the adsorption could be improved by adding $SiO_2$ to make its surface charge more positive and yield adsorption capacities for selenite and selenate of 32.7 and 11.3 mg/g respectively (Chan et al., 2009). Nano-manganese iron oxide ($MnFe_2O_4$) has also been shown to have a higher adsorption capacity for selenium oxy-anions than other naturally magnetic material (Gonzalez et al., 2010). The removal of selenate or selenite is pH-independent between pH 2 to 6 and occurs within five minutes of contact time. Recently, some other metals were identified to be very promising. Sun et al. (2015) found that the adsorption capacity of selenite and selenate followed the trend $CuFe_2O_4 > CoFe_2O_4 \gg MnFe_2O_4$, which is consistent with the order of hydroxyl group contents and surface charges on the bimetal oxide.

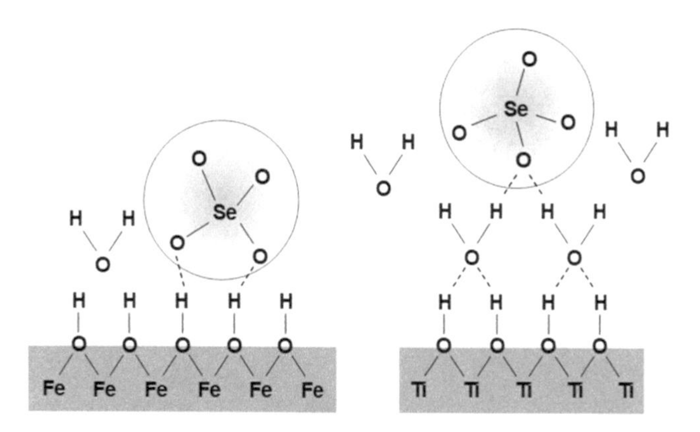


**FIGURE 8.5** Schematic interaction between selenite and surface of metal oxide. (Redrawn from Jordan et al., 2013.)

#### *8.3.1.4 Other Nanomaterials*

Other nanoparticles that can be utilized for adsorbing metal and organic ions are nano-zeolite, nano-silicon, and polymer–clay nanocomposites. Zeolite is commonly used in some applications such as separation, catalysis, ion exchange, and adsorption. Nano-zeolite (10–1000 nm) was found to be an effective adsorbent for the removal of chromium compounds (up to 97%) while natural zeolite (clinopitolite) could eliminate toxic metal only up to 55%. The use of nano-silicon resulted in cadmium removal in about 97% with a maximum adsorption capacity of 1,000 μmol/g at pH 7 (Mahmoud et al., 2011). Polymer–clay nanocomposites have also been studied for environmental applications such as sorbents for anionic pollutants. Chitosan–montmorillonite nanocomposites have been well studied for the adsorption of a vast array of anionic pollutants and are able to selectively adsorb selenate from contaminated waters (Bleiman & Mishael, 2010). The removal of selenium by the chitosan–montmorillonite nanocomposite was influenced by the polymer loading of the composite. The nanocomposites where two polymer layers were intercalated within the clay showed an increase in selenate removal.

### 8.3.2 Photooxidation Process

In general, the photooxidation process employs nanocatalysts of titanium dioxide nanoparticles ($TiO_2$) and photons from sunlight, especially ultra-violet (UV) rays. Photooxidation is an oxidation process on a particular polymer surface involving oxygen ($O_2$) and ozone ($O_3$). The process is done simply by placing nano-titanium dioxide inside the contaminated water to degrade the waste in open space. Titanium captures UV rays from the sun. Furthermore, the rays will make water ($H_2O$) react with oxygen to produce hydroxide radicals (OH*) and oxide radicals ($O^{2-}$). These radicals are organically destroyable into simple stable compounds, following the reaction:

$$mC_xH_yO_z + nOH^* + pO_2^- \rightarrow qCO_2 + sH_2O$$

where $C_xH_yO_z$ is an organic species.

Photooxidation with nanotechnology provides evidence that low organic waste concentration of 134 ppm can be reduced to 40 ppm in less than 1.5 hours. The higher the concentration of the waste the more time it will take to process it. For example, if the waste is increased up to 1,340 ppm, the time it takes to convert the waste to have concentration lower than 30 ppm is 24 hours. Although the process takes quite long, it does not produce secondary pollution or waste. The products are mostly water, nitrogen, and carbon dioxide. A smaller sized catalyst will cause more of the surface area of the active catalyst to radicalize water and oxygen. Decomposition of waste by using this technique has several advantages. One of them is that sunlight can be obtained freely.

The concept of photooxidation could also be used to manage the oxidation state of metal ions present in the wastewater, as shown in Figure 8.6.

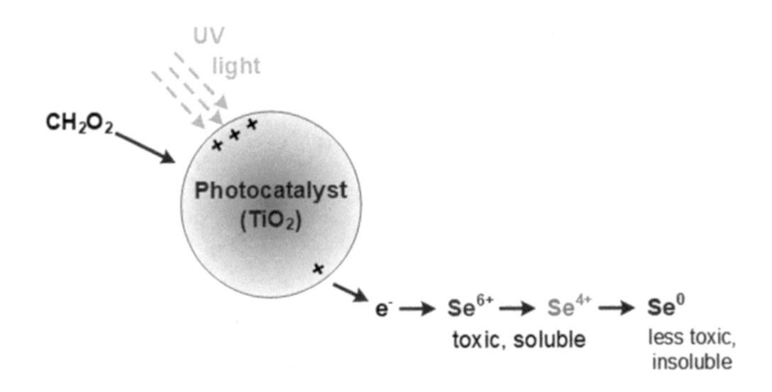


**FIGURE 8.6** Schematic of $Se^{6+}$ and $Se^{4+}$ photoreduction by $TiO_2$. (Redrawn from Holmes & Gu, 2016.)

In the case of selenite and selenate removal, the reduction of its metal oxidation state into $Se^0$ could reduce the toxicity of the selenium. Currently, the improvement of $TiO_2$ can be done by mixing it with zirconium (TiZr) which shows higher photocatalytic activity than standard n-$TiO_2$.

### 8.3.3 Disinfectant

Instead of an organic and non-organic chemical, the number of bacteria and other living organisms in the wastewater could be controlled to reach the acceptable or minimum level. New types of nanometallic particles have been developed to kill microbes, a process called oligodynamic disinfection; these particles have the ability to inactivate microorganisms at low concentrations. Various oligodynamic metals provide microbicidal, bactericidal, and viricidal properties; however, if the particle size is reduced to the nano-scale, these oligodynamic metals produce tremendous advantages in disinfection capacity due to their greater surface area, contact efficiency, and even better elution properties. Recently, it has been reported that the best nanometallic particles for water disinfection, from the strongest to weakest, are silver (Ag NPs), copper (Cu), zinc (Zn), titanium (Ti), and cobalt (Co). A combination of new oligodynamic materials is being developed, such as Ag deposited in titanium oxide and Ag-coated iron oxide has shown the reaction speed and high efficiency in killing bacteria. The perfect solubility of silver nitrate makes it easy to use in a variety of disinfection applications. Although various forms of silver have been used in a variety of disinfection applications, which include swimming pools and hospital hot water systems.

In addition to oligodynamic nonmetallic particles offering a good alternative disinfection system, there are some drawbacks to be considered. The rate of silver desorption is a crucial parameter that must be characterized before it is made. The U.S. EPA has issued a level of silver contamination in water, which is 100 ppb; therefore, it should be given more attention. However, since silver has extensive microbicidal properties, some nonpathogenic bacteria which are beneficial to the environment will be killed. Another nanomaterial that has a similar property is nano-ZnO. 2% of ZnO in the solution has been shown to have anti-bacterial effects, especially as it was observed primarily against *S. aureus* and *E. coli* bacteria.

Fullerenes are not commonly used in the disinfection of water, but some types of fullerenes have the potential to be applied, such as hydroxylated or fullerol C60. Compared with $TiO_2$, hydroxylated or fullerol C60 is a relatively nontoxic fullerenes product. Another obstacle in fullerol use in water treatment is the difficulty of immobilization, separation, and recycling of fullerol nanoparticles. The absence of an easy and inexpensive method to separate these small and light nanoparticles. However, fullerene power is attached to the surface can be taken into consideration. C60 encapsulated in polyvinyl pyrrolidone (PVP) produces antibacterial and photoactivity materials, although the nature of human toxicity has not been discovered. Functional groups in fullerene organic "enclosures" can facilitate the attachment of fullerenes to the surface of the material without the loss of antibacterial properties, a desirable property in disinfecting processes involving fixed beds, membranes, or surfaces.

## 8.4 Potential of Nanotechnology for Solving Air Pollution Problems

The recent report from the WHO stated that air pollution was mainly responsible for the death of around 7 million people during 2012. Air pollution led to several health effects such as inflammation, accelerated atherosclerosis, and altered cardiac functions. Other medical reports also mentioned that particulate matters (PM) that were inhaled by humans led to higher mortality rates, specifically for diabetes, chronic pulmonary, and patients with inflammatory diseases. Furthermore, there are possibilities for severe environmental damage caused by air pollution from fuel use. Other major sources of air pollution are metal mills, metal smelters, municipal incinerators, cement plants, and mining, farming, and petrochemical industries (Vaseashta et al., 2007; Shan et al., 2009; Mohamed, 2017).

The major pollutants in air pollution are nitrous oxide ($NO_x$), sulfur dioxide ($SO_2$), carbon monoxide (CO), and fine suspended PM (Indarto et al., 2008; Indarto, 2009, 2012b), and these pollutants triggered severe health problems for humans. There were several reports mentioned about acute exposure to air pollution that contained PM, $NO_x$, or VOCs that affected respiration and other health disorders (Vaseashta et al., 2007; Mohamed, 2017). The correlation between air pollution and human mortality has been studied at five cities in the United States: Watertown, Massachusetts; Harriman, Tennessee; St. Louis, Missouri; Steubenville, Ohio; and Portage, Wisconsin (Dockery et al., 1993). The data from that study is arranged and drawn in Figure 8.7 and it shows the relation between PM concentration on the city and the mortality rate. It can be seen in Figure 8.7 that there is a linear correlation between the PM concentration in air and the mortality rate (Dockery et al., 1993). Another air pollutant is greenhouse gases (GHGs), such as $CO_2$, $CH_4$, $N_2O$, and fluorinated gases. GHGs have the capability to remain in the atmosphere for a long time, which can trigger global warming.

We should also be concerned about indoor air pollutants. There was a study of VOCs that can be generated indoors, and which have several effects on human health. VOCs are the main cause of asthma problems in children, respiratory problems, and lung cancer (Lee et al., 2008; Ibrahim et al., 2016). Another source of indoor air pollutions is bioaerosols (aerosols that are generated biologically, such as fungi, bacteria, and viruses. Bioaerosols are easily spread in the indoor environment and can cause several diseases and trigger infections and allergies.

Innovation is needed to reduce air pollution, even for small concentrations of contaminants in the air. Nanotechnology with the nanomaterials as main components offer the solution for reducing air pollution.

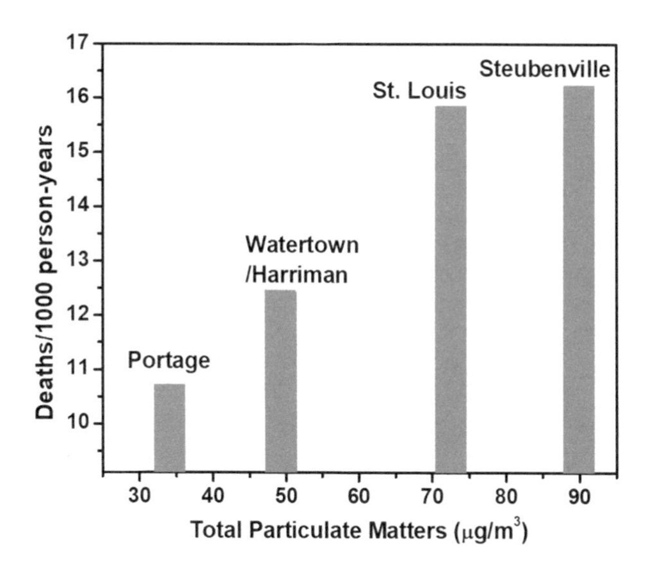


**FIGURE 8.7** The Effect of Air Pollution (Particulate Matters in air) on Mortality in 1979–1985 in the United States. (Drawn from data in Dockery et al., 1993.)

### 8.4.1 Adsorption Processes

Nanotechnology is one of the solutions to solve the air pollution problem. Nanotechnology offers the ability to develop nanomaterials with specific properties that can be utilized to solve the air pollution problem (Chirag, 2015; Mohamed, 2017). The contaminant and hazardous gas in air can be converted and degraded into non-toxic components by using nanomaterials.

There are two main routes for the application of nanomaterials to solve the air pollution problem. First is by the adsorption and filtration process, which applies nano-absorptive materials as active materials. Second is the catalytic degradation process which uses nano-catalyst materials.

The adsorption process can be utilized to adsorb contaminant and hazardous gas in the air by using nano-absorptive materials. The nano-adsorptive materials are commonly used for heavy metal adsorption in industrial smokestacks. Well-known nano-adsorptive materials that are installed in industrial smokestacks are nanostructured silica sorbents for lead (Pb) adsorption and carbon sorbents for mercury (Hg) adsorption (Kwon & Vidic, 2000; Booker & Boysen, 2005).

For the adsorption of pollutant gases, there are several examples of nano-adsorptive materials such as nanostructured Ca for $CO_2$ adsorption (Abanadez & Alvarez, 2003), alkali nanotubes (such as potassium titanate/K-Ti-NT and sodium nanotubes/Na-Ti-NT) for $CO_2$ adsorption, and nanostructured carbon for the adsorption of GHGs (Upendar et al., 2012). The nanostructured carbon materials have a superior ability to adsorb contaminants and hazardous gases in the air. The special properties of carbon nanostructures, such as a high number of active surface sites and high selectivity, are the main reason for excellent adsorption process (Gupta & Saleh, 2013; Wang et al., 2013). Furthermore, the addition of nanostructured carbon with a functional group or with other materials such as polymer could provide a higher number of active surface sites for adsorption (Vaseashta et al., 2007; Wang et al., 2013). The nanostructured carbons could allow for the addition of more than one functional group (OH, COOH, C=O), which enhances the selectivity and the stability and optimizes the adsorption ability (Gupta & Saleh, 2013). The nanostructured carbons that are commonly used for nano-absorptive materials are fullerenes, CNTs, graphene, and graphite (Mohamed, 2017). Nanostructured carbon has good contact with organic compounds through intermolecular forces such as hydrogen bonding electrostatic forces, $\pi$–$\pi$ interactions, van der Waals forces, electrostatic forces, and hydrophobic interactions (Mohamed, 2017). Another example of nanostructured carbon is activated coke from the coal industry (Shan et al., 2009). The activated coke is proven to be an adsorbent for the removal of sulfur oxides ($SO_x$). Furthermore, activated coke is proven to be an adsorbent for the removal of heavy metals and toxic trace materials.

For comparison, Table 8.2 shows several examples of nano-adsorptive materials.

### 8.4.2 Catalytic Degradation Processes

The catalytic degradation process using nano-catalyst materials can be applied for air pollutant treatment. The active surface is one of the most important properties for catalyst activity. There is the possibility

**TABLE 8.2**

Nano-Adsorptive Materials for Air Pollutants

| Nano-adsorptive Materials | Target Pollutant Gases | References |
|---|---|---|
| Nanostructured Ca | $CO_2$ | Abanadez & Alvarez (2003) |
| Alkali Nanotubes (K-Ti-NT and Na-Ti-NT) | $CO_2$ | Upendar et al. (2012) |
| Zinc nanocage ($Zn_{12}O_{12}$) | Carbon disulfide ($CS_2$) | Ghenaatian et al. (2013) |
| Activated coke | $SO_x$ | Shan et al. (2009) |
| Aligned-CNT | Aerosols | Yildiz & Bradford, (2013) |
| CNT-polymer | PM on air | Vaseashta et al. (2007) |
| CNTs deposited on quartz filters | VOCs | Amade et al. (2014) |
| CNT/NaClO | Isopropyl vapor | Hsu & Lu (2007) |
| Fullerene-like boron nitride nanocage | $N_2O$ | Esrafili (2017) |
| Graphene | $CO_2$ | Mishra & Ramaprabhu (2011) |

that decreasing the catalyst size (below 100 nm) leads to the optimizing of the utilization of the active surface area. In this process, air pollutants are converted into less dangerous/non-dangerous compounds. There are two types of catalytic degradation processes; the first is the common catalytic process and the second is the photo-catalytic process.

Common catalytic degradation processes that are well known are de-$NO_x$ converters and catalytic car exhaust emission reduction (Christensen et al., 2015). The de-$NO_x$ converter is commonly used in power plants or incineration plants. The nano-catalyst that is commonly used for the de-$NO_x$ converter is vanadium. In catalytic car exhaust emission reduction systems, a three-way catalyst has been used to convert CO into $CO_2$ and for $NO_x$ removal. The three-way catalyst consists of platinum (Pt), palladium (Pd), and rhodium (Rd). Pd and Pt promote the oxidation of CO and rhodium is used for $NO_x$ removal.

In the photocatalytic process for air pollutant treatment, semiconducting materials have been used. $TiO_2$ is a renowned material for use as a photocatalyst in this process. $TiO_2$ can reduce contaminants in the atmosphere such as VOCs, $NO_x$s, and other pollutants to become less toxic compounds (Shen et al., 2015). Furthermore, $TiO_2$ can oxidize CO to become $CO_2$, which is less toxic (Low et al., 2017). The mechanism of photocatalysis on the surface of $TiO_2$ had been proposed by Yang's group (Zhao & Yang, 2003). The first step is catalyst activation by the photon energy (hv) from UV light. In this step, both oxidizing ($h^+$) and reducing ($e^-$) agents are generated.

$$\mathrm{TiO_2} + hv \rightarrow h^+ + e^- \tag{8.1}$$

The scheme of oxidative (O) and reductive (R) reactions is shown in Figure 8.8, redrawn from Zhao & Yang, (2003). The mechanism of oxidative and reductive reactions expressed as:

Oxidative reaction:

$$\mathrm{OH^-} + h^+ \rightarrow \mathrm{OH^*} \tag{8.2}$$

Reductive reaction:

$$\mathrm{O_{2ads}} + e^- \rightarrow O^-_{2ads*} \tag{8.3}$$

By the oxidative reaction on the $TiO_2$ surface, the hydroxyl radical ($OH^*$) is formed. $OH^*$ is the primary oxidant during the photocatalysis oxidation, with the presence of oxygen. The role of oxygen is to prevent the recombination of hole–electron pairs. Complete photocatalysis oxidation is shown in reaction (4) below (Zhao & Yang, 2003):

$$\mathrm{OH^*} + \mathrm{pollutant} + \mathrm{O_2} \rightarrow \mathrm{products}\left(\mathrm{CO_2}, \mathrm{H_2O}, \mathrm{etc}\right) \tag{8.4}$$

Table 8.3 shows the catalytic degradation process for air pollutants (Shan et al., 2009; Christensen et al., 2015; Ibrahim et al., 2016). Some of the processes have been commercialized and others are still in the development stage.

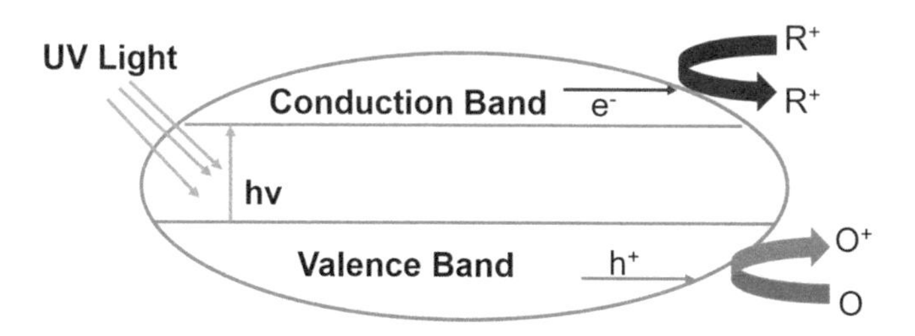


**FIGURE 8.8** The schematic of photocatalysis reaction on $TiO_2$ surface. (Redrawn from Zhao & Yang, 2003.)

**TABLE 8.3**

Applications of Nano-Catalysts in Catalytic Degradation Process for Air Pollutant

| Application/Product | Type of Nano-materials | Catalytic Process | Development Progress |
|---|---|---|---|
| Air pollution reduction by surface coating using nanomaterials | Commercial nanostructured $TiO_2$ | Photocatalysis process for removing $NO_x$ and VOC | Commercial |
| Catalyst for smoke stack of industry | Nanostructured vanadium (V) | Catalysis process for removing NOx | Commercial and applied for coal-fired power plants |
| Three-way catalyst for catalytic converter in diesel car | Nanostructured Pt, Pd and Rh | Catalysis process for removing $NO_{x,}$ unburnt hydrocarbon and oxidizing CO | Commercial and applied in diesel-fueled automobiles |
| Catalyst for hydrodesulfurization | Nanostructured cobalt-molybdenum (CoMo) on alumina | Catalysis process for breaking the bonds of S-O in $SO_2$ and breaking bond of C-S in $C_4H_4S$ | Commercial and applied in oil refinery |
| Catalyst for hydrodesulfurization | Molybdenum carbide ($MoC_xS_2$-$_x$) | Catalysis process for breaking the bonds of S-O in $SO_2$ and C-S in $C_4H_4S$ | Under development |
| Catalyst for $NO_x$ removal | Activated Coke | Catalysis process for removing $NO_x$ | Under development |
| Catalyst for $CH_4$ decomposition | Metallic nickel nanoparticles | Catalysis process for thermal decomposition of $CH_4$ to produce $H_2$ | Under development |

### 8.4.3 Production of Fuel from Air Pollutants

GHGs are the main contributor of air pollutants and $CO_2$ is the main component of GHGs (Whipple & Kenis, 2010; Nursanto et al., 2016; Nursanto et al., 2018). The trend of global carbon emissions based on carbon fuel is steadily increasing. To overcome this situation, $CO_2$ capture and transformation into valuable chemicals or carbon fuels and power from renewable energy resources should be developed. There are two common methods to utilize renewable energy sources, the first is by using it directly and the second is by using electricity that is generated by renewable energy sources. In regards to the photocatalytic process, only a few reports mention active and selective systems for photocatalytic $CO_2$. Additionally, the production rate of photocatalytic $CO_2$ is still not feasible from the point of view of cost. On the other hand, the electro-catalytic process or $CO_2$ electro-reduction has become a hot topic recently since it has several benefits compared to catalytic and photocatalytic process (Kondratenko et al., 2013). Furthermore, the $CO_2$ electro-reduction process has been implemented to produce formic acid in a large-scale process (Kondratenko et al., 2013).

$CO_2$ electro-reduction can be performed under ambient environmental conditions (temperature and pressure), powered by renewable energy sources (wind, solar, etc.), and the reaction is easily controlled (e.g. by varying working potential) (Whipple & Kenis, 2010; Lu et al., 2015). Furthermore, $CO_2$ electro-reduction as a direct electro-catalytic process can be performed in aqueous or non-aqueous electrolytes to transform $CO_2$ into valuable chemicals such as ethylene and formic acid, carbon fuels such as methane and methanol, and feedstock gases such as $H_2$ and CO for the Fischer–Tropsch process or hydrocarbon formation (Hori, 2008; Whipple & Kenis, 2010; Nursanto et al., 2016; Nursanto et al., 2018). Figure 8.9 shows the simple illustration for a $CO_2$ electro-reduction cell. In the anode part, the water oxidation reaction occurs and protons from the anode are passed through the ion-conducting membrane to the cathode. In the cathode part, $CO_2$ reduction occurs and transforms into reduced carbon forms.

## 8.5 Conclusion

Nanotechnology has been utilized in many applications, including in the field of environmental remediation. Several applications of nanotechnology for environmental remediation have been commercialized

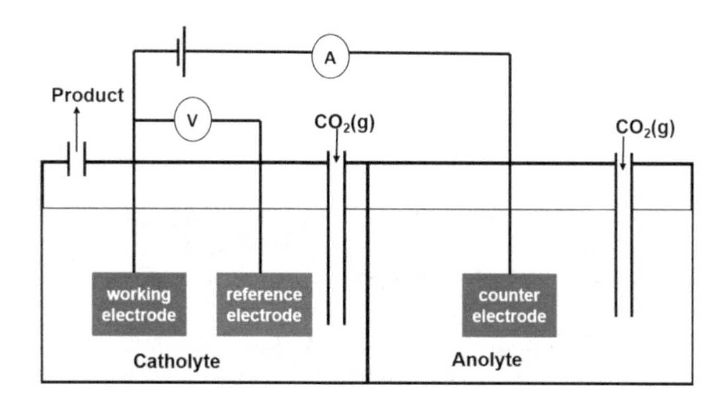


**FIGURE 8.9** $CO_2$ electro-reduction reactor.

and several are in the developing stage for commercialization. Many nanomaterials have been developed and shown to be able to remove major pollutants in soil, air, and water environments. Some materials, such as nZVI, graphene-based nanomaterial, NMO, and $TiO_2$, are even reliable for dealing with many pollutants and in different environments. Furthermore, the utilization of nanomaterials opens the possibility for future environmental remediation such as fuel production. The selection of materials for application must be based not only on performance but also on future waste handling and the economic point of view. The nanomaterial used for remediation must not be the next contributor to pollution.

## REFERENCES

Abanades, J.C., Alvarez, D. (2003) Conversion limits in the reaction of CO2 with lime. *Energy and Fuels, 17*, 308–315.

Amade, R., Hussain, S., Ocaña, I.R., Bertran, E. (2014) Growth and functionalization of carbon nanotubes on quartz filter for environmental applications. *Journal of Environmental Engineering & Ecological Science, 3*, 1–7.

Bleiman, N., Mishael, Y.G. (2010) Selenium removal from drinking water by adsorption to chitosan–clay composites and oxides: Batch and columns tests. *Journal of Hazardous Materials, 183*, 590–595.

Booker, R., Boysen, E. *Nanotechnology for Dummies*. Hoboken: Wiley Publishing Inc. (2005).

Boparai, H.K., Joseph, M., O'Carroll, D.M. (2011) Kinetics and thermodynamics of cadmium ion removal by adsorption onto nano zero valent iron particles. *Journal of Hazardous Materials, 186*, 458–465.

Cai, Y.Q., Cai, Y.E., Mou, S.F., Lu, Y.Q. (2005) Multi-walled carbon nanotubes as a solid-phase extraction adsorbent for the determination of chlorophenols in environmental water samples. *Journal of Chromatography A, 1081*, 245–247.

Cai, Y.Q., Jiang, G.B., Liu, J.F., Zhou, Q.X. (2003) Multiwalled carbon nanotubes as a solid-phase extraction adsorbent for the determination of bisphenol A, 4-n-nonylphenol, and 4-tert-octylphenol. *Analytical Chemistry, 75*, 2517–2521.

Chan, Y.T., Kuan, W.H., Chen, T.Y., Wang, M.K. (2009) Adsorption mechanism of selenate and selenite on the binary oxide systems. *Water Research, 43*, 4412–4420.

Chen, C.L., Wang, X.K. (2006) Adsorption of Ni(II) from aqueous solution using oxidized multiwall carbon nanotubes. *Industrial and Engineering Chemistry Research, 45*, 9144–9149.

Chirag, P.N. (2015) Nanotechnology: Future of environmental pollution control. *International Journal on Recent and Innovation Trend in Computing and Communication, 3*, 164–166.

Christensen, F.M., Brinch, A., Kjolhol, J.T., Mines, P.D., Schumacher, N., Jorgensen, T.H., Hummelsholj, R.M. (2015) Nano-enabled environmental products and technologies – Opportunities and drawbacks, The Danish Environmental Protection Agency, Copenhagen, DK. Available from: www.mst.dk.

Christian, F., Edith, Selly, Adityawarman, D., Indarto, A. (2013). Application of nanotechnologies in the energy sector: A brief and short review. *Frontiers in Energy, 7*, 6–18.

Dockery, D.W., Pope III, C.A., Xu, X., Spengler, J.D., Ware, J.H., Fay, M.E., Ferris, B.G., Speizer, F.E. (1993) An association between air pollution and mortality in six U.S. cities. *New England Journal of Medicine, 329*, 1753–1759.

EPA. *Cleaning Up the Nation's Waste Sites: Markets and Technology Trends*. EPA 542-R-04-015. Washington, DC: U.S. Environmental Protection Agency (2004).

Esrafili, M.D. (2017) N2O reduction over a fullerene-like boron nitride nanocage: A DFT study. *Physics Letters A*, *381*, 2085–2091.

Fu, F., Dionysiou, D.D., Liu, H. (2014) The use of zero-valent iron for groundwater remediation and wastewater treatment: A review. *Journal of Hazardous Materials*, *267*, 194–205.

Fugetsu, B., Satoh, S., Shiba, T., Mizutani, T., Lin, Y.B., Terui, N., Nodasaka, Y., Sasa, K., Shimizu, K., Akasaka, T., Shindoh, M., Shibata, K.I., Yokoyama, A., Mori, M., Tanaka, K., Sato, Y., Tohji, K., Tanaka, S., Nishi, N., Watari, F. (2004) Caged multiwalled carbon nanotubes as the adsorbents for affinity-based elimination of ionic dyes. *Environmental Science and Technology*, *38*, 6890–6896.

Fulekar, M.H. *Bioremediation Technology: Recent Advances*. New Delhi: Springer (2012).

Gao, Y., Li, Y., Zhang, L., Huang, H., Hu, J., Shah, S.M., Su, X.G. (2012) Adsorption and removal of tetracycline antibiotics from aqueous solution by graphene oxide. *Journal of Colloid and Interface Science*, *368*, 540–546.

Ghenaatian, H.R., Baei, M.T., Hashemian, S. (2013) Zn12O12 nano-cage as a promising adsorbent for CS2 capture. *Superlattices and Microstructures*, *58*, 198–204.

Gonzalez, C.M., Hernandez, J., Parsons, J.G., Gardea-Torresdey, J.L. (2010) A study of the removal of selenite and selenate from aqueous solutions using a magnetic iron/manganese oxide nanomaterial and ICP-MS. *Microchemical Journal*, *96*, 324–329.

Gupta, V.K., Saleh, T.A. (2013) Sorption of pollutants by porous carbon, carbon nanotubes and fullerene-an overview. *Environmental Science and Pollution Research International*, *20*, 2828–2843.

Handy, H., Santoso, A., Widodo, A., Palgunadi, J., Soerawidjaja, T.H., Indarto, A. (2014) H2S–CO2 separation using room temperature ionic liquid [BMIM][Br]. *Separation Science and Technology*, *49*, 2079–2084.

Ho, Y.C., Show, K.Y., Guo, X.X., Norli, I., Alkarki Abbas, F.M., Morad, M. (2012) Industrial discharge and their effect to the environment. Available from: http://www.intechopen.com/books/industrial-waste/industrial-emissions-and-their-effect-on-the-environment.

Holmes, A.B., Gu, F.X. (2016) Emerging nanomaterials for the application of selenium removal for wastewater treatment. *Environmental Science: Nano*, *3*, 982–996.

Hori, Y. *Electrochemical $CO_2$ Reduction on Metal Electrodes*. New York: Springer (2008).

Hsu, S., Lu, C. (2007) Modification of single-walled carbon nanotube for enhancing isopropyl alcohol vapor adsorption from water streams. *Separation Science and Technology*, *42*, 2751–2766.

Hwang, Y.H., Kim, D.G., Shin, H.S. (2011) Mechanism study of nitrate reduction by nano zero valent iron. *Journal of Hazardous Materials*, *185*, 1513–1521.

Ibrahim, R.K., Hayyan, M., AlSaadi, M.A., Hayyan, A., Ibrahim, S. (2016) Environmental application of nanotechnology: Air, soil, and water. *Environmental Science and Pollution Research International*, *23*, 13754–13788.

Indarto, A. (2009) Soot growth mechanisms from polyynes. *Environmental Engineering Science*, *26*, 1685–1691.

Indarto, A. (2012a) Decomposition of dichlorobenzene in a dielectric barrier discharge. *Environmental Technology*, *33*, 663–666.

Indarto, A. (2012b) Heterogeneous reactions of HONO formation from NO2 and HNO3: A review. *Research on Chemical Intermediates*, *38*, 1029–1041.

Indarto, A. (2016) Partial oxidation of methane to methanol with nitrogen dioxide in dielectric barrier discharge plasma: Experimental and molecular modeling. *Plasma Sources Science and Technology*, *25*, 025002.

Indarto, A., Choi, J.W., Lee, H., Song, H.K. (2008) Decomposition of greenhouse gases by plasma. *Environmental Chemistry Letters*, *6*, 215–222.

Jiang, L., Liu, Y., Zeng, G., Xiao, F., Hu, X., Hu, X., Wang, H., Li, T., Zhou, L., Tan, X. (2016) Removal of 17β-estradiol by few-layered graphene oxide nanosheets from aqueous solutions: External influence and adsorption mechanism. *Chemical Engineering Journal*, *284*, 93–102.

Jin, X., Yu, C., Li, Y., Qi, Y., Yang, L., Zhao, G., Hu, H. (2011) Preparation of novel nano-adsorbent based on organic inorganic hybrid and their adsorption for heavy metals and organic pollutants presented in water environment. *Journal of Hazardous Materials*, *186*, 1672–1680.

Jordan, N., Ritter, A., Foerstendorf, H., Scheinost, A.C., Weiß, S., Heim, K., Grenzer, J., Mücklich, A., Reuther, H. (2013) Adsorption mechanism of selenium(VI) onto maghemite. *Geochimica et Cosmochimica Acta*, *103*, 63–75.

Jouhara, H., Czajczyńska, D., Ghazal, H., Krzyżyńska, R., Anguilano, L., Reynolds, A.J., Spencer, N. (2017) Municipal waste management systems for domestic use. *Energy*, *139*, 458–506.

Keun Song, H.K., Choi, J.W., Lee, H., Indarto, A. (2005) Gliding arc plasma processing for decomposition of chloroform. *Toxicological and Environmental Chemistry*, *87*, 509–519.

Kondratenko, E.V., Mul, G., Baltrusaitis, J., Larrazábal, G.O., Pérez-Ramírez, J. (2013) Status and perspectives of CO2 conversion into fuels and chemicals by catalytic, photocatalytic and electrocatalytic processes. *Energy and Environmental Science*, *6*, 3112–3135.

Konicki, W., Aleksandrzak, M., Mijowska, E. (2017) Equilibrium, kinetic and thermodynamic studies on adsorption of cationic dyes from aqueous solutions using graphene oxide. *Chemical Engineering Research and Design*, *123*, 35–49.

Kwon, S., Vidic, R.D. (2000) Evaluation of two sulfur impregnation methods on activated carbon and bentonite for the production of elemental mercury sorbents. *Environmental Engineering Science*, *17*, 303–313.

Lee, B., Ji, J.H., Bae, G.N. (2008) Inactivation of S. epidermis, B. subtilis, and E. coli bacteria bioaerosols deposited on a filter utilizing airborne silver nanoparticles. *Journal of Microbiology and Biotechnology*, *18*, 176–182.

Li, Y.H., Zhu, Y.Q., Zhao, Y.M., Wu, D.H., Luan, Z.K. (2006) Different morphologies of carbon nanotubes effect on the lead removal from aqueous solution. *Diamond and Related Materials*, *15*, 90–94.

Liang, P., Ding, Q., Song, F. (2005) Application of multiwalled carbon nanotubes as solid phase extraction sorbent for preconcentration of trace copper in water samples. *Journal of Separation Science*, *28*, 2339–2343.

Liang, P., Liu, Y., Guo, L., Zeng, J., Lu, H.B. (2004) Multiwalled carbon nanotubes as solid-phase extraction adsorbent for the preconcentration of trace metal ions and their determination by inductively coupled plasma atomic emission spectrometry. *Journal of Analytical Atomic Spectrometry*, *19*, 1489.

Long, R.Q., Yang, R.T. (2001) Carbon nanotubes as superior sorbent for dioxin removal. *Journal of the American Chemical Society*, *123*, 2058–2059.

Low, J., Cheng, B., Yu, J. (2017) Surface modification and enhanced photocatalytic CO2 reduction performance of TiO2: A review. *Applied Surface Science*, *392*, 658–686.

Lu, C., Chung, Y.L., Chang, K.F. (2006) Adsorption thermodynamic and kinetic studies of trihalomethanes on multiwalled carbon nanotubes. *Journal of Hazardous Materials*, *138*, 304–310.

Lu, Q., Rosen, J., Jiao, F. (2015) Nanostructured metallic electrocatalysts for carbon dioxide reduction. *ChemCatChem*, *7*, 38–47.

Luo, S., Qin, P., Shao, J., Peng, L., Zeng, Q., Gu, J.D. (2013) Synthesis of reactive nanoscale zero valent iron using rectorite supports and its application for Orange II removal. *Chemical Engineering Journal*, *223*, 1–7.

Mahmoud, M.E., Yakout, A.A., Abdel-Aal, H., Osman, M.M. (2011) Enhanced biosorptive removal of cadmium from aqueous solutions by silicon dioxide nano-powder, heat inactivated and immobilized Aspergillus ustus. *Desalination*, *279*, 291–297.

Méndez-Paz, D., Omil, F., Lema, J.M. (2005) Anaerobic treatment of azo dye Acid Orange 7 under batch conditions. *Enzyme and Microbial Technology*, *36*, 264–272.

Mishra, A.K., Ramaprabhu, S. (2011) Carbon dioxide adsorption on graphene sheets. *AIP Advances*, *1*, 032152.

Mohamed, E.F. (2017) Nanotechnology: Future of environmental air pollution control. *Environmental Management and Sustainable Development*, *6*, 2.

Mossa Hosseini, S., Ataie-Ashtiani, B., Kholghi, M. (2011) Nitrate reduction by nano-Fe/Cu particles in packed column. *Desalination*, *276*, 214–221.

Moussavi, G., Hossaini, Z., Pourakbar, M. (2016) High-rate adsorption of acetaminophen from the contaminated water onto double-oxidized graphene oxide. *Chemical Engineering Journal*, *287*, 665–673.

Mu, Y., Jia, F., Ai, Z., Zhang, L. (2017) Iron oxide shell mediated environmental remediation properties of nano zero-valent iron. *Environmental Science: Nano*, *4*, 27–45.

Muñoz, J., Gallego, M., Valcárcel, M. (2005) Speciation of organometallic compounds in environmental samples by gas chromatography after flow preconcentration on fullerenes and nanotubes. *Analytical Chemistry*, *77*, 5389–5395.

Nanotechproject. (2009) Nanoremediation map. Available from: http://www.nanotechproject.org/inventories/remediation_map/ [accessed 7 July 2018].

Neumann, A., Kaegi, R., Voegelin, A., Hussam, A., Munir, A.K.M., Hug, S.J. (2013) Arsenic removal with composite iron matrix filters in Bangladesh: A field and laboratory study. *Environmental Science and Technology, 47*, 4544–4554.

Nursanto, E.B., Da Hye Won, D.H., Jee, M.S., Kim, H., Kim, N.K., Jung, K.D., Hwang, Y.J., Min, B.K. (2018) Facile and cost effective synthesis of oxide-derived silver catalyst electrode via chemical solution deposition for CO2 electroreduction. *Topics in Catalysis, 61*, 389–396.

Nursanto, E.B., Jeon, H.S., Kim, C., Jee, M.S., Koh, J.H., Hwang, Y.J., Min, B.K. (2016) Gold catalyst reactivity for CO2 electro-reduction: From nano particle to layer. *Catalysis Today, 260*, 107–111.

Pandey, A., Singh, P., Iyengar, L. (2007) Bacterial decolorization and degradation of azo dyes. *International Biodeterioration and Biodegradation, 59*, 73–84.

Qiu, X., Fang, Z., Yan, X., Gu, F., Jiang, F. (2012) Emergency remediation of simulated chromium (VI)-Polluted river by nanoscale zero-valent iron: Laboratory study and numerical simulation. *Chemical Engineering Journal, 193–194*, 358–365.

Salipira, K.L., Mamba, B.B., Krause, R.W., Malefetse, T.J., Durbach, S.H. (2007) Carbon nanotubes and cyclodextrin polymers for removing organic pollutants from water. *Environmental Chemistry Letters, 5*, 13–17.

Shan, G., Surampalli, R.Y., Tyagi, R.D., Zhang, T.C. (2009) Nanomaterials for environmental burden reduction, waste treatment, and nonpoint source pollution control: A review. *Frontiers of Environmental Science and Engineering in China, 3*, 249–264.

Shen, W., Zhang, C., Li, Q., Zhang, W., Cao, L., Ye, J. (2015) Preparation of titanium dioxide nanoparticle modified photocatalytic self-cleaning concrete. *Journal of Cleaner Production, 87*, 762–765.

Shen, Y., Chen, B.L. (2015) Sulfonated graphene nanosheets as a superb adsorbent for various environmental pollutants in water. *Environmental Science and Technology, 49*, 7364–7372.

Shimizu, A., Tokumura, M., Nakajima, K., Kawase, Y. (2012) Phenol removal using zero-valent iron powder in the presence of dissolved oxygen: Roles of decomposition by the Fenton reaction and adsorption/precipitation. *Journal of Hazardous Materials, 201–202*, 60–67.

Stanmore, B.R. (2004) The formation of dioxins in combustion systems. *Combustion and Flame, 136*, 398–427.

Sun, W., Pan, W., Wang, F., Xu, N. (2015) Removal of Se(IV) and Se(VI) by MFe2O4 nanoparticles from aqueous solution. *Chemical Engineering Journal, 273*, 353–362.

Tanboonchuy, V., Hsu, J.C., Grisdanurak, N., Liao, C.H. (2011) Gas-bubbled nano zero-valent iron process for high concentration arsenate removal. *Journal of Hazardous Materials, 186*, 2123–2128.

Upendar, K., Sri Hari Kumar, A., Lingaiah, N., Rama Rao, K.S., Sai Prasad, P.S. (2012) Low- temperature CO2 adsorption on alkali metal titanate nanotubes. *International Journal of Greenhouse Gas Control, 10*, 191–198.

Vaseashta, A., Vaclavikova, M., Vaseashta, S., Gallios, G., Roy, P., Pummakarnchana, O. (2007) Nanostructures in environmental pollution detection, monitoring, and remediation. *Science and Technology of Advanced Materials, 8*, 47–59.

Wang, J., Chen, B.L. (2015) Adsorption and coadsorption of organic pollutants and a heavy metal by graphene oxide and reduced graphene materials. *Chemical Engineering Journal, 281*, 379–388.

Wang, J., Chen, B.L., Xing, B.S. (2016) Wrinkles and folds of activated graphene nanosheets as fast and efficient adsorptive sites for hydrophobic organic contaminants. *Environmental Science and Technology, 50*, 3798–3808.

Wang, J.X., Jiang, D.Q., Gu, Z.Y., Yan, X.P. (2006) Multiwalled carbon nanotubes coated fibers for solid-phase microextraction of polybrominated diphenyl ethers in water and milk samples before gas chromatography with electron-capture detection. *Journal of Chromatography A, 1137*, 8–14.

Wang, S., Sun, H., Ang, H.M., Tadé, M.O. (2013) Adsorptive remediation of environmental pollutants using novel graphene-based nanomaterials. *Chemical Engineering Journal, 226*, 336–347.

Whipple, D.T., Kenis, P.J.A. (2010) Prospects of CO2 utilization via direct heterogeneous electrochemical reduction. *Journal of Physical Chemistry Letters, 1*, 3451–3458.

Widodo, A., Sujatnika, Y., Awali, D.R., Prakoso, T., Adhi, T.P., Soerawidjaja, T.H., Indarto, A. (2015) Thermal heat-free regeneration process using antisolvent for amine recovery. *Chemical Engineering and Processing: Process Intensification, 89*, 75–79.

Xu, J., Wang, L., Zhu, Y.F. (2012) Decontamination of bisphenol A from aqueous solution by graphene adsorption. *Langmuir, 28*, 8418–8425.

Yamani, J.S., Lounsbury, A.W., Zimmerman, J.B. (2014) Towards a selective adsorbent for arsenate and selenite in the presence of phosphate: Assessment of adsorption efficiency, mechanism, and binary separation factors of the chitosan-copper complex. *Water Research, 50*, 373–381.

Yang, K., Xing, B. (2007) Desorption of polycyclic aromatic hydrocarbons from carbon nanomaterials in water. *Environmental Pollution, 145*, 529–537.

Yang, K., Zhu, L., Xing, B. (2006) Adsorption of polycyclic aromatic hydrocarbons by carbon nanomaterials. *Environmental Science and Technology, 40*, 1855–1861.

Yang, S.T., Chen, S., Chang, Y., Cao, A., Liu, Y., Wang, H. (2011) Removal of methylene blue from aqueous solution by graphene oxide. *Journal of Colloid and Interface Science, 359*, 24–29.

Yildiz, O., Bradford, P.D. (2013) Aligned carbon nanotube sheet high efficiency particulate air filters. *Carbon, 64*, 295–304.

Yin, W., Wu, J., Li, P., Wang, X., Zhu, N., Wu, P., Yang, B. (2012) Experimental study of zero-valent iron induced nitrobenzene reduction in groundwater: The effects of pH, iron dosage, oxygen and common dissolved anions. *Chemical Engineering Journal, 184*, 198–204.

Yunus, I.S., Harwin, Kurniawan, A., Adityawarman, D., Indarto, A. (2012) Nanotechnologies in water and air pollution treatment. *Environmental Technology Reviews, 1*, 136–148.

Zhao, G., Jiang, L., He, Y., Li, J., Dong, H., Wang, X., Hu, W. (2011) Sulfonated graphene for persistent aromatic pollutant management. *Advanced Materials, 23*, 3959–3963.

Zhao, J., Yang, X. (2003) Photocatalytic oxidation for indoor air purification: A literature review. *Building and Environment, 38*, 645–654.

Zhou, Q.X., Ding, Y.J., Xiao, J.P. (2006a) Sensitive determination of thiamethoxam, Imidacloprid and acetamiprid in environmental water samples with solid-phase extraction packed with multiwalled carbon nanotubes prior to high-performance liquid chromatography. *Analytical and Bioanalytical Chemistry, 385*, 1520–1525.

Zhou, Q.X., Xiao, J.P., Wang, W.D. (2006b) Using multi-walled carbon nanotubes as solid phase extraction adsorbents to determine dichlorodiphenyltrichloroethane and its metabolites at trace level in water samples by high performance liquid chromatography with UV detection. *Journal of Chromatography A, 1125*, 152–158.

# 9

# *Non-Magnetic Metal Oxide Nanostructures and Their Application in Wastewater Treatment*

**Debanjan Guin and Chandra Shekhar Pati Tripathi**

**CONTENTS**



## 9.1 Introduction

The rapid growth in world population demands an increase in freshwater supplies. The freshwater supply to meet the expanding need is further restricted by water pollution. In the twenty-first century, water pollution is one of the biggest environmental problems all over the globe. Water pollution affects the entire biosphere with plants and organisms living in these bodies of water. With the spread of industrialization around the world, pollution has spread with it. In general, water pollution takes place mostly due to the release of industrial waste. Detergents, polychlorinated biphenyls (PCBs), and substances such as dioxins, which are the unwanted by-products of manufacturing and combustion processes, organic dyes, human trash, chemicals dumped by hospitals and other companies, and runoffs of chemical fertilizers and pesticides from farm fields cause water pollution. Surface water resources like oceans, lakes, and rivers are mostly affected due to pollution.

With frequent catastrophic oil spill accidents and oily wastewater from industries, oil–water separation has become one of the top environmental concerns and the treatment of oily wastewater has become a worldwide issue. It is reported that the massive Exxon Valdez oil spill accident (11 million gallons) in 1987 caused fatal damage to nearly 300,000 living species in the habitat, and the long-term ecological impact is still going on (Carson et al. 2003). A similar man-made disaster, the Deepwater Horizon oil spill (4.9 million barrels) took place in 2010 in the Gulf of Mexico (Brody et al. 2012). Researchers are currently investigating many direct and indirect solutions to address the grand challenge. Conventional techniques to clean oil from water include physical methods like centrifugation, air flotation, ultrasonic separation gravity separation, and vacuum suction. Chemical methods like degradation with chemical dispersants and *in situ* burning are another alternative. There are different convention biological processes also available. However, all of these approaches have been found to be expensive, relatively inefficient, and even to bring about secondary pollution. Due to these reasons, research in the area of oil–water separation has attracted a great deal of interest and has emerged as a rapidly growing area for the researchers and industrial community to resolve oil spills and resource recovery issues.

Therefore, the development of low cost and efficient functional materials for the effective treatment of oil-polluted water has been the focus of the research in this area. Recent advancements in nanotechnology enable us to develop affordable wastewater treatment, which allows us to overcome the existing challenges and economic utilization of the water sources to expand the water supply. In most cases non-magnetic nanomaterials with superhydrophobic and superoleophilic properties introduced as absorbents for filtration methods. It is reported that combining superhydrophobicity (water contact angles of >150°) with superoleophilicity (oil contact angles of <5°), surfaces with special wettability (special wettable materials) can be fabricated, which are required to separate oil from water, or water from oil.

Different nanosized semiconducting oxides like ZnO or $TiO_2$ are used as photocatalysts (under UV or visible light) for the removal of toxic organic pollutants. Nanosized silver is the most widely used material due to its low toxicity and microbial inactivation in water. Membranes made by nanomaterials such as carbon nanotubes (CNTs), zeolites, and graphene are widely used for desalination of water. Nowadays, functional silica nanoparticles (NPs) in the polymer matrix are used efficiently for the removal of oil from water. Hydrophobic ZnO NPs coated in mesh have been proven to be a robust inorganic system which is capable of handling the extreme heat and wear typical of oil extraction.

In this chapter, the application of a range of non-magnetic metal oxide nanostructures in addressing the environmental challenges such as oil–water separation, degradation of organic pollutants, and removal of toxic metal ions is discussed.

## 9.2 Oil–Water Separation

With the rapid growth in industrialization, wastewater management has become an important issue as increasing oil spill accidents are adversely affecting the environment. Due to the increasing pollution, there is a growing demand to find suitable methods in the field of oil–water separation, oil clean up, etc. (Fingas 1995, Pelletier and Siron 1999, Lessard and DeMarco 2000). As a result, efforts are being made to find simple methods for the effective treatment of oily water. Conventional methods applied for oil–water separation, such as bioremediation, chemical demulsifiers, air flotation, and adsorption, etc., suffer from the disadvantages of low separation efficiency, high energy cost, complex operational processes, and secondary pollution. Thus, it is a great challenge for scientists to develop new technologies for oil–water separation.

Research in the field of special wettable materials has gained increasing attention due to their application for oil–water separation. Wettable materials with opposite affinities toward water and oil are supposed to be the most efficient materials for selective oil–water separation. The wettability of a solid surface determines the wetting/dewetting properties when a liquid comes into contact with the solid surface. The surface architecture and surface chemistry are the two main factors which govern the wetting behavior of the solid surface. Materials with specific wettability (superhydrophobicity, superhydrophilicity, superoleophobicity, and superoleophilicity) have been designed and fabricated with the appropriate surface structure and composition for effective oil–water separation. Materials with superhydrophilicity and superoleophobicity can selectively absorb or filter water from oil–water mixtures and are known as "water-removing" types of materials. Similarly, superhydrophobic and superoleophilic materials can selectively filter or absorb oil from oil–water mixtures, and are usually called "oil-removing" types of materials.

Two different approaches, i.e. filtration and absorption with special wettable nanomaterials are used for oil–water separation (Figure 9.1). In the filtration process, only one phase is allowed to pass through a filter which is in the form of meshes or textile membranes. Copper and stainless steel (SS) are the most commonly used metallic meshes as the filtration substrates for oil–water separation due to their mechanical strength, high filtration flux, and large-scale production. The absorption method is a method during which one phase is selectively absorbed and other phase is repelled using three dimensional (3D) porous materials such as cotton, aerogels, sponges, and fibrous films. The controlled filtration/absorption process

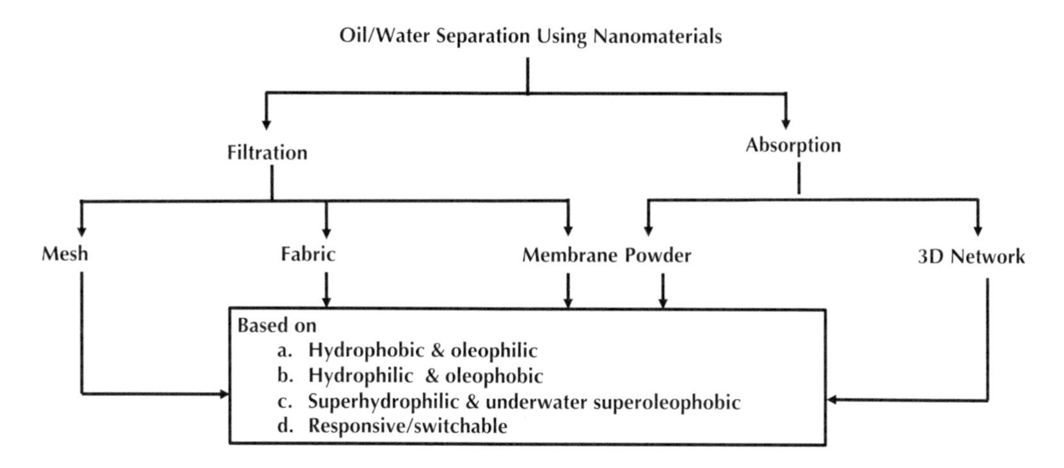


**FIGURE 9.1** Oil–water separation methods based on the filtration and absorption with special wettable materials.

is a relatively new area of research where the designed material shows controlled and switchable wettability via various means such as by the application of an external stimulus, e.g. a magnet or a change in pH, light, etc. (Xue et al. 2014, Ma et al. 2016, Gupta et al. 2017). In this section, we summarize the recent developments of highly efficient, less costly, and simple processes based on the filtration and absorption methods for oil–water separation, using nanomaterials with superwettability (Tian et al. 2014).

## 9.2.1 Oil–Water Separation Based on Filtration

### *9.2.1.1 Metal Oxide-Modified Meshes*

Metallic meshes are the class of substrate which has been studied most for oil–water separation. The surface of these metallic meshes has been modified accordingly for this purpose.

There are four kinds of special combinations of wetting properties, i.e. hydrophobic and oleophilic, hydrophilic and oleophobic, superhydrophilic and underwater superoleophobic, and responsive/switchable, that have been achieved on these metallic meshes. In the case of synthesis of materials that have hydrophobic and oleophilic, hydrophilic and oleophobic, superhydrophilic and superhydrophilic, and underwater superoleophobic properties, metal oxide polymer nanocomposites are used, whereas polymer hydrogels are used to get responsive/switchable property.

#### *9.2.1.1.1 Hydrophobic and Oleophilic Meshes*

For the preparation of metallic meshes that have hydrophobic and oleophilic properties, Feng et al. modified stainless steel mesh with polytetrafluroethylene (PTFE). PTFE is a hydrophobic and moderately oleophilic material. The combination of the low surface energy of PTFE and rough surface of the mesh resulted in a superhydrophobic and superoleophilic mesh. The PTFE-modified mesh exhibited special wettability with a water contact angle (WCA) of about 156° and oil contact angle (OCA) of nearly 0° due to the formation of micro and nanostructures on the surface (Feng et al. 2004). There are a number of superhydrophobic and superoleophilic polymers such as polyurethane sponge (Wang et al. 2014), pentaerythritol tetra(3-mercaptopropionate) (PETMP) (Chen et al. 2015), 2,4,6,8-tetramethyl-2,4,6,8-tetravi nylcyclotetrasiloxane (TMTVSi) (Sparks et al. 2013), and polydimethysiloxane (PDMS) (Su et al. 2012), which were used to modify the surface of metallic meshes and have been used efficiently for the filtration of oil–water mixtures. In order to obtain more hydrophobicity, different metal oxide NPs are introduced on the polymer-coated meshes which also reduce the surface smoothness of the meshes and directly influence the efficiency. To control the surface roughness, various groups have reported the growth of metal oxide crystals on metallic mesh substrates through chemical reactions.

Wang et al. fabricated a superhydrophobic and superoleophilic zinc oxide (ZnO)-coated SS mesh films. They used a three-step process to prepare the ZnO-coated mesh films. The ZnO-coated mesh films exhibited good selectivity, excellent recyclability, and a high level of ultraviolet (UV) stability. The morphology of ZnO was controlled by varying the reaction temperature and reaction time. The ZnO surfaces were modified with stearic acid molecules to increase the hydrophobicity of ZnO surfaces. When an oil drop was placed on the ZnO-coated mesh films with a pore size of 38 μm, the oil was captured and transported along the mesh film rather than passing through it. They also performed this oil capture phenomenon underwater with the ZnO-coated mesh film having a pore size of approximately 600 μm. The water contact angle and oil contact angle of ZnO-coated SS mesh films was of approximately 156° and 0°, respectively and remained almost unchanged after long-term UV irradiation. (Wang et al. 2012). Superhydrophobic aligned ZnO nanorods were grown on a metal mesh by other research groups. The special wettability and capillary effect increased separation efficiency and speed.

Different chemical methods have been adopted to increase the roughness of metallic meshes, but a common problem of these methods is to achieve uniformity on the surface. To overcome this problem, one of the easiest methods is to oxidize the metal surface by immersing it in an oxidizing agent. Dai et al. dipped Cu meshes into 1.0 M NaOH and 0.05 M $K_2S_2O_8$ at room temperature for 30 minutes to get a uniform rough $Cu(OH)_2$ surface on Cu meshes (Dai et al. 2014). The oxidized copper mesh was then immersed in a $2 \times 10^{-4}$ M octadecylphosphonic acid/tetrahydrofuran (ODPA/THF) solution to form an ODPA coating formed on the oxidized copper mesh. It is interesting that not only does ODPA-coating on the oxidized copper mesh produce an "oil-removing" type of oil–water separation mesh with superhydrophobicity and superoleophilicity with WCA and OCA of 158.9° and 0°, respectively, but straightforward oxidation of a copper substrate also produces a "water-removing" type oil–water separation mesh with underwater superoleophobicity. These types of simple but creatively designed devices allow the *in situ* cleanup of an oil spill in sea water.

#### *9.2.1.1.2 Hydrophilic and Oleophobic Meshes*

Though hydrophobic and oleophilic surfaces used for oil–water separation have proved to be a very promising solution, they are easily fouled and blocked due to their intrinsic oleophilic property (Xue et al. 2011). Thus, the development of novel materials with resistance to oil fouling have drawn the attention of researchers. The fabrication of hydrophilic–oleophobic metal meshes with durable surface coatings and high water flux is another alternative as an efficient filtration device for the cleanup of an oil layer in water. These materials selectively let the denser water pass through the filtrate while the oil is left behind. However, synthesis of these kinds of materials is not an easy task, as from a theoretical point of view it is difficult to prepare materials that simultaneously display hydrophilicity and oleophobicity. As water usually has a higher surface tension than those of oils it is difficult to prepare materials with both hydrophilic and oleophobic properties. Nevertheless, researchers have found several ways to design substrates with these unique hydrophilic–oleophobic wetting characteristics. Yang et al. reported, for the first time, a superhydrophilic and superoleophobic coating material that can be applied on almost any substrate. This special composite consisted of hydrophilic poly(diallyldimethylammonium chloride) (PDDA), oleophobic sodium perfluorooctanoate (PFO), and $SiO_2$ NPs (PDDA–PFO/$SiO_2$). The superhydrophilic and superoleophobic nanocomposite PDDA–PFO/$SiO_2$ was fabricated by spray coating NP–polymer suspensions onto the substrate. The nanocomposite-coated mesh exhibits water permeation and oil repellency behaviors. The water molecules were able to pass through the surfaces due to a water-induced molecular arrangement. However, the presence of a high surface concentration of fluorinated groups made the surface superoleophobic. The developed nanocomposite was able to overcome the problems of easy fouling and hard recycling limitations of previous methods. The oil–water separation was performed as shown in Figure 9.2. The contact angle measurements presented in the figure shows that the WCA gradually decreased from 165° to 0° within nine minutes, while the contact angle for hexadecane was 155° in that period. The water was allowed to pass through the mesh while the oil flowed over the mesh without any penetration (Yang et al. 2012). Chitosan-based superhydrophilic and superoleophobic coatings were fabricated by the coating process. The rough surface structure combined with the molecular rearrangement enabled the coating with WCA of 0° and a hexadecane contact angle of 157° ± 1°. The CTS-PFO/$SiO_2$ nanocomposite applied on the SS mesh for oil–water separation showed excellent water permeation and strong oil repellency behaviors, excellent anti-fouling capacity, high separation efficiency, and easy

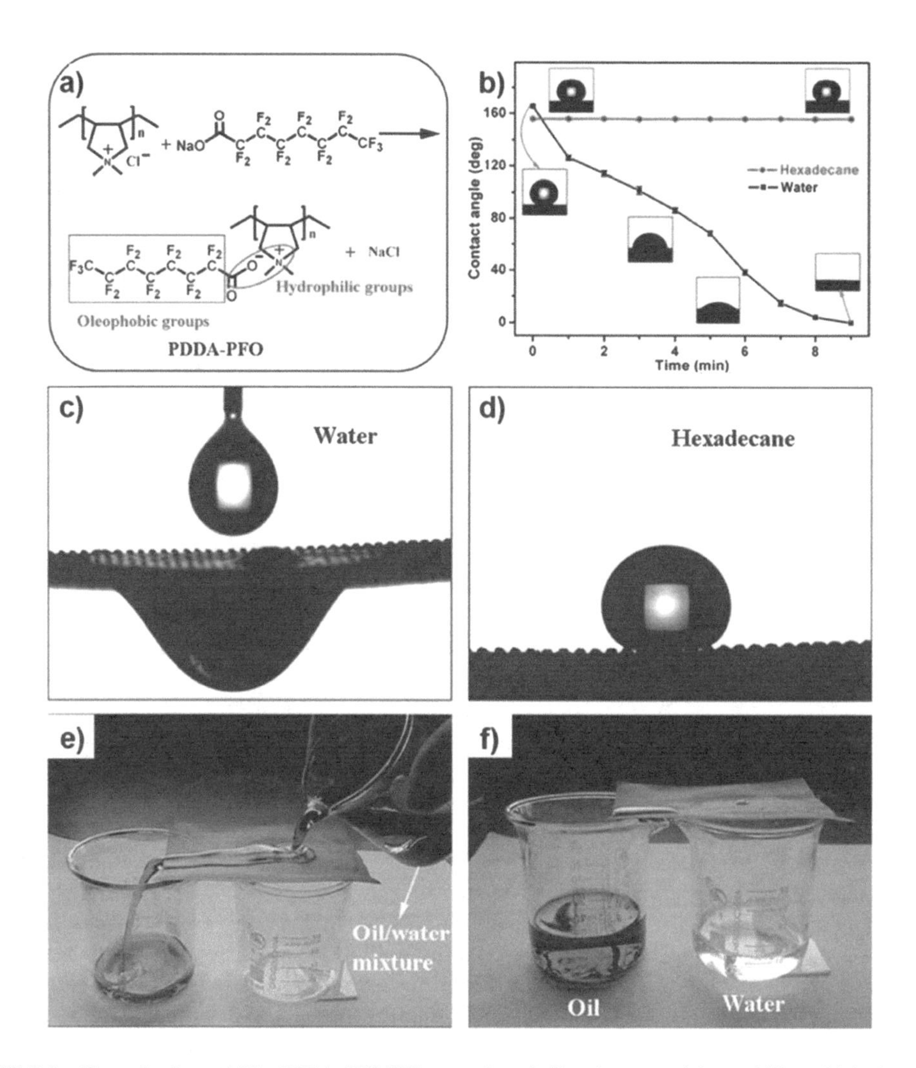


**FIGURE 9.2** (**See color insert.**) The PDDA–PFO/$SiO_2$-coated mesh film shows special wettability, with both superhydrophilic and superoleophobic properties. (a) Chemical reaction scheme used for the synthesis of PDDA–PFO. (b) Time dependence of contact angles for water and hexadecane on the PDDA–PFO/$SiO_2$ coating. (c) Water droplet spreading on and permeating through the mesh. (d) Shape of a hexadecane droplet on the mesh with a contact angle of 157°. (e and f) Oil–water separation experiment was performed on the PDDA–PFO/$SiO_2$-coated mesh. (Reproduced from Yang et al., *J. Mater. Chem.*, 2012b, 22, 2834–2837. Reprinted with permission from The Royal Society of Chemistry.)

recyclability (Yang et al. 2014). Polyelectrolyte–fluorosurfactant complex (PFC)-based separation material (PFC/$SiO_2$) showed similar behavior for oil–water separation (Yang et al. 2015).

### *9.2.1.1.3 Superhydrophilic and Underwater Superoleophobic Meshes*

To overcome the limitations of the above-mentioned methods, researchers recently proposed an innovative approach of constructing superhydrophilic and underwater superoleophobic surfaces using high-surface-energy materials to enhance the antifouling ability and separation efficiency of the commercial membranes. The surface of the metallic mesh is modified in such a way that the surface which is superhydrophilic in air will exhibit superoleophobicity underwater. Liu et al synthesized materials having these kinds of properties by the straightforward oxidation of Cu meshes with 0.05 m $K_2S_2O_8$ and 1.0 m NaOH. Synthesized $Cu(OH)_2$ with micro- and nanoscale hierarchical structures on a Cu mesh exhibits superhydrophilicity and superoleophilicity in air (WCA and OCA around 0°) and underwater superoleophobicity in water is around 166.2±1.38° (Liu et al. 2013). Zhang et al. adopted a similar approach to synthesized $Cu(OH)_2$ nanowires on the surface of a Cu mesh and with the changing morphology of $Cu(OH)_2$, the mesh could effectively separate not only layered oil–water mixtures but also oil-in-water emulsions (Zhang et al. 2013a). Gondal and his group synthesized synergistic superhydrophilic-underwater superoleophobic surfaces by spray coating of $TiO_2$ on stainless steel substrate for the study of gravity-driven oil–water separation. Using the coated mesh with pore sizes 50 and 100 µm, they achieved 99% separation efficiency. The adsorbed layer of water on the coated surface, the formation of a water-film between the individual wires of the mesh, and the strength of the underwater super-oleophobicity all contribute to this enhanced efficiency of oil–water separation (Gondal et al. 2014). The Bin Ding group has recently fabricated a novel microsphere/nanofiber composite membrane with intriguing superhydrophilicity and underwater super-oleophobicity via combined electrospinning and electro-spraying methods (Ge et al. 2017). The laser ablation method has also been used to create rough nanoscale structures on silica glass surfaces (Yong et al. 2015). These highly transparent as-prepared surfaces exhibit underwater superoleophobicity and ultralow oil adhesion.

##### *9.2.1.1.4 Meshes with Switchable Wettability*

The two main factors which control the surface wettability are chemical composition and geometrical structure. Recently, the change in the surface wettability with the application of external stimuli such as light irradiation, electric fields, thermal treatment, change in pH, etc. has been studied intensively (Krupenkin et al. 2004, Rosario et al. 2004, Xia et al. 2007). Polymers with responsive wettability can switch from wetting (superwetting) to antiwetting (superantiwetting), or vice versa. Therefore, they can act as "water-removing" or "oil-removing" depending on the external circumstances. Therefore, the responsive wettable material offers a significant advantage for controllable oil–water separation in various situations.

Wang et al. demonstrated for the first time the fabrication of superoleophobic $Ti/TiO_2$ surfaces by a combination of anodization and laser technology. These surfaces were fabricated by forming $TiO_2$ nanotube arrays on the micro-Ti structure via a simple electrochemical method. The superoleophobicity of the surface was achieved by modifying the titanium with perfluorosilane. The surfaces exhibited both superhydrophilicity and superoleophibicity to various liquids. As a typical photosensitive functional material, the wettability of $TiO_2$ can be reversibly switched between hydrophobic and hydrophilic or a highly amphiphilic state under UV light. Tunable oleophobicity was also achieved by manipulating the UV irradiation. It was observed that the UV illumination had an important effect on the wettability of the superoleophobic surface with the CA turning smaller. It is well-known that UV illumination on $TiO_2$ will generate hydroxyl groups, which will enhance the surface oleophilicity (Wang et al. 2010). Tian et al. reported the fabrication of a ZnO nanorod array-coated stainless steel mesh film, which showed excellent controllability and high separation efficiency of different types of oil–water mixtures in an oil–water–solid three-phase system. The prepared film exhibited switchable superhydrophobicity– superhydrophilicity, and underwater superoleophobicity at the special oil–water–solid three-phase interface. The contact angle (CA) measurement showed that the aligned ZnO nanorod array-coated mesh film behaves superhydrophobically with a WCA of 155° after storage in the dark, while the coated mesh film behaves superhydrophilically with a WCA of 0° under UV irradiation. The effect of the pore size of the mesh substrate on the wettability of the aligned ZnO nanorod array-coated mesh was also investigated. The suitable pore size of the mesh displaying superoleophobicity underwater that was obtained was smaller than 200 µm (Tian et al. 2012b).

#### ***9.2.1.2 Metal Oxide-Modified Fabrics/Textiles***

Similar to metallic meshes, textiles can also be modified with polymer and NPs to get superhydrophobic and superoleophilic properties. The textiles are advantageous over metallic meshes as they are cheaper, lightweight, flexible, and resistive against corrosion. Particularly, natural textiles such as cotton have added advantages like being eco-friendly, biodegradable, and reusable. To achieve superhydrophobicity, textiles are subjected to functionalization with suitable NPs. In most of the cases surface hydroxyl groups

(–OH) of textiles are chemically reacted with functional polymers or organic molecules to produce a functional surface and on that surface NPs can be incorporated.

Xue et al. prepared superhydrophobic cotton textiles with a dual-size hierarchical structure obtained by the complex coating of silica particles with functional groups on microscale natural cotton fibers followed by hydrophobization with stearic acid, 1*H*, 1*H*, 2*H*, 2*H*-perfluorodecyltrichlorosilane (PFTDS). Cotton textiles were dipped into an amine-functionalized silica particle solution and nipped by a padder. This process was repeated twice. The textiles were dried and again dipped into an epoxy-functionalized silica particle solution and nipped by the padder. This process was also repeated two times, and the textiles were dried. During drying, reactions occurred between the epoxy and amine groups, thus making the silica particles form a robust layer on the cotton fibers and leaving the outer surface of the fiber full of epoxy groups available for further surface grafting. Then, stearic acid, PFTDS, or their combination is grafted onto the rough surface to obtain a superhydrophobic property. The contact angle measurements showed the superhydrophobicity of the prepared textiles (Xue et al. 2008). Preparation of superhydrophobic cotton fabrics by the incorporation of silica NPs and subsequent hydrophobization with hexadecyltrimethoxysilane (HDTMS) was reported by Xu et al. They synthesized silica NPs via the sol–gel method with surfactant emulsification, using methyl trimethoxysilane (MTMS) as the precursor and ammonium hydroxide as the catalyst. Water shedding angle (WSA) technique and WCA measurements were used to evaluate the superhydrophobicity. The $SiO_2$ hydrosol with subsequent HDTMS-modified cotton fabric exhibited excellent superhydrophobicity with a WCA of 151.9° for a 5 μL droplet and WSA of 13° for a 15 μL droplet respectively. When the cotton surface with silica particles was modified with HDTMS, the $SiOCH_3$ groups in HDTMS were converted into Si–OH. HDTMS was hydrolyzed to form alkylsilanol. The dehydration reaction between the alkylsilanol and the hydroxyl group on the surface of $SiO_2$ particles resulted in a superhydrophobic rough surface (Xu et al. 2011).

Xu et al. reported an ammonia-responsive superamphiphobic coating, which turned out to be superhydrophilic and superoleophobic upon ammonia exposure. A heptadecafluorononanoic acid-modified $TiO_2$ sol (HFA-$TiO_2$) was mixed with silica NPs in ethanol to form the coating solution. The coating solution was applied to polyester fabric and polyurethane sponge by the dip-coating method. The coated fabric was superamphiphobic without exposure to ammonia vapor with the WCA and OCA being more than 150° but changed to superhydrophilic and superoleophobic upon exposure to ammonia vapor. In less than three seconds after the exposure, the WCA changed from 152° to 0° while the OCA remains unchanged during the complete process. The transition from superhydrophobic to superhydrophilic was due to the formation of ammonium carboxylate ions in the presence of ammonia based on the cleavage of titanium carboxylate coordination bonding. (Xu et al. 2015). Guo et al. fabricated a superhydrophobic polydopamine@$SiO_{2-}$ (PDA@$SiO_2$)-coated cotton fabric through a simple and inexpensive one-pot approach without requiring high temperature and toxic substances. The as-prepared fabric showed great resistance to mechanical abrasion, wear, and ultrasonic treatment, and had excellent superhydrophobicity stability toward UV irradiation, high temperature, and organic solvents immersion. The prepared superhydrophobic/superoleophilic fabric was used to treat oils on the water and separate the oil–water mixture with high efficiency (Guo et al. 2017).

Xiong et al. reported a responsive fabric with switchable wettability by fabricating the hierarchical ZnO nanorod structure on polyethylene terephthalate (PET) by atomic layer deposition (ALD) as well as hydrothermal treatment as shown in Figure 9.3. The membranes were able to separate any phase from the water–oil mixture simply via pre-wetting with the heavier phase in the mixture. During the separation process, the component with higher density, i.e. water, easily passes through the pre-wetted membrane, ultimately leaving just the other component with lower density, i.e. diesel oil. It provided a very simple method for *in situ* tuning the wettability without using any additional external stimuli (Xiong et al. 2015). In a water environment, a droplet of $CCl_4$ could maintain its spherical shape on the surface of modified PET fabric (Figure 9.4a). No residual $CCl_4$ was observed on the fabric after it was taken out of water (Figure 9.4b). This confirms the underwater oleophobicity of functionalized PET fabric. With a specific underwater oleophobicity, the functionalized PET fabrics could effectively separate water from an oil–water mixture through a pre-wetting process with water. In the diesel oil–water mixture, water stays at the bottom of the filtration cell due to its relatively larger density and since the functionalized PET fabric is pre-wetted by water beforehand, water easily penetrates through the fabric while blocking diesel oil.

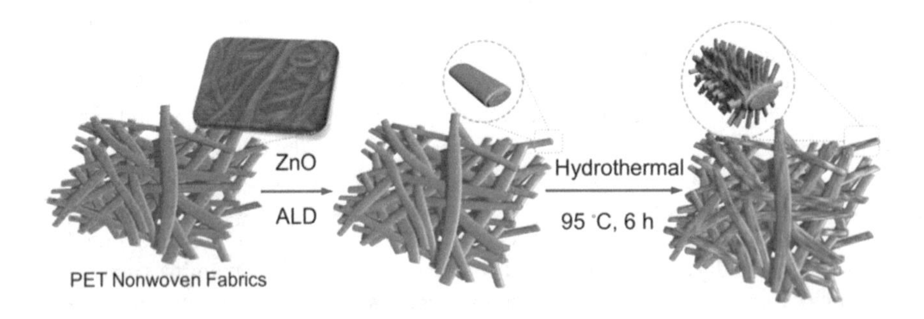


**FIGURE 9.3** Surface functionalization process of PET nonwoven fabrics with a large-field SEM image of the pristine nonwoven. (Reproduced from Xiong et al., *Journal of Membrane Science* 493 (2015) 478–485. Reprinted with permission from Elsevier B.V.)

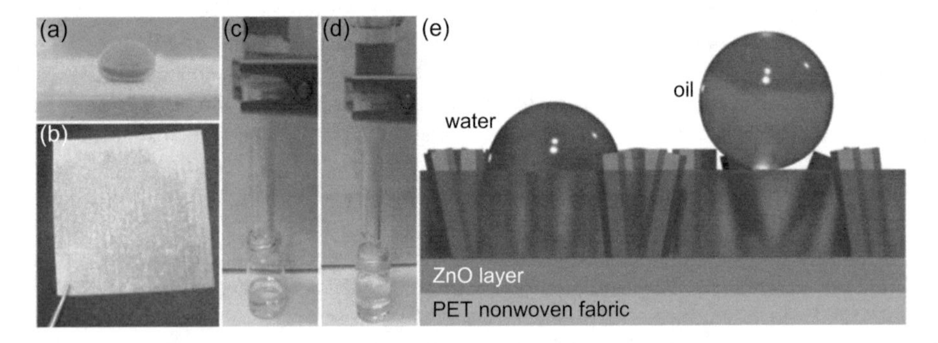


**FIGURE 9.4** **(See color insert.)** (a) Photograph of $CCl_4$ droplet, colored blue is placed in water on top of the functionalized PET fabric, (b) no residual $CCl_4$ left on the PET fabric, (c–d) diesel oil (dyed blue) is blocked in the filtration cell by water pre-wetted PET fabric, water is filtering through the fabric, and (e) the diagram showing the mechanism of water-permeable process: the blue rods/layer and the green layer represent ZnO nanostructure and pre-wetted liquid layer of water, respectively; oil droplet (brown) is staying on top of the water layer. (Reproduced from Xiong et al., *Journal of Membrane Science* 493 (2015) 478–485. Reprinted with permission from Elsevier B.V.)

Further addition of water into the filtration cell will lead to the continuous separation of water from the mixture only driven by the force of gravity (Figure 9.4c,d). The mechanism of permeation of water is shown in Figure 9.4e. As water comes into contact with ZnO nanostructures, a thin layer of water cushion is introduced since the surface of ZnO is rich with hydroxyl groups. As the oil droplets are added, water gets trapped into the roughened ZnO nanostructure. The as-formed composite interface shows an oleophobic property underwater, and these trapped water molecules will greatly reduce the contact area between oil droplets and the surface of the deposition layer.

## 9.2.2 Oil–Water Separation Based on Absorption

The superwetting materials fabricated for oil–water separation as discussed in the previous sections have proved efficient in oil–water separation. These materials could selectively filter oil due to their superhydrophilicity while completely repelling water due to their superhydrophobicity. However, the application of these materials in the place of oil spills is limited due to their low absorbency, because the surface oil should be first collected and then be filtered from top to bottom. Recent researchers have suggested that three-dimensional (3D) porous materials can be modified to offer combined superhydrophobic and superoleophilic properties for water-repelling and oil-absorbing applications. Nowadays, research efforts have been devoted to evolving advanced nanomaterial-based absorbents with special wetting properties, porous structure, and selective hydrophobicity and oleophilicity.

### *9.2.2.1 Metal Oxide Nanoparticle/Porous Material Composites*

The addition of NPs to a porous material has been found to an efficient method to increase hydrophobicity and the oleophilicity, which is required for the absorbance of oil. Su et al. fabricated superhydrophobic foam by coating the inner surface of polyurethane (PU) foam with a superhydrophobic film of nanosilica. The foam selectively absorbed oil from oil–water mixture without using any additional energy or chemical agent (Su 2009). A superhydrophobic/superoleophilic filter paper was prepared by treating filter paper with a mixture of hydrophobic silica NPs and polystyrene solution in toluene. The modified filter paper, when immersed in water, strongly repelled water and remained dry after being taken out. It selectively absorbed oil from an oil–water surface. In addition to oil, the modified filter paper was able to absorb a variety of nonpolar organic solvents (Wang et al. 2010). Tao et al. reported the fabrication of a new kind of highly efficient oil absorbent by grafting organic groups on the surface of hydrophobic monolithic hierarchically porous silica (MHS) from the sol–gel phase separation approach. The modified MHS was able to quickly capture the drops of oils or organic solvent and absorption in less than three seconds when it came into contact with oil on an oil–water surface. It exhibited uptake capacities up to eight times its weight for absorbing oil and other organic solvents (Tao et al. 2011). Ge et al. have demonstrated a very simple method to fabricate robust superhydrophobic polyurethane (PU) sponge by dipping PU sponges in a dispersion of hydrophobic $SiO_2$-NPs and polyfluorowax (PFW). PFW was used to increase the adhesion between the sponge and the metal oxide coating. The as-prepared sponge was applied for the removal of oil spills from the water surface and the investigated absorption capacities for a variety of oils and organic solvents. The high porosity and the flexible structure of the prepared sponge allowed them to be compressed to more than 75% volume reduction at low stress values (4–24 kPa) (Ge et al. 2015). Gao et al. successfully fabricated superhydrophobic and superoleophilic films on nickel surface for a facile ammonia-evaporation-induced method. Functionalized metal NPs and nanowire arrays were grown on Ni foam. The synthesized Ni foam possesses a high oil–water separation efficiency (99.6% for water and cyclohexane mixtures) and a very good cycle performance (Gao et al. 2014). Shuai et al. fabricated a novel superhydrophobic poly(-dimethylsiloxane) (PDMS)-$TiO_2$-coated polyurethane (PU) sponge by growing $TiO_2$ NPs on a PU sponge by the sol–gel method followed by polymerization of PDMS. The as-prepared PDMS-$TiO_2$-PU sponge exhibited excellent selectivity and a high absorption capacity for a range of oils and organic compounds. The regeneration of the absorption material was easily achievable simply by mechanical squeezing. The recyclability of the PU sponge was tested for more than 60 cycles and was found almost unchanged. (Shuai et al. 2015). Gao et al. have reported a superhydrophobic and superoleophilic molybdenum disulfide ($MoS_2$) sponge for highly efficient separation and absorption of oils and organic solvents from water. A perfect $MoS_2$ monolayer (without airborne contaminants) on $SiO_2$/Si substrates is intrinsically hydrophilic with a WCA of 70°, while multiple layer films (beyond three layers) are slightly hydrophobic (Gao et al. 2016). Zhang et al. have demonstrated a one-step solution immersion method for the bulk fabrication of superhydrophobic PU sponge by anchoring hydrophobic $SiO_2$ NPs onto a porous frame. The synthesized superhydrophobic and superoleophilic $SiO_2$/PU sponges had a WCA above 150° and an OCA below 5°. The superhydrophobicity of PU was found almost consistent with the original surface, with no change in the WCA even after 100 cycles. The prepared PU sponge showed absorption capacities between 14–27 times their own weight depending on the density and viscosity of the oils and solvents. (Zhang et al. 2017). Li et al. also fabricated superhydrophobic polyurethane (PU) sponges by dip-coating candle soot (CS) and $SiO_2$ NPs on sponge. It exhibited a high absorption capacity (up to 65 times of its own weight) and excellent separation efficiency for oil and other organic solvents (Li et al. 2017).

Recently, Kong et al. prepared a hydrophobic and oleophilic polyurethane ($Al_2O_3$/PUF) foam sponge in a three-step process. Hollow nanospheres of $Al_2O_3$ prepared by hydrothermal method were modified with γ-methacryloxypropyltrimethoxysilane to obtain hydrophobic $Al_2O_3$. One-step foaming technology was applied to fabricate hydrophobic and oleophilic polyurethane ($Al_2O_3$/PUF) foam sponge from the hydrophobic $Al_2O_3$ and polyurethane sponge. The synthetic route of hydrophobic and oleophilic polyurethane ($Al_2O_3$/PUF) foam sponge is shown in Scheme 9.1. The prepared $Al_2O_3$/PUF foam sponge showed a WCA of 144° and a very high absorption capacity for oil and other organic solvents. The absorption capacity of the foam sponge was investigated and found to be 37 g/g. The absorption of oil experiment

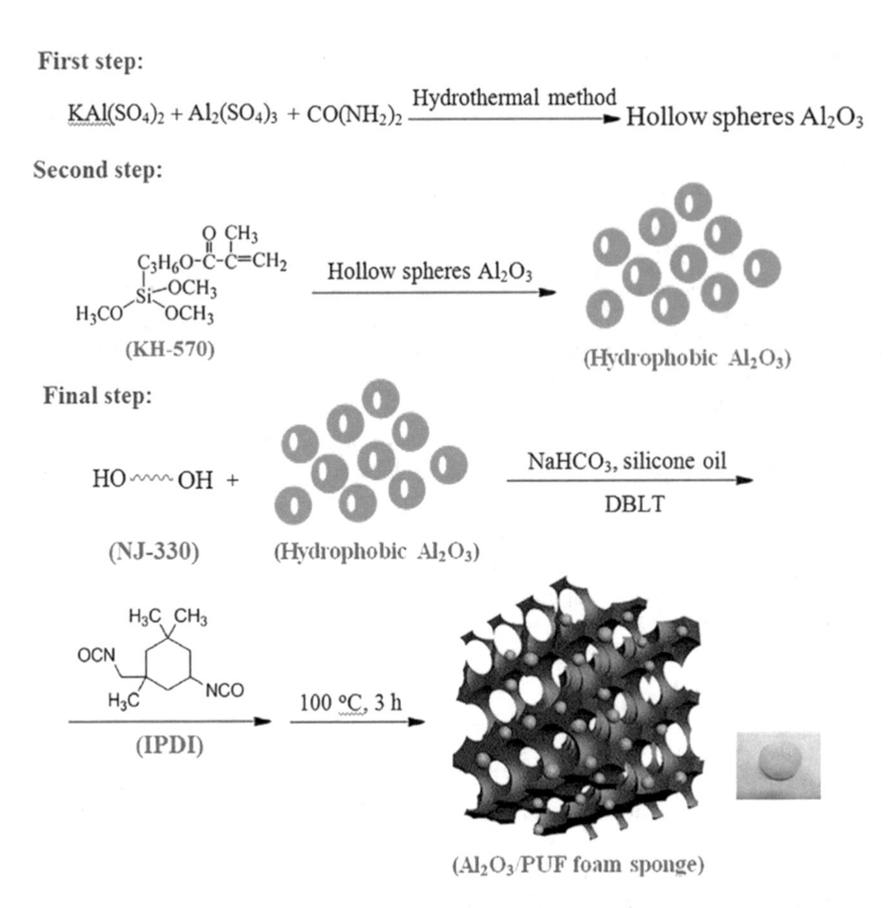


**SCHEME 9.1** The synthetic route of hydrophobic and oleophilic polyurethane ($Al_2O_3$/PUF) foam sponge. (Reproduced from Kong et al., *Journal of Industrial and Engineering Chemistry*, 58 (2018) 369–375. Reprinted with permission from the Elsevier B.V. on behalf of The Korean Society of Industrial and Engineering Chemistry.)

is shown in Figure 9.5. The oil was collected through a simple mechanical squeezing process. and it was reusable for up to 10 cycles while no change in the absorption capacity (Kong et al. 2018).

The absorption capacity is the key parameter used for the evaluation of the performances of the hydrophobic absorbent materials for oil–water separation. However, the limited absorption capacity of the porous absorbing materials is their biggest drawback. This requires the preparation of a huge amount of absorbing material for efficient absorption of oil from oil–water mixture. Therefore, it is important to overcome the capacity limitation of absorbent materials.

## 9.3 Removal of Organic Water Pollutants

The presence of organic contaminants in water due to intense agricultural and industrial activities is of great concern. There are various types of organic contaminants such as dyes, detergents, pesticides, volatile organic compounds (VOCs), and others. These contaminants can be toxic or carcinogenic even in a very small amount. The removal of these hazardous components from water resources has attracted significant attention from researchers. The conventional methods used for the treatment of water are adsorption of activated carbon, coagulation, sedimentation, filtration, chemical, and reverse osmosis. These commonly used processes do not completely eliminate waste. For example, activated charcoal is not able to remove organic pollutants at ppb levels. The reverse osmosis method used for water purification hasn't been very successful in removing small molecules. Additionally, these processes are also not economical

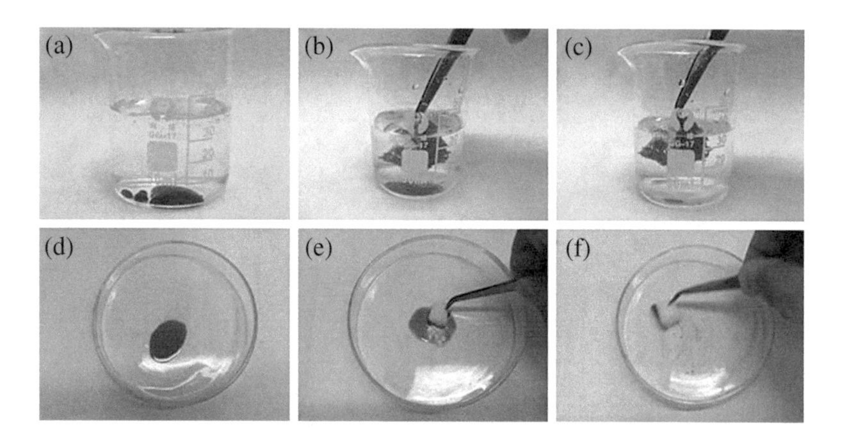

**FIGURE 9.5 (See color insert.)** A series of photos for the process of absorption and collection of lubricating oil (dyed with oil red) from the water. (Reproduced from Kong et al., *Journal of Industrial and Engineering Chemistry*, 58 (2018) 369–375. Reprinted with permission from the Elsevier B.V. on behalf of The Korean Society of Industrial and Engineering Chemistry.)

and can generate toxic secondary pollutants (Viessman and Hammer 1998, Chong et al. 2010). Due to increasing environmental awareness and strict regulations, these toxic contaminants have been of great concern worldwide. Thus, the development of an effective and economical method to remove these organic pollutants from water and to return it to safe levels is desirable. Recent research in the field of nanostructures of metal oxide semiconductors has made it possible to develop economically feasible and environmentally stable methods for the effective treatment of wastewater. The efficacy of metal oxide NPs in the decomposition of organic pollutants in water has been highlighted by many research groups. This section summarizes the recent progress in the application of non-magnetic nanostructures for the efficient removal of organic pollutants from water.

An efficient option to remove organic pollutants from water is via their degradation through advanced oxidation processes (AOP). This method includes photochemical degradation processes (e.g. UV/O3), photocatalysis (e.g. $TiO_2$/UV) and chemical oxidation processes (e.g. O3) (Poyatos et al. 2010). Among AOPs, $TiO_2$ NP-mediated photocatalysis has emerged as the most promising technology for the removal of organic pollutants. The first report of the application of $TiO_2$ in photocatalysis was reported by Frank and Bard for the oxidation of $CN^-$ and $SO_3^{2-}$ in an aqueous medium under sunlight (Frank and Bard 1977). The photocatalytic reduction of $CO_2$ by Inoue et al. attracted more interest than titania photocatalysis (Inoue et al. 1979). Photocatalysis using zinc oxide (ZnO) NPs has also attracted significant attention for the water purification as it is capable of removing chemical as well as biological contaminants.

### 9.3.1 $TiO_2$ Photocatalysis

Intensive studies in the field of photocatalysis were started when Akira Fujishima and Kenichi Honda discovered a new technology for hydrogen production by the photo-induced splitting of water on $TiO_2$ electrodes (Fujishima and Honda 1972). Photocatalysis is described as a process in which light is used to activate a substrate to facilitate photo reactions but with the catalyst remaining unconsumed. The photocatalytic process is classified into two categories, i.e. homogeneous and heterogeneous processes (Fujishima et al. 2000, Rajeshwar et al. 2008, Rehman et al. 2009). The heterogeneous photocatalysis process has been found to be technically superior for the degradation of various organic pollutants in wastewater. This process has several advantages over the competing processes. They are (i) complete mineralization, (ii) no waste disposal problem, (iii) low cost, and (iv) necessity of mild temperature and pressure conditions only. The basic principles underlying photocatalysis are already established and have been reported in the literature (Zhao and Yang 2003, Dung et al. 2005). When an electron from the valence band (VB) absorbs the photon energy $h\nu$, which is equal to or higher than the band gap of the semiconductor photocatalyst, it gets promoted to the conduction band (CB), and a photocatalytic reaction is initiated. Due to this reaction,

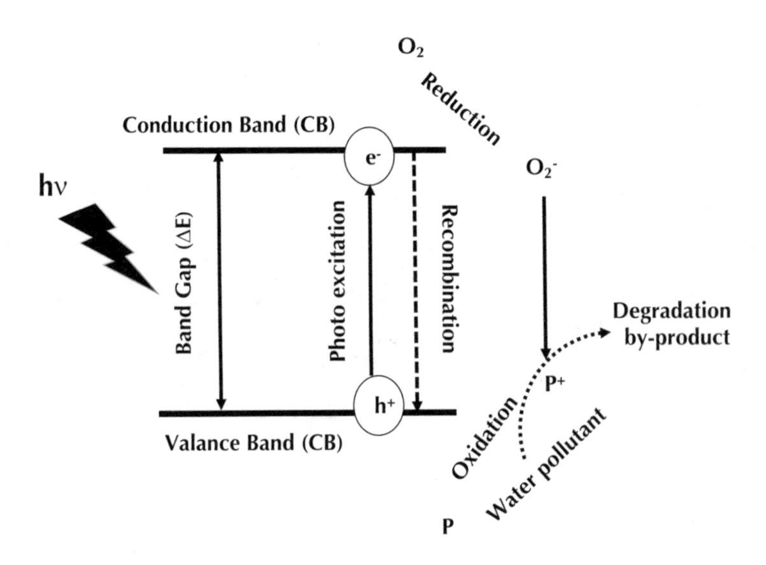


**FIGURE 9.6** Schematic illustration of $TiO_2$ photocatalytic mechanism.

a positively charged hole is created in the VB and a negatively charged electron in the CB of the photocatalyst. Thus, an electron–hole pair is created in this process and this allows the progress of the photocatalytic reaction through a series of chemical events (Gaya and Abdullah 2008). The generated electrons and holes migrate to the surface of the catalyst. The positively charged hole is used to produce hydroxyl radicals (OH*) by oxidizing the OH or water at the surface. This acts as an extremely powerful oxidant of organic pollutants. The photo-excited electron located in the CB is reduced to form the superoxide radical anion ($O_2$) upon reaction with oxygen and hydroperoxide radical (OOH) upon further reaction with $H^+$ (Zhan and Tian 1998). The recombination of the generated electron and hole is unwanted as it limits the efficiency of the photocatalytic process. It has been found that the OH ion is the most plentiful radical in the aqueous solution of $TiO_2$ and the reaction of OH radical with organic pollutants is the most crucial step in the degradation of organic pollutants (Lasa et al. 2005, Li Puma et al. 2008). Figure 9.6 shows the schematic representation of a semiconductor photocatalytic mechanism.

The following photoreactions on the photocatalyst surface have been widely proposed.

Photoexcitation:

$$TiO_2 / SC + h\nu \rightarrow e^- + h^+ \tag{9.1}$$

Oxygen ionosorption:

$$(O_2)_{ads} + e^- \rightarrow O_2 \tag{9.2}$$

Ionization of water:

$$H_2O \rightarrow OH^- + H^+ \tag{9.3}$$

Protonation of superoxides:

$$O_2^{*-} + H^+ \rightarrow HOO^* \tag{9.4}$$

The hydroperoxyl radical formed in (9.4) also has the same scavenging property as $O_2$, thus doubly prolonging the lifetime of the photohole:

$$HOO^* + e^- \rightarrow HOO^- \tag{9.5}$$

$$HOO^- + H^+ \rightarrow H_2O_2 \quad (9.6)$$

Both the oxidation and reduction take place at the surface of the photocatalyst (Figure 9.6). Recombination between the electron and the hole occurs unless oxygen is available to scavenge the electrons to form superoxides ($O_2^{*-}$), its protonated form the hydroperoxyl radical ($HOO^*$) and subsequently $H_2O_2$.

Titanium dioxide occurs in three different crystalline forms: anatase, rutile, and brookite. The anatase form of $TiO_2$ acts as the most photoactive phase because of its improved charge-carrier mobility and the higher number of surface hydroxyl groups (Pelaez et al. 2012). However, due to a relatively large band gap of 3.2 eV, the anatase form of $TiO_2$ is only activated with UV light ($\lambda \leq 387$ nm). This limits the utilization of a wide part of the solar energy spectrum. Researchers have made attempts to enhance the absorption range of $TiO_2$ under solar irradiation. Doping with metal and non-metal elements, surface modification, dye sensitization, and fabrication of composites have been some recent methods to improve the efficiency of $TiO_2$ under visible light (Umebayashi et al. 2002, Wodka et al. 2010).

The combination of noble metals like Ag, Au, Pt, and Pd with $TiO_2$ have also been studied extensively for their properties and their contribution toward visible light absorption by acting as an electron trap, promoting interfacial charge transfer, and therefore delaying recombination of the electron–hole pair (Li and Li 2001, Behar and Rabani 2006, Ishida and Haruta 2007, Seery et al. 2007, Zeng et al. 2007). Silver has gained particular attention due to its properties of enhancing the photocatalytic efficiency under visible light. Seery et al. reported the enhanced photocatalytic activity with Ag doping of $TiO_2$ (Seery et al. 2007). Nolan et al. reported that a very high absorbance was shown by silver NPs in the visible light region and further proposed a mechanism for the visible light absorbance by Ag–$TiO_2$ (Nolan et al. 2010). They also suggested that the surface plasmon resonance of the silver NPs and the surface oxidized Ag are responsible for the visible light responsiveness of $TiO_2$. Wodka et al. synthesized Ag–$TiO_2$ composites with a different concentration of silver NPs on the surface of $TiO_2$. They found that a very small amount (1 wt. %) of silver NPs greatly affected the photocatalytic performance of $TiO_2$ (Wodka et al. 2010). The synthesized composite also showed higher oxidation and reduction activity than the pure $TiO_2$. Zhao and his group fabricated $TiO_2/Al_2O_3$ composite membrane via the sol–gel method for photocatalytic removal of organic pollutants (Zhang et al. 2006a). They controlled the pore size of the membrane mainly by the sol properties and the immersion time. The $TiO_2/Al_2O_3$ composite membranes showed better photocatalytic activity. These composite membranes were found to have the multifunction of photocatalysis and separation simultaneously. In recent years, various groups have combined Ag and other materials to form composites for use in water treatment and self-cleaning cotton textiles (Wu et al. 2013, Xiao et al. 2015).

Doping by non-metal elements (C, N, B, etc.) has also been found to significantly increase the visible light-driven photocatalytic efficiency (Jing et al. 2013). Cao et al. were the first to report the application of carbon-doped $TiO_2$ single crystal nanorods in the photocatalytic degradation of methylene blue (MB), p-nitrophenol (PNP), and rhodamine B (RhB) (Shao et al. 2017). Similarly, other carbon materials such as CNTs, fullerene, graphene oxide (GO), and reduced GO-modified $TiO_2$ particles also provide a larger specific surface area similar to activated carbon and have shown excellent adsorptive and photocatalytic properties (Meng et al. 2011, Perera et al. 2012, Fotiou et al. 2013). Activated carbon was initially used as an inert support for $TiO_2$ in photodegradation reactions, but it is now concluded that activated carbon increases adsorption capability and can help to enrich organic pollutants around the catalyst which directly increases the photocatalytic efficiency (Fu et al. 2004, Foo and Hameed 2010,). Cozzoli et al. synthesized organic-molecule-capped $TiO_2$ NPs. They used oleic acid as a capping agent. Due to the capping of oleic acid, $TiO_2$ became highly soluble in organic non-polar solvents (hexane, toluene) and formed an optically transparent solution. It was observed that the oleic-acid-capped $TiO_2$ NPs are a more efficient photocatalyst than bare $TiO_2$ NPs for the phenol oxidation reaction under visible light irradiation (Cozzoli et al. 2003). Jiang et al. developed a simple surface modification protocol through a traditional organic reaction between the surface hydroxyls on $TiO_2$ NPs and the NCO groups of tolylene diisocyanate (TDI) (Jiang et al. 2007).

This TDI-capped $TiO_2$ showed higher efficiency not only for the absorption of visible light but also in its photocatalytic efficiency under visible light compared with unmodified $TiO_2$. Appropriate organic capping agents not only tune the band gap but they also increase the absorption of organic pollutants on the surface of a catalyst. Oh et al. (2007) prepared platinum-treated fullerene/TiO2 composites (Pt-fullerene/$TiO_2$) and investigated their photocatalytic effect for the degradation of organic dye (Meng et al. 2011). Nguyen-Phan et al. in 2011 prepared $TiO_2$/GO composites for the photodegradation of MB. The $TiO_2$/GO composite was prepared by a one-step colloidal blending method using commercial $TiO_2$, GO, and deionized water. The composite material showed excellent adsorptivity and photocatalytic activity under UV and visible light. The increase in the GO content resulted in higher degradation efficiencies of MB.

GO plays the roles of adsorbent, electron acceptor, and photosensitizer in order to accelerate photodecomposition. In the degradation of MB, GO first acts as an additional adsorbent for contaminant molecules that subsequently diffuse to the phase boundary or the interface to undergo effective decomposition. The higher the GO content, the much higher density of oxygen-containing functional groups, such as carboxyl, epoxy, and hydroxyl groups, leads to more ionic/electro interaction and therefore, the better adsorptivity (Figure 9.7). GO also plays the role of an electron acceptor that accelerates the interfacial electron-transfer process from $TiO_2$, strongly hindering the recombination of charge carriers and thus improving the photocatalytic activity (Nguyen-Phan et al. 2011).

To improve the photocatalytic efficiency of $TiO_2$ NPs, Jiang et al. assembled $TiO_2$ NPs on a graphene sheet via *in situ* depositing $TiO_2$ NPs on GO nanosheets by a liquid phase deposition, followed by calcination at 200°C. Due to the excellent enhancing effect, large surface area, much-increased adsorption capacity, and the strong electron-transfer ability of the thermally reduced GO, the photocatalytic properties of $TiO_2$ were enhanced and under optimal conditions, the photo-oxidative degradation rate of all the azo dyes (methyl orange [MeOr], methyl red, orange G, and acid orange 7) was observed at 7.4, higher than conventional commercial P25 $TiO_2$ (Jiang et al. 2011). Beyond the increasing surface-to-volume ratio, this carbon family may be tailored to enhance specificity toward adsorbents through the modification of their surface groups (Carp et al. 2004).

Liu and his group modified and functionalized $TiO_2$ nanotubes with Au NPs in such a way that they can be used as recyclable surface-enhanced Raman spectroscopy (SERS) substrates for multifold organic pollutants (detection of Rhodamine 6G [R6G] dye, herbicide 4-chlorophenol [4-CP], persistent organic pollutant [POP] dichlorophenoxyacetic acid [2,4-D], and organophosphate pesticide methyl-parathion [MP]) detection

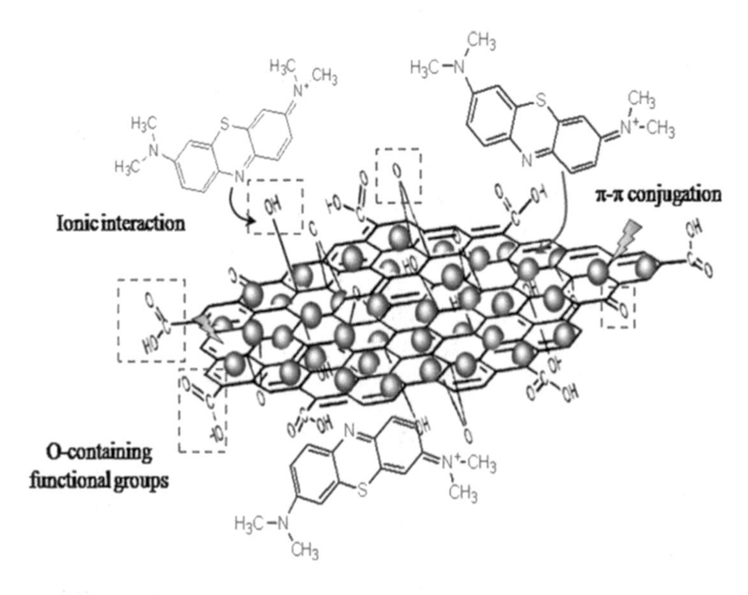


**FIGURE 9.7** Proposed interactions between $TiO_2$/graphene oxide composites and dye molecule that efficiently improve the photodegradation of methylene blue. (Reproduced from Nguyen-Phan et al., *Chemical Engineering Journal* 170 (2011) 226–232. Reprinted with permission from Elsevier B.V.)

as well as degradation to nontoxic materials under UV light. The $TiO_2$ nanotube array was synthesized on a ZnO template and the Au NPs were photodeposited on the $TiO_2$ nanotubes under UV light radiation (Li et al. 2010). A more detailed description of the work done on the application of a metal NP-based visible light catalyst in the treatment of contaminants that can be found elsewhere (Ayati et al. 2014, Fagan et al. 2016).

### 9.3.2 Zinc Oxide Photocatalysis

ZnO is an n-type semiconductor oxide with a band gap of 3.37 eV, binding energy of 60 meV, and deep violet absorption at room temperature. Due to its higher absorption efficiency in comparison to $TiO_2$, it has attracted attention for application as an alternative photocatalyst (Qiu et al. 2008, Yogendra et al. 2011). Lu et al. reported the synthesis of a newly structured ZnO hierarchical micro/nanoarchitecture by a facile solvothermal approach in an aqueous solution of ethylenediamine (EDA, $NH_2CH_2CH_2NH_2$) without self-assembled templates or matrices and investigates its photocatalytic performance. The fabricated ZnO nanostructures showed a strong structure-induced enhancement of photocatalytic performance and exhibit a significantly improved photocatalytic property and durability in the photodegradation of MeOr than that of other nanostructured ZnO, such as the powders of NPs, nanosheets, and nanoneedles. The special structural features of the micro/nanoarchitectured ZnO, i.e. a high surface-to-volume ratio with effective prevention from aggregation, leads to high photocatalytic activity, as observed in the decomposition of MeOr under UV light irradiation (Lu et al. 2008). Xu et al. reported the synthesis of various morphologies (cauliflower-like, truncated hexagonal conical, nanospherical, nanorods, tubular, and hourglass-like) of ZnO via a simple solvothermal method in solvents THF, decane, acetone, ethanol, water, and toluene, respectively, using zinc acetylacetonate as the zinc source. The morphologies of ZnO catalysts play an important role in the photocatalytic activity; this arises from differences in surface areas, polar planes, or oxygen vacancies. All of the ZnO samples showed enhanced activity for photocatalytic phenol degradation (Xu et al. 2009). Tian et al. demonstrated an effective route for easy, cost-effective, and large-scale production of ZnO photocatalysts by the direct calcination of zinc acetate at moderate temperature. The photocatalytic performance of the prepared ZnO was also evaluated. It was found that ZnO had superior photocatalytic activity and good reuse performance in the degradation of organic pollutants. This method would be important for the application of ZnO photocatalytic materials due to its advantages of simplicity, low cost, high production, and the excellent performance of the resulting products (Tian et al. 2012a).

Nanocomposites are preferable for photocatalysis, because of their higher light absorption, better suppression of photoinduced electron–hole pair recombination and increased charge separation. The enhanced photocatalytic activity of the nanocomposites is due to the reduction of the rate of recombination and an increase in the charge separation. According to inter-particle electron-transfer theory, the photoexcited electrons can be transferred between the CBs of coupled photocatalysts. This leads to the enhancement of charge separation and an increase in the lifetime of the charge carriers (Serpone et al. 1984). Nur et al. have also reported that a highly efficient photocatalyst can be developed by coupling two semiconductor metal oxides with different band gaps. Their proposed mechanism has suggested that the photogenerated electrons are driven further away and this allows efficient charge separation in the nanocomposites (Nur et al. 2007). For ZnO coupled with other semiconductors, $TiO_2$/ZnO, $SnO_2$/ZnO, $SnO_2$/ZnO/$TiO_2$, and $Co_3O_4$/ZnO are the most investigated materials for photocatalytic processes.

The photocatalytic activity of $TiO_2$/ZnO for the decolorization of MB was reported by Moradi et al. The results showed that the best photocatalytic activity of nanocomposite was obtained with a 50:10 M ratio of $TiO_2$:ZnO (Moradi et al. 2016). The synthesis of a $SnO_2$/ZnO/$TiO_2$ composite semiconductor by two simple methods, i.e. the sol–gel method and the solid-state method was published by Yang et al. Its photocatalytic activity was tested for the degradation of MeOr. They reported that the hybrid catalyst formed by the combination of three different band gap semiconductor oxides is able to extend the spectra response to the visible region, facilitate the transfer of the electrons, reduce the recombination probability, and increase the charge carrier lifetime, as a consequence of the enhancement of the photocatalytic activity for SnO/ZnO/$TiO_2$ samples (Yang et al. 2012a).

Huang et al. reported the synthesis of ZnO nanorod and nanotube arrays by the solvothermal process and the preparation of ZnO/$TiO_{2-x}$Ny heterojunction composites. The photocatalytic ability of the

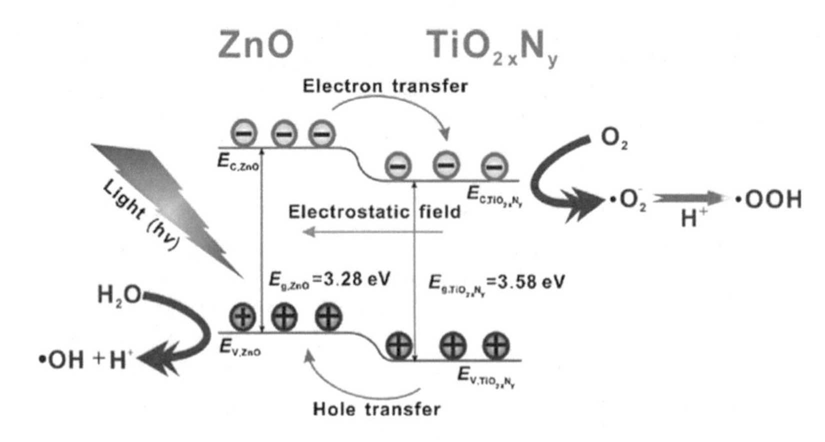


**FIGURE 9.8** Schematic illustration of the band structure and charge separation of the combined ZnO/$TiO_{2-x}N_y$ heterojunction. (Reproduced from Huang et al., *Applied Catalysis B: Environmental* 123–124 (2012) 9–17. Reprinted with permission from Elsevier B.V.)

prepared composites for the decomposition of NO was investigated. The ZnO nanotube arrays/$TiO_{2-x}N_y$ composites were found to have excellent photocatalytic activities in comparison to pure ZnO nanorods, nanotube, TiO2 (P25), and $TiO_{2-x}$Ny. The unique surface features and the heterojunction structure of the composites can promote an increase in the charge separation of the photogenerated electrons and holes within the nanostructures to enhance photocatalytic reaction (Huang et al. 2012). The mechanism of the separation of electron–hole pairs is shown in Figure 9.8.

## 9.4 Removal of Toxic Metal Ions from Water

The presence of heavy metals in the environment, even at trace levels, can be extremely harmful to living organisms. It is very well known that metal ions such as lead, copper, nickel, zinc, mercury, cadmium, and chromium released into the environment due to various natural and human activities have adverse effects on ecological systems and human health as they are not biodegradable and tend to accumulate (Naushad et al. 2016). Hence, the metal contaminants must be removed from water to prevent intake by organisms. As a result of environmental and health-related problems, drinking water and wastewater regulations have been toughened. The effective removal of these undesirable metals from water is thus a very important task before the researchers. Several methods have been proposed to remove metal contaminants from water which include chemical precipitation, electrochemical technology, membrane filtration, ion-exchange, and adsorption (Fu and Wang 2011, Fu et al. 2012, Zhang et al. 2013b, Shayeh et al. 2016). Due to its high efficiency, cost-effectiveness, and ease of operation, the adsorption treatment has gained considerable attention in the past few decades (Dabrowski 2001). Nanosized metal oxides (e.g. manganese oxides, aluminum oxides, magnesium oxides, titanium oxides, and cerium oxides) have been found to be promising absorbents for the removal of heavy metal ions from water (Coston et al. 1995). Recent studies have suggested that many nanosized metal oxides have been successful in the removal of toxic metal ions to the limit specified by the different environmental agencies (Deliyanni et al. 2009). However, the increased surface energy due to the reduction in size of the metal oxides has led to their poor stability. Thus, they become prone to agglomeration due to Van der Waals forces or other interactions and so their removal capacity is greatly decreased (Pradeep and Anshup 2009). It has been suggested by various research groups that the impregnation of nanosized metal oxides into porous supports to fabricate composites can lead to higher stability and better performances. In this section, we present the absorption of heavy metal ions by nanosized metal oxides and their composites.

Alumina ($Al_2O_3$) is the most commonly used adsorbent for heavy metals and the γ-alumina form has been found to be more adsorptive than α-alumina (Li et al. 2008). Nanosized alumina has gained extensive attention in recent years because of its large specific surface area, absence of internal diffusion resistance, and high surface binding energy. Experiments have confirmed that the adsorption performance of nanosized alumina is better for its higher surface area and smaller granularity (Türker 2007, Zhang et al. 2008). Nanosized $\gamma$-$Al_2O_3$ prepared by the sol–gel method has been used for separation of trace metal ions (Chang et al. 2003). Rahmani et al. reported that nanostructured $\gamma$-$Al_2O_3$ synthesized by ammonium acetate fuel is an effective adsorbent for the adsorption of $Pb^{2+}$, $Ni^{2+}$, and $Zn^{2+}$ (Rahmani et al. 2010). They demonstrated that the adsorption process was spontaneous and exothermic under natural conditions. The maximum capacities of adsorption for $Pb^{2+}$, $Ni^{2+}$, and $Zn^{2+}$ were 125, 83.33, and 58.82 mg $g^{-1}$, respectively. Poursani et al. synthesized nanostructured $\gamma$-$Al_2O_3$ by the sol–gel method. It was found to be an effective adsorbent for the adsorption of $Cr^{6+}$ and $Pb^{2+}$ ions from aqueous solutions under optimized conditions of pH (3 and 5 for Cr(VI) and Pb(II), respectively), adsorbent weight (3 g/L), contact time (4 h), and at room temperature (25°C) (Poursani et al. 2015). Hydrothermal synthesis of high surface area mesoporous $\gamma$-$Al_2O_3$ NPs by using the sodium salicylate as a template for the efficient removal of arsenic from contaminated water was reported (Patra et al. 2012). The highest absorption efficiency of the synthesized material for arsenic removal was found at pH 6.0. The adsorption studies also confirmed that the mesoporous alumina surface has a larger affinity for As(V) over As(III).

To overcome the poor adsorption of some heavy metal ions, chemical or physical modification of the sorbent surface with some organic compounds, especially chelating ones, is usually used to load the surface with some donor atoms such as oxygen, nitrogen, sulfur, and phosphorus (Hiraide et al. 1994, Manzoori et al. 1998, Ghaedi et al. 2008). The removal mechanism is changed when the functional group is immobilized on the surface of alumina. The target metal ions are removed both by adsorption on the surface of the alumina and by a surface attraction/chemical-bonding phenomenon on the newly added chemicals. A very common procedure to deposit an organic coating on an inorganic oxide such as alumina, iron oxide, and silica is to dissolve it in a proper medium and mix the solution with inorganic oxide particles for a period of time, followed by evaporation of the solvent and air drying of the adsorbent. A method was described for the immobilization of 2-mercaptonicotinic acid on the surface of three different alumina adsorbents. According to this report, selective separation and preconcentration processes of Pb(II) and Cu(II) from seawater samples were established (Mahmoud 2002). Pu et al synthesized $\gamma$-$Al_2O_3$ via the sol–gel method. $\gamma$-mercaptopropy-trimethoxysilane ($\gamma$-MPTMS) was then coated on $\gamma$-$Al_2O_3$ NPs by mixing the organic solution with inorganic oxide particles for a period of time, followed by evaporation of the solvent and air-drying the resultant adsorbent. This improved its selectivity toward Cu, Hg, Au, and Pd ions rather than other ions (Pu et al. 2004). Afkhami et al. reported the adsorption behavior of 2, 4-Dinitrophenlhydrazin (DNPH) coated $\gamma$-$Al_2O_3$ in the removal of metal ions. They found that the sorption process was greatly affected by parameters such as pH, contact time, and adsorbent dosage. The maximum adsorption capacity values of Cr(III), Cd(II), and Pb(II) ions in a mixture of six ion metals with modified alumina NPs (calculated by the Langmuir equation) were 100.0, 83.33, and 100.0 mg-$g^{-1}$, respectively (Afkhami et al 2010).

Two types of $\gamma$-$Al_2O_3$ nanofibers with different size and surface area synthesized via the hydrothermal method were functionalized with thiol and octyl groups by refluxing the toluene solution of 3-mercaptopropyltrimethoxysilane (MPTMS) and n-octyltriethoxysilane (OTES), respectively (Yang et al. 2010). These grafted nanofibers exhibited excellent sorption capacity. The thiol group-grafted nanofibers showed a high sorption capacity for $Pb^{2+}$ and $Cd^{2+}$. The octyl group-grafted nanofibers displayed superhydrophobicity and could efficiently adsorb highly diluted hydrophobic 4-nonylphenol from the aqueous solution. The high sorption efficiency of decorated nanofibers is mainly due to their unique structural features. The coated functional groups on the surface of the nanofibers improve the accessibility of sorption sites to the adsorbents. Also, large interconnected voids form in the aggregation of the randomly oriented nanofibers, which ensures that the contaminated solution flows through the sorption bed easily. The irregularly aggregated functionalized $\gamma$-$Al_2O_3$ nanofibers and a single functional group-grafted $\gamma$-$Al_2O_3$ nanofiber are shown in Figure 9.9.

Ceria is another rare earth metal oxide which is very commonly used as an adsorbent. The morphology, size, shape, and surface area greatly affect its adsorptive properties. In particular, ceria hollow nanosphere structures have several advantages in adsorption and catalysis. The hollow interior space effectively enhances the spatial dispersion, which results in not only higher surface area but also facile

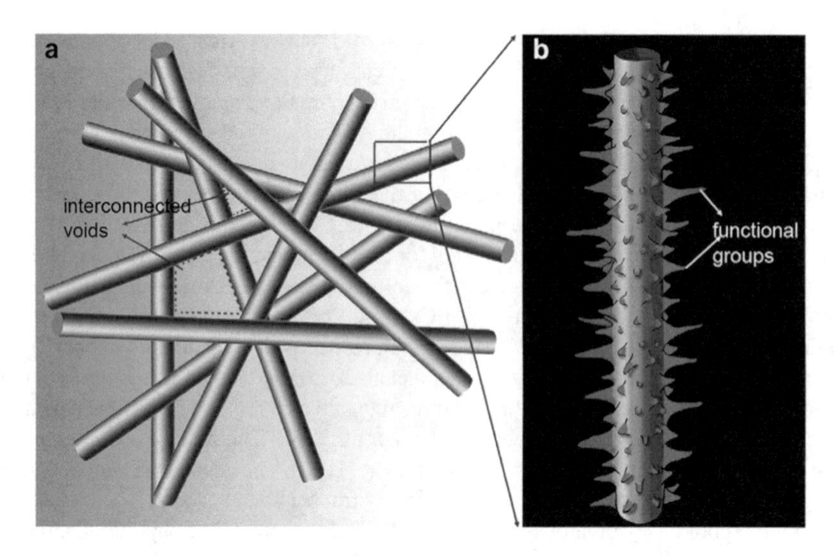


**FIGURE 9.9** The schematic diagrams of (a) irregularly aggregated functionalized $\gamma$-$Al_2O_3$ nanofibers and (b) a single functional group grafted $\gamma$-$Al_2O_3$ nanofiber. (Reproduced from Yang et al., *Water Research* 44 (2010) 741–750. Reprinted with permission from Elsevier B.V.)

mass transportation of molecules to the active sites. Cao et al. synthesized hollow $CeO_2$ NPs by the hydrothermal method and observed that these hollow $CeO_2$ NPs can efficiently remove 5.4 mg/g for Cr (VI) and 9.2 mg/g for Pb (II) in water, which was nearly 70 times higher than that of the commercial bulk ceria material (Cao et al. 2010). The application of $CeO_2$ NPs as an adsorbent for the removal of low amounts of dissolved Cr(VI) from pure water solutions was demonstrated by Recillas et al. (Recillas et al. 2010). The formation of the agglomerates of the $CeO_2$ nanoparticles with the addition of Cr(VI) solution due to the destabilization of the NP dispersion provides an easy way to remove the product in order to separate and re-use the $CeO_2$ NPs and to obtain a concentrated chromium solution by desorption. An exceptionally strong adsorption performance of hydrous cerium oxide NPs synthesized by a simple precipitation process for the removal of both As(III) and As(V) was reported by Li et al. (Li et al. 2012). The adsorption capacity of the prepared nanoparticles at the neutral pH was found to be 170 mg/g on As(III), and 107 mg/g on As(V). These NPs demonstrated a high adsorption capacity and extremely fast removal rate for arsenic without pre-oxidation or pH adjustment, which makes them very attractive for real field applications (Sun et al. 2012).

ZnO is known as an environmentally friendly material. The presence of hydroxyl groups on the surface of ZnO NPs makes them a promising candidate for the removal of heavy metal ions. In a study reported by Sheela et al., the application of ZnO nanoparticles for the removal of Zn(II), Cd(II), and Hg(II) ions from aqueous solution was investigated (Sheela et al. 2012). The ZnO nanoparticles were synthesized by the precipitation method. The adsorption of heavy metal ions on the ZnO was measured by the batch method. The effect of pH, contact time, concentration of metal ions, and temperature on adsorption was also studied. The maximum adsorption capacities for Zn(II), Cd(II), and Hg(II) were found to be 357, 387, and 714 mg/g for ions, respectively. The highest adsorption capacity of ZnO NPs for Hg(II) is due to the smallest hydrated ionic radii of Hg(II) in comparison to Zn and Cd. This allows them to move faster and reach the adsorption sites on the ZnO NPs. The mechanism of adsorption of heavy metal ions was attributed to both ion-exchange processes and the adsorption process. Wang et al. reported the mass fabrication of porous ZnO nanoplates based on solvothermal methods (Wang et al. 2010). Ethylene glycol was used to control the morphology and subsequent annealing. With the addition of ethylene glycol, the morphology evolved from nanosheets and plates to particles and finally microspheres. The fabricated ZnO nanoplates were porous with two terminal non-polar planes. The porous nanoplates have a pore diameter of 5–20 nm and a high specific surface area (147 $m^2g^{-1}$) showed strong and selective adsorption to cationic contaminants. It exhibited a very strong absorption efficiency of 1600 mg/g for the removal of Cu(II) ions (Wang et al. 2010). The adsorbed hydrated Cu(II) ions could partially hydrolyze, leading to the formation of Cu–O–Cu on the pore walls and hence multiplayer adsorption, exhibiting a Freundlich-type adsorptive behavior. It is worth noting that the adsorption behavior for each type of ZnO nanostructure was different. The adsorption behavior of ZnO nanoplates was best described using the Freundlich model which indicates that the adsorption occurs on a heterogeneous surface. Whereas the ZnO nanopowders were best fit to the Langmuir model, suggesting monolayer adsorption onto a uniform surface. Ma et al. reported the preparation of ZnO nanosheets via the hydrothermal method for $Pb^{2+}$ removal (Ma et al. 2010). The ZnO nanosheets showed good sorption capacities for $Pb^{2+}$ due to the surface hydroxyl groups while simultaneously forming new nanocomposites by doping $Pb^{2+}$ to act as secondary nanoadsorbents to avoid the secondary pollution of the regeneration procedure caused by conventional water treatment methods. Khan et al. reported the synthesis of ZnO nanosheets by the low-temperature stirring method and evaluated their efficiency for selective adsorption of Cd(II) in aqueous solution. Adsorption capacities of 97.36 mg $g^{-1}$ achieved for Cd(II) in aqueous solution for ZnO nanosheets (Khan et al. 2013).

The selectivity of $MnO_2$ nanoparticles towards toxic ions varies depending on their size and solid support. $MnO_2$ stabilized on sand support is selective for Cr(VI), Cd(II), and Pb(II) ions having an absorption capacity of 0.326 mmol/g, 0.111mmol/g, and 3.33 mg/g respectively. $MnO_2$ on silica support is able to absorb 0.030 mmol of Pb(II) and 0.396 mg/g of Mn (II) respectively. T. Pradeep and his group reported that $MnO_2$ nanoparticles stabilized on reduced GO can remove $Hg^{2+}$ ions with 100% efficiency (Sreeprasad et al. 2011). Due to the large surface area, high Young's modulus, good thermal conductivity, high-speed electron mobility, and electrocatalytic activities of GO, it can also be considered an excellent adsorbent for heavy metal ions in water. The selectivity and adsorption efficiency can be increased by the incorporation of nanomaterials, mainly metal oxides. Gohel et al. reported the synthesis of ZnO–GO and $TiO_2$–GO composites and their application in heavy metal ions removal. Composites have been used for the removal of different concentration of lead, cobalt, mercury, cadmium, and chromium (Gohel et al. 2017).

The oxide of titanium is chemically inert and has many applications ranging from anticorrosion, photocatalysis, photovoltaics, $H_2$ sensing, lithium batteries, and as an adsorbent for the removal of contaminants in polluted waters (Bavykin et al. 2006). Luo at al. first reported the application of $TiO_2$ for the removal and recovery of arsenic. The wastewater was acquired from a copper smelting industry in China (Luo et al. 2010). The $TiO_2$ was prepared by hydrolysis of titanyl sulfate. The BET surface area was 196 $m^2/g$, and the point of zero charge was 5.8. Batch experiments were employed for the absorption studies. They reported 21 successive treatment cycles using the regenerated $TiO_2$. Arsenic was recovered by pre-concentrating the extracted solutions. Since As(III) forms "inner-sphere bidentate binuclear complexes", it will bind to the OH surface sites on the $TiO_2$ (Pena et al. 2006). The results were confirmed using extended X-ray absorption fine structure spectroscopy (EXAFS), X-ray photoelectron spectroscopy (XPS), and surface complexation modeling. Engates et al. reported the effect of particle size, sorbent concentration, and exhaustion on the adsorption of Pb, Cd, Cu, Zn, and Ni to $TiO_2$ nanoparticles and $TiO_2$ anatase bulk particles (Engates and Shipley 2011). Adsorption and exhaustion studies were carried out for single- and multi-metal adsorption. Large adsorption capacities for the $TiO_2$ nanoparticles compared to their bulk counterpart were reported. The data correlated to the Langmuir isotherm model indicating monolayer adsorption on the surface of the $TiO_2$ nanoparticles. The exhaustion experiments showed that at pH 6, $TiO_2$ nanoparticles were exhausted after three cycles and at pH 8 after eight cycles. These results supported the possibility of $TiO_2$ nanoparticles as a potential remediation for heavy metal removal from contaminated waters. This group also reported the regeneration of $TiO_2$ nanoparticles for heavy metal removal (Hu and Shipley 2013). In brief, nanosized metal oxides have been found as an efficient adsorbent towards the removal of heavy metal ions. However, their agglomeration due to instability as small particles has hampered their application. An effective solution to this problem is the fabrication of hybrid adsorbents by coating or impregnating metal oxide nanoparticles into supports of larger sizes. The widely used supports include natural hosts such as bentonite, sand, metallic oxide materials such as $Al_2O_3$ membranes, porous manganese oxide complexes, and synthetic polymer hosts such as cross-linked ion-exchange resins (Hu et al. 2004, Zhang et al. 2006b, Eren et al. 2010, Lee et al. 2010, Ray and Shipley 2015).

## 9.5 Conclusions and Future Perspectives

Wastewater treatment is one of the major environmental concerns of the twenty-first century. Nanostructured non-magnetic metal oxides can be applied to wastewater treatment purposes. Metal oxides have numerous advantages compared with conventional materials and their bulk counterparts. As reflected in this chapter, the majority of research work done to date incorporating non-magnetic metal oxide nanostructures for environmental applications has been in the treatment of wastewater, whether it is in the area of oil–water separation, removal of organic pollutants and other toxic metal ions from water. The recent progress in many metal oxide nanostructure systems functionalized with organic/inorganic compounds has shown great promise in providing rapid separation and removal of water contaminants. However, a greater focus is needed on transforming the laboratory-scale research to real field applications where the conditions can vary greatly. While the chapter discussed the suitability of the metal oxides for removal of water pollutants, a number of questions need to be answered for future applications, including the effect of size on the application, issues with large-scale application, mechanisms behind the application, and the possible challenges and economic values of applying the nanostructures for wastewater treatment. The applications demonstrated and outlined in this chapter ensures that non-magnetic metal oxide nanostructures in improving the quality of our environment will continue to attract great interest.

## REFERENCES

Afkhami, A., Saber-Tehrani, M., Bagheri, H., 2010. Simultaneous removal of heavy-metal ions in wastewater samples using nano-alumina modified with 2, 4-dinitrophenylhydrazine. *Journal of Hazardous Materials*, *181*(1–3), pp.836–844.

Ayati, A., Ahmadpour, A., Bamoharram, F.F., Tanhaei, B., Mänttäri, M., Sillanpää, M., 2014. A review on catalytic applications of Au/TiO2 nanoparticles in the removal of water pollutant. *Chemosphere*, *107*, pp.163–174.

Bavykin, D.V., Friedrich, J.M., Walsh, F.C., 2006. Protonated titanates and TiO2 nanostructured materials: Synthesis, properties, and applications. *Advanced Materials*, *18*(21), pp.2807–2824.

Behar, D., Rabani, J., 2006. Kinetics of hydrogen production upon reduction of aqueous TiO2 nanoparticles catalyzed by Pd0, Pt0, or Au0 coatings and an unusual hydrogen abstraction; steady state and pulse radiolysis study. *The Journal of Physical Chemistry B*, *110*(17), pp.8750–8755.

Brody, T.M., Bianca, P.D., Krysa, J., 2012. Analysis of inland crude oil spill threats, vulnerabilities, and emergency response in the midwest United States. *Risk Analysis*, *32*(10), pp.1741–1749.

Cao, C.Y., Cui, Z.M., Chen, C.Q., Song, W.G., Cai, W., 2010. Ceria hollow nanospheres produced by a template-free microwave-assisted hydrothermal method for heavy metal ion removal and catalysis. *The Journal of Physical Chemistry C*, *114*(21), pp.9865–9870.

Carp, O., Huisman, C.L., Reller, A., 2004. Photoinduced reactivity of titanium dioxide. *Progress in Solid State Chemistry*, *32*, pp.33–177.

Carson, R.T., Mitchell, R.C., Hanemann, M., Kopp, R.J., Presser, S., Ruud, P.A., 2003. Contingent valuation and lost passive use: Damages from the Exxon Valdez oil spill. *Environmental and Resource Economics*, *25*(3), pp.257–286.

Chang, G., Jiang, Z.C., Peng, T.Y., Hu, B., 2003. Preparation of high-specific-surface-area nanometer-sized alumina by sol-gel method and study on adsorption behaviors of transition metal ions on the alumina powder with ICP-AES. *Acta Chimica Sinica-Chinese Edition*, *61*(1), pp.100–103.

Chen, Q., de Leon, A., Advincula, R.C., 2015. Inorganic–organic thiol–ene coated mesh for oil/water separation. *ACS Applied Materials and Interfaces*, *7*(33), pp.18566–18573.

Chong, M.N., Jin, B., Chow, C.W., Saint, C., 2010. Recent developments in photocatalytic water treatment technology: A review. *Water Research*, *44*(10), pp.2997–3027.

Coston, J.A., Fuller, C.C., Davis, J.A., 1995. Pb2+ and Zn2+ adsorption by a natural aluminum-and iron-bearing surface coating on an aquifer sand. *Geochimica et Cosmochimica Acta*, *59*(17), pp.3535–3547.

Cozzoli, P.D., Kornowski, A., Weller, H., 2003. Low-temperature synthesis of soluble and processable organic-capped anatase TiO2 nanorods. *Journal of the American Chemical Society*, *125*(47), pp.14539–14548.

Dąbrowski, A., 2001. Adsorption—From theory to practice. *Advances in Colloid and Interface Science*, *93*(1–3), pp.135–224.
Dai, C., Liu, N., Cao, Y., Chen, Y., Lu, F., Feng, L., 2014. Fast formation of superhydrophobic octadecylphosphonic acid (ODPA) coating for self-cleaning and oil/water separation. *Soft Matter*, *10*(40), pp.8116–8121.
Deliyanni, E.A., Peleka, E.N., Matis, K.A., 2009. Modeling the sorption of metal ions from aqueous solution by iron-based adsorbents. *Journal of Hazardous Materials*, *172*(2–3), pp.550–558.
Dung, N.T., Van Khoa, N., Herrmann, J.M., 2005. Photocatalytic degradation of reactive dye RED-3BA in aqueous TiO2 suspension under UV-visible light. *International Journal of Photoenergy*, *7*(1), pp.11–15.
Engates, K.E., Shipley, H.J., 2011. Adsorption of Pb, Cd, Cu, Zn, and Ni to titanium dioxide nanoparticles: Effect of particle size, solid concentration, and exhaustion. *Environmental Science and Pollution Research International*, *18*(3), pp.386–395.
Eren, E., Tabak, A., Eren, B., 2010. Performance of magnesium oxide-coated bentonite in removal process of copper ions from aqueous solution. *Desalination*, *257*(1–3), pp.163–169.
Fagan, R., McCormack, D.E., Dionysiou, D.D., Pillai, S.C., 2016. A review of solar and visible light active TiO2 photocatalysis for treating bacteria, cyanotoxins and contaminants of emerging concern. *Materials Science in Semiconductor Processing*, *42*, pp.2–14.
Feng, L., Zhang, Z., Mai, Z., Ma, Y., Liu, B., Jiang, L., Zhu, D., 2004. A super-hydrophobic and super-oleophilic coating mesh film for the separation of oil and water. *Angewandte Chemie*, *116*(15), pp.2046–2048.
Fingas, M., 1995. Oil spills and their cleanup. *Chemistry and Industry (London)*, *27*, pp.1005–1008.
Foo, K.Y., Hameed, B.H., 2010. Decontamination of textile wastewater via TiO2/activated carbon composite materials. *Advances in Colloid and Interface Science*, *159*(2), pp.130–143.
Fotiou, T., Triantis, T.M., Kaloudis, T., Pastrana-Martínez, L.M., Likodimos, V., Falaras, P., Silva, A.M.T., Hiskia, A., 2013. Photocatalytic degradation of microcystin-LR and off-odor compounds in water under UV-A and solar light with a nanostructured photocatalyst based on reduced graphene oxide–TiO2 composite. Identification of intermediate products. *Industrial and Engineering Chemistry Research*, *52*(39), pp.13991–14000.
Frank, S.N., Bard, A.J., 1977. Heterogeneous photocatalytic oxidation of cyanide and sulfite in aqueous solutions at semiconductor powders. *The Journal of Physical Chemistry*, *81*(15), pp.1484–1488.
Fu, F., Wang, Q., 2011. Removal of heavy metal ions from wastewaters: A review. *Journal of Environmental Management*, *92*(3), pp.407–418.
Fu, F., Xie, L., Tang, B., Wang, Q., Jiang, S., 2012. Application of a novel strategy—Advanced Fenton-chemical precipitation to the treatment of strong stability chelated heavy metal containing wastewater. *Chemical Engineering Journal*, *189–190*, pp.283–287.
Fu, P., Luan, Y., Dai, X., 2004. Preparation of activated carbon fibers supported TiO2 photocatalyst and evaluation of its photocatalytic reactivity. *Journal of Molecular Catalysis A: Chemical*, *221*(1–2), pp.81–88.
Fujishima, A., Honda, K., 1972. Electrochemical photolysis of water at a semiconductor electrode. *Nature*, *238*(5358), p.37–38.
Fujishima, A., Rao, T.N., Tryk, D.A., 2000. Titanium dioxide photocatalysis. *Journal of Photochemistry and Photobiology C: Photochemistry Reviews*, *1*(1), pp.1–21.
Gao, R., Liu, Q., Wang, J., Liu, J., Yang, W., Gao, Z., Liu, L., 2014. Construction of superhydrophobic and superoleophilic nickel foam for separation of water and oil mixture. *Applied Surface Science*, *289*, pp.417–424.
Gao, X., Wang, X., Ouyang, X., Wen, C., 2016. Flexible superhydrophobic and superoleophilic MoS2 sponge for highly efficient oil-water separation. *Scientific Reports*, *6*, p.27207.
Gaya, U.I., Abdullah, A.H., 2008. Heterogeneous photocatalytic degradation of organic contaminants over titanium dioxide: A review of fundamentals, progress and problems. *Journal of Photochemistry and Photobiology C: Photochemistry Reviews*, *9*(1), pp.1–12.
Ge, B., Men, X., Zhu, X., Zhang, Z., 2015. A superhydrophobic monolithic material with tunable wettability for oil and water separation. *Journal of Materials Science*, *50*(6), pp.2365–2369.
Ge, J., Zhang, J., Wang, F., Li, Z., Yu, J., Ding, B., 2017. Superhydrophilic and underwater superoleophobic nanofibrous membrane with hierarchical structured skin for effective oil-in-water emulsion separation. *Journal of Materials Chemistry A*, *5*(2), pp.497–502.

Ghaedi, M., Niknam, K., Shokrollahi, A., Niknam, E., Rajabi, H.R., Soylak, M., 2008. Flame atomic absorption spectrometric determination of trace amounts of heavy metal ions after solid phase extraction using modified sodium dodecyl sulfate coated on alumina. *Journal of Hazardous Materials*, *155*(1–2), pp.121–127.

Gohel, V.D., Rajput, A., Gahlot, S., Kulshrestha, V., 2017, December. Removal of toxic metal ions From potable water by graphene oxide composites. *Macromolecular Symposia 376*(1), p.1700050.

Gondal, M.A., Sadullah, M.S., Dastageer, M.A., McKinley, G.H., Panchanathan, D., Varanasi, K.K., 2014. Study of factors governing oil–water separation process using TiO2 films prepared by spray deposition of nanoparticle dispersions. *ACS Applied Materials and Interfaces*, *6*(16), pp.13422–13429.

Gupta, R.K., Dunderdale, G.J., England, M. W., Hozumi, A., 2017. Oil/water separation techniques: a review of recent progresses and future directions. *Journal of Material Chemistry* A 5, pp.16025–16058.

Guo, F., Wen, Q., Peng, Y., Guo, Z., 2017. Simple one-pot approach toward robust and boiling-water resistant superhydrophobic cotton fabric and the application in oil/water separation. *J. Mater. Chem. A*, *5*(41), pp.21866–21874.

Hiraide, M., Sorouradin, M.H., Kawaguchi, H., 1994. Immobilization of dithizone on surfactant-coated alumina for preconcentration of metal ions. *Analytical Sciences*, *10*(1), pp.125–127.

Hu, J., Shipley, H.J., 2013. Regeneration of spent TiO2 nanoparticles for Pb (II), Cu (II), and Zn (II) removal. *Environmental Science and Pollution Research International*, *20*(8), pp.5125–5137.

Hu, P.Y., Hsieh, Y.H., Chen, J.C., Chang, C.Y., 2004. Characteristics of manganese-coated sand using SEM and EDAX analysis. *Journal of Colloid and Interface Science*, *272*(2), pp.308–313.

Huang, Y., Wei, Y., Wu, J., Guo, C., Wang, M., Yin, S., Sato, T., 2012. Low temperature synthesis and photocatalytic properties of highly oriented ZnO/$TiO_{2-x}N_y$ coupled photocatalysts. *Applied Catalysis B: Environmental*, *123–124*, pp.9–17.

Inoue, T., Fujishima, A., Konishi, S., Honda, K., 1979. Photoelectrocatalytic reduction of carbon dioxide in aqueous suspensions of semiconductor powders. *Nature*, *277*(5698), pp.637–638.

Ishida, T., Haruta, M., 2007. Gold catalysts: Towards sustainable chemistry. *Angewandte Chemie*, *46*(38), pp.7154–7156.

Jiang, D., Xu, Y., Hou, B., Wu, D., Sun, Y., 2007. Synthesis of visible light-activated TiO2 photocatalyst via surface organic modification. *Journal of Solid State Chemistry*, *180*(5), pp.1787–1791.

Jiang, G., Lin, Z., Chen, C., Zhu, L., Chang, Q., Wang, N., Wei, W., Tang, H., 2011. TiO2 nanoparticles assembled on graphene oxide nanosheets with high photocatalytic activity for removal of pollutants. *Carbon*, *49*(8), pp.2693–2701.

Jing, L., Zhou, W., Tian, G., Fu, H., 2013. Surface tuning for oxide-based nanomaterials as efficient photocatalysts. *Chemical Society Reviews*, *42*(24), pp.9509–9549.

Khan, S.B., Rahman, M.M., Marwani, H.M., Asiri, A.M., Alamry, K.A., 2013. An assessment of zinc oxide nanosheets as a selective adsorbent for cadmium. *Nanoscale Research Letters*, *8*(1), p.377.

Kong, L., Li, Y., Qiu, F., Zhang, T., Guo, Q., Zhang, X., Yang, D., Xu, J., Xue, M., 2018. Fabrication of hydrophobic and oleophilic polyurethane foam sponge modified with hydrophobic A12O3 for oil/water separation. *Journal of Industrial and Engineering Chemistry*, *58*, pp.369–375.

Krupenkin, T.N., Taylor, J.A., Schneider, T.M., Yang, S., 2004. From rolling ball to complete wetting: The dynamic tuning of liquids on nanostructured surfaces. *Langmuir*, *20*(10), pp.3824–3827.

Lasa, H., Serrano, B., Salaices, M., 2005 *Photocatalytic Reaction Engineering*, Springer, New York.

Lee, S.M., Kim, W.G., Laldawngliana, C., Tiwari, D., 2010. Removal behavior of surface modified sand for Cd (II) and Cr (VI) from aqueous solutions. *Journal of Chemical and Engineering Data*, *55*(9), pp.3089–3094.

Lessard, R.R., DeMarco, G., 2000. The significance of oil spill dispersants. *Spill Science and Technology Bulletin*, *6*(1), pp.59–68.

Li, J., Shi, Y., Cai, Y., Mou, S., Jiang, G., 2008. Adsorption of di-ethyl-phthalate from aqueous solutions with surfactant-coated nano/microsized alumina. *Chemical Engineering Journal*, *140*(1–3), pp.214–220.

Li, J., Zhao, Z., Kang, R., Zhang, Y., Lv, W., Li, M., Jia, R., Luo, L., 2017. Robust superhydrophobic candle soot and silica composite sponges for efficient oil/water separation in corrosive and hot water. *Journal of Sol-Gel Science and Technology*, *82*(3), pp.817–826.

Li, R., Li, Q., Gao, S., Shang, J.K., 2012. Exceptional arsenic adsorption performance of hydrous cerium oxide nanoparticles: Part A. Adsorption capacity and mechanism. *Chemical Engineering Journal*, *185–186*, pp.127–135.

Li, X., Chen, G., Yang, L., Jin, Z., Liu, J., 2010. Multifunctional Au-Coated TiO2 Nanotube Arrays as Recyclable SERS Substrates for Multifold Organic Pollutants Detection. *Advanced Functional Materials, 20*(17), pp.2815–2824.

Li, X.Z., Li, F.B., 2001. Study of Au/Au3+-TiO2 photocatalysts toward visible photooxidation for water and wastewater treatment. *Environmental Science and Technology, 35*(11), pp.2381–2387.

Li Puma, G.L., Bono, A., Krishnaiah, D., Collin, J.G., 2008. Preparation of titanium dioxide photocatalyst loaded onto activated carbon support using chemical vapor deposition: A review paper. *Journal of Hazardous Materials, 157*(2–3), pp.209–219.

Liu, N., Chen, Y., Lu, F., Cao, Y., Xue, Z., Li, K., Feng, L., Wei, Y., 2013. Straightforward oxidation of a copper substrate produces an underwater superoleophobic mesh for oil/water separation. *ChemPhysChem, 14*(15), pp.3489–3494.

Lu, F., Cai, W., Zhang, Y., 2008. ZnO hierarchical micro/nanoarchitectures: Solvothermal synthesis and structurally enhanced photocatalytic performance. *Advanced Functional Materials, 18*(7), pp.1047–1056.

Luo, T., Cui, J., Hu, S., Huang, Y., Jing, C., 2010. Arsenic removal and recovery from copper smelting wastewater using TiO2. *Environmental Science and Technology, 44*(23), pp.9094–9098.

Ma, Q., Cheng, H., Fane, A. G., Wang, R., Zhang, H., 2016. Recent development of advanced materials with special wettability for selective oil/water separation. *Small, 12*, pp.2186–2202.

Ma, X., Wang, Y., Gao, M., Xu, H., Li, G., 2010. A novel strategy to prepare ZnO/PbS heterostructured functional nanocomposite utilizing the surface adsorption property of ZnO nanosheets. *Catalysis Today, 158*(3–4), pp.459–463.

Mahmoud, M.E., 2002. Study of the selectivity characteristics incorporated into physically adsorbed alumina phases. II. Mercaptonicotinic acid and potential applicatssions as selective stationary phases for separation, extraction, and preconcentration of lead (II) and copper (II). *Journal of Liquid Chromatography and Related Technologies, 25*(8), pp.1187–1199.

Manzoori, J.L., Sorouraddin, M.H., Shabani, A.M.H., 1998. Determination of mercury by cold vapour atomic absorption spectrometry after preconcentration with dithizone immobilized on surfactant-coated alumina. *Journal of Analytical Atomic Spectrometry, 13*(4), pp.305–308.

Meng, Z.D., Zhu, L., Choi, J.G., Chen, M.L., Oh, W.C., 2011. Effect of Pt treated fullerene/TiO2 on the photocatalytic degradation of MO under visible light. *Journal of Materials Chemistry, 21*(21), pp.7596–7603.

Moradi, S., Aberoomand-Azar, P., Raeis-Farshid, S., Abedini-Khorrami, S., Givianrad, M.H., 2016. The effect of different molar ratios of ZnO on characterization and photocatalytic activity of TiO2/ZnO nanocomposite. *Journal of Saudi Chemical Society, 20*(4), pp.373–378.

Naushad, M., Ahamad, T., Sharma, G., Al-Muhtaseb, A.H., Albadarin, A.B., Alam, M.M., ALOthman, Z.A., Alshehri, S.M., Ghfar, A.A., 2016. *Chemical Engineering Journal, 300*, pp.306–316.

Nguyen-Phan, T.D., Pham, V.H., Shin, E.W., Pham, H.D., Kim, S., Chung, J.S., Kim, E.J., Hur, S.H., 2011. The role of graphene oxide content on the adsorption-enhanced photocatalysis of titanium dioxide/graphene oxide composites. *Chemical Engineering Journal, 170*(1), pp.226–232.

Nolan, N.T., Seery, M.K., Hinder, S.J., Healy, L.F., Pillai, S.C., 2010. A systematic study of the effect of silver on the chelation of formic acid to a titanium precursor and the resulting effect on the anatase to rutile transformation of TiO2. *The Journal of Physical Chemistry C, 114*(30), pp.13026–13034.

Nur, H., Misnon, I.I., Wei, L.K., 2007. Stannic oxide-titanium dioxide coupled semiconductor photocatalyst loaded with polyaniline for enhanced photocatalytic oxidation of 1-octene. *International Journal of Photoenergy, 2007*, 1–6.

Patra, A.K., Dutta, A., Bhaumik, A., 2012. Self-assembled mesoporous γ-Al2O3 spherical nanoparticles and their efficiency for the removal of arsenic from water. *Journal of Hazardous Materials, 201–202*, pp.170–177.

Pelaez, M., Nolan, N.T., Pillai, S.C., Seery, M.K., Falaras, P., Kontos, A.G., Dunlop, P.S.M., Hamilton, J.W.J., Byrne, J.A., O'shea, K., Entezari, M.H., Dionysiou, D.D., 2012. A review on the visible light active titanium dioxide photocatalysts for environmental applications. *Applied Catalysis B: Environmental, 125*, pp.331–349.

Pelletier, É., Siron, R., 1999. Silicone-based polymers as oil spill treatment agents. *Environmental Toxicology and Chemistry: An International Journal, 18*(5), pp.813–818.

Pena, M., Meng, X., Korfiatis, G.P., Jing, C., 2006. Adsorption mechanism of arsenic on nanocrystalline titanium dioxide. *Environmental Science and Technology, 40*(4), pp.1257–1262.

Perera, S.D., Mariano, R.G., Vu, K., Nour, N., Seitz, O., Chabal, Y., Balkus Jr, K.J., 2012. Hydrothermal synthesis of graphene-TiO2 nanotube composites with enhanced photocatalytic activity. *ACS Catalysis*, *2*(6), pp.949–956.

Poursani, A.S., Nilchi, A., Hassani, A.H., Shariat, M., Nouri, J., 2015. A novel method for synthesis of nano-γ-Al 2 O 3: study of adsorption behavior of chromium, nickel, cadmium and lead ions. *International Journal of Environmental Science and Technology*, *12*(6), pp.2003–2014.

Poyatos, J.M., Muñio, M.M., Almecija, M.C., Torres, J.C., Hontoria, E., Osorio, F., 2010. Advanced oxidation processes for wastewater treatment: State of the art. *Water, Air, and Soil Pollution*, *205*(1–4), pp. 187–204.

Pradeep, T., Anshup, 2009. Noble metal nanoparticles for water purification: A critical review. *Thin Solid Films*, *517*(24), pp.6441–6478.

Pu, X., Jiang, Z., Hu, B., Wang, H., 2004. γ-MPTMS modified nanometer-sized alumina micro-column separation and preconcentration of trace amounts of Hg, Cu, Au and Pd in biological, environmental and geological samples and their determination by inductively coupled plasma mass spectrometry. *Journal of Analytical Atomic Spectrometry*, *19*(8), pp.984–989.

Qiu, R., Zhang, D., Mo, Y., Song, L., Brewer, E., Huang, X., Xiong, Y., 2008. Photocatalytic activity of polymer-modified ZnO under visible light irradiation. *Journal of Hazardous Materials*, *156*(1–3), pp.80–85.

Rahmani, A., Mousavi, H.Z., Fazli, M., 2010. Effect of nanostructure alumina on adsorption of heavy metals. *Desalination*, *253*(1–3), pp.94–100.

Rajeshwar, K., Osugi, M.E., Chanmanee, W., Chenthamarakshan, C.R., Zanoni, M.V.B., Kajitvichyanukul, P., Krishnan-Ayer, R., 2008. Heterogeneous photocatalytic treatment of organic dyes in air and aqueous media. *Journal of Photochemistry and Photobiology C: Photochemistry Reviews*, *9*(4), pp.171–192.

Ray, P.Z., Shipley, H.J., 2015. Inorganic nano-adsorbents for the removal of heavy metals and arsenic: A review. *RSC Advances*, *5*(38), pp.29885–29907.

Recillas, S., Colón, J., Casals, E., González, E., Puntes, V., Sánchez, A., Font, X., 2010. Chromium VI adsorption on cerium oxide nanoparticles and morphology changes during the process. *Journal of Hazardous Materials*, *184*(1–3), pp.425–431.

Rehman, S., Ullah, R., Butt, A.M., Gohar, N.D., 2009. Strategies of making TiO2 and ZnO visible light active. *Journal of Hazardous Materials*, *170*(2–3), pp.560–569.

Rosario, R., Gust, D., Garcia, A.A., Hayes, M., Taraci, J.L., Clement, T., Dailey, J.W., Picraux, S.T., 2004. Lotus effect amplifies light-induced contact angle switching. *The Journal of Physical Chemistry B*, *108*(34), pp.12640–12642.

Seery, M.K., George, R., Floris, P., Pillai, S.C., 2007. Silver doped titanium dioxide nanomaterials for enhanced visible light photocatalysis. *Journal of Photochemistry and Photobiology A: Chemistry*, *189*(2–3), pp.258–263.

Serpone, N., Borgarello, E., Grätzel, M., 1984. Visible light induced generation of hydrogen from H 2 S in mixed semiconductor dispersions; improved efficiency through inter-particle electron transfer. *Journal of the Chemical Society, Chemical Communications*, *6*(6), pp.342–344.

Shao, J., Sheng, W., Wang, M., Li, S., Chen, J., Zhang, Y., Cao, S., 2017. In situ synthesis of carbon-doped TiO2 single-crystal nanorods with a remarkably photocatalytic efficiency. *Applied Catalysis B: Environmental*, *209*, pp.311–319.

Shayeh, J.S., Siadat, S.O.R., Sadeghnia, M., Niknam, K., Rezaei, M., Aghamohammadi, N., 2016. Advanced studies of coupled conductive polymer/metal oxide nano wire composite as an efficient supercapacitor by common and fast Fourier electrochemical methods. *Journal of Molecular Liquids*, *220*, pp.489–494.

Sheela, T., Nayaka, Y.A., Viswanatha, R., Basavanna, S., Venkatesha, T.G., 2012. Kinetics and thermodynamics studies on the adsorption of Zn (II), Cd (II) and Hg (II) from aqueous solution using zinc oxide nanoparticles. *Powder Technology*, *217*, pp.163–170.

Shuai, Q., Yang, X., Luo, Y., Tang, H., Luo, X., Tan, Y., Ma, M., 2015. A superhydrophobic poly (dimethylsiloxane)-TiO2 coated polyurethane sponge for selective absorption of oil from water. *Materials Chemistry and Physics*, *162*, pp.94–99.

Sparks, B.J., Hoff, E.F., Xiong, L., Goetz, J.T., Patton, D.L., 2013. Superhydrophobic hybrid inorganic–organic thiol-ene surfaces fabricated via spray-deposition and photopolymerization. *ACS Applied Materials and Interfaces*, *5*(5), pp.1811–1817.

Sreeprasad, T.S., Maliyekkal, S.M., Lisha, K.P., Pradeep, T., 2011. Reduced graphene oxide–metal/metal oxide composites: Facile synthesis and application in water purification. *Journal of Hazardous Materials, 186*(1), pp.921–931.

Su, C., 2009. Highly hydrophobic and oleophilic foam for selective absorption. *Applied Surface Science, 256*(5), pp.1413–1418.

Su, C., Xu, Y., Zhang, W., Liu, Y., Li, J., 2012. Porous ceramic membrane with superhydrophobic and superoleophilic surface for reclaiming oil from oily water. *Applied Surface Science, 258*(7), pp.2319–2323.

Sun, W., Li, Q., Gao, S., Shang, J.K., 2012. Exceptional arsenic adsorption performance of hydrous cerium oxide nanoparticles: Part B. Integration with silica monoliths and dynamic treatment. *Chemical Engineering Journal, 185–186*, pp.136–143.

Tao, S., Wang, Y., An, Y., 2011. Superwetting monolithic SiO2 with hierarchical structure for oil removal. *Journal of Materials Chemistry, 21*(32), pp.11901–11907.

Tian, C., Zhang, Q., Wu, A., Jiang, M., Liang, Z., Jiang, B., Fu, H., 2012a. Cost-effective large-scale synthesis of ZnO photocatalyst with excellent performance for dye photodegradation. *Chemical Communications, 48*(23), pp.2858–2860.

Tian, D., Zhang, X., Tian, Y., Wu, Y., Wang, X., Zhai, J., Jiang, L., 2012b. Photo-induced water–oil separation based on switchable superhydrophobicity–superhydrophilicity and underwater superoleophobicity of the aligned ZnO nanorod array-coated mesh films. *Journal of Materials Chemistry, 22*(37), pp.19652–19657.

Tian, Y., Su, B., Jiang, L., 2014. Interfacial material system exhibiting superwettability. *Advanced Materials, 26*(40), pp.6872–6897.

Türker, A.R., 2007. New sorbents for solid-phase extraction for metal enrichment. *CLEAN – Soil, Air, Water, 35*(6), pp.548–557.

Umebayashi, T., Yamaki, T., Itoh, H., Asai, K., 2002. Analysis of electronic structures of 3d transition metal-doped TiO2 based on band calculations. *Journal of Physics and Chemistry of Solids, 63*(10), pp.1909–1920.

Viessman, W., Jr, Hammer, M.J., 1998 *Water Supply and Pollution Control*, Addison-Wesley Longman, Menlo Park, CA.

Wang, C.F., Tzeng, F.S., Chen, H.G., Chang, C.J., 2012. Ultraviolet-durable superhydrophobic zinc oxide-coated mesh films for surface and underwater–oil capture and transportation. *Langmuir, 28*(26), pp.10015–10019.

Wang, F., Lei, S., Li, C., Ou, J., Xue, M., Li, W., 2014. Superhydrophobic Cu mesh combined with a superoleophilic polyurethane sponge for oil spill adsorption and collection. *Industrial and Engineering Chemistry Research, 53*(17), pp.7141–7148.

Wang, S., Li, M., Lu, Q., 2010a. Filter paper with selective absorption and separation of liquids that differ in surface tension. *ACS Applied Materials and Interfaces, 2*(3), pp.677–683.

Wang, X., Cai, W., Lin, Y., Wang, G., Liang, C., 2010b. Mass production of micro/nanostructured porous ZnO plates and their strong structurally enhanced and selective adsorption performance for environmental remediation. *Journal of Materials Chemistry, 20*(39), pp.8582–8590.

Wodka, D., Bielańska, E., Socha, R.P., Elzbieciak-Wodka, M., Gurgul, J., Nowak, P., Warszyński, P., Kumakiri, I., 2010. Photocatalytic activity of titanium dioxide modified by silver nanoparticles. *ACS Applied Materials and Interfaces, 2*(7), pp.1945–1953.

Wu, D., Wang, L., Song, X., Tan, Y., 2013. Enhancing the visible-light-induced photocatalytic activity of the self-cleaning TiO2-coated cotton by loading Ag/AgCl nanoparticles. *Thin Solid Films, 540*, pp.36–40.

Xia, F., Ge, H., Hou, Y., Sun, T., Chen, L., Zhang, G., Jiang, L., 2007. Multiresponsive surfaces change between superhydrophilicity and superhydrophobicity. *Advanced Materials, 19*(18), pp.2520–2524.

Xiao, G., Zhang, X., Zhang, W., Zhang, S., Su, H., Tan, T., 2015. Visible-light-mediated synergistic photocatalytic antimicrobial effects and mechanism of Ag-nanoparticles@ chitosan–TiO2 organic–inorganic composites for water disinfection. *Applied Catalysis B: Environmental, 170–171*, pp.255–262.

Xiong, S., Kong, L., Huang, J., Chen, X., Wang, Y., 2015. Atomic-layer-deposition-enabled nonwoven membranes with hierarchical ZnO nanostructures for switchable water/oil separations. *Journal of Membrane Science, 493*, pp.478–485.

Xu, L., Hu, Y.L., Pelligra, C., Chen, C.H., Jin, L., Huang, H., Sithambaram, S., Aindow, M., Joesten, R., Suib, S.L., 2009. ZnO with different morphologies synthesized by solvothermal methods for enhanced photocatalytic activity. *Chemistry of Materials, 21*(13), pp.2875–2885.

Xu, L., Zhuang, W., Xu, B., Cai, Z., 2011. Fabrication of superhydrophobic cotton fabrics by silica hydrosol and hydrophobization. *Applied Surface Science*, *257*(13), pp.5491–5498.

Xu, Z., Zhao, Y., Wang, H., Wang, X., Lin, T., 2015. A Superamphiphobic coating with an ammonia-triggered transition to superhydrophilic and superoleophobic for oil–water separation. *Angewandte Chemie*, *54*(15), pp.4527–4530.

Xue, C.H., Jia, S.T., Zhang, J., Tian, L.Q., Chen, H.Z., Wang, M., 2008. Preparation of superhydrophobic surfaces on cotton textiles. *Science and Technology of Advanced Materials*, *9*(3), p.035008.

Xue, Z., Cao, Y., Liu, N., Feng, L., Jiang, L., 2014. Special wettable materials for oil/water separation. *Journal of Materials Chemistry A*, *2*(8), pp.2445–2460.

Xue, Z., Wang, S., Lin, L., Chen, L., Liu, M., Feng, L., Jiang, L., 2011. A novel superhydrophilic and underwater superoleophobic hydrogel-coated mesh for oil/water separation. *Advanced Materials*, *23*(37), pp.4270–4273.

Yang, D., Paul, B., Xu, W., Yuan, Y., Liu, E., Ke, X., Wellard, R.M., Guo, C., Xu, Y., Sun, Y., Zhu, H., 2010. Alumina nanofibers grafted with functional groups: A new design in efficient sorbents for removal of toxic contaminants from water. *Water Research*, *44*(3), pp.741–750.

Yang, G., Yan, Z., Xiao, T., 2012a. Preparation and characterization of SnO2/ZnO/TiO2 composite semiconductor with enhanced photocatalytic activity. *Applied Surface Science*, *258*(22), pp.8704–8712.

Yang, J., Song, H., Yan, X., Tang, H., Li, C., 2014. Superhydrophilic and superoleophobic chitosan-based nanocomposite coatings for oil/water separation. *Cellulose*, *21*(3), pp.1851–1857.

Yang, J., Yin, L., Tang, H., Song, H., Gao, X., Liang, K., Li, C., 2015. Polyelectrolyte-fluorosurfactant complex-based meshes with superhydrophilicity and superoleophobicity for oil/water separation. *Chemical Engineering Journal*, *268*, pp.245–250.

Yang, J., Zhang, Z., Xu, X., Zhu, X., Men, X., Zhou, X., 2012b. Superhydrophilic–superoleophobic coatings. *Journal of Materials Chemistry*, *22*(7), pp.2834–2837.

Yogendra, K., Naik, S., Mahadevan, K.M., Madhusudhana, N., 2011. A comparative study of photocatalytic activities of two different synthesized ZnO composites against Coralene Red F3BS dye in presence of natural solar light. *International Journal of Environmental Sciences and Research*, *1*(1), pp.11–15.

Yong, J., Chen, F., Yang, Q., Du, G., Shan, C., Bian, H., Farooq, U., Hou, X., 2015. Bioinspired transparent underwater superoleophobic and anti-oil surfaces. *Journal of Materials Chemistry A*, *3*(18), pp.9379–9384.

Zeng, Y., Wu, W., Lee, S., Gao, J., 2007. Photocatalytic performance of plasma sprayed Pt-modified TiO2 coatings under visible light irradiation. *Catalysis Communications*, *8*(6), pp.906–912.

Zhan, H., Tian, H., 1998. Photocatalytic degradation of acid azo dyes in aqueous TiO2 suspension I. The effect of substituents. *Dyes and Pigments*, *37*(3), pp. 231–239.

Zhang, F., Zhang, W.B., Shi, Z., Wang, D., Jin, J., Jiang, L., 2013a. Nanowire-haired inorganic membranes with superhydrophilicity and underwater ultralow adhesive superoleophobicity for high-efficiency oil/water separation. *Advanced Materials*, *25*(30), pp.4192–4198.

Zhang, H., Quan, X., Chen, S., Zhao, H., Zhao, Y., 2006a. Fabrication of photocatalytic membrane and evaluation its efficiency in removal of organic pollutants from water. *Separation and Purification Technology*, *50*(2), pp.147–155.

Zhang, L., Huang, T., Zhang, M., Guo, X., Yuan, Z., 2008. Studies on the capability and behavior of adsorption of thallium on nano- Al2O3. *Journal of Hazardous Materials*, *157*(2–3), pp.352–357.

Zhang, Q., Wang, N., Zhao, L., Xu, T., Cheng, Y., 2013b. Polyamidoamine dendronized hollow fiber membranes in the recovery of heavy metal ions. *ACS Applied Materials and Interfaces*, *5*(6), pp.1907–1912.

Zhang, S., Cheng, F., Tao, Z., Gao, F., Chen, J., 2006b. Removal of nickel ions from wastewater by Mg(OH)2/MgO nanostructures embedded in Al2O3 membranes. *Journal of Alloys and Compounds*, *426*(1–2), pp.281–285.

Zhang, X., Zhi, D., Zhu, W., Sathasivam, S., Parkin, I.P., 2017. Facile fabrication of durable superhydrophobic SiO2/polyurethane composite sponge for continuous separation of oil from water. *RSC Advances*, *7*(19), pp.11362–11366.

Zhao, J., Yang, X., 2003. Photocatalytic oxidation for indoor air purification: A literature review. *Building and Environment*, *38*(5), pp.645–654.

# 10

# *Nanohybrid Graphene Oxide for Advanced Wastewater Treatment*

**Efstathios V. Liakos, Ilias T. Sarafis, Athanasios C. Mitropoulos, and George Z. Kyzas**

**CONTENTS**



## 10.1 Introduction

Graphene is considered to be the "hottest" material of recent years due to its numerous properties (mechanical, optical, environmental, etc). In 1986 Boehm et al. [1] described in details the structure of graphite having a single atomic sheet [1]. The major turn in graphene history happened during the 2000s, where it was proved that 2-D crystals (like graphene) did not have thermodynamic stability suggesting their non-existence in room temperature (conditions) [2]. Particularly, a graphene sheet is thermodynamically unstable if its size is less than about 20 nm (graphene is the least stable structure until about 6,000 atoms) and becomes the most stable fullerene (as within graphite) only for molecules larger than 24,000 atoms [3].

The pioneer of graphene science is Konstantin Novoselov who successfully isolated and characterized with various techniques an exfoliated graphene mono-layer [4]; A.K. Geim and K.S. Novoselov were awarded in 2010 with the highest honor (the Nobel Prize) for their impact on graphene science. But what is graphene? The reply was clear and given by IUPAC: Graphene is a carbon layer (single) of graphite, having a structure/nature similar or analogous with an aromatic hydrocarbon (polycyclic) of quasi infinite size [5]. That means that graphene is a flat mono-layer of hybridized $sp^2$ atoms of carbon, which are densely packed into an ordered two-dimension honeycomb network [6]. A hexagonal unit cell of graphene comprises two equivalent sub-lattices of carbon atoms, joined together by sigma ($\sigma$) bonds with a carbon–carbon bond length of 0.142 nm [7]. Each carbon atom in the lattice has a $\pi$-orbital that contributes to a delocalized network of electrons, making graphene sufficiently stable compared to other nanosystems [8]. The applicability of graphene is based on an advantageous network provided by this material: the combination of high three-dimensional aspect ratio and large specific surface area, superior mechanical stiffness and flexibility, remarkable optical transmittance, exceptionally high electronic and thermal conductivities, impermeability to gases, as well as many other supreme properties. Due to all of the above, Novoselov characterized it as a miracle material [9].

One of the most important applications of graphene in wastewater treatment is the use of the oxidized form of graphene (graphene oxide). The large-scale production of functionalized graphene at low cost should result in good adsorbents for water purification [10]. This is due to the two-dimensional layer structure, large surface area, pore volume and presence of surface functional groups in these materials; the inorganic nanoparticles also prevent adsorbent aggregation. Water, as it is known (i.e. from several good handbooks), can be treated and purified by multiple techniques, such as desalination, filtration, membranes, flotation, adsorption, disinfection, sedimentation. Certainly, adsorption holds advantages over other methods (various of which will be shown in the following), such as for example ease of operation and comparatively low cost. Adsorption is the surface phenomenon where pollutants are adsorbed on the surface of a material (adsorbent) via physical and/or chemical forces. It depends on many factors such as temperature, solution pH, concentration of pollutants, contact time, particle size, temperature, nature of the adsorbate and adsorbent etc. Apart from its use in adsorption, graphene oxide (GO) can be used as a supplement in membrane technology and especially nanofiltration. So, in this chapter the use of graphene oxide is analyzed for wastewater treatment and many examples for membranes (anti-fouling properties) and adsorbents are given.

## 10.2 Synthesis Procedures

Before analyzing further, it is mandatory to give a definition of graphene composites. Graphene composites are considered to be all graphene-based materials which have been modified (grafting with reactive groups, functionalizations with polymers, complexes with other sources, etc.).

GO, which is considered to be the most known graphene composite material, is the result of the chemical exfoliation of graphite. It is a highly oxidized form of graphene, consisting of numerous and different-type oxygen functionalities. Many theories have been developed in the past for the determination of the exact chemical structure of GO [11, 12]. This is mainly because of the complexity of the material (including sample-to-sample variability), and of course its amorphous, berthollide character i.e., non-stoichiometric atomic composition [13]. The Lerf–Klinowski model describes a theory according to which, the carbon plane in GO is decorated with hydroxyl and epoxy(1,2-ether) functional groups [14]. The consideration for the existence of some carbonyl groups is correct, most likely as carboxylic acids along the sheet edges but also as organic carbonyl defects within the sheet [15, 16]. The synthesis of GO is based on three preparation methods: (i) Brodie's [17], (ii) Staudenmaier's [18], or Hummers' method [19]. The major part of all methods is the chemical exfoliation of graphite using an oxidizing agent in the presence of mineral acid. Two methods (Brodie's and Staudenmaier's methods) apply a combination of $KClO_4$ with $HNO_3$ in order to oxidize graphite. Hummers' method uses the addition of graphite to potassium permanganate and $H_2SO_4$. The oxidation of graphite breaks up the π-conjugation of the stacked graphene sheets into nanoscale graphitic sp2 domains surrounded by highly disordered oxidized domains ($sp^3$ C\C) as well as defects of carbon vacancies [20]. The GO sheets produced consist of phenol, hydroxyl, and epoxy groups mainly at the basal plane and carboxylic acid groups at the edges [21], and can thus readily exfoliate to form a stable, light brown colored, single layer suspension in water [20].

Next, some major synthesis procedures are analyzed in detail to describe the nanohybrid graphene oxide preparation.

### 10.2.1 Synthesis of Graphene Oxide

In their study, Suresh Kumar et al. [22] used graphite for the synthesis of GO. Specifically, 3 g of graphite flakes were dissolved in a mixture of $H_2SO_4/H_3PO_4$ (360/40 mL) and then 18 g of $KMnO_4$ were added slowly under continuous stirring for 12 h at 50°C. After this process, in order to exfoliate GO into single layers, the mixture was cooled down to room temperature and it was mixed with a solution ~400 mL ice water with concentration 3 mL of 30% hydrogen peroxide followed by sonication for 0.5 h. In addition, the diluted mixture was centrifuged at 10,000 rpm for 15 min. Multiple washings (30% HCI) were made with water in order to remove the residues from the solid particles. Then, the experimental process has a vacuum drying process at room temperature for 12 h [22].

The next step is the preparation of Mn $Fe_2O_4$ nanoparticles. Briefly, 100 mL of deionized water was used to dilute 0.845 g of $MnSO_4$ $H_2O$ and 2.7 g of $FeCl_3{\cdot}6H_2O$ (the molar ratio of Mn:Fe in the mixture was 1:2). The mixture was then continuously stirred, and heated at temperature 80°C. To slightly modify the pH of the mixture to 10.5, drops of 8 M NaOH were inserted slowly to the same temperature for 5 min and then the mixture was cooled down to room temperature. To separate the blackish precipitates, a magnetical process was used. The obtained blackish precipitates were then washed with an excess of water to remove the unreacted quantity and then followed by washing with propanol. Finally, the obtained blackish precipitates were dried at room temperature for 24 h [22].

The final step is the synthesis of GO–$MnFe_2O_4$ nanohybrids. So, in the final experimental process, 0.5 g of GO were inserted into 400 mL of water and then dispersed by ultrasonication for 5 min. 0.845 g of $MnSO_4$ $H_2O$ and 2.7 g $FeCl_3{\cdot}6H_2O$ were added to the colloidal GO mixture and after this, this solution was stirred for 30 min (until the increase of the mixture temperature at 80°C). To increase the pH of the solution, similarly 8 M NaOH was added dropwise and heated up to the same temperature with the mixture. The experimental reaction process was then continued for 5 min and then the yielded mixture cooled down to room temperature. A magnetical process was then used to separate the nanohybrids particles. The nanohybrids particles were washed with an excess of $H_2O$ and propanol and dried for 24 h at room temperature [22].

### 10.2.2 Preparation of GO and Reduced GO Mixture

In their study, Lee et al. [23] synthesized neat GO from natural graphite according to a modified Hummers' method. Potassium permanganate, nitric acid, sulfuric acid, and hydrogen peroxide were used as reagents to oxidize graphite to GO. 0.12 g of graphite flakes were added to a mixture of concentrated $H_2SO_4/HNO_3$ (6 mL:0.132 mL) and 0.72 g potassium permanganate was then inserted gradually to the mixture under stirring for 2 h to temperature 35–45°C. After this step, the mixture was heated up to 100°C and stirred for 30 min. After that, 42 mL of water and 1.2 mL hydrogen peroxide were added. The mixture was cooled down to room temperature to remove the acidic supernatant from the mixture and then centrifuged at 13,000 rpm for 15 min. After the removal of the acidic supernatant by centrifugation process, distilled water was inserted into the mixture to dilute the acidic remnant from GO. Afterward, to redisperse the pellets, the mixture of GO pellets and distilled water was vortex-stirred for 1 min. This process was done to obtain a nearly neutral aqueous mixture and was continuously repeated with centrifugation and vortex, alternatively. The graphite oxide pellets were then added to N-methyl pyrrolidone (NMP) and followed by a sonication process. A tip sonicator was used for the exfoliation of graphite oxide to GO. Ultrasonication was also performed in an ice water bath for 1 h. Subsequently, the preparation of reduced graphene oxide (rGO), was achieved by reducing dispersed GO in the resultant homogeneous GO mixture using hydrazine ($N_2H_4$, Sigma-Aldrich) in the ratio of 0.7 mg/mg of GO. Then, the mixture was stirred for 10 h at 80°C [23].

Then, to prepare the respective membranes, NMR was used for the exfoliation of graphene oxide by the sonication process. To obtain various GO concentrations (0.02, 0.05, 0.14, 0.20, and 0.39 wt.%), a NMR mixture of polysulfone (PSf) (15 wt.% PSf and 85 wt.% of NMR) was used in order to disperse the nanoplatelets of GO. The solution was then stirred to 60°C and kept at room temperature overnight. To remove the bubbles in the solutions, the mixture was sonicated for 1 h. Then an Elcometer 3570 (Micrometric Film Applicator) was used to cast the polymer mixture on a non-woven polyester fabric. Thence, to achieve an entire liquid–liquid de-mixing, a water bath was used for 24 h for the immersion of the produced membranes. The yielded PSf/GO membranes have ratio GO:PSf 0.16, 0.32, 0.92, 1.30, and 2.60 wt.% [23].

### 10.2.3 Synthesis of $Fe_3O_4$/rGO

In another study, Wang et al. [24] used 120 mg of graphite oxide, undergoing an ultrasonication process for 2 h to disperse in deionized water. The yielded exfoliated GO nanosheets were stuck to $Fe_3O_4$ nanoparticles with the application of a post-oxidation method. A solution of 1 M $FeCl_3{\cdot}6H_2O$ was added to GO and extra-sonicated for 1 h. Thence, with the use of a dropping funnel, a $KBH_4$ solution (1.65 M) was injected to the dispersion solution of GO under vigorous stirring. The reaction was finished after 4 h and the mixture was cooled down to room temperature. The synthesized black solution was filtered and washed several times with water and ethanol to remove residual acid and dissociative Fe(II), To get $Fe_3O_4$/rGO, the yielded solid with the addition of anhydrous ethanol was soaked for 60 min. Finally, the yielded solid residues were dried at 60°C in vacuum atmosphere [24].

The synthesis of polypyrrole (PPy)-decorated $Fe_3O_4$/rGO was completed by using ammonium persulfate (APS) as the oxidant Ppy was embedded on the $Fe_3O_4$/rGO composite surface. The process was conducted by *in situ* polymerization under non-acidic conditions. In order to form dispersion solution, the magnetic $Fe_3O_4$/rGO nanoparticles were first sonicated for 3 h into 100 mL hexadecyltrimethylammonium bromide (CTAB) solution. 1 mL pyrrole monomer was inserted and dissolved into this solution. Then, 20 mL of APS solution was dropped slowly, after cooling to 273 K and stirred at this temperature for 8 h. The yielded product was then washed with water and anhydrous ethanol and finally dried at 60°C in a vacuum oven. The modification process of the Ppy-$Fe_3O_4$/rGO composite is schematically illustrated in Figure 10.1. [24].

### 10.2.4 Preparation of Graphene Oxide

Zhibin Wu et al. [25] also followed the modified Hummers method for the preparation of GO from graphite powder (particle size ≤ 30 μm). Briefly, 1 g of sodium nitrate and 2 g of graphite were placed in a 250 mL beaker. During the stirring process in an ice bath, 46 mL of $H_2SO_4$ (98%) were added. 1 g of sodium nitrate and 6 g potassium permanganate were slowly dropped under vigorous stirring (283 K) into the suspension. After vigorous stirring for 2 h in the ice-bath, the mixture was stirred for 30 min to 303 K. As the reaction process continued, the color of the solution gradually transformed to brownish paste. The new yielded paste was dissolved into 92 mL of ultrapure water under vigorous agitation for 30 min to 368 K. The consequence of this was to change the color of the suspension to bright yellow. 10 mL of hydrogen peroxide (30 wt.%) was added to the solution to terminate the reaction process and was then stirred to room temperature for 2 h. This process was achieved, when the suspension temperature was 333 K. The precipitate after the centrifugation process was washed repeatedly with 5% hydrochloric acid, in order to remove residual metal ions, and then washed with deionized water to remove the sulfate ions. Finally, the yielded precipitate-GO was sonicated and dried at 338 K under vacuum [25].

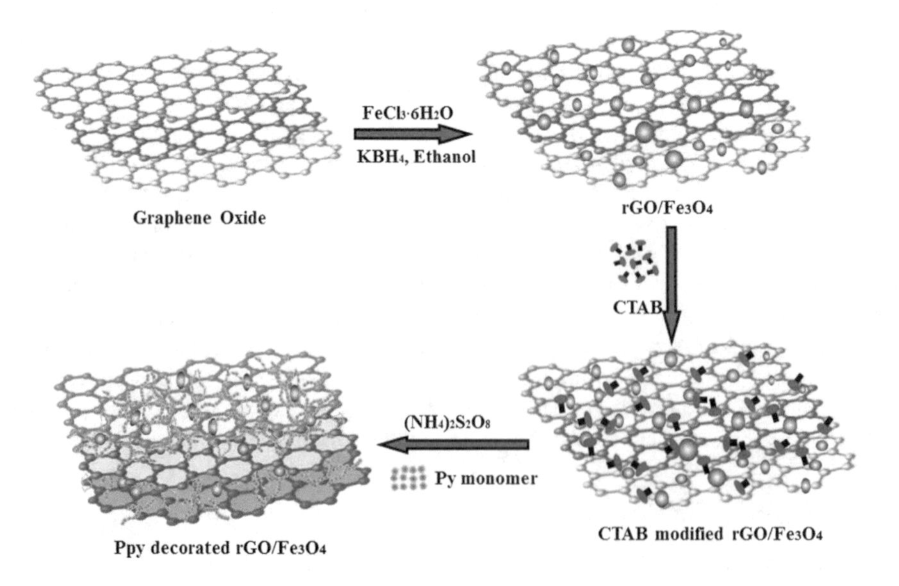


**FIGURE 10.1 (See color insert.)** Schematic illustration of the ternary composites preparation. (Reprinted with permission from Hou Wang et al. [24]. © (2015) Elsevier.)

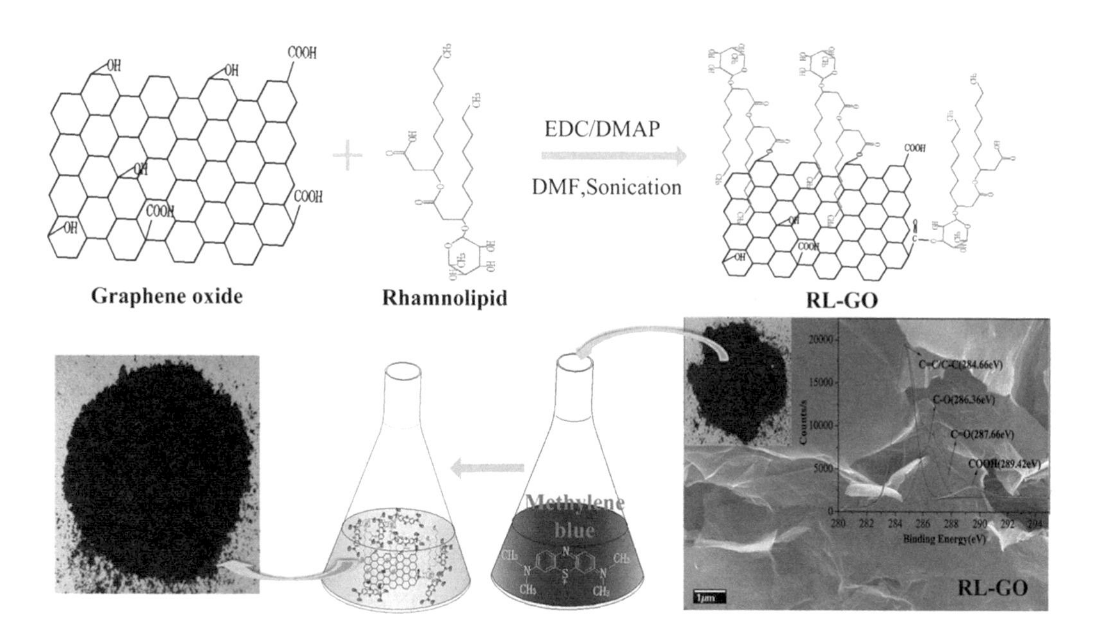


**FIGURE 10.2** **(See color insert.)** Schematic depiction of the formation of RL–GO and application for removal of MB. (Reprinted with permission from Zhibin Wu et al. [25]. © (2014) Elsevier.)

As presented in Figure 10.2, the synthesized RL–GO hybrid composite was prepared by one-step ultrasonication process. 200 mg of GO was dissolved into 100 mL of dimethylformamide (DMF) and followed by 1 h sonication. Then, 600 mg of rhamnolipid was added to the GO suspension and sonicated under vigorous stirring until complete dissolution. After that, 1 g of N-(3-dimethylaminopropyl-N-ethylcarbodiimide) hydrochloride and 200 mL of 4-(dimethylamino) pyridine were added to the suspension. The reaction process under stirring and ultrasonication was allowed to progress for over 3 h. Then, the procedure continued under vigorous stirring, by adding methanol, to the precipitation of the suspension. A black solid precipitate was yielded by the centrifugation process. The process contained washing of black solid precipitate five times with anhydrous ethanol and then twice with ultrapure water. Finally, freeze-drying under vacuum was used on the yielded rhamnolipid-functionalized GO composite. The obtained GO composite was suitable for adsorption applications [25].

### 10.2.5 Synthesis of Magnetic Chitosan Functionalized with Graphene Oxide

A very promising approach regarding the use of nanohybrid GO in wastewater treatment is the combination with magnetic chitosan by Fan et al. [26]. 25 mL of double distilled water, 1.7312 g of $FeCl_3$ 6 $H_2O$ and 0.6268 g of Fe $Cl_2$ 4 $H_20$ were inserted to ammonia solution, which was purged with N and stirred in a bath of water for 3 h at 90°C. With magnetic separation, the magnetic particles used in the chitosan coating were yielded. In order to give a final content of 1.5% (w/v), 0.3 g chitosan was diluted in 30 mL 3% of acetic solution. A four-neck rounded bottom flask was used in order to add the chitosan mixture and 0.1 g magnetic particles. After that, 2.0 mL neat glutaraldehyde was inserted into the reaction flask to blend with the mixture and was stirred for 2 h at 60°C. The obtained precipitate was purged with petroleum ether, ethanol, and distilled water until the pH was approximately 7, and then was dried at 50°C in a vacuum oven. The yielded product was magnetic chitosan [26].

GO was synthesized from purged natural graphite according to the modified Hummers' method. Briefly, potassium permanganate and natural graphite were agitated for 12 h at 60°C with the mixed acid ($HNO_3$:$H_2SO_4$= 1:9). After that, it was put in the mixed liquor hydrogen peroxide and agitated for 1 h, in order to yield a bright yellow material. This obtained bright yellow material was pureed with 2 M HCI in order to remove bisulfate ions and then extra washed with an abundant amount of water in order to make the solution neutral. In addition, the GO was obtained by a centrifuging process and was then dried in a vacuum desiccator [26].

So, for the functionalized final material to be prepared, a special procedure must be followed. Ultrapure water was used to sonicate GO for 3 h, in order to obtain a GO dispersion. Then in order to activate the carboxyl groups of GO, a mixture of 0.05 M NHS and 0.05 M EDC was inserted into the GO dispersion under continuous stirring for 2 h. Then, in order to maintain the pH of the obtained solution at 7.0 dissolved sodium hydroxide was used. After that, the activated GO mixture and 0.1 g of magnetic chitosan were inserted in a flask and dispersed by ultrasonic dispersion in distilled water for 10 min, and the blended solutions were agitated for 2 h at 60°C. The yielded precipitate was washed with 2% (w/v) NaOH and distilled water in turn until pH was about 7.0. Furthermore, the yielded material was collected and separated with a magnet and it was dried at 50°C in a vacuum oven so as to obtain the final MCGO product. Figures 10.3 and 10.4 present the synthesis of magnetic chitosan and GO and their application, respectively [26].

## 10.3 Characterizations

Some AFM and SEM results from a prepared GO hybrid material indicate that the average flake size was about 2 μm. The average thickness results of the GO flake were ~1 nm as presented in Figure 10.5a. The average size was ~6 nm as presented in Figure 10.5b and refers to the NP grown on the surface of GO as measured by AFM. Furthermore, XRD patterns of GO were confirmed using Cu Ka radiation ($\lambda = 1{,}542$ Å). Figure 10.6a presents the GO-XRD pattern with a diffraction peak at scattering angle $2\theta = 9.4°$ which attributed to (001) plane with an interlayer separation ~9.5 Å. Figure 10.6b presents the XRD pattern from the NP [22].

The Debye–Scherrer equation was used to calculate the average particle size, which seems to be ~11 nm (corresponding to (311) line). Figure 10.6c displays the XRD pattern of the GONH, revealing both the NPs and the GO flakes diffraction peaks. It presents that during the experimental preparation of the

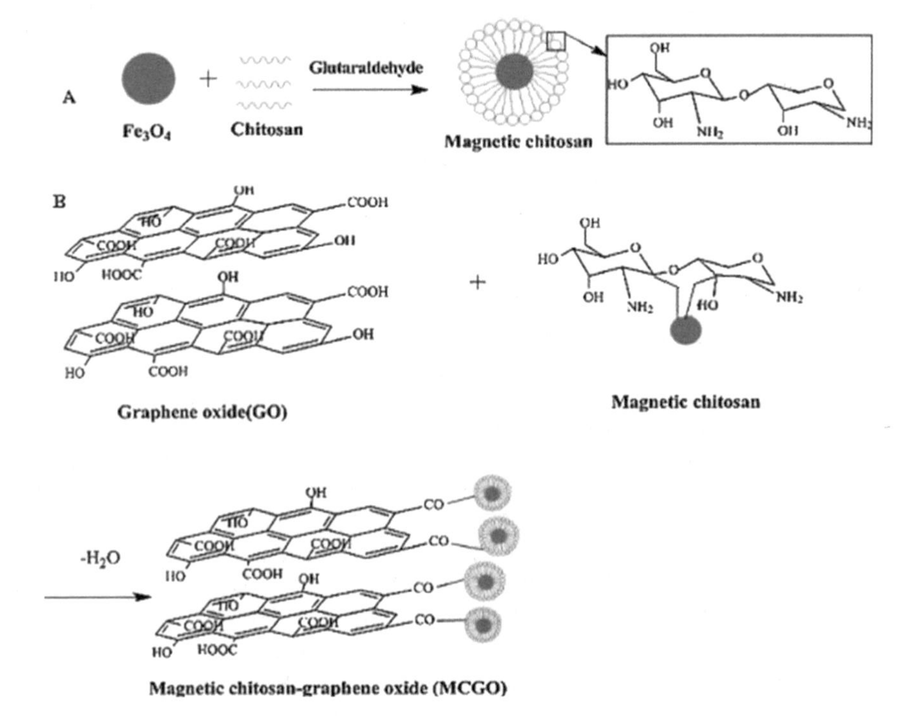


**FIGURE 10.3** Schematic depiction of the formation of (a) magnetic chitosan and (b) MCGO. (Reprinted with permission from Lulu Fan et al. [26]. © (2012) Elsevier.)

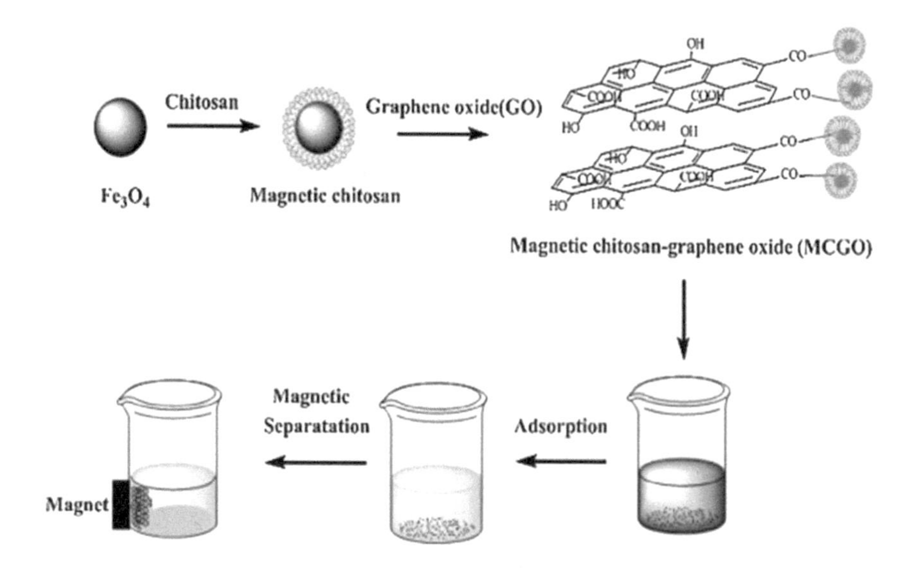


**FIGURE 10.4** Synthesis of MCGO and their application for adsorption of Methylene blue (MB) with the use of an external magnetic field. (Reprinted with permission from Lulu Fan et.al [26]. © (2012) Elsevier.)

nanohybrids, because of the partial reduction of the GO had as a sequence, there is a decrease of the peak due to the (001) reflection plane. The diffraction peaks corresponding to $-Fe_2O_3$ and $-MnO_2$, on the surface of the NPs were absent in the XRD pattern, and that confirms the formation of NPs and GONH. XRD analysis pictures displayed that the average size of the nanoparticles was ~7.5 nm.

The peak of GO, in the XRD pattern of the nanohybrids, was reduced very much due to the fact, that the NP grown on the epiphany of GO averts its restacking. Furthermore, the fact that the size of the NPs was decreased could be attributed to the reason that one side of the NPs growth was blocked when grown *in situ* onto the graphene surface. Figure 10.6c presents the typical SEM surface micrograph of the synthesized nanohybrid. The coverage of the NPs on graphene was confirmed to be uniform for different samples [22].

Figure 10.7a presents the FTIR spectrum of GO, where the characteristic absorption peaks were observed at 1236, 1046, 1415, 1620 and 1729 $cm^{-1}$, which can be attributed to the epoxy C–O stretching

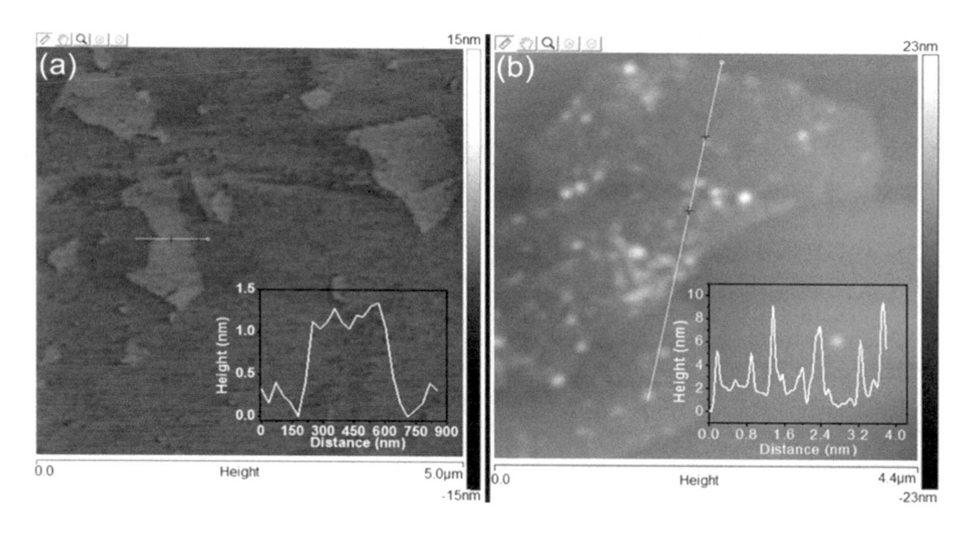


**FIGURE 10.5** AFM image of (a) GO and (b) GONH. (Reprinted with permission from Suresh Kumar et al. [22]. © (2014) American Chemical Society.)

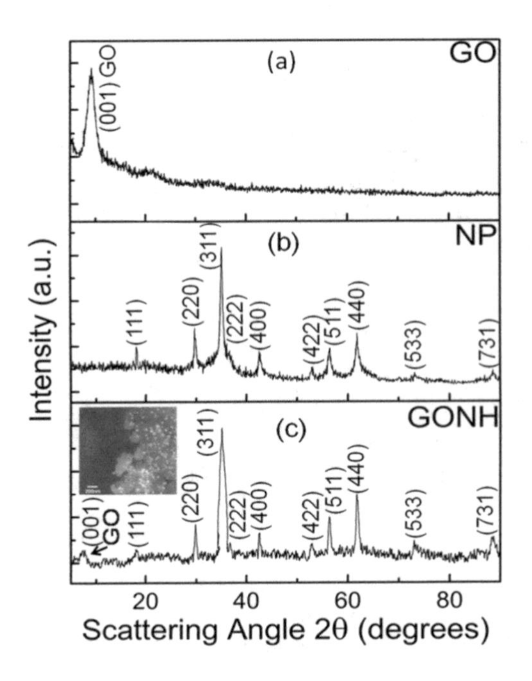


**FIGURE 10.6** XRD pattern of graphene oxide (a) NP; (b) GONH; (c) and typical FESEM image of GONH (inset of [c]). (Reprinted with permission from Suresh Kumar et al. [22], © (2014) American Chemical Society.)

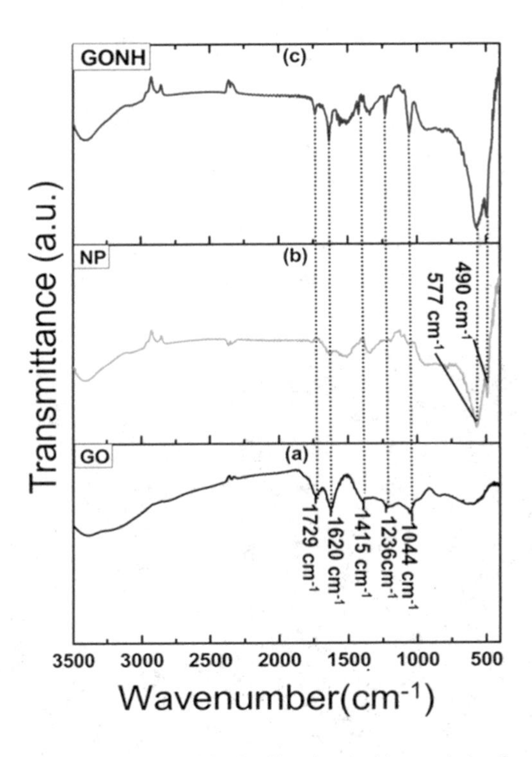


**FIGURE 10.7** FTIR spectra of (a) GO; (b) NP; (c) GONH. (Reprinted with permission from Suresh Kumar et al. [22]. © (2014) American Chemical Society.)

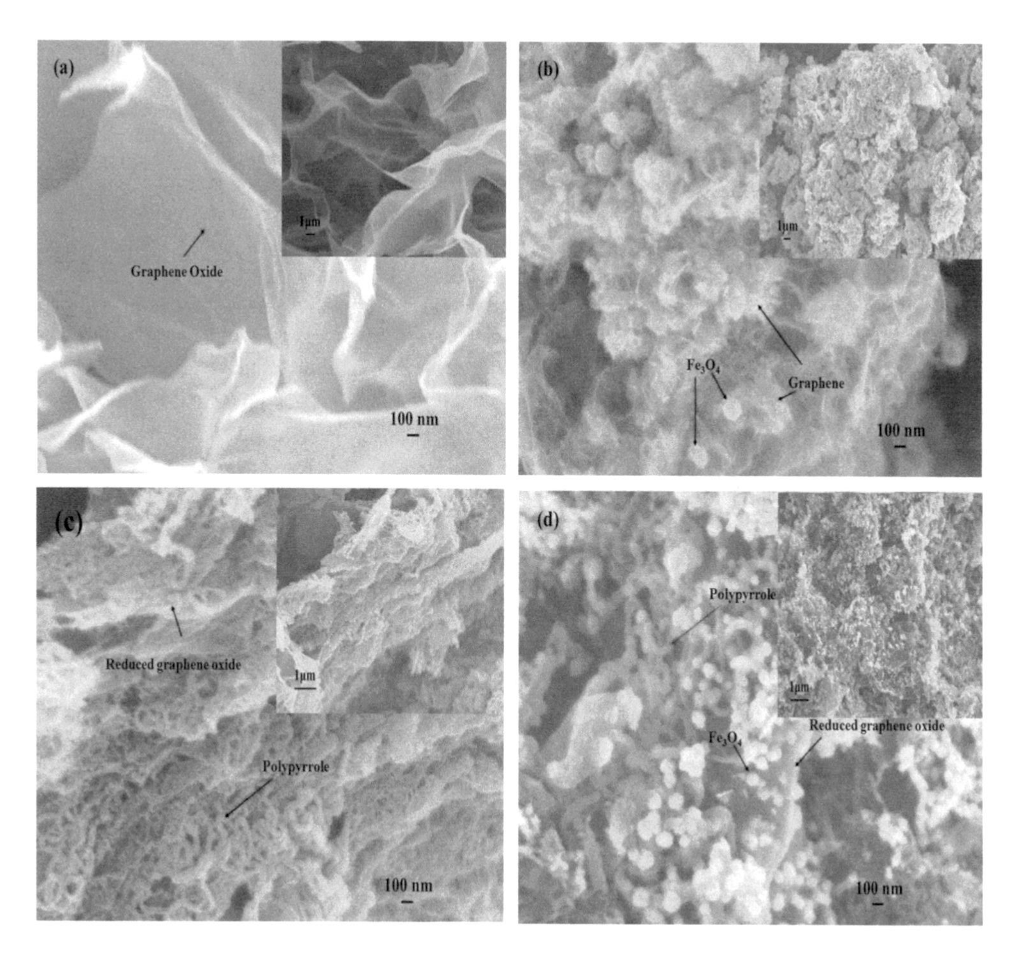


**FIGURE 10.8** SEM images of (a) GO, (b) $Fe_3O_4$/rGO, (c) rGO/Ppy and (d) Ppy-$Fe_3O_4$/rGO composites. The inset is the corresponding lower magnification images. (Reprinted with permission from Hou Wang et al. [24]. © (2015) Elsevier.)

vibrations, alkoxy C–O stretching, O–H deformation, and due to adsorbed water molecules, C=C in-plane stretching vibrations or C=O stretching, respectively.

All these measurements, clearly confirm the formation of GO and the appearance into the graphene skeleton of oxygenated functionalities, which have been used in the experimental process for the growth of magnetic NPs. Furthermore, the pH of the mixture affects the charge to the –OH and –COOH groups. Figure 10.7b displays the absorption peaks of NP at 577 and 490 $cm^{-1}$ due to manganese ferrite (metal–O stretching vibrations). FTIR measurements confirm the formation of $MnFe_2O_4$ NPs. Figure 10.7c presents the FTIR spectrum of the GONH and presents both the characteristic peaks of NP and GO. These results confirm the successful preparation of the nanohybrids [22].

Some interesting microscopic images can also be a tool for the examination of the surface of nanohybrid graphenes. Figure 10.8 depicts the low and high magnification images of GO, $Fe_3O_4$/rGO, Ppy/rGO, and Ppy-$Fe_3O_4$/rGO. As it can be presented in Figure 10.8a, the prepared GO was sheet-like in shape morphology, with a slick surface and single layer structure with wrinkled edges. The formation of loosely packed $Fe_3O_4$ nanoparticles onto rGO sheets, as displayed in Figure 10.8b, was achieved while a large number of granular particles attach on the rGO surface after the addition of $Fe^{3+}$. The diameter of the spherical $Fe_3O_4$ nanoparticles was 50–80 nm. Figure 10.8c presents the rough surface morphology of the Ppy decorated in the rGO sheets. In Figure 10.8d the ternary hybrids are observed [24].

## 10.4 Adsorption Application of Nanohybrid Graphene Oxides

### 10.4.1 Adsorption Isotherms

It is important to shape the most proper adsorption equilibrium connection in the endeavor to find new materials to access a perfect adsorption framework with [27], which is key for steady prediction of adsorption factors and correlation (mostly quantitative) of material's behavior for different adsorbent systems (or for different experimental parameters) [28, 29].

Explaining the phenomenon through which the preservation (or release) or mobility of a substance from the aqueous porous media or aquatic environments to a solid-phase at a persistent temperature and pH takes place, in broad-spectrum, an adsorption isotherm is an invaluable curve [30, 31]. The mathematical association which establishes a significant role towards the modeling analysis, operational design and applicable practice of the adsorption systems is normally represented by plotting a graph between solid-phase and its residual concentration [32].

When the concentration of the solute remains unchanged as a result of zero net transfer of solute adsorbed and desorbed from sorbent surface, a condition of equilibrium is achieved. These associations between the equilibrium concentration of the adsorbate in the solid and liquid phase at persistent temperature are defined by the equilibrium sorption isotherms. Linear, favorable, strongly favorable, irreversible and unfavorable are some of the isotherm shapes that may form. Understandings of the mechanism of adsorption and the surface properties, along with the extent of the affinity of the adsorbents, are delivered by the physicochemical parameters accompanied by the fundamental thermodynamic suppositions [33].

In the past, some basic adsorption equations were described in detail revealing their diversity (Langmuir (L), Freundlich (F), Langmuir–Freundlich (L–F), Brunauer–Emmett–Teller, Redlich–Peterson (R–P), Dubinin–Radushkevich, Temkin, Toth, Koble–Corrigan, Sips, Khan, Hill, Flory–Huggins and Radke–Prausnitz isotherm) [34]. Although thermodynamics are very important in adsorption theory explanations, the first successful approach was the adsorption dynamics (kinetics); thermodynamics were then explained in the second stage. Thermodynamics correlate the stage of dynamic equilibrium (having both adsorption and desorption movements/rates) [35, 36]. The differentiation in the physical explanations of adsorption parameters for each model separately shows that the single approach was not adequate [37]. Nowadays, the most widely used models are L, F, and L–F (see Table 10.1).

The uptake of each pollutant in equilibrium ($Q_e$ in mg/g) is estimated using the mass balance equation [38]:

$$Q_e = \frac{(C_0 - C_e)V}{m} \tag{10.1}$$

**TABLE 10.1**

Widely Used Isotherm Equations in Adsorption Works

| Isotherm | Equation | Ref |
|---|---|---|
| Langmuir | $Q_e = \frac{Q_m K_L C_e}{1 + K_L C_e}$ | [39] |
| Freundlich | $Q_e = K_F (C_e)^{1/n_f}$ | [40] |
| Langmuir–Freundlich | $Q_e = \frac{Q_m K_{LF} C_e^{1/n}}{1 + K_{LF} C_e^{1/n}}$ | [38] |

$Q_m$ (mg/g) is the maximum amount of adsorption; $K_L$ (L/mg) is the L adsorption equilibrium constant; ($K_F$ ($mg^{1-1/n}$ $L^{1/n}$/g) is the F constant representing the adsorption capacity; $n_f$ (dimensionless) is the F constant depicting the adsorption intensity; $K_{LF}$ ($(L/mg)^{1/n}$) is the L–F constant; n (dimensionless) is the L–F heterogeneity constant.

where $C_0$ and $C_e$ in mg/L express the initial and equilibrium concentration of pollutants; *V* in L is the adsorbate volume; *m* in g expresses the adsorbent's mass.

Although the L and F isotherms were firstly introduced about 90 years ago, they still remain the two most commonly used adsorption isotherm equations. Their success undoubtedly reflects their ability to fit a wide variety of adsorption data quite well. The L model represents chemisorption on a set of well-defined localized adsorption sites, having the same adsorption energies independent of surface coverage and no interaction between adsorbed molecules. L isotherm assumes mono-layer coverage of adsorbate onto adsorbent. F isotherm gives an expression encompassing the surface heterogeneity and the exponential distribution of active sites and their energies. This isotherm does not predict any saturation of the adsorbent surface; thus infinite surface coverage is predicted, indicating physisorption on the surface. One of the most promising extensions to L and F isotherms is the L–F equation, which is a general purpose isotherm for heterogeneous surfaces. The L–F isotherm is essentially F isotherm with a suitable asymptotic property in the sense that it approaches a maximum at high concentrations. Moreover, L–F trends to the L isotherm, when the "heterogeneity parameter" (b) is set to unity.

### 10.4.2 Examples

A recent example of nanohybrid GO for adsorption application was given by Kumar et al. [22]. Aqueous mixtures of metal ions Pb(II), As(III) and As(V), were prepared with different concentrations, and treated with NPs and GONH. The degree of surface charge ionization and speciation of the adsorbate is affected by the pH of the mixture which finally is a very important factor. In the case of Pb(II), As(V), and As(III), the adsorption data of NPs and GONH are revealed in Figure 10.9a with concentration 100 mg/L. The pH values showed that the maximum adsorption was for As(III) at pH = 6.5, As(V) at pH=4 and Pb(II) at pH=5. At the initial and final pH values there was a small change [22].

After finding the optimum pH, the adsorption on different quantities of metal ions was investigated. In order to find the maximum adsorption $Q_m$, the equilibrium data were fitted to the L adsorption isotherm. Figure 10.9b presents the measurements for As(V). The adsorbent results were $Q_m$ = 136 mg/g for NP and 207 mg/g for GONH. As(III) showed $Q_m$ = 97 mg/g for NP and 146 mg/g for GONH [22]. The maximum adsorption capacity for Pb(II) by adding NP and GONH adsorbents was 488 and 673 mg/g, respectively.

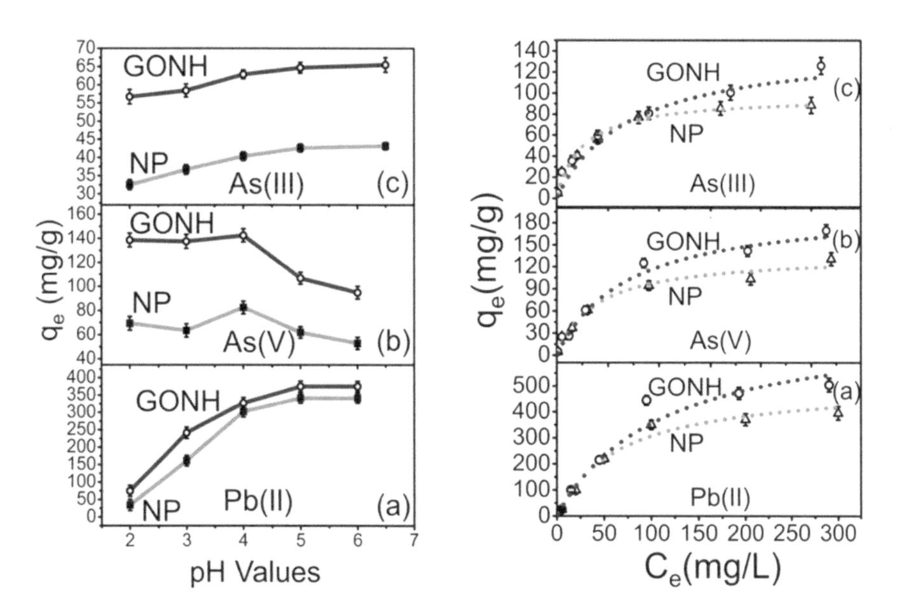


**FIGURE 10.9** (a) Effect of pH at different pH values with initial concentration of 100 mg/L for (a) Pb(II), (b) As(V), and (c) As(III); (b) Langmuir adsorption isotherm for (a) Pb(II), (b) As(V), and (c) As(III) with varied initial heavy metal ion concentration ranging from 0 to 400 mg/L pH values were kept at 5, 4 and 6.5 for Pb(II), As(V) and As(III), respectively. (Reprinted with permission from Suresh Kumar et al. [22]. © (2014) American Chemical Society.)

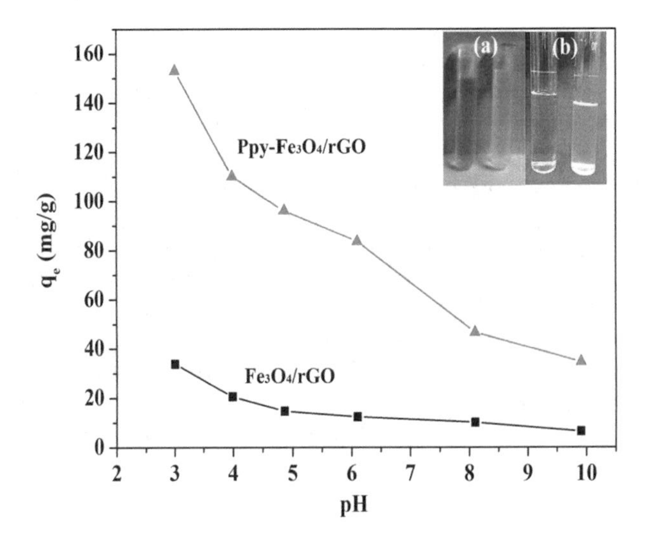


**FIGURE 10.10 (See color insert.)** Cr (VI) adsorption and the pH effect for $Fe_3O_4$/rGO and Ppy-$Fe_3O_4$/rGO. C (Cr(VI)) initial = 48.4 mg/L, $m/V$ = 0.25 g/L, T = 303 K. In the inserted photograph presented: (a) the Cr(VI) removal ability of Ppy-$Fe_3O_4$/rGO and (b) the chemical experiment for $SO_4^{-2}$ and after $Ba^{2+}$ injection. (Reprinted with permission from Hou Wang et al. [24]. © (2015) Elsevier.)

On the metal solution chemistry (i.e., redox, hydrolysis, reactions, polymerization, and coordination) the pH of the solution has a strong influence, as well as the ionic state of the functional groups on the surface of the adsorbent. Figure 10.10 displays the Cr(VI) adsorption for the initial pH in aqueous solution. The removal capacity of Ppy-$Fe_3O_4$/rGO composite which has a lower surface area is higher compared with only $Fe_3O_4$/rGO nanoparticles. Thus, it is determined that during the adsorption process the surface area is not the crucial factor. The Cr(VI) adsorption happens mainly through electrostatic attraction for $Fe_3O_4$/rGO composite. In addition, the Cr (VI) removal by the Ppy-$Fe_3O_4$/rGO composite, during the adsorption process, besides electrostatic interaction, may still be included in ion exchange and chemical reduction process [24].

One very interesting finding is the effect of adsorbent's dosage. So, Figure 10.11 depicts the effects of the RL–GO (or GO) dosage on the removal of MB. It was observed that when the dosage of RL–GO (or GO) increased, the removal percentage of MB increased, too. Moreover, it was observed that when the adsorbent dose was increased, the removal percentage increased slightly. It also could be observed that, when the adsorbent dose was increased, the removal capacity of MB decreased. This happened because

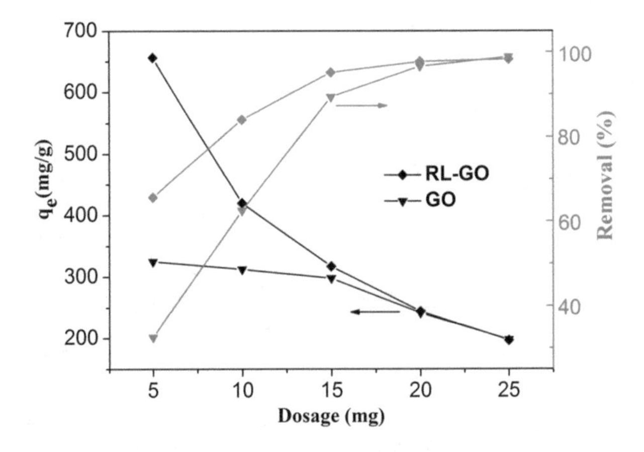


**FIGURE 10.11** Effect of adsorbent dosage on the adsorption capacity of MB. (Reprinted with permission from Zhibin Wu et al. [25]. © (2014) Elsevier.)

**TABLE 10.2**

Adsorption Kinetic Parameters of MB onto RL–GO (pH Value, 7.0; RL–GO dose, 400 mg/L, Temperature 298 K)

| | Pseudo-First Order Kinetic | | | | Pseudo-Second Order Kinetic | | | |
|---|---|---|---|---|---|---|---|---|
| $C_0$ (mg/L) | $q_e$, exp (mg/g) | $k_1$ (1/*h*) | $q_e$, cal (mg/g) | $R^2$ (–) | $q_e$, exp (mg/g) | $k_2$ (g/mg.h) | $q_e$, cal (mg/g) | $R^2$ (–) |
| 100 | 242 | 0.19 | 55.31 | 0.688 | 242 | 0.0208 | 241.55 | 0.999 |
| 150 | 305 | 0.28 | 152.56 | 0.985 | 305 | 0.0068 | 309.60 | 0.999 |
| 200 | 365 | 0.25 | 127.13 | 0.926 | 365 | 0.0093 | 364.96 | 0.999 |

a higher adsorbent dose provides a large excess of the active sites, which have as a result a lower utility of the sites at a specific concentration of MB solution. Meanwhile, it was observed under the experimental conditions that the obtained RL–GO have superior properties compared to GO in MB adsorption capacity. This can be confirmed by the measurements from XRD and XPS. The results indicate that RL–GO have a bigger layer spacing and contained more oxygen to the functional groups compared with GO for MB adsorption [25].

Also, in the present study, in order to analyze the kinetics of MB adsorption onto RL–GO, the intra-particle diffusions, Boyd's film-diffusion, the pseudo-first-order, and pseudo-second-order models were used, with the purpose of examining three different initial MB concentrations. During the adsorption process, the pseudo-first-order model and its linearized-integral form were widely used. Table 10.2 presents the adsorption kinetic parameters of MB onto RL–GO [25].

The plots for MB adsorption are presented in Figure 10.12.

## 10.5 Applications of Nanohybrid Graphene Oxide Membranes

Wu et al. [41] studied the anti-fouling property of a $SiO_2$-GO/PSf hybrid membrane in order to examine the long term performance of hybrid membranes, the anti-fouling experiment was performed and operated three cycles. The anti-fouling properties of membranes were achieved by calculation of FRR, Rt, Rr and Rir values. The first cycle of membrane fouling and washing was employed in order to estimate the values of anti-fouling properties. Figure 10.13a presents the value of FRR for pure PSf membrane. Figure 10.13b presents that the value of Rir is as high as half of the total fouling. FRR value augments to 72%, when doping PSf membrane with $SiO_2$-GO and the value of Rir in maximum fouling simultaneously reduced sharply to one third. Finally, the $SiO_2$-GO/PSf hybrid membrane presents good performance stability through long time operation and the better anti-fouling ability, compared with the other hybrid membranes from GO and $SiO_2$ [41].

In another study, Zhu et al. [42] investigated the adsorption and desorption ability of PVDF/GO/LiCl nanohybrid membranes. In order to make the solution with dye for testing, 10 mg/L Rhodamine B particles were diluted in distilled water. The PVDF/GO/LiCl membrane surface functional groups consist mainly of hydroxyl and carboxyl groups, because of GO and LiCl synergistic effects. Figure 10.14 shows that the nanohybrid membranes of PVDF/GO/LiCl at pH 8 changed the surface zeta potential from −13.47 to −21.49 mV.

The Rhodamine B-ethanol solutions first became stable, and then the absorbance values were confirmed. Calculations confirmed that the rates of decolorization (DR, %) of all type of nanohybrid membranes exceeded 85%. DRs of M1–M5 nanohybrid membranes after 20 cycles still exceeded 80%. These results confirm that Rhodamine B adsorbed and desorbed better with nanohybrid membranes of PVDF/GO/LiCl, and could be used for dye removal. Figure 10.15 displays images of PVDF/GO/LiCl nanohybrid membranes [42].

Figures 10.16 and 10.17 display the flux decline curves at 100 kPa and flux recovery ratio of the M0–M5 PVDF/GO/LiCl nanohybrid membranes, respectively [42].

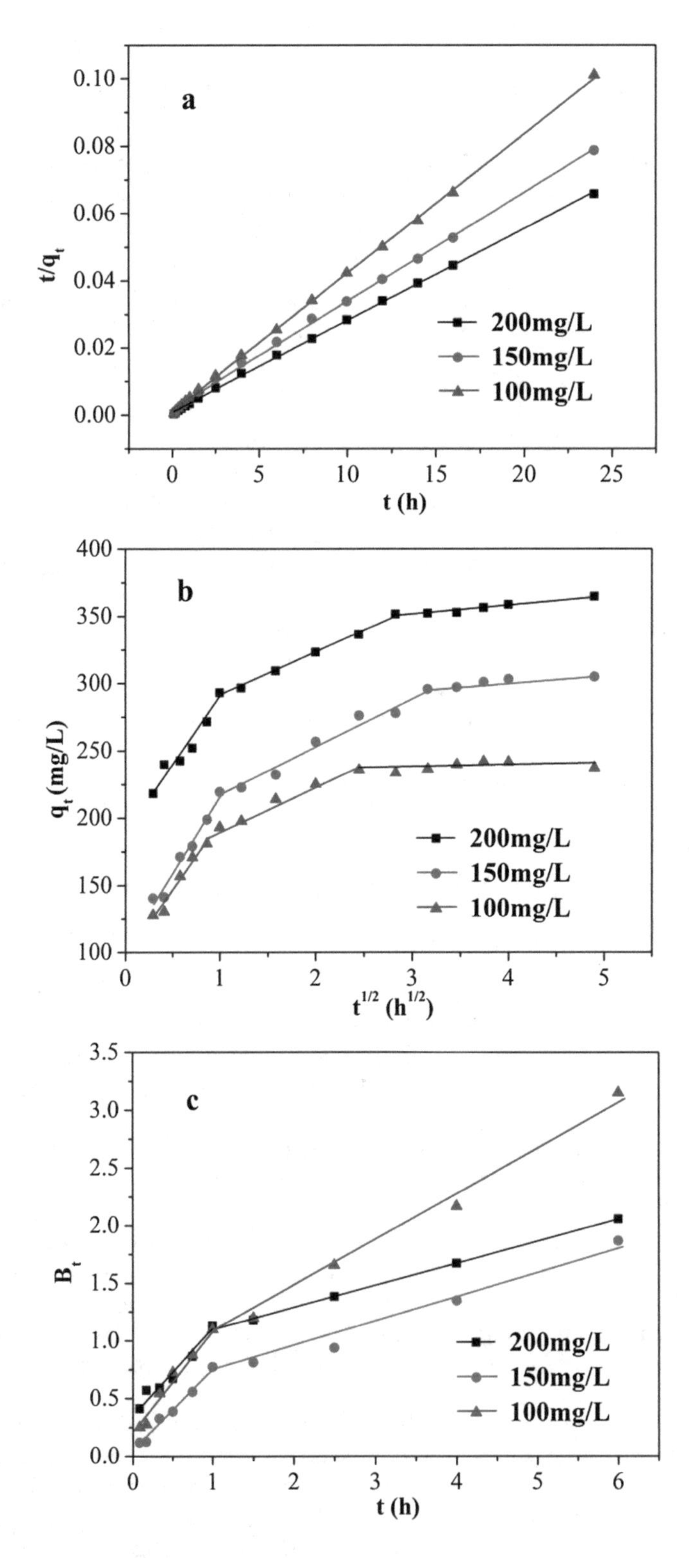


**FIGURE 10.12** (a) Pseudo-second-order plots for MB adsorption (b) Intra-particle diffusion plots for MB adsorption (c) Boyd plots for MB adsorption. (Reprinted with permission from Zhibin Wu et al. [25]. © (2014) Elsevier.)

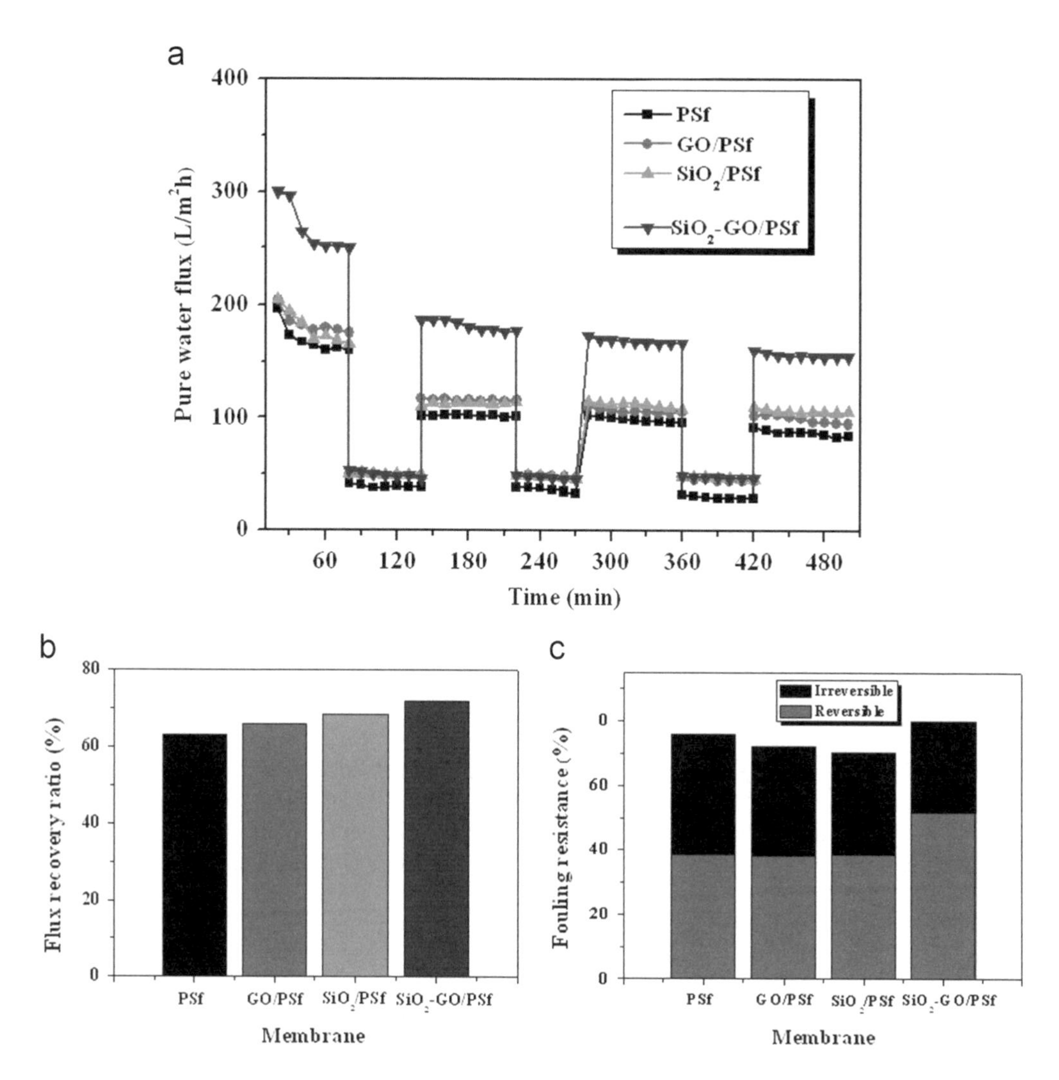


**FIGURE 10.13 (See color insert.)** Membranes with different inorganic materials and the time-dependent fluxes during an anti-fouling experiment with BSA filtration (1 mg/mL, pH=7.4) at 0.2 MPa, (b) FRR of the prepared membranes and (c) Fouling resistance of the prepared membranes. (Reprinted with permission from Huiqing Wu et al. [41]. © (2014) Elsevier.)

Peng et al. [43] prepared a reduced GO composite nanohybrid membrane. In order to obtain an acceptable rejection ratio at a high permeation flux, the composite films were fabricated with various mass ratios of $SiO_2$ to GO, and the rejection ratios and permeation fluxes while removing SDS/diesel oil/$H_2O$ emulsion and MB solution were measured.

As presented in Figure 10.18, the M1 PDA (GO/$SiO_2$ (mass ratio) – 2 mg/0.67 mg immersed into PDA for 24 h) membrane reveals at the pure water flux, a sharp decrease to 133.2 L $m^{-2}$ $h^{-1}$ compared with pure PVDF membrane with pure water flux 1389.1 L $m^{-2}$ $h^{-1}$. Occasionally, M1PDA membrane revealed an ultra-high removal ratio for diesel and MB, which is 99.2 and 99.8%, respectively. It's worth noticing that a visible change of rejection ratio is not observed with more content of $SiO_2$, while the rejection ratio of MB remained 98% with a mass ratio of $SiO_2$:GO 4:3.

As presented in Figure 10.19, the retention capacity of MB dye was limited, with a feed solution at a low pH value, but at a respectively high pH value the composite membrane is revealed to have a better ability to remove MB and could be explained by the functional groups of Si –OH, –COOH, –OH, in the membrane as receptors for hydrogen ions, with the charge of the composite membrane being negative at higher pH value. It's worth noting that the repulsion between the positive charge of the membrane and positive charge of MB lead to the low removal ratio of MB [43].

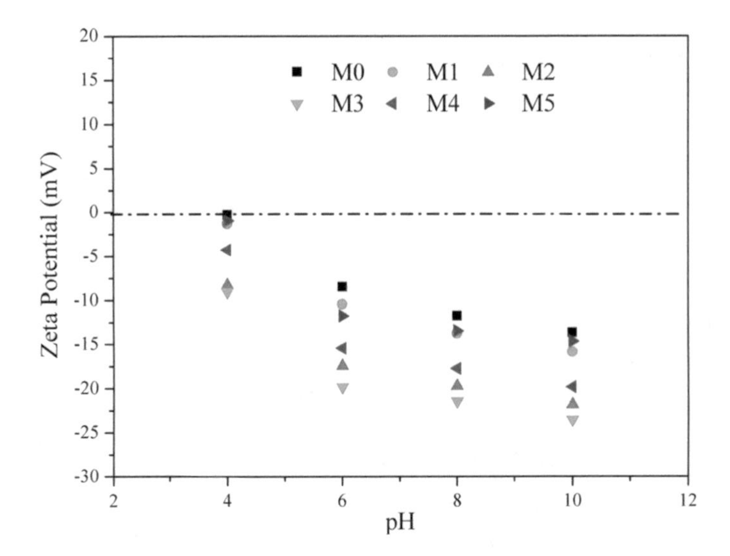


**FIGURE 10.14** **(See color insert.)** Zeta potential of PVDF/GO/LiCl nanohybrid membranes. (Reprinted with permission from Zhenya Zhu et al. [42]. © (2017) Elsevier.)

Moreover, Wang et al. [44] achieved the dye removal using GO and polyelectrolyte complexes (PECs). For the enhancement of the quality of the dye product, a well-known vital process is dye desalting and purification. Furthermore, when the dye molecules have impurities, one of the most promising technologies to clean them from impurities is nanofiltration (NF). Table 10.3 presents the retention of various dyes with the GO/PECs membranes. For instance, after incorporating GO, the retention of Congo red increased from 91.5% to 99.5% and this result is beneficial for the improvement of the membrane selectivity [44].

Table 10.4 shows that when increasing the processing temperature increased the permeate flux and decreased the water content in the permeate. The diffusion coefficients and the increases in the water vapor pressure on the feed side caused the increase of flux. Because the apparent activation energy of water ($E_p$ = 13.12 kJ mol$^{-1}$) was much lower than that of ethanol ($E_p$ = 28.13 kJmol$^{-1}$) a decrease in selectivity was revealed. The Arrhenius law was used in order to calculate the apparent activation energy [44].

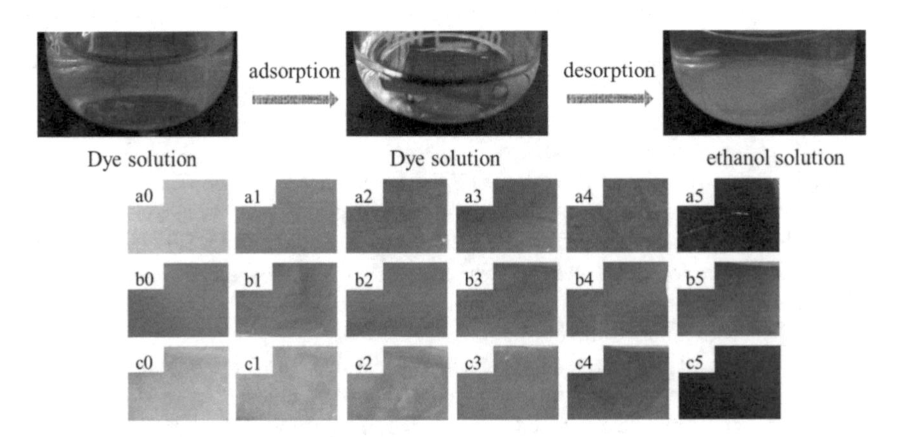


**FIGURE 10.15** **(See color insert.)** The images of PVDF/GO/LiCl nanohybrid membranes (a0–a5: pristine membranes, b0–b5: after immersing the Rhodamine B solution for 36 h, c0–c5: after decoloration by ethanol alcohol). (Reprinted with permission from Zhenya Zhu et al. [42]. © (2017) Elsevier.)

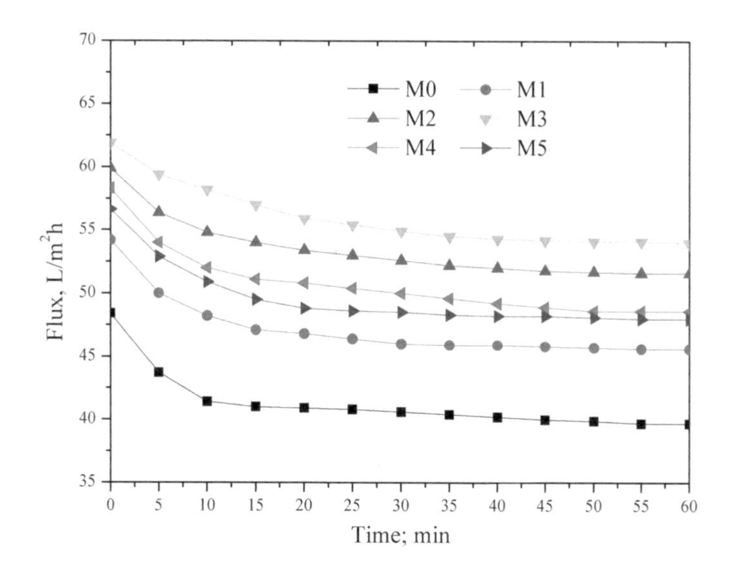


**FIGURE 10.16** **(See color insert.)** M0–M5 PVDF/GO/LiCl nanohybrid membranes and the flux decline curves at 100 kPa. (Reprinted with permission from Zhenya Zhu et al. [42]. © (2017) Elsevier.)

## 10.6 Conclusions

The most interesting findings can be summarized as: in the case of GONH, it was observed that for concentration 100 mg/L of As(V), Pb(II) and As(III) the greatest treatment was achieved at optimum adsorbent pH 4, 5 and 6.5, respectively. The results for As(V) were $Q_m = 136$ mg/g for NP and 207 mg/g for GONH, while As(III) presented $Q_m = 97$ mg/g for NP and 146 mg/g for GONH. Finally, for GONH

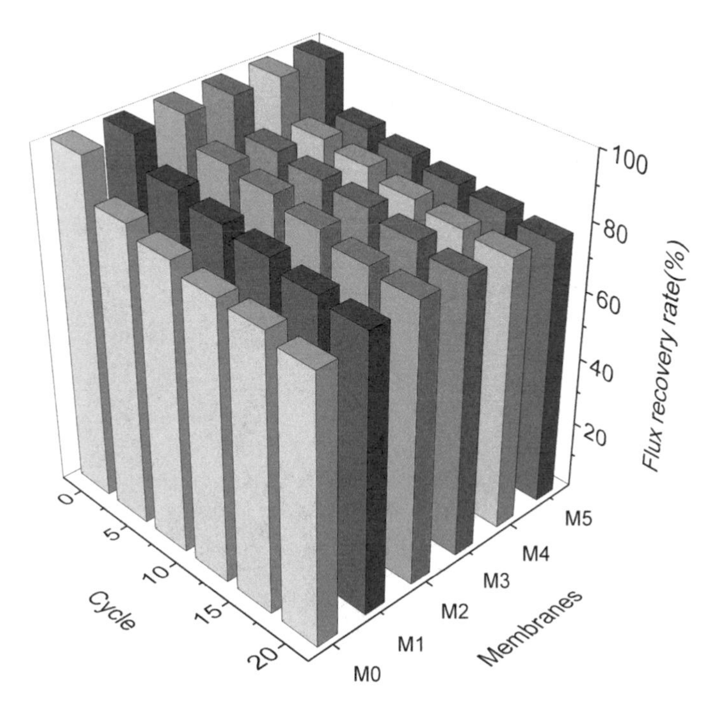


**FIGURE 10.17** Flux recovery ratio of PVDF/GO/LiCl nanohybrid membranes. (Reprinted with permission from Zhenya Zhu et al. [42]. © (2016) Elsevier.)

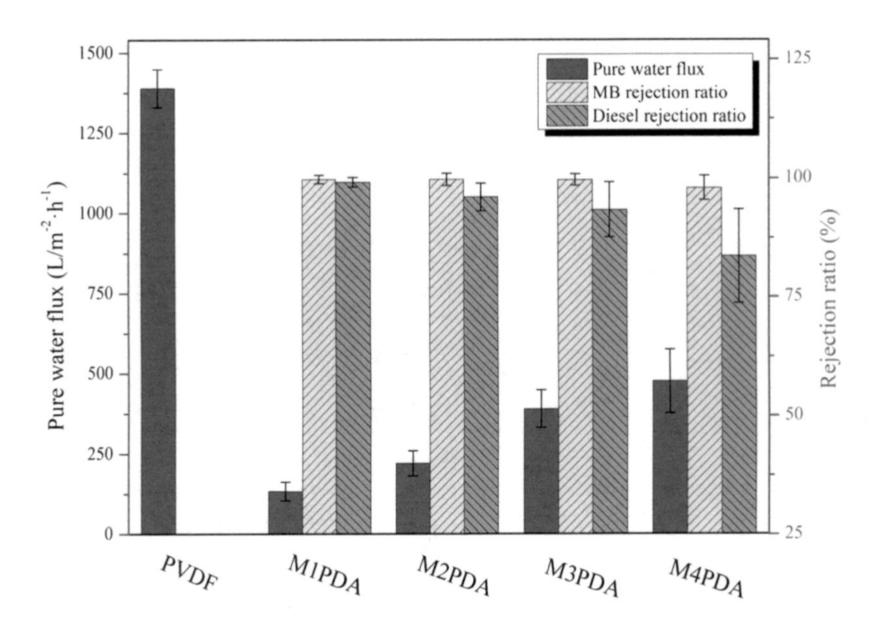


**FIGURE 10.18** Pure water flux and removal ratio under 0.09 MPa (membrane M1 PDA has ($GO/SiO_2$) (mass ratio) – 2 mg/0.67 mg and immersed into PDA for 24 h, membrane M2 PDA has ($GO/SiO_2$) (mass ratio) – 2 mg/1.34 mg and immersed into PDA for 24 h, membrane M3 PDA has ($GO/SiO_2$) (mass ratio) – 2 mg/2 mg and immersed into PDA for 24 h, membrane M4 PDA has ($GO/SiO_2$) (mass ratio) – 2 mg/2.67 mg and immersed into PDA for 24 h). (Reprinted with permission from Yixin Peng et al. [43]. © (2018) Elsevier.)

the maximum adsorption capacity was observed at 488 mg/g for NP and 673 mg/g for GONH. The Ppy-$Fe_3O_4$/rGO composite revealed better Cr(VI) adsorption with $q_e$=155 mg/g compared to $Fe_3O_4$/rGO composite with $q_e$=36 mg/g. The better adsorption capacity of Ppy-$Fe_3O_4$/rGO composite could be attributed to the electrostatic interaction, ion exchange and chemical reduction process. The above are some indicative (and not only those explained in this chapter) examples of how big the impact of nanohybrid in decontamination technology is.

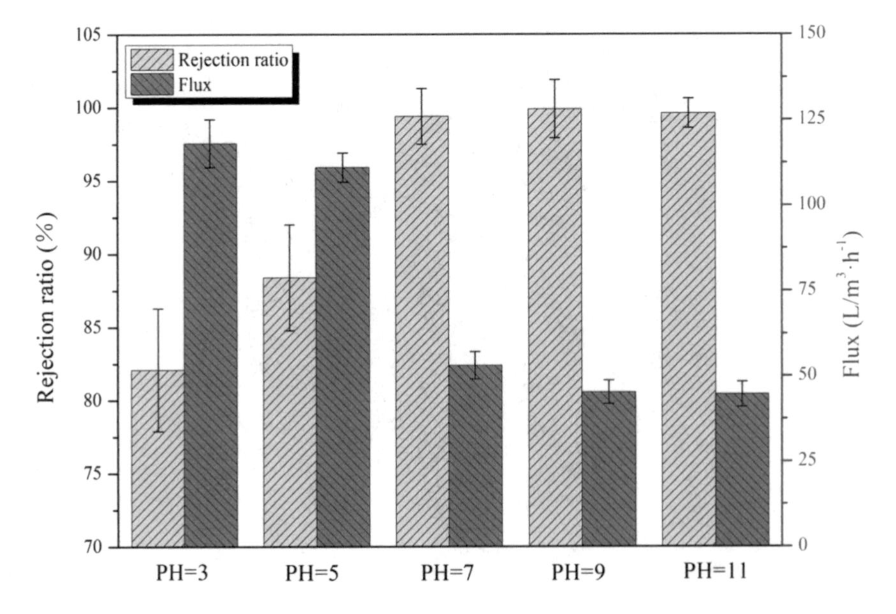


**FIGURE 10.19** MB removal ratio and flux by M1 PDA at different pH value. (Reprinted with permission from Yixin Peng et al. [43]. © (2018) Elsevier.)

**TABLE 10.3**

Dye Retention with Different Membranes

| Membrane | Dye Molecule | Retention (%) | Permeance ($kg/m^2$ h MPa) | Pressure (MPa) | Ref |
|---|---|---|---|---|---|
| (NaSS-AC)/PS | Acid red | 96 | 50–58 | 0.4 | [45] |
| CMCNa/PP | Congo red | 99.9 | 77.5 | 0.08 | [46] |
| CMCNa/PP | Sunset yellow | 82.2 | 86.3 | 0.08 | [46] |
| PES-TA | Methyl green | 98 | 20 | 0.5 | [47] |
| Sulfonated PES | Reactive orange 16 | 92.1 | 32 | 2 | [48] |
| PEI/PAA/PVA/GA[a] | Congo red | $91.5 \pm 0.4$ | $8.1 \pm 0.3$ | 0.5 | [44] |
| PEI/PAA/PVA/GA[a] | Methyl blue | $87.3 \pm 0.6$ | $8.5 \pm 0.2$ | 0.5 | [44] |
| PEI/PAA/PVA/GA[a] | Methyl orange | $83.4 \pm 0.7$ | $8.9 \pm 0.2$ | 0.5 | [44] |
| (PEI-modified GO)/PAA/PVA/GA[b] | Congo red | $99.5 \pm 0.1$ | $8.4 \pm 0.3$ | 0.5 | [44] |
| (PEI-modified GO)/PAA/PVA/GA[b] | Methyl blue | $99.3 \pm 0.1$ | $8.7 \pm 0.2$ | 0.5 | [44] |
| (PEI-modified GO)/PAA/PVA/GA[b] | Methyl orange | $87.6 \pm 0.3$ | $9.6 \pm 0.5$ | 0.5 | [44] |

[a] Hydrolysis conditions for PAN support membrane: hydrolysis temperature 65°C, hydrolysis time 30 min; preparative conditions: 30 min filtration time, 0.25 wt.% PEI aqueous solution, 0.05 wt.% PAA aqueous solution, 0.5 wt.% PVA aqueous solution, 3 wt.% glutaraldehyde aqueous solution.

[b] Hydrolysis conditions for PAN support membrane: hydrolysis temperature 65°C, hydrolysis time 30 min; preparative conditions: 30 min filtration time, 0.25wt.% PEI aqueous solution with 0.1 g/L GO, 0.05 wt.% PAA aqueous solution, 0.5 wt.% PVA aqueous solution, 3 wt.% glutaraldehyde aqueous solution.

**TABLE 10.4**

Pervaporation Performance of Different Membranes

| Membrane | Feed Solution | Water Content in Permeate (wt.%) | Flux (g/($m^2$ h)) | T (°C) | Ref |
|---|---|---|---|---|---|
| Chitosan/$TiO_2$ | 90% Ethanol/water | 96.8 | 278 | 80 | [49] |
| Zeolite-filled chitosan | 90% Ethanol/water | 99.7 | 16 | 30 | [49] |
| ZSM-5 incorporated in polyimide | 90% Isopropanol/water | 99.55 | 936 | 60 | [50] |
| PVA-zeolite 4A | 76.43% Ethanol/water | 99.55 | 936 | 60 | [51] |
| PSS-$ZrO_2$/PDDA-$ZrO_2$ | 95% Ethanol/water | 99.9 | 340 | 50 | [52] |
| PEI/PAA/PVA/GA | 95% Ethanol/water | $91.78 \pm 0.3$ | $229 \pm 17$ | 50 | [44] |
| ((PEI-modified GO)/PAA)$_1$/PVA/GA[a] | 95% Ethanol/water | $94.4 \pm 0.13$ | $368 \pm 15$ | 80 | [44] |
| ((PEI-modified GO)/PAA)$_1$/PVA/GA[a] | 95% Ethanol/water | $95.4 \pm 0.1$ | $268 \pm 10$ | 50 | [44] |
| ((PEI-modified GO)/PAA)$_1$/PVA/GA[a] | 95% Ethanol/water | $96.9 \pm 0.1$ | $196 \pm 13$ | 40 | [44] |
| ((PEI-modified GO)/PAA)$_1$/PVA/GA[a] | 95% Ethanol/water | $98.1 \pm 0.1$ | $156 \pm 4$ | 50 | [44] |

[a] Hydrolysis conditions for PAN support membrane: hydrolysis temperature 65°C, hydrolysis time 30 min; preparative conditions: 30 min filtration time, 0.25wt.% PEI aqueous solution with 0.1 g/L GO, 0.05 wt.% PAA aqueous solution, 0.5 wt.% PVA aqueous solution, 3 wt.% glutaraldehyde aqueous solution.

## REFERENCES

1. Boehm, H.P., R. Setton, and E. Stumpp, Nomenclature and terminology of graphite intercalation compounds. *Carbon*, 1986. **24**(2): p. 241–245.
2. Nemes-Incze, P., et al., Anomalies in thickness measurements of graphene and few layer graphite crystals by tapping mode atomic force microscopy. *Carbon*, 2008. **46**(11): p. 1435–1442.
3. Shenderova, O.A., V.V. Zhirnov, and D.W. Brenner, Carbon nanostructures. *Critical Reviews in Solid State and Materials Sciences*, 2002. **27**(3–4): p. 227–356.
4. Novoselov, K.S., et al., Electric field effect in atomically thin carbon films. *Science*, 2004. **306**(5696): p. 666–669.
5. IUPAC, Recommended terminology for the description of carbon as A solid (IUPAC Recommendations 1995). *Pure and Applied Chemistry*, 1995. **67**: p. 491.
6. Ivanovskii, A.L., Graphene-based and graphene-like materials. *Russian Chemical Reviews*, 2012. **81**(7): p. 571–605.
7. Avouris, P., and C. Dimitrakopoulos, Graphene: Synthesis and applications. *Materials Today*, 2012. **15**(3): p. 86–97.
8. Zhu, Y., et al., Graphene and graphene oxide: Synthesis, properties, and applications. *Advanced Materials*, 2010. **22**(35): p. 3906–3924.
9. Novoselov, K.S., et al., A roadmap for graphene. *Nature*, 2012. **490**(7419): p. 192–200.
10. Kemp, K.C., et al., Environmental applications using graphene composites: Water remediation and gas adsorption. *Nanoscale*, 2013. **5**(8): p. 3149–3171.
11. Compton, O.C., et al., Crumpled graphene nanosheets as highly effective barrier property enhancers. *Advanced Materials*, 2010. **22**(42): p. 4759–4763.
12. Compton, O.C., and S.T. Nguyen, Graphene oxide, highly reduced graphene oxide, and graphene: Versatile building blocks for carbon-based materials. *Small*, 2010. **6**(6): p. 711–723.
13. Dreyer, D.R., et al., The chemistry of graphene oxide. *Chemical Society Reviews*, 2010. **39**(1): p. 228–240.
14. Lerf, A., et al., Structure of graphite oxide revisited. *The Journal of Physical Chemistry B*, 1998. **102**(23): p. 4477–4482.
15. Pei, S., and H.M. Cheng, The reduction of graphene oxide. *Carbon*, 2012. **50**(9): p. 3210–3228.
16. Pei, Z., et al., Adsorption characteristics of 1,2,4-trichlorobenzene, 2,4,6-trichlorophenol, 2-naphthol and naphthalene on graphene and graphene oxide. *Carbon*, 2013. **51**(1): p. 156–163.
17. Brodie, B.C., On the atomic weight of graphite. *Philosophical Transactions of the Royal Society of London*, 1859. **149**: p. 249–259.
18. Staudenmaier, L., Verfahren zur Darstellung der graphitsaure. *Berichte Der Deutschen Chemischen Gesellschaft*, 1898. **31**(2): p. 1481–1487.
19. Hummers Jr, W.S., and R.E. Offeman, Preparation of graphitic oxide. *Journal of the American Chemical Society*, 1958. **80**(6): p. 1339.
20. Krishnan, D., et al., Energetic graphene oxide: Challenges and opportunities. *Nano Today*, 2012. **7**(2): p. 137–152.
21. Singh, V., et al., Graphene based materials: Past, present and future. *Progress in Materials Science*, 2011. **56**(8): p. 1178–1271.
22. Kumar, S., et al., Graphene oxide-MnFe2O4 magnetic nanohybrids for efficient removal of lead and arsenic from water. *ACS Applied Materials and Interfaces*, 2014. **6**(20): p. 17426–17436.
23. Lee, J., et al., Graphene oxide nanoplatelets composite membrane with hydrophilic and antifouling properties for wastewater treatment. *Journal of Membrane Science*, 2013. **448**: p. 223–230.
24. Wang, H., et al., Facile synthesis of polypyrrole decorated reduced graphene oxide-$Fe_3O_4$ magnetic composites and its application for the Cr(VI) removal. *Chemical Engineering Journal*, 2015. **262**: p. 597–606.
25. Wu, Z., et al., Adsorptive removal of methylene blue by rhamnolipid-functionalized graphene oxide from wastewater. *Water Research*, 2014. **67**: p. 330–344.
26. Fan, L., et al., Fabrication of novel magnetic chitosan grafted with graphene oxide to enhance adsorption properties for methyl blue. *Journal of Hazardous Materials*, 2012. **215–216**: p. 272–279.
27. Srivastava, V.C., et al., Adsorptive removal of phenol by bagasse fly ash and activated carbon: Equilibrium, kinetics and thermodynamics. *Colloids and Surfaces A: Physicochemical and Engineering Aspects*, 2006. **272**(1–2): p. 89–104.

28. Gimbert, F., et al., Adsorption isotherm models for dye removal by cationized starch-based material in a single component system: Error analysis. *Journal of Hazardous Materials*, 2008. **157**(1): p. 34–46.
29. Ho, Y.S., J.F. Porter, and G. McKay, Equilibrium isotherm studies for the sorption of divalent metal ions onto peat: Copper, nickel and lead single component systems. *Water, Air, and Soil Pollution*, 2002. **141**(1–4): p. 1–33.
30. Allen, S.J., G. McKay, and J.F. Porter, Adsorption isotherm models for basic dye adsorption by peat in single and binary component systems. *Journal of Colloid and Interface Science*, 2004. **280**(2): p. 322–333.
31. Limousin, G., et al., Sorption isotherms: A review on physical bases, modeling and measurement. *Applied Geochemistry*, 2007. **22**(2): p. 249–275.
32. Ncibi, M.C., Applicability of some statistical tools to predict optimum adsorption isotherm after linear and non-linear regression analysis. *Journal of Hazardous Materials*, 2008. **153**(1–2): p. 207–212.
33. Bulut, E., M. Özacar, and İ.A. Şengil, Adsorption of malachite green onto bentonite: Equilibrium and kinetic studies and process design. *Microporous and Mesoporous Materials*, 2008. **115**(3): p. 234–246.
34. Malek, A., and S. Farooq, Comparison of isotherm models for hydrocarbon adsorption on activated carbon. *AIChE Journal*, 1996. **42**(11): p. 3191–3201.
35. De Boer, J.H., *The Dynamical Character of Adsorption*, 1953. Oxford, UK: Oxford University Press.
36. Myers, A.L., and J.M. Prausnitz, Thermodynamics of mixed-gas adsorption. *AIChE Journal*, 1965. **11**(1): p. 121–127.
37. Ruthven, D.M., *Principles of Adsorption and Adsorption Processes*, 1984. Hoboken, NJ: John Wiley & Sons.
38. Tien, C., *Adsorption Calculations and Modeling*, 1994. Boston, MA: Butterworth-Heinemann.
39. Langmuir, I., The constitution and fundamental properties of solids and liquids. Part I. Solids. *The Journal of the American Chemical Society*, 1916. **38**(2): p. 2221–2295.
40. Freundlich, H., Over the adsorption in solution. *Zeitschrift für Physiologische Chemie*, 1906. **57**: p. 385–470.
41. Wu, H., B. Tang, and P. Wu, Development of novel SiO2–GO nanohybrid/polysulfone membrane with enhanced performance. *Journal of Membrane Science*, 2014. **451**: p. 94–102.
42. Zhu, Z., et al., Preparation and characteristics of graphene oxide-blending PVDF nanohybrid membranes and their applications for hazardous dye adsorption and rejssection. *Journal of Colloid and Interface Science*, 2017. **504**: p. 429–439.
43. Peng, Y., et al., A novel reduced graphene oxide-based composite membrane prepared via a facile deposition method for multifunctional applications: Oil/water separation and cationic dyes removal. *Separation and Purification Technology*, 2018. **200**: p. 130–140.
44. Wang, N., et al., Self-assembly of graphene oxide and polyelectrolyte complex nanohybrid membranes for nanofiltration and pervaporation. *Chemical Engineering Journal*, 2012. **213**: p. 318–329.
45. Akbari, A., et al., New UV-photografted nanofiltration membranes for the treatment of colored textile dye effluents. *Journal of Membrane Science*, 2006. **286**(1–2): p. 342–350.
46. Yu, S., et al., Application of thin-film composite hollow fiber membrane to submerged nanofiltration of anionic dye aqueous solutions. *Separation and Purification Technology*, 2012. **88**: p. 121–129.
47. Zhang, Q., et al., Positively charged nanofiltration membrane based on cardo poly(arylene ether sulfone) with pendant tertiary amine groups. *Journal of Membrane Science*, 2011. **375**(1–2): p. 191–197.
48. Van der Bruggen, B., et al., Mechanisms of retention and flux decline for the nanofiltration of dye baths from the textile industry. *Separation and Purification Technology*, 2001. **22–23**(1–2): p. 519–528.
49. Sun, H., et al., Surface-modified zeolite-filled chitosan membranes for pervaporation dehydration of ethanol. *Applied Surface Science*, 2008. **254**(17): p. 5367–5374.
50. Mosleh, S., et al., Zeolite filled polyimide membranes for dehydration of isopropanol through pervaporation process. *Chemical Engineering Research and Design*, 2012. **90**(3): p. 433–441.
51. Huang, Z., et al., Multilayer poly(vinyl alcohol)-zeolite 4A composite membranes for ethanol dehydration by means of pervaporation. *Separation and Purification Technology*, 2006. **51**(2): p. 126–136.
52. Zhang, G., J. Li, and S. Ji, Self-assembly of novel architectural nanohybrid multilayers and their selective separation of solvent-water mixtures. *AIChE Journal*, 2012. **58**(5): p. 1456–1464.

# 11

# *Titanium Dioxide-Based Nanohybrids as Photocatalysts for Removal and Degradation of Industrial Contaminants*

**Hafeez Anwar, Iram Arif, Uswa Javeed, and Yasir Javed**

**CONTENTS**



## 11.1 Introduction

Pollution-free water, air and land are becoming an important challenge for the survival of human society on the earth in this age of industrialization. To maintain quality of life, it is necessary to minimize the pollution level from the earth's atmosphere that originates from untreated hazardous disposal, the expulsion of materials into water and the discharge of polluted gases into the air. This may create chronic harm to the earth's environmental balance, which is vital for the life on earth (Chong et al., 2010, Schwarzenbach et al., 2010, Pelaez et al., 2012. Approximately 3.2 million people die worldwide every year due to contaminated water, imperfect sanitation and poor hygiene (Liu et al., 2012). This indicates that the purity or impurity of water directly influences the quality of life not only of humans but also of other living beings.

Water pollution is generated only if the load of pollution overreaches the intrinsic regenerative capacity of the water's origin (Schwarzenbach et al., 2010). The main sources of water pollution are toxic materials released from industry and from agricultural fertilizers, which highly influence water quality (Chong et al., 2010, Gupta et al., 2012). A series of studies have proved that industrial, especially textile, dyes are the major organic compounds which increase ecological pollution threats. All over the world, about 15–20% dye compounds are converted into textile waste during the production process and are freed as drainage into the green environment (Konstantinou and Albanis, 2004, Akpan and Hameed, 2009). More than 0.7 million tons of organic-based fabricated dyes are produced every year specifically for the manufacturing of plastics, food, leather goods, textile, electronics and industrial painting (Chong et al., 2010). In the last century, millions of colored chemical compounds have been produced, and approximately 10,000 of these were industrially fabricated (Konstantinou and Albanis, 2004, Akpan and Hameed, 2009). About 21–377 $m^3$ of water per ton of dye are used by the textile industry during production processes such as finishing and dyeing (Schwarzenbach et al., 2010, Gupta et al., 2012). Dyes and pigments are further divided into different types on the basis of the variety of colorants or additives, which creates variation in colors due to the change in the range of visible light absorption (Pagga and Brown, 1986, Rajeshwar et al., 2008). The different delineations between dyes and pigments are as follows: dyes can be soluble or partly soluble organics because most of them are extracted from animals and carbon-based plants (Pagga and Brown, 1986). These are colored components which form suspensions in the medium, whereas the pigments are not soluble compounds that have no chemical affinity for substrates to be colored (Pagga and Brown, 1986). Thousands of different types of pigments and dyes, mostly azo dyes, are used by industries. Furthermore, dyes are also categorized in terms of function or structure, or in terms of base, acid, reactive, dispersive, direct, cationic, anionic and so on.

With the passage of time, various effective methods, technologies and treatments have been introduced to treat industrial wastewater but lower-cost and less time-consuming procedures still attract the attention of researchers (Gupta et al., 2012). To clean water of industrial contaminations, photocatalysis is an effective method. The credit of discovering the phenomenon of photocatalysis goes to Fujishima and Honda, who used water splitting under UV irradiation in the presence of titanium dioxide ($TiO_2$) (Fujishima and Honda, 1972, Fujishima et al., 2000). After this discovery, a series of studies were carried out by chemical engineers, chemists and physicists to understand the typical procedure and enhance the efficiency of photocatalysis (Fujishima and Honda, 1972, Fujishima et al., 2000). These studies are mostly relevant to energy storage and the renewal of energy.

"Photocatalyst" is a combination of two terms: photo, meaning light, and catalyst, a suitable compound that subordinates the free activation enthalpy of the concerned reaction. Therefore, the photocatalyst can be elucidated so as to increase the rate of a photoreaction in the presence of a catalyst. The sensitized photoreaction can occur in two different ways when the catalyst absorbs the light: by energy transfer, by establishing an activated state of the reactant of interest, which could be easily oxidized in its ground state, or by electron ($e^-$) transfer, with the electron playing the role of an acceptor or donor. In this chapter, we mainly stress the importance and the requirements of the photocatalytic process for industry, followed by the properties of nanohybrids affecting the photocatalysis process. We start with the fundamental and basic mechanism of the photocatalysis process.

### 11.1.1 Basic Mechanism of Photocatalysis

When light of the required energy falls on the sensitizer, the photocatalyst oxidizes the organic molecules and an electron ($e^-$) from the valence band transfers to the higher energy band by creating a hole in the valence band. The transfer of the electron ($e^-$) into the conduction band creates a large number of electrons that are equally oxidized or reduced respectively which can be involved in redox activity. The oxidation of organic molecules is the most dominant feature of photocatalysis. For the oxidization of an organic molecule, the required redox potential is maintained by the location of the valence band and the organic molecule's redox potential with respect to a standard electrode. When the organic molecule will become negative redox potential rather than the redox level of photo generated hole, it may produce a cation radical of organic molecule by reducing a hole. This reaction leads to product formation. Secondly, when the hole reduces by water or absorbs $OH^-$ ions, $\bullet OH$ and other radicals are formed, with the power to oxidize organic matter. The electron is captured by the adsorbed oxygen from the air (Linsebigler et al., 1995). The mechanism of photocatalysis is shown in Figure 11.1.

Photocatalysis is considered as an advanced oxidation technology (AOT) in environmental chemistry (Lee and Park, 2013). The idea of AOT was introduced by Glaze and collaborators (Munter, 2001). They explained the principles of $TiO_2$ photocatalysis AOT as a procedure which encompasses the generation

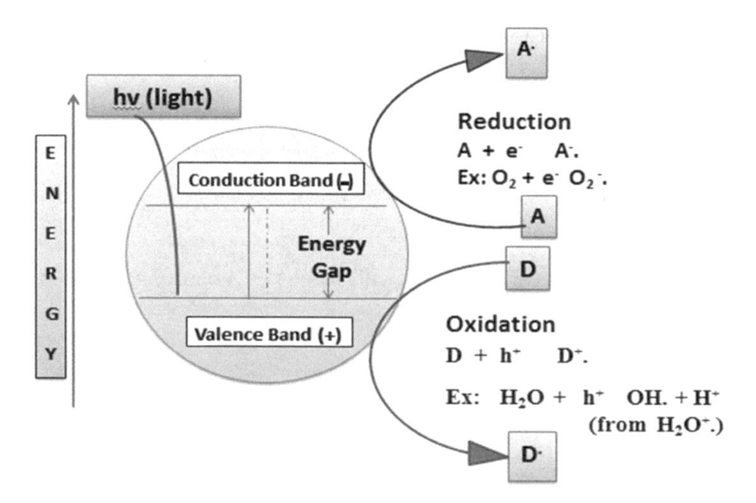


**FIGURE 11.1** Mechanism of photocatalysis.

and utilization of influential redox transient species, predominantly the hydroxyl radical. This radical can be generated during different processes and is highly effective in the oxidation of various species, especially for dyes. Amongst the non-photochemical AOT, one could differentiate among oxidation with $O_3/H_2O_2$, $O_3/OH^-$, electrochemical oxidation, Fenton's processes, plasma, radiolysis and ultrasonic treatments. In photochemical processes, one can determine the oxidation in supercritical and subcritical water, whereas in photo-Fenton's processes, heterogeneous photocatalysis will come after photoanalysis of water in UVV, $UV/O_3$, $UV/H_2O_2/O_3$. After flourine, the main radical which is able to react with every organic compound is hydroxyl (Gligorovski et al., 2015). Hydroxyl is a very strong oxidizing agent which reacts 106–1012 times quicker than other agents such as $O_3$ (Litter, 2005, Meichtry et al., 2007, Litter, 2009).

The rate and efficiency of the mineralization of an organic compound by the photocatalysis procedure mainly depends on factors such as the size, structure, shape and surface of the catalyst, light intensity, pH, reaction temperature, the concentration of wastewater and the amount of the catalyst (Fujishima et al., 2000, Wang et al., 2007, Rajeshwar et al., 2008, Rehman et al., 2009, Sasidharan et al., 2011).

The structure of the catalyst plays an important role in achieving the best photocatalyst response. For example, the higher efficiency of spherical-shaped zinc oxide (ZnO) samples was observed by Saravanan et al.; these samples have large surface areas compared to rod-shaped ZnO (Saravanan et al., 2013). Similarly, nanosized $TiO_2$ showed better recycling and water purification ability as compared to the bulk form (Cernuto et al., 2011, Han et al., 2014, Saravanan et al., 2017). Therefore, it concludes that, as the size of the catalyst decreases, the surface-to-volume ratio increases, which helps a larger number of molecules to accumulate on the surface of the catalyst. This leads to higher charge transfer and a higher number of active sites for reaction, so high photocatalytic activity can be achieved by using nanosized catalytic materials (Saravanan et al., 2013, Khoa et al., 2015).

There are basically two types of photocatalysis, i.e., homogenous photocatalysis and heterogeneous photocatalysis. In the first type, the same phase carries both the reactants and the photocatalysts. Generally, the photo-Fenton and ozone systems are included in homogeneous photocatalysis. •OH is the reactive species, which has been used for different reactions. In the second type of photocatalysis, catalysis has the catalyst in a different phase from the reactants. There are different types of reactions existing in the category of heterogeneous photocatalysis: metal deposition, mild or total oxidations, gaseous pollutant removal, dehydrogenation, water detoxification, and so on. There are reviews of heterogeneous photocatalysis topics having more than 1100 references which could be performed in different mediums: aqueous solutions, pure organic liquid phases and pure organic liquids. Typical heterogeneous catalysis can be divided into five different and independent steps, i.e., reactant transfer to the surface in the liquid phase, at least one of the reactants adsorbed, reaction in the adsorbed phase, products desorbed and products removed from the interface region. the photocatalytic process takes place in the phase of adsorption. Conventional catalysis is different from heterogeneous photocatalysis due to the activation mode, which is thermal for conventional and photonic for heterogeneous (Kaan et al., 2012, Ibhadon and Fitzpatrick, 2013).

## 11.2 The Importance and Requirements of the Photocatalytic Process for Industries

There are some methods which are used conventionally for the treatment of wastewater, such as flocculation, coagulation, reverse osmosis and adsorption process by using a large number of adsorbing materials (Kishimoto et al., 2007). One of the most important treatments for the removal and degradation of contaminants from wastewater is adsorption. This method is not only economical but preferable due to its capacity, efficiency and high applications (Mattson and Mark, 1971, Cheremisinoff and Ellerbusch, 1978). Researchers working on this technique have applied different absorbent materials for the degradation and removal of various pollutants (Lee and Low, 1989, Cowan et al., 1991, Groffman et al., 1992, Pollard et al., 1992, Kesraoui-Ouki et al., 1993, Rodda et al., 1993, Periasamy and Namasivayam, 1994, Srivastava et al., 1997, Gupta et al., 1998, Jain and Sikarwar, 2006).

Semiconductor photocatalysis has become a popular and interesting treatment over conventional techniques of wastewater treatment from the previous 25 years in the fields of energy generation and ecological harm. Since the invention of water splitting to oxygen and hydrogen by UV-irradiated $TiO_2$ after Fujishima and Honda's (Fujishima and Honda, 1972) efforts at applying different semiconductor photocatalysts (Fujishima et al., 2000), great attention has been fixated on its ecological uses such as water and air decontamination treatment by the application of a semiconductor photocatalyst. Light sources and semiconductor photocatalysts play an important role in the process of photocatalysis.

Semiconductor materials could be extensively utilized for the photocatalysis process (Egerton, 1997). The great advantage of the photocatalysis process is due to its supportive electronic structure, adsorption properties and charge transport properties. The electronic structure of semiconductors always has an empty conduction band and a filled valence band. Due to the influence of the high energy of the conduction band, electrons move from atom to atom freely whereas the valence band is always at low energy because its orbits are completely occupied by electrons. When light strikes the semiconductor material, the semiconductor is excited due to photon energy and electrons transfer to the empty conduction band from the filled lower energy level. $TiO_2$ as a photocatalyst is getting more attention due to its safety, low cost and high photocatalytic activity under either solar irradiation or UV light (Hagen, 2015). Moreover, due to the photocatalytic activity of $TiO_2$, it is used as an additive in medicines or food and as an electrode for solar cells for the degradation of organic contaminants. Secondly, the use of solar energy is still limited due to the photo-inefficiency of $TiO_2$ catalysts; only 5.0% of the solar spectrum can be used. This is why the up gradation of nominative $TiO_2$ and its optimization is required. This photocatalyst could be used at a commercial level in photocatalytic water treatment processes.

A comparison of the advantages and disadvantages of the photocatalytic system and conventional water treatment will help to further highlight the importance of the photocatalytic system. Biological treatment techniques are highly reliable and high load operation can be processed through this system, but it is difficult to make this system stable for minimizing the sludge problem due to the requirement of a large number of operating managers. Congulation/precipitation is highly efficient but at the same time it has high level of sludge problem that makes it difficult to maintain smooth operation. Fenton technology has wide coverage, is easy to maintain and shows effective colored and discoloration of wastewater, but it is very costly and iron salt is needed to remove the equipment. The main advantages of photocatalytic AOT are that non-biodegradable wastewater treatment is feasible, the small cost required for operations and installation, having no sludge treatment cost, unmanned and unskilled operation is possible, easy pre-processing is possible directly to the industrial water treatment, a small area is required for treatment and a change in water quantity and quality could provide facilitation for operation. In the photocatalysis process, a UV-lamp is used for the activation of catalyst $TiO_2$ but its life time is very low, though this is a bearable disadvantage of this process. These advantages over conventional techniques make the photocatalytic process more attractive and important (Lee and Park, 2013).

## 11.3 Characterization Tools for Photocatalysts

The properties of the material and the reaction conditions affect the photocatalytic activity of photocatalysts. The quality of the photocatalyst, i.e., the crystallinity, defects and surface area are critical factors that affect the photocatalytic activity (Shinde et al., 2013). Crystallinity affects the photocatalytic activity of the material as it reduces the recombination losses (Carp et al., 2004). Therefore, various methods were co-opted to enhance the photocatalytic activity, such as structure optimization, the preparation of nanocomposites in combination with photosensitive materials, doping and surface modification (Yang et al., 2015, Luo et al., 2017). In addition to this, precise characterization of the structure and photocatalytic activity also plays a vital role, as it not only provides the information about mechanisms but also helps to achieve high photocatalytic activity.

The most important technique used for the structural analysis of photocatalysts is X-ray diffraction (XRD). It has become the fundamental technique for analyzing the relationship between photocatalytic activity and the crystal phase of various photocatalysts like $TiO_2$ and its nanohybrids (Ohno et al., 2004, Teoh et al., 2005, Zhou et al., 2018). Studies showed that both anatase–brookite and anatase–rutile mixtures of a definite percentage display higher photocatalytic activity as compared to pure anatase (Kawahara et al., 2002, Miyagi et al., 2004). Scanning tunneling microscopy (STM) and high-resolution transmission electron microscopy (HRTEM) help to investigate the effects of these structures on the photocatalytic activity of the material and the distribution of active sites on the surface of a photocatalyst (Yu et al., 2007, Xiang et al., 2012). HRTEM, in combination with electron energy-loss spectroscopy (EELS), was also used for the analysis of the atomic structures of photocatalysts like oxygen states, lattices and band gaps and so on. (Zhang et al., 2013).

Material defects also affect the photocatalytic activity of the photocatalysts. There are two kinds of defects where recombination occurs, i.e., bulk and surface defects. Both of these affect the adsorption and the surface reactivity of the material, which is directly related to the photo-activation of the material (Mattioli et al., 2008). The major defect in $TiO_2$ is that of oxygen vacancy and this was investigated by using first-principles calculations (Cho et al., 2006, Mattioli et al., 2008). The characterization of these defects is mostly carried out by photoluminescence spectroscopy (PL) (Yuwono et al., 2006, Li et al., 2009, Wu et al., 2009, Wang et al., 2010a, He et al., 2012, Cao et al., 2015), STM (Gong et al., 2006, Lee et al., 2011, Naldoni et al., 2012, Scheiber et al., 2012) and electron paramagnetic resonance spectroscopy (EPR) (Berger et al., 2005, Priebe et al., 2013, Muñoz-Batista et al., 2014). Higher surface area provides more active sites for the reaction which leads towards the enhancement of photocatalytic activity (Sakthivel et al., 2004).

There are various techniques used to determine the surface area of the material. Brunauer–Emmett–Teller (BET) is mostly used for the characterization of the surface area of the material. The main limitations for the use of BET is the need for a vacuum, baking and outgassing of materials, and the sample must be in solid form (powder) (Walton and Snurr, 2007). In the photocatalytic process, the material is dispersed in the liquid, so the measurement of the surface area of the sample will be the aggregation size instead of actual particle surface area. BET results show the maximum surface area available but results can be much less than the actual available surface area due to dispersion in the liquid (Djurišić et al., 2014). BET results cannot guarantee higher photocatalytic activity (Yasui et al., 2012, Hong et al., 2013, Liu et al., 2013).

One of the major challenges in photocatalysis is ultrafast charge recombination and its separation. Time-resolved spectroscopies are an effective tool for this purpose. Time-resolved infrared (IR), time-resolved PL and transient absorption spectroscopy (TAS) are the main techniques used to study the mechanism of photocatalytic activity (Shi et al., 2007, Tang et al., 2008, Wang et al., 2010b, Cowan et al., 2011).

## 11.4 $TiO_2$ as Photocatalyst

$TiO_2$ is one of the most preferred catalysts for the degradation of dyes in industrial wastewater (Mondal and Sharma, 2014, Elango and Sivaperuman, 2015). $TiO_2$ is used as a photocatalyst in dye wastewater treatment mainly because of its high oxidizing ability, good chemical stability, non-toxicity and long-term photo-stability (Han et al., 2009, Skocaj et al., 2011, Hoffeditz et al., 2016). When it absorbs the light corresponding to its band gap, i.e., mostly UV irradiations, electron–hole pairs are generated due to oxidation (Fujishima et al., 2000). The holes are generated due to the irradiation diffusion into the surface of $TiO_2$ and react with water to form hydroxyl groups (•OH) as shown in Figure 11.2. These holes and hydroxyl groups (•OH) interact with organic molecules on the surface of $TiO_2$. On the other hand, electrons present in conduction bands react with oxygen molecules present in the air and generate super oxide anions ($O_2^-$•) (Nakata and Fujishima, 2012).

The anatase form of $TiO_2$ is most photocatalytic as compared to rutile and brookite (Luttrell et al., 2014). The anatase is more suitable for photocatalytic activity because it has a high degree of hydroxylation, adsorption power, stability characteristics and optimal position of conduction band (Kalathil et al., 2013, Khan et al., 2014, Gnanasekaran et al., 2015). The morphology is also a very important factor which affects the final degradation efficiency (Wang et al., 2007, Wang et al., 2008c, Saravanan et al., 2011). To make it more efficient as a photocatalyst, $TiO_2$ is combined with other materials to widen its band gap. It is a great challenge to tailor a good photocatalyst from $TiO_2$ that can efficiently harness the energy from natural sunlight, which consists of no more than 5% UV light and 45% visible light (Shen et al., 2011). $TiO_2$ has a bandgap of 3.2 eV and it only absorbs ultraviolet light. (Dette et al., 2014).

### 11.4.1 Structure of $TiO_2$

$TiO_2$ has a closely packed tetragonal and orthorhombic crystalline structure (Pawar et al., 2011). It is a polymorphic material having three polymorphous systems, namely anatase, rutile and brookite. These three different symmetries show stability according to their nanosize (Fujishima et al., 2008). Rutile and anatase have tetragonal symmetry with space group $P4_2/mnm$ and $I4_1/amd$ (Howard et al., 1991) while brookite has an orthorhombic symmetry exhibiting the Pbca space group (Di Paola et al., 2013).

In these structures, each titanium atom is paired with six oxygen atoms at equal distances and each oxygen atom is paired with three atoms of titanium (Khataee and Mansoori, 2012). Due to the relative spacing, the octahedral structure is different in each type. Rutile has three faces but (110) and (100) are extremely low in energy and are very effective and polycrystalline. The extra characteristic of (110) is that it shows good stability at high temperatures (Mizukoshi and Masahashi, 2010, Zhao et al., 2017). $TiO_2$ atoms which are exposed show very low electronic density. (001) does not show good stability at high temperatures (Diebold, 2003). In (001) phase, two rows of $O_2$ are exchanged with a single row of titanium exposed atoms. These rows are not axial but equatorial in type (Xia et al., 2013). The structures of $TiO_2$ and crystal images are given in Figure 11.3.

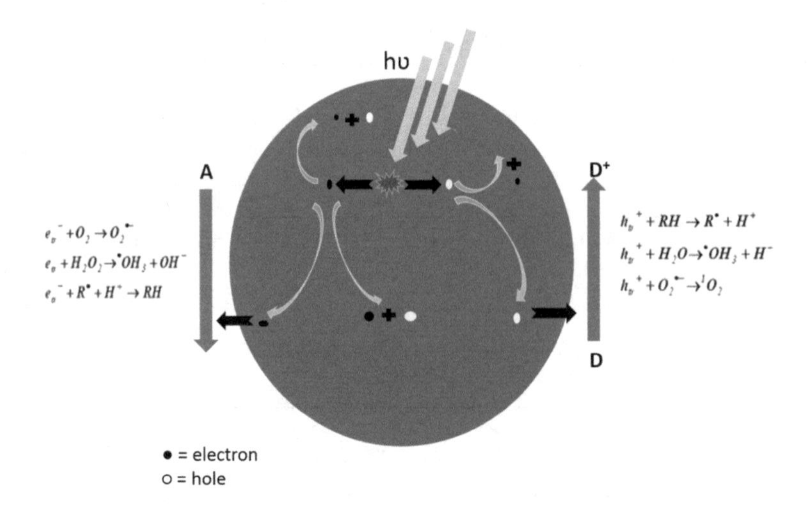


**FIGURE 11.2** Processes occurring on bare $TiO_2$ particles after UV excitation.

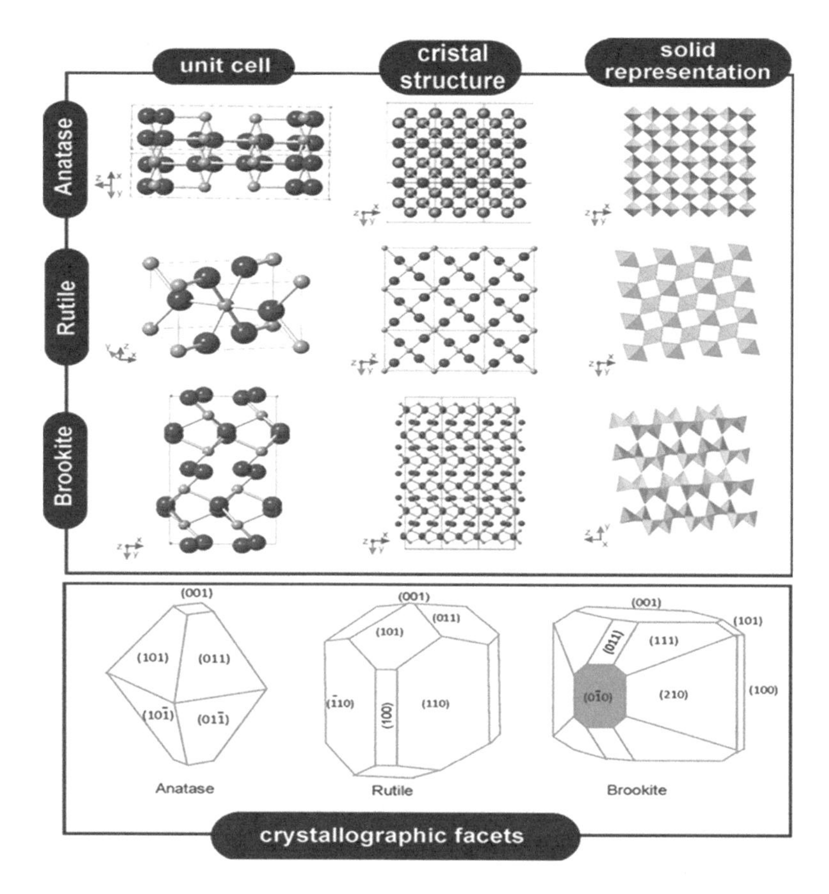


**FIGURE 11.3** **(See color insert.)** Structures of $TiO_2$ and crystal images. (Adapted from Pimentel et al., 2016.)

## 11.4.2 $TiO_2$-Based Nanohybrids as Photocatalysts

The degradation of organic pollutants is mostly carried out by photocatalysis and $TiO_2$ is a promising photocatalyst due to its stability and good photocatalytic activity but it only has range in the UV region. To enhance its range for photocatalysis, it is doped/combined with various materials using different approaches. Different $TiO_2$-based nanohybrids have been reported in the past few decades that enhance the photocatalytic activity of $TiO_2$. A more detailed discussion of these nanohybrids is given in the following section.

### *11.4.2.1 $TiO_2$/CdSe Nanohybrids*

The tunable band gaps of cadmium selenide (CdSe) quantum dots (QDs) help to enhance the range of light energy from the visible to the infrared range of the spectrum (Prasad et al., 2017) and have an efficient carrier multiplication ability when exposed to visible light (McGuire et al., 2008). When these nanostructures are deposited on the surface of $TiO_2$, they can inject the carriers to produce photocurrent under visible light. Therefore, CdSe nanohybrids with $TiO_2$ have widened the range for photocatalytic activity as compared to pristine $TiO_2$ (Hassan et al., 2014). The schematic diagram of the mechanism of photocatalytic activity of $TiO_2$/CdSe nanohybrids is shown in Figure 11.4.

Hosseini et al. prepared $TiO_2$/CdSe and $TiO_2$/CdSe in combination with cadmium sulfide (CdS) nanohybrids by using a hydrothermal technique and deposited them on the surface of reduced graphene oxide (rGO) via an *in situ* fixation technique and used them for the oxidation of aromatic alcohols. Results showed that the $TiO_2$/CdSe/CdS showed higher photocatalytic activity, up to 87% under visible irradiation,

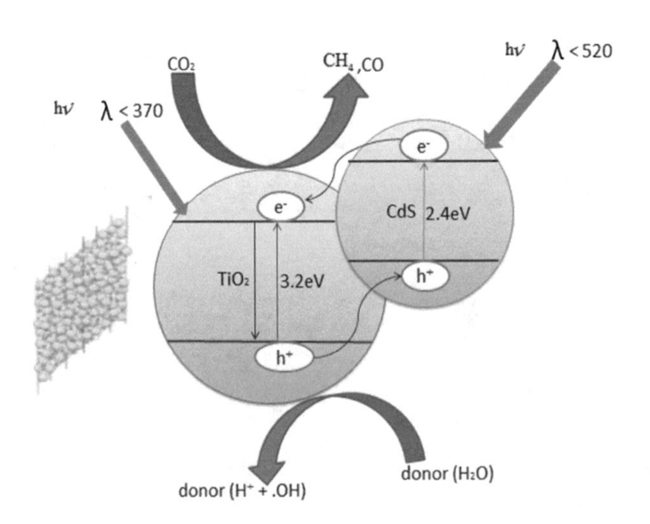


**FIGURE 11.4** Mechanism of photocatalytic activity of $TiO_2$/CdSe nanohybrids. (Redrawn from Ahmad Beigi et al., 2014.)

as compared to $TiO_2$/CdSe. It was also concluded that the combination of CdS and CdSe enhanced the light sensitivity which caused higher product of photo-oxidation (Hosseini and Mohebbi, 2018). Ho et al. also reported CdSe- and CdSe-sensitized $TiO_2$ nanohybrids synthesized by using an ultrasound-driven approach and used them for the degradation of 4-chlorophenol under visible irradiation. In these nanohybrids, CdSe acted as a photosensitizer and helped to prevent recombination. Results showed that the 4-chlorophenol degraded by up to 68% after 8 hours irradiation. To check the mineralization of the dye, total organic carbon analysis was carried out which showed a 22% decrease in total organic carbon (TOC) during the degradation of 4-chlorophenol (Ho and Yu, 2006). Shih-Chen Lo et al. also reported the photodegradation of 4-chlorophenol using $TiO_2$/CdSe nanohybrids. Sol–gel and coupled photocatalysis were used for the synthesis of $TiO_2$/CdSe nanohybrids. The specific surface area of material was 7.0 $m^2g^{-1}$ and the deposition proportion of CdSe was 28%. Results showed higher photodegradation efficiency of 4- chlorophenol was achieved at high pH values (Lo et al., 2004). Yang et al. also synthesized $TiO_2$/CdSe nanohybrids and used them for the degradation of anthracene-9-carboxylic acid (ACA) under monochromatic green light of 550 nm wavelength. The well-dispersed CdSe nanoparticles were deposited on the inner and outer surfaces of 4 μm-long $TiO_2$ nanotubes using the electrodeposition technique. The voltage played a vital role during the formation of the CdSe nanoparticles and provided active sites for the CdSe crystals on the $TiO_2$ nanotubes. Results showed that 5–15 nm sized particles generated more electron-hole pairs as compared to 25–50 nm sized nanoparticles. It was concluded that with an increase in particle size, photocatalytic activity decreases (Yang et al., 2010). Wang et al. prepared a series of $TiO_2$/CdSe heterostructures and used them for the degradation of $CO_2$ in the presence of water. Results showed that these structures can only be used for the reduction of $CO_2$ under visible irradiation ($\lambda > 420$ nm). IR spectroscopy and gas chromatography were also used to investigate the gas-phase reactants and products, respectively. Gas chromatographic analysis showed that the $CH_4$ was the primary product and $CH_3OH$, $H_2$ and CO were detected as secondary products (Wang et al., 2009). Mir et al. reported the photodegradation of methylene blue (MB) dye using $TiO_2$/CdSe nanoparticles mixture (w/w ratio 50:1–9). The results showed 67% degradation of the dye in 1 hour and the prepared photocatalyst could remove MB dye from wastewater in the presence, as well as in the absence, of UV irradiation (Mir et al., 2017).

#### 11.4.2.2 $TiO_2$/CNT Nanohybrids

Carbon nanotubes (CNTs) combined with various metal oxides have gained the attention of researchers due to remarkable photocatalytic activity for the degradation of industrial dyes in wastewater (Saleh, 2011, Saleh, 2013). CNTs help to control the morphology of $TiO_2$ particles as a dispersing agent (Kamil et al., 2014). Nitrate and fluoride traces in water can cause severe health issues and $TiO_2$/CNT-based systems were reported as photocatalysts to remove these contaminants from water (Suriyaraj and Selvakumar,

2016). Xie et al. exhibited a novel technique for the coating of $TiO_2$ on the surface of CNT to form $TiO_2$/CNTs nanohybrids and used them for the degradation of rhodamine B (RhB) under visible irradiation. SEM and TEM results showed that the CNT were uniformly entangled along with $TiO_2$ on the whole surface which enhanced the specific surface area for photocatalytic activity. The average crystallite size of $TiO_2$ nanoparticles was approximately 3.5 nm. The SEM results also showed some agglomeration of $TiO_2$ nanoparticles. Energy-dispersive X-ray spectroscopy (EDX) analysis also confirmed the formation of $TiO_2$/CNT nanohybrids as shown in Figure 11.5.

The $TiO_2$/CNTs nanohybrids showed much better photocatalytic results as compared to commercially available $TiO_2$ nanomaterials, as shown in Figure 11.6. Commercially available $TiO_2$ nanoparticles and P25 adsorbed only 11.30% of RhB dye whereas $TiO_2$/CNT nanohybrids adsorbed approximately 34.60% under the same conditions in the dark. Under UV irradiation, the decolorization rate for P25, $TiO_2$ and $TiO_2$/CNT nanohybrids were 55.3%, 42.8% and 92.3%, respectively, which showed that $TiO_2$/CNT nanohybrids are a better photocatalyst, as shown in Figure 11.6 b (Xie et al., 2012).

Miribangul et al. also synthesized $TiO_2$/CNT nanohybrids by varying the amount of CNT (0.1–0.5 wt.%) and characterized them by using TEM, powder X-ray diffraction (PXRD) and UV–Vis absorption spectroscopy. These were used for the degradation of Sudan (I) to measure the photocatalytic efficiency of prepared nanohybrids under UV–Vis irradiation. Nanohybrids with 0.30 wt. % of CNT showed best results as compared to other hybrids. Results showed that the reaction temperature affects the reaction rate and half-life (in terms of reaction). As temperature increases, half-life decreases, whereas reaction rate increases constantly (Miribangul et al., 2016). Kamil et al. prepared the $TiO_2$/MWCNT by sol–gel technique and simple evaporation method. The Bismarck brown R dye was degraded and its photocatalytic efficiency was found. Results showed that the 0.5% w/w ratio of MWCNT in nanohybrids exhibits the highest degradation of the Bismarck brown R dye. (Kamil et al., 2014). Abbas et al. prepared $TiO_2$/CNT nanohybrids and used them to remove the E.coli bacteria from water in the absence of light. Various samples were prepared by varying the amount of CNT in the nanohybrids (Abbas et al., 2016).

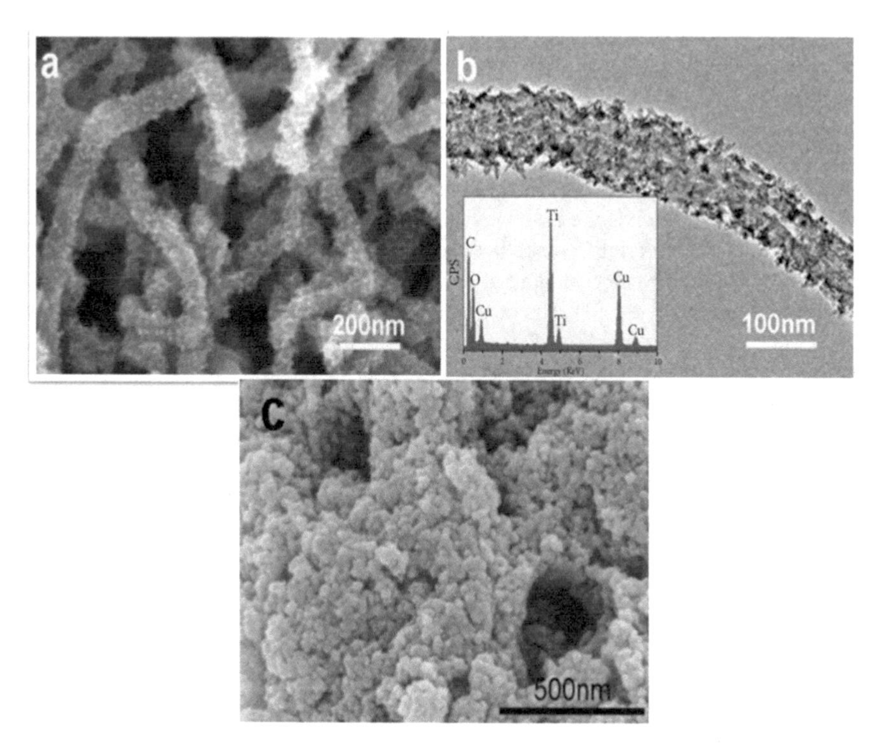


**FIGURE 11.5** (a) Typical SEM image, (b) TEM image and EDX pattern (inset) of as-prepared CNT/$TiO_2$ heterostructures. (c) SEM image of the as-prepared $TiO_2$ nanoparticles. (Adapted from Xie et al., 2012.)

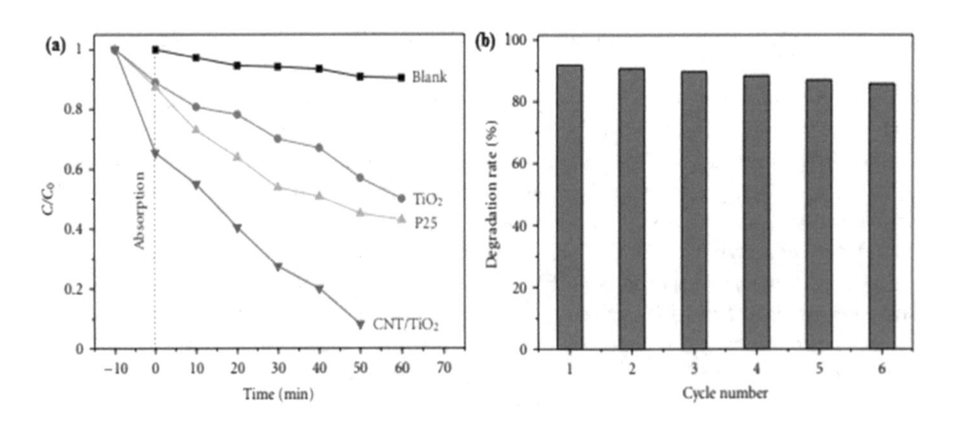


**FIGURE 11.6 (See color insert.)** (a) The photocatalytic degradation of RhB in the absence of any photocatalysts (the blank test) and in the presence of different photocatalysts. (b) Six cycles of degradation of RhB using CNT/$TiO_2$ nanohybrids as the photocatalyst. (Adapted from Xie et al., 2012.)

Vilamiki et al. also used these nanohybrids for the inactivation of *Staphylococcus aureus* and *Escherichia coli* in the presence of visible light (Koli et al., 2016). Zouzellka et al. reported the synthesis of $TiO_2$/MWCNT nanohybrids and used them for the degradation of 4-chlorophenol. Results showed that the presence of MWCNT not only reduced the recombination but also enhanced the photocatalytic activity. Deposition of thin layers of MWCNT affects the photocatalytic activity as it increased for a two-fold layer as compared to single layer (Zouzelka et al., 2016).

#### *11.4.2.3 $TiO_2$/ZnO Nanohybrids*

In the past decade, $TiO_2$/ZnO nanohybrids have gained the attention of researchers as photocatalysts (Liao et al., 2008). $TiO_2$/ZnO showed better photocatalytic activity when compared to individual $TiO_2$ or ZnO. Different chemical methods were used for the synthesis of $TiO_2$/ZnO nanocomposites like hydrothermal technique and sol–gel. (Wang et al., 2008b, Xu et al., 2011, Rusu et al., 2016). The difference between the band gap of ZnO and $TiO_2$ is very small. When light falls on the surface of the material, it interacts with the ZnO and excites the electron from its valence band to the conduction band, leaving a hole behind. Due to the potential difference between ZnO and $TiO_2$, this electron jumps from the conduction band of ZnO to the conduction band of $TiO_2$. As a result, a hole from the valence band of $TiO_2$ transfers into the valence band of ZnO. The recombination rate reduces due to the presence of a potential barrier of ZnO and $TiO_2$ heterojunction. These electron–hole pairs lead to redox reactions which cause the degradation of organic dyes as shown in Figure 11.7 (Xiao et al., 2014). Xiao also reported the synthesis of $TiO_2$ nanotubes combined with ZnO by using a two-step anodization in combination with pyrolysis. The results showed that the ZnO was uniformly grafted on the surface of $TiO_2$ nanotubes. These heterostructures showed better photocatalytic activity as compared to the pristine $TiO_2$ (Xiao, 2012).

Xiao et al. also reported the nanostructure of $TiO_2$ nanotubes and ZnO nanorods prepared by using a two-step anodization in combination with electrochemical deposition. These nanohybrids showed enhanced photocatalytic activity due to the interfacial integration of $TiO_2$ nanotubes with ZnO nanorods towards the organic pollutants under UV irradiation (Xiao et al., 2014). Araújoa et al. also prepared the $TiO_2$/ZnO hierarchical nanostructures by using a two-step method in combination with the hydrothermal technique and used them for the degradation of RhB dye under visible irradiation. Their results showed that 90% of the RhB dye was degraded after 1 hour and 10 minutes (Kwiatkowski et al., 2015, Araújo et al., 2016). Core–brush nanohybrids of $TiO_2$/ZnO were prepared on the glass substrate by using the magnetron sputtering method and aqueous solution growth. These nanohybrids were used for the degradation of bromo-pyrogallol red (Br-PGR) dye under both visible and UV irradiation and it showed excellent results due to the small band gap, low recombination rate, high density active sites and wide irradiation

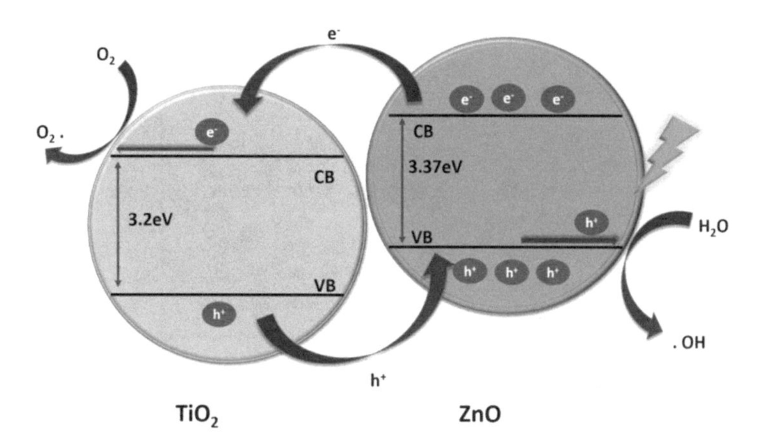


**FIGURE 11.7** Schematic diagram illustrating the charge-transfer process in the ZnO/TNTs heterostructure when used as a photocatalyst under UV light irradiation. (Reproduced with permission from Xiao, 2012.)

range. The photocatalytic kinetics were analyzed using the Langmuir–Hinshelwood model. $TiO_2$ thin film and $TiO_2$/ZnO nanohybrids were both investigated for the degradation of dye under visible and UV irradiation. Results showed that the Br-PGR dye was degraded about 80% in the presence of $TiO_2$/ZnO nanohybrids as compared to the $TiO_2$, i.e., 32% under UV irradiation. In visible light, it degraded only 40.5% for $TiO_2$/ZnO nanohybrids, whereas for $TiO_2$, there was no obvious degradation of Br-PGR dye, as shown in Figure 11.8 (Yan et al., 2012).

$TiO_2$/ZnO mixed nanohybrids were also prepared by using hydrothermal and precipitation techniques, with titanium tetrachloride and zinc nitrate used as precursors. These were used for the photodegradation of Procion Red MX-5B dye. A better photocatalytic activity was observed compared to individual ZnO and $TiO_2$ (William IV et al., 2008). Nanohybrids of varying ZnO/$TiO_2$ molar ratios have been prepared from their precursors and annealed at 600°C and 900°C. These nanohybrids were used for the degradation of azo and brilliant golden yellow (BGY) dyes under solar irradiation. Results showed that the molar ratio of ZnO/$TiO_2$ nanohybrids, initial dye concentration, pH and photocatalyst loading affect the photocatalytic activity of ZnO/$TiO_2$ nanohybrids. ZnO/$TiO_2$ nanohybrids with a molar ratio of 1:1

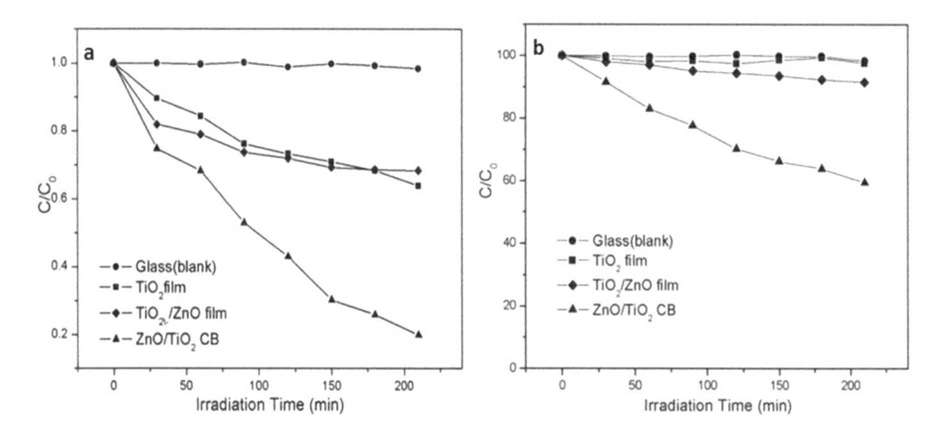


**FIGURE 11.8** Photodecomposition of Br-PGR under UV light (250 nm) and visible irradiation respectively (a) photodegradation rate of Br-PGR by different photocatalysts under UV light irradiation (b) photodegradation rate of Br-PGR by different photocatalysts under UV light irradiation. (Reproduced with the permission from Yan et al., 2012.)

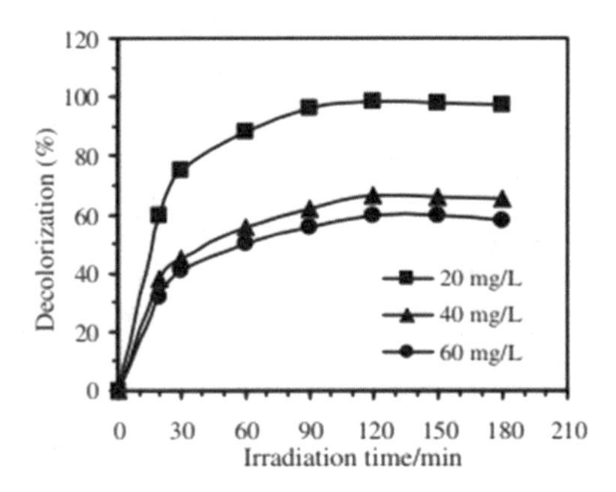


**FIGURE 11.9** Effect of initial dye concentration on decolorization of BGY. (Adapted from Habib et al., 2013.)

or 3:1 prepared at 900°C showed 98% decolorization of dyes. It was also observed that as the amount of ZnO increased, the photocatalytic activity also increased, compared to the amount of $TiO_2$. In addition to this, photocatalytic activity increased as the amount of photocatalyst increased up to 8 g/L; after that, it started decreasing, which may be due to the blockage of light diffusion into the material. Similarly, pH also affects the photocatalytic activity and the degradation of dyes increases as pH increases from 6.5 to 7 but after that, it started decreasing. Initial dye concentration also has an impact on the decolorization of the dyes, as shown in Figure 11.9. It showed that decolorization decreased with increase in the initial dye concentration for BGY dye (Habib et al., 2013).

#### *11.4.2.4 $TiO_2$/CS Nanohybrids*

Biocompatible polymers have gained the attention of researchers in the past few decades due to their eco-friendly and biomedical applications. These polymers can decompose easily in the environment (Wu and Wu, 2006). Chitosan (CS) has proven itself a good host material for metal oxides in the synthesis of organic/inorganic nanohybrids (Huang et al., 2004). CS has two functional groups and good adsorbing ability (Moradi Dehaghi et al., 2014), which can be used for the removal of microbes from wastewater (Xiao et al., 2015). CS and $TiO_2$ can easily interact with the negatively charged cell surfaces by electrostatic interaction and their nanohybrids are reported to remove the various entities from wastewater (Ozerin et al., 2006, Raut et al., 2016). A schematic diagram of the crosslinking of CS with $TiO_2$ nanostructure is given in Figure 11.10.

Haldorai et al. reported the high photocatalytic activity of $TiO_2$/CS nanohybrids against MB dye. $TiO_2$/CS nanohybrids were prepared by using a chemical precipitation technique and the results showed 90% degradation of the MB dye in the UV–Vis region and that it can be reused for the degradation (Haldorai and Shim, 2014). $TiO_2$/CS/feldspar nanohybrids were also reported for the degradation of Black 1 (AB1) dye under and without UV irradiation. The results showed that the degradation of AB1 dye increases under UV irradiation up to 97%, the reaction was exothermic and experimental data was best fitted to the Freundlich model (Yazdani et al., 2014). The photodegradations of three reactive dyes of RY 17, RR 120 and RB 220 using $TiO_2$/CS nanohybrids were reported. These nanohybrids were synthesized by using a solution casting technique and Arquad T50 HFP was used as a surfactant. These $TiO_2$/CS nanohybrids films with crosslinks had low efficiency as compared to non-crosslinked samples. The experimental data of dye removal efficiencies was best fit to the Langmuir–Hinshelwood model, as shown in Figure 11.11. For all dyes at their initial concentrations (10–100 mg $L^{-1}$), dye removal efficiency was higher under UV conditions as compared to conditions without UV light. The sorption-to-photocatalysis ratio of these three dyes was 70.7:29.3, 78.5:21.5 and 92.2:7.8, respectively. Langmuir isotherm showed that $q_{max}$

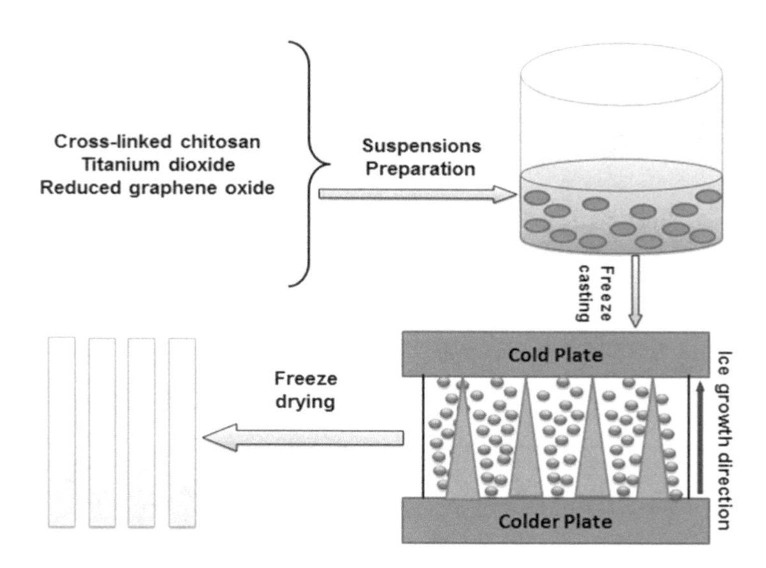


**FIGURE 11.10** Preparation of $TiO_2$/CS nanostructures.

of RR120, RY17, and RB220 were 46.8, 427.1 and 229.1 mg-dye.$g^{-1}$-chitosan-$TiO_2$ film, respectively (Norranattrakul et al., 2013).

$TiO_2$/CS nanohybrids were also used for the degradation of three different kinds of dyes, i.e., MB, Reactive Red 2 (RR) and RhB. A kinetic study was carried out using the Langmuir–Hinshelwood model. The results showed that 90.9%, 94.8% and 80.2% of MB, RR and RhB, respectively, were degraded under UV irradiation. This increased to 95.6%, 99.9% and 85.04% for MB, RR and RhB, respectively, when $TiO_2$/CS nanohybrids were combined with $H_2O_2$ oxidant. Photocatalytic activity was significantly affected by substrate concentration, dosage, $H_2O_2$ concentration and light intensity (Farzana and Meenakshi, 2014).

### *11.4.2.5 $TiO_2$/PANI Nanohybrids*

Polyaniline (PANI)/$TiO_2$ nanohybrids are mostly used as photocatalysts due to their better range in UV and visible light, stability and recycling ability. PANI is not only a good donor but also acts as an acceptor of electrons in photocatalysis (Wetterskog et al., 2014). PANI/$TiO_2$ nanohybrids were synthesized by a chemical polymerization technique. Using the PANI/$TiO_2$ nanohybrids film as photocatalysts, results showed that 67.1% and 83.2% of RhB was degraded under sunlight and UV irradiation within 120 min respectively (Jinzhang et al., 2007). Wei et al. prepared PANI/$TiO_2$ nanohybrids using a hydrothermal method and their photocatalytic properties were investigated by the photodegradation of gaseous acetone under UV ($\lambda = 254$ nm) and visible light irradiation ($\lambda > 400$ nm). In fact, the photocatalytic results of these nanohybrids were superior to those of pure $TiO_2$ and PANI samples (Wei et al., 2011). Li et al. prepared a new type of macroporous PANI/$TiO_2$ nanohybrid by *in situ* oxidative polymerization that showed excellent photocatalytic activity and regeneration ability under visible light for degradation of organic wastewater, as shown in Figure 11.12 (Li et al., 2015). Radoičić et al. developed a novel photocatalytic system based on carbonized PANI/$TiO_2$ nanohybrids by using a simple bottom-up approach. Photocatalytic degradations of MB and RhB were carried out that showed excellent photocatalytic activity. (Radoičić et al., 2017).

### *11.4.2.6 $TiO_2$/Polypyrrole Nanohybrids*

Polypyrrole also belongs to the family of conducting polymers. It gains the attention of researchers due to its environmental stability, easy synthesis and high conductivity (Lu and Lin, 2003). Polypyrrole is an important part of organic–inorganic nanostructures. $TiO_2$/Polypyrrole nanohybrids have gained

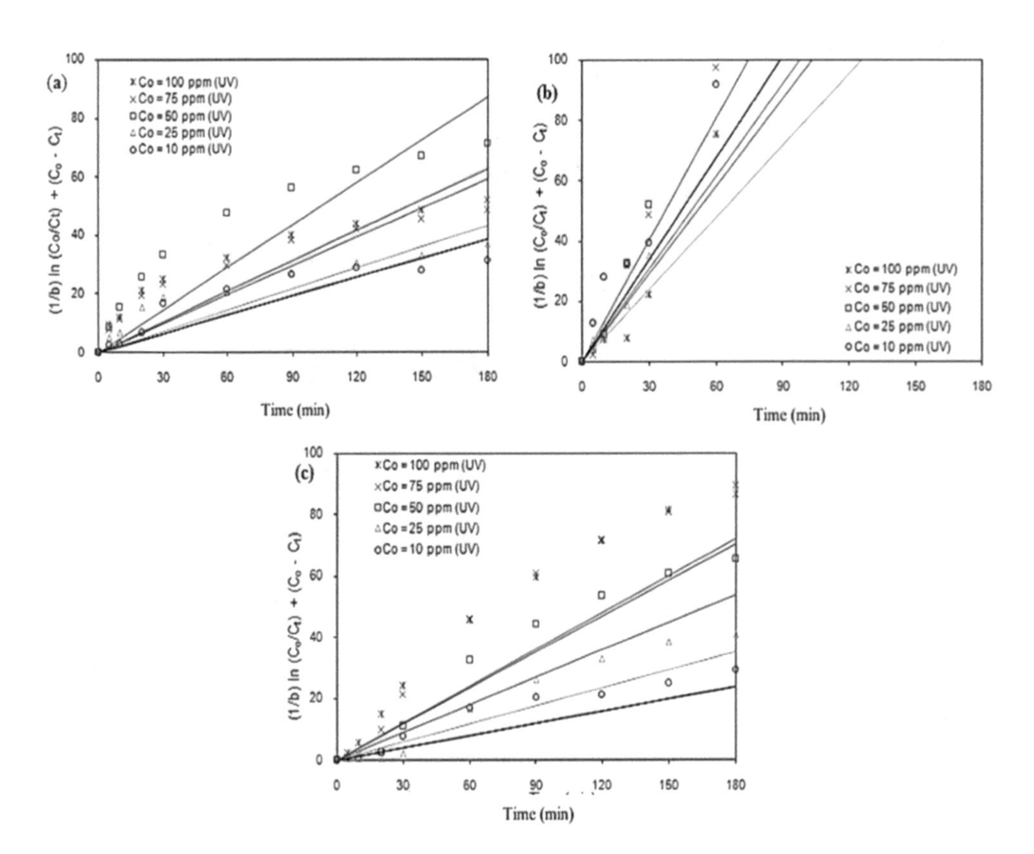


**FIGURE 11.11** **(See color insert.)** Effect of $TiO_2$/CS films on dye removal for various types and concentration of dye using Langmuir–Hinshelwood model, (a) RR 120, (b) RY 17, and (c) RB 220. (Adapted from Norranattrakul et al., 2013.)

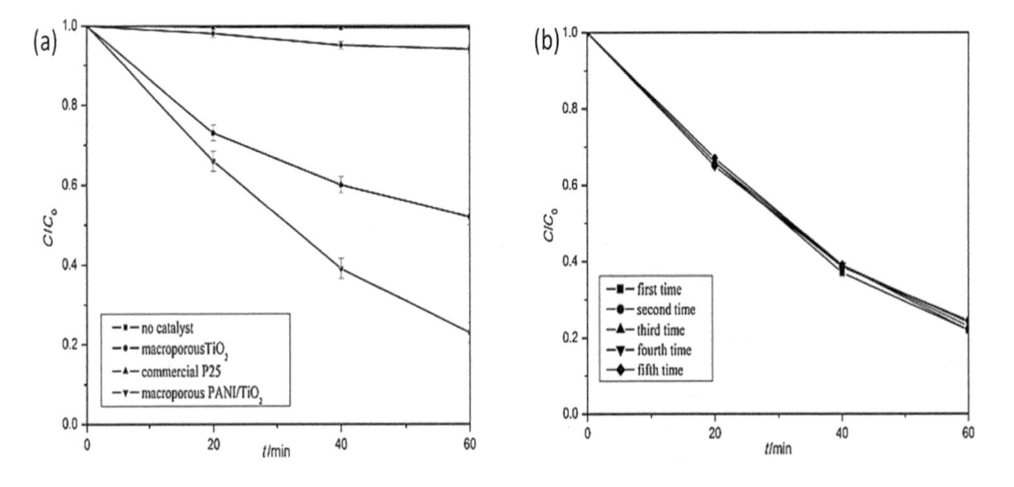


**FIGURE 11.12** (a) Degradation curves of MO solution by various samples under visible light irradiation (b) Degradation curves of MO solution by regenerated PANI/$TiO_2$ composites under visible light irradiation. (Adapted from Li et al., 2015.)

importance due to their photocatalytic applications. These nanostructures have narrow band gaps, which makes them better at absorbing visible light compared to their individual constituents. Polypyrrole and $TiO_2$ interact by a covalent bond with Polypyrrole behaving as donor and $TiO_2$ behaving as acceptor (Ullah et al., 2017). Wang et al. prepared a series of $TiO_2$/Polypyrrole nanohybrids by varying the amount of both $TiO_2$ and Polypyrrole and used them for the photocatalytic degradation of methyl orange under solar irradiation. An *in situ* deposition oxidative polymerization technique was used for the synthesis of $TiO_2$/Polypyrrole nanohybrids in which ferric chloride was used as an oxidant. Results showed that the prepared samples have better photocatalytic activity towards methyl orange compared to $TiO_2$ nanoparticles (Wang et al., 2008a). Li et al. also synthesized novel hierarchically macroporous $TiO_2$/Polypyrrole nanohybrids by polymerization of a pyrrole monomer with macro/mesoporous $TiO_2$. Complete hybridization was realized since the dual porous structure of the sensitizing material was efficient at infiltrating the pyrrole interior and introducing UV photons that penetrated the porous network for local polymerization. Collective performance favors the fabrication of hybrids with a well-adhered interface between the polymer monomer and the inorganic sensitizing material. Mechanisms of polymerization and interface interaction were discussed in terms of semiconductor photocatalysis. The photoresponse of these nanohybrids was shifted towards the visible light spectrum (2.67–2.92 eV), indicating a narrow band gap compared to the 3.18 eV of pure $TiO_2$. The sample of Polypyrrole/TiO2-24 showed higher photoactivity and photocurrent w.r.t other nanohybrids and the $TiO_2$ under visible light irradiation. It may be due to its ordered macrospores, uniform Polypyrrole sensitizer layer and relatively less hole–electron recombination rate. (Li et al., 2014). Denga et al. also synthesized the $TiO_2$/Polypyrrole nanohybrids by using surface molecular imprinting technique in which methyl orange was used for photodegradation. Results showed that the prepared nanohybrids have better selectivity and adsorption capability compared to the control Polypyrrole/$TiO_2$ nanocomposites (Deng et al., 2012).

Polypyrrole/$TiO_2$ nanohybrids were also reported for the degradation of MB dye. Thermogravimetric analysis (TGA) results showed better response compared to simple Polypyrrole or $TiO_2$, whereas X-ray photoelectron spectroscopy (XPS) analysis showed better doping capability for prepared Polypyrrole/$TiO_2$ nanohybrids compared to Polypyrrole or $TiO_2$. Langmuir isotherm and pseudo-second-order models were successfully employed to explain the adsorption of MB dye. Hydrogen bonding, electrostatic and hydrophobic interaction played vital roles in the degradation of the dye and the dye was completely adsorbed within 30 minutes, which indicated that the prepared composites can be considered potential absorbents for MB dye (Li et al., 2013). He et al. also prepared Polypyrrole/$TiO_2$ nanohybrids by a surface molecular imprinting technique and degradation of RhB under visible irradiation was carried out. Results showed better stability and reusability of nanohybrids compared to pure $TiO_2$, as shown in Figure 11.13 (He et al., 2014).

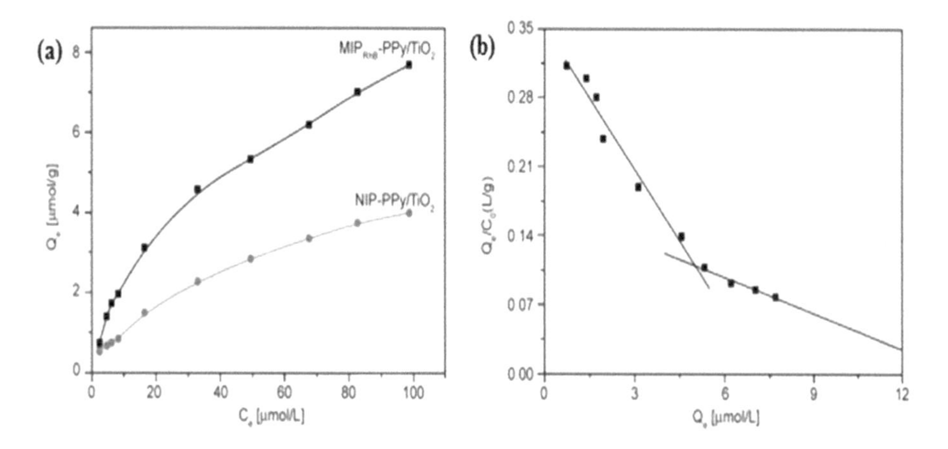


**FIGURE 11.13** (a) Adsorption isotherms of MIPRhB-PolyPyrrole/$TiO_2$ and NIP-PolyPyrrole/$TiO_2$ for RhB (b) Scatchard plot of the adsorption of MIPRhB-PolyPyrrole/$TiO_2$ for RhB. (Adapted from He et al., 2014.)

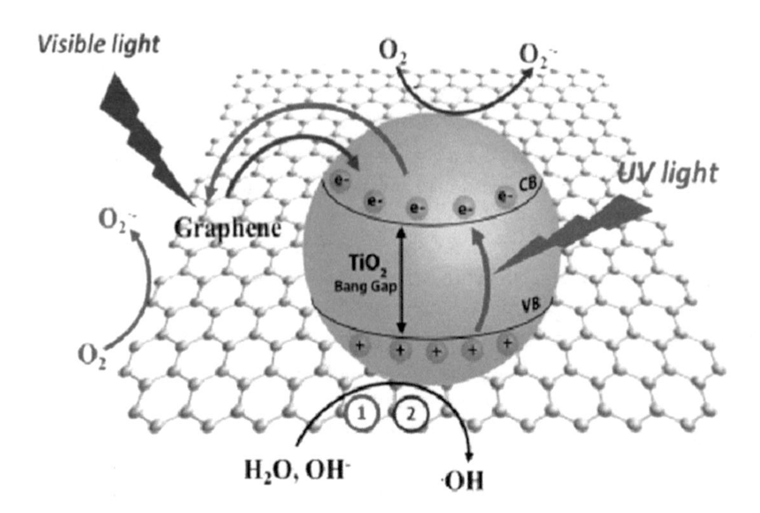


**FIGURE 11.14 (See color insert.)** Mechanisms of UV and visible light activation of $TiO_2$ with graphene. (Adapted from Giovannetti et al., 2017.)

### *11.4.2.7 $TiO_2$/Graphene Nanohybrids*

Graphene has gained the attention of scientists due to its applications in various fields (Novoselov, 2011). It is the elementary part of all the graphitic forms of carbon and $sp^2$ hybrid bonds of a single atomic layer of carbon atoms that are arranged in honeycomb structure (Papageorgiou et al., 2017). The role of graphene in the photocatalytic activity of $TiO_2$/GO nanohybrids is not completely understood. When light falls on the surface of the photocatalyst, it creates electron–hole pairs. These electrons are injected into the graphene because it has a greater positive Fermi level compared to $TiO_2$. The electrons present in the conduction band of $TiO_2$ are also injected into the graphene due to the difference in the work function of graphene compared to $TiO_2$. Graphene scavenges these electrons by dissolved oxygen, thereby reducing recombination and enhancing electron transportation, which leads to higher photocatalytic activity. The schematic diagram of the activation of $TiO_2$ with graphene is shown in Figure 11.14.

Liu et al. synthesized $TiO_2$/GO nanohybrids by assembling $TiO_2$ nanorods on the surface of graphene oxide sheets at the toluene/water interface and used them for the degradation of MB dye. These prepared nanohybrids were easily dispersed in water. Results showed that the photodegradation of the MB is much better in the second cycle compared to the first cycle due to the reduction of graphene under UV irradiation (Liu et al., 2010). Zhang et al. synthesized $TiO_2$/rGO and $SnO_2$/rGO nanohybrids by a direct redox reaction between reactive $Ti^{3+}$ and $Sn^{2+}$ cations and graphene oxide and used them for the degradation of RhB under visible irradiation. During the reaction, graphene oxide reduced to rGO. $TiO_2$ and $SnO_2$ were deposited on the surface of the rGO. Results showed that the photodegradation of dye is much higher than the commercially available $TiO_2$ due to the presence of rGO (Zhang et al., 2011). Zhang et al. also synthesized $TiO_2$/Graphene nanohybrids via a one-step hydrothermal technique and used them for the photodegradation of MB dye. These prepared nanohybrids showed excellent absorptivity for the MB dye compared to bare and $TiO_2$/CNT nanohybrids, as shown in Figure 11.15 (Zhang et al., 2010).

## 11.5 Conclusion and Future Perspectives

Photocatalysis is the most widely used eco-friendly method to degrade or completely eliminate organic contaminants in industrial wastewater. The photocatalytic activity of any photocatalyst depends upon a number of factors, such as the quality of the photocatalyst, which includes crystallinity, and defects, and the available surface area of the photocatalyst. All photocatalysts are basically semiconductors. Several metal oxide semiconductors have been investigated as potential photocatalysts for dye wastewater

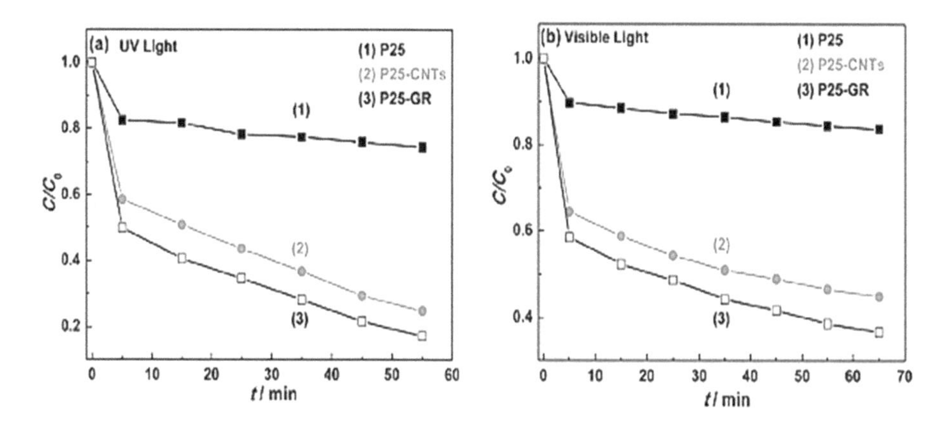


**FIGURE 11.15** **(See color insert.)** Photodegradation of MB under (a) UV light and (b) visible light (400 nm) over (1) P25, (2) P25/CNTs, and (3) P25/GR photocatalysts, respectively. (Reproduced with the permission from Zhang et al., 2010.)

treatment. $TiO_2$ and its nanohybrids are the most widely used metal oxides to treat industrial wastewater. There are different approaches and methods to synthesize $TiO_2$-based nanohybrids and the selection of any of the preparation methods affects their physical and chemical properties such as structural, morphological, optical and electrical properties. These nanohybrids exploit the maximum degradation capability of industrial dyes as these provide modification of the band gap of pristine $TiO_2$ towards the visible light portion of the spectrum. On a final note, although the application of $TiO_2$-based nanohybrids as photocatalysts is making rapid progress, there is nevertheless a lot of work which needs to be done in terms of facile synthesis methods, characterization with suitable analytical tools to get full control of desired properties and hence, efficient use for degrading industrial dyes in the future.

## Acknowledgments

H. Anwar is grateful to higher education commission (HEC), Pakistan for funding under the project SRGP-266.

## REFERENCES

Abbas, N., Shao, G.N., Haider, M.S., Imran, S.M., Park, S.S., Jeon, S.J. & Kim, H.T., 2016. Inexpensive sol-gel synthesis of multiwalled carbon nanotube-TiO2 hybrids for high performance antibacterial materials. *Materials Science and Engineering: C, Materials for Biological Applications*, 68, 780–788.

Ahmad Beigi, A.A., Fatemi, S. & Salehi, Z., 2014. Synthesis of nanocomposite CdS/TiO2 and investigation of its photocatalytic activity for CO2 reduction to CO and CH4 under visible light irradiation. *Journal of CO2 Utilization*, 7, 23–29.

Akpan, U.G. & Hameed, B.H., 2009. Parameters affecting the photocatalytic degradation of dyes using TiO2-based photocatalysts: A review. *Journal of Hazardous Materials*, 170, 520–529.

Araújo, E.S., Da Costa, B.P., Oliveira, R.A.P., Libardi, J., Faia, P.M. & De Oliveira, H.P., 2016. TiO2/ZnO hierarchical heteronanostructures: Synthesis, characterization and application as photocatalysts. *Journal of Environmental Chemical Engineering*, 4, 2820–2829.

Berger, T., Sterrer, M., Diwald, O., Knözinger, E., Panayotov, D., Thompson, T.L. & Yates, J.T., 2005. Light-induced charge separation in anatase TiO2 particles. *The Journal of Physical Chemistry B*, 109, 6061–6068.

Cao, S., Chen, C., Zhang, J., Zhang, C., Yu, W., Liang, B. & Tsang, Y., 2015. MnOx quantum dots decorated reduced graphene oxide/TiO2 nanohybrids for enhanced activity by a UV pre-catalytic microwave method. *Applied Catalysis B: Environmental*, 176–177, 500–512.

Carp, O., Huisman, C.L. & Reller, A., 2004. Photoinduced reactivity of titanium dioxide. *Progress in Solid State Chemistry*, 32, 33–177.

Cernuto, G., Masciocchi, N., Cervellino, A., Colonna, G.M. & Guagliardi, A., 2011. Size and shape dependence of the photocatalytic activity of TiO2 nanocrystals: A total scattering Debye function study. *Journal of the American Chemical Society*, 133, 3114–3119.

Cheremisinoff, P.N. & Ellerbusch, F., 1978. *Carbon Adsorption Handbook*. Ann Arbor Science Publishers.

Cho, E., Han, S., Ahn, H.-S., Lee, K.-R., Kim, S.K. & Hwang, C.S., 2006. First-principles study of point defects in rutile $TiO_{2-x}$. *Physical Review B*, 73, 193202.

Chong, M.N., Jin, B., Chow, C.W. & Saint, C., 2010. Recent developments in photocatalytic water treatment technology: A review. *Water Research*, 44, 2997–3027.

Cowan, A.J., Barnett, C.J., Pendlebury, S.R., Barroso, M., Sivula, K., GräTzel, M., Durrant, J.R. & Klug, D.R., 2011. Activation energies for the rate-limiting step in water photooxidation by nanostructured α-Fe2O3 and TiO2. *Journal of the American Chemical Society*, 133, 10134–10140.

Cowan, C.E., Zachara, J.M. & Resch, C.T., 1991. Cadmium adsorption on iron oxides in the presence of alkaline-earth elements. *Environmental Science and Technology*, 25, 437–446.

Deng, F., Li, Y., Luo, X., Yang, L. & Tu, X., 2012. Preparation of conductive polypyrrole/TiO2 nanocomposite via surface molecular imprinting technique and its photocatalytic activity under simulated solar light irradiation. *Colloids and Surfaces A: Physicochemical and Engineering Aspects*, 395, 183–189.

Dette, C., PéRez-Osorio, M.A., Kley, C.S., Punke, P., Patrick, C.E., Jacobson, P., Giustino, F., Jung, S.J. & Kern, K., 2014. TiO2 anatase with a bandgap in the visible region. *Nano Letters*, 14, 6533–6538.

Di Paola, A., Bellardita, M. & Palmisano, L., 2013. Brookite, the least known TiO2 photocatalyst. *Catalysts*, 3, 36–73.

Diebold, U., 2003. The surface science of titanium dioxide. *Surface Science Reports*, 48, 53–229.

Djurišić, A.B., Leung, Y.H. & Ching Ng, A.M., 2014. Strategies for improving the efficiency of semiconductor metal oxide photocatalysis. *Materials Horizons*, 1, 400–410.

Egerton, T.A., 1997. Titanium compounds, inorganic. In: C. Ley, ed., *Kirk-Othmer Encyclopedia of Chemical Technology*. John Wiley & Sons.

Elango, S. & Sivaperuman, K., 2015. Sol-gel mediated synthesis of tri-doped TiO2 nanoparticles towards application of photo catalysis and its kinetic study. *International Journal of ChemTech Research*, 8, 588–597.

Farzana, M.H. & Meenakshi, S., 2014. Synergistic effect of chitosan and titanium dioxide on the removal of toxic dyes by the photodegradation technique. *Industrial and Engineering Chemistry Research*, 53, 55–63.

Fujishima, A. & Honda, K., 1972. Electrochemical photolysis of water at a semiconductor electrode. *Nature*, 238, 37–38.

Fujishima, A., Rao, T.N. & Tryk, D.A., 2000. Titanium dioxide photocatalysis. *Journal of Photochemistry and Photobiology C: Photochemistry Reviews*, 1, 1–21.

Fujishima, A., Zhang, X. & Tryk, D.A., 2008. TiO2 photocatalysis and related surface phenomena. *Surface Science Reports*, 63, 515–582.

Giovannetti, R., Rommozzi, E., Zannotti, M. & D'amato, C.A., 2017. Recent advances in graphene based TiO2 nanocomposites (GTiO2Ns) for photocatalytic degradation of synthetic dyes. *Catalysts*, 7, 305.

Gligorovski, S., Strekowski, R., Barbati, S. & Vione, D., 2015. Environmental implications of hydroxyl radicals (• OH). *Chemical Reviews*, 115, 13051–13092.

Gnanasekaran, L., Hemamalini, R. & Ravichandran, K., 2015. Synthesis and characterization of TiO2 quantum dots for photocatalytic application. *Journal of Saudi Chemical Society*, 19, 589–594.

Gong, X.-Q., Selloni, A., Batzill, M. & Diebold, U., 2006. Steps on anatase TiO2(101). *Nature Materials* 5, 665–670.

Groffman, A., Peterson, S. & Brookins, D., 1992. Removing lead from wastewater using zeolite. *Water Environment and Technology*, 4, 54–59.

Gupta, V., Sharma, S., Yadav, I. & Mohan, D., 1998. Utilization of bagasse fly ash generated in the sugar industry for the removal and recovery of phenol and p-nitrophenol from wastewater. *The Journal of Chemical Technology & Biotechnology: International Research in Processing, Environmental and Clean Technology*, 71, 180–186.

Gupta, V.K., Ali, I., Saleh, T.A., Nayak, A. & Agarwal, S., 2012. Chemical treatment technologies for wastewater recycling—An overview. *RSC Advances*, 2, 6380–6388.

Habib, M.A., Shahadat, M.T., Bahadur, N.M., Ismail, I.M.I. & Mahmood, A.J., 2013. Synthesis and characterization of ZnO-TiO2 nanocomposites and their application as photocatalysts. *International Nano Letters*, 3, 5.

Hagen, J., 2015. *Industrial Catalysis: A Practical Approach*. John Wiley & Sons.

Haldorai, Y. & Shim, J., 2014. Novel chitosan-TiO2 nanohybrid: Preparation, characterization, antibacterial, and photocatalytic properties. *Polymer Composites*, 35, 327–333.

Han, F., Kambala, V.S.R., Srinivasan, M., Rajarathnam, D. & Naidu, R., 2009. Tailored titanium dioxide photocatalysts for the degradation of organic dyes in wastewater treatment: A review. *Applied Catalysis A: General*, 359, 25–40.

Han, G., Wang, L., Pei, C., Shi, R., Liu, B., Zhao, H., Yang, H. & Liu, S., 2014. Size-dependent optical properties and enhanced visible light photocatalytic activity of wurtzite CdSe hexagonal nanoflakes with dominant {0 0 1} facets. *Journal of Alloys and Compounds*, 610, 62–68.

Hassan, Y., Chuang, C.-H., Kobayashi, Y., Coombs, N., Gorantla, S., Botton, G.A., Winnik, M.A., Burda, C. & Scholes, G.D., 2014. Synthesis and optical properties of linker-free TiO2/CdSe nanorods. *The Journal of Physical Chemistry C*, 118, 3347–3358.

He, M.Q., Bao, L.L., Sun, K.Y., Zhao, D.X., Li, W.B., Xia, J.X. & Li, H.M., 2014. Synthesis of molecularly imprinted polypyrrole/titanium dioxide nanocomposites and its selective photocatalytic degradation of rhodamine B under visible light irradiation. *Express Polymer Letters*, 8, 850–861.

He, Z., Que, W., Chen, J., Yin, X., He, Y. & Ren, J., 2012. Photocatalytic degradation of methyl orange over nitrogen–fluorine codoped TiO2 nanobelts prepared by solvothermal synthesis. *ACS Applied Materials and Interfaces*, 4, 6816–6826.

Ho, W. & Yu, J.C., 2006. Sonochemical synthesis and visible light photocatalytic behavior of CdSe and CdSe/TiO2 nanoparticles. *Journal of Molecular Catalysis A: Chemical*, 247, 268–274.

Hoffeditz, W.L., Son, H.J., Pellin, M.J., Farha, O.K. & Hupp, J.T., 2016. Engendering long-term air and light stability of a TiO2-supported porphyrinic dye via atomic layer deposition. *ACS Applied Materials and Interfaces*, 8, 34863–34869.

Hong, Y., Tian, C., Jiang, B., Wu, A., Zhang, Q., Tian, G. & Fu, H., 2013. Facile synthesis of sheet-like ZnO assembly composed of small ZnO particles for highly efficient photocatalysis. *Journal of Materials Chemistry A*, 1, 5700–5708.

Hosseini, F. & Mohebbi, S., 2018. Photocatalytic oxidation based on modified titanium dioxide with reduced graphene oxide and CdSe/CdS as nanohybrid materials. *Journal of Cluster Science*, 29, 289–300.

Howard, C.J., Sabine, T.M. & Dickson, F., 1991. Structural and thermal parameters forrutile and anatase. *Acta Crystallogr aphica B*, 47, 462–468.

Huang, H., Yuan, Q. & Yang, X., 2004. Preparation and characterization of metal–chitosan nanocomposites. *Colloids and Surfaces B, Biointerfaces*, 39, 31–37.

Ibhadon, A.O. & Fitzpatrick, P., 2013. Heterogeneous photocatalysis: Recent advances and applications. *Catalysts*, 3, 189–218.

Jain, R. & Sikarwar, S., 2006. Photocatalytic and adsorption studies on the removal of dye Congo red from wastewater. *International Journal of Environment and Pollution*, 27, 158–178.

Jinzhang, G., Shengying, L., Wu, Y., Guohu, Z., Lili, B. & Li, S., 2007. Preparation and photocatalytic activity of PANI/TiO2 composite film. *Rare Metals*, 26, 1–7.

Kaan, C.C., Aziz, A.A., Ibrahim, S., Matheswaran, M. & Saravanan, P., 2012. Heterogeneous photocatalytic oxidation an effective tool for wastewater treatment – A review. In: M. Kumarasamy, ed., *Studies on Water Management Issues*. InTech.

Kalathil, S., Khan, M.M., Ansari, S.A., Lee, J. & Cho, M.H., 2013. Band gap narrowing of titanium dioxide (TiO2) nanocrystals by electrochemically active biofilms and their visible light activity. *Nanoscale*, 5, 6323–6326.

Kamil, A.M., Hussein, F.H., Halbus, A.F. & Bahnemann, D.W., 2014. Preparation, characterization, and photocatalytic applications of MWCNTs/TiO2 composite. *International Journal of Photoenergy*, 2014, 475713.

Kawahara, T., Konishi, Y., Tada, H., Tohge, N., Nishii, J. & Ito, S., 2002. A patterned TiO2(anatase)/TiO2(rutile) bilayer-type photocatalyst: Effect of the anatase/rutile junction on the photocatalytic activity. *Angewandte Chemie*, 114, 2935–2937.

Kesraoui-Ouki, S., Cheeseman, C. & Perry, R., 1993. Effects of conditioning and treatment of chabazite and clinoptilolite prior to lead and cadmium removal. *Environmental Science and Technology*, 27, 1108–1116.

Khan, M.M., Ansari, S.A., Pradhan, D., Ansari, M.O., Lee, J. & Cho, M.H., 2014. Band gap engineered TiO2 nanoparticles for visible light induced photoelectrochemical and photocatalytic studies. *Journal of Materials Chemistry A*, 2, 637–644.

Khataee, A.R. & Mansoori, G.A., 2012. *Nanostructured Titanium Dioxide Materials: Properties, Preparation and Applications*. World Scientific.

Khoa, N.T., Kim, S.W., Yoo, D.H., Cho, S., Kim, E.J. & Hahn, S.H., 2015. Fabrication of Au/graphene-wrapped ZnO-nanoparticle-assembled hollow spheres with effective photoinduced charge transfer for photocatalysis. *ACS Applied Materials and Interfaces*, 7, 3524–3531.

Kishimoto, N., Morita, Y., Tsuno, H. & Yasuda, Y., 2007. Characteristics of electrolysis, ozonation, and their combination process on treatment of municipal wastewater. *Water Environment Research*, 79, 1033–1042.

Koli, V.B., Dhodamani, A.G., Raut, A.V., Thorat, N.D., Pawar, S.H. & Delekar, S.D., 2016. Visible light photo-induced antibacterial activity of TiO2-MWCNTs nanocomposites with varying the contents of MWCNTs. *Journal of Photochemistry and Photobiology A: Chemistry*, 328, 50–58.

Konstantinou, I.K. & Albanis, T.A., 2004. TiO2-assisted photocatalytic degradation of azo dyes in aqueous solution: Kinetic and mechanistic investigations: A review. *Applied Catalysis B: Environmental*, 49, 1–14.

Kwiatkowski, M., Bezverkhyy, I. & Skompska, M., 2015. ZnO nanorods covered with a TiO2 layer: Simple sol–gel preparation, and optical, photocatalytic and photoelectrochemical properties. *Journal of Materials Chemistry A*, 3, 12748–12760.

Lee, C.K. & Low, K.S., 1989. Removal of copper from solution using moss. *Environmental Technology Letters*, 10, 395–404.

Lee, J., Sorescu, D.C. & Deng, X., 2011. Electron-induced dissociation of CO2 on TiO2(110). *Journal of the American Chemical Society*, 133, 10066–10069.

Lee, S.-Y. & Park, S.-J., 2013. TiO2 photocatalyst for water treatment applications. *Journal of Industrial and Engineering Chemistry*, 19, 1761–1769.

Li, G.-S., Zhang, D.-Q. & Yu, J.C., 2009. A new visible-light photocatalyst: CdS quantum dots embedded mesoporous TiO2. *Environmental Science and Technology*, 43, 7079–7085.

Li, J., Feng, J. & Yan, W., 2013. Excellent adsorption and desorption characteristics of polypyrrole/TiO2 composite for methylene blue. *Applied Surface Science*, 279, 400–408.

Li, S., Du, C., Zhao, D., Zheng, J., Liu, H. & Wang, Y., 2015. Preparation and application of a new-type of ordered macroporous PANI/TiO2 photocatalyst. *Chemistry Letters*, 44, 568–570.

Li, X., Jiang, G., He, G., Zheng, W., Tan, Y. & Xiao, W., 2014. Preparation of porous PPyTiO2 composites: Improved visible light photoactivity and the mechanism. *Chemical Engineering Journal*, 236, 480–489.

Liao, D.L., Badour, C.A. & Liao, B.Q., 2008. Preparation of nanosized TiO2/ZnO composite catalyst and its photocatalytic activity for degradation of methyl orange. *Journal of Photochemistry and Photobiology A: Chemistry*, 194, 11–19.

Linsebigler, A.L., Lu, G. & Yates Jr, J.T., 1995. Photocatalysis on TiO2 surfaces: Principles, mechanisms, and selected results. *Chemical Reviews*, 95, 735–758.

Litter, M.I., 2005. Introduction to photochemical advanced oxidation processes for water treatment. In: P. Boule, D.W. Bahnemann, P.K.J. Robertson, eds., *Environmental Photochemistry Part II*. Springer, 325–366.

Litter, M.I., 2009. Treatment of chromium, mercury, lead, uranium, and arsenic in water by heterogeneous photocatalysis. *Advances in Chemical Engineering*, 36, 37–67.

Liu, J., Bai, H., Wang, Y., Liu, Z., Zhang, X. & Sun, D.D., 2010. Self-assembling TiO2 nanorods on large graphene oxide sheets at a two-phase interface and their anti-recombination in photocatalytic applications. *Advanced Functional Materials*, 20, 4175–4181.

Liu, L., Johnson, H.L., Cousens, S., Perin, J., Scott, S., Lawn, J.E., Rudan, I., Campbell, H., Cibulskis, R., Li, M., Mathers, C., Black, R.E. & Child Health Epidemiology Reference Group of WHO and UNICEF, 2012. Global, regional, and national causes of child mortality: An updated systematic analysis for 2010 with time trends since 2000. *The Lancet*, 379, 2151–2161.

Liu, T.-J., Wang, Q. & Jiang, P., 2013. Morphology-dependent photo-catalysis of bare zinc oxide nanocrystals. *RSC Advances*, 3, 12662–12670.

Lo, S.C., Lin, C.F., Wu, C.H. & Hsieh, P.H., 2004. Capability of coupled CdSe/TiO2 for photocatalytic degradation of 4-chlorophenol. *Journal of Hazardous Materials*, 114, 183–190.

Lu, S.-Y. & Lin, I.-H., 2003. Characterization of polypyrrole-CdSe/CdTe nanocomposite films prepared with an all electrochemical deposition process. *The Journal of Physical Chemistry B*, 107, 6974–6978.
Luo, C., Ren, X., Dai, Z., Zhang, Y., Qi, X. & Pan, C., 2017. Present perspectives of advanced characterization techniques in TiO2-based photocatalysts. *ACS Applied Materials and Interfaces*, 9, 23265–23286.
Luttrell, T., Halpegamage, S., Tao, J., Kramer, A., Sutter, E. & Batzill, M., 2014. Why is anatase a better photocatalyst than rutile? - Model studies on epitaxial TiO(2) films. *Scientific Reports*, 4, 4043.
Mattioli, G., Filippone, F., Alippi, P. & Amore Bonapasta, A.A., 2008. Ab initio study of the electronic states induced by oxygen vacancies in rutile and anatase TiO2. *Physical Review B*, 78, 241201.
Mattson, J.S. & Mark, H.B., 1971. *Activated Carbon: Surface Chemistry and Adsorption from Solution*. M. Dekker.
Mcguire, J.A., Joo, J., Pietryga, J.M., Schaller, R.D. & Klimov, V.I., 2008. New aspects of carrier multiplication in semiconductor nanocrystals. *Accounts of Chemical Research*, 41, 1810–1819.
Meichtry, J.M., Lin, H.J., De La Fuente, L., Levy, I.K., Gautier, E.A., Blesa, M.A. & Litter, M.I., 2007. Low-cost TiO2 photocatalytic technology for water potabilization in plastic bottles for isolated regions. Photocatalyst fixation. *Journal of Solar Energy Engineering*, 129, 119–126.
Mir, I.A., Singh, I., Birajdar, B. & Rawat, K., 2017. A facile platform for photocatalytic reduction of methylene blue dye by CdSe-TiO2 nanoparticles. *Water Conservation Science and Engineering*, 2, 43–50.
Miribangul, A., Ma, X., Zeng, C., Zou, H., Wu, Y., Fan, T. & Su, Z., 2016. Synthesis of TiO2/CNT composites and its photocatalytic activity toward sudan (I) degradation. *Photochemistry and Photobiology*, 92, 523–527.
Miyagi, T., Kamei, M., Mitsuhashi, T., Ishigaki, T. & Yamazaki, A., 2004. Charge separation at the rutile/anatase interface: A dominant factor of photocatalytic activity. *Chemical Physics Letters*, 390, 399–402.
Mizukoshi, Y. & Masahashi, N., 2010. Photocatalytic activities and crystal structures of titanium dioxide by anodization: Their dependence upon current density. *Materials Transactions*, 51, 1443–1448.
Mondal, K. & Sharma, A., 2014. Photocatalytic oxidation of pollutant dyes in wastewater by TiO2 and ZnO nano-materials—A mini-review. In: A. Misra & J.R. Bellare, eds., *Nanoscience & Technology for Mankind*. The National Academy of Sciences India, 36–72.
Moradi Dehaghi, S.M., Rahmanifar, B., Moradi, A.M. & Azar, P.A., 2014. Removal of permethrin pesticide from water by chitosan–zinc oxide nanoparticles composite as an adsorbent. *Journal of Saudi Chemical Society*, 18, 348–355.
MuñOz-Batista, M.J., GóMez-Cerezo, M.N., Kubacka, A., Tudela, D. & Fernández-García, M.2014. Role of interface contact in CeO2–TiO2 photocatalytic composite materials. *ACS Catalysis*, 4, 63–72.
Munter, R., 2001. Advanced oxidation processes–current status and prospects. *Proceedings of the Estonian Academy of Sciences, Chemistry*, 50, 59–80.
Nakata, K. & Fujishima, A., 2012. TiO2 photocatalysis: Design and applications. *Journal of Photochemistry and Photobiology C: Photochemistry Reviews*, 13, 169–189.
Naldoni, A., Allieta, M., Santangelo, S., Marelli, M., Fabbri, F., Cappelli, S., Bianchi, C.L., Psaro, R. & Dal Santo, V., 2012. Effect of nature and location of defects on bandgap narrowing in black TiO2 nanoparticles. *Journal of the American Chemical Society*, 134, 7600–7603.
Norranattrakul, P., Siralertmukul, K. & Nuisin, R., 2013. Fabrication of chitosan/titanium dioxide composites film for the photocatalytic degradation of dye. *Journal of Metals, Materials and Minerals*, 23, 9–22.
Novoselov, K.S., 2011. Nobel lecture: Graphene: Materials in the flatland. *Reviews of Modern Physics*, 83, 837–849.
Ohno, T., Akiyoshi, M., Umebayashi, T., Asai, K., Mitsui, T. & Matsumura, M., 2004. Preparation of S-doped TiO2 photocatalysts and their photocatalytic activities under visible light. *Applied Catalysis A: General*, 265, 115–121.
Ozerin, A.N., Zelenetskii, A.N., Akopova, T.A., Pavlova-Verevkina, O.B., Ozerina, L.A., Surin, N.M. & Kechek'yan, A.S., 2006. Nanocomposites based on modified chitosan and titanium oxide. *Polymer Science Series A*, 48, 638–643.
Pagga, U. & Brown, D., 1986. The degradation of dyestuffs: Part II Behaviour of dyestuffs in aerobic biodegradation tests. *Chemosphere*, 15, 479–491.
Papageorgiou, D.G., Kinloch, I.A. & Young, R.J., 2017. Mechanical properties of graphene and graphene-based nanocomposites. *Progress in Materials Science*, 90, 75–127.

Pawar, S.G., Chougule, M.A., Patil, S.L., Raut, B.T., Godse, P.R., Sen, S. & Patil, V.B., 2011. Room temperature ammonia gas sensor based on polyaniline-TiO2 Nanocomposite. *IEEE Sensors Journal*, 11, 3417–3423.

Pelaez, M., Nolan, N.T., Pillai, S.C., Seery, M.K., Falaras, P., Kontos, A.G., Dunlop, P.S.M., Hamilton, J.W.J., Byrne, J.A., O'shea, K., Entezari, M.H. & Dionysiou, D.D., 2012. A review on the visible light active titanium dioxide photocatalysts for environmental applications. *Applied Catalysis B: Environmental*, 125, 331–349.

Periasamy, K. & Namasivayam, C., 1994. Process development for removal and recovery of cadmium from wastewater by a low-cost adsorbent: Adsorption rates and equilibrium studies. *Industrial and Engineering Chemistry Research*, 33, 317–320.

Pollard, S.J.T., Fowler, G.D., Sollars, C.J. & Perry, R., 1992. Low-cost adsorbents for waste and wastewater treatment: A review. *Science of the Total Environment*, 116, 31–52.

Prasad, S., Alhesseny, H.S., Alsalhi, M.S., Devaraj, D. & Masilamai, V., 2017. A high power, frequency tunable colloidal quantum dot (cdse/zns) laser. *Nanomaterials*, 7, 29.

Priebe, J.B., Karnahl, M., Junge, H., Beller, M., Hollmann, D. & Brückner, A., 2013. Water reduction with visible light: Synergy between optical transitions and electron transfer in Au-TiO(2) catalysts visualized by in situ EPR spectroscopy. *Angewandte Chemie – International Edition*, 52, 11420–11424.

Radoičić, M., Ćirić-Marjanović, G., Spasojević, V., Ahrenkiel, P., Mitrić, M., Novaković, T. & Šaponjić, Z., 2017. Superior photocatalytic properties of carbonized PANI/TiO 2 nanocomposites. *Applied Catalysis B: Environmental*, 213, 155–166.

Rajeshwar, K., Osugi, M.E., Chanmanee, W., Chenthamarakshan, C.R., Zanoni, M.V.B., Kajitvichyanukul, P. & Krishnan-Ayer, R., 2008. Heterogeneous photocatalytic treatment of organic dyes in air and aqueous media. *Journal of Photochemistry and Photobiology C: Photochemistry Reviews*, 9, 171–192.

Raut, A.V., Yadav, H.M., Gnanamani, A., Pushpavanam, S. & Pawar, S.H., 2016. Synthesis and characterization of chitosan-TiO2: Cu nanocomposite and their enhanced antimicrobial activity with visible light. *Colloids and Surfaces B, Biointerfaces*, 148, 566–575.

Rehman, S., Ullah, R., Butt, A.M. & Gohar, N.D., 2009. Strategies of making TiO2 and ZnO visible light active. *Journal of Hazardous Materials*, 170, 560–569.

Rodda, D.P., Johnson, B.B. & Wells, J.D., 1993. The effect of temperature and pH on the adsorption of copper (II), Lead (II), and zinc (II) onto goethite. *Journal of Colloid and Interface Science*, 161, 57–62.

Rusu, E., Ursaki, V., Gutul, T., Vlazan, P. & Siminel, A. 2016. Characterization of TiO 2 nanoparticles and ZnO/TiO2 composite obtained by hydrothermal methoded. In: *3rd International Conference on Nanotechnologies and Biomedical Engineering*. Springer, 93–96.

Sakthivel, S., Shankar, M.V., Palanichamy, M., Arabindoo, B., Bahnemann, D.W. & Murugesan, V., 2004. Enhancement of photocatalytic activity by metal deposition: Characterisation and photonic efficiency of Pt, Au and Pd deposited on TiO2 catalyst. *Water Research*, 38, 3001–3008.

Saleh, T.A., 2011. The influence of treatment temperature on the acidity of MWCNT oxidized by HNO3 or a mixture of HNO3/H2SO4. *Applied Surface Science*, 257, 7746–7751.

Saleh, T.A., 2013. The role of carbon nanotubes in enhancement of photocatalysis. In: S. Suzuki, ed., *Syntheses and Applications of Carbon Nanotubes and Their Composites*. InTech.

Saravanan, R., Gracia, F. & Stephen, A., 2017. Basic principles, mechanism, and challenges of photocatalysis. In: M.M. Khan, D. Pradhan & Y. Sohn, eds., *Nanocomposites for Visible Light-induced Photocatalysis*. Springer, 19–40.

Saravanan, R., Shankar, H., Prakash, T., Narayanan, V. & Stephen, A., 2011. ZnO/CdO composite nanorods for photocatalytic degradation of methylene blue under visible light. *Materials Chemistry and Physics*, 125, 277–280.

Saravanan, R., Thirumal, E., Gupta, V., Narayanan, V. & Stephen, A., 2013. The photocatalytic activity of ZnO prepared by simple thermal decomposition method at various temperatures. *Journal of Molecular Liquids*, 177, 394–401.

Sasidharan, S., Chen, Y., Saravanan, D., Sundram, K.M. & Yoga Latha, L., 2011. Extraction, isolation and characterization of bioactive compounds from plants' extracts. *African Journal of Traditional, Complementary, and Alternative Medicines*, 8, 1–10.

Scheiber, P., Fidler, M., Dulub, O., Schmid, M., Diebold, U., Hou, W., Aschauer, U. & Selloni, A., 2012. (Sub) Surface mobility of oxygen vacancies at the TiO2 anatase (101) surface. *Physical Review Letters*, 109, 136103.

Schwarzenbach, R.P., Egli, T., Hofstetter, T.B., Von Gunten, U. & Wehrli, B., 2010. Global water pollution and human health. *Annual Review of Environment and Resources*, 35, 109–136.

Shen, C.S.H., Ta, Y.W., Ching, J.J. & Yang, T.C., 2011. Recent developments of metal oxide semiconductors as photocatalysts in advanced oxidation processes (AOPs) for treatment of dye waste-water. *Journal of Chemical Technology and Biotechnology*, 86, 1130–1158.

Shi, J., Chen, J., Feng, Z., Chen, T., Lian, Y., Wang, X. & Li, C., 2007. Photoluminescence characteristics of TiO2 and their relationship to the photoassisted reaction of water/methanol mixture. *The Journal of Physical Chemistry C*, 111, 693–699.

Shinde, S.S., Bhosale, C.H. & Rajpure, K.Y., 2013. Kinetic analysis of heterogeneous photocatalysis: Role of hydroxyl radicals. *Catalysis Reviews*, 55, 79–133.

Skocaj, M., Filipic, M., Petkovic, J. & Novak, S., 2011. Titanium dioxide in our everyday life; is it safe? *Radiology and Oncology*, 45, 227–247.

Srivastava, S., Tyagi, R., Pal, N. & Mohan, D., 1997. Process development for removal of substituted phenol by carbonaceous adsorbent obtained from fertilizer waste. *Journal of Environmental Engineering*, 123, 842–851.

Suriyaraj, S.P. & Selvakumar, R., 2016. Advances in nanomaterial based approaches for enhanced fluoride and nitrate removal from contaminated water. *RSC Advances*, 6, 10565–10583.

Tang, J., Durrant, J.R. & Klug, D.R., 2008. Mechanism of photocatalytic water splitting in TiO2. Reaction of water with photoholes, importance of charge carrier dynamics, and evidence for four-hole chemistry. *Journal of the American Chemical Society*, 130, 13885–13891.

Teoh, W.Y., Mädler, L., Beydoun, D., Pratsinis, S.E. & Amal, R., 2005. Direct (one-step) synthesis of TiO2 and Pt/TiO2 nanoparticles for photocatalytic mineralisation of sucrose. *Chemical Engineering Science*, 60, 5852–5861.

Ullah, H., Tahir, A.A. & Mallick, T.K., 2017. Polypyrrole/TiO2 composites for the application of photocatalysis. *Sensors and Actuators B: Chemical*, 241, 1161–1169.

Walton, K.S. & Snurr, R.Q., 2007. Applicability of the BET method for determining surface areas of microporous metal–organic frameworks. *Journal of the American Chemical Society*, 129, 8552–8556.

Wang, C., Thompson, R.L., Baltrus, J. & Matranga, C., 2009. Visible light photoreduction of CO2 using CdSe/Pt/TiO2 heterostructured catalysts. *The Journal of Physical Chemistry Letters*, 1, 48–53.

Wang, D., Wang, Y., Li, X., Luo, Q., An, J. & Yue, J., 2008a. Sunlight photocatalytic activity of polypyrrole–TiO2 nanocomposites prepared by 'in situ'method. *Catalysis Communications*, 9, 1162–1166.

Wang, D., Zhao, H., Wu, N., El Khakani, M.A. & Ma, D., 2010a. Tuning the charge-transfer property of PbS-quantum dot/TiO2-nanobelt nanohybrids via quantum confinement. *The Journal of Physical Chemistry Letters*, 1, 1030–1035.

Wang, H., Xie, C., Zhang, W., Cai, S., Yang, Z. & Gui, Y., 2007. Comparison of dye degradation efficiency using ZnO powders with various size scales. *Journal of Hazardous Materials*, 141, 645–652.

Wang, N., Li, X., Wang, Y., Hou, Y., Zou, X. & Chen, G., 2008b. Synthesis of ZnO/TiO2 nanotube composite film by a two-step route. *Materials Letters*, 62, 3691–3693.

Wang, X., Feng, Z., Shi, J., Jia, G., Shen, S., Zhou, J. & Li, C., 2010b. Trap states and carrier dynamics of TiO 2 studied by photoluminescence spectroscopy under weak excitation condition. *Physical Chemistry Chemical Physics*, 12, 7083–7090.

Wang, Y., Li, X., Lu, G., Chen, G. & Chen, Y., 2008c. Synthesis and photo-catalytic degradation property of nanostructured-ZnO with different morphology. *Materials Letters*, 62, 2359–2362.

Wei, J., Zhang, Q., Liu, Y., Xiong, R., Pan, C. & Shi, J., 2011. Synthesis and photocatalytic activity of polyaniline–TiO2 composites with bionic nanopapilla structure. *Journal of Nanoparticle Research*, 13, 3157–3165.

Wetterskog, E., Agthe, M., Mayence, A., Grins, J., Wang, D., Rana, S., Ahniyaz, A., Salazar-Alvarez, G. & Bergström, L., 2014. Precise control over shape and size of iron oxide nanocrystals suitable for assembly into ordered particle arrays. *Science and Technology of Advanced Materials*, 15, 055010.

William IV, L., Kostedt, I., Ismail, A.A. & Mazyck, D.W., 2008. Impact of heat treatment and composition of ZnO–TiO2 nanoparticles for photocatalytic oxidation of an azo dye. *Industrial and Engineering Chemistry Research*, 47, 1483–1487.

Wu, T.-M. & Wu, C.-Y., 2006. Biodegradable poly (lactic acid)/chitosan-modified montmorillonite nanocomposites: Preparation and characterization. *Polymer Degradation and Stability*, 91, 2198–2204.

Wu, Z., Dong, F., Zhao, W., Wang, H., Liu, Y. & Guan, B., 2009. The fabrication and characterization of novel carbon doped TiO2 nanotubes, nanowires and nanorods with high visible light photocatalytic activity. *Nanotechnology*, 20, 235701.

Xia, Y., Zhu, K., Kaspar, T.C., Du, Y., Birmingham, B., Park, K.T. & Zhang, Z., 2013. Atomic structure of the anatase TiO2(001) surface. *The Journal of Physical Chemistry Letters*, 4, 2958–2963.

Xiang, Q., Yu, J. & Jaroniec, M., 2012. Synergetic effect of MoS2 and graphene as cocatalysts for enhanced photocatalytic H2 production activity of TiO2 nanoparticles. *Journal of the American Chemical Society*, 134, 6575–6578.

Xiao, F.X., 2012. Construction of highly ordered ZnO–TiO2 nanotube arrays (ZnO/TNTs) heterostructure for photocatalytic application. *ACS Applied Materials and Interfaces*, 4, 7055–7063.

Xiao, F.X., Hung, S.F., Tao, H.B., Miao, J., Yang, H.B. & Liu, B., 2014. Spatially branched hierarchical ZnO nanorod-TiO 2 nanotube array heterostructures for versatile photocatalytic and photoelectrocatalytic applications: Towards intimate integration of 1D–1D hybrid nanostructures. *Nanoscale*, 6, 14950–14961.

Xiao, G., Su, H. & Tan, T., 2015. Synthesis of core–shell bioaffinity chitosan–TiO2 composite and its environmental applications. *Journal of Hazardous Materials*, 283, 888–896.

Xie, Y., Qian, H., Zhong, Y., Guo, H. & Hu, Y., 2012. Facile low-temperature synthesis of carbon nanotube/nanohybrids with enhanced visible-light-driven photocatalytic activity. *International Journal of Photoenergy*, 2012, 682138.

Xu, X., Wang, J., Tian, J., Wang, X., Dai, J. & Liu, X., 2011. Hydrothermal and post-heat treatments of TiO2/ZnO composite powder and its photodegradation behavior on methyl orange. *Ceramics International*, 37, 2201–2206.

Yan, X., Zou, C., Gao, X. & Gao, W., 2012. ZnO/TiO 2 core–brush nanostructure: Processing, microstructure and enhanced photocatalytic activity. *Journal of Materials Chemistry*, 22, 5629–5640.

Yang, L., Luo, S., Liu, R., Cai, Q., Xiao, Y., Liu, S., Su, F. & Wen, L., 2010. Fabrication of CdSe nanoparticles sensitized long TiO2 nanotube arrays for photocatalytic degradation of anthracene-9-carbonxylic acid under green monochromatic light. *The Journal of Physical Chemistry C*, 114, 4783–4789.

Yang, Z., Wang, B., Cui, H., An, H., Pan, Y. & Zhai, J., 2015. Synthesis of crystal-controlled TiO2 nanorods by a hydrothermal method: Rutile and brookite as highly active photocatalysts. *The Journal of Physical Chemistry C*, 119, 16905–16912.

Yasui, M., Katagiri, K., Yamanaka, S. & Inumaru, K., 2012. Molecular selective photocatalytic decomposition of alkylanilines by crystalline TiO 2 particles and their nanocomposites with mesoporous silica. *RSC Advances*, 2, 11132–11137.

Yazdani, M., Bahrami, H. & Arami, M., 2014. Feldspar/titanium dioxide/chitosan as a biophotocatalyst hybrid for the removal of organic dyes from aquatic phases. *Journal of Applied Polymer Science*, 131(10), 1–9.

Yu, J., Wang, G., Cheng, B. & Zhou, M., 2007. Effects of hydrothermal temperature and time on the photocatalytic activity and microstructures of bimodal mesoporous TiO2 powders. *Applied Catalysis B: Environmental*, 69, 171–180.

Yuwono, A.H., Zhang, Y., Wang, J., Zhang, X.H., Fan, H. & Ji, W., 2006. Diblock copolymer templated nanohybrid thin films of highly ordered TiO2 nanoparticle arrays in PMMA matrix. *Chemistry of Materials*, 18, 5876–5889.

Zhang, H., Lv, X., Li, Y., Wang, Y. & Li, J., 2010. P25-graphene composite as a high performance photocatalyst. *ACS Nano*, 4, 380–386.

Zhang, J., Xiong, Z. & Zhao, X.S., 2011. Graphene–metal–oxide composites for the degradation of dyes under visible light irradiation. *Journal of Materials Chemistry*, 21, 3634–3640.

Zhang, L., Miller, B.K. & Crozier, P.A., 2013. Atomic level in situ observation of surface amorphization in anatase nanocrystals during light irradiation in water vapor. *Nano Letters*, 13, 679–684.

Zhao, H., Pan, F. & Li, Y., 2017. A review on the effects of TiO 2 surface point defects on CO 2 photoreduction with H 2 O. *Journal of Materiomics*, 3, 17–32.

Zhou, F., Yan, C., Sun, Q. & Komarneni, S., 2018. TiO2/sepiolite nanocomposites doped with rare earth ions: Preparation, characterization and visible light photocatalytic activity. *Microporous and Mesoporous Materials*.

Zouzelka, R., Kusumawati, Y., Remzova, M., Rathousky, J. & Pauporté, T., 2016. Photocatalytic activity of porous multiwalled carbon nanotube-TiO2 composite layers for pollutant degradation. *Journal of Hazardous Materials*, 317, 52–59.

# 12

## *Dendrimer-Based Hybrid Nanomaterials for Water Remediation: Adsorption of Inorganic Contaminants*

**Herlys Viltres, Oscar F. Odio, and Edilso Reguera**

**CONTENTS**



## 12.1 Introduction

Nanomaterials are currently on the cutting edge of materials science research and are gradually finding applications in our daily life, including life sciences, energy, and environmental remediation. Currently, the surface assembly of functional inorganic nanostructures with dendrimers is emerging as a powerful and challenging strategy for developing sophisticated hybrid nanomaterials with excellent performance and amazing applications. Dendrimers are novel synthetic polymers that possess highly branched structures with unique three-dimensional molecular configurations and large numbers of reactive end groups (Chou and Lien, 2011, Han et al., 2012, Tomalia, 2005a, Tomalia, 2005b). Hence, they have become one of the research hotspot areas for host–guest chemistry. Their size, unique shape, open interior cavities, particle-like topography with numerous end-groups, and exciting properties such as high solubility, high reactivity, and low viscosity (Patel and Patel, 2013, Tomalia et al., 1985) make them widely applicable in a range of fields such as water treatment (Wang et al., 2015), separating agents (Ulaszewska et al., 2012), drug-delivery systems (Jędrych et al., 2014), gene therapy (Wang et al., 2010), catalysts (Karakhanov et al., 2015, Natarajan and Jayaraman, 2011), electronic applications, and chemical sensors (Sánchez-Navarro and Rojo, 2012, Satheeshkumar et al., 2015).

This chapter covers recent advances in the use of dendrimer-containing hybrid nanomaterials for water remediation, with emphasis on contamination by heavy metals and radioactive elements. The rest of this section introduces some aspects of water pollution and dendrimer chemistry. Later sections deal with available methodologies for the synthesis of the hybrid nanostructures and representative applications of these nanoplatforms in the removal of contaminants, including several efforts devoted to elucidate the adsorption mechanism. Finally, some considerations and future perspectives are provided.

### 12.1.1 Water Pollution

Water resources are crucial for all life forms. Keeping this natural resource free of pathogens and toxic chemicals is essential for human health and the maintenance of Earth ecosystems. Nowadays, there exist several anthropogenic activities responsible for the contamination of water resources, such as the rapid growth of population, urbanization, and industrialization (Elimelech and Phillip, 2011). According to the World Health Organization, 1.2 billion people worldwide are still without access to clean drinking water, while 2.6 billion do not have access to basic sanitation (Kurniawan et al., 2012, Pereira et al., 2009). Even at low concentrations, environmental pollutants represent serious threats to freshwater sources, public health, and living organisms (Kurniawan et al., 2006, Kurniawan et al., 2012, Repo et al., 2010, Teijon et al., 2010). This chapter is focused on two inorganic pollutant families: heavy metals and radioactive isotopes.

In the field of environmental pollution, heavy metals mainly refer to metals or metalloids ions that show significant toxicity to living organisms, such as mercury (Hg), cadmium (Cd), lead (Pb), chromium (Cr), zinc (Zn), copper (Cu), cobalt (Co), nickel (Ni), tin (Sn), arsenic (As), selenium (Se), thallium (Tl), antimony (Sb), and so on (Lee et al., 2015, Liu, 2014, Maleki et al., 2015). The major sources of heavy metals are the effluents from chemical factories of fuels, batteries, plastics, paintings, fertilizers, and also from metallurgical and mining processes (Ebrahimi et al., 2013, Hayati et al., 2016, Hua et al., 2012). Water contamination by heavy metal ions has become a severe environmental problem in recent decades. Owing to their resistance, non-biodegradability, high toxicity, and cumulative effects in biological systems, such metal ions signify a serious threat to human and aquatic ecosystems, even if they appear in low concentrations, due to their toxic and carcinogenic effects (Hayati et al., 2016, Hayati et al., 2017, Liu, 2014, Maleki et al., 2015). Precious metals (e.g. gold [Au], silver [Ag], platinum [Pt], and palladium [Pd]) constitute a subfamily within the heavy metals. Not only can they be toxic but they are also limited resources widely used in a variety of advanced applications such as electronic devices and catalysis. Thus, it is important to develop proper methodologies for their recovery and reuse due to their high economic impact.

Much of the research work in the field of nuclear waste management is still centered on the separation/preconcentration of several isotopes of uranium (U) (Alijani et al., 2015, Sengupta et al., 2016), thorium (Th) (Chen et al., 2007, Sengupta et al., 2016), plutonium (Pu) (Gupta et al., 2016, Kumar et al., 2016), and cesium (Cs) (Thammawong et al., 2013) due to their great importance in nuclear applications and to the irreversible damage that they provoke in living organisms (Taylor, 1989). An important alpha-emitting long-lived radionuclide is $^{237}Np$ (neptunium), which requires proper mitigation steps in the case of nuclear fallout since exposure can induce bone, lung, and liver cancers. The separation of these elements requires special conditions that will be discussed in section 12.3.3.

### 12.1.2 Properties and Synthesis of Dendrimers

Dendrimers were first presented by the group of Fritz Vögtle in 1978 (Buhleier et al., 1978). They consist of repetitively branched molecules with three-dimensional morphology (Behbahani et al., 2014, Kumar et al., 2017b, Zhang et al., 2014). Each macromolecule is constituted by elementary building blocks called *dendrons*, which are connected according to a tree-like pattern around a multifunctional core (Felder-Flesch, 2016). They are routinely synthesized as nanostructures with tunable physical and chemical properties that may be designed and controlled as a function of their size, shape, surface chemistry, and interior void space, which in turn depends on the nature of the core, the type and number of branches, and the terminal functional groups (Felder-Flesch, 2016, Sun et al., 2016).

One of the most intensively studied dendrimers in the last decades is poly(amidoamine) (PAMAM), mainly because of its low toxicity, low cost, and accessibility. Lower generations (G1 and G2) comprise flexible molecules with starlike shapes and no appreciable inner regions, while medium-sized generations (G3–G5) exhibit both starlike and spherelike shapes with an internal space separated from the outer shell of the dendrimer. Contrarily, larger generations (> G6) present a clear spherelike shape with large empty inner regions and dense surfaces due to the structure of the outer shell (Gröhn et al., 2000, Kumar et al., 2017b). PAMAM consists of three basic units: an ethylenediamine (EDA) core, repeating units with secondary amide and tertiary amine functions, and terminal primary amine groups (Figure 12.1A). Their synthesis follows a divergent approach and is accomplished by a serial repetition of two reactions: Michael addition of amine groups to the double bond of methyl acrylate (MA), followed by amidation of the resulting methyl ester with EDA molecules. Therefore, each complete reaction sequence results in a new generation with an increase in the dendrimer diameter of about 1 nm. Due to the high density of functional nitrogen groups, the PAMAM dendrimers and their derivatives could display a strong binding affinity for toxic heavy metal ions in aqueous solutions, (Rether and Schuster, 2003) making them favored adsorbents in water purification.

Dendritic structures are mostly synthesized by two classic approaches: divergent and convergent.

- *Divergent dendrimer synthesis (inside out)*: the first dendrimers were prepared following this strategy by the groups of Denkewalter (Denkewalter et al., 1981), Tomalia (Tomalia et al., 1985), and Newkome (Newkome et al., 1985). The construction of the dendrimer takes place in a stepwise manner starting from the core and building up the macromolecule towards the periphery using two basic operations: (1) coupling of the monomer and (2) deprotection or transformation of the monomer end-group (activation). This permits to create a new reactive surface functionality and the further coupling of a new monomer (Boas et al., 2006). Hence, the first generation G1 is obtained by the coupling of a ligand unit on the central core (Figure 12.1A). The repetition of such sequence allows the synthesis of higher generations Gn. (Felder-Flesch, 2016.)
- *Convergent growth (outside in)*: this strategy was first introduced by Fréchet (Hawker and Frechet, 1990). The growth of the dendrimers starts from the periphery toward the core

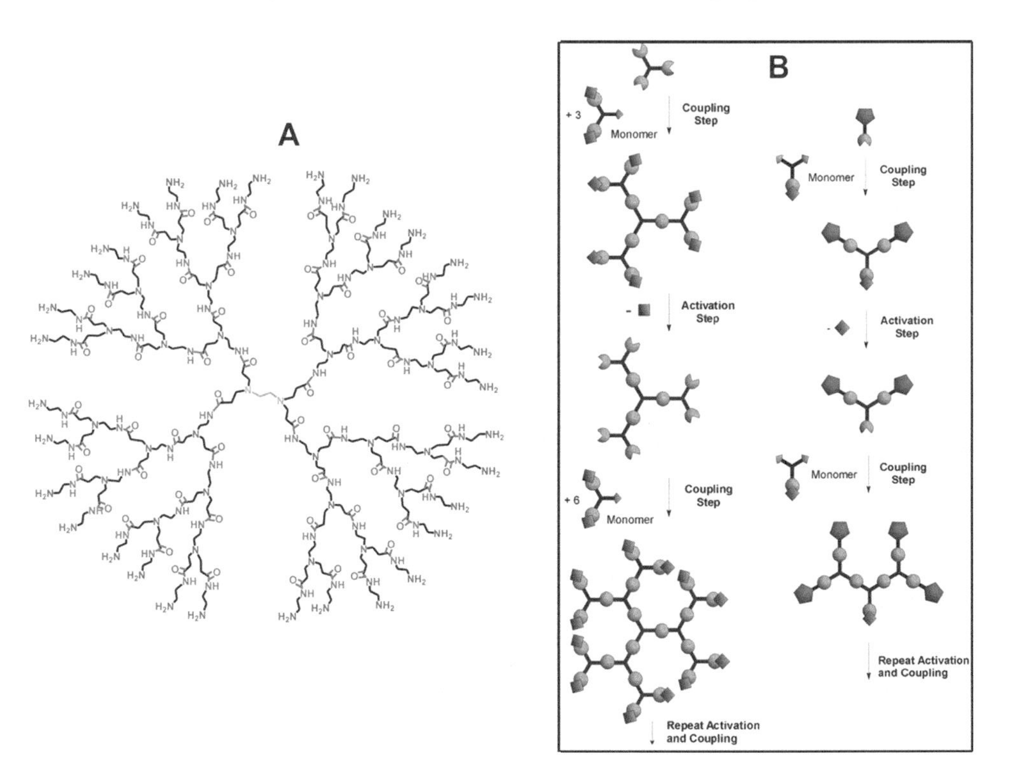


**FIGURE 12.1 (See color insert.)** (**A**) Structure of PAMAM-G4 dendrimer. (**B**) Schematic representation of divergent (*left*) and convergent (*right*) strategies for dendrimer synthesis. (Adapted with permission from Grayson and Frechet, 2001, American Chemical Society.)

(Figure 12.1B). Only a small number of reactive sites are involved in each reaction step, thus giving rise to few secondary reactions and to rather defectless structures. However, one issue relies on the central position of the reactive groups: for high generations, the focal point is more and more isolated and buried by branches, which leads to reactivity decrease (Felder-Flesch, 2016, Grayson and Frechet, 2001).

## 12.2 Synthesis of Dendrimer-Containing Nanohybrids

### 12.2.1 Dendrimer/Dendron Growth from Core Nanomaterials

One of the most popular strategies for the synthesis of hybrid nanomaterials comprises the divergent growth of dendrimers/dendrons from the inorganic nanomaterial containing the initial dendritic core (Chou and Lien, 2011, Kim et al., 2016, Kim and Park, 2017, Liu et al., 2008, Zhou et al., 2013). In this methodology, nanomaterials cores are first functionalized in order to bear amine groups on their surface; from these functions, alternate reactions with MA and EDA or other organic ligands with the appropriate terminal group give the desired generations. The resulting core/shell hybrid structures generally have good monodispersity and abundant surface functional groups. The divergent approach is more effective since it allows the high yield formation of clean structures with empty internal cavities, while the convergent approach likely means that inorganic cores could be trapped in the cavities.

Either the nature of the nanomaterial core or the amine-containing ligand can vary in accordance with the desired application. For example, Pan et al. (2005) modified the surface of iron oxide nanoparticles (NPs) with (3-aminopropyl)trimethoxysilane (APTMS) through direct surface silanization to obtain PAMAM-G4 dendrons; the hybrid nanostructures showed excellent water dispersity. The resulting dendrons can be further derivatized in order to obtain different terminal functions like carboxylic (Yen et al., 2017) or thiourea groups (Niu et al., 2016). In some instances, the method can be improved by using $SiO_2$-coated NPs prior to amine silanization, as was demonstrated by Liang et al. (2017) The $SiO_2$ shell not only avoids the aggregation of magnetic NPs, but also increases the incorporation of APTMS molecules as starting points for PAMAM growth, which gives nanomaterials with denser dendritic content (Figure 12.2). The system could be designed even more complex; thus, in a recent work, Jiang et al. (2018) reported the synthesis of magnetic hydrazide-functionalized PAMAM dendrons embedded with $TiO_2$ NPs for the high performance enrichment of phosphopeptides. Also, other functional inorganic materials like graphene oxide (GO) can be decorated with dendritic structures by this strategy to obtain magnetic GO functionalized with PAMAM molecules (mGO-PAMAM) (Einollahi Peer et al., 2018, Ma et al., 2017b). By taking advantage of GO functional groups, EDA is directly incorporated

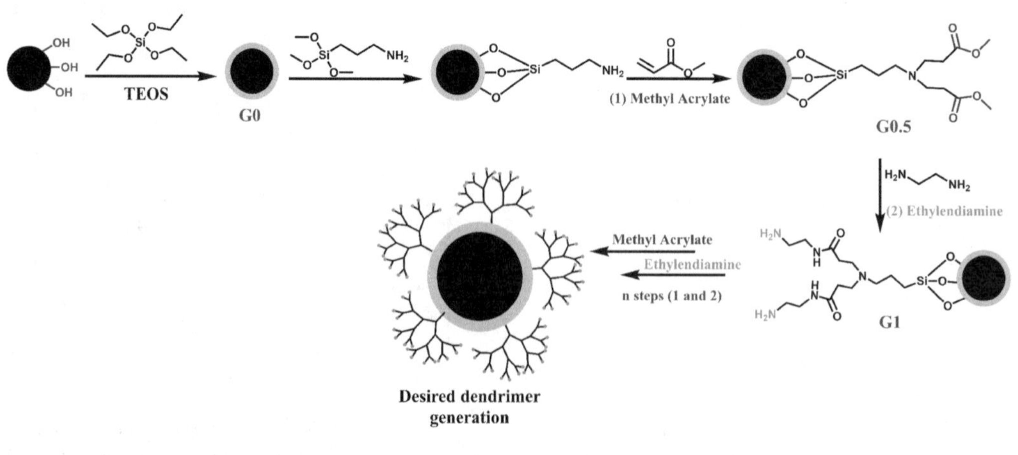

**FIGURE 12.2** Schematic representation of the synthesis of $Fe_3O_4$@$SiO_2$-PAMAM NPs through the divergent growth of PAMAM dendrons at the surface of APTMS-functionalized $SiO_2$ shell.

into the carbonaceous material as a prior step for dendron growth. Either EDA or tris(2-aminoethyl) amine (TAEA) () have been used during the formation of subsequent generations. In some instances, primary amines are not the starting points for dendritic growth, e.g. Pourjavadi et al. (2016) developed G3-PAMAM dendrons from the ester groups of poly(methyl acrylate) chains coating $Fe_3O_4$ NPs.

### 12.2.2 Dendrimer-Assembled Nanomaterials

In this strategy, pre-synthesized NPs and dendrimers are assembled together through specific chemical or physical interactions, which include covalent bonds (Yu et al., 2009), electrostatic interaction (Shi et al., 2007), and hydrogen bonding (Xu et al., 2012). Electrostatic interactions can be exploited in a facile approach called layer-by-layer (LbL) deposition, in which prefabricated structures bearing opposite charges are taken together. This method was employed in the synthesis of iron oxide NPs covered with a bilayer of polystyrene sulfonate (PSS) and PAMAM-G5 dendrimers can be obtained (Wang et al., 2007). First, as-synthesized magnetite NPs with positive charges at the surface was assembled with negatively charged PSS, followed by the deposition of PAMAN dendrimers previously treated in order to acquire positive charges. The resulting hybrid nanoplatform was subjected to an acetylation reaction to neutralize the remaining terminal amine groups of the dendrimers.

Recently, coordinated and covalent assemblies have been the preferred choice. Kong and co-workers (Kong et al., 2017) developed a simple approach to synthesize PAMAM-modified $MoS_2$ nanoflakes (Figure 12.3A). PAMAM-G5 dendrimers were coupled with lipoic acid (LA). Then, PAMAM-LA was anchored to the $MoS_2$ nanoflakes through the linkage between LA disulfide bonds and $MoS_2$ surface. The resulting PAMAM-LA-$MoS_2$ NPs present improved stability compared to pristine aggregated $MoS_2$ nanoflakes because of the dendritic coating (Figure 12.3B). In other paper, GO flakes were successfully modified with PAMAM by activating the carboxyl groups in GO with carbonyl chlorides (Nonahal et al., 2018). Also, several coupling strategies like EDC/NHS (Hayati et al., 2016) and click chemistry (Desmecht et al., 2018) have been used for assembling PAMAM onto carbon nanotubes (CNT) walls. Moreover, Mohammadi and co-workers (2018) coupled Au NPs with thiol-terminated PAMAM dendrimers due to Au-S coordination. Finally, a recent paper reports the design of a novel nanocomposite formed with magnetite nanoflowers as cores and polydopamine (PDA) and PAMAM layers as successive shells (Sun et al., 2018). PDA shell is formed by self-polymerization of dopamine (DP) in the presence of the NPs. At alkaline pH, the resulting surface PDA quinone moieties are allowed to react with the dendrimer free amine groups to form the second shell. The final $Fe_3O_4$@PDA@PAMAM nanohybrid still retains suitable magnetic properties for further applications.

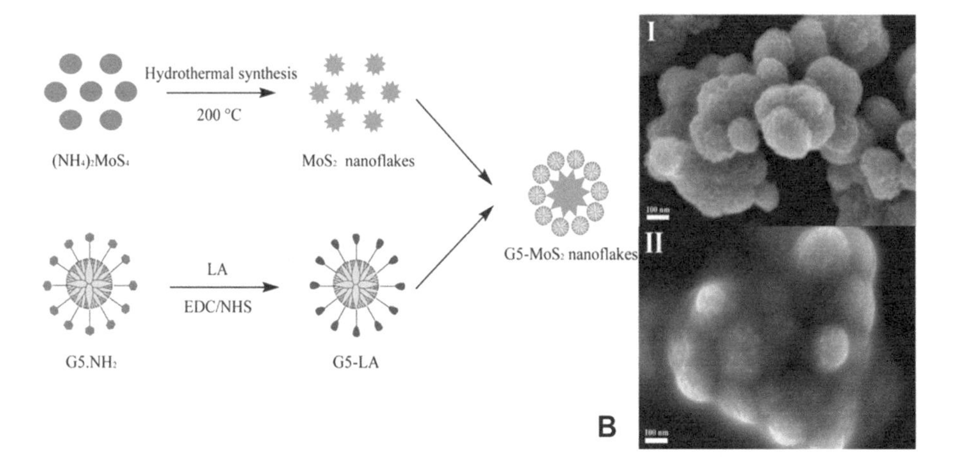


**FIGURE 12.3** (**A**) Schematic representation of the synthesis of PAMAMG5-LA-$MoS_2$ nanoflakes. (**B**) FESEM images of $MoS_2$ (I) and PAMAMG5-LA-$MoS_2$ nanoflakes (II). (Adapted with permission from Kong et al., 2017, American Chemical Society.)

### 12.2.3 Dendron-Assembled Nanomaterials

Dendrons with various functional groups can be incorporated into several materials mainly in two ways: a direct grafting process and a ligand exchange reaction. In this strategy, dendrons are first synthesized in a step-by-step manner similar to the conventional divergent approach used to synthesize dendrimers; then, they are incorporated into the material surface mainly in two ways: direct grafting process or ligand exchange reaction.

Woo et al. (2007) reported a study on the noncovalent functionalization of single-walled carbon nanotubes (SWCNTs) by three dendrons with different terminal groups. Dendrons with benzyl and carboxyl groups were successfully attached to the carbon surface through $\pi$–$\pi$ interactions and strong van der Waals interactions; contrarily, dendrons with methyl ester functions as terminal groups were incapable of being grafted onto the SWCNT surface. CNTs have also been dendron-conjugated through click chemistry, in particular, the copper(I)-catalyzed azide-alkyne cycloaddition (CuAAC) (Battigelli et al., 2013, Clavé and Campidelli, 2011). This versatile reaction is simple, fast and renders good yields. For instance, Battigelli et al. (2013) prepared CNTs with acetylene moieties at the surface, which were easily coupled to dendrons bearing protected guanidinium groups at the periphery and one azide group at the apex (Figure 12.4). The resulting positive-charged dendronized CNTs were tested as siRNA carriers.

Basly et al. (2013) reported the design of iron oxide NPs functionalized with small hydrophilic dendrons containing a phosphonic acid group at the dendron apex. When initial hydrophobic NPs coated

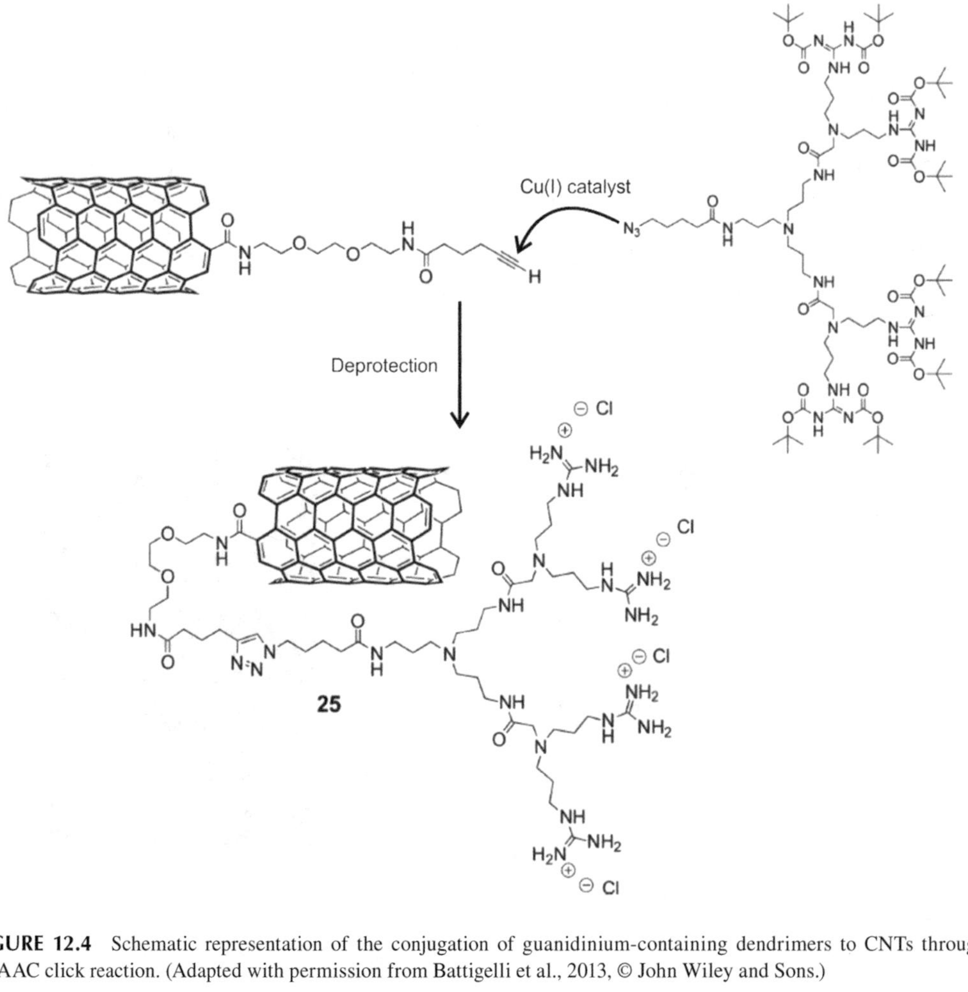

**FIGURE 12.4** Schematic representation of the conjugation of guanidinium-containing dendrimers to CNTs through CuAAC click reaction. (Adapted with permission from Battigelli et al., 2013, © John Wiley and Sons.)

with oleic acid molecules were used, dendron incorporation occurred through a ligand exchange process that stabilized the NPs in polar media. Contrarily, when naked surface NPs were employed, dendrons were directly grafted. In both cases, grafting rates were similar, but the anchoring seemed to be stronger when naked NPs were used, which can be related to the geometry of the iron-phosphonate complexes at the NP surface. In another study, Xu and coworkers (Xu et al., 2014) developed a novel multifunctional supramolecular hybrid dendrimer (SHD). It comprised dual-functionalized low generation peptide dendrons (PDs) self-assembled onto fluorescent CdSe/ZnS quantum dots (QDs). The PDs were modified with arginine moieties as peripheral groups for improved biocompatibility, while LA was incorporated into the dendron core for organic–inorganic coordination. *In situ* UV irradiation triggers the rupture of the LA disulfide bridge leading to the formation of free thiol moieties that self-assemble on the QD surface.

### 12.2.4 Dendrimer-Entrapped Nanomaterials

The unique internal structure of dendrimers allows their use as templates to entrap different metal and semiconductor nanomaterials. Such nanohybrids are typically produced in a stepwise procedure by complexation of the metal ions followed by their controlled reduction to form the inorganic nanostructure. Host–guest structures are only achieved by fine tuning of the experimental conditions, depending on the dendrimer architecture and the size of the NPs, which in turn is a function of the metal:dendrimer ratio, the dendrimer concentration, and the rate of metal reduction. In a pioneer paper, the group of Amis (Gröhn et al., 2000) made a systematic study on the fabrication of Au NPs using PAMAM dendrimers as polymeric templates. They noted that lower PAMAM generations (up to G4) are incapable of templating the Au colloids, but they tend to stabilize the colloids by surrounding the NP surface with multiple molecular dendrimers (see Section 12.2.5). However, for larger generations (G6–G9) both the internal free space and the chain flexibility inside the structure allow for the host of one Au NP, whose diameter is precisely controlled by the dendrimer generation. Interestingly, PAMAM-G10 rather hosts several smaller Au NPs due to the decrease in chain flexibility, leading to a different morphology of the hybrid nanomaterial. The work of Priyam and coworkers (Priyam et al., 2010) also supports the same conclusions about the influence of dendrimer generation in the morphology of nanohybrids.

By studying the template synthesis of semiconductor HgTe NPs employing PAMAM dendrimers, they demonstrated that multiple G5 dendrimers are needed for colloid stabilization, while G7 dendrimers effectively encapsulate the semiconductor NPs. Nevertheless, when the Hg(II):G7 ratio is lowered enough, some semiconductor NPs are formed outside the dendrimer internal cavities and are only surface stabilized by the external functional groups. Remarkably, they noted that HgTe-PAMAMG7 nanocapsules tend to self-organize by forming a necklace-type array due to the interpenetration of the branches from neighboring dendrimers, which is favored by the chain flexibility of PAMAMG7 and changes the optical properties (Figure 12.5). Other reports applying this synthetic strategy with different dendrimers and NPs can be found elsewhere (Desmecht et al., 2018, Knecht et al., 2008, Liu et al., 2014).

### 12.2.5 Dendrimer-Stabilized Nanomaterials

In dendrimer-stabilized nanomaterials, the inorganic nanostructure is coated with multiple dendrimers anchored to the NP surface. Dendrimers are generally of medium and low generations, while NPs tend to be larger than in Section 12.2.4. According to the literature, multiple types of dendrimers can be used to stabilize metal or metal oxide NPs. The stabilized nanohybrids are prepared through the *in situ* formation of the NPs in the presence of the dendrimers.

In an interesting paper, Strable et al. (2001) reported the synthesis and stabilization of maghemite NPs in the presence of carboxylated PAMAM-G4.5 dendrimers. The inorganic phase is formed by oxidative hydrolysis of Fe(II) ions at the dendrimer–solution interface under controlled and mild conditions, leading to NPs with a narrow size distribution in the 20–30 nm range. The work shows that these dendrimers not only effectively control the formation of well-defined iron oxide NPs but also limit interparticle aggregation, which results in high colloidal stability. Further studies with smaller generations of the same carboxylated dendrimer also enabled effective stabilization of NPs; contrarily, neither amine- nor

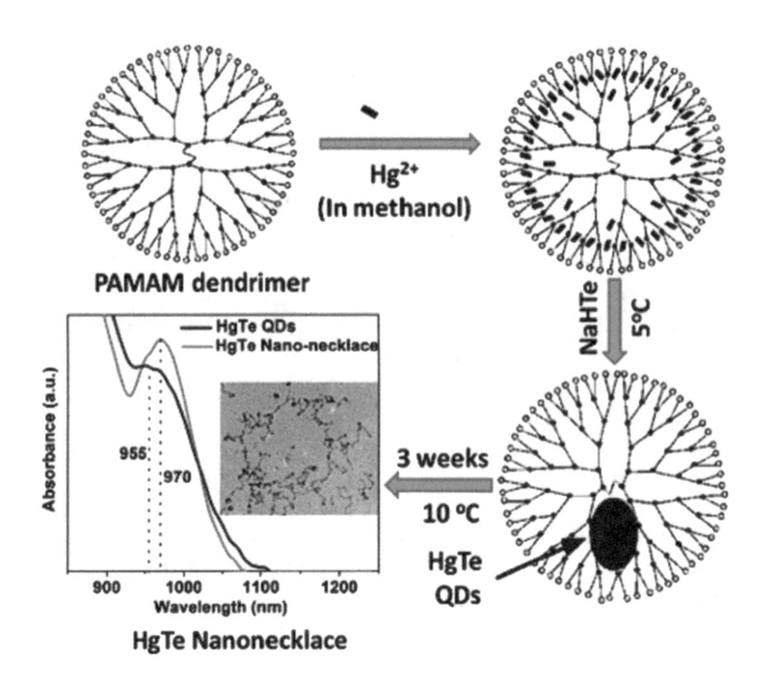


**FIGURE 12.5** Scheme of the synthesis of HgTe QDs trapped by PAMAM dendrimers; TEM image shows the self-organization of the hybrid QDs into a necklace-type array, which leads to changes in the UV vis spectrum. (Reproduced with permission from Priyam et al., 2010, American Chemical Society.)

hydroxyl-terminated G5 dendrimers were capable of stabilizing the NPs, which showed the important role of electrostatic interactions in NP stabilization (Stanicki et al., 2017, Strable et al., 2001).

More recently, Yuan et al. (2013) fabricated near-monodisperse 12.1 nm Ag NPs stabilized with PAMAN-G5. Dendrimers act as complex agents for Ag(I) ions and stabilizing macromolecules after the formation of colloidal Ag NPs via $NaBH_4$ reduction. In a similar work, Liu and coworkers (2013) reported the synthesis and stabilization of 5.5 nm Au NPs by using PAMAM-G2 dendrimers without the need for additional reducing agents. Also, Pd NPs stabilized with low generation glycodendrimers have been synthesized by the same route (Figure 12.6A) (Gatard et al., 2014). Here, glycodendrimers coordinate Pd(II) ions through the intradendritic triazole ligands prior to $NaBH_4$ reduction. Transmission electron microscopy (TEM) imaging shows the formation of ultra-small (average diameter: 2.3 nm) Pd-stabilized NPs with good monodispersity owing to a low concentration of the metal precursor used during the process (Figure 12.6B). Other recent examples comprising more complex experimental setups can be found elsewhere (Domracheva, 2018, Zhao et al., 2018).

## 12.3 Applications for Aqueous Contaminant Removal: Adsorption Mechanisms

The three-dimensional shapes and the presence of multiple functional groups in dendrimers make them very useful for attracting ions and molecules, which can be adsorbed either inside the cavities or at the periphery. Adsorption mechanisms typically comprise electrostatic attraction, chemical interaction, and physical adsorption. The nature of the interactions depends mainly on the chemistry of the functional groups, the dendritic porosity, the kind of the target species, and the solution pH. However, as far as water remediation technologies are concerned, dendrimers by themselves have limited practical use due to their high water solubility, which makes difficult their rapid and facile separation from the treated water. In order to overcome this key issue, dendrimers can be wisely combined with several functional nanomaterials, thereby increasing their range of applicability since they offer many advantages: high and reactive surface area, great porosity, enhanced chemical, thermic, and mechanical resistance, facile

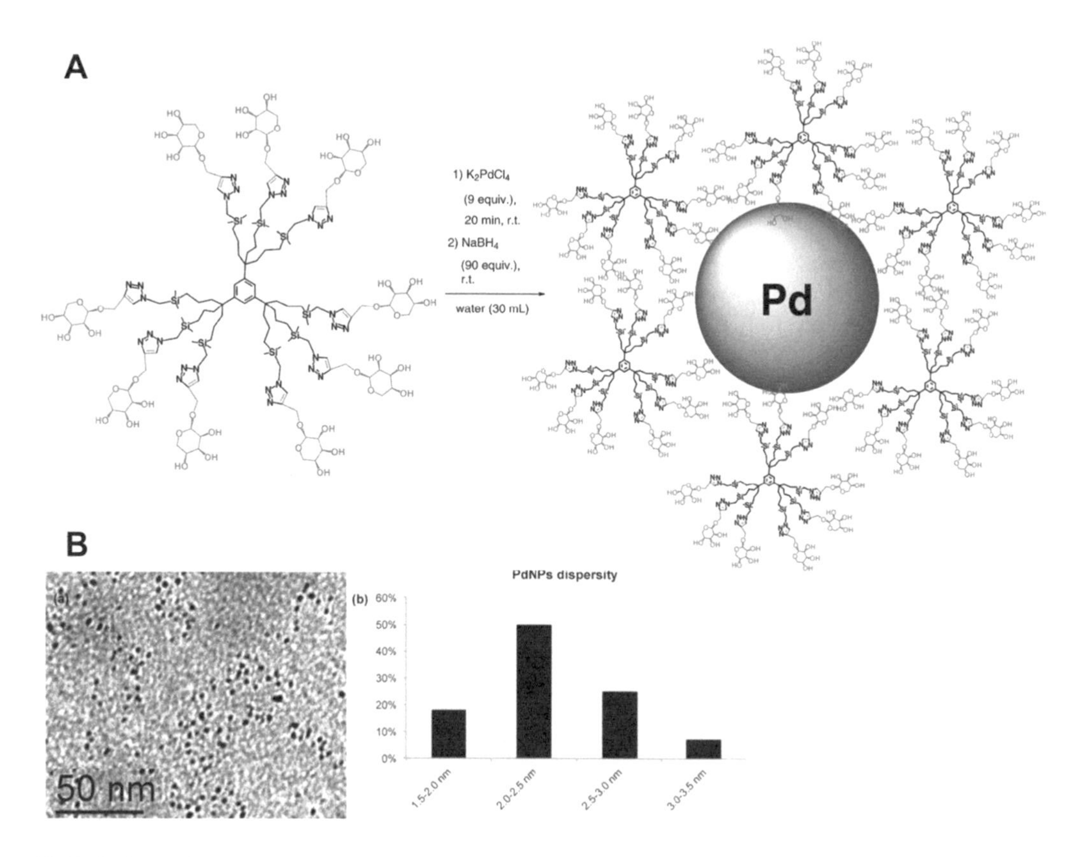


**FIGURE 12.6** (**A**) Schematic illustration of the synthesis of Pd NPs stabilized with glycodendrimers. (**B**) TEM image and size distribution histogram of Pd/dendrimer NPs. (Adapted with permission from Gatard et al., 2014, © 2014 John Wiley and Sons.)

separation and reuse, and environmentally friendly processes. The most common functional materials used in water decontamination technologies are $SiO_2$ and $TiO_2$ NPs, carbonaceous materials like CNT and GO, and the family of spinel ferrite NPs, which allow for magnetic separation of the loaded sorbents (Odio and Reguera, 2017).

Among the available dendrimer-based sorbents for water treatment, PAMAM is very frequently used for the removal of heavy metals from wastewater because it contains numerous cavities and is easy to functionalize with a range of chemical groups that can selectively chelate metal ions from solutions (Ma et al., 2009, Maiti et al., 2005, Rether and Schuster, 2003). Furthermore, its excellent solubility in water, as well as its reactive amine or ester groups (in the case of intermediate generations) on the periphery, allows for the preparation of polychelatogens with selective complexation of heavy-metal ions (Rether and Schuster, 2003). After adsorption, sorbent separated from remediated water can be chemically treated to detach the sequestered metals; the resulting metal concentrates are destined to reuse or disposal. At the same time, this procedure enables a new adsorption-recovery cycle with the same sorbent (Song et al., 2017b).

In this section, we present a concise overview of the applications of dendrimer-based hybrid nanoadsorbents for aqueous contaminant removal. In some cases, we review several attempts to elucidate possible adsorption mechanisms; such information is extremely useful because detailed knowledge of the sorbent–sorbate interactions at atomic and molecular levels allows for the improvement and optimization of remediation processes.

### 12.3.1 Heavy Metals

In an interesting report, Niu et al. (2016) studied the adsorption behavior of a series of $SiO_2$-supported sulfur-capped PAMAM dendrons toward Hg(II) ions. G0, G1, and G2 PAMAM dendrons were modified in order to exhibit terminal methylthiurea moieties. It was confirmed that adsorption capacity increased with increasing dendron generation up to 337 mg/g at optimum pH 6. Adsorption isotherms fitted to the Langmuir model and the process was found to take place by a chemical mechanism. Also, selectivity experiments showed that for the three adsorbents Hg(II) ions bind with high preference respect to harder Pearson acids like Fe(III), Ni(II), Zn(II), and Cd(II) ions, which suggests that sulfur atoms from thiourea moieties play an important role in the adsorption process. Indeed, density functional theory (DFT) calculations revealed that G0 dendrons interact with Hg(II) through the S atom in a mono-coordinated manner, whereas G1 behaves as a pentadentate ligand through the both terminal S atoms, the N atom of the internal tertiary amine group, and two O atoms from the amide functions (Figure 12.7). Natural bond orbital (NBO) analysis confirmed a charge transfer process from ligand to metal ions.

In other work, G3-PAMAM-graft-poly(methyl acrylate) magnetic nanocomposites were designed for the magnetic removal of Pb(II) ions (Pourjavadi et al., 2016). Maximum adsorption capacity ($Q_{max}$) of 310 mg/g was reached at pH 5–6. Adsorption decreases with decreasing pH due to the protonation of primary and tertiary amines; such positive charges make difficult the approach and penetration of Pb(II) cations into the dendron arms. the isotherm followed the Langmuir model and showed chemical interactions upon adsorption. Moreover, DFT calculations of several Pb-PAMAM configurations along the dendron segments predicted high adsorption free energies. The most likely structure comprised the coordination of Pb with the N atom of the tertiary amine and two O atoms from amide groups; this 3-coordinated complex also shows the less positive charge over the metal atom, which is in accordance with a charge transfer mechanism from the ligand to the metal. Similar calculations with Cd instead of Pb also yielded the same conclusions. These strong interactions between Pb(II) and PAMAM-MNC adsorbent resulted in high adsorption capacity of adsorbent, which is consistent with experimental observations. Another attempt to achieve the magnetic recovery of Pb(II) ions by dendron-functionalized GO was presented by Ma et al. (2017a) In this case, amine-based dendrons were prepared with maleic acid moieties as terminals in the intermediate generations. Pb(II) adsorption tests were performed with either amine-terminated or carboxyl-terminated dendrons. For the two sets, it was noted that the higher the generation, the higher the ion uptake. Also, amine-terminated dendrons were always more effective than the parent carboxyl-terminated. In an attempt to shed light on the adsorption mechanism, X-ray photoelectron spectroscopy (XPS) measurements were performed; based on the analysis of high resolution spectra for Pb 4f and N 1s orbitals, authors confirmed that Pb(II) adsorption comprises the formation of

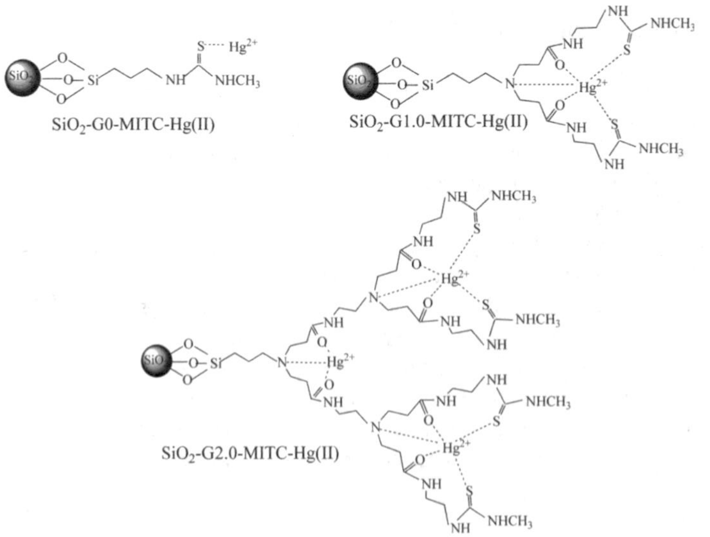

**FIGURE 12.7** Scheme of the proposed chelating mechanism of Hg(II) with methylthiourea derivatives of PAMAM dendrimers. (Reproduced with permission from Niu et al., 2016, American Chemical Society.)

amino-complexes. The XPS technique was also employed in a similar work dealing with Hg(II) uptake by magnetic GO-PAMAM dendrons (Ma et al., 2017b). In this case, the analysis of the high resolution spectra suggested that during adsorption Hg(II) is reduced to Hg(I), but neither the resulting amino-complexes nor the electron source were clarified.

The critical effect of pH on adsorption systems governed by electrostatic interactions is clearly demonstrated in the work of Xiao and coworkers (2016) Authors study the adsorption of selenite ($HSeO_3^-$) and selenate ($SeO_4^{2-}$) anions by using different generations (G0–G4) of PAMAM-functionalized GO. It was noted that for Se(VI) species, increasing the acidity of the solution markedly enhances the adsorption capacity of the GO-G4 composites (Figure 12.8A); such behavior is consistent with strong electrostatic attractions between the double-charged anion and the increasingly protonated dendron amines. In contrast, Se(IV) species are more insensible to lowering the solution pH; this effect is the combination of weaker electrostatic attractions and the likely formation of inner-sphere complexes between Se(IV) and hydroxyl groups at the surface of GO. These conclusions are also supported by comparing the adsorption capacity of different dendron generations (Figure 12.8B). Se(VI) species are more retained as dendron growth due to increasing number of amine groups, while Se(IV) species follow the same trend in a more discrete fashion except for a decline in G4 generation, which was explained by the authors claiming that large dendrons might hinder the access of selenite to the GO surface.

The adsorption capacities of metal cations with medium Pearson hardness can be sensibly enhanced by increasing the nitrogen functions of the dendritic ligand. For example, Jiryaei et al. (2017) designed a melamine-based dendrimer amine-modified magnetic NPs (MDA-$Fe_3O_4$) for the efficient removal of Pb(II) ions. Adsorption capacity was near 350 mg/g at an optimum pH of 5. Isotherm fitted well to the Freundlich model, which is consistent with multiple adsorption sites. In another approach following this idea, Hayati et al. (2017) fabricated a novel material based on a fifth generation of poly(propyleneimine) (PPI-G5) dendrimer linked to $SiO_2$ NPs for the removal of several divalent cations. The high number of amine groups gives maximum adsorption uptakes at neutral pH in the range of 440–500 mg/g; excellent fit to Langmuir-type behavior suggests homogeneity of the binding sites. By converting mg/g to mmol/g, it is noted that $Q_{max}$ increases in the order Pb(II) < Cu(II) ≈ Ni(II) ≈ Co(II), which confirmed the preferential affinity of aminated ligands for divalent cations of the first transition row. Similar conclusions have been supported by the work of Einollahi Peer et al. (2018) Authors tested $Fe_3O_4$-GO nanosheets grafted with PAMAM-G2 dendrons for the magnetic recovery of Cd(II), Pb(II), and Cu(II) ions. Maximum molar adsorption capacities follow the order Pb(II) < Cd(II) < Cu(II). It is interesting to note that Cu(II) adsorption was best fitted to the Langmuir model, while Pb(II) and Cd(II) rather obeyed

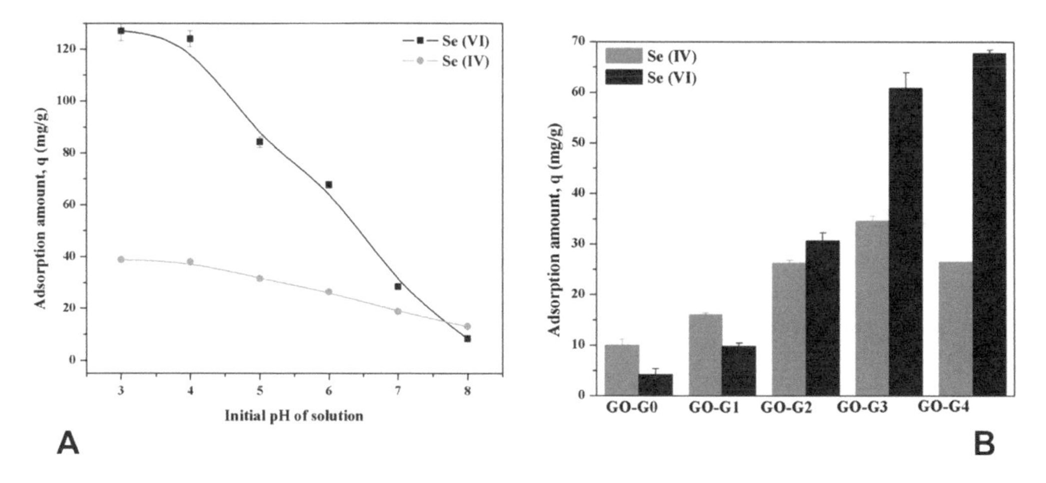


**FIGURE 12.8** (**A**) Adsorption behavior of Se (IV) and Se(VI) species over the surface of GO-PAMAMG4 composites as a function of initial pH. (**B**) Adsorption behavior at pH 6 of Se(IV) and Se(VI) species over the surface of GO-PAMAMGn composites as a function of PAMAM dendrimer generation. (Adapted from Xiao et al., 2016, with permission from Elsevier.)

the Freundlich-type. These differences suggest that Cu(II) ions are solely linked to the dendritic amine moieties, while Cd(II) and Pb(II) also anchor to the functional groups of the GO surface. For the last two mentioned systems it was found that the treatment of metal-loaded nanohybrids in an acidic medium displays great desorption performance; after five cycles, nanomaterials still retain excellent adsorption capacities. Other amine-based dendritic nanomaterials ($Fe_3O_4$@PDA@PAMAM) have been tested for magnetic separation of Cu(II) ions, but with less efficiency compared to the nanoplatforms mentioned above (Sun et al., 2018).

Recently, branched poly(ethyleneimine) [b-PEI] dendrimers have been employed in the capture of Pb(II) and Cr(VI) species from wastewater. Hydrophilic b-PEI was first grafted to mesoporous magnetic clusters (MMC) to allow for facile adsorbent separation (Lee et al., 2018). Such a hybrid nanoplatform was able to remove Pb(II) ions with a $Q_{max}$ of 216 mg/mol. Adsorption follows the typical pH dependence due to the progressive protonation of amine groups, being optimum at pH 6. The high selectivity of the adsorbent toward Pb(II) in the presence of alkaline and alkaline-earth cations evidenced the chelating nature of the interactions between amines and Pb(II) ions. In order to expand the removal capabilities for other contaminants, hybrid nanocomposites were modified by converting the dendron amines to permanent quaternary ammonium groups. The resulting permanent positive PEI dendrimers (p-PEI) were able to uptake oxyanions like chromates with high efficiency ($Q_{max}$=334 mg/g) through electrostatic attraction. However, the active sites are poorly chromate selective in the presence of $NO_3^-$ due to anion size differences. Isotherm studies showed that MMC-b-PEI-Pb(II) system is well-fitted by the Langmuir model, while MMC-p-PEI-Cr(VI) system is better explained by the Freundlich model. In the last case, it is apparent that quaternary ammonium groups are not evenly distributed along the iron oxide surface. Moreover, reusability studies also differ for both hybrid nanomaterials, since MMC-b-PEI conserves its adsorption capacity after three cycles, but under the same conditions, MMC-p-PEI loses almost a third. Amine-based dendrimers have been also used for the treatment of other highly toxic species-forming oxyanions like As(III) with promising results (Hayati et al., 2018).

Finally, we mention the use of dendritic hybrid nanomaterials for an environmental application not related to water remediation. In an experimental study complemented with computational calculations, Song and coworkers (2017b) tested silica-gel-supported PAMAM dendrimers of several half and complete generations (G0.5 – G3.0) for the removal of Co(II) ions present in fuel ethanol. Results indicated that for half generations an increase in ester moieties corresponds to an increase in Co(II) adsorption. However, for complete generations the $Q_{max}$ is achieved by G2; the decrease of G3 capacity was devoted to a more crowded dendritic surface, which hampers the diffusion of the cations into the dendron cavities needed to adopt a stable complex configuration. Further DFT optimizations showed that in half generations Co(II) is preferentially forming a quadri-coordinated complex with two amine N atoms and two carbonyl O atoms from ester groups. In the case of complete generations, the preferred Pb-PAMAM complex configuration entails penta-coordination in which an external primary amine group also participates; such result disagrees with similar DFT calculations previously cited, in which the more stable Pb-PAMAM complex was tridentate (Pourjavadi et al., 2016). NBO analysis suggested the occurrence of ligand to metal charge transfer with prevalent electron donation from internal N atoms. Similar strategies have been adopted in order to extract other heavy metals from fuel ethanol [Hg(II) and Ag(I)] (Song et al., 2017a) and waste dimethyl sulfoxide (DMSO) [Cd(II) and Fe(III)] (Zhu et al., 2018).

### 12.3.2 Precious Metals

So far, there are few reports with promising results that make use of dendrimer-based hybrid nanomaterials to trap precious metals. For instance, Zhang and coworkers (2015) designed a highly branched diethylenetriamine (DETA)-based dendrimer grown at the surface of $SiO_2$ particles for the recovery of Au(III) ions. The high density of amine groups gives a $Q_{max}$ value of 418 mg/g at room temperature and pH 2.5. At this pH value, Au(III) exists as an $AuCl_4^-$ complex and dendron amine groups are fully protonated; therefore, high values of $Q_{max}$ are devoted to electrostatic attractions between the solute and the sorbent surface. This also explains why the sorbent exhibits high Au(III) selectivity in binary systems containing several cations that remain electropositive at this pH. Based on thermodynamic data, authors have also suggested the formation of inner-sphere complexes in which the N lone pair coordinates to the

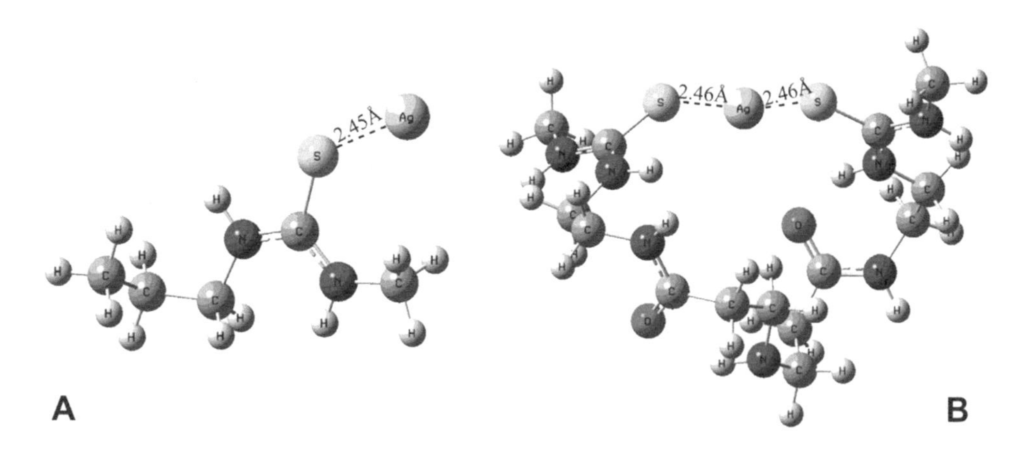


**FIGURE 12.9** **(See color insert.)** DFT optimized geometries of the complexes formed by the interaction of Ag(I) with PAMAM-G0 (**A**) and G1 (**B**) generations. (Adapted from Zhang et al., 2018, with permission from Elsevier.)

Au atoms. After adsorption, gold is separated from the sorbent by treatment with thiourea; the material still exhibited very good adsorption performance after three regeneration cycles.

In a different approach, Zhang et al. (2018) performed a theoretical and experimental study concerning the adsorption of Ag(I) by silica-supported PAMAM dendrimers bearing methylthiourea moieties as terminal functions. Given that Ag(I) is a soft Pearson acid, thiourea groups are introduced in order to enhance the adsorption capacity due to metal–sulfur interactions. As a result, $Q_{max}$ increased from G0 to G2 up to 140 mg/g at pH 6. In addition, the adsorbents are highly selective for Ag(I) in the presence of harder cations like Fe(II), Zn(II), Cd(II), and Pb(II). Isotherms are well fitted to the Langmuir model while thermodynamic data showed a strong chemisorption process. Further, DFT calculations were used to predict the most likely Ag-dendron adducts; in the case of G0 dendrons, which only exhibits one terminal group, Ag(I) is forming a monodentate complex through the S atom (Figure 12.9A). However, for dendrons with multiple functional groups, the most stable adduct configuration entails a bidentate complex with two S atoms (Figure 12.9B), though tridentate and pentadentate coordination involving amine N atoms and carbonyl O atoms are also possible. Further, NBO analysis showed that in all cases ligand to metal charge transfer processes are dominated by the S lone pair electron donation to the Ag 5s orbital.

### 12.3.3 Radioactive Elements

As in the case of precious metals, scarce reports deal with the application of dendrimer-based material for the water removal of radioisotopes. In the special case of this application, adsorbing materials are exposed to high energy particles (α and β) and γ-radiation; hence, they must display radiolytic stability in order to avoid performance losses because of the degradation of the adsorbing functions. In addition, nuclear wastes are often very acidic media due to high concentrations (1–10 M) of $HNO_3$, which requires the use of materials with chemical stability under these conditions. Another important issue is the need for rapid and efficient separation of the radiotoxic metal from the sorbent after water treatment, aiming to avoid sorbent damage and metal radioactivity losses before the next application.

In recent studies, Sengupta et al. have tested multi-walled CNTs functionalized with G1 and G2 PAMAM generations for the adsorption and recover of radiotoxic Pu(IV), (Kumar et al., 2017b) Am(III)(Kumar et al., 2017a) and Np(V)(Sengupta et al., 2018) species in acidic water. Experimental results showed that both materials present good selectivity in simulated nuclear waste water and high adsorption capacities, and metal back recovery was efficient with several stripping agents like oxalic acid and EDTA. In all cases, the G2 dendrimer presented higher $Q_{max}$ value (see Table 12.1) and was radiolytically more stable, though both materials do not present important performance loses up to 500

**TABLE 12.1**

Recently Reported Nanohybrid Materials for the Uptake of Several Heavy Metal Pollutants

| Nanohybrid Material | Pollutant | Interaction | $Q_{max}$ (mg/g) | Reference |
|---|---|---|---|---|
| $SiO_2$-PAMAMG2(sulfur) | Hg(II) | Complexation | 337 | Niu et al. (2016) |
| $Fe_3O_4$-PAMAM-*g*-poly(methyl acrylate) | Pb(II) | Complexation | 310 | Pourjavadi et al. (2016) |
| GO-PAMAMG4 | Se(IV) | Complex. + Elect. | 61 | Xiao et al. (2016) |
| | Se(VI) | Electrostatic | 78 | |
| MMC-b-PEI | Pb(II) | Complexation | 216 | Lee et al. (2018) |
| MMC-p-PEI | Cr(VI) | Electrostatic | 334 | |
| $SiO_2$-PPIG5 | Pb(II) | Complexation | 471 | Hayati et al. (2017) |
| | Ni(II) | | 438 | |
| | Cu(II) | | 460 | |
| | Co(II) | | 503 | |
| MDA-$Fe_3O_4$ | Pb(II) | Complexation | 333 | Jiryaei Sharahi and Shahbazi (2017) |
| $Fe_3O_4$-GO-PAMAMG3.0 | Hg(II) | Complexation + Reduction | 114 | Ma et al. (2017b) |
| $Fe_3O_4$-GO-PMAAMG3.0 | Pb(II) | Complexation | 181 | Ma et al. (2017a) |
| $Fe_3O_4$@PDA@PAMAM | Cu(II) | Physisorption | 97 | Sun et al. (2018) |
| $SiO_2$-DETAG2 | Au(III) | Complex. + Elect. | 418 | Zhang et al. (2015) |
| $Fe_3O_4$-GO-PAMAMG3 | Pd(IV) | Complexation | 3.6 | Yen et al. (2017) |
| | Au(III) | | 3.6 | |
| | Pd(II) | | 2.8 | |
| | Ag(I) | | 2.8 | |
| $SiO_2$-PAMAMG2(sulfur) | Ag(I) | Complexation | 140 | Zhang et al. (2018) |
| $Fe_3O_4$-GO-PAMAMG3.0 | Cd(II) | Complex. + Elect. | 449 | Einollahi Peer et al. (2018) |
| | Pb(II) | | 340 | |
| | Cu(II) | | 354 | |
| MWCNT-PAMAMG2 | $Pu^{4+}$ | Complexation | 92.48 | Kumar et al. (2017b) |
| MWCNT-PAMAMG2 | $Am^{3+}$ | Complexation | 97.62 | Kumar et al. (2017a) |
| MWCNT–PAMAMG2 | Np(V)) | Complexation | 59 | Sengupta et al. (2018) |
| $SiO_2$-PAMAMG6 | U(VI) | Complexation | 303 | Shaabana et al. (2018) |

kGy. Adsorption thermodynamics reflect chemisorption processes. Moreover, for the three ions and with independence of the dendrimer generation, it was observed that adsorption increases with the concentration of $HNO_3$, which suggests that $NO_3^-$ anions favor complex formation. Due to the high acidic conditions, adsorption does not seem to occur through amine functions but through the amide carbonyl groups. Indeed, recent DFT studies dealing with the complexation of nitrate-coordinated Am(III) and U(VI) species by a single amidoamine arm, it was showed that both N amine and O amide atoms participate in ion complexes; however, the protonation of amine terminal groups effectively weakens the coordination strength since amidoamine behaves as a monodentate ligand through the carbonyl oxygen (Deb et al., 2018). Further, luminescence studies with Eu(III), which is an inactive surrogate of Am(III), gave new insights about the nature of the interactions between this ion and dendronic amidoamine functions (Kumar et al., 2017a). Such investigations suggested that G2 dendrimers form more symmetrical and less covalent complexes than G1, but in both cases, no water molecules are present in the metal coordination sphere.

PAMAM have been also used for U(VI) adsorption in batch and fixed-bed column modes (Shaaban et al., 2018) G1–G6 dendrons were grown over $SiO_2$ and the resulting hybrid materials were tested is less harsh conditions. Uptake capacities increased with dendrimer generation up to 303 mg/g for G6 at pH 4.5. At lower pH values adsorption decreases due to amine protonation.

## 12.4 Final Remarks and Perspectives

Dendrimer-based hybrid nanomaterials have the potential to become the next generation of adsorbents. This chapter was intended to provide a concise picture of the current state of the art about the design and use of these hybrid nanostructures for the adsorption and recovery of important water contaminants like heavy metals and radioactive elements. Dendrimers exhibit a highly branched architecture with open internal cavities and multiple functional groups that make them excellent candidates for many applications in different fields. Their effective combination with novel functional nanomaterials has opened new attractive scenarios and possibilities, in particular for water treatment processes, which require sorbents with special characteristics in order to be efficient, cheap, simple, and environmentally friendly. By means of several synthetic methodologies, it is possible to design a wide range of hybrid nanomaterial and tune key adsorption properties like grafting density, dendrimer size and shape, and the number and nature of functional groups. Within dendrimers, amine-based structures have been so far the preferred choice, especially PAMAM and their derivatives. In the case of functional materials, silica gel, CNTs, GO, and iron oxide magnetic NPs are customary.

In the last five years, these hybrid materials have been tested for the adsorption and recovery of heavy metals, especially Pb(II), Hg(II), Cu(II), and other divalent cations. Up to now, there have been reports of several nanoplatforms with promising adsorption capabilities, but the topic is far from being exhausted. First, we consider that the potentials of dendrimers have not been yet exploited, especially the possibility of being properly designed with specific functional groups that could enhance the uptake and selectivity of a given metal; second, the range of species is still limited, in particular, for highly toxic elements like As and Cr. Moreover, in the near future, it would be desirable to include other functional materials like metallic and mixed ferrite NPs in the design of dendronic hybrid materials with novel properties.

Adsorption studies comprise mechanistic insights; the information is used for improving material performance and the efficiency of the process. In this sense, efforts have been centered in thermodynamic and kinetic analysis, but in many instances, these phenomenological approaches have not been corroborated with spectroscopic data and surface characterization techniques, which provide local information at the atomic and molecular level. In addition, theoretical studies remain a valuable tool in order to understand surface interactions and predict the adsorption behavior for the successful removal of water contaminants.

## REFERENCES

Alijani, H., Beyki, M. H. & Mirzababaei, S. N. 2015. Adsorption of UO22+ ions from aqueous solution using amine functionalized MWCNT: Kinetic, thermodynamic and isotherm study. *Journal of Radioanalytical and Nuclear Chemistry*, 306, 165–173.

Basly, B., Popa, G., Fleutot, S., Pichon, B. P., Garofalo, A., Ghobril, C., Billotey, C., Berniard, A., Bonazza, P., Martinez, H., Felder-Flesch, D. & Begin-Colin, S. 2013. Effect of the nanoparticle synthesis method on dendronized iron oxides as MRI contrast agents. *Dalton Transactions*, 42, 2146–2157.

Battigelli, A., Wang, J. T.-W., Russier, J., Da Ros, T., Kostarelos, K., Al-Jamal, K. T., Prato, M. & Bianco, A. 2013. Ammonium and guanidinium dendron–carbon nanotubes by amidation and click chemistry and their use for siRNA delivery. *Small*, 9, 3610–3619.

Behbahani, M., Gorji, T., Mahyari, M., Salarian, M., Bagheri, A. & Shaabani, A. 2014. Application of polypropylene amine dendrimers (POPAM)-grafted MWCNTs hybrid materials as a new sorbent for solid-phase extraction and trace determination of gold (III) and palladium (II) in food and environmental samples. *Food Analytical Methods*, 7, 957–966.

Boas, U., Christensen, J. B. & Heegaard, P. M. 2006. *Dendrimers in Medicine and Biotechnology: New Molecular Tools*. Royal Society of Chemistry.

Buhleier, E., Wehner, W. & Vögtle, F. 1978. 'CASCADE'-and 'nonSKID–CHAIN–LIKE' syntheses of molecular cavity topologies. *ChemInform*, 9, 155–158.

Chen, C., Li, X., Zhao, D., Tan, X. & Wang, X. 2007. Adsorption kinetic, thermodynamic and desorption studies of Th (IV) on oxidized multi-wall carbon nanotubes. *Colloids and Surfaces A: Physicochemical and Engineering Aspects*, 302, 449–454.

Chou, C.-M. & Lien, H.-L. 2011. Dendrimer-conjugated magnetic nanoparticles for removal of zinc (II) from aqueous solutions. *Journal of Nanoparticle Research*, 13, 2099–2107.

Clavé, G. & Campidelli, S. 2011. Efficient covalent functionalisation of carbon nanotubes: The use of "click chemistry". *Chemical Science*, 2, 1887–1896.

Deb, A. K. S., Pahan, S., Dasgupta, K., Panja, S., Debnath, A. K., Dhami, P. S., Ali, S. M., Kaushik, C. P. & Yadav, J. S. 2018. Carbon nano tubes functionalized with novel functional group- amido-amine for sorption of actinides. *Journal of Hazardous Materials*, 345, 63–75.

Denkewalter, R. G., Kolc, J. & Lukasavage, W. J. 1981. Macromolecular highly branched homogeneous compound based on lysine units. Google Patents.

Desmecht, A., Steenhaut, T., Pennetreau, F., Hermans, S. & Riant, O. 2018. Synthesis and catalytic applications of multi-walled carbon nanotube-Polyamidoamine dendrimer hybrids. *Chemistry - A European Journal*, 24, 12992–13001.

Domracheva, N. 2018. Multifunctional properties of γ-Fe2O3 Nanoparticles Encapsulated Into Liquid-Crystalline Poly (propylene imine) Dendrimer. In: N. Domracheva, M. Caporali & E. Rentschler, eds., *Novel Magnetic Nanostructures*. Elsevier.

Ebrahimi, R., Maleki, A., Shahmoradi, B., Daraei, H., Mahvi, A. H., Barati, A. H. & Eslami, A. 2013. Elimination of arsenic contamination from water using chemically modified wheat straw. *Desalination and Water Treatment*, 51, 2306–2316.

Einollahi Peer, F. E., Bahramifar, N. & Younesi, H. 2018. Removal of Cd (II), Pb (II) and Cu (II) ions from aqueous solution by polyamidoamine dendrimer grafted magnetic graphene oxide nanosheets. *Journal of the Taiwan Institute of Chemical Engineers*, 87, 225–240.

Elimelech, M. & Phillip, W. A. 2011. The future of seawater desalination: Energy, technology, and the environment. *Science*, 333, 712–717.

Felder-Flesch, D. 2016. *Dendrimers in Nanomedicine*. Pan Stanford.

Gatard, S., Salmon, L., Deraedt, C., Ruiz, J., Astruc, D. & Bouquillon, S. 2014. Palladium nanoparticles stabilized by glycodendrimers and their application in catalysis. *European Journal of Inorganic Chemistry*, 2014, 4369–4375.

Grayson, S. M. & Fréchet, J. M. 2001. Convergent dendrons and dendrimers: from synthesis to applications. *Chemical Reviews*, 101, 3819–3868.

Gröhn, F., Bauer, B. J., Akpalu, Y. A., Jackson, C. L. & Amis, E. J. 2000. Dendrimer templates for the formation of gold nanoclusters. *Macromolecules*, 33, 6042–6050.

Gupta, N. K., Sengupta, A., Boda, A., Adya, V. C. & Ali, S. M. 2016. Oxidation state selective sorption behavior of plutonium using N, N-dialkylamide functionalized carbon nanotubes: Experimental study and DFT calculation. *RSC Advances*, 6, 78692–78701.

Han, K. N., Yu, B. Y. & Kwak, S.-Y. 2012. Hyperbranched poly (amidoamine)/polysulfone composite membranes for Cd (II) removal from water. *Journal of Membrane Science*, 396, 83–91.

Hawker, C. J. & Frechet, J. M. J. 1990. Preparation of polymers with controlled molecular architecture. A new convergent approach to dendritic macromolecules. *Journal of the American Chemical Society*, 112, 7638–7647.

Hayati, B., Maleki, A., Najafi, F., Daraei, H., Gharibi, F. & Mckay, G. 2016. Synthesis and characterization of PAMAM/CNT nanocomposite as a super-capacity adsorbent for heavy metal (Ni2+, Zn2+, As3+, CO2+) removal from wastewater. *Journal of Molecular Liquids*, 224, 1032–1040.

Hayati, B., Maleki, A., Najafi, F., Daraei, H., Gharibi, F. & Mckay, G. 2017. Adsorption of Pb2+, Ni2+, Cu2+, CO2+ metal ions from aqueous solution by PPI/SiO2 as new high performance adsorbent: Preparation, characterization, isotherm, kinetic, thermodynamic studies. *Journal of Molecular Liquids*, 237, 428–436.

Hayati, B., Maleki, A., Najafi, F., Gharibi, F., Mckay, G., Gupta, V. K., Puttaiah, S. H. & Marzban, N. 2018. Heavy metal adsorption using PAMAM/CNT nanocomposite from aqueous solution in batch and continuous fixed bed systems. *Chemical Engineering Journal*, 346, 258–270.

Hua, M., Zhang, S., Pan, B., Zhang, W., Lv, L. & Zhang, Q. 2012. Heavy metal removal from water/wastewater by nanosized metal oxides: A review. *Journal of Hazardous Materials*, 211–212, 317–331.

Jędrych, M., Borowska, K., Galus, R. & Jodłowska-Jędrych, B. 2014. The evaluation of the biomedical effectiveness of poly (amido) amine dendrimers generation 4.0 as a drug and as drug carriers: A systematic review and meta-analysis. *International Journal of Pharmaceutics*, 462, 38–43.

Jiang, D., Li, X., Lv, X. & Jia, Q. 2018. A magnetic hydrazine-functionalized dendrimer embedded with TiO2 as a novel affinity probe for the selective enrichment of low-abundance phosphopeptides from biological samples. *Talanta*, 185, 461–468.

Jiryaei Sharahi, F. & Shahbazi, A. 2017. Melamine-based dendrimer amine-modified magnetic nanoparticles as an efficient Pb (II) adsorbent for wastewater treatment: Adsorption optimization by response surface methodology. *Chemosphere*, 189, 291–300.

Karakhanov, E. A., Maksimov, A. L., Zakharian, E. M., Kardasheva, Y. S., Savilov, S. V., Truhmanova, N. I., Ivanov, A. O. & Vinokurov, V. A. 2015. Palladium nanoparticles encapsulated in a dendrimer networks as catalysts for the hydrogenation of unsaturated hydrocarbons. *Journal of Molecular Catalysis A: Chemical*, 397, 1–18.

Kim, H. R., Jang, J. W. & Park, J. W. 2016. Carboxymethyl chitosan-modified magnetic-cored dendrimer as an amphoteric adsorbent. *Journal of Hazardous Materials*, 317, 608–616.

Kim, K.-J. & Park, J.-W. 2017. Stability and reusability of amine-functionalized magnetic-cored dendrimer for heavy metal adsorption. *Journal of Materials Science*, 52, 843–857.

Knecht, M. R., Weir, M. G., Frenkel, A. I. & Crooks, R. M. 2008. Structural rearrangement of bimetallic alloy PdAu nanoparticles within dendrimer templates to yield core/shell configurations. *Chemistry of Materials*, 20, 1019–1028.

Kong, L., Xing, L., Zhou, B., Du, L. & Shi, X. 2017. Dendrimer-modified MoS2 nanoflakes as a platform for combinational gene silencing and photothermal therapy of tumors. *ACS Applied Materials and Interfaces*, 9, 15995–16005.

Kumar, P., Sengupta, A., Deb, A. K. S. & Ali, S. M. 2017a. Poly (amidoamine) Dendrimer functionalized carbon nanotube for efficient sorption of trivalent f-elements: A Comparison Between 1st and 2nd Generation. *Chemistry Select*, 2, 975–985.

Kumar, P., Sengupta, A., Deb, A. K. S., Dasgupta, K. & Ali, S. M. 2016. Sorption behaviour of Pu 4+ and PuO 2 2+ on amido amine-functionalized carbon nanotubes: Experimental and computational study. *RSC Advances*, 6, 107011–107020.

Kumar, P., Sengupta, A., Singha Deb, A. K., Dasgupta, K. & Ali, S. M. 2017b. Understanding the sorption behavior of Pu4+ on poly (amidoamine) dendrimer functionalized carbon nanotube: Sorption equilibrium, mechanism, kinetics, radiolytic stability, and back-extraction studies. *Radiochimica Acta*, 105, 677–688.

Kurniawan, T. A., Chan, G. Y. S., Lo, W.-H. & Babel, S. 2006. Physico–chemical treatment techniques for wastewater laden with heavy metals. *Chemical Engineering Journal*, 118, 83–98.

Kurniawan, T. A., Sillanpää, M. E. T. & Sillanpää, M. 2012. Nanoadsorbents for remediation of aquatic environment: Local and practical solutions for global water pollution problems. *Critical Reviews in Environmental Science and Technology*, 42, 1233–1295.

Lee, M. Y., Lee, J. H., Chung, J. W. & Kwak, S. Y. 2018. Hydrophilic and positively charged polyethylenimine-functionalized mesoporous magnetic clusters for highly efficient removal of Pb (II) and Cr (VI) from wastewater. *Journal of Environmental Management*, 206, 740–748.

Lee, S.-J., Park, J. H., Ahn, Y.-T. & Chung, J. W. 2015. Comparison of heavy metal adsorption by peat moss and peat moss-derived biochar produced under different carbonization conditions. *Water, Air, and Soil Pollution*, 226, 9.

Liang, X., Ge, Y., Wu, Z. & Qin, W. 2017. DNA fragments assembled on polyamidoamine-grafted core-shell magnetic silica nanoparticles for removal of mercury (II) and methylmercury (I). *Journal of Chemical Technology and Biotechnology*, 92, 819–826.

Liu, H., Guo, J., Jin, L., Yang, W. & Wang, C. 2008. Fabrication and functionalization of dendritic poly (amidoamine)-immobilized magnetic polymer composite microspheres. *The Journal of Physical Chemistry. B*, 112, 3315–3321.

Liu, H., Wang, H., Xu, Y., Guo, R., Wen, S., Huang, Y., Liu, W., Shen, M., Zhao, J., Zhang, G. & Shi, X. 2014. Lactobionic acid-modified dendrimer-entrapped gold nanoparticles for targeted computed tomography imaging of human hepatocellular carcinoma. *ACS Applied Materials and Interfaces*, 6, 6944–6953.

Liu, H., Xu, Y., Wen, S., Chen, Q., Zheng, L., Shen, M., Zhao, J., Zhang, G. & Shi, X. 2013. Targeted tumor computed tomography imaging using low-generation dendrimer-stabilized gold nanoparticles. *Chemistry*, 19, 6409–6416.

Liu, X. 2014. *Dendrimer Modified Magnetite Particles for Heavy Metal (Copper (II), Cobalt (II), Zinc (II), Nickel (II)) Removal from Water.* University of Delaware.

Ma, F., Qu, R., Sun, C., Wang, C., Ji, C., Zhang, Y. & Yin, P. 2009. Adsorption behaviors of Hg (II) on chitosan functionalized by amino-terminated hyperbranched polyamidoamine polymers. *Journal of Hazardous Materials*, 172, 792–801.

Ma, Y. X., Kou, Y. L., Xing, D., Jin, P. S., Shao, W. J., Li, X., Du, X. Y. & La, P. Q. 2017a. Synthesis of magnetic graphene oxide grafted polymaleicamide dendrimer nanohybrids for adsorption of Pb (II) in aqueous solution. *Journal of Hazardous Materials*, 340, 407–416.

Ma, Y., Xing, D., Shao, W., Du, X. & La, P. 2017b. Preparation of polyamidoamine dendrimers functionalized magnetic graphene oxide for the adsorption of Hg (II) in aqueous solution. *Journal of Colloid and Interface Science*, 505, 352–363.

Maiti, P. K., Çağın, T., Lin, S. & Goddard, W. A. 2005. Effect of Solvent and pH on the Structure of PAMAM Dendrimers. *Macromolecules*, 38, 979–991.

Maleki, A., Pajootan, E. & Hayati, B. 2015. Ethyl acrylate grafted chitosan for heavy metal removal from wastewater: Equilibrium, kinetic and thermodynamic studies. *Journal of the Taiwan Institute of Chemical Engineers*, 51, 127–134.

Mohammadi, S., Salimi, A. & Qaddareh, S. H. 2018. Amplified FRET based CA15-3 immunosensor using antibody functionalized luminescent carbon-dots and AuNPs-dendrimer aptamer as donor-acceptor. *Analytical Biochemistry*, 557, 18–26.

Natarajan, B. & Jayaraman, N. 2011. Synthesis and studies of Rh (I) Catalysts within and across poly (alkyl aryl ether) dendrimers. *Journal of Organometallic Chemistry*, 696, 722–730.

Newkome, G. R., Yao, Z., Baker, G. R. & Gupta, V. K. 1985. Micelles. Part 1. Cascade molecules: A new approach to micelles. A [27]-arborol. *The Journal of Organic Chemistry*, 50, 2003–2004.

Niu, Y., Yang, J., Qu, R., Gao, Y., Du, N., Chen, H., Sun, C. & Wang, W. 2016. Synthesis of silica-gel-supported sulfur-capped PAMAM dendrimers for efficient Hg(II) adsorption: Experimental and DFT study. *Industrial and Engineering Chemistry Research*, 55, 3679–3688.

Nonahal, M., Rastin, H., Saeb, M. R., Sari, M. G., Moghadam, M. H., Zarrintaj, P. & Ramezanzadeh, B. 2018. Epoxy/PAMAM dendrimer-modified graphene oxide nanocomposite coatings: Nonisothermal cure kinetics study. *Progress in Organic Coatings*, 114, 233–243.

Odio, O. F. & Reguera, E. 2017. Nanostructured spinel ferrites: Synthesis, functionalization, nanomagnetism and environmental applications. In: M. S. Seehra, ed., *Magnetic Spinels-Synthesis, Properties and Applications*. InTech.

Pan, B.-F., Gao, F. & Ao, L.-M. 2005. Investigation of interactions between dendrimer-coated magnetite nanoparticles and bovine serum albumin. *Journal of Magnetism and Magnetic Materials*, 293, 252–258.

Patel, H. & Patel, P. 2013. Dendrimer applications–a review. *International Journal of Pharmacy and Biological Sciences*, 4, 454–463.

Pereira, L. S., Cordery, I. & Iacovides, I. 2009. *Coping with Water Scarcity: Addressing the Challenges*. Springer Science & Business Media.

Pourjavadi, A., Abedin-Moghanaki, A. & Hosseini, S. H. 2016. Synthesis of poly (amidoamine)-graft-poly (methyl acrylate) magnetic nanocomposite for removal of lead contaminant from aqueous media. *International Journal of Environmental Science and Technology*, 13, 2437–2448.

Priyam, A., Blumling, D. E. & Knappenberger, K. L. 2010. Synthesis, characterization, and self-organization of dendrimer-encapsulated HgTe quantum dots. *Langmuir*, 26, 10636–10644.

Repo, E., Warchol, J. K., Kurniawan, T. A. & Sillanpää, M. E. T. 2010. Adsorption of Co (II) and Ni (II) by EDTA-and/or DTPA-modified chitosan: Kinetic and equilibrium modeling. *Chemical Engineering Journal*, 161, 73–82.

Rether, A. & Schuster, M. 2003. Selective separation and recovery of heavy metal ions using water-soluble N-benzoylthiourea modified PAMAM polymers. *Reactive and Functional Polymers*, 57, 13–21.

Sánchez-Navarro, M. & Rojo, J. 2012. Synthetic strategies to create dendrimers: Advantages and drawbacks. In: J. M. de la Fuente & V. Grazu, eds., *Frontiers of Nanoscience*. Elsevier.

Satheeshkumar, C., Ravivarma, M., Rajakumar, P., Ashokkumar, R., Jeong, D.-C. & Song, C. 2015. Synthesis, photophysical and electrochemical properties of stilbenoid dendrimers with phenothiazine surface group. *Tetrahedron Letters*, 56, 321–326.

Sengupta, A., Deb, A. K. S., Gupta, N. K., Kumar, P., Dasgupta, K. & Ali, S. M. 2018. Evaluation of 1st and 2nd generation of poly (amidoamine) dendrimer functionalized carbon nanotubes for the efficient removal of neptunium. *Journal of Radioanalytical and Nuclear Chemistry*, 315, 331–340.

Sengupta, A., Sk., J., Boda, A. & Ali, S. M. 2016. An amide functionalized task specific carbon nanotube for the sorption of tetra and hexa valent actinides: Experimental and theoretical insight. *RSC Advances*, 6, 39553–39562.

Shaaban, A., Khalil, A., Lasheen, T., Nouh, E. S. A. & Ammar, H. 2018. Polyamidoamine dendrimers modified silica gel for uranium (VI) removal from aqueous solution using batch and fixed-bed column methods. *Desalination and Water Treatment*, 102, 197–210.

Shi, X., Thomas, T. P., Myc, L. A., Kotlyar, A., Baker, J. R. 2007. Synthesis, characterization, and intracellular uptake of carboxyl-terminated poly (amidoamine) dendrimer-stabilized iron oxide nanoparticles. *Physical Chemistry Chemical Physics*, 9, 5712–5720.

Song, X., Niu, Y., Qiu, Z., Zhang, Z., Zhou, Y., Zhao, J. & Chen, H. 2017a. Adsorption of Hg (II) and Ag (I) from fuel ethanol by silica gel supported sulfur-containing PAMAM dendrimers: Kinetics, equilibrium and thermodynamics. *Fuel*, 206, 80–88.

Song, X., Niu, Y., Zhang, P., Zhang, C., Zhang, Z., Zhu, Y. & Qu, R. 2017b. Removal of Co (II) from fuel ethanol by silica-gel supported PAMAM dendrimers: Combined experimental and theoretical study. *Fuel*, 199, 91–101.

Stanicki, D., Vander Elst, L., Muller, R. N., Laurent, S., Felder-Flesch, D., Mertz, D., Parat, A., Begin-Colin, S., Cotin, G. & Greneche, J.-M. 2017. Iron-oxide nanoparticle-based contrast agents. In: V. C. Pierre & M. J. Allen, eds., *Contrast Agents for MRI*. Royal Society of Chemistry.

Strable, E., Bulte, J. W. M., Moskowitz, B., Vivekanandan, K., Allen, M. & Douglas, T. 2001. Synthesis and characterization of soluble iron oxide– dendrimer composites. *Chemistry of Materials*, 13, 2201–2209.

Sun, W., Mignani, S., Shen, M. & Shi, X. 2016. Dendrimer-based magnetic iron oxide nanoparticles: Their synthesis and biomedical applications. *Drug Discovery Today*, 21, 1873–1885.

Sun, Y., Li, D., Yang, H. & Guo, X. 2018. Fabrication of Fe3O4@ Polydopamine@ Polyamidoamine core–shell nanocomposites and its application for Cu (II) adsorption. *New Journal of Chemistry*.

Taylor, D. M. 1989. The biodistribution and toxicity of plutonium, americium and neptunium. *The Science of the Total Environment*, 83, 217–225.

Teijon, G., Candela, L., Tamoh, K., Molina-Díaz, A. & Fernández-Alba, A. R. 2010. Occurrence of emerging contaminants, priority substances (2008/105/CE) and heavy metals in treated wastewater and groundwater at Depurbaix facility (Barcelona, Spain). *The Science of the Total Environment*, 408, 3584–3595.

Thammawong, C., Opaprakasit, P., Tangboriboonrat, P. & Sreearunothai, P. 2013. Prussian blue-coated magnetic nanoparticles for removal of cesium from contaminated environment. *Journal of Nanoparticle Research*, 15, 1689.

Tomalia, D. A. 2005a. Birth of a new macromolecular architecture: Dendrimers as quantized building blocks for nanoscale synthetic polymer chemistry. *Progress in Polymer Science*, 30, 294–324.

Tomalia, D. A. 2005b. The dendritic state. *Materials Today*, 8, 34–46.

Tomalia, D. A., Baker, H., Dewald, J., Hall, M., Kallos, G., Martin, S., Roeck, J., Ryder, J. & Smith, P. 1985. A new class of polymers: Starburst-dendritic macromolecules. *Polymer Journal*, 17, 117–132.

Ulaszewska, M. M., Hernando, M. D., Ucles, A., Rosal, R., Rodríguez, A., Garcia-Calvo, E. & Fernández-Alba, A. R. 2012. Chemical and ecotoxicological assessment of dendrimers in the aquatic environment. In: M. Farré & D. Barceló, eds., *Comprehensive Analytical Chemistry*. Elsevier.

Wang, P., Ma, Q., Hu, D. & Wang, L. 2015. Removal of Reactive Blue 21 onto magnetic chitosan microparticles functionalized with polyamidoamine dendrimers. *Reactive and Functional Polymers*, 91–92, 43–50.

Wang, P., Zhao, X. H., Wang, Z. Y., Meng, M., Li, X. & Ning, Q. 2010. Generation 4 polyamidoamine dendrimers is a novel candidate of nano-carrier for gene delivery agents in breast cancer treatment. *Cancer Letters*, 298, 34–49.

Wang, S. H., Shi, X., VanAntwerp, M., Cao, Z., Swanson, S. D., Bi, X. & Baker, J. R. 2007. Dendrimer-functionalized iron oxide nanoparticles for specific targeting and imaging of cancer cells. *Advanced Functional Materials*, 17, 3043–3050.

Woo, S., Lee, Y., Sunkara, V., Cheedarala, R. K., Shin, H. S., Choi, H. C. & Park, J. W. 2007. "Fingertip"-guided noncovalent functionalization of carbon nanotubes by dendrons. *Langmuir*, 23, 11373–11376.

Xiao, W., Yan, B., Zeng, H. & Liu, Q. 2016. Dendrimer functionalized graphene oxide for selenium removal. *Carbon*, 105, 655–664.

Xu, X., Jian, Y., Li, Y., Zhang, X., Tu, Z. & Gu, Z. 2014. Bio-inspired supramolecular hybrid dendrimers self-assembled from low-generation peptide dendrons for highly efficient gene delivery and biological tracking. *ACS Nano*, 8, 9255–9264.

Xu, X., Yuan, H., Chang, J., He, B. & Gu, Z. 2012. Cooperative hierarchical self-assembly of peptide dendrimers and linear polypeptides into nanoarchitectures mimicking viral capsids. *Angewandte Chemie*, 51, 3184–3187.

Yen, C. H., Lien, H. L., Chung, J. S. & Yeh, H. D. 2017. Adsorption of precious metals in water by dendrimer modified magnetic nanoparticles. *Journal of Hazardous Materials*, 322, 215–222.

Yu, J., Zhao, H., Ye, L., Yang, H., Ku, S., Yang, N. & Xiao, N. 2009. Effect of surface functionality of magnetic silica nanoparticles on the cellular uptake by glioma cells in vitro. *Journal of Materials Chemistry*, 19, 1265–1270.

Yuan, X., Wen, S., Shen, M. & Shi, X. 2013. Dendrimer-stabilized silver nanoparticles enable efficient colorimetric sensing of mercury ions in aqueous solution. *Analytical Methods*, 5, 5486–5492.

Zhang, F., Wang, B., He, S. & Man, R. 2014. Preparation of graphene-oxide/polyamidoamine dendrimers and their adsorption properties toward some heavy metal ions. *Journal of Chemical and Engineering Data*, 59, 1719–1726.

Zhang, P., Niu, Y., Qiao, W., Xue, Z., Bai, L. & Chen, H. 2018. Experimental and DFT investigation on the adsorption mechanism of a silica gel-supported sulfur-capped PAMAM dendrimers for Ag (I). *Journal of Molecular Liquids*, 263, 390–398.

Zhang, Y., Qu, R., Sun, C., Ji, C., Chen, H. & Yin, P. 2015. Improved synthesis of silica-gel-based dendrimer-like highly branched polymer as the Au (III) adsorbents. *Chemical Engineering Journal*, 270, 110–121.

Zhao, L., Ling, Q., Liu, X., Hang, C., Zhao, Q., Liu, F. & Gu, H. 2018. Multifunctional triazolylferrocenyl Janus dendron: Nanoparticle stabilizer, smart drug carrier, and supramolecular nanoreactor. *Applied Organometallic Chemistry*, 32, e4000.

Zhou, S. L., Li, J., Hong, G. B. & Chang, C. T. 2013. Dendrimer modified magnetic nanoparticles as adsorbents for removal of dyes. *Journal of Nanoscience and Nanotechnology*, 13, 6814–6819.

Zhu, Y., Niu, Y., Li, H., Ren, B., Qu, R., Chen, H. & Zhang, Y. 2018. Removal of Cd (II) and Fe (III) from DMSO by silica gel supported PAMAM dendrimers: Equilibrium, thermodynamics, kinetics, and mechanism. *Ecotoxicology and Environmental Safety*, 162, 253–260.

# 13

# *Carbonaceous and Polysaccharide-Based Nanomaterials: Synthesis and Their Importance in Environmental Applications*

**Jaise Mariya George, Ragam N. Priyanka, and Beena Mathew**

**CONTENTS**

The environment is everything that isn't me.

**Albert Einstein**

## 13.1 Introduction

Nanostructured materials are materials with a microstructure and a characteristic length scale which is on the order of a few (typically 1–100) nanometers. Nanomaterials have gained importance in technological progress due to their tunable physicochemical characteristics such as melting point, electrical and thermal conductivity, magnetic property, catalytic activity, wettability, light absorption and scattering, resulting in enhanced performance over their bulk counterparts. Nanoscale formation of materials from the macrolevel also changes the magnetic properties.[1] Quantum confinement is the change of electronic and optical properties when the material sampled is of sufficiently small size – typically 10 nanometers or less. When the size of a material is as small as the de Broglie wavelength of an electron, their magnetic and optical properties deviate substantially from those of the bulk material.

Carbonaceous nanomaterials and nanopolysaccharides have several environmental benign applications. The catalytic and adsorption properties of fullerenes, carbon nanotubes (CNT), graphene and nanopolysaccharides are used for environmental remediation. Nanotechnology has extensive applications in medicine and technology, but the adverse effects of nanomaterials should be handled properly.

## 13.2 Carbon-Based Nanomaterials for Environmental Applications

### 13.2.1 Introduction Carbon-Based Nanomaterials

Carbon and its allotropes occupy the supreme position among all the elements with their versatile structures, properties, compounds, and applications. The reason behind the wide acceptance of carbon is its basic electronic structure. Carbon has six electrons distributed as $1s^2\, 2s^2\, 2p^2$ where the 2s and 2p orbitals have only a narrow energy difference. This permits the shift of one electron from the 2s orbital to the empty 2p orbital which offers the element a choice to hybridize to either an sp, $sp^2$ or $sp^3$ type. Thus, carbon has many allotropes with extreme varieties of physical, chemical, electrical and optical properties such as graphite, fullerenes, nanotubes, diamond, coke and so forth. The properties are fine-tuned by modifying synthesis precursors along with reaction conditions like heat and pressure (Figure 13.1).[2,3]

Graphite is the most stable and commonly found allotrope of carbon. It has a planar structure with multiple layers of carbon atoms arranged at the corners of hexagons like in a honeycomb with a separation of 0.142 nm. Weak π-interactions are present between each such layer which holds them together. It is an

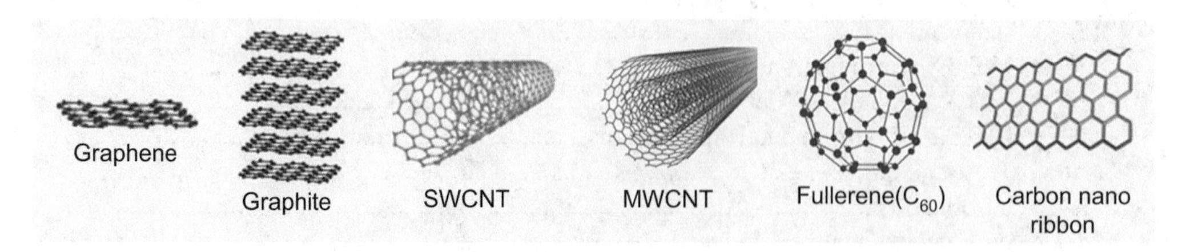


**FIGURE 13.1** Carbon-based nanomaterials.

electrical conductor due to the possible delocalization of $\pi$ electrons. Graphene is a collection of fewer than 10 layers of graphite. In graphite, graphene sheets are arranged in parallel with a distance of 0.34 nm between the layers. In 2010, Giem and Novoselov received the Nobel Prize for separating a graphene layer of single atom thickness from graphite, which they achieved in 2004. They used Scotch tape to exfoliate graphene flakes from graphite. This two-dimensional graphene has interesting thermal, electronic and mechanical properties which makes it a favorite choice for applications in electrical and optical devices, and it is stronger than steel. Buckyballs (1D), CNTs (2D) and graphite (3D) can be effectively prepared from quality graphene sheets. Graphene can be synthesized via different techniques. The mechanical cleavage method produces graphene sheets without many defects but the process is not suitable to be scaled up. The graphene sheets produced by epitaxial growth have fewer defects but are discontinuous. The chemical vapor deposition (CVD) method is not affordable, whereas the total organic synthesis method renders graphene with many defects. Thus, the chemical synthesis of graphene is found to be the most economical route to acquiring quality graphene sheets in satisfactory amounts. The thermal decomposition of silicon carbide is used for the large-scale production of graphene. The nature of synthesized graphene is obtained by various characterization techniques such as X-ray diffraction (XRD), scanning electron microscopy (SEM), transmission electron microscopy (TEM) and Raman spectroscopy. Graphene oxide is synthesized from graphite, most popularly by Hummer's method.[4] To a solution of graphite containing sodium nitrate and sulphuric acid, potassium permanganate is added to obtain graphene oxide.

Fullerenes were discovered in 1985 and Kroto, Curl and Smalley shared the Nobel Prize for this discovery in 1996. Major examples include $C_{60}$, Buckminsterfullerene or Buckyball, $C_{70}$, $C_{100}$, and $C_{400}$. It is formed out of only carbon atoms arranged in the shape of a hollow sphere or ellipsoid. They are mostly synthesized using methods such as arc discharge and laser ablation, as well as CVD. In the arc discharge method, an electric arc is applied between the graphite electrodes placed in an inert environment like helium gas under low pressure to vaporize carbon as fullerene at the cathode. But here we obtain fullerenes of low purity, approximating to 15%. Higher order fullerenes can be obtained by the laser ablation method, where graphite rods are exposed to extremely high temperatures by a dual pulsed laser and vaporized to produce fullerenes or CNTs. One of the most popular methods used for nanocarbon synthesis is CVD, which allow the combustion of hydrocarbon fuels such as methane, ethylene, acetylene or CO under low pressure to obtain most pure products. In addition, this method is much preferred since it uses the least amount of energy.[3,5]

CNTs are a cylindrical form of fullerenes which consist of carbon atoms arranged at the corners of a hexagon and has pentagonal units where they have a curve. They have extreme superconductivity and tensile strength. Depending upon the atom thickness of the wall of the nanotubes, they have been further classified into single-walled (SWCNT), double-walled (DWCNT) as well as multi-walled (MWCNT) carbon nanotubes. SWCNTs, DWCNTs and MWCNTs have also been synthesized by the same methods used for fullerenes, but generally by using metal catalysts along with the carbon source. The arc evaporation method provides the highest structural quality CNTs by applying very high temperature. In CVD, at a lower temperature and relatively low cost, more quantities of CNTs of desired length and diameters can be obtained by tuning the temperature and pressure conditions. The deciding factors for the formation of SWCNT or MWCNT are conditions such as pressure and heat, as well as the composition and size of the catalyst and the carbon source. Thus, CVD is the method used for the commercial preparation of various kinds of CNTs.[3,5]

Graphene nanoribbons (GNR) and graphene oxide nanoribbons (GONR) are recently developed carbon-based materials with fascinating properties and applications. GNRs are thin graphene strips less than 50 nm in width. GONRs are obtained as intermediate compounds during the preparation of GNRs. In addition to the exciting properties of common carbonaceous nanomaterials, GNRs possess an elevated area normalized edge–plane structure. They can be synthesized from carbon sources like graphene and CNTs by methods like CVD, sonochemical, lithography and plasma etching. Recent years have witnessed the exclusive use of GNRs and GONRs in the fields of optoelectronics, batteries, solar cells, sensors and nanoelectronics.[6]

In addition to the major allotropes of carbon in nanodimensions discussed above, there are many more forms of carbonaceous nanomaterials known to have interesting properties and valuable applications. These include the torus which is a doughnut-shaped CNT, nanobuds which are combinations of fullerenes and CNTs, peapods which have fullerene enclosed inside CNT, cup-stacked CNTs, carbon megatubes, carbon onions, fluorescent carbon quantum dots, carbon nanospheres, fibers, bucky papers and so forth.[3,5]

### 13.2.2 Modification of Carbonaceous Nanomaterials

The structure and composition of carbonaceous nanomaterials favor hydrophobic interactions with external species, limiting their application to mostly organic pollutants. To achieve the removal of a larger group of hydrophilic pollutants present in various water bodies such as toxic metal ions, it is necessary to introduce suitable groups onto the surface of carbonaceous nanomaterials. Modifications are carried out in accordance with the type of application and the nature of the target pollutant.

CNTs have been elaborately explored by modifying their surface to fit for several applications. A variety of modifications is reported at the tips and the side walls of CNTs. Some of the non-magnetic modifications are discussed below.[7]

#### *13.2.2.1 Modification by Oxidation*

Surface modification by oxidation is a common method used for carbonaceous nanomaterials to introduce hydrophilic groups like –OH, –C=O and –COOH groups. The nanomaterials are refluxed in the presence of inorganic acids such as $HNO_3$ or $H_2SO_4$ and an oxidizing agent like $KMnO_4$, $H_2O_2$ or NaOCl. These functionalized materials can be used directly or any specific guest species can be grafted onto the surface additionally to make them suitable for desired applications. In addition to these liquid-phase oxidation techniques, CNT surfaces can be oxidized under $O_2$ gas by heating. But sometimes, it may burn SWCNTs which have a thin layer of carbon atoms. It is also known that upon oxidation under $O_2$ or $CO_2$ gas, the tube caps of CNTs are etched away, which allows the peeling of outer layers, increasing the surface area drastically. The oxidation of CNTs has the additional advantage of leaving the surface clean, making more sites available. Once CNTs are sonicated in the presence of acid, it allows for reducing the length of the nanotubes and introduces hydrophilic functional groups to them. Both these approaches are helpful for proper solubility of CNTs in an aqueous medium, increasing its application in wastewater samples. As an example, we can observe the work of Datsyuk et al.[8] where the oxidation of MWCNT was carried out with different reagents. In this method, $HNO_3$ treatment provided maximum functionalization and nanotube shortening compared to the $H_2SO_4$–$H_2O_2$ mixture. Also, they used $NH_4OH$ and $H_2O_2$ to conduct basic oxidation and reported that it resulted in MWCNTs with improved structural integrity by the removal of metal oxides and amorphous carbon impurity. Microwave irradiation is one of the fastest methods of acid functionalization, completing the reaction in less than one hour (Figure 13.2).

#### *13.2.2.2 Alkali-Mediated Modification*

Alkali-mediated surface activation of carbon materials was also found to be proficient at augmenting surface area and pore volume. Ma et al.[9] heated MWCNTs in the presence of KOH for an hour at 750°C under an inert atmosphere; these were then washed and dried to obtain alkali-modified MWCNT. They evaluated the effectiveness of the modified MWCNTs for adsorption of both cationic and anionic dyes from water and found significant activity. Secondary forces such as π–π, hydrogen bonding and

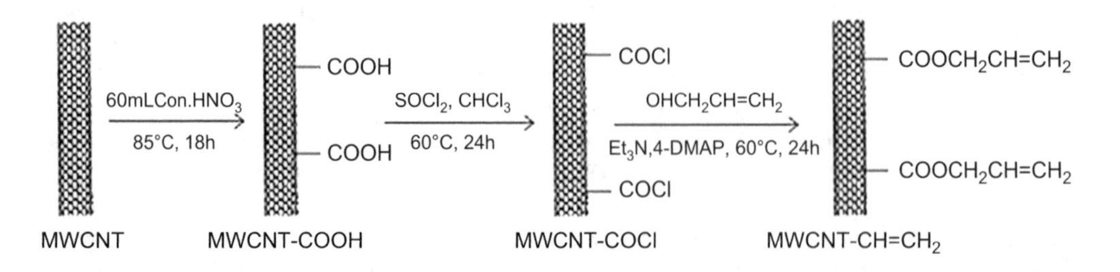


**FIGURE 13.2** Acid modification on MWCNTs. (Reproduced from Sebastian et al.[13] With permission from Taylor & Francis.)

electrostatic interactions act behind the improved adsorption. Similar MWCNTs were also reported to be capable in the removal of ethylbenzene, toluene and m-xylene from water samples.

#### *13.2.2.3 Non-Magnetic Oxides*

Oxides of transition metals such as Ag, Fe, Ti, Ce, Zr and bimetallic Pd–Fe were used to modify CNT surfaces to improve their adsorption efficiency towards water pollutants such as arsenic, cadmium, copper, fluoride and 2,4-dichlorophenol. In the synthesis, metal oxides are precipitated onto the surface of functionalized CNTs in an alkaline medium. Examples include the incorporation of photo-reactive titania to MWCNT and the introduction of non-magnetic $Fe_2O_3$ onto MWCNTs. They were effectively used for the catalytic degradation of organic contaminants like o-chlorophenol and resorcinol by the Fenton reaction. 2,4-dinitrophenol and 2,6-dinitro-p-cresol were degraded under sunlight by a $TiO_2$–MWCNT system.

#### *13.2.2.4 Specific Compounds*

The incorporation of specific chemicals has been found to increase the activity of CNTs towards specific contaminants. For instance, Bandaru et al.[10] used an ethanol solution of cysteamine hydrochloride to modify SWCNT–COOH to introduce –SH groups on the nanosurface, which led to higher dispersibility in the water medium. They found a threefold increase of adsorption of $Hg^{2+}$ ions than the unmodified SWCNTs. Another modification of MWCNTs with a diglycol amide derivative carried out by Deb et al.[11] achieved actinide ion adsorption. Similarly, MWCNTs modified with tannic acid displayed admirable adsorption of rare earth metals. Some of the other reported groups incorporated to CNTs include ethylenediamine, cyclodextrin, amino compounds and iodide/sulfur.

#### *13.2.2.5 Polymers*

Another class of compounds used to modify the surface of CNTs for functional applications is polymers. Dendrimer-modified MWCNT–COOH were proved efficient at adsorbing dye solutions. In another report, Shao et al.[12] grafted poly (methyl methacrylate) (PMMA) which is a known adsorbent of organic contaminants onto the surface of CNTs by a plasma technique. The obtained PMMA-modified CNTs efficiently removed 4,4′-dichlorinated biphenyl from the aqueous solutions since the CNTs could solve the sedimentation of PMMA in water. MWCNT-based sensors and structure-specific sorbents were developed through molecular imprinting technology (Figure 13.3).[13–15]

#### *13.2.2.6 Carbon–Carbon Hybrids*

Similar to the incorporation of other chemical moieties to CNT, carbon nanomaterials such as graphene are also used to make hybrids for specifically targeted applications to improve the efficiency of bare

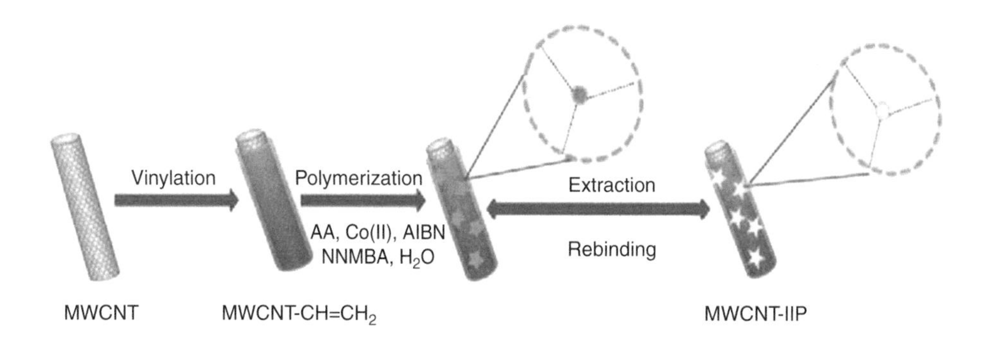


**FIGURE 13.3** MWCNT based molecular imprinted polymer as a sorbent for Co(II) ion extraction. (Reproduced from Sebastian et al.[13] With permission from Taylor & Francis.)

CNTs. Ai and Jiang prepared a self-assembled hybrid of graphene–CNT from pre-exfoliated graphene oxide and MWCNTs.[16] Whereas Sui et al. obtained graphene–CNT hybrid aerogels by a green method starting from graphene oxide and MWCNTs and utilized them for the adsorption of both non-polar organic compounds and metal ions.[17] Dichiara et al. constructed paper-containing hybrids of CNTs and graphene to remove organic pollutants and metal ions.[18,19] The hybrid adsorbent paper showed improved adsorption efficiency than bare CNTs and activated carbon for prepared hybrids towards aqueous $Cu^{2+}$ ions by 50% and fourfold respectively.[18] Moreover, a hybrid of SWCNTs and graphene nanoplatelets was 25% better than both the individual components at organic pollutant (diquat dibromide, 2,4-dichlorophenoxyacetic acid and 1-pyrenebutyric acid) adsorption.[19]

### 13.2.3 Environmental Applications of Carbonaceous Nanomaterials

#### *13.2.3.1 Adsorption of Environmental Pollutants*

Adsorption is the physical adhesion of atoms, molecules or ions on the surface of the solid. The significant application of this phenomenon is the fast and convenient removal of pollutants from water bodies such as heavy metal ions, dyes and pesticides, to name a few. An adsorbent is a material which can adsorb desired target species. Adsorption is a surface phenomenon and thus the larger the surface area of the adsorbent, the more efficient adsorption occurs. Thus, nano-adsorbents are very relevant since they are made of nanomaterials which possess high surface-to-volume ratio in comparison with their bulk counterparts (Figure 13.4).[20]

Carbonaceous nanosorbents are biocompatible and biodegradable. They are less toxic and less costly. They have a high surface to volume ratio and controlled pore size distribution. Carbon-based nanomaterials are effective in broad pH range and are observed as consistent with isotherms such as BET, Freundlich or Langmuir. They are effective sorbents owing to their high equilibrium rates and great sorbent capacity compared to activated carbon. Higher rates are ascribed to the polarizability of π electrons which occurs upon π–π electron-donor–acceptor (EDA) interactions through aromatic sorbates. These interactions reduce the heterogeneity of adsorption energies and lack intermediate mechanisms such as

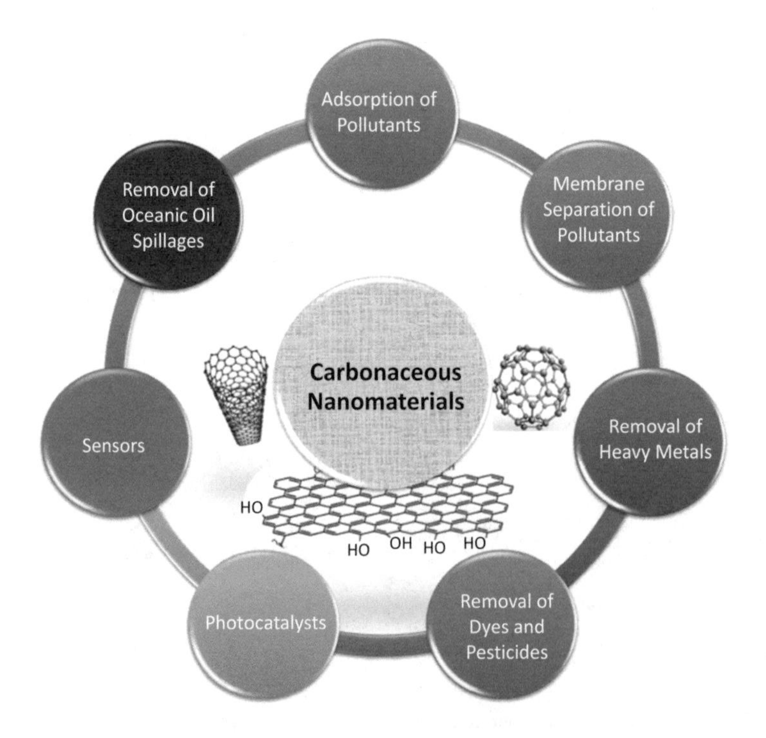


**FIGURE 13.4** Environmental applications of carbonaceous nanomaterials.

pore diffusion during the adsorption. Yang et al.[21] strengthened this conclusion by comparing different carbon-based nanomaterials such as MWCNTs, SWCNT and fullerenes for the sorption of polycyclic aromatic hydrocarbons (PAHs) like pyrene, phenanthrene and naphthalene.[2]

The adsorption mechanism of carbonaceous sorbents with organic sorbates, such as PAHs, trihalomethanes, naphthalene, and so on, is mainly hydrophobic and non-specific. Conversely, the adsorption of inorganic sorbates such as metal contaminants occurs via complexation reactions. Hence, the major feature influencing the sorption efficacy is not just the total surface area but the richness of functional groups available on the surface for effective complexation. Factors like pH and divalent cation concentration of the solution also can affect the extent of sorption reasonably.[2]

Fullerene-based sorbents have been proved significant for pollutants like heavy metals, PAHs, organic and organometallic compounds.[22] Ballesteros et al. showed the utility of $C_{60}$ fullerene for adsorbing organic pollutants such as PAHs, phenols, amines and M-methyl carbamates.[23] Additionally, they were successful in the sorption of organometallic compounds like organolead and metallocenes better than conventional sorbents such as silica gel 100. Samonin et al.[24] showed that the incorporation of $C_{60}$ fullerene to activated carbon resulted in sorption enhancement towards metal cations by 2.5–5-fold. They exemplified this by studying adsorption behavior towards the metals silver, lead and copper. The utility of fullerenes towards heavy metal sorption was explored by Alekseeva et al. by comparing the efficacy of fullerene with polystyrene and a fullerene–polystyrene composite.[25] They found that the adsorption behavior was enhanced upon introducing fullerene and the absorption behavior towards metals followed the order: Zn (II) < Ni (II) < Cd (II) < Cu (II).

Carbon-based materials like fullerenes and nanotubes are easily susceptible to modification which makes them popular in applications like adsorption. Functionalisation allows for the targeting of the specific adsorption of micropollutants, permitting the removal of even low concentrations of pollutants. It has been reported that –COOH or –OH functionalized CNTs displayed improved adsorption to polar and low molecular weight compounds when compared to the activated carbon sorbent. This effect is attributed to the presence of hydrophilic functional groups present on the CNT surface.[2]

Among different carbonaceous nanomaterials, CNTs possess a significant position in the field of pollutant adsorption owing to their hollow and layered structure and small size. Thus, they can provide their surface and interstitial space, as well as their internal cavity, as adsorption sites. This renders them with superior adsorption efficacy for the removal of environmental contaminants. Depending upon the nature of the sorbates, CNTs are reported to display different sorption mechanisms due to interactions such as hydrogen bonding, electrostatic, hydrophobic and $\pi$–$\pi$ type. For heavy metal adsorption, complexation reactions between functional groups on the CNT surface and metal ions along with their physical adsorption and electrostatic interaction comes into their chief role.[26] Song et al.[27] compared the adsorption capacity of MWCNT towards hydrophobic phenanthrene and hydrophilic $Cd^{2+}$ and obtained sorption of 162.1 mg/g and 11.18, respectively. They reasoned the hydrophobic and $\pi$–$\pi$ interaction between aromatic phenanthrene and the graphene sheets of CNTs as the factor causing much-improved adsorption towards the hydrocarbon.

In order to achieve superior adsorption activity, surface modifications were conducted on the surface of the CNTs. Several methods were followed for surface modification including acid treatment, grafting specific functional groups and loading metals or microorganisms. Li et al.[28] introduced active functional groups like –OH, –COOH, –C=O on the CNT surfaces to electrostatically attract hydrophilic metal ions. They employed oxidizing agents such as $KMnO_4$, $HNO_3$ and $H_2O_2$ and obtained enhanced metal adsorption for cadmium ions from 1.1 to 11, 5.1 and 2.6 mg/g, respectively. In another report, Shao et al.[12] grafted chitosan onto the surface of MWCNT. An enhanced number of amino and hydroxyl groups made MWCNTs good adsorbents for copper, lead and $UO_2^+$ ions from water samples by effective chelation with the cations.

Hadavifar et al.[29] functionalized MWCNTs with amino and thiol groups by using ethylenediamine, cyanuric chloride and sodium 2-mercaptoethanol sequentially. They studied mercury (II) ion adsorption from wastewater samples and found more efficiency for thiol functionalization than for amino. Toxic pollutants like lead, cadmium and 1,2-dichlorobenzene were removed by CNTs as per the reports of various groups.[7]

Additionally, carbonaceous nanomaterials are also employed as efficient scaffolds for other materials with inherent sorption skill to obtain improved sorption capacity. They showed obvious improvement in absorption of pollutant cations in solution. For instance, CNTs decorated with $CeO_2$ showed effective removal of chromium and arsenate ions, while the use of amorphous alumina allowed removal of fluoride ions. Introducing polypyrrole resulted in the separation of perchlorate from the solution. Several studies have focused on metal preconcentration, oxidation or removal by other material embedded CNTs.[2]

### *13.2.3.2 Membrane Separation of Pollutants*

Membrane separation is a physical method of separating solutes of a solution through mechanical filtration. It is a very significant step in wastewater treatment. The methods employ a membrane with a porous thin-layered structure, limiting the passage of particles like metals, salt, bacteria, viruses and so on. Membrane-mediated water purification is relatively simple, less costly and efficient. The method has high partition effectiveness and it is eco-friendly.[20]

Various types of separation processes are reported to be used for water treatment. Microfiltration, ultrafiltration, nanofiltration, diafiltration, forward osmosis, reverse osmosis, membrane bioreactors, membrane distillation, electrodialysis and capacitive deionization (CDI) are examples to name a few. Serious issues encountered in common membrane filtration process are membrane biofouling and energy consumption. Energy consumption leads to a decline in membrane efficiency as well as a rise in maintenance cost. This influences the quality and flux of obtained water.[20]

Separation membranes employing the fascinating properties of CNTs are obtained which can decrease the energy requirements and the cost of the method reasonably. CNTs can improve the efficiency of membranes when used as filler in the membrane matrix. CNTs can alleviate the fouling of membranes by avoiding the takeoff between selectivity and permeability. CNT membrane filters can provide high stability, flexibility and large surface area resulting in very successful filtration of biological and chemical contaminants in wastewater.[7,20]

CNTs have been explored as functional materials of fabricating membranes. The utilization of hollow structured CNTs as functional materials to fabricate membranes is due to the high adsorption affinity. Also, they can allow swift flow of water through their hydrophobic and smooth walls to bring about faster filtration. Upon the passage of wastewater through the CNT-incorporated membrane, the pollutants will be adsorbed to the surface of CNT fixed on the membrane.[26]

Wang et al.[30] arranged a membrane incorporated with MWCNT and polyvinylidene fluoride and proved it to be effective at separating personal care products and pharmaceuticals from water. The polyvinylidene fluoride membrane fixes MWCNTs firmly so that no CNTs can enter the water and lead to ingestion by humans. The resultant membrane needed lesser pressure than that used in the reverse osmosis technique. Employing the exceptional electrical conductivity of CNTs, the CNT incorporated membranes were used as electrochemical filters. The filters electrochemically degraded organic pollutants and were successful in deactivating bacteria and fungi. The pollutant is degraded either by direct electrochemical oxidation or by the action of generated reactive oxygen species (ROS). Bisphenol-A was reportedly removed by Bakr and Rahaman employing an electrochemical CNT filter and Liu et al. were successful in degrading tetracycline by similar means.[30]

### *13.2.3.3 Removal of Heavy Metals*

Heavy metals are a major threat to nature since once introduced they cannot be degraded biologically. They exist forever and continue to accumulate and contaminate water, soil, and air. They cause solemn threats to the life and health of humans and terrestrial and aquatic plants and animal life, even at minute concentrations. Minamata is a lethal disease caused by mercury overdose. Heavy metals have relatively high density and are toxic in lower concentrations. Most common heavy metals are Hg(II), As(III)/(V), Pb(II), Cd(II) and Cr(VI).[31]

In the past few decades, a lot of studies have been devoted to appropriate removal of heavy metals in water. Important methods for heavy metal removal include filtration, adsorption, reduction, precipitation, electrochemical removal and ion exchange. Out of these known methods, adsorption holds significance since it is simple, manageable at a wide range of pH, feasible to perform at the industrial level and relatively less costly. Furthermore, efficiency and selectivity can be well tuned by functionalization.[31]

Silica, $TiO_2$, zeolite, chitosan and polymers are the major conventional adsorbents. The scientific world is always engaged in the search for the best material for the purpose, and carbonaceous nanomaterials such as CNTs and graphene hold a noteworthy place in heavy metal removal. The interesting properties of carbon nanomaterials such as large surface areas, great pore volumes, tunable surfaces, scalable production, and the presence of oxygen-containing functional groups enables CNTs and graphene to work better than conventional heavy metal sorbents. CNTs and graphene are reported to remove toxic heavy metal contaminants such as $Hg^{2+}$, $As^{3+/5+}$, $Pb^{2+}$, $Cd^{2+}$, $U^{6+}$, $Cr^{4+}$ and $Co^{2+}$.[32]

Graphene, owing to its large specific surface area, inherent oxygen-containing functional groups and significant stability is much explored for heavy metal removal. Deng et al.[33] and Li et al.[34] reported the utility of graphene for the effective sorption of Pb (II) ions from water. Lujaneine et al.[35] carried out adsorption of cobalt, copper, lead and nickel ions using graphene and found that sorption efficiency is pH dependent.[31] Table 13.1 gives a summary of the graphene-based compounds employed in heavy metal adsorption.

CNTs require acid- or base-mediated introduction of functionalities but graphene is rich in oxygen-containing functionalities, inviting interactions from cations. Nanocomposites and hybrids of CNTs and graphene with multiple functionalities have opened a wide area of scientific research for environmental applications. They are arising as smarter materials providing high capacity for the fast and easy separation and removal of heavy metal ions in water with notable selectivity and reasonable reusability. Hence, modification on the surface of carbonaceous nanomaterials is extremely desirable. Modifications are generally carried out using chemical methods like oxidation and deposition as well as hydrothermal, sol–gel, electrochemical and microemulsion methods. Chemical processes are more desired owing to the possibility of incorporating multiple functional groups in simple and selective methods involving a reducing agent, supportive, additive or stabilizer. Functionalisation is used to enhance the number of available oxygen-containing groups such as –OH, –COOH and –C=O and to introduce new functionalities like amino groups or thiols. This will result in the enhancement of interactions between heavy metal ions and nanomaterial surface which ultimately result in their efficient removal.[32] Table 13.2 gives the reports on CNT-based adsorption of heavy metals.

#### *13.2.3.4 Removal of Organic Dyes and Pesticides*

Dyes are substances which can impart color to materials and are widely used in industries such as textiles, paper production, cosmetics, paints, food and so on. Unfortunately, these dye residues are being dumped into nature, polluting water and soil, which in turn reaches living beings and causes undesirable effects. Along with minimizing dye wastes, the effective remediation of nature from dye wastes is also of great concern. Carbonaceous nanomaterials such as CNTs, graphene and graphene oxide have superior activity in the removal of dye residues over conventional materials like activated carbon.

**TABLE 13.1**

Graphene-based Nano Materials for Heavy Metal Removal

| Nanomaterials | Heavy Metal Ion | $q_m$ (mg/g) | Reference |
|---|---|---|---|
| Graphene nanosheets (GNS) | $Pb^{2+}$ | 22.4 | 36 |
| GNS-500 | $Pb^{2+}$ | 35.2 | |
| GNS-cetyltrimethylammonium bromide | $Cr^{6+}$ | 21.57 | 37 |
| $SiO_2$-GNS | $Pb^{2+}$ | 113.6 | 38 |
| $MnO_2$-GNS | $Hg^{2+}$ | 10.8 | 39 |
| Graphene oxide (GO) | $Cr^{6+}$ | 43.72 | 40 |
| GO | $Pb^{2+}$ | 35.6 | 41 |
| GO-EDTA | $Pb^{2+}$ | 479 | 42 |
| GO-Dimethylaminoethyl methacrylate copolymer | $Cr^{6+}$ | 82.4 | 43 |
| Graphene sand composite | $Cr^{6+}$ | 2859.38 | 44 |
| $TiO_2$-GO | $Pb^{2+}$ | 65.6 | 41 |

**TABLE 13.2**

CNT-Based Nanomaterials for Heavy Metal Removal

| Nanomaterial | Heavy Metal Ion | $q_m$ (mg/g) | Reference |
|---|---|---|---|
| Carbon nanotube (CNT) | $Pb^{2+}$ | 17.44 | 45 |
| CNT ($HNO_3$) | $Cr^{3+}$ | 0.5 | 46 |
| CNT ($HNO_3$) | $Pb^{2+}$ | 49.95 | 45 |
| Single-walled CNT (SWCNT) | $Ni^{2+}$ | 9.22 | 47 |
| Multi-walled CNT (MWCNT) | $Ni^{2+}$ | 7.53 | |
| MWCNT ($HNO_3$) | $Pb^{2+}$ | 97.08 | 48 |
| SWCNT (NaOCl) | $Zn^{2+}$ | 43.66 | 49 |
| MWCNT (NaOCl) | $Zn^{2+}$ | 32.68 | |
| MWCNT | $Hg^{2+}$ | 71.10 | 50 |
| MWCNT-OH | | 78.90 | |
| MWCNT-COOH | | 134.00 | |
| MWCNT-$NH_2$ | | 205.00 | |

From the literature, we can observe that functionalized CNTs such as oxidized MWCNTs are efficient at adsorbing organic dyes such as methylene blue and methyl orange from wastewater. Sadegh et al.[51] showed that an adsorbent synthesized by CNT modification, namely, MWCNT–COOH–cysteamine, could significantly enhance the removal of Amido black 10B. SWCNTs adsorbed Blue 29 dye efficiently, owing to their large surface area.[7]

Pesticide residues from agricultural land are finally accumulated in water bodies and soil. Carbon nanomaterials such as CNTs are found to actively remove these pollutants. Deng et al.[33] observed higher diuron removal by oxidized MWCNTs via adsorption due to the increased surface area and pore volume and this was efficient at basic pH ranges. It was found that the use of amino-functionalized CNTs decreased the uptake of chlordane and p,p′-dichlorodiphenyldichloroethylene pesticides by lettuce plants, indicating higher adsorption efficiency obtained by functionalization. The efficiency of semiconducting and metallic SWCNTs was compared towards pyrenebutyric acid, diquat dibromide and 2,4-dichlorophenoxyacetic acid. The activity was higher for semiconducting SWCNTs which can be due to the absence of much electron density, leading to better adsorption (Figure 13.5).[7]

#### *13.2.3.5 Photocatalysis*

Photocatalysis is the method used for the removal of toxic organic contaminants such as dyes and pesticides from water by the absorption of light. This process can enhance the biodegradability of wastes in water bodies, industries and houses. Active studies are contributing to synthesizing stable, non-toxic and inexpensive photocatalysts. Efforts are being made towards synthesizing photocatalysts with higher absorption of light and lower electron–hole recombination, to make more active species to catalyze the degradation of pollutants.[20]

Photocatalysts based on graphene reportedly deactivated bacteria due to their capacity to reduce oxygen to hydrogen dioxide. However, further improvements in the study are not available. CNTs are promising as photocatalysts since they can generate electron-hole pair upon exposure to light owing to the presence of semiconductor nature and low energy gap between the valence and conduction bands. They are mesoporous in nature and have high mechanical strength and suitable chemical, electrical and thermal features to support great applications. In addition, the $sp^2$ carbon network allows the longtime travel of electrons and thus reduces electron–hole recombination. The free electrons and holes are the particles which can generate ROS or radicals. The generated hydroxide and superoxide radicals are highly efficient oxidizing agents which can degrade the organic pollutants present in water.

Photocatalysts such as $TiO_2$ which are fabricated on carbonaceous nanomaterials like CNTs and graphene are known to degrade several organic pollutants. Wang et al. synthesized $TiO_2$–MWCNT for the

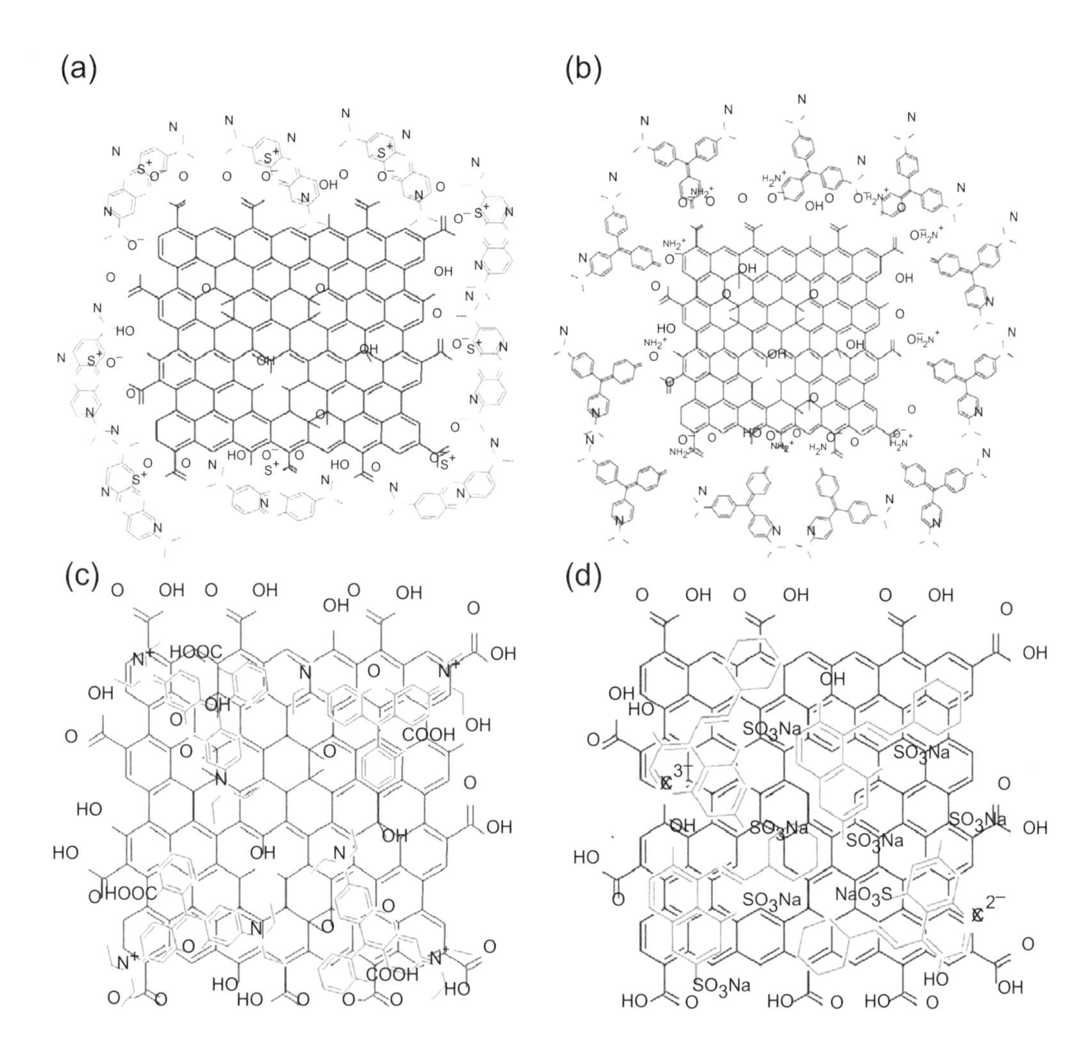


**FIGURE 13.5** **(See color insert.)** Schematic representation of exfoliated graphene oxide interaction with organic dyes (a) methylene blue, (b) methyl violet, (c) rhodamine B, and (d) orange G. (Reproduced from Ramesha et al.[52] With permission from Elsevier.)

degradation of 2,4-dinitrophenol and 2,6-dinitro-p-cresol.[53,54] A $TiO_2$–CNT nanocomposite was synthesized by several groups and they found enhanced photocatalytic activity compared to bare TiO2. Several organic compounds, such as phenol, benzene and azo dyes like methyl orange and procion red MX-5B were successfully degraded by such composites.[55–57] Zhang et al. employed a $TiO_2$–graphene nanocomposite for gas-phase photocatalytic reduction of benzene and methylene blue.[58]

### *13.2.3.6 Sensors*

Conventional sensors are not sufficiently competent to monitor and detect contaminants in water bodies. An efficient sensor should possess qualities such as rapid response, sensitivity, selectivity, reliability and accuracy. Thus, in recent years, nanomaterials have been used to overcome problems in the field of sensing and monitoring water pollutants. They provide a fast response with three- to four-order enhanced sensitivity to targets in comparison to thin film-based sensors.[20]

Carbon nanomaterials are shown to have good fluorescence features, making them very useful in the optical sensing of target pollutants. CNTs are also great candidates for sensor application since their large specific area and electrical conductivity makes them suitable for detecting pollutants electrochemically.

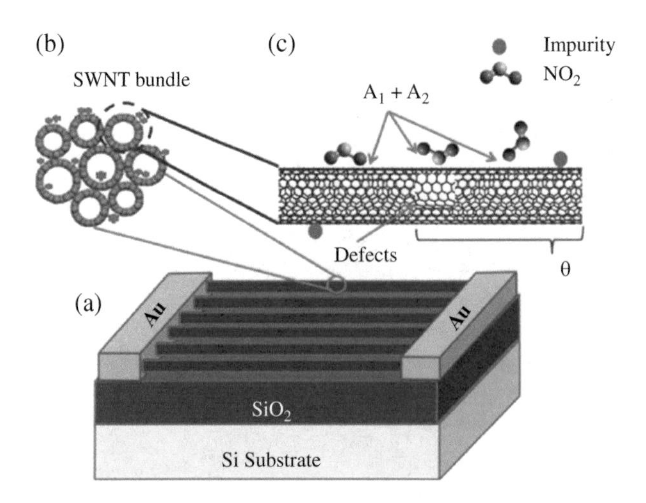


**FIGURE 13.6 (See color insert.)** Schematic representation of SWCNT-gas sensor for $NO_2$ monitoring. (Reproduced from Kumar et al.[59] With permission from Elsevier.)

Fascinating properties of CNTs such as their non-metallic nature, electronic transport, minute size per amount of material and thermal power can be exploited for environmental sensing applications. CNTs are applied as optical sensors, resistive sensors, gas ionization sensors, chemical field effect transistors, capacitive sensors and thermoelectric sensors. They can be used to make CNT devices and to modify electrode surfaces, and electrochemical sensing can be achieved. Kumar et al.[59] constructed SWCNT-based gas sensor by electrophoresis for monitoring $NO_2$ levels. The sensor was found to act quickly, selectively and sensitively. Another report reveals the utility of MWCNT arrays grown on a silica substrate as anodes in ionization sensors for gas detection (Figure 13.6).[7]

CNT nanowire-based sensors have contributed greatly to the field of sensors. Any change in analyte composition and concentration can be easily detected by observing current fluctuation since the adsorption of the charged analyte on to the CNT surface can significantly perturb its conductance. The uncovered large surface area of SWCNTs gives them efficient sensing ability with a fast and sensitive response and low detection limits. In a gas sensor developed by Kong et al.[60] using SWCNT, the electrical resistance of the sensor was changed by three times upon sensing $NO_2$ or $NH_3$ at the parts-per-million (ppm) level. Label-free electrical detection was provided by the direct adsorption of proteins or nucleic acid targets to the surface of the CNT electrode array. Some studies employ the conductive properties of CNTs to amplify signal pathways in both recognition and transduction events. Detection limits achieved by many of these methods are very high, in the order of $10^{-18}$ M or $10^{-21}$ M.[7]

For instance, Safavi et al.[61] obtained a selective Hg(II) sensor by adsorbing cold mercury vapor onto SWCNTs in industrial wastewater. Sensing was monitored by impedance studies and obtained a low detection limit of 0.64 μg $mL^{-1}$ Hg (II) in different samples of wastewater. In another case, de Oliveira et al.[62] employed CNT-based voltammetric sensors for sensing anthraquinone dye in wastewater. They coated sulphuric acid-activated MWCNTs on to glassy carbon electrodes and were able to detect anthraquinone in concentration as low as $2.7 \times 10^{-9}$ mol $L^{-1}$. Hence, one can understand that carbonaceous nanomaterials have shown good sensitivity and excellent stability for real sample analysis.[7]

Fullerenes also were studied for their sensor applications. For instance, Serrano and Gallego reported that $C_{60}$ or Buckminsterfullerene is effectively detecting volatile organic compounds such as toluene, benzene, ethylbenzene and xylene.[63] Graphene and graphene-based nanomaterials such as graphene oxide and its reduced form are extensively explored for their gas sensor activity. It was reported that their adsorption capacity and electron mobility, as well as good signal-to-noise ratio, are the factors that make them competent.[64]

(GNRs and GONRs are recent additions to the carbonaceous material-based sensors. They are stable and biocompatible and have exciting electronic and catalytic properties. GNRs were reported to detect biomolecules such as aptamers and antibodies, while GONRs detected adenosine triphosphate and methylene blue. The fabrication of GNRs and GONRs with several biomolecules, magnetic and non-magnetic materials are reportedly offering versatile sensing applications. MWCNT–GONR composites were reported as carbaryl biosensors while GONR–polyethylene glycol–antibody biomatrix was shown effectively detecting chloramphenicol.[6]

#### 13.2.3.7 Removal of Oceanic Oil Spillages

Relative to land, oil spillage in water bodies is not easily manageable since it spreads so rapidly. Hence, oil removal also should be very rapid. Carbon-based materials such as CNTs are known to be effective at removing the pollution of seawater due to oil spillage. CNTs possess highlighted properties such as large surface area, remarkable hydrophobicity, porosity, recyclability and chemical inertness. These properties offer CNTs a rapid oil adsorption capacity. CNTs are used for adsorbing oil spillages in the form of sponges, foam or by being vertically aligned on solid supports.[7]

## 13.3 Nanostructured Polysaccharides

### 13.3.1 Nanocellulose

#### 13.3.1.1 Introduction to Nanocellulose

Cellulose nanomaterials are naturally occurring, with unique structural, mechanical and optical properties, and are considered sustainable nanomaterials due to their ready availability, biocompatibility and biodegradability. Nanocellulose, which is a type of polysaccharide, is an individualized cellulose nanofiber. Cellulose is widely distributed in nature by storage in the plant body. The abundant reserves of nanocellulose became the main reason for its wide use.

Cellulose is a high molecular weight natural homopolymer, composed of α-1,4-linked anhydro-D-glucose units. The sugar units are connected by joining the H and –OH group together with the elimination of water, resulting in a disaccharide, namely cellobiose. In nanosized cellulose, the distinct properties arise due to its surface-to-volume ratio. Different chemical modifications have been attempted on cellulose nanofibers by the natural advantage of an abundance of hydroxyl groups at the surface. Those groups can be functionalized by introducing functional groups for different target functions.[65]

Different types of nanocellulose exist and are given a variety of names. These names mainly include homogenized cellulose pulps that are commonly designated as nanofibrillated cellulose (NFC) or microfibrillated cellulose (MFC), acid hydrolyzed cellulose whiskers described as cellulose nanocrystal (CNC) or nanocrystalline cellulose (NCC) and bacterially synthesized cellulose or bacterial cellulose (BC).[66]

#### 13.3.1.2 Synthesis Methods

Nanocrystalline cellulose (NCC) and MFC or NFC were fabricated from wood pulp fibers which are used as a common raw material. In 1983, Herrick et al.[67] and Turbak et al.[68] were carried out revolutionary work in the synthesis of cellulose nanofibers. The successful synthesis involved repeated treatments of cellulose wood pulp/water suspensions by a mechanical homogenizer.

Usually, the synthesis of NFCs is carried out by a number of methods such as mechanical treatment including that by cryogenic grinders or microfluidizers, ultrasonic homogenizers and high-pressure homogenizers, electrospinning and steam explosion. To overcome the difficulties of high energy consumption due to fiber delamination during the nanofibrillation process, pre-treatments such as enzymatic/mechanical pre-treatment[69] and oxidation[70] or carboxymethylation were carried out and followed by mechanical treatment. The diameters were in the range of 5–60 nm.

Strong acid hydrolysis of cellulosic fibers with hydrochloric or sulfuric acid can be used for the manufacture of NCC/cellulose nanowhiskers. Properties of cellulose such as source, agitation, hydrolysis time and temperature will affect the dimensions of the nanocellulose. The average diameter and typical length vary from 5 to 70 nm and 100 and 250 nm, respectively. Ionic hydrolysis is an alternate way to synthesize NCC using an ionic liquid, 1-butyl-3-methylimidazolium hydrogen sulfate ($bmimHSO_4$) as a catalyst.[71]

Bacterial nanocelluloses (BNCs) are mainly derived from sugars and alcohols as the raw materials with low molecular weight. BNCs with a diameter of 20–100 nm are mainly synthesized using *Gluconacetobacter xylinus* or *Acetobacter xylinum* that come under the category of gram-negative acetic acid bacteria. Both bacterial nanocellulose and plant-based nanocellulose have silmiar molecular formulae, but their nanofiber structure is different significantly.

#### *13.3.1.3 Environmental Applications*

Nanocellulose plays a vital role in wastewater treatment. It integrates fundamental cellulose features such as hydrophilicity, wide functionalization ability, and the ease of fabrication of variable semi-crystalline fiber textures owing to nanodimensional properties like quantum size effects, high aspect ratio, large surface area as well as chemical accessibility. Nanocellulose has higher adsorptive capacity than the microsized cellulose.

Nanocellulose consists of a large number of hydroxyl functionalities. It has some adverse effects, even though these reactive –OH groups will help the modification of nanocellulose. This improves their adsorption behavior in wastewater treatment. The functionalized nanocellulose imparts high adsorption capacity towards toxic water contaminants.

The sorption of hazardous toxins like heavy metal ions can achieved by carboxyl-functionalized nanocellulose. The $-COO^-$ groups provide strengths to nanocellulose and enhance the adsorption of heavy metals from water. The scavenging of uranyl ($UO_2^{2+}$) ions of about 167 mg $g^{-1}$ was improved in nanofiber cellulose (NFC) due to the presence of negatively charged carboxylate groups. Graft co-polymerization through acrylic or maleic or itaconic acid was conducted to increase the concentration of $-COO^-$ groups in NFC obtained from rice straw. The sorption of heavy metal impurities like $Ni^{2+}$, $Cr^{3+}$, $Cd^{2+}$ and $Pb^{2+}$ can be achieved by the ion exchange process between $-COO^-$ groups and metal ions.

Succinylated nanocelluloses were obtained by the functionalization through succinic anhydride. These kinds of nanocellulose modifications participate in the adsorption of various heavy metal ions from water resources.

The incorporation of amino functionality avails the bioactive group ($NH^{3+}$). Amine functionalities have a chelating effect on heavy metals and dyes which increases the sorption capacity of the nanocellulose materials. Polyvinylamine modified cellulose nanocomposite is used for the removal of anionic dyes, such as congo red 4BS, acid red GR and reactive light yellow K-4G. This chemisorption was found to be pseudo second-order kinetics. NFC was modified with aminopropyltriethoxysilane (APS) for the uptake of $Cu^{2+}$, $Ni^{2+}$ and $Cd^{2+}$. The synergistic effect between the amino group on APS and the –OH group on the cellulose improves the adsorption capacity. The decontamination of heavy metals in water is enhanced by the incorporation of amino functional groups in nanostructured cellulose. Figure 13.7 shows the amination modification on cellulose nanowhiskers for dye removal.

Cellulose-based nanocomposites are novel candidates for the remediation of wastewater treatment, with large adsorption efficiency and spontaneous decontamination of water pollutants. These nanohybrid materials are biodegradable, biocompatible and environmentally friendly and exhibit properties such as flexibility, toughness, low density and high adsorption capacity. Different types of cellulose-based nanocomposites such as cellulose/clay nanocomposites, cellulose/polymer nanocomposites and cellulose/silsesquioxane nanocomposites, have been studied for wastewater treatment.

#### *13.3.1.4 Mechanism*

The mechanism of adsorption of nanocellulose and heavy metal ions are discussed here. The possible mechanism involve through a solid-phase and liquid-phase interaction. The solid phase contains the nanocellulose and liquid phase involves the dissolved metal ions to be adsorbed. The uptake of heavy metal ions by nanocellulose involves a number of processes such chemical adsorption, chelation, ion exchange complexation reaction, adsorption on surface and pores, entrapment in inter and intra-fibrillar capillaries, adsorption by physical forces, and spaces of the structural nanocellulose framework due to concentration gradient and diffusion through cell wall and membrane.

The mechanism involved in the binding of nanocellulose crystals with organic dyes is (Rhodamine dye) due to the hydrogen bonding and electrostatic attraction. The electrostatic interaction mechanism

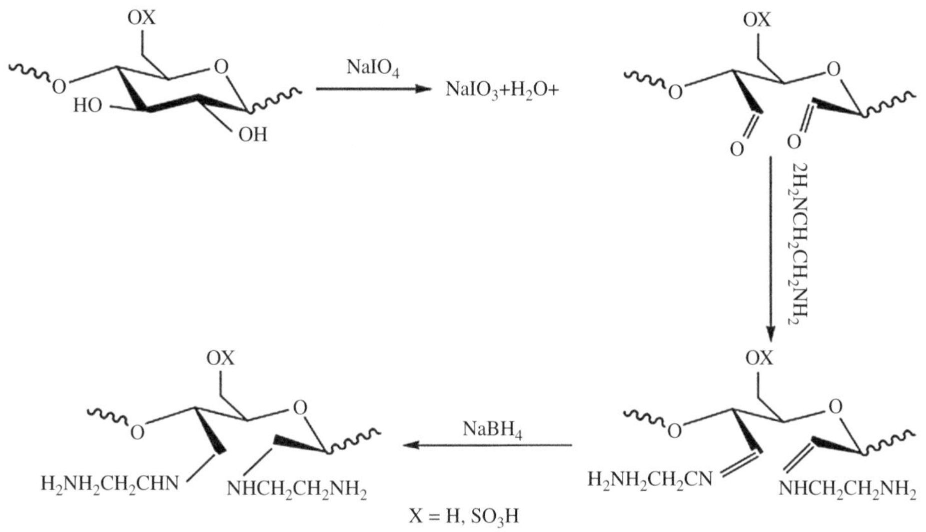

OX $NaIO_4$ $NaIO_3$+$H_2O$+ OX

$2H_2NCH_2CH_2NH_2$

OX $NaBH_4$ OX

$H_2NH_2CH_2CHN$ $NHCH_2CH_2NH_2$ $H_2NH_2CH_2CN$ $NHCH_2CH_2NH_2$

X = H, $SO_3H$

**FIGURE 13.7** Sodium periodate oxidation and amination reaction of cellulose nanowhiskers. (Reproduced from Jin et al.[72] With permission from Springer Nature.)

between cellulose-$O^-$ and dyes on the membrane surface were depicted in Figure 13.8. The strong ionic interaction occurs between the highly negative charged nanocelluloses with positively charged dyes.

## 13.3.2 Nanochitosan

### *13.3.2.1 Introduction*

In 1811, Henri Braconnot, a French professor of natural history, discovered chitosan, a chitin bio-polymer. The de-acetylated product of chitin, chitosan was synthesized by Professor C. Rouget in 1859.[74]

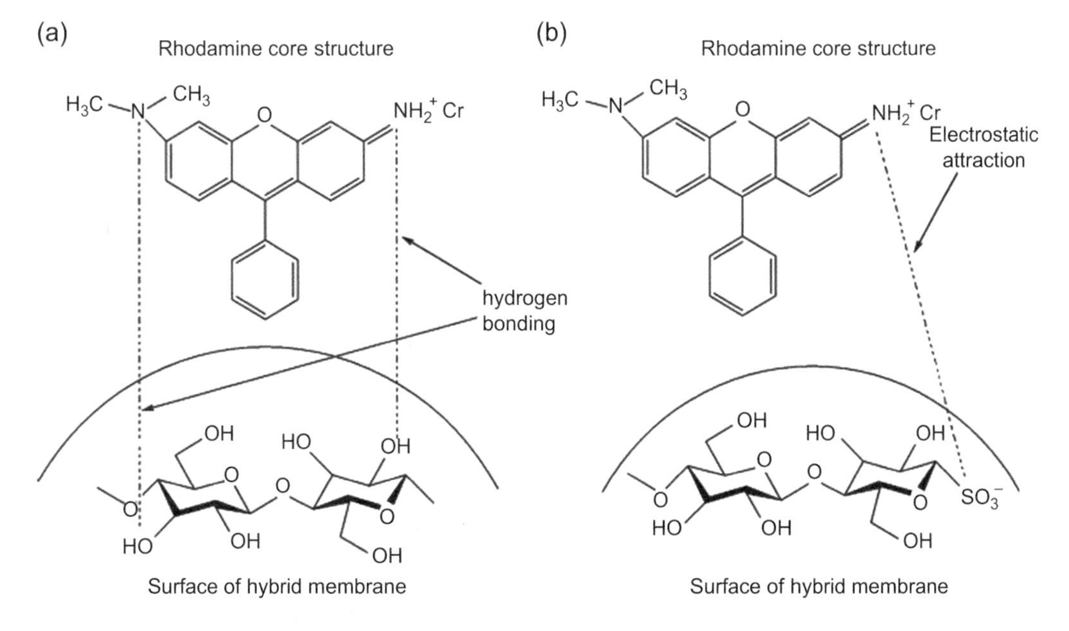


**FIGURE 13.8** A schematic representation for the mechanisms of binding of the dyes with NCCs via (a) hydrogen bonding and (b) electrostatic attraction. (Reproduced from Zoheb et al.[73] With permission from Elsevier.)

The macromolecular organization such as chitosan and modified chitosan are some biosorbents for environmental application. Chitosan is a natural non-toxic biopolymer of polyamino-saccharide, synthesized from the deacetylation of chitin, consisting of unbranched chains of -(1, 4)2-acetoamido-2-deoxy-D-glucose. Chitosan can be chemically modified and it contains a large number of chelating amino groups, and is more useful than chitin.[75] Chitosan is a biocompatible and biodegradable material with effective antibacterial properties and environmental friendliness. The presence of amino and hydroxyl groups by virtue of its structure made nanochitosan into an attractive candidate for the sorption of heavy metals from wastewater. Rather than the usual chitosan, nanochitosan sorbents have enhanced application in separation process owing to high specific surface area, small size and quantum size effect that could make it show higher abilities for metal ions.[76]

#### *13.3.2.2 Synthesis Methods*

Chitin and chitosan can be easily modified into gels, beads, nanoparticles, membranes, nanofibers, scaffolds and sponges.[77] The synthesis of chitosan nanoparticles is carried out mainly by the ionic gelation of chitosan and tripolyphosphate. Chitosan particles with an average size of 40 to 100 nm have been prepared by this method. Ionic gelation method is widely used among researchers due to its simplicity and ease of preparation of nanoparticles. The electrostatic interactions between the positively charged amino group of chitosan and anionic cross-linking agent like sodium tripolyphosphate attributed the formation of chitosan nanoparticles. The general procedure involves the addition of polyethylene glycol (PEG) crosslinking agent into a solution of chitosan polymer in acetic acid having a stabilizer such as Tween 80. Other methods for the fabrication of nanochitosan are the self-assembling method, reverse micellar method, emulsification solvent diffusion method, micro-emulsion method, emulsion-droplet coalescence technique, premix membrane emulsification method, spray drying and the desolvation method. Chitosan nanofibers can be synthesized by different methods including chemical vapor deposition method, sol–gel method, thermal oxidation method, electrospinning method, etc. The suitable candidate for the manufacturing of chitosan nanofibers is the electro-spinning method.

#### *13.3.2.3 Environmental Applications*

The ionotropic gelation of chitosan and tripolyphosphate cross-linking agent were used for the synthesis of nanochitosan particles and tested for $Cd^{2+}$ and $Pb^{2+}$ adsorption. Chitosan nanofibers made by electrospinning were used for the adsorption of $Pb^{2+}$ and $Cu^{2+}$ ions.

Chitosan nanofibers synthesized by electrospinning are potential candidates for the adsorption of $Pb^{2+}$ and $Cu^{2+}$ due to higher surface area per unit mass, huge porosity with fine pores and easy exposure of functional groups to metal ions. Nanochitosan can be used for the removal of arsenate ions. Chitosan nanoparticles are the active choice for the adsorption of hazardous dyes because of their higher surface area. Nanochitosan was used for the adsorption of anthraquinone type Acid Green 27 (AG27) dye. The proposed adsorption mechanism stated that the electrostatic interactions between the positively charged amino group on the chitosan and the anionic dye were attributed to the adsorption. Chitosan nanoparticles have high adsorption capacity over a set of dyes than unmodified chitosan.

The adsorption capacity and selectivity of nanochitosan can be improved by its amination functionalization. The incorporation of $-NH_2$ group on chitosan nanoparticles effectively was carried out by modification agents such as ethylenediamine, hexanediamine, diethylenetriamine, etc.

Chitosan–hydroxyapatite nanocomposites, chitosan–montmorillonite nanocomposites, polyethylene oxide (PEO)/chitosan, nanofiber membrane molecular chitosan/polyacrylic acid etc. are applicable for the sorption of both heavy metal and organic dyes from wastewater.

The mechanism involved in the sorption of metal ions by nanochitosan heavy metal ions are ion-exchange, electrostatic interaction, and formation of ion-pairs, and metal chelation. Dye removal mechanism of nanochitosan can take place by various methods such as ion-exchange, chemical bonding hydrogen bonds, van der Waals force, hydrophobic attractions, aggregation mechanisms physical adsorption and dye interactions occur simultaneously.

### 13.3.3 Nanostarch

#### *13.3.3.1 Introduction*

Starch is a natural, renewable and biodegradable polymer and the major energy reserve of higher plants. It is the second most abundant biomass material in nature. Starch consists of two glucosidic macromolecules amylose, with weight percentage range between 72 and 82% and amylopectins range from 18 to 28%. Amylose is a linear/slightly branched (1→4)-α-D-glucan and amylopectin is a highly branched macromolecule consisting of (1→4)-α-D-glucan short chains linked through α-(1→6) linkages.[78] The biosynthesized starches are semicrystalline granules consists of densely packed polysaccharides with a small quantity of water. The combination of crystalline and amorphous lamellae regions form the inner structure of starch, which together give the crystalline and amorphous growth rings.[79] Various characterization methods were used for the analysis of the intrinsic structure of starch. The methods include nuclear magnetic resonance (NMR) spectroscopy, small angle X-ray microfocus scattering and X-ray microfocus diffraction. Studies revealed that starch granules possessed an annular structure of alternant crystalline and semicrystalline layers.

#### *13.3.3.2 Synthesis Methods*

The structure and morphology of starch nanocrystals defined by the botanical origin of starch and the comparative amount of amylose and amylopectin. The size and yield of starch nanocrystals depend upon the conditions of hydrolysis through the extraction process (such as the type of acid, acid concentration, time and temperature). The low hydrolytic temperature from 35°C to 45°C used to prevent the demolition of the starch crystalline structure and gelatinization. During the hydrolysis process, acid type and extent of hydrolysis are quite different. Hydrochloric acid (HCl) hydrolysis needs more than 15 days while only 5–7 days are needed for sulfuric acid ($H_2SO_4$) hydrolysis.

The conditions of hydrolysis for the extraction of starch nanocrystals with HCl[80] and $H_2SO_4$[81] and its effects on morphology and yield were studied. Starch nanoplatelets were formed during HCl hydrolysis in low yield and the long treatment time prevents its application as a nanofiller in nanomaterials. For the large scale production of starch nanocrystals, $H_2SO_4$ hydrolysis is an adaptable method.

However, Kim and Lim studied alternative ways to obtain starch nanocrystals by regeneration.[82] Regeneration is another method to obtain starch nanocrystals complex formation (or co-crystallisation) with other components. Starch dissolved in an aqueous DMSO solution was heated for 1 h and stirred for 24 h and filtered through a 10 µm pore size polytetrafluoroethylene (PTFE) membrane into n-butanol (3 days at 70°C). Maximum starch precipitation was 6.78% of the initial amount. The complex formation involved amylose rather than amylopectin. During the complex formation, a large portion was an amorphous matrix, so it needed enzymatic hydrolysis for selective removal. Starch nanoparticles displayed spherical or oval shape with diameters in the range 10–20 nm. Combined enzymatic and acid hydrolysis and ultrasound methods were also used for the synthesis of starch nanoparticles.

#### *13.3.3.3 Environmental Applications*

Owing to the biodegradability and biocompatibility of starch nanoparticles, they can be applied in any field, even though they have weak colloidal stability and reduced functionality. Modification on starch nanoparticles by sodium hypochlorite increases the active chlorine, carbonyl and carboxyl contents. Oxidation modification of starch nanoparticles by this method forms negatively charged surface and adsorbs positively charged heavy metal ions such as $Pb^{2+}$ and $Cu^{2+}$ in aqueous solution.[83] Starch nanoparticles and its derivatives are useful for biodegradable food packaging carriers and have a wide environmental impact. But enhanced oxygen transfer rates at elevated humidity conditions limit its applications.

## 13.4 Nanotoxicity

### 13.4.1 Introduction

Nanoparticles have innumerable applications in human life nowadays such as electronic equipment, medicine, diagnostics, drug delivery, textiles, food and personal care applications. This emerging trend of using nanoparticles in everyday life sometimes causes an adverse impact on health and the environment. The toxic effect of nanoparticles is called nanotoxicity and the study of nanotoxicity is called nanotoxicology. The linear relationship between the surface area and size also affects the toxicity. Compared to the bulk, the toxic level is higher than material with a smaller size. Due to the small size of these materials, they can easily penetrate cell membranes and biological barriers causing cell damage.

Different methods of preparation were executed for various morphologically relevant (spheres, rods, wires, prisms, cubes, films and particles) nanomaterials which differ in kinetics and environmental transport. In the inhibition of *Escherichia coli*, the high atom density of triangular nanoplates of AgNPs show greater activity than corresponding spherical or rod-shaped nanoparticles.[84,85]

### 13.4.2 Carbon Nanomaterial-Induced Toxicity

#### *13.4.2.1 Carbon Nanotubes*

CNTs are extensively studied and applied in many fields. During *in vivo* studies in mice, these nanosized materials can be inhaled and can migrate to other body parts.[86] The inhaled CNTs are able to penetrate into cells and can accumulate in cytoplasm. This leads to immunologic toxicity, lung insult and adverse cardiovascular problems. The direct injection or implantations of CNTs on mice results in asbestos-like pathogenic behaviors. This also includes inflammation and development of lesions known as granulomas.[87] Moreover, the injection of CNTs into the tail vein of mice leads a reversible damage to the tests.[88] Several arguments suggested to propose the mechanism of nanotoxicity of these materials and one of them is the generation of ROS.[89] The inactive pristine CNTs adsorbed/assimilated by living bodies will be occupied at the active site or hydrophobic core of the protein. This interaction between the protein and CNTs results in the loss of protein function and may be the initial step to nanotoxicity.

CNTs can bind with proteins and lead to conformational changes (partial unfolding) and a decrease in enzymatic activity. Depending upon the strength of the interaction between proteins and CNTs, two potential mechanisms are suggested rather than the chemical reactions among CNTs and proteins. In strong interaction between the protein and CNTs, the ternary structure of the functional domain of proteins may get destroyed. In weak interactions, CNTs may decrease the target binding of the incoming ligand into the corresponding binding site.[90]

#### *13.4.2.2 Graphene*

Graphene and its derivatives have versatile applications in biomedical and engineering fields. Both the toxic and non-toxic effects were observed in *in vitro* and *in vivo* studies on graphene compounds. Graphene exhibits comparatively lower toxicity than the CNTs and the fact that toxicity is found to be shape and size selective. While CNTs have lower toxicity at lower concentrations, but the toxicity of graphene is higher at lower concentrations. i.e., toxicity and concentrations of graphene are inversely proportional.[91] Compared with graphene, functionalized graphene are found to be less toxic. Because graphene cannot be degradable, but its biocompatibility and solubility can be enhanced on functionalization with hydrophilic polymers such as polyethylene glycol (PEG), which will reduce its hazardous nature. Administration of GO in mice leads to acute toxicity and lung granuloma death and pulmonary edema fibrosis.

The treatment of graphene (>500 μg/mL) with plants such as tomato, cabbage, lettuce and red spinach showed significant inhibition of growth, the number and size of leaves and biomass level, in a dose-dependent manner. The potential adverse effect of graphenes on plants may mainly depend on the plant species, graphene dose and exposure time.[92]

Many reports have suggested that oxidative stress in a target cell by the generation of ROS is one of the mechanisms for the development of toxicity.[93] Antioxidant enzymes such as glutathione peroxidase or superoxide dismutase are helpful to reduce and eliminate ROS. Small and biocompatible materials capped graphene nanomaterials tend to form stable colloidal dispersion and are more suitable to be removed/excreted from the application site.

### 13.4.3 Quantum Dots-Induced Toxicity

High quality quantum dots (QDs) have tremendous applications in biology and medicine. The intriguing area of research comprises the clinical applications of QDs to enhance biocompatibility and applicability in therapeutics. The heavy metal content present in the quantum dots makes them a potential hazard. The most frequently used QDs are CdSe–ZnS, CdTe–CdSe and CdTe–ZnS core–shell nanocrystals for bioimaging and therapeutic researches.

The intracellular uptake of QDs can interrupt the oxidative balance of the cell and generate oxidative stress. This leads to the formation of superoxide ($O^{2-}$), peroxide radicals ($ROO^-$), hydroxyl radicals ($HO^-$), hydrogen peroxide ($H_2O_2$) and singlet oxygen, collectively known as ROS which adversely affect cellular functions. Cd-based QDs also cause neurotoxicity in which the heavy metal content can kill neuronal cells and affect the signal transmission from the brain to the nervous system. However, no abnormal behavior or tissue damage was observed in small animal models (mice and rats) over a period of few months after the systematic administration of QDs.[94]

### 13.4.4 Metal–Metal Oxide Nanomaterial-Induced Toxicity

Nanometals, such as nano-Ag (AgNPs), nano-Au (AuNPs), nano-copper (CuNPs), nano-nickel (NiNPs), nano-cobalt, (CoNPs), nano-aluminum (AlNPs), and other nanoparticles and metal oxide nanoparticles such as nano-ZnO (ZnONPs), nano-$TiO_2$ ($TiO_2$ NPs) and nano-CuO (CuONPs) have also been extensively studied.[95] Among these, AgNPs, AuNPs, ZnONPs and $TiO_2$NPs are the most commonly used nanoparticles in industries, medicine, drug delivery and health care products. Nevertheless, several reports have indicated the toxicity of these metal–metal oxide nanoparticles. As suggested earlier, different nanomaterials have different toxic potency. The toxic level of three metal oxide nanoparticles, such as CuONPs, CdONPs and $TiO_2$NPs were studied.[96] Among the three metal oxide nanoparticles, CuONPs exhibited more cytoxicity and DNA damage and lead to the formation of 8-hydroxy-20-deoxysuanosine (8-OHdG), while $TiO_2$NPs were the least, without producing a substantial amount of 8-OHdG.

Nickel oxide nanoparticles (NiONPs) can induce oxidative stress, cellular ROS, apoptosis/necrosis, lipid peroxidation, and mitochondrial dysfunction in tomato seedling roots.[97] The activities of the anti-oxidative enzymes such as glutathione, catalase, and SOD were all improved. The dissolution of Ni ions from NiONPs was responsible for the cause of cell death, via triggering the mitochondrion-dependent intrinsic apoptotic pathway. The release of $Zn^{2+}$ from dissolved ZnONPs was hazardous to the aquatic organisms. Thus, dissolution of ZnONPs play a vital role in inducing toxicity and lead to disruption of cellular zinc homeostasis, mitochondrial damage and cell death.[98]

Wide medicinal applications of silver nanoparticles (AgNPs) have been discovered.[99,100] AgNPs can develop induced mutation, genotoxicity, apoptosis and oxidative stress mediated by ROS formation in cultured cells, animal tissues and mouse lymphoma cells. In the presence of water, surface oxidation results in the release of $Ag^+$ ions from the suface of AgNPs.[101] Several factors such as the size of AgNPs, temperature, the sulfur concentrations present in AgNPs, oxygen, pH and light determines the rate of release of $Ag^+$ ions. The interaction between the $Ag^+$ ions with molecular oxygen creates superoxide radicals and oxidative stress, resulting in apoptosis and the response of stress-related genes-expression.[102,103] The main reason for the AgNP-induced toxicity is due to the formation of reactive species of $Ag^+$ ions.

Metal–metal oxide nanoparticles with redox characteristic properties can increase the generation of ROS. These nanoparticles act as catalyst in ROS production through Fenton or Fenton-like reactions, or the Haber–Weiss cycle reaction giving hydroxyl radicals.[104–106]

Fenton-like reaction

$$Cu^+ + H_2O_2 \rightarrow Cu^{2+} + {}^\bullet OH + OH^-$$

$$Ag + H_2O_2 \rightarrow Ag^+ + {}^\bullet OH + OH^-$$

The generation of oxidative stress mediated by ROS formation induce the toxicity through free-radical mechanisms. The formation of highly reactive and short lived species cause difficulties in elucidating the

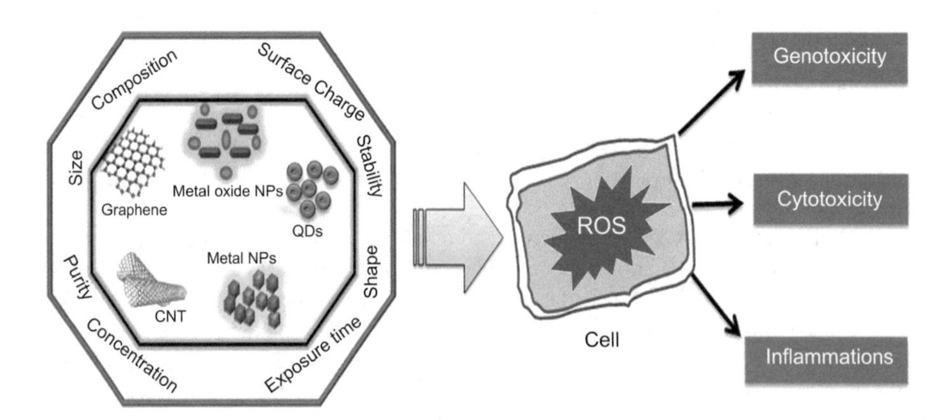


**FIGURE 13.9** Schematic representation of nanotoxicity-induced different nanoforms and their properties on generating reactive oxygen species (ROS).

proper mechanism of toxicity induced by nanoparticles. The direct isolation of ROS, with a short life of about $10^{-9}$ seconds is difficult without derivatization for structural identification. Commonly ESR spin trapping techniques are used to identify the formation of ROS from chemical reactions. The schematic representation (Figure 13.9) gives the idea about the toxicity generated by nanoparticles.

Proper caution is needed for the safe handling of nanoparticles. Suitable care is implemented to ensure the personal health and safety of the people who are involved in the manufacturing and application of nanomaterials. The consumer should be aware of its effect on the environment for better future.

## 13.5 Conclusions and Future Perspectives

The whole chapter discussed carbonaceous, polysaccharide nanomaterials and their environmental applications. Carbon-based nanomaterials such as graphene, fullerene, CNTs and so forth possess a large volume of applications as nanomaterials which are of environmental concern. Nanopolysaccharides are very effective in the adsorption of heavy metals and hazardous pollutants in wastewater. The nanotoxicity of various materials which are applicable in technology and medicine, evolve several impact on health and enviorment. The handling, production and use of these materials should be taken carefully.

## REFERENCES

1. R. Singh, *J. Magn. Magn. Mater.*, 2013, **346**, 58–73.
2. M. S. Mauter, M. Elimelech, *ACS Environ. Sci. Technol. Sci. Technol.*, 2008, **42**, 5843–5859.
3. P. S. Karthik, A. L. Himaja, S. P. Singh, *Carbon Lett.*, 2014, **15**, 219–237.
4. Q. Zheng, J. K. Kim Synthesis, structure, and properties of graphene and graphene oxide. In: *Graphene for Transparent Conductors*, Springer, Newyork, NY, 2015.
5. S. Potnis, *Resonance*, 2017, **22**, 257–268.
6. U. Rajaji, *Int. J. Electrochem. Sci.*, 2018, 6643–6654.
7. B. Sarkar, S. Mandal, Y. F. Tsang, P. Kumar, K. H. Kim, Y. S. Ok, *Sci. Total Environ.*, 2018, **612**, 561–581.
8. V. Datsyuk, M. Kalyva, K. Papagelis, J. Parthenios, D. Tasis, A. Siokou, I. Kallitsis, C. Galiotis, *Carbon N. Y.*, 2008, **46**, 833–840.
9. J. Ma, F. Yu, L. Zhou, L. Jin, M. Yang, J. Luan, Y. Tang, H. Fan, Z. Yuan, J. Chen, *ACS Appl. Mater. Interfaces*, 2012, **4**, 5749–5760.
10. N. M. Bandaru, N. Reta, H. Dalal, A. V. Ellis, J. Shapter, N. H. Voelcker, *J. Hazard. Mater.*, 2013, **261**, 534–541.
11. A. K. S. Deb, P. Ilaiyaraja, D. Ponraju, B. Venkatraman, *J. Radioanal. Nucl. Chem.*, 2012, **291**, 877–883.
12. D. Shao, J. Hu, Z. Jiang, X. Wang, *Chemosphere*, 2011, **82**, 751–758.

13. M. Sebastian, B. Mathew, *J. Macromol. Sci. Part A Pure Appl. Chem.*, 2018, **0**, 1–11.
14. M. Sebastian, B. Mathew, *J. Mater. Sci.*, 2018, **53**, 3557–3572.
15. M. Sebastian, B. Mathew, *Int. J. Polym. Anal. Charact.*, 2018, **23**, 18–28.
16. L. Ai, J. Jiang, *Chem. Eng. J.*, 2012, **192**, 156–163.
17. Z. Sui, Q. Meng, X. Zhang, R. Ma, B. Cao, *J. Mater. Chem*, 2012, **22**, 8767–8771.
18. A. B. Dichiara, M. R. Webber, W. R. Gorman, R. E. Rogers, *ACS Appl. Mater. Interfaces*, 2015, **7**, 15674–15680.
19. A. B. Dichiara, T. J. Sherwood, J. Benton-Smith, J. C. Wilson, S. J. Weinstein, R. E. Rogers, *Nanoscale*, 2014, **6**, 6322–6327.
20. O. K. Bishoge, L. Zhang, S. L. Suntu, H. Jin, A. A. Zewde, Z. Qi, *J. Environ. Sci. Heal. A*, 2018, **4529**, 1–18.
21. F. H. Yang, A. J. Lachawiec, R. T. Yang, *J. Phys. Chem. B*, 2006, **110**, 6236–6244.
22. K. Yang, L. Zhu, B. Xing, *Environ. Sci. Technol.*, 2006, **40**, 1855–1861.
23. E. Ballesteros, M. Gallego, M. Valcárcel, *J. Chromatogr. A*, 2000, **869**, 101–110.
24. V. V. Samonin, V. Y. Nikonova, M. L. Podvyaznikov, *Prot. Met.*, 2008, **44**, 190–192.
25. O. V. Alekseeva, N. A. Bagrovskaya, A. V. Noskov, *Russ. J. Appl. Chem.*, 2015, **88**, 436–441.
26. B. Song, P. Xu, G. Zeng, J. Gong, P. Zhang, H. Feng, Y. Liu, X. Ren, *Rev. Environ. Sci. Bio Technol.*, 2018, 17, 571–590.
27. B. Song, G. Zeng, J. Gong, P. Zhang, J. Deng, C. Deng, J. Yan, P. Xu, C. Lai, C. Zhang, M. Cheng, *Chemosphere*, 2017, **172**, 449–458.
28. Y. H. Li, S. Wang, Z. Luan, J. Ding, C. Xu, D. Wu, *Carbon N. Y.*, 2003, **41**, 1057–1062.
29. M. Hadavifar, N. Bahramifar, H. Younesi, Q. Li, *Chem. Eng. J.*, 2014, **237**, 217–228.
30. Y. Wang, J. Zhu, H. Huang, H. H. Cho, *J. Memb. Sci.*, 2015, **479**, 165–174.
31. A. E. Burakov, E. V. Galunin, I. V. Burakova, A. E. Kucherova, S. Agarwal, A. G. Tkachev, V. K. Gupta, *Ecotoxicol. Environ. Saf.*, 2018, **148**, 702–712.
32. J. Xu, Z. Cao, Y. Zhang, Z. Yuan, Z. Lou, X. Xu, X. Wang, *Chemosphere*, 2018, **195**, 351–364.
33. X. Deng, L. Lü, H. Li, F. Luo, *J. Hazard. Mater.*, 2010, **183**, 923–930.
34. F. Li, X. Wang, T. Yuan, R. Sun, *J. Mater. Chem. A*, 2016, **4**, 11888–11896.
35. G. Lujaniene, S. Semcuk, I. Kulakauskaite, K. Mazeika, *J. Radioanal. Nucl. Chem.*, 2016, **307**, 2267–2275.
36. Z. H. Huang, X. Zheng, W. Lv, M. Wang, Q. H. Yang, F. Kang, *Langmuir*, 2011, **27**, 7558–7562.
37. Y. Wu, H. Luo, H. Wang, C. Wang, J. Zhang, Z. Zhang, *J. Colloid Interface Sci.*, 2013, **394**, 183–191.
38. L. Hao, H. Song, L. Zhang, X. Wan, Y. Tang, Y. Lv, *J. Colloid Interface Sci.*, 2012, **369**, 381–387.
39. T. S. Sreeprasad, S. M. Maliyekkal, K. P. Lisha, T. Pradeep, *J. Hazard. Mater.*, 2011, **186**, 921–931.
40. S. Yang, L. Li, Z. Pei, C. Li, J. Lv, J. Xie, B. Wen, S. Zhang, *Colloids Surf. A Physicochem. Eng. Asp.*, 2014, **457**, 100–106.
41. Y. C. Lee, J. W. Yang, *J. Ind. Eng. Chem.*, 2012, **18**, 1178–1185.
42. C. J. Madadrang, H. Y. Kim, G. Gao, N. Wang, J. Zhu, H. Feng, M. Gorring, M. L. Kasner, S. Hou, *ACS Appl. Mater. Interfaces*, 2012, **4**, 1186–1193.
43. H. L. Ma, Y. Zhang, L. Zhang, L. Wang, C. Sun, P. Liu, L. He, X. Zeng, M. Zhai, *Radiat. Phys. Chem.*, 2016, **124**, 159–163.
44. R. Dubey, J. Bajpai, A. K. Bajpai, *J. Water Proc. Eng.*, 2015, **5**, 83–94.
45. Y. Li, S. Wang, J. Wei, X. Zhang, C. Xu, Z. Luan, D. Wu, B. Wei, *Chem. Phys. Lett.*, 2002, **357**, 263–266.
46. M. A. Atieh, O. Y. Bakather, B. S. Tawabini, A. A. Bukhari, M. Khaled, M. Alharthi, M. Fettouhi, F. A. Abuilaiwi, *J. Nanomater.*, 2010, **2010**, 1–9.
47. X. W. Zhao, L. Wang, D. Liu, *J. Chem. Technol. Biotechnol.*, 2007, **82**, 1115–1121.
48. Y.-H. Li, J. Ding, Z. Luan, Z. Di, Y. Zhu, C. Xu, D. Wu, B. Wei, *Carbon N. Y.*, 2003, **41**, 2787–2792.
49. C. Lu, H. Chiu, *Chem. Eng. Sci.*, 2006, **61**, 1138–1145.
50. D. Zhang, Y. Yin, J. Liu, *Chem. Speciat. Bioavailab.*, 2017, **29**, 161–169.
51. H. Sadegh, K. Zare, B. Maazinejad, R. Shahryari-Ghoshekandi, I. Tyagi, S. Agarwal, V. K. Gupta, *J. Mol. Liq.*, 2016, **215**, 221–228.
52. G. K. Ramesha, A. V. Kumara, H. B. Muralidhara, S. Sampath, *J. Colloid Interface Sci.*, 2011, **361**, 270–277.
53. H. Wang, H. L. Wang, W. F. Jiang, *Chemosphere*, 2009, **75**, 1105–1111.

54. H. Wang, H. L. Wang, W. F. Jiang, Z. Q. Li, *Water Res.*, 2009, **43**, 204–210.
55. G. An, W. Ma, Z. Sun, Z. Liu, B. Han, S. Miao, Z. Miao, K. Ding, *Carbon N. Y.*, 2007, **45**, 1795–1801.
56. Y.-J. Xu, Y. Zhuang, X. Fu, *J. Phys. Chem. C*, 2010, **114**, 2669–2676.
57. Y. Yu, J. C. Yu, C. Y. Chan, Y. K. Che, J. C. Zhao, L. Ding, W. K. Ge, P. K. Wong, *Appl. Catal. B Environ.*, 2005, **61**, 1–11.
58. Y. Zhang, Z. R. Tang, X. Fu, Y. J. Xu, *ACS Nano*, 2010, **4**, 7303–7314.
59. D. Kumar, I. Kumar, P. Chaturvedi, A. Chouksey, R. P. Tandon, P. K. Chaudhury, *Mater. Chem. Phys.*, 2016, **177**, 276–282.
60. J. Kong, N. R. Franklin, C. Zhou, M. G. Chapline, S. Peng, K. Cho, H. Dai, *Science*, 2000, **287**, 622–625.
61. A. Safavi, N. Maleki, M. M. Doroodmand, *J. Hazard. Mater.*, 2010, **173**, 622–629.
62. R. de Oliveira, F. Hudari, J. Franco, M. Zanoni, *Chemosensors*, 2015, **3**, 22–35.
63. A. Serrano, M. Gallego, *J. Sep. Sci.*, 2006, **29**, 33–40.
64. N. Joshi, T. Hayasaka, Y. Liu, H. Liu, O. N. Oliveira, L. Lin, *Microchim. Acta*, 2018, 185. DOI:10.1007/s00604-018-2750-5.
65. B. M. Cherian, A. L. Leao, S. F. Souza, S. Thomas, L. A. Pothan, M. Kottaisamy, *Cellulose Fibers: Bio- and Nano-Polymer Composites*, Springer, Berlin, Heidelberg, 2011, pp. 539–587.
66. I. Siró, D. Plackett, *Cellulose*, 2010, **17**, 459–494.
67. F. W. Herrick, R. L. Casebier, J. K. Hamilton, K. R. Sandberg, *J. Appl. Polym. Sci. Appl. Polym. Symp.* (United States).
68. A. F. Turbak, F. W. Snyder, K. R. Sandberg, *J. Appl. Polym. Sci. Appl. Polym. Symp.*, 1983, **37**, 815–827.
69. M. Pääkko, M. Ankerfors, H. Kosonen, A. Nykänen, S. Ahola, M. Österberg, J. Ruokolainen, J. Laine, P. T. Larsson, O. Ikkala, T. Lindström, *Biomacromolecules*, 2007, **8**, 1934–1941.
70. T. Saito, S. Kimura, Y. Nishiyama, A. Isogai, *Biomacromolecules*, 2007, **8**, 2485–2491.
71. Z. Man, N. Muhammad, A. Sarwono, M. A. Bustam, M. Vignesh Kumar, S. Rafiq, *J. Polym. Environ.*, 2011, **19**, 726–731.
72. L. Jin, W. Li, Q. Xu, Q. Sun, *Cellulose*, 2015, **22**, 2443–2456.
73. Z. Karim, A. P. Mathew, M. Grahn, J. Mouzon, K. Oksman, *Carbohydr. Polym.*, 2014, **112**, 668–676.
74. A. Bhatnagar, M. Sillanpää, *Adv. Colloid Interface Sci.*, 2009, **152**, 26–38.
75. R. Jayakumar, D. Menon, K. Manzoor, S. V. Nair, H. Tamura, *Carbohydr. Polym.*, 2010, **82**, 227–232.
76. L. Qi, Z. Xu, *Colloids Surf. A Physicochem. Eng. Asp.*, 2004, **251**, 183–190.
77. S. Kim, *Chitin, Chitosan, Oligosaccharides and Their Derivatives: Biological Activities and Applications*, CRC Press, Boca Raton, FL, 2010, pp. 3–37.
78. M. Paris, H. Bizot, J. Emery, J. Y. Buzaré, A. Buléon, *Carbohydr. Polym.*, 1999, **39**, 327–339.
79. P. J. Jenkins, R. E. Comerson, A. M. Donald, W. Bras, G. E. Derbyshire, G. R. Mant, A. J. Ryan, *J. Polym. Sci. B Polym. Phys.*, 1994, **32**, 1579–1583.
80. J. L. Putaux, S. Molina-Boisseau, T. Momaur, A. Dufresne, *Biomacromolecules*, 2003, **4**, 1198–1202.
81. H. Angellier, L. Choisnard, S. Molina-Boisseau, P. Ozil, A. Dufresne, *Biomacromolecules*, 2004, **5**, 1545–1551.
82. J. Y. Kim, S. T. Lim, *Carbohydr. Polym.*, 2009, **76**, 110–116.
83. Q. Liu, F. Li, H. Lu, M. Li, J. Liu, S. Zhang, Q. Sun, L. Xiong, *Food Chem.*, 2018, **242**, 256–263.
84. J. R. Morones, J. L. Elechiguerra, A. Camacho, K. Holt, J. B. Kouri, J. T. Ramírez, M. J. Yacaman, *Nanotechnology*, 2005, **16**, 2346–2353.
85. J. Wetzel, S. Herrmann, L. S. Swapna, D. Prusty, A. T. John Peter, M. Kono, S. Saini, S. Nellimarla, T. W. Wong, L. Wilcke, O. Ramsay, A. Cabrera, L. Biller, D. Heincke, K. Mossman, T. Spielmann, C. Ungermann, J. Parkinson, T. W. Gilberger, *J. Biol. Chem.*, 2015, **290**, 1712–1728.
86. J. P. Ryman-rasmussen, M. F. Cesta, A. R. Brody, J. K. Shipley-Phillips, J. I. Everitt, E. W. Tewksbury, O. R. Moss, B. A. Wong, D. E. Dodd, M. E. Andersen, J. C. Bonner, *Nat. Nanotechnol.*, 2009, **4**, 747–751.
87. N. Kong, M. R. Shimpi, O. Ramström, M. Yan, *Carbohydr. Res.*, 2015, 405, 33–38.
88. Y. Bai, Y. Zhang, J. Zhang, Q. Mu, W. Zhang, E. R. S. E. Snyder, B. Yan, *Nat. Nanotechnol.*, 2011, **5**, 683–689.
89. A. Nel, T. Xia, L. Mädler, N. Li, *Science.*, 2006, **311**, 622–627.
90. G. Zuo, S. G. Kang, P. Xiu, Y. Zhao, R. Zhou, *Small*, 2013, **9**, 1546–1556.
91. L. Yan, F. Zhao, S. Li, Z. Hu, Y. Zhao, *Nanoscale*, 2011, **3**, 362–382.
92. P. Begum, R. Ikhtiari, B. Fugetsu, *Carbon N. Y.*, 2011, **49**, 3907–3919.

93. V. C. Sanchez, A. Jachak, R. H. Hurt, A. B. Kane, *Chem. Res. Toxicol.*, 2012, **25**, 15–34.
94. K. T. Yong, W. C. Law, R. Hu, L. Ye, L. Liu, M. T. Swihart, P. N. Prasad, *Chem. Soc. Rev.*, 2013, **42**, 1236–1250.
95. O. Bondarenko, K. Juganson, A. Ivask, K. Kasemets, M. Mortimer, A. Kahru, *Arch. Toxicol.*, 2013, **87**, 1181–1200.
96. X. Zhu, E. Hondroulis, W. Liu, C. Z. Li, *Small*, 2013, **9**, 1821–1830.
97. M. Faisal, Q. Saquib, A. A. Alatar, A. A. Al-Khedhairy, A. K. Hegazy, J. Musarrat, *J. Hazard. Mater.*, 2013, **250–251**, 318–332.
98. N. M. Franklin, N. J. Rogers, S. C. Apte, G. E. Batley, G. E. Gadd, P. S. Casey, *Environ. Sci. Technol.*, 2007, **41**, 8484–8490.
99. R. Vijayan, S. Joseph, B. Mathew, *Part. Sci. Technol.*, 2018, **0**, 1–11.
100. R. Vijayan, S. Joseph, B. Mathew, *Artif. Cells Nanomed. Biotechnol.*, 2018, **46**, 861–871.
101. N. Mei, Y. Zhang, Y. Chen, X. Guo, W. Ding, S. F. Ali, S. A. Biris, P. Rice, M. M.Moore, T. Chen, *Environ. Mol. Mutagen.*, 2012, **53**, 409–419.
102. E. Navarro, F. Piccapietra, B. Wagner, F. Marconi, R. Kaegi, N. Odzak, L. Sigg, R. Behra, *Environ. Sci. Technol.*, 2008, **42**, 8959–8964.
103. L. K. Limbach, P. Wick, P. Manser, R. N. Grass, A. Bruinink, W. J. Stark, *Environ. Sci. Technol.*, 2007, **41**, 4158–4163.
104. L. Gonzalez, D. Lison, M. Kirsch-Volders, *Nanotoxicology*, 2009, **3**, 61–71.
105. B. Wang, J. J. Yin, X. Zhou, I. Kurash, Z. Chai, Y. Zhao, W. Feng, *J. Phys. Chem. C*, 2013, **117**, 383–392.
106. W. He, Y. T. Zhou, W. G. Wamer, M. D. Boudreau, J. J. Yin, *Biomaterials*, 2012, **33**, 7547–7555.

# 14

## *Role of Nanoclay Polymers in Agriculture: Applications and Perspectives*

**Allah Ditta**

**CONTENTS**



## 14.1 Introduction

The world population is increasing day by day, demanding more food to be produced from the limited resources available. The increasing demand for food is being fulfilled through intensive farming, which otherwise requires more fertilizers and pesticides to be applied. This extensive and unavoidable use of these amendments is causing environmental pollution in the form of greater utilization of fossil fuels and eutrophication through extensive use of fertilizers, thereby producing more pollution and causing environmental threats. There is a dire need to produce alternatives to the conventional amendments in the form of fertilizers and pesticides which do little harm to the environment and have more capacity to produce optimum crop yields, ensuring soil health and quality. In this regard, nanotechnology in the form of nano-fertilizers and nano-pesticides could be a great step towards the achievement of sustainable development goals (Ditta 2012; Ditta et al. 2015; Ditta and Arshad 2016).

In nanotechnology, we deal with the particle with at least one dimension being 100 nm or less and the size range for nanoparticles in a colloidal particulate system ranges from 10–100 nm (Nakache et al. 1999; Auffan et al. 2009). These particles have more surface area to volume ratio, therefore they have more absorption capacity in the plant and animal cells (Ditta 2012, Figure 14.1). This capacity for enhanced absorption by the cells tackles the problem of wastage of any nutrient due to inefficient absorption by the

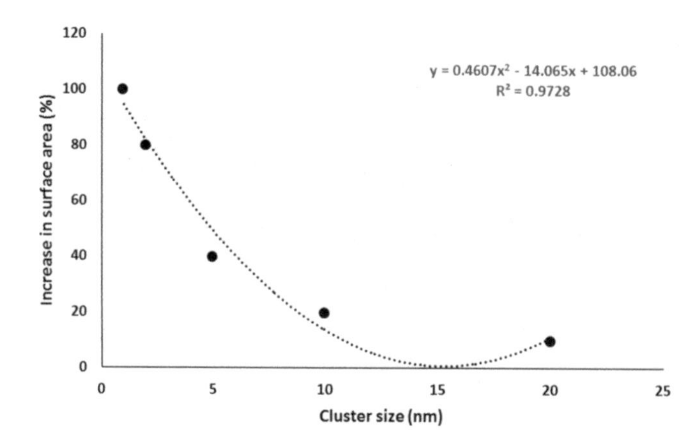


**FIGURE 14.1** Relationship between cluster size (nm) and surface area (%) (Reproduced from Ditta 2012. With permission.)

target like in the form of phosphatic fertilizers which have not more than 20% fertilizer use efficiency. With the passage of time, the recent development in nanotechnology has led us to explore the potential of combined forms of nutrients in the form of nanohybrids. Nanohybrids are materials with a size ranging from 10–100 nm and formed through the combination of two materials. These materials have been prepared on the principle of cost-effectiveness which is only possible through the efficient utilization of the materials being used, e.g. nanoclay polymers, nano-encapsulated pesticides and nano-superabsorbents.

In agriculture, polymers like nanoclay polymers act as carrier materials and have been successfully utilized for controlled release of fertilizers and pesticides. These polymers have been successfully used in the delivery of pesticides with ingredients encapsulated in the polymers, thereby ensuring the slow release of pesticides with efficient absorption. Similarly, micro- and macro-nutrients have also been encapsulated or combined with these polymers and the combined product helps ensure a slow release of the nutrients under consideration, thereby increasing nutrient use efficiency. Moreover, nanoclay polymers have also been utilized for crop improvement through the efficient incorporation of genetic material into the plant cell, thereby generating new varieties with the potential to efficiently utilize nutrients and pesticides and helping in the control of pathogens. All above-mentioned applications in the field of agriculture are possible due to their unique properties like small size and large surface area to volume ratio Figure 14.1. In the following sections, a summarized view of what these nanohybrids are, their characteristics, factors affecting their properties and the applications of nanohybrids in various fields of agriculture is provided. Moreover, future perspectives about their applications and research needs is also discussed.

## 14.2 Nanocomposites

These are formed by the combination of two different materials by changing their composition and structure via exfoliation in the polymer matrix, acid and salt-induced modifications and via mechanically and thermally induced changes for improving catalytic activities at the nanoscale.

### 14.2.1 Types of Nanocomposites

Nanocomposites have been formulated and successfully utilized in agriculture; the details about each type are discussed below.

#### *14.2.1.1 Intercalated Nanocomposites*

In these types of nanocomposites, the polymer chains are penetrated into the interlayer region of the clay and a multilayer structure with alternate polymer/inorganic layers having an interlayer distance of a few nanometers is formed (Weiss et al. 2006).

#### 14.2.1.2 Exfoliated Nanocomposites

During exfoliated nanocomposite formation, there is extensive polymer penetration and clay layers are delaminated and randomly dispersed in a polymer matrix, which makes them the best to exhibit superior properties due to the optimal interaction between clay and polymer. These types of nanocomposites have exceptional properties with their nano-size and large surface area which reduces their quantity required as fillers during the formation of nanocomposites in comparison to conventional composites.

#### 14.2.1.3 Modified Clays

Similarly, there are two types of modified clays (Basak et al. 2012). The details of each modified clay are given below.

##### *14.2.1.3.1 Pillared Layered Clays*

In this type of modified clay, the lamellae of 2:1 clays are propped apart with modified nano-sized pillars of organic cations while charge-balancing cations are exchanged with polymeric hydroxyl cations. These clays have a large surface area and high adsorption potential with solids like microporous zeolites, silica, alumina, etc. After the formation of composites, such types of clays would ultimately increase the water absorption potential of the soil when applied under field conditions.

##### *14.2.1.3.2 Organoclays*

These are organically modified phyllosilicates which are formed by exchanging the interlayer cations with organocation, e.g. quaternary alkylammonium ions. These types of clays at nano-size have a large surface area and high absorption potential for oil in the water which ultimately would enhance the water absorption potential of a composite containing such a type of clay. During the formation of organoclays, native exchangeable inorganic cations are replaced with organic cations which make them organophilic.

### 14.2.2 Preparation of Nanoclay Polymer Composite

Nanoclay polymer composites are prepared according to the standard procedure by Liang and Liu (2007). The chemicals required for the preparation are as follows: a source of ammonia (urea and diammonium phosphate (DAP) granule fertilizer), acrylic acid, acryl amide, the cross-linker (N, N′-methylene bisacrylamide) and ammonium persulfate. Briefly, acrylic acid and acryl amide are dissolved in distilled water and then neutralized (neutralization degree = 60%) with the sources of ammonia listed above in a three-necked flask with a condenser, a thermometer and an opening for nitrogen gas to be added in the reaction mixture. In the three-necked flask, clay is added and dispersed by placing the flask on a magnetic stirrer with a heating control. Under a nitrogen atmosphere, N, N′-methylene bisacrylamide is added to the mixture solution and stirred on the magnetic stirrer at room temperature for 30 minutes. After a radical initiator, vigorous stirring with a gradual increase in temperature (70°C) is started and ammonium persulfate is added to the mixture. The resulting product after polymerization is washed several times with distilled water, dried at 100°C to a constant weight and screened.

Similarly, encapsulated urea formaldehyde fertilizer with superabsorbent and moisture preservation properties is prepared from the 37% formaldehyde solution and urea granules following standard procedures (Guo et al. 2005). Accordingly, the dried granules along with a certain amount of tetrachloride, polyethylene glycol octyl phenyl ether Span-80 are added before the cross-linker and ammonium persulfate. The temperature of the mixture is raised to 65°C and cross-linker and ammonium persulfate are added to get the desired superabsorbent. However, DAP and urea were loaded by the immersion of reweighed dry gels into the aqueous solution of each fertilizer for 20 h to reach swelling equilibrium (Sarkar et al. 2012). Thereafter, the swollen gels were dried at 60°C for six days to get and screen the final dried products.

## 14.3 Factors Affecting Water Absorbency of Nanocomposites

Water absorbency is one of the most critical and important characteristics that should be considered while preparing a nanocomposite with the objective to use it as a superabsorbent under water stress conditions. The materials with the highest water absorbency have the ability to supply water in a slow and steady manner. Scientists around the world have devised some key factors that can affect the water absorbency of nanocomposites. The details about each are discussed below.

### 14.3.1 Time

Earlier, it was found that water absorbency (*QH2O*) of nanocomposites varies with time. The materials having high water absorbency within a shorter period of time are need and could be utilized under arid and water stress conditions (Hedrick and Mowry 1952). For example, a starch-graft polyacrylamide/kaolinite composite showed the highest *QH2O* (4000 g H2O) within an hour (Wu et al. 2000).

### 14.3.2 Amount of Cross-Linker

Another important factor that affects the water absorption capacity of the nanocomposites is the amount of cross-linker (N, N′-methylene bisacrylamide) added. According to Wu et al. (2000), an inverse relationship exists between *QH2O* and the amount of cross-linker in the range of 0.3–0.02%. The reason behind this inverse relationship was suggested to be the decrease in the network space due to the formation of additional networks with the increasing concentration of cross-linker (0.3%) and water absorption in the composites get less space, therefore there is a decrease in the water absorption capacity of the composite. As when a cross-linker is added, polymer chains are produced which increase with the increasing amount of cross-linker added. Similarly, under low concentration of cross-linker (0.02%), there is less copolymerization between starch molecules and the acryl amide monomer which decreases the formation of a cross-linked network of superabsorbents and there is more space for water to be entrapped and, ultimately, more water absorbency by the composite under consideration. Other researchers have also found that cross-concentration in the range of 0.06–0.12% (Singh et al. 2011) and 0.06–0.75% (Guo et al. 2005) was optimum for maximum absorption capacity of the nanocomposite under consideration. They also found that concentration of the cross-linker below 0.065 significantly improved/increased the water absorption capacity of the nanocomposite.

### 14.3.3 Degree of Neutralization

The degree of neutralization shows the alkalinity or acidity of the material under consideration. In other words, it is the amount of acid or base required for complete neutralization of that substance. There is a direct relationship between the degree of neutralization and water absorbency of the nanocomposites. It has been found that water absorbency increased with the degree of neutralization, up to 90% degree of neutralization, and decreases gradually beyond this limit (Guo et al. 2005). However, with the application of montmorillonite (MMT) clay (7%), cross-linker (0.15%) and the degree of neutralization (80%), maximum absorption capacity was noted (Liu et al. 2006). Similarly, in another study, it was found that with the application of sepiolite hydrogel, water absorbency of the nanocomposite was significantly enhanced in comparison with sepiolite free hydrogel (Zhang et al. 2005). During this experiment, the water absorbency was estimated with the help of *Qw* (distilled water absorbency) and *Qp* (physiological saline water absorbency) parameters and the results showed that sepiolite hydrogel application at the rate of 15%, there was an increase of 11.6% and 14.5% in *Qw* and *Qp*, respectively. The sepiolite hydrogel used in the experiment was composed of N, N′-methylene bisacrylamide (0.02% w/w), potassium persulfate (0.41% w/w) and sepiolite (16.2% w/w). With the addition of these mentioned chemicals, a degree of neutralization of 85% was obtained at 65–80°C. Overall, sepiolite hydrogel synthesized in the laboratory showed maximum water absorbance of 830 g $H_2O$ $g^{-1}$ hydrogel and saline absorbance of 98 g $H_2O$ $g^{-1}$ hydrogel.

## 14.4 Structural Analysis of Nanocomposites

After the incorporation of various clays into the polymer network, the absorption bands at about 1030 $cm^{-1}$ were weakened and were attributed to Si–O stretching of clays as revealed through Fourier transform infrared (FTIR) spectra analysis (Sarkar et al. 2012). Moreover, it was also observed that the absorption bands of –OH stretching of various clays ranging from 3400–3700 $cm^{-1}$ disappeared. The reason behind band disappearance was suggested to be due to the graft copolymerization between –OH groups on kaolin and the monomers during polymerization. It has also been reported that a poly (acrylic acid)/mica superabsorbent composite could also be synthesized by the grafting of acrylic acid onto mica (Lin et al. 2001). A similar reason regarding the disappearance of the band suggested that the –OH group can react with acrylamide and kaolinite particle can chemically form a bond with the polymer chains which results in the formation of a starch graft acryl amide/kaolinite composite. From the above discussion, it has been suggested that the interaction between the nanoclay and polymers had some influence on the physicochemical properties of resulting superabsorbent nanocomposites.

Liu et al. (2006) conducted another experiment to find out the structure, spectral signature and micrographs of superabsorbent nanocomposites using FTIR spectroscopy, X-ray diffraction (XRD) and scanning electron microscopy (SEM). It was observed that there were successful intercalation and bonding between an acrylic acid monomer and montmorillonite layers. After polymerization, exfoliation and dispersion of montmorillonite layers at the nanoscale occurred. It was also found that the water absorption capacity of the nanocomposite formed and charge density are dependent on the proper ratio of carboxy and carboxylate groups. In the nanocomposite formed, the carboxyl group is responsible for water absorption while the carboxylate group has infiltration pressure which varies with the degree of neutralization (45–80%). Under a higher degree of neutralization, the ratio is imbalanced, which significantly reduces the water absorption capacity of the nanocomposite under consideration.

Zhang et al. (2005) conducted another experiment regarding FTIR spectra of sepiolite, sepiolite/poly (AA-co-AM) composite, and sepiolite-free poly (AA-co-AM) and observed strong absorption peaks of asymmetric and symmetric R-COONa groups at 1560 and 1410 $cm^{-1}$, respectively. Moreover, the absorption peak of CONH2 appeared at 1672 $cm^{-1}$. The peaks at 1675 and 1020 $cm^{-1}$ were attributed to the presence of C=O stretching and the Si–O group, respectively. During polymerization, esterization of carboxylic acid with silanol and the grafting of AA and/or AM on the sepiolite surface occurred which resulted in the disappearance of absorption bands of Si–O–H groups at 877 and 3689 $cm^{-1}$. Two possible mechanisms of the grafting process were suggested: 1) before radical polymerization, –OH groups in sepiolite may react with acrylic acid (AA) and 2) there might be the production of free radicals by the reaction of OH-groups and radicals and these produced radicals were utilized during graft polymerization with acrylic acid (AA) and acrylamide (AM) branches on sepiolite backbone.

## 14.5 Applications of Nanohybrids in Agriculture

Nanohybrids are well-renowned for their slow and steady supply of materials encapsulated like nutrients, pesticides etc. and effectively being used in agriculture. These have been utilized for efficient supply of water under water stress conditions, the slow and steady supply of nutrients and pesticides encapsulated in nanocarriers (Figure 14.2). Details about nanohybrid applications in various fields are discussed in the following sections.

### 14.5.1 Water Absorbency and Releasing Capacity of the Soil

Water absorbency is the most important feature of the nanocomposites, e.g. slow release membrane-encapsulated urea fertilizer with superabsorbent (Guo et al. 2005). Moreover, these nanocomposites with excellent water absorption capacity are economical and could be utilized at commercial scale for sustainable production of horticultural and agricultural crops plants. Recently, in a research study, a decrease of 15.5 and 22% in the evaporation of water from the soil on the 12th and 21st days were

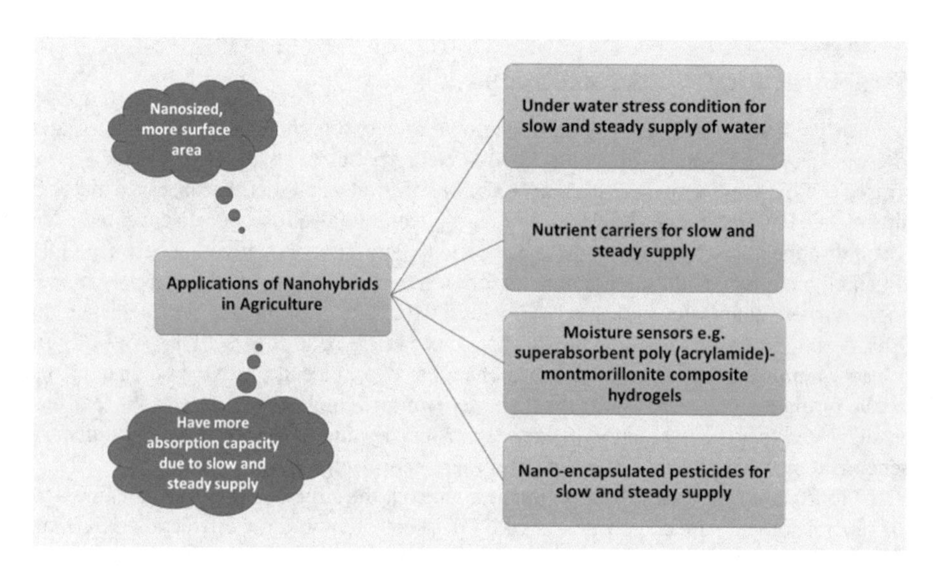


**FIGURE 14.2** Applications of nanohybrids in agriculture.

observed with the application of slow-release membrane-encapsulated urea fertilizer with superabsorbent and moisture preservation (SMUSMP). Similarly, more moisture retention in the soil was noted with the application of a nanoclay polymer composite (NCPC) as compared with the control and with the application of farmyard manure (FYM) (Jatav et al. 2013). In the control, the water release ratio of soil on the 15th and 30th days reached 62.3 and 90.2 wt% while with the application of FYM, the ratios were 59.6 and 82.6 wt% on the same days as stated above. In the case of NCPC application, the ratios of water release from the soil were 55.7 and 78.6 wt%, on the 15th and 30th days, respectively. For 50% water evaporation from the soil, the time required in the case of the control, FYM and NCPC was 12.2, 13.5 and 14.6 days, respectively, which shows an improvement in the water retention/absorption capacity of the soil. Moreover, the data regarding the soil moisture desorption curve showed that the volumetric contents of soil after 20 days were more with the application of NCPC (21.4%) in comparison with control (9.8%) and with the application of FYM (11.4), respectively. Overall, it was concluded that NCPC can result in a reduction of irrigation frequency not only under normal conditions but also under rainfed and drought stress conditions by improving the drought stress tolerance ability of the crop plants (Mukhopadhyay and De 2014).

### 14.5.2 Superabsorbent-Nano-Polymers for Water Stress Management

Water stress poses a great threat to food security and is one the important and crucial abiotic stresses, hindering about 50% of the global agriculture (Vinocur and Altman 2005; Krasensky and Jonak 2012). There have been many approaches employed in order to cope with water stress like the use of water stress tolerant genotypes, fertilizers usage and certain chemicals as absorbents of soil moisture for longer periods of time. In this regard, nanotechnology has also proposed a number of solutions to cope with environmental water stress e.g. nano-superabsorbent polymers. These being very small in size offer more surface area to hold more water which could be utilized for longer periods of time, thereby increasing the water use efficiency of crop plants and ultimately the optimum production of crops (Lokhande and Varadarajan 1992; Nge et al. 2004). Moreover, moisture availability for longer periods of time would reduce the amount of irrigation water required especially under arid and desert condition and increase the fertilizer use efficiency with optimum production of crop plants. Also, these nanohybrids become cost-effective when compared with the conventional amendments being used, as these offer great surface area to volume ratio which increases their absorption capacity and ultimately more utilization efficiency for optimum crop production. These absorbents have been successfully utilized in agriculture, horticulture and the health sector (Sakiyama et al. 1993).

### 14.5.3 Moisture Sensors

With the advancement of research and development, superabsorbent poly (acrylamide)-montmorillonite composite hydrogels have been successfully utilized as novel moisture sensor and proved better in performance in comparison to other sensors (Gao et al. 2001). Accordingly, water vapor absorption behavior of superabsorbent polymers and superabsorbent composites with clays was investigated through thermogravimetric and calorimetric methods. The absorption capacity of the studied materials was dependent on the relative humidity of the gas phase in a non-linear fashion. Overall, the absorption capacity of the superabsorbent composite with clay was superior to simple superabsorbent polymer and the former was suggested to have the potential to be used as a moisture sensor for under water stress conditions.

### 14.5.4 Slow and Sustainable Release of Nutrients

Slow release membrane-encapsulated fertilizers with superabsorbent can also serve as a slow and sustainable source of nutrients and can contribute in the achievement sustainable development goals (SDGs). It has been found that the nutrient release capacity of conventional chemical fertilizers treated and encapsulated with polymers had a slow and steady release for longer periods of times as compared to untreated ones which released about 90% of the contained nutrients within the first five days (Wu et al. 2008). The reason behind this quick release was suggested due to the dissolution of nutrients from the untreated chemical fertilizers in water and a fast release was observed. However, an opposite situation was suggested in case of slow release encapsulated fertilizers that the nutrient dissolved in soil solution get adsorbed on the surfaces of polymers added and also get entrapped due to high absorption capacity of the polymers, being superabsorbents. In this way, the absorbed nutrients in the soil solution and adsorbed nutrients will be released slowly and would act as a slow and steady source of nutrients for longer period of times.

The slow and steady release of nutrients is dependent on the nature of material being used as adsorbent. For example, the impact of polymers prepared from different types of clays (polymer synthesized from kaolinitic, micaceous and montmorillonite clay) was tested in an experiment and it was found that polymer synthesized from kaolinitic clay showed the maximum release of nutrients (90%) within 48 hours of application while 70% of the nutrients were released with the application of polymer synthesized from montmorillonite (Sarkar et al. 2012). Moreover, it was also found that the nutrient release was independent of amorphous aluminosilicate. Earlier, it was found that the encapsulation of fertilizers like urea with polymers and nanoclays i.e. nanoclay polymer composite (NCPC) showed a slow release of nitrogen in comparison to uncoated urea which showed a 90% release of nitrogen on the fifth day (Guo et al. 2005). In case of encapsulated urea, only 61% of nitrogen was released on the 30th day after application. Therefore, it was suggested NCPC can serve as a slow and steady source of nitrogen under rainfed cropping and drought stress conditions.

### 14.5.5 Nano-Encapsulated Pesticides

Nanocarriers have been successfully utilized for an efficient and slow and steady delivery of pesticides to control pests in crop plants (Khot et al. 2012). The slow and steady delivery has been made possible through various mechanisms including the encapsulation and entrapment via weak ionic attachments. These mechanisms not only provide a cushion against environmental degradation but also reduce the cost of application as a small amount is required due to efficient absorption. Moreover, the environmental concerns related with chemical run off are also reduced which ensures environmental stability. Processes like encapsulation and the entrapment of pesticides into the nanocarriers could be accomplished by taking into the account that it ensures their anchorage with plant roots and maximally with the rhizospheric soil (Johnston 2010). In future, there should be a focus on understanding the molecular and conformation mechanisms involved with the delivery of pesticides to the structural modification of nanoscale carriers involved.

Various researchers around the world have reported the efficacy of nano-encapsulated pesticides (Table 14.1). Recent development has been succeeded in the formulation of pesticides with the ability to

**TABLE 14.1**

Applications of Nano-encapsulated Pesticides/Herbicides in the Form of Nanohybrids for Sustainable Agriculture

| Nano-pesticides | Effect and Properties Related with the Control of Pests | Reference |
|---|---|---|
| 2,4-dichlorophenoxyacetate (2,4-D) was developed by the virtue of the formation of organic–inorganic nanohybrid material and its ion exchange property | Release of 2,4-D anions from the lamella of Zn–Al layered double hydroxide (ZAD) was controlled by the first order kinetic up to 12 h regardless of the structure of resulting controlled release formulation. | Hussein et al. (2005) |
| Polyhydroxybutyrate-co-hydroxyvalerate microspheres (PHBV-MS) encapsulated atrazine | Decreased genotoxicity and proved biodegradable herbicide release system | Grillo et al. (2010) |
| Pre-gelation of alginate then complexation between alginate and chitosan and then with paraquat | Soil sorption of paraquat, either free or associated with the nanoparticles, was dependent on organic matter contents. Also, association of paraquat with alginate/chitosan nanoparticles alters the release profile of the herbicide, as well as its interaction with the soil. | Silva et al. (2011) |
| Pesticide formulations were designed by combining wheat gluten, ethofumesate (model pesticide) and three montmorillonites (MMT) using a bi-vis extrusion process | Ethofumesate release was slowed down for all wheat gluten based-formulations as compared to the commercial product and this effect was increased in the presence of hydrophobic MMTs, due to a higher affinity for ethofumesate than for wheat gluten. Contrarily, hydrophilic MMT, proved ineffective to slow down its release despite the tortuous pathway achieved through a well-exfoliated structure. | Chevillard et al. (2012) |
| Azadirachtin encapsulated sodium alginate | Controlled release | Jerobin et al. (2012) |
| Emulsion of surfactants/oil/water and glyphosate | The nano-emulsion formulation showed lower ED50 i.e. 0.40 kg a.e./ha in controlling the weed than Roundup® was 0.48 kg a.e./ha. alleviating the negative effect of pesticide formulations into environment. | Jiang et al. (2012) |
| Cross-linked Chitosan-saponin, Chitosan-Cu, Chitosan/tripolyphosphate | Cu-chitosan nanoparticles were found most effective at 0.1% concentration and showed 89.5, 63.0 and 60.1% growth inhibition of *A. alternata*, *M. phaseolina* and *R. solani*, respectively in *in vitro* model. At the same concentration, Cu-chitosan nanoparticles also showed maximum of 87.4% inhibition rate of spore germination of *A. alternata*. Chitosan nanoparticles showed the maximum growth inhibitory effects (87.6%) on *in vitro* mycelial growth of *M. phaseolina* at 0.1% concentration. | Saharan et al. (2013) |
| Chitosan/tripolyphosphate encapsulated paraquat | Cytotoxicity and genotoxicity assays showed that the nanoencapsulated herbicide was less toxic than the pure compound, indicating its potential to control weeds while at the same time reducing environmental impacts. | Grillo et al. (2014) |
| Alginate Imidacloprid Emulsion | *In vitro* cytotoxicity results revealed that conc. loaded-pesticide nano-formulation was very less toxic than original pesticide. Also, significantly reduced sucking pest (leaf hoppers) population up to 15th day in which its population was found to be in range of 7–1 per three leaves while in other insecticidal treatment the leafhopper population recorded to be above threshold level. | Kumar et al. (2014) |

(*Continued*)

**TABLE 14.1 (CONTINUED)**

Applications of Nano-encapsulated Pesticides/Herbicides in the Form of Nanohybrids for Sustainable Agriculture

| Nano-pesticides | Effect and Properties Related with the Control of Pests | Reference |
|---|---|---|
| Polyacetic acid-polyethylene glycol-polyacetic acid encapsulated imidacloprid | Essential dosage of pesticide and environmental risk decreased significantly and indicated good performance for this formulation. | Memarizadeh et al. (2014) |
| Carboxymethyl chitosan encapsulated methomyl | Control efficacy of methomyl-loaded nanocapsules against the armyworm larvae was significantly superior to the original and was maintained 100% over seven days. | Sun et al. (2014) |
| Chitosan encapsulated Imazapic and Imazapyr | Encapsulation of the herbicides improved their mode of action and reduced their toxicity. | Maruyama et al. (2016) |
| The herbicide glyphosate (GLY) or 2,4-dichlorophenoxyacetic acids (2,4-D) was intercalated in the interlayer region of a Zn-Al-layered double hydroxide (LDH) to obtain LDH-GLY or LDH-2,4-D | LDH-GLY or LDH-2,4-D hybrid composite reduced the maximum 2,4-D or GLY contamination and retarded herbicides leaching through the soil. | Phuong et al. (2017) |
| Cypermethrin loaded calcium alginate nanocarriers | Proved a promising and safe candidate for sustained and slow release of cypermethrin and can reduce environment pollution caused by its excessive use | Patel et al. (2018) |

deliver a slow and steady supply of active ingredients through the increased solubility (due to their small size and large surface area), permeability, specificity and stability under harsh environmental conditions (Bhattacharyya et al. 2016).

In this regard, microencapsulation has been successfully carried out by Syngenta, Switzerland and this company has launched some of the products like Karate ZEON, Subdue MAXX, Ospray's Chyella, Penncap-M with nano-encapsulated pesticides in Australia and other countries (Gouin 2004). Stability of nano-encapsulated pesticides have been achieved through the protection of materials from premature degradation and this property has made the availability of active ingredients in the nano-encapsulated pesticides for longer periods of time. In addition to the supply of active ingredients for longer periods of time, it has also reduced pesticide exposure to human and other living organisms due to the usage of more pesticides with less efficiency and ultimately has led to the sustainability of our environment (Nuruzzaman et al. 2016). In the future, it must be kept in mind that the nano-delivery system of pesticides, like encapsulation, must be carried out through the use of non-toxic carrier materials so that it could contribute towards environmental sustainability through reduced negative impacts on our ecosystem (de Oliveira et al. 2014; Kah and Hofmann 2014; Bhattacharyya et al. 2016; Grillo et al. 2016).

## 14.6 Conclusion and Future Perspectives

From the above discussion, it is concluded that nanocomposites have unique properties when synthesized regarding their role as superabsorbents. Being superabsorbents, these serve as a slow and steady source of water which could be very helpful under rainfed and drought stress agriculture. These have also served as slow release nutrient sources and could provide nutrients like nitrogen to the crop plants for a longer period of time. In future, the research should be focused on other nutrients like phosphorus and others. In order to attain, the sustainable development goals set by the United Nations, we need to focus on the toxic effects of polymers used during the preparation of nanocomposites. In this way, we would make a world with zero poverty and no hunger along with a clean and sustainable environment.

## REFERENCES

Auffan, M., J. Rose, J.-Y. Bottero, G.V. Lowry, J.-P. Jolivet, M.R. Wiesner. 2009. Towards a definition of inorganic nanoparticles from an environmental, health and safety perspective. *Nature Nanotechnology* 4:634–641.

Basak, B.B., S. Pal, S.C. Datta. 2012. Use of modified clays for retention and supply of water and nutrients. *Current Science* 102:1272–1279.

Bhattacharyya, A., P. Duraisamy, M. Govindarajan, A.A. Buhroo, R. Prasad. 2016. Nano-biofungicides: Emerging trend in insect pest control. In: R. Prasad (Ed.), *Advances and Applications through Fungal Nanobiotechnology*. Springer International Publishing , pp. 307–319.

Chevillard, A., H. Angellier-Coussy, V. Guillard, N. Gontard, E. Gastaldi. 2012. Controlling pesticide release via structuring agropolymer and nanoclays based materials. *Journal of Hazardous Materials* 205–206:32–39.

de Oliveira, J.L., E.V. Campos, M. Bakshi, P.C. Abhilash, L.F. Fraceto. 2014. Application of nanotechnology for the encapsulation of botanical insecticides for sustainable agriculture: Prospects and promises. *Biotechnology Advances* 32:1550–1561.

Ditta, A. 2012. How nanotechnology is helpful in agriculture? *Advances in Natural Sciences: Nanoscience and Nanotechnology* 3:033002.

Ditta, A., M. Arshad. 2016. Applications and perspectives of using nanomaterials for sustainable plant nutrition. *Nanotechnology Reviews* 5:209–229.

Ditta, A., M. Arshad, M. Ibrahim. 2015. Nanoparticles in sustainable agricultural crop production: Applications and perspectives. In: M.H. Siddiqui, M.H. Al-Whaibi, F. Mohammad (Eds.), *Nanotechnology and Plant Sciences - Nanoparticles and Their Impact on Plants*. Springer, Switzerland, pp. 55–75.

Gao, D., R.B. Heimann, J. Lerchner, J. Seidel, G. Wolf. 2001. Development of a novel moisture sensor based on superabsorbent poly (acrylamide)-montmorillonite composite hydrogels. *Journal of Materials Science* 36:4567–4571.

Gouin, S. 2004. Microencapsulation: Industrial appraisal of existing technologies and trends. *Trends in Food Science and Technology* 15:330–347.

Grillo, R., P.C. Abhilash, L.F. Fraceto. 2016. Nanotechnology applied to bio-encapsulation of pesticides. *Journal of Nanoscience and Nanotechnology* 16:1231–1234.

Grillo, R., N.F.S. de Melo, R. de Lima, R.W. Lourenço, A.H. Rosa, L.F. Fraceto. 2010. Characterization of atrazine-loaded biodegradable poly(hydroxybutyrate-co-hydroxyvalerate) microspheres. *Journal of Polymers and the Environment* 18:26–32.

Grillo, R., et al. 2014. Chitosan/tripolyphosphate nanoparticles loaded with paraquat herbicide: An environmentally safer alternative for weed control. *Journal of Hazardous Materials* 278:163–171.

Guo, M., M. Liu, F. Zhan, L. Wu. 2005. Preparation and properties of a slow-release membrane-encapsulated urea fertilizer with superabsorbent and moisture preservation. *Industrial and Engineering Chemistry Research* 44:4206–4211.

Hedrick, R.M., D.T. Mowry. 1952. Effect of synthetic polyelectrolytes on aggregation, aeration, and water relationships of soil. *Soil Science* 73:427–442.

Hussein, M.Z., A.H. Yahaya, Z. Zainal, L.H. Kian. 2005. Nanocomposite-based controlled release formulation of an herbicide, 2,4- dichlorophenoxyacetate incapsulated in zinc-aluminum-layered double hydroxide. *Science and Technology of Advanced Materials* 6:956–962.

Jatav, G.K., R. Mukhopadhyay, N. De. 2013. Characterization of swelling behavior of nano clay composite. *International Journal of Innovative Research in Science, Engineering and Technology* 2:1560–1563.

Jerobin, J., R.S. Sureshkumar, C.H. Anjali, A. Mukherjee, N. Chandrasekaran. 2012. Biodegradable polymer-based encapsulation of neem oil nanoemulsion for controlled release of Aza-A. *Carbohydrate Polymers* 90:1750–1756.

Jiang, L.C., et al. 2012. Green nano-emulsion intervention for water-soluble glyphosate isopropylamine (IPA) formulations in controlling *Eleusine indica* (E. indica). *Pesticide Biochemistry and Physiology* 102:19–29.

Johnston, C.T. 2010. Probing the nanoscale architecture of clay minerals. *Clay Minerals* 45:245–279.

Kah, M., T. Hofmann. 2014. Nanopesticides research: Current trends and future priorities. *Environment International* 63:224–235.

Khot, L.R., S. Sankaran, J.M. Maja, R. Ehsani, E.W. Schuster. 2012. Applications of nanomaterials in agricultural production and crop protection: A review. *Crop Protection* 35:64–70.

Krasensky, J., C. Jonak. 2012. Drought, salt, and temperature stress-induced metabolic rearrangements and regulatory networks. *Journal of Experimental Botany* 63:1593–1608.

Kumar, S., G. Bhanjana, A. Sharma, M.C. Sidhu, N. Dilbaghi. 2014. Synthesis, characterization and on field evaluation of pesticide loaded sodium alginate nanoparticles. *Carbohydrate Polymers* 101:1061–1067.

Liang, R., M. Liu. 2007. Preparation of poly(acrylic acid-co-acrylamide)/kaolin and release kinetics of urea from it. *Journal of Applied Polymer Science* 106:3007–3015.

Lin, J., J. Wu, Z. Yang, M. Pu. 2001. Synthesis and properties of poly (acrylic acid)/mica superabsorbent nanocomposite. *Macromolecular Rapid Communications* 22:422–424.

Liu, P., L. Li, N. Zhou, J. Zhang, S. Wei, J. Shen. 2006. Synthesis and Properties of a poly(acrylic acid)/ montmorillonite superabsorbent nanocomposite. *Journal of Applied Polymer Science* 102:5725–5730.

Lokhande, H.T., P.V. Varadarajan. 1992. A new Guargum-based superabsorbent polymer synthesized using gamma radiation as a soil additive. *Bioresource Technology* 42:119–122.

Maruyama, C.R., et al. 2016. Nanoparticles based on chitosan as carriers for the combined herbicides imazapic and imazapyr. *Scientific Reports* 6:23854.

Memarizadeh, N., M. Ghadamyari, M. Adeli, K. Talebi. 2014. Preparation, characterization and efficiency of nanoencapsulated imidacloprid under laboratory conditions. *Ecotoxicology and Environmental Safety* 107:77–83.

Mukhopadhyay, R., N. De. 2014. Nano clay polymer composite: Synthesis, characterization, properties and application in rainfed agriculture. *Global Journal of Bio-Science and Biotechnology* 3:133–138.

Nakache, E., N. Poulain, F. Candau, A.M. Orecchioni, J.M. Irache. 1999. Biopolymer and polymer nanoparticles and their biomedical applications. In: H.S. Nalwa editor. *Handbook of Nanostructured Materials and Nanotechnology*. Academic Press, New York, pp. 577–635.

Nge, T.T., N. Hori, A. Takemura, H. Ono. 2004. Swelling behavior of chitosan/poly (acrylic acid) complex. *Journal of Applied Polymer Science* 92:2930–2940.

Nuruzzaman, M., M.M. Rahman, Y. Liu, R. Naidu. 2016. Nanoencapsulation, nano-guard for pesticides: A new window for safe application. *Journal of Agricultural and Food Chemistry* 64:1447–1483.

Patel, S., J. Bajpai, R. Saini, A.K. Bajpai, S. Acharya. 2018. Sustained release of pesticide (cypermethrin) from nanocarriers: An effective technique for environmental and crop protection. *Process Safety and Environmental Protection* 117:315–325.

Phuong, N.T.K., H.N.N. Ha, N.T.P. Dieu, B.T. Huy. 2017. Herbicide/Zn-Al-layered double hydroxide hybrid composite: Synthesis and slow/controlled release properties. *Environmental Science and Pollution Research* 24:19386–19392.

Saharan, V., A. Mehrotra, R. Khatik, P. Rawal, S.S. Sharma, A. Pal. 2013. Synthesis of chitosan-based nanoparticles and their in vitro evaluation against phytopathogenic fungi. *International Journal of Biological Macromolecules* 62:677–683.

Sakiyama, T., C.H. Chu, T. Fujii, T. Yano. 1993. Preparation of a polyelectrolyte complex gel from chitosan and κ-carrageenan and its pH-sensitive swelling. *Journal of Applied Polymer Science* 50:2021–2025.

Sarkar, S., S.C. Datta, D.R. Biswas. 2012. Synthesis and characterization of nanoclay-polymer composites from soil clay with respect to their water-holding capacities and nutrient – Release behavior. *Journal of Applied Polymer Science* 131:39951 (1–8).

Silva, MdS., D.S. Cocenza, R. Grillo, N.F.Sd Melo, P.S. Tonello, L.Cd Oliveira, D.L. Cassimiro, A.H. Rosa, L.F. Fraceto. 2011. Paraquat-loaded alginate/chitosan nanoparticles: Preparation, characterization and soil sorption studies. *Journal of Hazardous Materials* 190:366–374.

Singh, A., D.J. Sarkar, A.K. Singh, R. Parsad, A. Kumar, B.S. Parmar. 2011. Studies on novel nanosuperabsorbent composites: Swelling behavior in different environments and effect on water absorption and retention properties of sandy loam soil and soil-less medium. *Journal of Applied Polymer Science* 120:1448–1458.

Sun, C., et al. 2014. Encapsulation and controlled release of hydrophilic pesticide in shell cross-linked nanocapsules containing aqueous core. *International Journal of Pharmaceutics* 463:108–114.

Vinocur, B., A. Altman. 2005. Recent advances in engineering plant tolerance to abiotic stress: Achievements and limitations. *Current Opinion in Biotechnology* 16:123–132.

Weiss, J., P. Takhistov, D.J. McClements. 2006. Functional materials in food nanotechnology. *Journal of Food Science* 71:R107–R116.

Wu, J., J. Lin, M. Zhou, C. Wei. 2000. Synthesis and properties of starch-graft-polyacrylamide/clay superabsorbent composite. *Macromolecular Rapid Communications* 21:1032–1034.

Wu, L., M. Liu, R. Rui Liang. 2008. Preparation and properties of a double-coated slow-release NPK compound fertilizer with superabsorbent and water retention. *Bioresource Technology* 99:547–554.

Zhang, F., Z. Guo, H. Gao, Y. Li, L. Ren, L. Shi, L. Wang. 2005. Synthesis and properties of sepiolite/poly (acrylic acid-co-acrylamide) nanocomposites. *Polymer Bulletin* 55:419–428.

# Index